De Gruyter Textbook

Close-Range Photogrammetry and 3D Imaging

Close-Range Photogrammetry and 3D Imaging

2nd edition

Edited by
Thomas Luhmann, Stuart Robson,
Stephen Kyle and Jan Boehm

De Gruyter

ISBN 978-3-11-030269-1
e-ISBN 978-3-11-030278-3

Library of Congress Cataloging-in-Publication Data

A CIP catalog record for this book has been applied for at the Library of Congress.

Bibliographic information published by the Deutsche Nationalbibliothek

The Deutsche Nationalbibliothek lists this publication in the Deutsche Nationalbibliografie; detailed bibliographic data are available in the Internet at http://dnb.dnb.de.

Printing and binding: Hubert & Co. GmbH & Co. KG, Göttingen
♾ Printed on acid-free paper

Printed in Germany

www.degruyter.com

Preface

The first edition of "Close Range Photogrammetry" was published in 2006 as a translated and extended version of the original German book "Nahbereichsphotogrammetrie". It was well received by the large international community of photogrammetrists, metrologists and computer vision experts. This success was further recognised by the International Society of Photogrammetry and Remote Sensing (ISPRS) which awarded the authors the then newly inaugurated Karl Kraus Medal for excellence in authorship (2010).

Seven years have passed since the first English-language edition of the book and a new edition was required to reflect the rapid changes in computer and imaging technologies. This second edition is now entitled "Close-Range Photogrammetry and 3D Imaging". Three-dimensional information acquired from imaging sensors is widely used and accepted and so the title emphasises 3D imaging as a key measurement technology. The field of photogrammetry and optical 3D metrology is still growing, especially in areas that have no traditional link to photogrammetry and geodesy. However, whilst 3D imaging methods are established in many scientific communities, photogrammetry is still an engineering-driven technique where quality and accuracy play an important role.

It is the expressed objective of the authors to appeal to non-photogrammetrists and experts from many other fields in order to transfer knowledge and avoid re-invention of the wheel. The structure of the book therefore assumes different levels of pre-existing knowledge, from beginner to scientific expert. For this reason the book also presents a number of fundamental techniques and methods in mathematics, adjustment techniques, physics, optics, image processing and others. Although this information may also be found in other textbooks, the objective here is to create a closer link between different fields and present a common notation for equations and parameters.

The new edition is printed in four colours which enhances the information in figures and graphics. Numerous technical details have been updated and new sections added, for example on range cameras and point cloud processing. There are also additional numerical examples and more derivations of equations. The chapter on imaging technology has been completely re-structured to accommodate new digital sensors and remove sections on analogue cameras. The chapter on analytical data processing has been updated by a section on panoramic photogrammetry whilst the previous section on line photogrammetry has been removed. As a result, chapters on systems, measurement solutions and applications have also been modified in response to today's available technologies. Finally, the section on quality, accuracy and project planning has been updated to reflect new standards and guidelines.

Readers will not now find references to literature within the text. The extent of potentially relevant literature is both very large and increasing, so it is no longer possible to ensure full coverage. We have therefore decided to summarise relevant literature in a separate chapter. For each listed publication, cross-references are added that link to the related book sections. We hope that readers, who increasingly use the Internet as a literature search tool, will find this solution acceptable.

For the new edition the authoring team has changed. Our respected colleague Ian Harley, co-author of edition 1, unfortunately died unexpectedly in 2011. Despite the loss, the group was fortunate to find another colleague from University College London, Dr. Jan Boehm. The book is therefore still backed by an internationally known team of experts in photogrammetry, metrology and image analysis.

The authors aim to list any errors and updates at www.close-range-photogrammetriy.com and invite all readers to report mistakes directly to them in support of this.

The authors would also like to express their gratitude to the many generous colleagues who have helped complete the work and in particular to Folkmar Bethmann, Holger Broers, Henri Eisenbeiss, Clive Fraser, Heidi Hastedt, Christian Jepping, Thomas Kersten, Hans-Gerd Maas, Fabio Remondino, Dirk Rieke-Zapp, Mark Shortis and Hero Weber. In addition we thank all the companies, universities and institutes that have provided illustrative material and other valuable technical information. The book would also not have been possible without the professional approach to the translation task taken by our publisher, de Gruyter. Finally, of course, we would like to thank our families and colleagues for their patience and support during many months of translation, writing and editing.

Oldenburg/London, August 2013

Thomas Luhmann, Stuart Robson, Stephen Kyle, Jan Boehm

Content

1 Introduction

1.1 Overview

1.1.1 Content

Chapter 1 provides an overview of the fundamentals of photogrammetry, with particular reference to close-range measurement. After a brief discussion of the principal methods and systems, typical areas of applications are presented. The chapter ends with a short historical review of close-range photogrammetry.

Chapter 2 deals with mathematical basics. These include the definition of some important coordinate systems and the derivation of geometric transformations which are needed for a deeper understanding of topics presented later. In addition, a number of geometrical elements important for object representation are discussed. The chapter concludes with a summary of least squares adjustment and statistics.

Chapter 3 is concerned with photogrammetric image acquisition for close-range applications. After an introduction to physical basics and the principles of image acquisition, geometric fundamentals and imaging models are presented. There follow discussions of digital imaging equipment as well as specialist areas of image recording. Laser-based and 3D measuring systems are also briefly outlined. The chapter ends with a summary of targeting and illumination techniques.

Analytical methods of image orientation and object reconstruction are presented in Chapter 4. The emphasis here is on bundle triangulation. The chapter also presents methods for dealing with single, stereo and multiple image configurations based on measured image coordinates, and concludes with a review of panorama and multi-media photogrammetry.

Chapter 5 brings together many of the relevant methods of digital photogrammetric image processing. In particular, those which are most useful to dimensional analysis and three-dimensional object reconstruction are presented. A section on 3D point clouds has been added to the current edition.

Photogrammetric systems developed for close-range measurement are discussed in Chapter 6. They are classified into systems designed for single image, stereo image and multiple image processing. Interactive and automatic, mobile and stationary systems are considered, along with systems for surface reconstruction and the measurement of dynamic processes.

Chapter 7 discusses imaging project planning and quality criteria for practical measurement tasks. After an introduction to network planning and optimisation, the chapter concentrates on strategies for camera calibration.

Finally, Chapter 8 uses case studies and examples to demonstrate the potential for close-range photogrammetry in fields such as architecture and heritage conservation, the construction industry, manufacturing industry and medicine.

1.1.2 References

Relevant literature is directly referenced within the text in cases where it is highly recommended for the understanding of particular sections. In general, however, further reading is presented in Chapter 9 which provides an extensive list of thematically ordered literature. Here each chapter in the book is assigned a structured list of reference texts and additional reading. Efforts have been made to suggest reference literature which is easy to access. In addition, the reader is advised to make use of conference proceedings, journals and the webpages of universities, scientific societies and commercial companies for up-to-date information.

1.2 Fundamental methods

1.2.1 The photogrammetric process

Photogrammetry encompasses methods of image measurement and interpretation in order to derive the shape and location of an object from one or more photographs of that object. In principle, photogrammetric methods can be applied in any situation where the object to be measured can be photographically recorded. The primary purpose of a photogrammetric measurement is the three-dimensional reconstruction of an object in digital form (coordinates and derived geometric elements) or graphical form (images, drawings, maps). The photograph or image represents a store of information which can be re-accessed at any time.

Fig. 1.1: Photogrammetric images

Fig. 1.1 shows examples of photogrammetric images. The reduction of a three-dimensional object to a two-dimensional image implies a loss of information. In the first place, object areas which are not visible in the image cannot be reconstructed from it. This not only includes hidden parts of an object such as the rear of a building, but also regions which cannot be interpreted due to lack of contrast or size limitations, for example individual bricks in a building façade. Whereas the position in space of each point on the object may be defined by three coordinates, there are only two coordinates available to define the position of its image. There are geometric changes caused by the shape of the object, the relative positioning of camera and object, perspective imaging and optical lens defects. Finally, there are also radiometric (colour) changes since the reflected electromagnetic radiation recorded in the image is affected by the transmission media (air, glass) and the light-sensitive recording medium (film, electronic sensor).

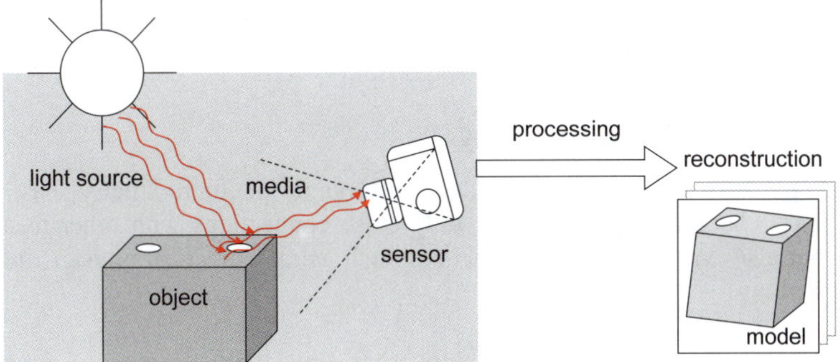

Fig. 1.2: From object to image

For the reconstruction of an object from images it is therefore necessary to describe the optical process by which an image is created. This includes all elements which contribute to this process, such as light sources, properties of the surface of the object, the medium through which the light travels, sensor and camera technology, image processing, and further processing (Fig. 1.2).

Methods of image interpretation and measurement are then required which permit the image of an object point to be identified from its form, brightness or colour distribution. For every image point, values in the form of radiometric data (intensity, grey value, colour value) and geometric data (position in image) can then be obtained. This requires measurement systems with the appropriate geometric and optical quality.

From these measurements and a mathematical transformation between image and object space, the object can finally be modelled.

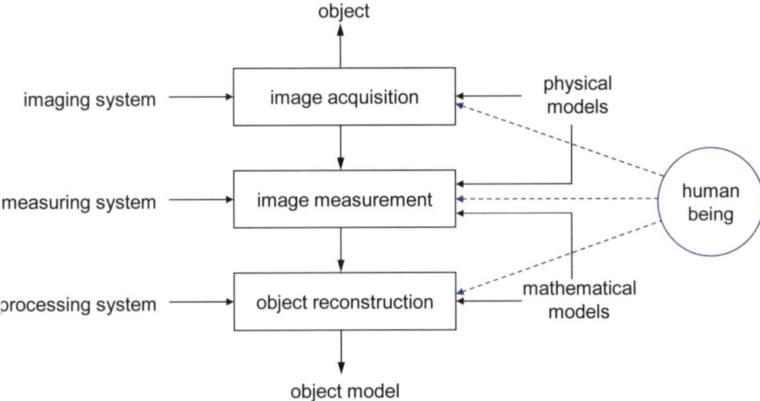

Fig. 1.3: The photogrammetric process: from object to model

Fig. 1.3 simplifies and summarises this sequence. The left hand side indicates the principal instrumentation used whilst the right hand side indicates the methods involved. Together with the physical and mathematical models, human knowledge, experience and skill play a significant role. They determine the extent to which the reconstructed model corresponds to the imaged object or fulfils the task objectives.

1.2.2 Aspects of photogrammetry

Because of its varied application areas, close-range photogrammetry has a strong interdisciplinary character. There are not only close connections with other measurement techniques but also with fundamental sciences such as mathematics, physics, information sciences or biology.

Close-range photogrammetry also has significant links with aspects of graphics and photographic science, for example computer graphics and computer vision, digital image processing, computer aided design (CAD), geographic information systems (GIS) and cartography.

Traditionally, there are further strong associations of close-range photogrammetry with the techniques of surveying, particularly in the areas of adjustment methods and engineering surveying. With the increasing application of photogrammetry to industrial metrology and quality control, links have been created in other directions, too.

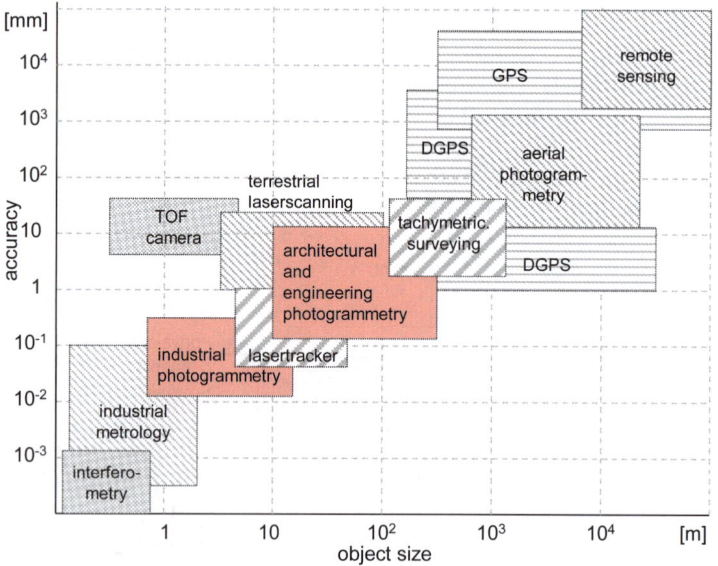

Fig. 1.4: Relationship between measurement methods and object size and accuracy[1]

[1] Unsharp borders indicating typical fields of applications of measuring methods

Fig. 1.4 gives an indication of the relationship between size of measured object, required measurement accuracy and relevant technology. Although there is no hard-and-fast definition, it may be said that close-range photogrammetry applies to objects ranging from 0.5 m to 200 m in size, with accuracies under 0.1 mm at the smaller end (manufacturing industry) and around 1cm at the larger end (architecture and construction industry).

Optical methods using light as the information carrier lie at the heart of non-contact 3D measurement techniques. Measurement techniques using electromagnetic waves may be subdivided in the manner illustrated in Fig. 1.5. Techniques based on light waves are as follows:

- Triangulation techniques:

  Photogrammetry (single, stereo and multiple imaging), angle measuring systems (theodolites), indoor GPS, structured light (light section procedures, fringe projection, phase measurement, moiré topography), focusing methods, shadow methods, etc.

- Interferometry:

  Optically coherent time-of-flight measurement, holography, speckle interferometry, coherent radar

- Time-of-flight measurement:

  Distance measurement by optical modulation methods, pulse modulation, etc.

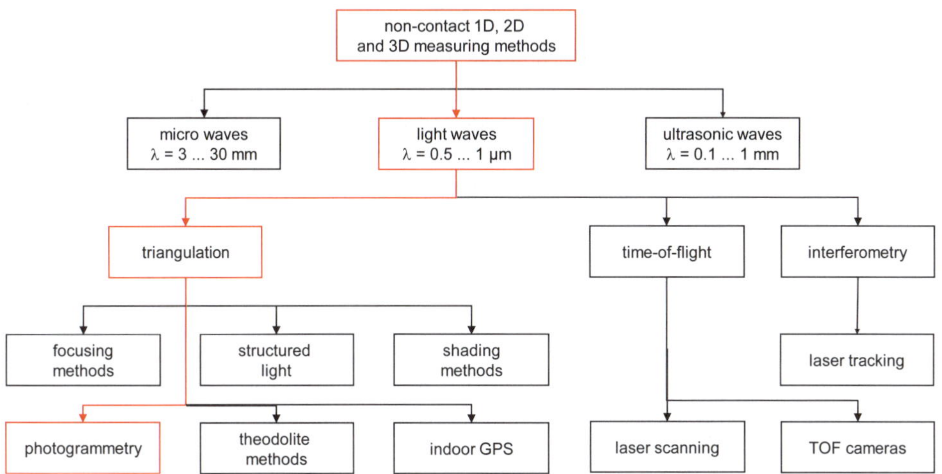

Fig. 1.5: Non-contact measuring methods

The clear structure of Fig. 1.5 is blurred in practice since multi-sensor and hybrid measurement systems utilise different principles in order to combine the advantages of each.

Photogrammetry can be categorised in a multiplicity of ways:

- By camera position and object distance:
  - – Satellite photogrammetry: processing of remote sensing and satellite images, $h >$ ca. 200 km
  - – Aerial photogrammetry: processing of aerial photographs, $h >$ ca. 300 m
  - – Terrestrial photogrammetry: measurements from a fixed terrestrial location
  - – Close-range photogrammetry: imaging distance $d <$ ca. 300 m
  - – Macro photogrammetry: image scale > 1 (microscope imaging)
  - – Mobile mapping: data acquisition from moving vehicles, d <ca. 100 m

- By number of measurement images:
  - – Single-image photogrammetry: single-image processing, mono-plotting, rectification, orthophotos
  - – Stereo photogrammetry: dual image processing, stereoscopic measurement
  - – Multi-image photogrammetry: n images where n>2, bundle triangulation

- By method of recording and processing:
  - – Plane table photogrammetry: graphical evaluation (until ca. 1930)
  - – Analogue photogrammetry: analogue cameras, opto-mechanical measurement systems (until ca. 1980)
  - – Analytical photogrammetry: analogue images, computer-controlled measurement
  - – Digital photogrammetry: digital images, computer-controlled measurement
  - – Videogrammetry: digital image acquisition and measurement
  - – Panorama photogrammetry: panoramic imaging and processing
  - – Line photogrammetry: analytical methods based on straight lines and polynomials

- By availability of measurement results:
  - – Offline photogrammetry: sequential, digital image recording, separated in time or location from measurement
  - – Online photogrammetry: simultaneous, multiple, digital image recording, immediate measurement
  - – Real-time photogrammetry: recording and measurement completed within a specified time period particular to the application

- By application or specialist area:
  - – Architectural photogrammetry: architecture, heritage conservation, archaeology
  - – Engineering photogrammetry: general engineering (construction) applications
  - – Industrial photogrammetry: industrial (manufacturing) applications
  - – Forensic photogrammetry: applications to diverse legal problems
  - – Biostereometrics: medical applications
  - – Multi-media photogrammetry: recording through media of different refractive indices
  - – Shape from stereo: stereo image processing (computer vision)
  - – Structure from motion: multi-image processing (computer vision)

1.2.3 Image-forming model

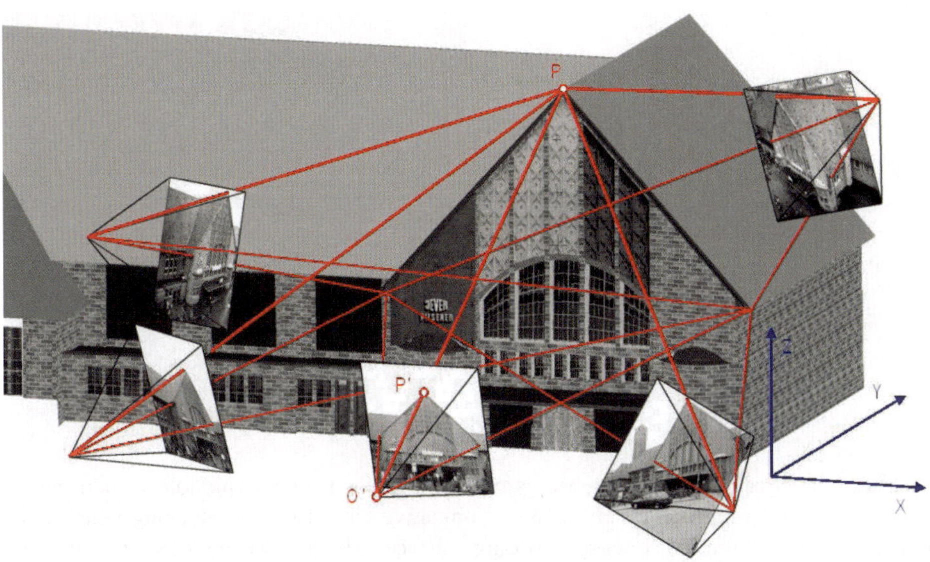

Fig. 1.6: Principle of photogrammetric measurement

Photogrammetry is a three-dimensional measurement technique which uses *central projection imaging* as its fundamental mathematical model (Fig. 1.6). Shape and position of an object are determined by reconstructing bundles of rays in which, for each camera, each image point P', together with the corresponding *perspective centre* O', defines the spatial direction of the ray to the corresponding object point P. Provided the imaging geometry within the camera and the location of the imaging system in object space are known, then every image ray can be defined in 3D object space.

From the intersection of at least two corresponding (homologous), spatially separated image rays, an object point can be located in three dimensions. In stereo photogrammetry two images are used to achieve this. In multi-image photogrammetry the number of images involved is, in principle, unlimited.

The *interior orientation* parameters describe the internal geometric model of a camera. Photogrammetric usage, deriving from German, applies the word to groups of camera parameters. Exterior (extrinsic) orientation parameters incorporate this angular meaning but extend it to include position. Interior (intrinsic) orientation parameters, which include a distance, two coordinates and a number of polynomial coefficients, involve no angular values. The use of the terminology here underlines the connection between two very important, basic groups of parameters.

With the model of the pinhole camera as its basis (Fig. 1.7), the most important reference location is the perspective centre O, through which all image rays pass. The interior orientation defines the position of the perspective centre relative to a reference system fixed in the camera (image coordinate system), as well as departures from the ideal central projection (image distortion). The most important parameter of interior orientation is the principal distance, c, which defines the distance between image plane and perspective centre (see section 3.3.2).

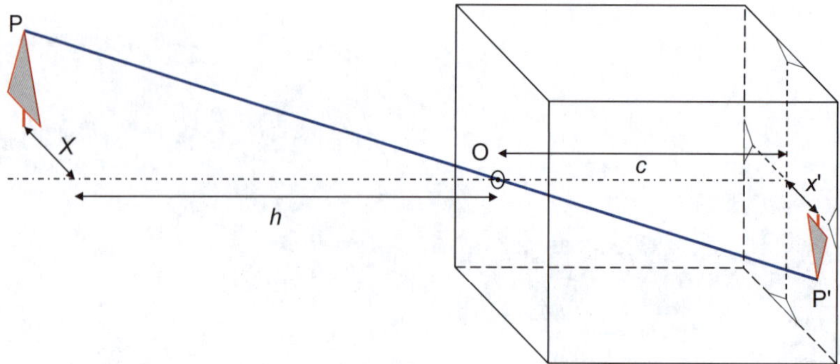

Fig. 1.7: Pinhole camera model

A real and practical photogrammetric camera will differ from the pinhole camera model. The necessity of using a relatively complex objective lens, a camera housing which is not built for stability and an image recording surface which may be neither planar nor perpendicular to the optical axis of the lens will all give rise to departures from the ideal imaging geometry. The interior orientation, which will include parameters defining these departures, must be determined by calibration for every camera.

A fundamental property of a photogrammetric image is the *image scale* or *photo scale*. The photo scale factor m defines the relationship between the object distance, h, and principal distance, c. Alternatively it is the relationship between a distance, X, parallel to the image plane in the object, and the corresponding distance in image space, x':

$$m = \frac{h}{c} = \frac{X}{x'} \tag{1.1}$$

The photo scale is in every case the deciding factor in resolving image details, as well as the photogrammetric measurement accuracy, since any measurement error in the image is multiplied in the object space by the scale factor (see section 3.3.1). Of course, when dealing with complex objects, the scale will vary throughout the image and a nominal or average value is usually quoted.

The *exterior orientation* parameters specify the spatial position and orientation of the camera in a global coordinate system. The exterior orientation is described by the coordinates of the perspective centre in the global system and three suitably defined angles expressing the rotation of the image coordinate system with respect to the global system (see section 4.2.1). The exterior orientation parameters are calculated indirectly, after measuring image coordinates of well identified object points.

Every measured image point corresponds to a spatial direction from projection centre to object point. The length of the direction vector is initially unknown, i.e. every object point lying on the line of this vector generates the same image point. In other words, although every three-dimensional object point transforms to a unique image point for given orientation parameters, a unique reversal of the projection is not possible. The object point can be located on the image ray, and thereby absolutely determined in object space, only by intersecting the ray with an additional known geometric element such as a second spatial direction or an object plane.

Every image generates a spatial *bundle of rays*, defined by the imaged points and the perspective centre, in which the rays were all recorded at the same point in time. If all the bundles of rays from multiple images are intersected as described above, a dense network is created. For an appropriate imaging configuration, such a network has the potential for high geometric strength. Using the method of *bundle triangulation* any number of images (ray bundles) can be simultaneously oriented, together with the calculation of the associated three-dimensional object point locations (Fig. 1.6, Fig. 1.8, see section 4.4).

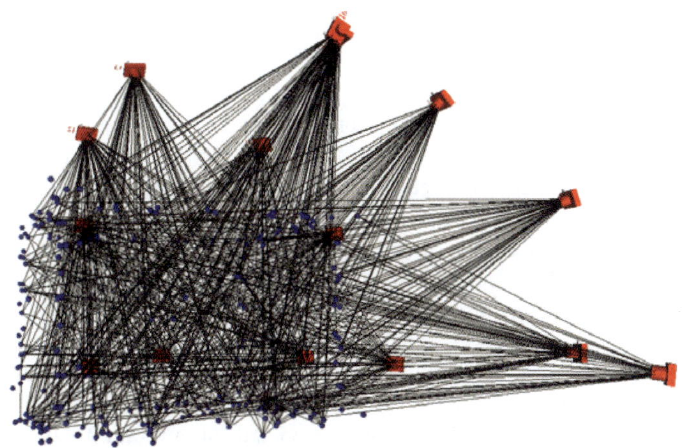

Fig. 1.8: Bundle of rays from multiple images

1.2.4 Photogrammetric systems and procedures

1.2.4.1 Analogue systems

Analogue systems which record images on film still exist and are still in use, but are now largely superseded by digital systems. This edition places its emphasis on digital systems and processing. Analogue systems will only be mentioned in passing, for example, where they illuminate an explanation or remain logically part of an overview.

1.2.4.2 Digital systems

The photogrammetric procedure has changed fundamentally with the development of digital imaging systems and digital image processing (Fig. 1.9). Digital image recording and processing offer the possibility of a fast, closed data flow from taking the images to presenting the results. Complete measurement of an object can therefore be made directly on site. Two general procedures are distinguished here. Offline photogrammetry uses a single camera with measurement results generated after all images have first been recorded and then evaluated together. Online photogrammetry records simultaneously using at least two cameras, with immediate generation or results. If the result is delivered within a certain process-specific time period, the term real-time photogrammetry is commonly used.

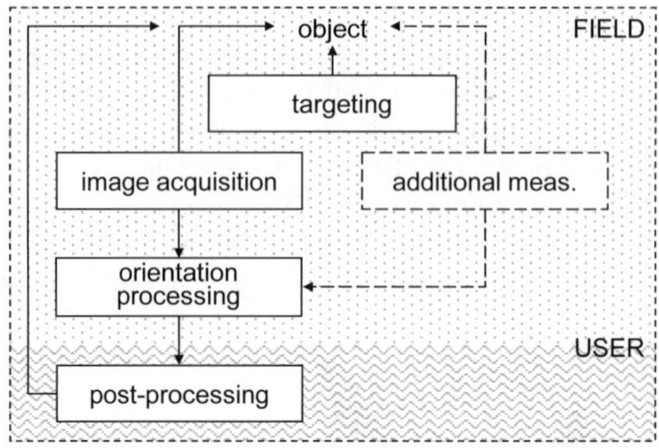

Fig. 1.9: Digital photogrammetric system

Automation and short processing cycles enable a direct integration with other processes where decisions can be made on the basis of feedback of the photogrammetric results. Digital systems are therefore critical to the application of photogrammetry in complex real-time processes, in particular industrial manufacturing and assembly, robotics and medicine where feedback with the object or surroundings takes place.

When imaging scenes with purely natural features, without the addition of artificial targets, the potential for automation is much lower. An intelligent evaluation of object structures and component forms demands a high degree of visual interpretation which is conditional on a corresponding knowledge of the application and further processing requirements. However, even here simple software interfaces, and robust techniques of image orientation and camera calibration, make it possible for non-expert users to carry out photogrammetric recording and analysis.

1.2.4.3 Recording and analysis procedures

Fig. 1.10 shows the principal procedures in close-range photogrammetry which are briefly summarised in the following sections.

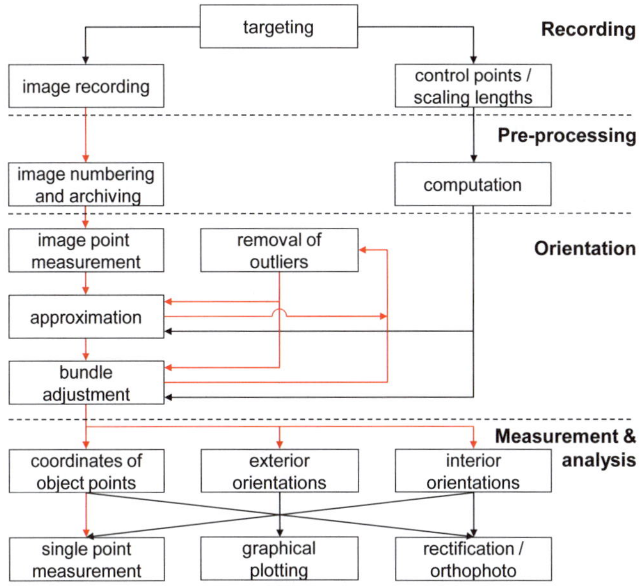

Fig. 1.10: Recording and analysis procedures (red - can be automated)

1. Recording

   a) Targeting[1]:
   target selection and attachment to object features to improve automation and increase the accuracy of target measurement in the image

   b) Determination of control points or scaling lengths:
   creation of a global object coordinate system by definition of reference (control) points and/or reference lengths (scales)

   c) Image recording:
   analogue or digital image recording of the object with a photogrammetric system

2. Pre-processing

   a) Numbering and archiving:
   assigning photo numbers to identify individual images and archiving or storing the photographs

   b) Computation:
   calculation of reference point coordinates and/or distances from survey observations (e.g. using network adjustment)

3. Orientation

   a) Measurement of image points:
   identification and measurement of reference and scale points
   identification and measurement of tie points

[1] Also increasingly known as signalizing, particularly to highlight the use of artificial targets.

b) Approximation:
calculation of approximate (starting) values for unknown quantities to be calculated by the bundle adjustment

c) Bundle adjustment:
adjustment program which simultaneously calculates parameters of both interior and exterior orientation as well as the object point coordinates which are required for subsequent analysis

d) Removal of outliers:
detection and removal of gross errors which mainly arise during measurement of image points

4. Measurement and analysis

a) Single point measurement:
creation of three-dimensional object point coordinates for further numerical processing

b) Graphical plotting:
production of scaled maps or plans in analogue or digital form.(e.g. hard copies for maps and electronic files for CAD models or GIS)

c) Rectification/Orthophoto:
generation of transformed images or image mosaics which remove the effects of tilt relative to a reference plane (rectification) and/or remove the effects of perspective (orthophoto)

This sequence can, to a large extent, be automated (connections in red in Fig. 1.10). Provided that the object features are suitably marked and identified using coded targets, initial values can be calculated and measurement outliers (gross errors) removed by robust estimation methods.

1.2.5 Photogrammetric products

In general, photogrammetric systems supply three-dimensional object coordinates derived from image measurements. From these, further elements and dimensions can be derived, for example lines, distances, areas and surface definitions, as well as quality information such as comparisons against design and machine control data. The direct determination of geometric elements such as straight lines, planes and cylinders is also possible without explicit calculation of point coordinates. In addition the recorded image is an objective data store which documents the state of the object at the time of recording. The visual data can be provided as corrected camera images, orthophotos or graphical overlays (Fig. 1.11). Examples of graphical presentation are shown in Fig. 1.12 and Fig. 1.13.

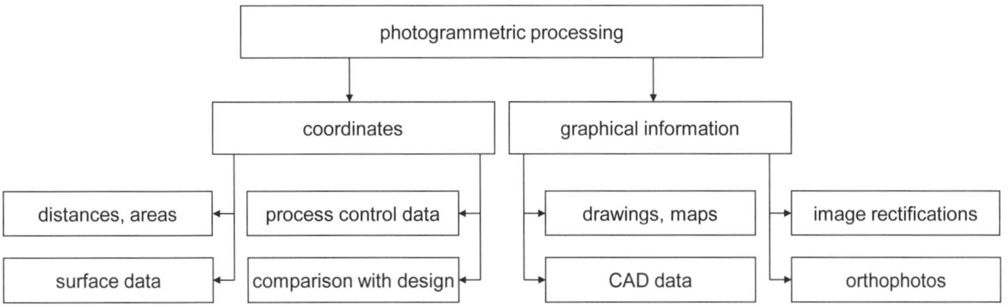

Fig. 1.11: Typical photogrammetric products

Fig. 1.12: Measurement image overlaid with part of the photogrammetrically generated CAD data

Fig. 1.13: Cylindrical projection of CAD data

1.3 Application areas

Much shorter imaging ranges, typically from a few centimetres to a few hundred metres, and alternative recording techniques, differentiate close-range photogrammetry from its aerial and satellite equivalents.

The following comments, based on ones made by Thompson as long ago as 1963, identify applications in general terms by indicating that photogrammetry is potentially useful when:

- the object to be measured is difficult to access;
- the object is not rigid and its instantaneous dimensions are required;
- it is not certain that measurement will be required at all, or even what measurements are required (i.e. the data is preserved for possible later evaluation);
- the object is very small;
- the use of direct measurement would influence the measured object or disturb events around it;
- real-time results are required;
- the simultaneous recording and the measurement of a very large number of points is required.

The following specific application areas (with examples) are amongst the most important in close-range photogrammetry:

- Automotive, machine and shipbuilding industries::
  - inspection of tooling jigs
  - reverse engineering of design models
  - manufacturing control
  - optical shape measurement
  - recording and analysing car safety tests
  - robot calibration
  - driver assistance systems
  - measurement of ship sections
  - shape control of ship parts

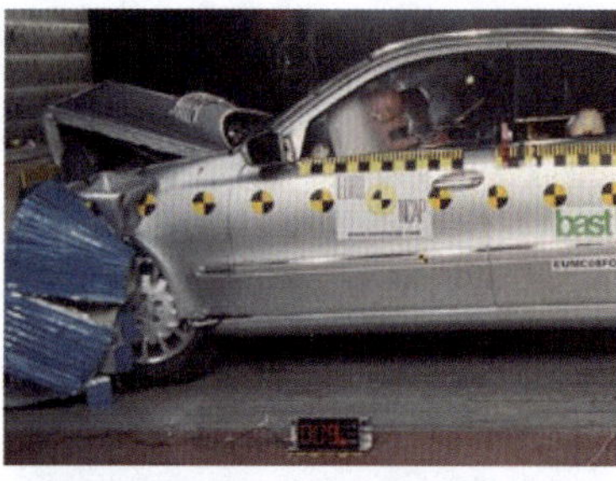

Fig. 1.14: Car safety test

- Aerospace industry:
  - measurement of parabolic antennae and mirrors
  - control of component assembly
  - inspection of tooling jigs
  - space simulations

Fig. 1.15: Parabolic mirror

- Architecture, heritage conservation, archaeology:
  - facade measurement
  - historic building documentation
  - deformation measurement
  - reconstruction of damaged buildings
  - mapping of excavation sites
  - modelling monuments and sculptures
  - 3D city models and texturing

Fig. 1.16: Building record

- Engineering:
  - as-built measurement of process plants
  - measurement of large civil engineering sites
  - deformation measurements
  - pipework and tunnel measurement
  - mining
  - evidence documentation
  - road and railway track measurement
  - wind power systems

Fig. 1.17: Engineering

- Medicine and physiology:
  - dental measurements
  - spinal deformation
  - plastic surgery
  - neuro surgery
  - motion analysis and ergonomics
  - microscopic analysis
  - computer-assisted surgery (navigation)

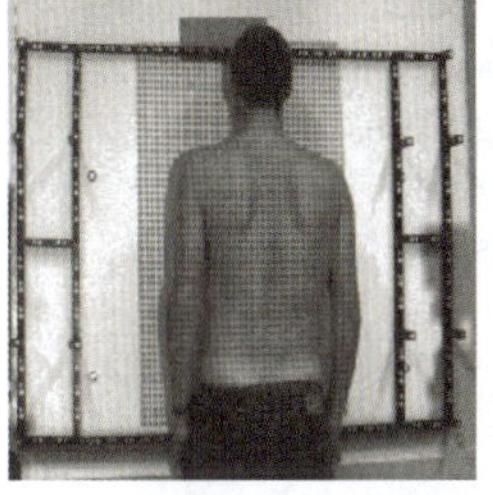

Fig. 1.18: Spinal analysis

- Police work and forensic analysis:
  - accident recording
  - scene-of-crime measurement
  - legal records
  - measurement of individuals

Fig. 1.19: Accident recording

- Animation and movie/film industries
  - body shape recording
  - motion analysis (of actors)
  - 3D movies
  - virtual reality (VR)

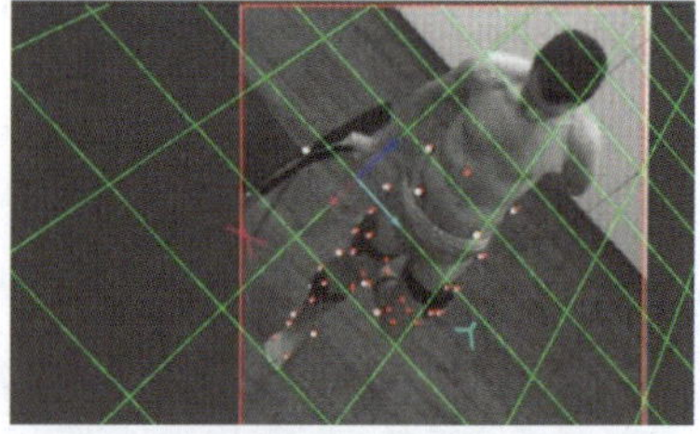

Fig. 1.20: Motion capture

- Information systems:
  - building information systems
  - facility management
  - production planning
  - image databases
  - internet applications (digital globes)

Fig. 1.21: Pipework measurement

- Natural sciences:
  - liquid flow measurement
  - wave topography
  - crystal growth
  - material testing
  - glacier and soil movements
  - etc.

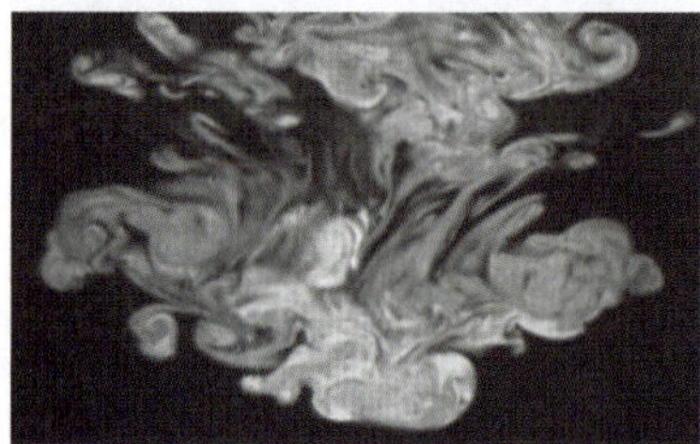

Fig. 1.22: Flow measurement

In general, similar methods of recording and analysis are used for all application areas of close-range photogrammetry and the following features are shared:

- powerful image recording systems
- freely chosen imaging configuration with almost unlimited numbers of pictures
- photo orientation based on the technique of bundle triangulation
- visual and digital analysis of the images
- presentation of results in the form of 3D models, 3D coordinate files, CAD data, photographs or drawings

Industrial and engineering applications make special demands of the photogrammetric technique:

- limited recording time on site (no significant interruption of industrial processes)
- delivery of results for analysis after only a brief time
- high accuracy requirements
- traceability of results to standard unit of dimension, the Metre
- proof of accuracy attained

1.4 Historical development

It comes as a surprise to many that the history of photogrammetry is almost as long as that of photography itself and that, for at least the first fifty years, the predominant application of photogrammetry was to close range, architectural measurement rather than to topographical mapping. Only a few years after the invention of photography during the 1830s and 1840s by Fox Talbot in England, by Niepce and Daguerre in France, and by others, the French military officer Laussedat began experiments in 1849 into measuring from perspective views by working on the image of a façade of the Hotel des Invalides. Admittedly Laussedat, usually described as the first photogrammetrist, was in this instance using a camera lucida for he did not obtain photographic equipment until 1852.

Fig. 1.23 shows an early example of Laussedat's work for military field mapping by "metrophotographie". In fact it was not a surveyor, but an architect, the German Meydenbauer, who coined the word “photogrammetry”. As early as 1858 Meydenbauer used photographs to draw plans of the cathedral of Wetzlar and by 1865 he had constructed his “great photogrammeter”, a forerunner of the phototheodolite.

Fig. 1.23: Early example of photogrammetric field recording, about 1867 (source: Laussedat 1899)

Fig. 1.24: One of the first photogrammetric cameras, by Brunner,1859 (Gruber 1930)

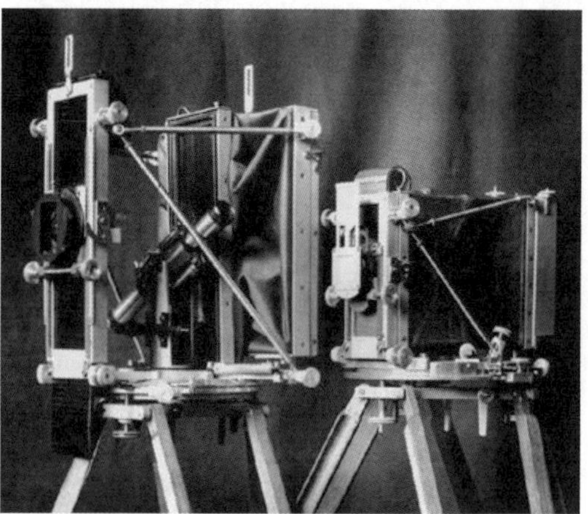

Fig. 1.25: Metric cameras by Meydenbauer (ca. 1890); left: 30x30 cm^2, right: 20x20 cm^2 (Albertz 2009)

Meydenbauer used photography as an alternative to manual methods of measuring façades. For this he developed his own large-format, glass-plate cameras (see Fig. 1.25) and, between 1885 and 1909, compiled an archive of around 16 000 metric[1] images of the most important Prussian architectural monuments. This represents a very early example of cultural heritage preservation by photogrammetry.

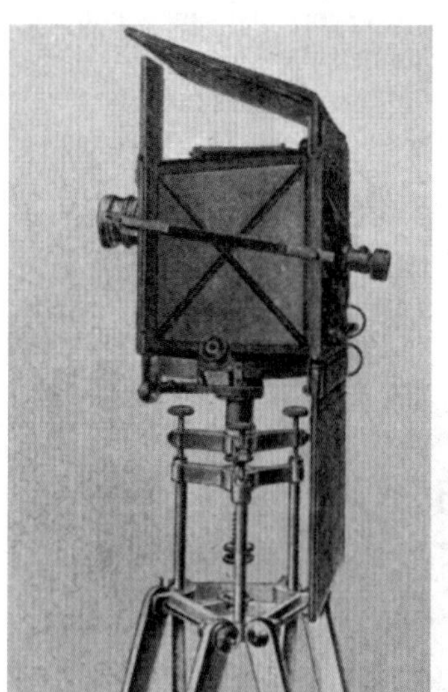

Fig. 1.26: Phototheodolite by Finsterwalder (1895) and Zeiss Jena 19/1318 (ca. 1904)

[1] A "metric" camera is defined as one with known and stable interior orientation.

The phototheodolite, as its name suggests, represents a combination of camera and theodolite. The direct measurement of orientation angles leads to a simple photogrammetric orientation. A number of inventors, such as Porro and Paganini in Italy, in 1865 and 1884 respectively, and Koppe in Germany, 1896, developed such instruments (Fig. 1.26).

Horizontal bundles of rays can be constructed from terrestrial photographs, with two or more permitting a point-by-point survey using intersecting rays. This technique, often called *plane table photogrammetry*, works well for architectural subjects which have regular and distinct features. However, for topographic mapping it can be difficult identifying the same feature in different images, particularly when they were well separated to improve accuracy. Nevertheless, despite the early predominance of architectural photogrammetry, mapping was still undertaken. For example, in the latter part of the 19th century, Paganini mapped the Alps, Deville the Rockies and Jordan the Dachel oasis, whilst Finsterwalder developed analytical solutions.

The development of stereoscopic measurement around the turn of the century was a major breakthrough in photogrammetry. Following the invention of the stereoscope around 1830, and Stolze's principle of the floating measuring mark in 1893, Pulfrich in Germany and Fourcade in South Africa, at the same time but independently[1], developed the stereocomparator which implemented Stolze's principle. These enabled the simultaneous setting of measuring marks in the two comparator images, with calculation and recording of individual point coordinates (Fig. 1.27).

Fig. 1.27: Pulfrich's stereocomparator (Zeiss, 1901)

Photogrammetry then entered the era of *analogue computation*, very different to the numerical methods of surveying. Digital computation was too slow at that time to compete with continuous plotting from stereo instruments, particularly of contours, and analogue computation became very successful for a large part of the 20th century.

In fact, during the latter part of the 19th century much effort was invested in developing stereoplotting instruments for the accurate and continuous plotting of topography. In

[1] Pulfrich's lecture in Hamburg announcing his invention was given on 23rd September 1901, while Fourcade delivered his paper in Cape Town nine days later on 2nd October 1901.

Germany, Hauck proposed a device and in Canada Deville claimed "the first automatic plotting instrument in the history of photogrammetry". Deville's instrument had several defects, but they inspired many developers such as Pulfrich and Santoni to overcome them.

In Germany, conceivably the most active country in the early days of photogrammetry, Pulfrich's methods were very successfully used in mapping. This inspired von Orel in Vienna to design an instrument for the "automatic" plotting of contours, which lead to the Orel-Zeiss Stereoautograph in 1909. In England, F. V. Thompson anticipated von Orel in the design of the Vivian Thompson stereoplotter and subsequently the Vivian Thompson Stereoplanigraph (1908). This was described by E. H. Thompson (1974) as "the first design for a completely automatic and thoroughly rigorous photogrammetric plotting instrument".

The rapid development of aviation, which began shortly after this, was another decisive influence on the course of photogrammetry. Not only is the Earth, photographed vertically from above, an almost ideal subject for the photogrammetric method, but also aircraft made almost all parts of the Earth accessible at high speed. In the first half, and more, of the 20th century these favourable circumstances allowed impressive development in photogrammetry, with tremendous economic benefit in air survey. On the other hand, the application of stereo photogrammetry to the complex surfaces relevant to close-range work was impeded by far-from-ideal geometry and a lack of economic advantage.

Although there was considerable opposition from surveyors to the use of photographs and analogue instruments for mapping, the development of stereoscopic measuring instruments forged ahead in very many countries during the period between the First World War and the early 1930s. Meanwhile, non-topographic use was sporadic for the reasons that there were few suitable cameras and that analogue plotters imposed severe restrictions on principal distance, on image format and on disposition and tilts of cameras. Instrumentally complex systems were being developed using optical projection (for example Multiplex), opto-mechanical principles (Zeiss Stereoplanigraph) and mechanical projection using space rods (for example Wild A5, Santoni Stereocartograph), designed for use with aerial photography. By 1930 the Stereoplanigraph C5 was in production, a sophisticated instrument able to use oblique and convergent photography. Even if makeshift cameras had to be used at close range, experimenters at least had freedom in the orientation and placement of these cameras and this considerable advantage led to some noteworthy work.

As early as 1933 Wild stereometric cameras were being manufactured and used by Swiss police for the mapping of accident sites, using the Wild A4 Stereoautograph, a plotter especially designed for this purpose. Such stereometric cameras comprise two identical metric cameras fixed to a rigid base of known length such that their axes are coplanar, perpendicular to the base and, usually, horizontal[1] (Fig. 3.35a, see section 4.3.1). Other manufacturers have also made stereometric cameras (Fig. 1.29) and associated plotters (Fig. 1.31) and a great deal of close-range work has been carried out with this type of equipment. Initially glass plates were used in metric cameras in order to provide a flat image surface without significant mechanical effort (see example in Fig. 1.28, Fig. 1.30). From the 1950s, film was increasingly used in metric cameras which were then equipped with a mechanical film-flattening device.

[1] This is sometimes referred to as the "normal case" of photogrammetry.

Fig. 1.28: Zeiss TMK 6 metric camera

Fig. 1.29: Zeiss SMK 40 and SMK 120 stereometric cameras

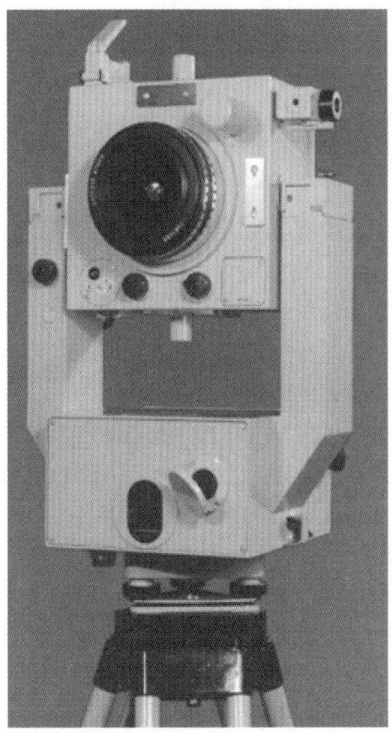

Fig. 1.30: Jenoptik UMK 1318

Fig. 1.31: Zeiss Terragraph stereoplotter

The 1950s were the start of the period of *analytical photogrammetry*. The expanding use of digital, electronic computers in that decade shifted interest from prevailing analogue methods to a purely analytical or numerical approach to photogrammetry. While analogue computation is inflexible, in regard to both input parameters and output results, and its accuracy is limited by physical properties, a numerical method allows virtually unlimited accuracy of computation and its flexibility is limited only by the mathematical model on which it is based. Above all, it permits over-determination which may improve precision, lead to the detection of gross errors and provide valuable statistical information about the measurements and the results. The first analytical applications were to photogrammetric triangulation. As numerical methods in photogrammetry improved, the above advantages, but above all their flexibility, were to prove invaluable at close range.

Subsequently stereoplotters were equipped with devices to record model coordinates for input to electronic computers. Arising from the pioneering ideas of Helava (1957), computers were incorporated in stereoplotters themselves, resulting in *analytical stereoplotters* with fully numerical reconstruction of the photogrammetric models. Bendix/OMI developed the first analytical plotter, the AP/C, in 1964 and, during the following two decades, analytical stereoplotters were produced by the major instrument companies and others (example in Fig. 1.32). While the adaptability of such instruments has been of advantage in close-range photogrammetry, triangulation programs with even greater flexibility were soon to be developed, which were more suited to the requirements of close-range work.

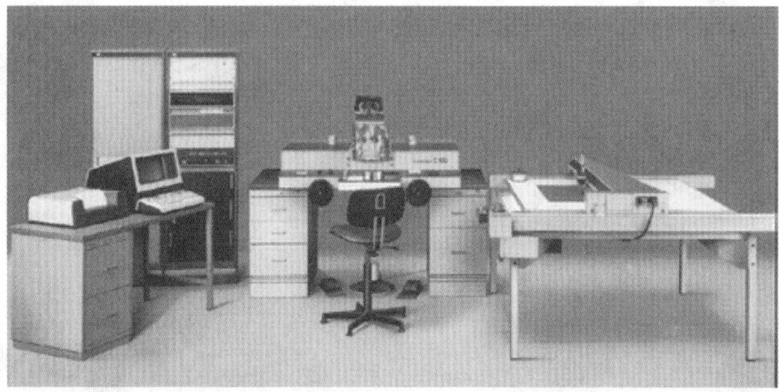

Fig. 1.32: Analytical Stereoplotter Zeiss Planicomp (ca. 1980)

Analytical photogrammetric triangulation is a method, using numerical data, of point determination involving the simultaneous orientation of all the photographs and taking all inter-relations into account. Work on this line of development, for example by the Ordnance Survey of Great Britain, had appeared before World War II, long before the development of electronic computers. Analytical triangulation required instruments to measure photo coordinates. The first stereocomparator designed specifically for use with aerial photographs was the Cambridge Stereocomparator designed in 1937 by E. H. Thompson. By 1955 there were five stereocomparators on the market and monocomparators designed for use with aerial photographs also appeared.

In the 1950s many mapping organisations were also experimenting with the new automatic computers, but it was the ballistic missile industry which gave the impetus for the development of the bundle method of photogrammetric triangulation. This is commonly known simply as the bundle adjustment and is today the dominant technique for triangulation in close-range photogrammetry. Seminal papers by Schmid (1956-57, 1958) and Brown (1958) laid the foundations for theoretically rigorous block adjustment. A number of bundle adjustment programs for air survey were developed and became commercially available, such as those by Ackermann et al. (1970) and Brown (1976). Programs designed specifically for close-range work have appeared since the 1980s, such as STARS (Fraser & Brown 1986), BINGO (Kruck 1983), MOR (Wester-Ebbinghaus 1981) or CAP (Hinsken 1989).

The importance of bundle adjustment in close-range photogrammetry can hardly be overstated. The method imposes no restrictions on the positions or the orientations of the cameras, nor is there any necessity to limit the imaging system to central projection. Of equal or greater importance, the parameters of interior orientation of all the cameras may be included as unknowns in the solution. Until the 1960s many experimenters appear to have given little attention to the calibration[1] of their cameras. This may well have been because the direct calibration of cameras focused for near objects is usually much more difficult than that of cameras focused for distant objects. At the same time, the inner orientation must usually be known more accurately than is necessary for vertical aerial photographs because the geometry of non-topographical work is frequently far from ideal. In applying the standard methods of calibration in the past, difficulties arose because of the finite distance of the targets, either real objects or virtual images. While indirect, numerical methods to overcome this difficulty were suggested by Torlegård (1967) and others, bundle adjustment now removes this concern. For high precision work, it is no longer necessary to use metric cameras which, while having the advantage of known and constant interior orientation, are usually cumbersome and expensive. Virtually any camera can now be used. Calibration via bundle adjustment is usually known as *self-calibration* (see section 4.4). Many special cameras have been developed to extend the tools available to the photogrammetrist. One example promoted by Wester-Ebbinghaus (1981) was a modified professional photographic camera with an inbuilt réseau, an array of engraved crosses on a glass plate which appear on each image (see Fig. 1.33).

Fig. 1.33: Rolleiflex SLX semi-metric camera (ca. 1980)

Fig. 1.34: Rollei MR2 multi-image restitution system (ca. 1990)

The use of traditional stereo photogrammetry at close ranges has declined. As an alternative to the use of comparators, multi-photo analysis systems which use a digitizing pad as a measuring device for photo enlargements (e.g. Rollei MR2, 1986, Fig. 1.34) have been widely used for architectural and accident recording.

[1] In photogrammetry, unlike computer vision, "calibration" refers only to interior orientation. Exterior orientation is not regarded as part of calibration.

Fig. 1.35: Partial-metric camera GSI CRC-1 (ca. 1986)

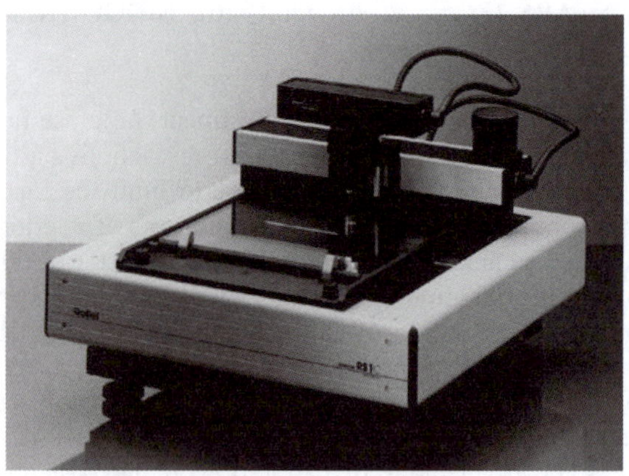

Fig. 1.36: Réseau-Scanner Rollei RS1 (ca. 1986)

Since the middle of the 1980s, the use of opto-electronic image sensors has increased dramatically. Advanced computer technology enables the processing of digital images, particularly for automatic recognition and measurement of image features, including pattern correlation for determining object surfaces. Procedures in which both the image and its photogrammetric processing are digital are often referred to as *digital photogrammetry*. Initially, standard video cameras were employed. These generated analogue video signals which could be digitised with resolutions up to 780 x 580 picture elements (pixel) and processed in real time (*real-time photogrammetry, videogrammetry)*. The first operational online multi-image systems became available in the late 1980s (example in Fig. 1.37). Automated precision monocomparators, in combination with large format réseau cameras, were developed for high-precision, industrial applications, e.g. by Fraser and Brown (1986) or Luhmann and Wester-Ebbinghaus (1986). Analytical plotters were enhanced with video cameras to become analytical correlators, used for example in car body measurement (Zeiss Indusurf 1987, Fig. 1.38). Closed procedures for simultaneous multi-image processing of grey level values and object data based on least squares methods were developed, e.g. by Förstner (1982) and Gruen (1985).

Fig. 1.37: Online multi-image system Mapvision (1987)

Fig. 1.38: Zeiss Indusurf (1987)

Fig. 1.39: POM online system with digital rotary table (1990)

Fig. 1.40: Réseau-scanning camera Rollei RSC (1990)

The limitations of video cameras in respect of their small image format and low resolution led to the development of scanning cameras which enabled the high resolution recording of static objects to around 6000 x 4500 pixels. In parallel with this development, electronic theodolites were equipped with video cameras to enable the automatic recording of directions to targets (Kern SPACE). With the Leica/Rollei system POM (Programmable Optical Measuring system, Fig. 1.39) a complex online system for the measurement of automotive parts was developed which used réseau-scanning cameras (Fig. 1.40) and a rotary table for all-round measurements.

Fig. 1.41: Still-video camera Kodak DCS 460 (ca. 1996)

Fig. 1.42: GSI VSTARS online industrial measurement system (ca. 1991)

Digital cameras with high resolution, which can provide a digital image without analogue signal processing, have been available since the beginning of the 1990s. Resolutions ranged from about 1000 x 1000 pixels (e.g. Kodak Megaplus) to over 4000 x 4000 pixels. Easily portable still video cameras could store high resolution images directly in the camera (e.g. Kodak DCS 460, Fig. 1.41). They have led to a significant expansion of photogrammetric measurement technology, particularly in the industrial field. *Online photogrammetric systems* (Fig. 1.42) have been brought into practical use, in addition to *offline systems*, both as mobile systems and in stationary configurations. Coded targets allowed the fully automatic identification and assignment of object features and orientation of the image sequences. Surface measurement of large objects were now possible with the development of pattern projection methods combined with photogrammetric techniques.

Fig. 1.43: PHIDIAS-MS multi-image analysis system (Phocad, 1994)

Interactive digital stereo systems (e.g. Leica/Helava DSP, Zeiss PHODIS) have existed since around 1988 (Kern DSP-1). They have replaced analytical plotters, but they are rarely employed for close-range use. Interactive, graphical multi-image processing systems are of more importance here as they offer processing of freely chosen image configurations in a CAD environment (e.g. Phocad PHIDIAS, Fig. 1.43). Easy-to-use, low-cost software packages (e.g. EOS PhotoModeler, Fig. 1.44, or Photometrix iWitness, Fig. 6.14) provide object reconstruction and creation of virtual 3D models from digital images without the need for a deep understanding of photogrammetry. Since around 2010 computer vision algorithms (interest operators, structure from motion approaches) have become very popular and provide fully automated 3D modelling for arbitrary imagery without any pre-knowledge or on-site measurements (Fig. 1.45).

A trend in close-range photogrammetry is now towards the integration or embedding of photogrammetric components in application-oriented hybrid systems. This includes links to such packages as 3D CAD systems, databases and information systems, quality analysis and control systems for production, navigation systems for autonomous robots and vehicles, 3D visualisation systems, internet applications, 3D animations and virtual reality. Another trend is the increasing use of methods from computer vision, such as projective geometry or pattern recognition, for rapid solutions which do not require high accuracy.

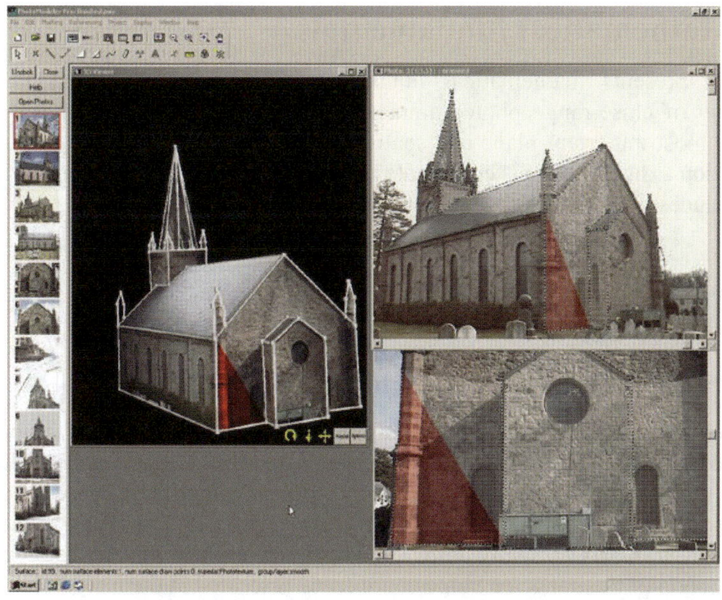

Fig. 1.44: Multi-image analysis system PhotoModeler (EOS, 2008)

Fig. 1.45: Automatic multi-image matching software (structure from motion) VisualSFM

Close-range photogrammetry is today a well-established, universal 3D measuring technique, routinely applied in a wide range of interdisciplinary fields. There is every reason to expect its continued development long into the future.

2 Mathematical fundamentals

This chapter presents mathematical fundamentals which are essential for a deeper understanding of close-range photogrammetry. After defining some common coordinate systems, the most important plane and spatial coordinate transformations are summarised. An introduction to homogeneous coordinates and graphical projections then follows and the chapter concludes with the basic theory of least-squares adjustment.

2.1 Coordinate systems

2.1.1 Image and camera coordinate systems

The *image coordinate system* defines a two-dimensional, image-based reference system of right-handed rectangular Cartesian coordinates, x'y'. In a film camera its physical relationship to the camera is defined by reference points, either *fiducial marks* or a *réseau*, which are projected into the acquired image (see section 3.3.2.1). For a digital imaging system, the sensor matrix normally defines the image coordinate system (see section 2.1.2). Usually the origin of the image or frame coordinates is located at the image centre.

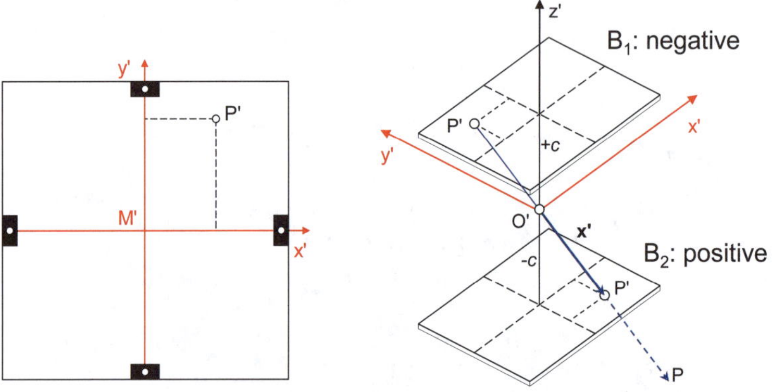

Fig. 2.1: Image and camera coordinate system

The relationship between the plane image and the camera, regarded as a spatial object, can be established when the image coordinate system is extended by the z' axis normal to the image plane, preserving a right-handed system (see Fig. 2.1). This 3D coordinate system will be called the *camera coordinate system* and its origin is located at the perspective centre O'. This axis coincides approximately with the optical axis. The origin of this 3D camera coordinate system is located at the perspective centre O'. The image position B_1 corresponds to a location in the physically acquired image, which is the image negative. For a number of mathematical calculations it is easier to use the corresponding image position B_2, in the equivalent positive image (see Fig. 2.1). Here the vector of image coordinates **x'** points in the same direction as the vector to the object point P. In this case the principal distance must be defined as a negative value leading to the three-dimensional image vector **x'**:

$$\mathbf{x}' = \begin{bmatrix} x' \\ y' \\ z' \end{bmatrix} = \begin{bmatrix} x' \\ y' \\ -c \end{bmatrix} \qquad (2.1)$$

Thus the image vector **x'** describes the projection ray, with respect to the image coordinate system, from the image point to the object point. The spatial position of the perspective centre in the image coordinate system is given by the parameters of *interior orientation* (see section 3.3.2).

2.1.2 Pixel and sensor coordinate system

The *pixel coordinate system* is designed for the storage of data defined by the rows and columns of a digital image. It is a left-handed system, *u,v*, with its origin in the upper left element (Fig. 2.2). The digital image can be viewed as a two-dimensional matrix which, in the case of multiple stored channels (e.g. colour channels), can also be defined as multi-dimensional (see also section 5.1.3). A digital image only has a relationship to the physical image sensor in the camera when the pixel coordinate system directly corresponds to the sensor coordinate system or the corner point coordinates of an image detail are stored.

Pixel coordinates *u,v* can be converted to image coordinates *x',y'* using pixel separations Δx_S, Δy_S and the sensor dimensions s'_x, s'_y . This shifts the origin to the centre of the sensor (centre of image) and converts to a right-handed system (see section 3.3.2.1 and eqn. 3.49).

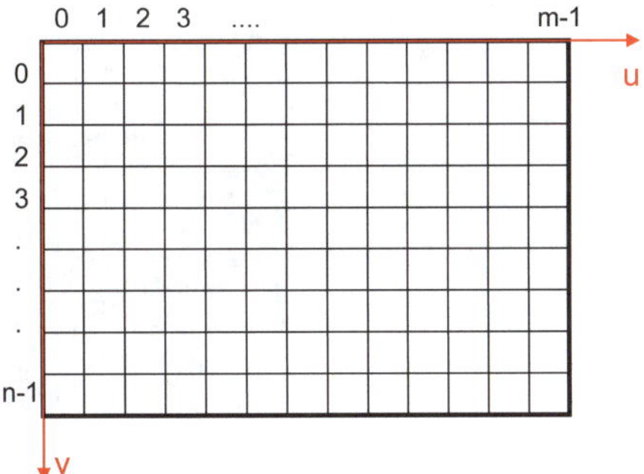

Fig. 2.2: Pixel coordinate system

2.1.3 Model coordinate system

The spatial Cartesian *model coordinate system* xyz is used to describe the relative position and orientation of two or more images (image coordinate systems). Normally its origin is at the perspective centre of one of the images. In addition, the model coordinate system may be parallel to the related image coordinate system (see section 4.3.3).

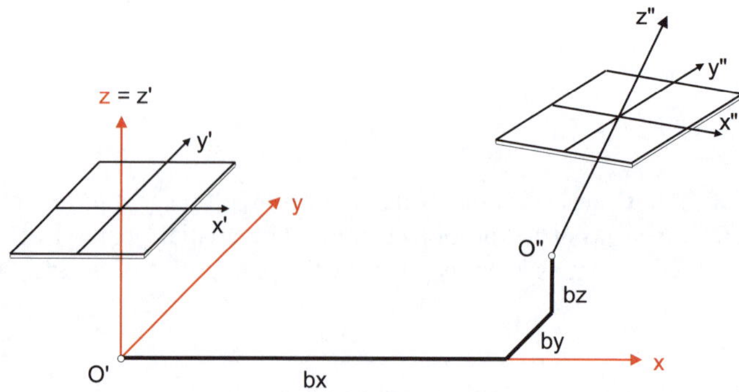

Fig. 2.3: Model coordinate system

2.1.4 Object coordinate system

The term *object coordinate system*, also known as the world coordinate system, is here used for every spatial Cartesian coordinate system XYZ that is defined by reference points of the object. For example, national geodetic coordinate systems (X=easting, Y=northing, Z=altitude, origin at the equator) are defined by geodetically measured reference points[1]. Another example is the local object or workpiece coordinate system of a car body that is defined by the constructional axes (X = longitudinal car axis, Y = front axle, Z = height, origin at centre of front axle).

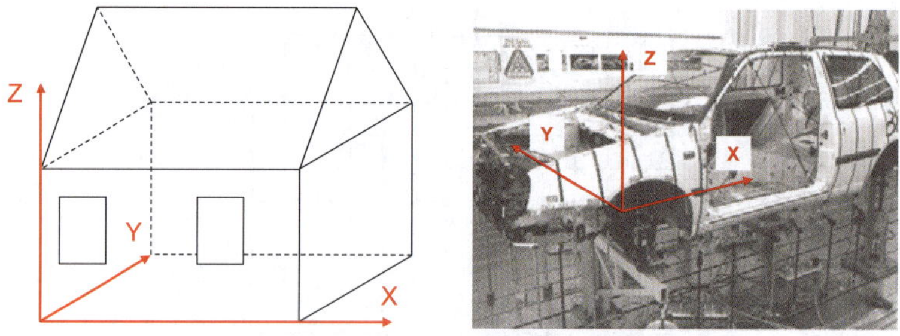

Fig. 2.4: Object coordinate systems

A special case of three-dimensional coordinate system is an arbitrarily oriented one used by a 3D measuring system such as a camera or a scanner. This is not directly related to any superior system or particular object but if, for instance, just one reference scale is given (Fig. 2.5), then it is still possible to measure spatial object coordinates.

The definition of origin, axes and scale of a coordinate system is also known as the *datum*.

[1] National systems of geodetic coordinates which use the geoid as a reference surface are equivalent to a Cartesian coordinate system only over small areas.

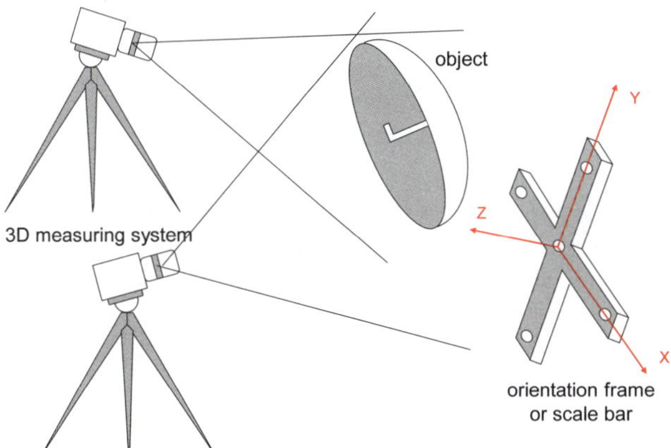

Fig. 2.5: 3D instrument coordinate system

2.2 Coordinate transformations

2.2.1 Plane transformations

2.2.1.1 Similarity transformation

The *plane similarity transformation* is used for the mapping of two plane Cartesian coordinate systems (Fig. 2.6). Generally a 4-parameter transformation is employed which defines two translations, one rotation and a scaling factor between the two systems. Angles and distance proportions are maintained.

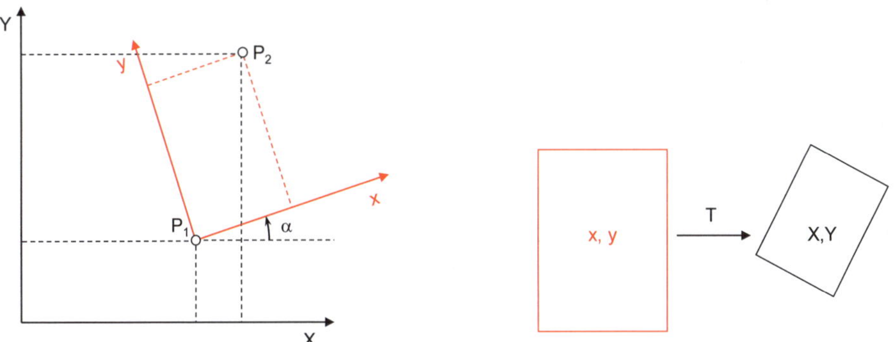

Fig. 2.6: Plane similarity transformation

Given a point P in the xy source system, the XY coordinates in the target system are:

$$
\begin{aligned}
X &= a_0 + a_1 \cdot x - b_1 \cdot y \\
Y &= b_0 + b_1 \cdot x + a_1 \cdot y
\end{aligned}
\tag{2.2}
$$

or

$$\begin{aligned} X &= a_0 + m \cdot (x \cdot \cos\alpha - y \cdot \sin\alpha) \\ Y &= b_0 + m \cdot (x \cdot \sin\alpha + y \cdot \cos\alpha) \end{aligned} \tag{2.3}$$

Here a_0 and b_0 define the translation of the origin, α is the rotation angle and m is the global scaling factor. In order to determine the four coefficients, a minimum of two identical points is required in both systems. With more than two identical points the transformation parameters can be calculated by an over-determined least-squares adjustment.

In matrix notation (2.2) is expressed as:

$$\begin{gathered} \mathbf{X} = \mathbf{A} \cdot \mathbf{x} + \mathbf{a} \\ \begin{bmatrix} X \\ Y \end{bmatrix} = \begin{bmatrix} a_1 & -b_1 \\ b_1 & a_1 \end{bmatrix} \cdot \begin{bmatrix} x \\ y \end{bmatrix} + \begin{bmatrix} a_0 \\ b_0 \end{bmatrix} \end{gathered} \tag{2.4}$$

or in non-linear form with $a_0 = X_0$ and $b_0 = Y_0$

$$\begin{gathered} \mathbf{X} = m \cdot \mathbf{R} \cdot \mathbf{x} + \mathbf{X}_0 \\ \begin{bmatrix} X \\ Y \end{bmatrix} = m \cdot \begin{bmatrix} \cos\alpha & -\sin\alpha \\ \sin\alpha & \cos\alpha \end{bmatrix} \cdot \begin{bmatrix} x \\ y \end{bmatrix} + \begin{bmatrix} X_0 \\ Y_0 \end{bmatrix} \end{gathered} \tag{2.5}$$

R is the rotation matrix corresponding to rotation angle α. This is an orthogonal matrix having orthonormal column (or row) vectors and it has the properties:

$$\mathbf{R}^{-1} = \mathbf{R}^T \qquad \text{and} \qquad \mathbf{R}^T \cdot \mathbf{R} = \mathbf{I}$$

For the reverse transformation of coordinates from the target system into the source system, the transformation equations (2.5) are re-arranged as follows:

$$\begin{gathered} \mathbf{x} = \frac{1}{m} \cdot \mathbf{R}^{-1} \cdot (\mathbf{X} - \mathbf{X}_0) \\ \begin{bmatrix} x \\ y \end{bmatrix} = \frac{1}{m} \cdot \begin{bmatrix} \cos\alpha & \sin\alpha \\ -\sin\alpha & \cos\alpha \end{bmatrix} \cdot \begin{bmatrix} X - X_0 \\ Y - Y_0 \end{bmatrix} \end{gathered} \tag{2.6}$$

or explicitly with the coefficients of the forward transformation:

$$\begin{aligned} x &= \frac{a_1(X - a_0) + b_1(Y - b_0)}{a_1^2 + b_1^2} \\ y &= \frac{a_1(Y - b_0) - b_1(X - a_0)}{a_1^2 + b_1^2} \end{aligned} \tag{2.7}$$

2.2.1.2 Affine transformation

The *plane affine transformation* is also used for the mapping of two plane coordinate systems (Fig. 2.7). This 6-parameter transformation defines two displacements, one rotation, one shearing angle between the axes and two separate scaling factors.

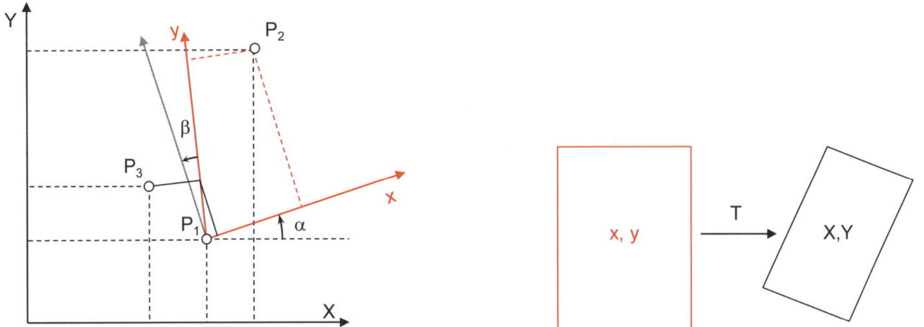

Fig. 2.7: Plane affine transformation

For a point P in the source system, the XY coordinates in the target system are given by:

$$
\begin{aligned}
X &= a_0 + a_1 \cdot x + a_2 \cdot y \\
Y &= b_0 + b_1 \cdot x + b_2 \cdot y
\end{aligned}
\tag{2.8}
$$

or in non-linear form with $a_0=X_0$ and $b_0=Y_0$

$$
\begin{aligned}
X &= X_0 + m_X \cdot x \cdot \cos\alpha - m_Y \cdot y \cdot \sin(\alpha+\beta) \\
Y &= Y_0 + m_X \cdot x \cdot \sin\alpha + m_Y \cdot y \cdot \cos(\alpha+\beta)
\end{aligned}
\tag{2.9}
$$

The parameters a_0 and b_0 (X_0 and Y_0) define the displacement of the origin, α is the rotation angle, β is the shearing angle between the axes and m_X, m_Y are the scaling factors for x and y. In order to determine the six coefficients, a minimum of three identical points is required in both systems. With more than three identical points, the transformation parameters can be calculated by over-determined least-squares adjustment.

In matrix notation the affine transformation can be written as:

$$
\mathbf{X} = \mathbf{A} \cdot \mathbf{x} + \mathbf{a}
$$

$$
\begin{bmatrix} X \\ Y \end{bmatrix} = \begin{bmatrix} a_1 & a_2 \\ b_1 & b_2 \end{bmatrix} \cdot \begin{bmatrix} x \\ y \end{bmatrix} + \begin{bmatrix} a_0 \\ b_0 \end{bmatrix}
$$

or (2.10)

$$
\begin{bmatrix} X \\ Y \end{bmatrix} = \begin{bmatrix} m_X \cdot \cos\alpha & -m_Y \cdot \sin(\alpha+\beta) \\ m_X \cdot \sin\alpha & m_Y \cdot \cos(\alpha+\beta) \end{bmatrix} \cdot \begin{bmatrix} x \\ y \end{bmatrix} + \begin{bmatrix} X_0 \\ Y_0 \end{bmatrix}
$$

A is the affine transformation matrix. For transformations with small values of rotation and shear, the parameter a_1 corresponds to the scaling factor m_X and the parameter b_2 to the scaling factor m_Y.

For the reverse transformation from coordinates in the target system to coordinates in the source system, eqn. 2.10 is re-arranged as follows:

$$\mathbf{x} = \mathbf{A}^{-1} \cdot (\mathbf{X} - \mathbf{a}) \tag{2.11}$$

or explicitly with the coefficients with the original, forward transformation:

$$\begin{aligned} x &= \frac{a_2(Y-b_0) - b_2(X-a_0)}{a_2 b_1 - a_1 b_2} \\ y &= \frac{b_1(X-a_0) - a_1(Y-b_0)}{a_2 b_1 - a_1 b_2} \end{aligned} \tag{2.12}$$

2.2.1.3 Polynomial transformation

Non-linear deformations (Fig. 2.8) can be described by polynomials of degree n:

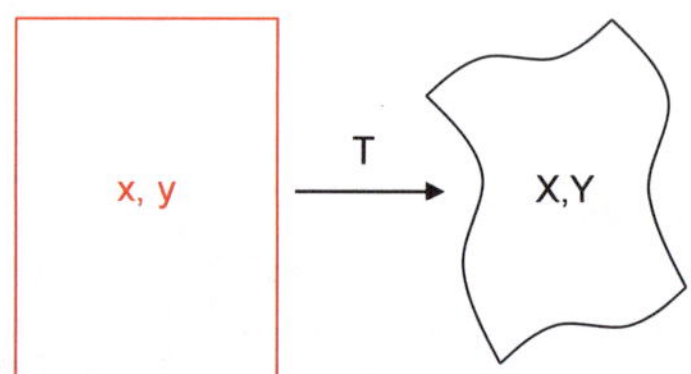

Fig. 2.8: Plane polynomial transformation

In general, the transformation model can be written as:

$$\begin{aligned} X &= \sum_{j=0}^{n} \sum_{i=0}^{j} a_{ji} \cdot x^{j-i} \cdot y^{i} \\ Y &= \sum_{j=0}^{n} \sum_{i=0}^{j} b_{ji} \cdot x^{j-i} \cdot y^{i} \end{aligned} \tag{2.13}$$

where n = degree of polynomial

A polynomial with $n = 2$ is given by:

$$\begin{aligned} X &= a_{00} + a_{10} \cdot x + a_{11} \cdot y + a_{20} \cdot x^2 + a_{21} \cdot x \cdot y + a_{22} \cdot y^2 \\ Y &= b_{00} + b_{10} \cdot x + b_{11} \cdot y + b_{20} \cdot x^2 + b_{21} \cdot x \cdot y + b_{22} \cdot y^2 \end{aligned} \tag{2.14}$$

The polynomial with $n = 1$ is identical to the affine transformation (2.8). In general, the number of coefficients required to define a polynomial transformation of degree n is

$u = (n+1) \cdot (n+2)$. In order to determine the u coefficients, a minimum of $u/2$ identical points is required in both systems.

2.2.1.4 Bilinear transformation

The *bilinear transformation* is similar to the affine transformation but extended by a mixed term:

$$\begin{aligned} X &= a_0 + a_1 \cdot x + a_2 \cdot y + a_3 \cdot x \cdot y \\ Y &= b_0 + b_1 \cdot x + b_2 \cdot y + b_3 \cdot x \cdot y \end{aligned} \tag{2.15}$$

In order to determine the eight coefficients, a minimum of four identical points is required.

The bilinear transformation can be used in the unconstrained transformation and interpolation of quadrilaterals, for example in réseau grids or digital surface models.

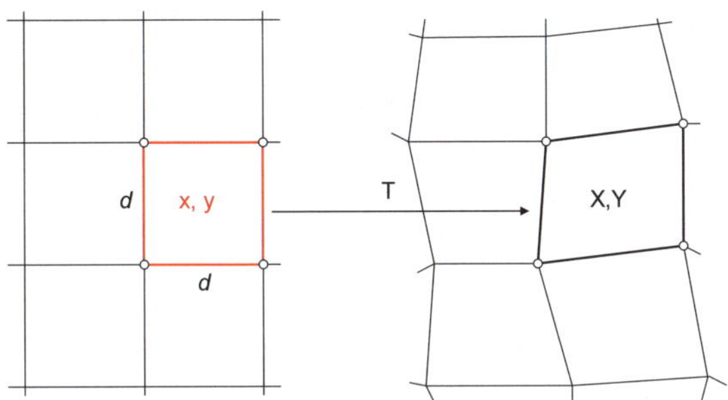

Fig. 2.9: Bilinear transformation

For the transformation of a square with side length d (Fig. 2.9), the coefficients can be calculated as follows:

$$\begin{bmatrix} a_0 \\ a_1 \\ a_2 \\ a_3 \end{bmatrix} = \mathbf{A} \cdot \begin{bmatrix} x_1 \\ x_2 \\ x_3 \\ x_4 \end{bmatrix} \quad \text{and} \quad \begin{bmatrix} b_0 \\ b_1 \\ b_2 \\ b_3 \end{bmatrix} = \mathbf{A} \cdot \begin{bmatrix} y_1 \\ y_2 \\ y_3 \\ y_4 \end{bmatrix} \tag{2.16}$$

$$\text{where} \quad \mathbf{A} = \begin{bmatrix} 1 & 0 & 0 & 0 \\ -1/d & 1/d & 0 & 0 \\ -1/d & 0 & 1/d & 0 \\ 1/d^2 & -1/d^2 & -1/d^2 & 1/d^2 \end{bmatrix}$$

2.2.1.5 Projective transformation

The *plane projective transformation* maps two plane coordinate systems using a central projection. All projection rays are straight lines through the perspective centre (Fig. 2.10).

The transformation model is:

$$
\begin{aligned}
X &= \frac{a_0 + a_1 \cdot x + a_2 \cdot y}{1 + c_1 \cdot x + c_2 \cdot y} \\
Y &= \frac{b_0 + b_1 \cdot x + b_2 \cdot y}{1 + c_1 \cdot x + c_2 \cdot y}
\end{aligned}
\tag{2.17}
$$

The system of equations 2.17 is not linear. By multiplying by the denominator and rearranging, the following linear form can be derived. This is suitable as an observation equation in an adjustment procedure.

$$
\begin{aligned}
a_0 + a_1 x + a_2 y - X - c_1 x X - c_2 y X &= 0 \\
b_0 + b_1 x + b_2 y - Y - c_1 x Y - c_2 y Y &= 0
\end{aligned}
\tag{2.18}
$$

In order to determine the eight coefficients, four identical points are required where no three may lay on a common straight line. With more than four points, the system of equations can be solved by adjustment (see calculation scheme in section 4.2.6). For the derivation of (2.17) the spatial similarity transformation can be used (see section 2.2.2.2).

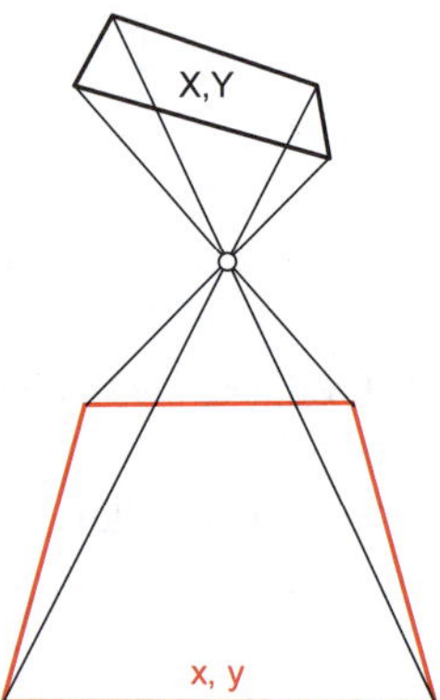

Fig. 2.10: Plane projective transformation

The reverse transformation can be calculated by re-arrangement of equations 2.17:

$$
\begin{aligned}
x &= \frac{a_2 b_0 - a_0 b_2 + (b_2 - b_0 c_2) X + (a_0 c_2 - a_2) Y}{a_1 b_2 - a_2 b_1 + (b_1 c_2 - b_2 c_1) X + (a_2 c_1 - a_1 c_2) Y} \\
y &= \frac{a_0 b_1 - a_1 b_0 + (b_0 c_1 - b_1) X + (a_1 - a_0 c_1) Y}{a_1 b_2 - a_2 b_1 + (b_1 c_2 - b_2 c_1) X + (a_2 c_1 - a_1 c_2) Y}
\end{aligned}
\tag{2.19}
$$

In this form the equations again express a projective transformation. By substitution of terms the following form is derived:

$$x = \frac{a'_0 + a'_1 X + a'_2 Y}{1 + c'_1 X + c'_2 Y}$$
$$y = \frac{b'_0 + b'_1 X + b'_2 Y}{1 + c'_1 X + c'_2 Y} \tag{2.20}$$

where

$$a'_0 = \frac{a_2 b_0 - a_0 b_2}{N} \qquad b'_0 = \frac{a_0 b_1 - a_1 b_0}{N} \qquad c'_1 = \frac{b_1 c_2 - b_2 c_1}{N}$$
$$a'_1 = \frac{b_2 - b_0 c_2}{N} \qquad b'_1 = \frac{b_0 c_1 - b_1}{N} \qquad c'_2 = \frac{a_2 c_1 - a_1 c_2}{N}$$
$$a'_2 = \frac{a_0 c_2 - a_2}{N} \qquad b'_2 = \frac{a_1 - a_0 c_1}{N} \qquad N = a_1 b_2 - a_2 b_1$$

The plane projective transformation preserves rectilinear properties and intersection points of straight lines. In contrast, angles, length and area proportions are not invariant. An additional invariant property of the central projection are the *cross ratios* of distances between points on a straight line. They are defined as follows:

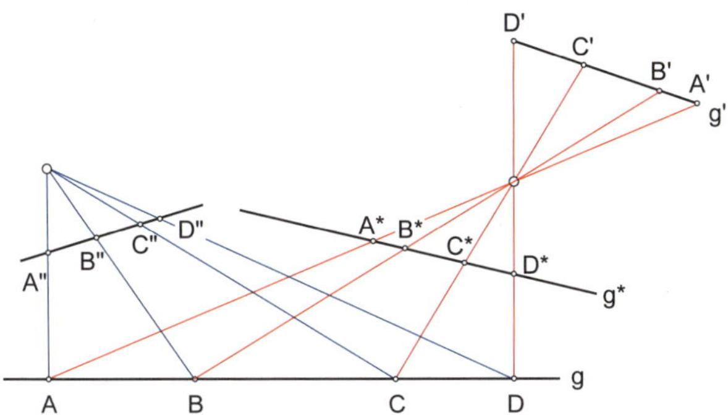

Fig. 2.11: Cross ratios

$$\lambda = \frac{\overline{AB}}{\overline{BC}} \div \frac{\overline{AD}}{\overline{CD}} = \frac{\overline{A^*B^*}}{\overline{B^*C^*}} \div \frac{\overline{A^*D^*}}{\overline{C^*D^*}} = \frac{\overline{A'B'}}{\overline{B'C'}} \div \frac{\overline{A'D'}}{\overline{C'D'}} = \frac{\overline{A''B''}}{\overline{B''C''}} \div \frac{\overline{A''D''}}{\overline{C''D''}} \tag{2.21}$$

The cross ratios apply to all straight lines that intersect a bundle of perspective rays in an arbitrary position (Fig. 2.10).

The plane projective transformation is applied to single image analysis, e.g. for rectification or coordinate measurement in single images (see section 4.2.6).

Example 2.1:

Given 8 points in the source and target coordinate systems with the following plane coordinates:

Nr.	x	y	X	Y
1	−12.3705	−10.5075	0	0
2	−10.7865	15.4305	0	5800
3	8.6985	10.8675	4900	5800
4	11.4975	−9.5715	4900	0
5	7.8435	7.4835	4479	4580
6	−5.3325	6.5025	1176	3660
7	6.7905	−6.3765	3754	790
8	−6.1695	−0.8235	1024	1931

These correspond to the image and control point coordinates in Fig. 5.50.

The plane transformations described in section 2.2.1.1 to section 2.2.1.5 then give rise to the following transformation parameters:

Coeff.	4-param transf.	6-param transf.	Bilinear transf.	Projective transf.	Polynomial 2nd order
a_0	2524.3404	2509.3317	2522.4233	2275.9445	2287.8878
a_1	237.2887	226.9203	228.0485	195.1373	230.9799
a_2		7.3472	11.5751	−11.5864	16.4830
a_3			2.1778		2.9171
a_4					2.2887
a_5					−0.0654
b_0	2536.0460	2519.9142	2537.9164	2321.9622	2348.9782
b_1	5.6218	21.5689	23.1202	−9.0076	28.0384
b_2		250.1298	255.9436	222.6108	250.7228
b_3			2.9947		−0.2463
b_4					3.5332
b_5					2.5667
c_1				−0.0131	
c_2				−0.0097	
s_0 [mm]	369.7427	345.3880	178.1125	3.1888	38.3827

The standard deviation s_0 indicates the spread of the transformed points in the XY system. It can be seen that the projective transformation has the best fit, with the 2nd order polynomial as second best. The other transformations are not suitable for this particular distribution of points.

2.2.2 Spatial transformations

2.2.2.1 Spatial rotations

Rotation matrix using trigonometric functions

For plane transformations, rotations take effect about a single point. In contrast, spatial rotations are performed successively about the three axes of a spatial coordinate system. Consider a point P in the source system xyz which is rotated with respect to the target system XYZ. Using trigonometric functions, individual rotations about the three axes of the target system are defined as follows:

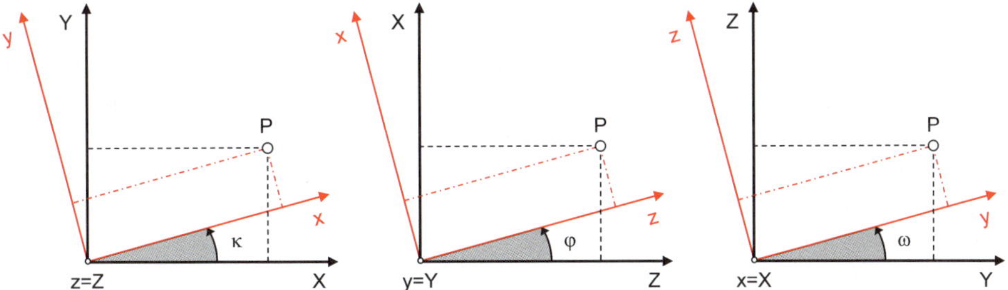

Fig. 2.12: Definition of spatial rotation angles

1. Rotation about Z-axis:

A Z-axis rotation is conventionally designated by angle κ. This is positive in an anticlockwise direction when viewed down the positive Z axis towards the origin. From eqn. 2.5, this results in the following point coordinates in the target system XYZ:

$$
\begin{aligned}
X &= x \cdot \cos\kappa - y \cdot \sin\kappa \\
Y &= x \cdot \sin\kappa + y \cdot \cos\kappa \\
Z &= z
\end{aligned}
\qquad \text{or} \qquad
\begin{gathered}
\mathbf{X} = \mathbf{R}_\kappa \cdot \mathbf{x} \\
\begin{bmatrix} X \\ Y \\ Z \end{bmatrix} = \begin{bmatrix} \cos\kappa & -\sin\kappa & 0 \\ \sin\kappa & \cos\kappa & 0 \\ 0 & 0 & 1 \end{bmatrix} \cdot \begin{bmatrix} x \\ y \\ z \end{bmatrix}
\end{gathered}
\tag{2.22}
$$

2. Rotation about Y-axis:

The corresponding rotation about the Y-axis is designated by rotation angle φ. This results in the following XYZ target point coordinates:

$$
\begin{aligned}
X &= x \cdot \cos\varphi + z \cdot \sin\varphi \\
Y &= y \\
Z &= -x \cdot \sin\varphi + z \cdot \cos\varphi
\end{aligned}
\qquad \text{or} \qquad
\begin{gathered}
\mathbf{X} = \mathbf{R}_\varphi \cdot \mathbf{x} \\
\begin{bmatrix} X \\ Y \\ Z \end{bmatrix} = \begin{bmatrix} \cos\varphi & 0 & \sin\varphi \\ 0 & 1 & 0 \\ -\sin\varphi & 0 & \cos\varphi \end{bmatrix} \cdot \begin{bmatrix} x \\ y \\ z \end{bmatrix}
\end{gathered}
\tag{2.23}
$$

3. Rotation about X-axis:

Finally the X axis rotation is designated by angle ω, which results in XYZ values:

$$
\begin{aligned}
X &= x \\
Y &= y \cdot \cos\omega - z \cdot \sin\omega \\
Z &= y \cdot \sin\omega + z \cdot \cos\omega
\end{aligned}
\qquad \text{or} \qquad
\begin{gathered}
\mathbf{X} = \mathbf{R}_\omega \cdot \mathbf{x} \\
\begin{bmatrix} X \\ Y \\ Z \end{bmatrix} = \begin{bmatrix} 1 & 0 & 0 \\ 0 & \cos\omega & -\sin\omega \\ 0 & \sin\omega & \cos\omega \end{bmatrix} \cdot \begin{bmatrix} x \\ y \\ z \end{bmatrix}
\end{gathered}
\tag{2.24}
$$

The given rotation matrices are orthonormal, i.e.

$$\mathbf{R} \cdot \mathbf{R}^T = \mathbf{R}^T \cdot \mathbf{R} = \mathbf{I} \qquad \mathbf{R}^{-1} = \mathbf{R}^T \qquad \text{and} \qquad \det(\mathbf{R}) = 1 \tag{2.25}$$

The complete rotation **R** of a spatial coordinate transformation can be defined by the successive application of 3 individual rotations, as defined above. Only certain combinations of these 3 rotations are possible and these may be applied about either the fixed axial directions of the target system or the moving axes of the source system. If a general rotation is defined about *moving axes* in the order ω φ κ, then the complete rotation is given by:

$$\mathbf{X} = \mathbf{R} \cdot \mathbf{x} \tag{2.26}$$

where

$$\mathbf{R} = \mathbf{R}_\omega \cdot \mathbf{R}_\varphi \cdot \mathbf{R}_\kappa \tag{2.27}$$

and

$$\mathbf{R} = \begin{bmatrix} r_{11} & r_{12} & r_{13} \\ r_{21} & r_{22} & r_{23} \\ r_{31} & r_{32} & r_{33} \end{bmatrix}$$
$$= \begin{bmatrix} \cos\varphi\cos\kappa & -\cos\varphi\sin\kappa & \sin\varphi \\ \cos\omega\sin\kappa + \sin\omega\sin\varphi\cos\kappa & \cos\omega\cos\kappa - \sin\omega\sin\varphi\sin\kappa & -\sin\omega\cos\varphi \\ \sin\omega\sin\kappa - \cos\omega\sin\varphi\cos\kappa & \sin\omega\cos\kappa + \cos\omega\sin\varphi\sin\kappa & \cos\omega\cos\varphi \end{bmatrix}$$

If the rotation is alternatively defined about *fixed* axes in the order ω φ κ, then the rotation matrix is given by:

$$\mathbf{R}^* = \mathbf{R}_\kappa \cdot \mathbf{R}_\varphi \cdot \mathbf{R}_\omega \tag{2.28}$$

This is mathematically equivalent to applying the same rotations about moving axes but in the reverse order.

From eqn. 2.26 the inverse transformation which generates the coordinates of a point P in the *rotated* system xyz from its XYZ values is therefore given by:

$$\mathbf{x} = \mathbf{R}^T \cdot \mathbf{X} \tag{2.29}$$

where

$$\mathbf{R}^T = \mathbf{R}_\kappa^T \cdot \mathbf{R}_\varphi^T \cdot \mathbf{R}_\omega^T \tag{2.30}$$

Note that in this inverse transformation, the individually inverted rotation matrices are multiplied in the reverse order.

From the matrix coefficients $r_{11} \dots r_{33}$ in eqn. 2.27, the individual rotation angles can be calculated as follows:

$$\sin\varphi = r_{13} \qquad \tan\omega = -\frac{r_{23}}{r_{33}} \qquad \tan\kappa = -\frac{r_{12}}{r_{11}} \tag{2.31}$$

Eqn. 2.31 shows that the determination of φ is ambiguous due to solutions for sin φ in two quadrants. In addition, there is no unique solution for the rotation angles if the second rotation (φ in this case) is equal to 90° or 270°. (Cosine φ in r_{11} and r_{33} then causes division by zero).

A simple solution to this ambiguity problem is to alter the order of rotation. In the case that the secondary rotation is close to 90°, the primary and secondary rotations can be exchanged, leading to the new order φ ω κ. This procedure is used in close-range photogrammetry when the viewing direction of the camera is approximately horizontal (see Fig. 2.13 and also section 4.2.1.2). The resulting rotation matrix is then given by:

$$\mathbf{R}_{\varphi\omega\kappa} = \mathbf{R}_{\varphi} \cdot \mathbf{R}_{\omega} \cdot \mathbf{R}_{\kappa} \tag{2.32}$$

where

$$\mathbf{R}_{\varphi\omega\kappa} = \begin{bmatrix} r_{11} & r_{12} & r_{13} \\ r_{21} & r_{22} & r_{23} \\ r_{31} & r_{32} & r_{33} \end{bmatrix}$$

$$= \begin{bmatrix} \cos\varphi\cos\kappa + \sin\varphi\sin\omega\sin\kappa & -\cos\varphi\sin\kappa + \sin\varphi\sin\omega\cos\kappa & \sin\varphi\cos\omega \\ \cos\omega\sin\kappa & \cos\omega\cos\kappa & -\sin\omega \\ -\sin\varphi\cos\kappa + \cos\varphi\sin\omega\sin\kappa & \sin\varphi\sin\kappa + \cos\varphi\sin\omega\cos\kappa & \cos\varphi\cos\omega \end{bmatrix}$$

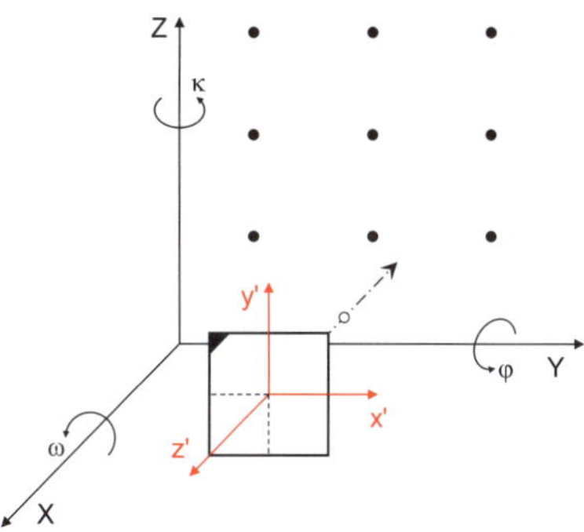

Fig. 2.13: Image configuration where ω = 0°, φ = 90° and κ = 90°

Example 2.2:

Referring to Fig. 2.13, an image configuration is shown where the primary rotation $\omega = 0°$, the secondary rotation $\varphi = 90°$ and the tertiary rotation $\kappa = 90°$. In this case the $\mathbf{R}_{\varphi\omega\kappa}$ reduces to

$$\mathbf{R}_{\varphi\omega\kappa} = \begin{bmatrix} 0 & 0 & 1 \\ 1 & 0 & 0 \\ 0 & 1 & 0 \end{bmatrix}$$

This rotation matrix represents an exchange of coordinate axes. The first row describes the transformation of the X axis. Its x, y and z elements are respectively 0, 0 and 1, indicating a transformation of X to z. Correspondingly, the second row shows Y transforming to x and the third row transforms Z to y.

The exchange of rotation orders is not a suitable solution for arbitrarily oriented images (see Fig. 3.36 and Fig. 4.50). Firstly, the rotation angles of images freely located in 3D space are not easy to visualise. Secondly, ambiguities cannot be avoided, which leads to singularities when calculating orientations.

Rotation matrix using algebraic functions

The ambiguities for trigonometric functions (above) can be avoided when a rotation matrix with algebraic functions is used. The three independent rotations are described by four algebraic parameters (*quaternions*) $a \ldots d$.

$$\mathbf{R}^T = \begin{bmatrix} d^2 + a^2 - b^2 - c^2 & 2(ab - cd) & 2(ac + bd) \\ 2(ab + cd) & d^2 - a^2 + b^2 - c^2 & 2(bc - ad) \\ 2(ac - bd) & 2(bc + ad) & d^2 - a^2 - b^2 + c^2 \end{bmatrix} \tag{2.33}$$

Implicitly, this rotation matrix contains a common scaling factor:

$$m = a^2 + b^2 + c^2 + d^2 \tag{2.34}$$

Using the constraint $m = 1$, an orthogonal rotation matrix with three independent parameters is obtained.

The geometric interpretation of this rotation matrix is not easy. However, using the rotation matrix of eqn. 2.27 the transformation of the four coefficients into standard rotation angles can be performed as follows:

$$\begin{aligned} \cos\varphi \cdot \sin\kappa &= 2(dc - ab) \\ \cos\varphi \cdot \cos\kappa &= d^2 + a^2 - b^2 - c^2 \\ \cos\varphi \cdot \sin\omega &= 2(da - bc) \\ \cos\varphi \cdot \cos\omega &= d^2 - a^2 - b^2 + c^2 \\ \sin\varphi &= 2(ac + bd) \end{aligned} \tag{2.35}$$

In summary, a rotation matrix with algebraic functions offers the following benefits:

- no use of trigonometric functions,
- simplified computation of the design matrix and faster convergence in adjustment systems,
- no singularities,
- faster computation by avoiding power series for internal trigonometric calculations.

Rotation matrix with direction cosines

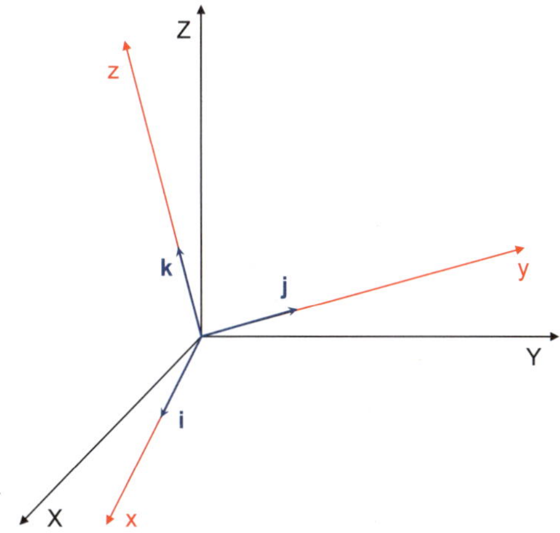

Fig. 2.14: Direction cosines

The spatial rotation matrix can be regarded as a matrix of *direction cosines* of the angles δ between the original and the rotated coordinate axes. The unit vectors **i**,**j**,**k** are defined in the direction of the rotated axes (Fig. 2.14).

$$\mathbf{R} = \begin{bmatrix} \cos\delta_{xX} & \cos\delta_{yX} & \cos\delta_{zX} \\ \cos\delta_{xY} & \cos\delta_{yY} & \cos\delta_{zY} \\ \cos\delta_{xZ} & \cos\delta_{yZ} & \cos\delta_{zZ} \end{bmatrix} = \begin{bmatrix} \mathbf{i} & \mathbf{j} & \mathbf{k} \end{bmatrix} \tag{2.36}$$

Differential rotation matrix for small rotations

For differential rotations the rotation matrix (2.27) reduces to

$$\mathbf{dR} = \begin{bmatrix} 1 & -d\kappa & d\varphi \\ d\kappa & 1 & -d\omega \\ -d\varphi & d\omega & 1 \end{bmatrix} \tag{2.37}$$

Normalisation of rotation matrices

If the coefficients of a rotation matrix are not explicitly derived from three rotational values, but instead are the result of a calculation process (e.g. the determination of exterior orientation or a spatial similarity transformation) then the matrix can show departures from orthogonality and orthonormality. Possible causes are systematic errors in the input data or limits to computational precision. In this case, the matrix can be orthonormalised by methods such as the Gram-Schmidt procedure or the following similar method:

With the initial rotation matrix (to be orthonormalised):

$$\mathbf{R} = \begin{bmatrix} r_{11} & r_{12} & r_{13} \\ r_{21} & r_{22} & r_{23} \\ r_{31} & r_{32} & r_{33} \end{bmatrix} = \begin{bmatrix} \mathbf{u} & \mathbf{v} & \mathbf{w} \end{bmatrix} \tag{2.38}$$

create direction vectors which have unit length (unit vectors), and are mutually orthogonal, and which form the new (orthonormal) matrix as follows:

$$\mathbf{u'} = \frac{\mathbf{u}}{|\mathbf{u}|} \qquad \mathbf{s} = \mathbf{v} - \frac{\mathbf{v} \cdot \mathbf{u'}}{\mathbf{u'}} \qquad \mathbf{v'} = \frac{\mathbf{s}}{|\mathbf{s}|} \qquad \mathbf{w'} = \mathbf{u'} \times \mathbf{v'}$$

$$\mathbf{R'} = \begin{bmatrix} \mathbf{u'} & \mathbf{v'} & \mathbf{w'} \end{bmatrix} \quad \text{: orthonormalised rotation matrix} \tag{2.39}$$

Example 2.3:

A rotation matrix **R** is defined by angles $\omega = 35°$, $\varphi = 60°$, $\kappa = 30°$ according to eqn. 2.27. In this example, the values of the coefficients after the third decimal place are subject to computational error (see also example 2.4):

$$\mathbf{R} = \begin{bmatrix} 0.433273 & 0.844569 & -0.324209 \\ -0.248825 & 0.468893 & 0.855810 \\ 0.876000 & -0.284795 & 0.409708 \end{bmatrix} \quad \text{and} \quad \det(\mathbf{R}) = 1.018296$$

which, when multiplied by its transpose, does not result in a unit matrix:

$$\mathbf{R}^T\mathbf{R} = \begin{bmatrix} 1.017015 & -0.000224 & 0.005486 \\ -0.000224 & 1.014265 & 0.010784 \\ 0.005486 & 0.010784 & 1.005383 \end{bmatrix} \quad \text{and} \quad \det(\mathbf{R}^T\mathbf{R}) = 1.036927$$

The matrix orthonormalised according to (2.39) is given by:

$$\mathbf{R'} = \begin{bmatrix} 0.429633 & 0.8387032 & -0.334652 \\ -0.246735 & 0.465529 & 0.849944 \\ 0.868641 & -0.282594 & 0.406944 \end{bmatrix} \quad \text{and} \quad \det(\mathbf{R'}) = 1.000000$$

The three column vectors are now orthogonal to one another in pairs and all have unit length.

Comparison of coefficients

The spatial rotation defined in

$$\mathbf{X} = \mathbf{R} \cdot \mathbf{x}$$

depends on the nine coefficients $r_{11} \ldots r_{33}$ of **R**. See, for example, the rotation order $\omega\ \varphi\ \kappa$ about rotated axes which defines **R** in eqn. 2.27. If the identical transformation result is to be achieved by a rotation matrix **R'** using a different rotation order, the coefficients of **R'** must be equal to those of **R**:

$$\mathbf{R} = \mathbf{R}'$$

If the rotation angles ω',φ',κ' of rotation matrix **R'** are to be calculated from the explicitly given angles ω,φ,κ of **R**, this can be achieved by a comparison of matrix coefficients and a subsequent reverse calculation of the trigonometric functions.

Example 2.4:

Given the rotation matrix of eqn. 2.27 defined by angles $\omega = 35°$, $\varphi = 60°$, $\kappa = 30°$, determine the rotation angles ω',φ',κ' belonging to the equivalent rotation matrix **R'** defined by eqn. 2.32:

1. Evaluate the coefficients $r_{11} \ldots r_{33}$ of **R** by multiplying out the individual rotation matrices in the order $\mathbf{R} = \mathbf{R}_\omega \cdot \mathbf{R}_\varphi \cdot \mathbf{R}_\kappa$, substituting the given values of $\omega\ \varphi\ \kappa$:

$$\mathbf{R} = \begin{bmatrix} 0.433013 & -0.250000 & 0.866025 \\ 0.839758 & 0.461041 & -0.286788 \\ -0.327576 & 0.851435 & 0.409576 \end{bmatrix}$$

2. Write the coefficients $r'_{11} \ldots r'_{33}$ of **R'** in trigonometric form by multiplying the individual rotation matrices in the order $\mathbf{R}' = \mathbf{R}_\varphi \cdot \mathbf{R}_\omega \cdot \mathbf{R}_\kappa$. Assign to each coefficient the values from **R**, i.e. $r'_{11} = r_{11}$, $r'_{12} = r_{12}$, and so on.
3. Calculate the rotation angles ω',φ',κ' of **R'** by solution of trigonometric equations:
   $\omega' = 16.666°$ $\varphi' = 64.689°$ $\kappa' = 61.232°$

2.2.2.2 Spatial similarity transformation

Mathematical model

The *spatial similarity transformation* is used for the shape-invariant mapping of a three-dimensional Cartesian coordinate system xyz into a corresponding target system XYZ. Both systems can be arbitrarily rotated, shifted and scaled with respect to each other. It is important to note that the rectangularity of the coordinate axes is preserved. This transformation is therefore a special case of the general affine transformation which requires 3 scaling factors and 3 additional shearing parameters for each coordinate axis - a total of 12 parameters.

The spatial similarity transformation, also known as a 3D Helmert transformation, is defined by 7 parameters, namely 3 translations to the origin of the xyz system (vector $\mathbf{X_0}$ defined by X_0,Y_0,Z_0), 3 rotation angles ω,φ,κ about the axes XYZ (implied by orthogonal rotation matrix **R**) and one scaling factor m (Fig. 2.15). The 6 parameters for translation and rotation correspond to the parameters of exterior orientation (see section 4.2.1). Parameters are applied in the order rotate - scale - shift and the transformation function for a point P(x,y,z), defined by vector **x**, is given by:

$$\mathbf{X} = \mathbf{X_0} + m \cdot \mathbf{R} \cdot \mathbf{x} \tag{2.40}$$

or

$$\begin{bmatrix} X \\ Y \\ Z \end{bmatrix} = \begin{bmatrix} X_0 \\ Y_0 \\ Z_0 \end{bmatrix} + m \cdot \begin{bmatrix} r_{11} & r_{12} & r_{13} \\ r_{21} & r_{22} & r_{23} \\ r_{31} & r_{32} & r_{33} \end{bmatrix} \cdot \begin{bmatrix} x \\ y \\ z \end{bmatrix}$$

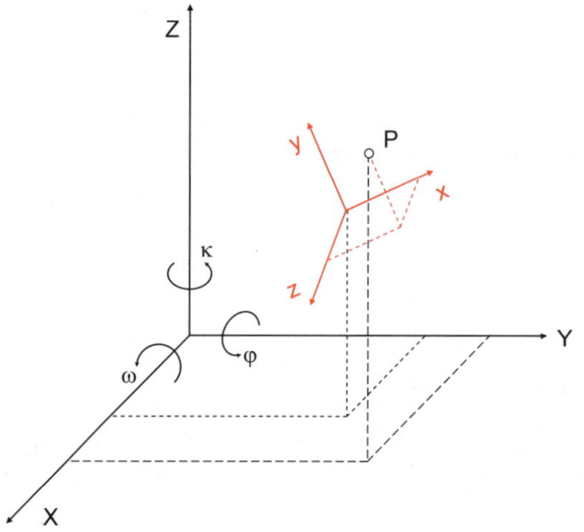

Fig. 2.15: Spatial similarity transformation

In order to determine the seven parameters, a minimum of seven observations is required. These observations can be derived from the coordinate components of at least three spatially distributed reference points (control points). They must contain at least 2 X, 2 Y and 3 Z components[1] and they must not lie on a common straight line in object space.

The spatial similarity transformation is of fundamental importance to photogrammetry for two reasons. Firstly, it is a key element in the derivation of the collinearity equations, which are the fundamental equations of analytical photogrammetry (see section 4.2.2). Secondly, it is used for the transformation of local 3D coordinates (e.g. model coordinates, 3D measuring machine coordinates) into an arbitrary superior system (e.g. object or world coordinate system), for instance in the case of absolute orientation (see section 4.3.5) or bundle adjustment (see section 4.4).

[1] It is assumed that the viewing direction is approximately parallel to the Z axis. For other image orientations appropriately positioned minimum control information is required.

There are simplified solutions for a transformation between two systems that are approximately parallel. In the general case both source and target system have an arbitrary relative orientation, i.e. any possible translation and rotation may occur. The calculation of transformation parameters then requires linearisation of the system of equations defined by the similarity transformation (2.40). Sufficiently accurate initial values are then required in order to determine the unknown parameters (see below).

The system of equations is normally over-determined and the solution is performed by least-squares adjustment (see section 2.4). This derives an optimal fit between both coordinate systems. According to eqn. 2.40 every reference point defined in both systems generates up to three equations:

$$
\begin{aligned}
X &= X_0 + m \cdot (r_{11} \cdot x + r_{12} \cdot y + r_{13} \cdot z) \\
Y &= Y_0 + m \cdot (r_{21} \cdot x + r_{22} \cdot y + r_{23} \cdot z) \\
Z &= Z_0 + m \cdot (r_{31} \cdot x + r_{32} \cdot y + r_{33} \cdot z)
\end{aligned}
\tag{2.41}
$$

By linearizing the equations at approximate parameter values, corresponding correction equations are built up. Any reference point with defined X, Y and Z coordinates (full reference point) provides three observation equations. Correspondingly, reference points with fewer coordinate components generate fewer observation equations but they can still be used for parameter estimation. Thus a transformation involving 3 full reference points already provides 2 redundant observations. The 3-2-1 method (see section 4.4.3), used in industrial metrology, is based on 6 observations, does not derive a scale change, and therefore results in zero redundancy.

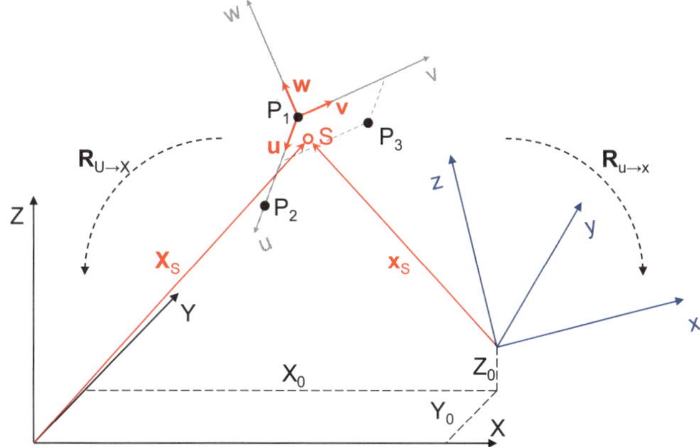

Fig. 2.16: Calculation of approximate values for 3D similarity transformation

Approximate values

In order to calculate approximate values of the translation and rotation parameters of the similarity transformation, an intermediate coordinate system is formed. This is derived from 3 reference points P_1,P_2,P_3, defined in an intermediate system uvw and known in both the target system XYZ and the source system xyz (Fig. 2.16). The purpose at this stage is to calculate the parameters which transform the reference points from intermediate system uvw to coordinate systems XYZ and xyz.

$$\begin{aligned} \mathbf{P}_{XYZ} &= \mathbf{R}_{u\to X} \cdot \mathbf{P}_{uvw} + \mathbf{T}_{u\to X} \\ \mathbf{P}_{xyz} &= \mathbf{R}_{u\to x} \cdot \mathbf{P}_{uvw} + \mathbf{T}_{u\to x} \end{aligned} \tag{2.42}$$

Solving both equations for $\mathbf{P}_{uvw}$ and re-arranging:

$$\mathbf{R}^T_{u\to X} \cdot (\mathbf{P}_{XYZ} - \mathbf{T}_{u\to X}) = \mathbf{R}^T_{u\to x} \cdot (\mathbf{P}_{xyz} - \mathbf{T}_{u\to x})$$

and finally for the coordinates of a point in system XYZ:

$$\begin{aligned} \mathbf{P}_{XYZ} &= \mathbf{R}_{u\to X} \cdot \mathbf{R}^T_{u\to x} \cdot \mathbf{P}_{xyz} + \mathbf{T}_{u\to X} - \mathbf{R}_{u\to X} \cdot \mathbf{R}^T_{u\to x} \cdot \mathbf{T}_{u\to x} \\ &= \mathbf{R}^0_{x\to X} \cdot \mathbf{P}_{xyz} + (\mathbf{T}_{u\to X} - \mathbf{R}^0_{x\to X} \cdot \mathbf{T}_{u\to x}) \end{aligned} \tag{2.43}$$

Here matrices $\mathbf{R}_{u\to X}$ and $\mathbf{R}_{u\to x}$ describe the rotation of each system under analysis with respect to the intermediate system. The vectors $\mathbf{T}_{u\to X}$ and $\mathbf{T}_{u\to x}$ describe the corresponding translations. The expression in brackets describes the translation between systems XYZ and xyz:

$$\mathbf{X}^0_{x\to X} = \mathbf{T}_{u\to X} - \mathbf{R}^0_{x\to X} \cdot \mathbf{T}_{u\to x} \tag{2.44}$$

To calculate the required parameters, the u axis of the intermediate system is constructed through P_1 and P_2 and the uv plane through P_3 (corresponds to the 3-2-1 method). From the local vectors defined by the reference points $\mathbf{P}_i(X_i,Y_i,Z_i)$, $i = 1\ldots3$, normalised direction vectors are calculated. Here vectors **u**,**v**,**w** are derived from the coordinates of $\mathbf{P}_i$ in the source system xyz, while **U**,**V**,**W** are calculated from the target system coordinates XYZ:

$$\begin{aligned} \mathbf{U} &= \frac{\mathbf{P}_2 - \mathbf{P}_1}{|\mathbf{P}_2 - \mathbf{P}_1|} & \mathbf{u} &= \frac{\mathbf{p}_2 - \mathbf{p}_1}{|\mathbf{p}_2 - \mathbf{p}_1|} \\ \mathbf{W} &= \frac{\mathbf{U} \times (\mathbf{P}_3 - \mathbf{P}_1)}{|\mathbf{U} \times (\mathbf{P}_3 - \mathbf{P}_1)|} & \mathbf{w} &= \frac{\mathbf{u} \times (\mathbf{p}_3 - \mathbf{p}_1)}{|\mathbf{u} \times (\mathbf{p}_3 - \mathbf{p}_1)|} \\ \mathbf{V} &= \mathbf{W} \times \mathbf{U} & \mathbf{v} &= \mathbf{w} \times \mathbf{u} \end{aligned} \tag{2.45}$$

Vector **u** is a unit vector on the u axis, **w** is perpendicular to the uv plane and **v** is perpendicular to **u** and **w**. These 3 vectors directly define the rotation matrix from uvw to XYZ (see eqn. 2.36):

$$\mathbf{R}_{U\to X} = [\mathbf{U} \quad \mathbf{V} \quad \mathbf{W}] \qquad \mathbf{R}_{u\to x} = [\mathbf{u} \quad \mathbf{v} \quad \mathbf{w}] \tag{2.46}$$

In the same way rotation matrix $\mathbf{R}_{xyz}$ is defined by the reference point coordinates in system xyz. The approximate rotation matrix from the xyz to the XYZ system is obtained from successive application of the above two matrices as follows:

$$\mathbf{R}^0_{x\to X} = \mathbf{R}_{U\to X} \cdot \mathbf{R}^T_{u\to x} \tag{2.47}$$

Using the centroid of the reference points in both coordinate systems, approximate values for the translation parameters of the similarity transformation can be calculated:

$$\mathbf{X}_S = \begin{bmatrix} X_S \\ Y_S \\ Z_S \end{bmatrix} = \mathbf{T}_{u\to X} \qquad \text{: centroid in XYZ system}$$

$$\mathbf{x}_S = \begin{bmatrix} x_S \\ y_S \\ z_S \end{bmatrix} = \mathbf{T}_{u\to x} \qquad \text{: centroid in xyz system} \tag{2.48}$$

According to (2.44) the translation can then be calculated:

$$\mathbf{X}^0_{x\to X} = \mathbf{X}_S - \mathbf{R}^0_{x\to X} \cdot \mathbf{x}_S \tag{2.49}$$

The approximate scale factor can be calculated from the point separations:

$$m^0 = \frac{|\mathbf{P}_2 - \mathbf{P}_1|}{|\mathbf{p}_2 - \mathbf{p}_1|} = \frac{\sqrt{(X_2 - X_1)^2 + (Y_2 - Y_1)^2 + (Z_2 - Z_1)^2}}{\sqrt{(x_2 - x_1)^2 + (y_2 - y_1)^2 + (z_2 - z_1)^2}} \tag{2.50}$$

Example 2.5:

5 points are known in the source and target systems and have the following 3D coordinates:

Nr.	x	y	z	X	Y	Z
1	110.0	100.0	110.0	153.435	170.893	150.886
2	150.0	280.0	100.0	98.927	350.060	354.985
3	300.0	300.0	120.0	214.846	544.247	319.103
4	170.0	100.0	100.0	179.425	250.771	115.262
5	200.0	200.0	140.0	213.362	340.522	252.814

Approximate values, calculated using points 1, 2 and 3 as above, are:

Rotation:

$$\mathbf{R}^0_{x\to X} = \begin{bmatrix} 0.432999 & -0.249986 & 0.866036 \\ 0.839769 & 0.461019 & -0.286790 \\ -0.327565 & 0.851450 & 0.409552 \end{bmatrix}$$

Scale factor:

$$m^0 = 1.5$$

Translation:

$$\mathbf{X}^0_{x\to X} = \begin{bmatrix} -23.41 \\ 10.49 \\ 9.63 \end{bmatrix}$$

2.2.2.3 Homogeneous coordinate transformations

Homogeneous coordinates

Graphical transformations are transformations and projections used in computer graphics and projective geometry. In this field *homogeneous coordinates*[1] are often used to form these functions.

$$\boldsymbol{x}_h = \begin{bmatrix} x_h \\ y_h \\ z_h \\ w \end{bmatrix} \tag{2.51}$$

These include the important special case $w = 1$ for Cartesian coordinates x,y,z:

$$\boldsymbol{x} = \begin{bmatrix} x_h/w \\ y_h/w \\ z_h/w \\ w/w \end{bmatrix} = \begin{bmatrix} x \\ y \\ z \\ 1 \end{bmatrix} \tag{2.52}$$

Using homogeneous coordinates, all coordinate transformations and translations, as well as axonometric and central projections, can be formulated in a unified way in any combination. They are therefore perfectly suited to calculations in computer graphics and CAD systems. The photogrammetric equations can also be elegantly expressed in homogeneous coordinates (see section 4.2.4.2).

General transformations

The general linear transformation of homogeneous coordinates is given by:

$$\boldsymbol{X} = \boldsymbol{T} \cdot \boldsymbol{x} \tag{2.53}$$

where $\boldsymbol{T}$ is the transformation or projection matrix[2].

$$\boldsymbol{T} = \left[\begin{array}{ccc|c} a_{11} & a_{12} & a_{13} & a_{14} \\ a_{21} & a_{22} & a_{23} & a_{24} \\ a_{31} & a_{32} & a_{33} & a_{34} \\ \hline a_{41} & a_{42} & a_{43} & a_{44} \end{array}\right] = \left[\begin{array}{c|c} \underset{3,3}{\mathbf{T}_{11}} & \underset{1,3}{\mathbf{T}_{12}} \\ \hline \underset{3,1}{\mathbf{T}_{21}} & \underset{1,1}{\mathbf{T}_{22}} \end{array}\right] \tag{2.54}$$

The result of this transformation always results in a new homogeneous coordinate vector. The four sub-matrices contain information as follows:

[1] Homogeneous vectors and matrices are represented by bold, italic text.

[2] Note that $\boldsymbol{T}$ is a homogeneous matrix whilst the four sub-matrices are not.

$\mathbf{T}_{11}$: scaling, reflection in a line, rotation
$\mathbf{T}_{12}$: translation
$\mathbf{T}_{21}$: perspective
$\mathbf{T}_{22}$: homogeneous scaling (factor w)

Scaling or reflection about a line is performed by the factors s_X, s_Y, s_Z:

$$\boldsymbol{T}_S = \left[\begin{array}{ccc|c} s_X & 0 & 0 & 0 \\ 0 & s_Y & 0 & 0 \\ 0 & 0 & s_Z & 0 \\ \hline 0 & 0 & 0 & 1 \end{array}\right] \quad \text{: scaling, reflection in a line} \tag{2.55}$$

A spatial rotation results if $\mathbf{T}_{11}$ is replaced by the rotation matrix derived in section 2.2.2.1:

$$\boldsymbol{T}_R = \left[\begin{array}{ccc|c} r_{11} & r_{12} & r_{13} & 0 \\ r_{21} & r_{22} & r_{23} & 0 \\ r_{31} & r_{32} & r_{33} & 0 \\ \hline 0 & 0 & 0 & 1 \end{array}\right] \quad \text{: spatial rotation} \tag{2.56}$$

Translation by a vector x_T,y_T,z_T is performed by the projection matrix:

$$\boldsymbol{T}_T = \left[\begin{array}{ccc|c} 1 & 0 & 0 & x_T \\ 0 & 1 & 0 & y_T \\ 0 & 0 & 1 & z_T \\ \hline 0 & 0 & 0 & 1 \end{array}\right] \quad \text{: translation} \tag{2.57}$$

Combined transformations $\boldsymbol{T}_1$, $\boldsymbol{T}_2$ etc. can be created by sequential multiplication of single projection matrices as follows:

$$\boldsymbol{X} = \boldsymbol{T} \cdot \boldsymbol{x} = \boldsymbol{T}_n \cdot \ldots \cdot \boldsymbol{T}_2 \cdot \boldsymbol{T}_1 \cdot \boldsymbol{x} \tag{2.58}$$

In general, the multiplication order may not be changed because the projections are not necessarily commutative.

The reverse transformation is given by:

$$\boldsymbol{x} = \boldsymbol{T}^{-1} \cdot \boldsymbol{X} = \boldsymbol{T}_1^{-1} \cdot \boldsymbol{T}_2^{-1} \cdot \ldots \cdot \boldsymbol{T}_n^{-1} \cdot \boldsymbol{X} \tag{2.59}$$

This inversion is only possible if the projection matrix is not singular, as is the normal case for the transformation of one 3D system into another. However, if the vector $\boldsymbol{x}$ is projected onto a plane, the projection matrix does become singular. The original coordinates cannot then be calculated from the transformed plane coordinates $\boldsymbol{X}$.

Using homogenous coordinates, the spatial similarity transformation of eqn. 2.40 is given by ($m = s_x = s_y = s_z$):

$$
\begin{aligned}
\boldsymbol{X} &= \boldsymbol{T}_T \cdot \boldsymbol{T}_S \cdot \boldsymbol{T}_R \cdot \boldsymbol{x} \\
\begin{bmatrix} X \\ Y \\ Z \\ 1 \end{bmatrix} &= \begin{bmatrix} 1 & 0 & 0 & X_0 \\ 0 & 1 & 0 & Y_0 \\ 0 & 0 & 1 & Z_0 \\ 0 & 0 & 0 & 1 \end{bmatrix} \cdot \begin{bmatrix} m & 0 & 0 & 0 \\ 0 & m & 0 & 0 \\ 0 & 0 & m & 0 \\ 0 & 0 & 0 & 1 \end{bmatrix} \cdot \begin{bmatrix} r_{11} & r_{12} & r_{13} & 0 \\ r_{21} & r_{22} & r_{23} & 0 \\ r_{31} & r_{32} & r_{33} & 0 \\ 0 & 0 & 0 & 1 \end{bmatrix} \cdot \begin{bmatrix} x \\ y \\ z \\ 1 \end{bmatrix} \\
&= \begin{bmatrix} mr_{11} & mr_{12} & mr_{13} & X_0 \\ mr_{21} & mr_{22} & mr_{23} & Y_0 \\ mr_{31} & mr_{32} & mr_{33} & Z_0 \\ 0 & 0 & 0 & 1 \end{bmatrix} \cdot \begin{bmatrix} x \\ y \\ z \\ 1 \end{bmatrix}
\end{aligned}
\tag{2.60}
$$

Projections

For the projection of a 3D object into an image or drawing plane, it is common to distinguish between axonometric and perspective transformations.

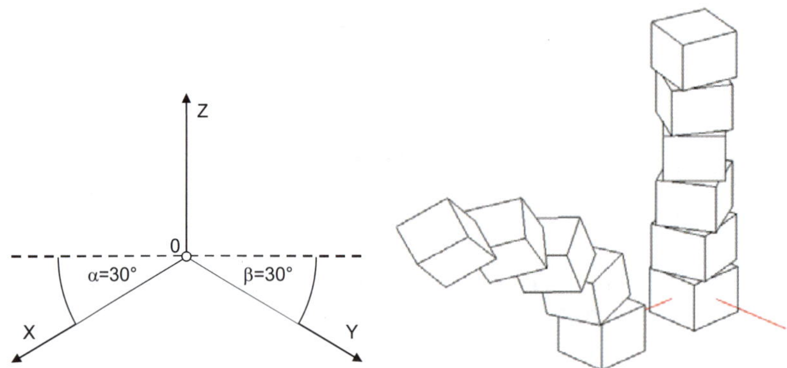

Fig. 2.17: Isometric projection

For axonometric projections the object is projected onto the desired plane using a parallel projection. An example is the isometric projection widely used in CAD technology. The projection matrix for the isometric projection (Fig. 2.17) is given by:

$$
\boldsymbol{T}_I = \begin{bmatrix} -\cos(30°) & \cos(30°) & 0 & 0 \\ -\sin(30°) & -\sin(30°) & 1 & 0 \\ 0 & 0 & 0 & 0 \\ 0 & 0 & 0 & 1 \end{bmatrix} \quad : \text{isometry} \tag{2.61}
$$

where the transformed Z coordinate is discarded in the visualisation.

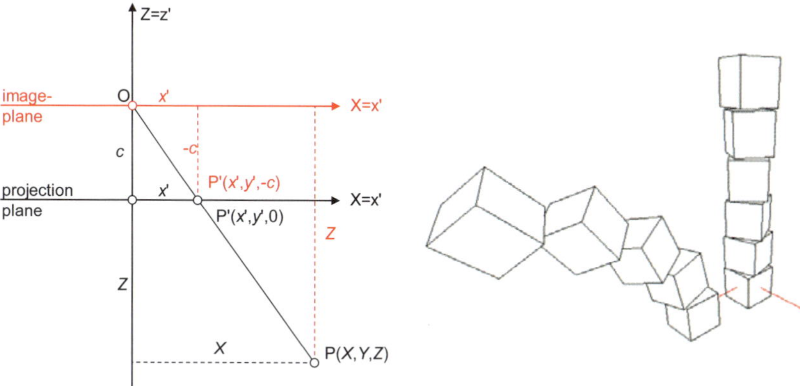

Fig. 2.18: Central projection

The central projection is modelled firstly for the following special case. The projection plane is oriented normal to the viewing direction Z with the distance c to the perspective centre at O. Referring to Fig. 2.18, the following ratios can be derived[1].

$$\frac{x'}{X} = \frac{y'}{Y} = \frac{c}{Z+c} \tag{2.62}$$

and further rearranged to give x' and y':

$$x' = \frac{X}{\frac{Z}{c}+1} \qquad y' = \frac{Y}{\frac{Z}{c}+1} \tag{2.63}$$

If the perspective centre moves to infinity (focal length c becomes infinite), the denominator becomes 1 and the central projection changes to a parallel projection. Without affecting validity, the image coordinate system can then be shifted to the perspective centre (red position in Fig. 2.18), which leads to the following projection equations:

$$x' = \frac{-c}{Z} \cdot X = \frac{1}{m} \cdot X \qquad y' = \frac{-c}{Z} \cdot Y = \frac{1}{m} \cdot Y \tag{2.64}$$

In matrix form the transformation (2.64) is given by:

$$\bar{\boldsymbol{x}} = \boldsymbol{T}_Z \cdot \boldsymbol{X}$$

$$\begin{bmatrix} \bar{x} \\ \bar{y} \\ \bar{z} \\ w \end{bmatrix} = \left[\begin{array}{ccc|c} 1 & 0 & 0 & 0 \\ 0 & 1 & 0 & 0 \\ 0 & 0 & 1 & 0 \\ \hline 0 & 0 & -1/c & 0 \end{array}\right] \cdot \begin{bmatrix} X \\ Y \\ Z \\ 1 \end{bmatrix} = \begin{bmatrix} X \\ Y \\ Z \\ -Z/c \end{bmatrix} \tag{2.65}$$

[1] Here the usual notations for image coordinates x',y' and c are used.

and for the resulting homogenous coordinates after division by $m = -Z/c$:

$$\boldsymbol{x}' = \boldsymbol{T}_S^{-1} \cdot \bar{\boldsymbol{x}}$$

$$\begin{bmatrix} x' \\ y' \\ z' \\ 1 \end{bmatrix} = \begin{bmatrix} -c/Z & 0 & 0 & 0 \\ 0 & -c/Z & 0 & 0 \\ 0 & 0 & -c/Z & 0 \\ 0 & 0 & 1/Z & 0 \end{bmatrix} \cdot \begin{bmatrix} X \\ Y \\ Z \\ 1 \end{bmatrix} = \begin{bmatrix} -c \cdot X/Z \\ -c \cdot Y/Z \\ -c \\ 1 \end{bmatrix} = \begin{bmatrix} X/m \\ Y/m \\ -c \\ 1 \end{bmatrix} \tag{2.66}$$

It is obvious that rows 3 and 4 of the transformation matrix $\boldsymbol{T}_S$ are linearly dependent and the matrix cannot be inverted. It is therefore not possible to calculate 3D object coordinates from 2D image coordinates in this case.

If the above mentioned special case is extended to an arbitrary exterior orientation of the image plane (position and orientation in space), the transformation of object coordinates into image coordinates can be performed by the following matrix operation with respect to (2.60):

$$\boldsymbol{x}' = \boldsymbol{T}_S^{-1} \cdot \boldsymbol{T}_Z \cdot \boldsymbol{T}_R^{-1} \cdot \boldsymbol{T}_T^{-1} \cdot \boldsymbol{X}$$

$$\begin{bmatrix} x' \\ y' \\ z' \\ 1 \end{bmatrix} = \begin{bmatrix} \frac{r_{11}}{m} & \frac{r_{21}}{m} & \frac{r_{31}}{m} & \frac{-(r_{11}X_0 + r_{21}Y_0 + r_{31}Z_0)}{m} \\ \frac{r_{12}}{m} & \frac{r_{22}}{m} & \frac{r_{32}}{m} & \frac{-(r_{12}X_0 + r_{22}Y_0 + r_{32}Z_0)}{m} \\ \frac{r_{13}}{m} & \frac{r_{23}}{m} & \frac{r_{33}}{m} & \frac{-(r_{13}X_0 + r_{23}Y_0 + r_{33}Z_0)}{m} \\ \frac{-r_{13}}{c \cdot m} & \frac{-r_{23}}{c \cdot m} & \frac{-r_{33}}{c \cdot m} & \frac{r_{11}X_0 + r_{21}Y_0 + r_{31}Z_0}{c \cdot m} \end{bmatrix} \cdot \begin{bmatrix} X \\ Y \\ Z \\ 1 \end{bmatrix} \tag{2.67}$$

2.3 Geometric elements

The geometric reconstruction of a measured object is the major goal of a photogrammetric process. This section therefore gives a short summary of geometric elements and their mathematical definition. It distinguishes between planar elements, spatial elements and surface descriptions that are the basic result of a photogrammetric measurement. For a detailed description of the methods of analytical geometry, the reader should refer to specialist literature on geometry and 3D computer graphics.

Except in very few cases, photogrammetric methods are based on measurement of discrete object points. Geometric elements such as straight lines, planes, cylinders etc. are normally calculated in a post-processing step using the measured 3D points. For over-determined solutions, least-squares fitting methods are used. Computed geometric elements can then either be combined or intersected in order to create additional geometric elements such as the intersection line between two planes. Alternatively, specific dimensions can be derived from them, such as the distance between two points (Fig. 2.19).

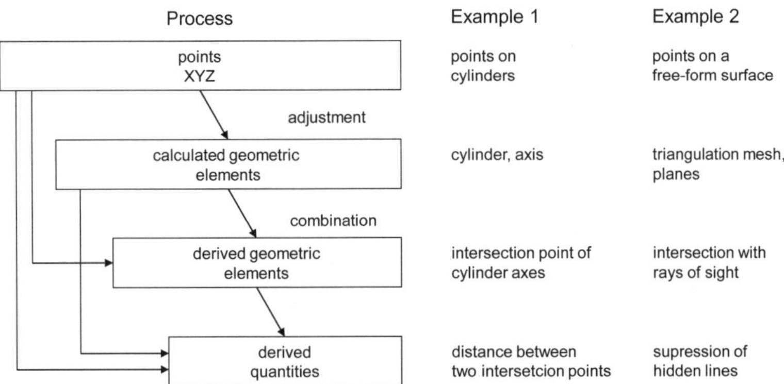

Fig. 2.19: Calculation progress for geometric elements

In addition to the determination of regular geometric shapes, the determination and visualisation of arbitrary three dimensional surfaces (free-form surfaces) is of increasing importance. This requires a basic knowledge of different ways to represent 3D surfaces, involving point grids, triangle meshing, analytical curves etc.

Many of these calculations are embedded in state-of-the-art 3D CAD systems or programs for geometric quality analysis. CAD and photogrammetric systems are therefore often combined. However, geometric elements may also be directly employed in photogrammetric calculations, e.g. as conditions for the location of object points (see section 4.3.2.3). In addition, some evaluation techniques enable the direct calculation of geometric 3D elements without the use of discrete points (e.g. contour method, section 4.4.7.2).

2.3.1 Analytical geometry in the plane

2.3.1.1 Straight line

Parametric form

The straight line g between two points P_1 and P_2 (Fig. 2.20) is to be determined. For all points $P(x,y)$ belonging to g, the proportional relationship

$$\frac{y-y_1}{x-x_1}=\frac{y_2-y_1}{x_2-x_1} \tag{2.68}$$

leads to the parametric form of the straight line:

$$\mathbf{x} = \mathbf{x}_1 + t \cdot (\mathbf{x}_2 - \mathbf{x}_1)$$
$$\begin{bmatrix} x \\ y \end{bmatrix} = \begin{bmatrix} x_1 \\ y_1 \end{bmatrix} + t \cdot \begin{bmatrix} x_2 - x_1 \\ y_2 - y_1 \end{bmatrix} \tag{2.69}$$

Point $P_1(x_1,y_1)$ is defined at $t = 0$ and point $P_2(x_2,y_2)$ at $t = 1$.

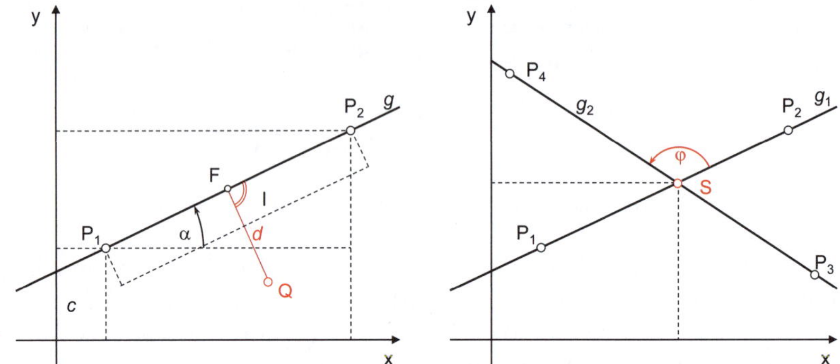

Fig. 2.20: Definition of straight lines

The distance d of a point $Q(x,y)$ from the straight line is defined by:

$$d = \frac{(y_2 - y_1)(x - x_1) - (x_2 - x_1)(y - y_1)}{l} \tag{2.70}$$

with

$$l = \sqrt{(x_2 - x_1)^2 + (y_2 - y_1)^2}$$

The foot $F(x,y)$ of the perpendicular from Q to the line is given by:

$$\begin{bmatrix} x_L \\ y_L \end{bmatrix} = \begin{bmatrix} x_1 \\ y_1 \end{bmatrix} + s \cdot \begin{bmatrix} y_1 - y_2 \\ x_2 - x_1 \end{bmatrix} \tag{2.71}$$

where

$$s = \frac{(y_1 - y_P)(x_2 - x_1) - (x_1 - x_P)(y_2 - y_1)}{l^2}$$

Analytical form

The analytical form of a straight line

$$A \cdot x + B \cdot y + C = 0 \tag{2.72}$$

leads to the following relations (Fig. 2.20):

$$y = m \cdot x + c$$

where

$$m = \tan\alpha = -\frac{A}{B} = \frac{y_2 - y_1}{x_2 - x_1} \qquad \text{: slope} \tag{2.73}$$

$$c = -\frac{C}{B} \qquad \text{: intersection point on y axis}$$

The (perpendicular) distance d to point Q(x,y) is given by:

$$d = \frac{A \cdot x + B \cdot y + C}{\sqrt{A^2 + B^2}} \tag{2.74}$$

Intersection of two straight lines

Given two straight lines g_1 and g_2, their point of intersection S is derived from (2.72) as:

$$\begin{aligned} x_S &= \frac{B_1 \cdot C_2 - B_2 \cdot C_1}{A_1 \cdot B_2 - A_2 \cdot B_1} \\ y_S &= \frac{C_1 \cdot A_2 - C_2 \cdot A_1}{A_1 \cdot B_2 - A_2 \cdot B_1} \end{aligned} \tag{2.75}$$

Alternatively, from (2.69) two equations for parameters t_1 und t_2 are obtained:

$$\begin{aligned} (x_2 - x_1) \cdot t_1 + (x_3 - x_4) \cdot t_2 &= x_3 - x_1 \\ (y_2 - y_1) \cdot t_1 + (y_3 - y_4) \cdot t_2 &= y_3 - y_1 \end{aligned} \tag{2.76}$$

The point of intersection is obtained by substituting t_1 or t_2 into the original straight line equations:

$$\begin{gathered} \mathbf{x}_S = \mathbf{x}_1 + t_1 \cdot (\mathbf{x}_2 - \mathbf{x}_1) = \mathbf{x}_3 + t_2 \cdot (\mathbf{x}_4 - \mathbf{x}_3) \\ \begin{bmatrix} x_S \\ y_S \end{bmatrix} = \begin{bmatrix} x_1 \\ y_1 \end{bmatrix} + t_1 \cdot \begin{bmatrix} x_2 - x_1 \\ y_2 - y_1 \end{bmatrix} = \begin{bmatrix} x_3 \\ y_3 \end{bmatrix} + t_2 \cdot \begin{bmatrix} x_4 - x_3 \\ y_4 - y_3 \end{bmatrix} \end{gathered} \tag{2.77}$$

The angle between both lines is given by:

$$\tan\varphi = \frac{A_1 \cdot B_2 - A_2 \cdot B_1}{A_1 \cdot A_2 - B_1 \cdot B_2} = \frac{m_2 - m_1}{m_1 \cdot m_2 + 1} \tag{2.78}$$

Alternatively, if both lines are defined by their direction vectors

$$\mathbf{a} = \mathbf{x}_2 - \mathbf{x}_1 \qquad \text{and} \qquad \mathbf{b} = \mathbf{x}_4 - \mathbf{x}_3$$

then the angle between them can be found from the scalar product of both vectors:

$$\cos\varphi = \frac{\mathbf{a}\cdot\mathbf{b}}{|\mathbf{a}|\cdot|\mathbf{b}|} = \frac{\mathbf{a}^T\mathbf{b}}{|\mathbf{a}|\cdot|\mathbf{b}|} \tag{2.79}$$

The scalar product is zero if the lines are mutually perpendicular.

Regression line

The generalised regression line, which is a best-fit to a set of points, is the straight line which minimises the sum of squared distances d_i of all points P_i from the line (Fig. 2.21). For n point coordinates with equal accuracy in the x and y directions, the criterion is expressed as:

$$d_1^2 + d_2^2 + \ldots + d_n^2 = \sum_{i=1}^{n} d_i^2 \rightarrow \min \tag{2.80}$$

The regression line passes through the centroid of the points

$$x_0 = \frac{1}{n}\sum_{i=1}^{n} x_i \qquad \text{and} \qquad y_0 = \frac{1}{n}\sum_{i=1}^{n} y_i \tag{2.81}$$

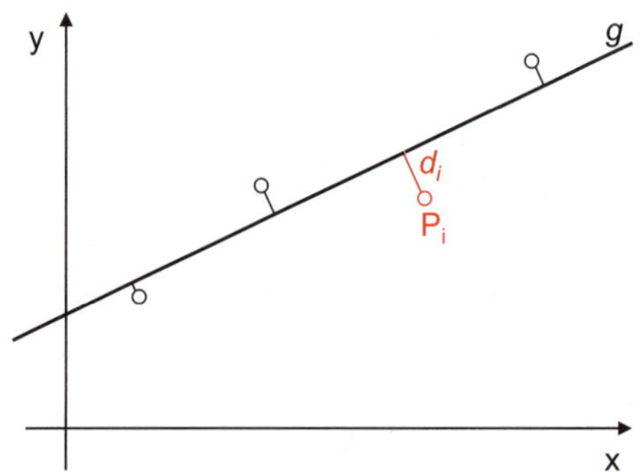

Fig. 2.21: Regression line

One point on the straight line is therefore directly given. The direction of the line is defined by

$$\tan 2\varphi = \frac{2\sum(x_i - x_0)(y_i - y_0)}{\sum(y_i - y_0)^2 - (x_i - x_0)^2} \tag{2.82}$$

Alternatively, the direction of the line can be expressed by the direction vector (a,b) which is equal to the eigenvector of the maximum eigenvalue of matrix **B**:

$$\mathbf{B} = \mathbf{A}^T \cdot \mathbf{A}$$

$$\text{where} \quad \mathbf{A} = \begin{bmatrix} x_1 - x_0 & y_1 - y_0 \\ \vdots & \vdots \\ x_n - x_0 & y_n - y_0 \end{bmatrix} \tag{2.83}$$

Without restriction, the optimisation principle based on minimum quadratic distances according to (2.80) can be applied to regression lines in space as well as other best-fit elements.

2.3.1.2 Circle

From the generalised equation for second order curves (conic sections)

$$Ax^2 + 2Bxy + Cy^2 + 2Dx + 2Ey + F = 0 \tag{2.84}$$

the special cases of circle and ellipse are of major interest in close-range photogrammetry. For a circle with centre (x_M, y_M) and radius r, the equation is typically written as:

$$(x - x_M)^2 + (y - y_M)^2 = r^2 \tag{2.85}$$

This can be re-arranged in the form:

$$x^2 + y^2 - 2x_M x - 2y_M y + x_M^2 + y_M^2 - r^2 = 0 \tag{2.86}$$

This can be further re-arranged as:

$$x^2 + y^2 + 2D'x + 2E'y + F' = 0 \tag{2.87}$$

This is equivalent to eqn. 2.84 with $A = C = 1$ and $B = 0$. There are effectively only three independent, unknown parameters, D', E' and F' and the circle can therefore be defined with a minimum of three points. The linear form of (2.87) can be used directly to solve for D', E' and F' in a least-squares solution where there are more than three points.

By comparing eqn. 2.87 with eqn. 2.86, the radius and centre of the circle can be further derived as follows:

$$r = \sqrt{D'^2 + E'^2 - F'} \qquad x_M = -D' \qquad y_M = -E' \tag{2.88}$$

Alternatively, the distance of any point $P_i(x_i, y_i)$ from the circumference is given by:

$$d_i = r_i - r = \sqrt{(x_i - x_M)^2 + (y_i - y_M)^2} - r \tag{2.89}$$

The non-linear eqn. 2.89 can also be used as an observation equation after linearisation. The best-fit circle is obtained by least-squares minimisation of all point distances d_i. With initial approximate values for the centre coordinates and radius, the design matrix **A** consists of the derivatives

$$\frac{\partial d_i}{\partial x_M} = -\frac{x_i - x_M}{r_i} \qquad \frac{\partial d_i}{\partial y_M} = -\frac{y_i - y_M}{r_i} \qquad \frac{\partial d_i}{\partial r} = -1 \tag{2.90}$$

Although the linear approach offers a direct solution without the requirement for initial parameter values, the non-linear approach directly generates the geometrically meaningful parameters of circle centre coordinates and circle radius. For over-determined data, the two solutions will generate slightly different circles because different parameters are used in their respective optimisations. In this case it may be advantageous to use the linear solution to find initial estimates for circle centre and radius and then apply the non-linear solution to optimise these estimates.

2.3.1.3 Ellipse

The determination of ellipse parameters, in particular the centre coordinates, is an important part of the measurement of circular targets which are projected as ellipses in the central perspective image (see section 5.4.2.5). As a good approximation, the calculated ellipse centre corresponds to the required centre of the circular target (see section 3.6.1.2 for restrictions).

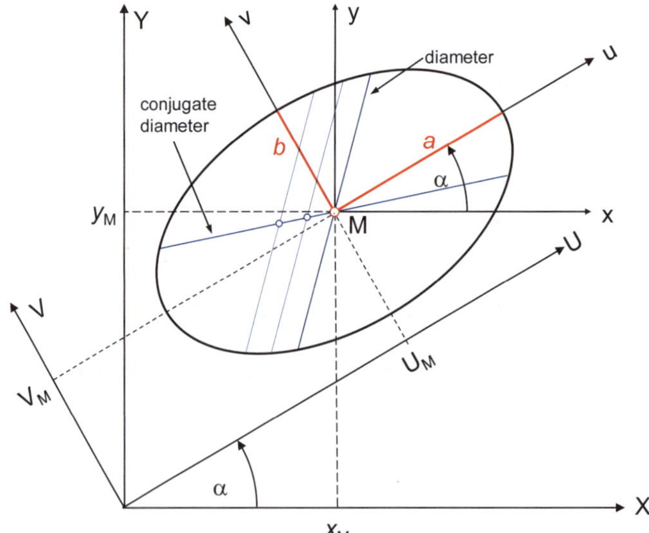

Fig. 2.22: Geometry of an ellipse

A simple method for the determination of the ellipse centre is based on the geometry of ellipse diameters. Ellipse diameters are chords that are bisected by the ellipse centre. A conjugate diameter is defined by the straight line through the mid-point of all chords which are parallel to a given diameter. A given diameter and its conjugate intersect at the ellipse centre (see Fig. 2.22). A possible implementation of this technique is presented in section 5.4.2.5.

For a full determination of ellipse parameters, a similar approach to the calculation of circle parameters can be applied. The approach is again based on fitting measured points to the generalised equation for second order curves (conic section), which is repeated here:

$$Ax^2 + 2Bxy + Cy^2 + 2Dx + 2Ey + F = 0 \tag{2.91}$$

This equation provides a direct linear solution for the unknown parameters A, B, C, D, E, F by substituting measured values for x,y in order to create an observation equation. However, as in the case of a circle where only 3 of these parameters are required, for an ellipse only 5 are required, as explained below. This requires a minimum of 5 measured points in the image which generate 5 observation equations.

The following analysis indicates a suitable modification of the generalised conic, as well as a derivation of the ellipse parameters (major and minor axes, centre position and rotation angle) from the generalised equation parameters.

The simple form of a non-rotated ellipse, with semi-major axis a, semi-minor axis b and centre at the origin of the coordinate axes, is illustrated by the uv system in Fig. 2.22 and given by:

$$\frac{u^2}{a^2} + \frac{v^2}{b^2} = 1 \tag{2.92}$$

The axial rotation α must be applied in order to transform the equation from the uv and UV systems to the xy and XY systems respectively. In this process, it is convenient to use terms c and s where $c = \cos\alpha$ and $s = \sin\alpha$ as follows:

$$\begin{aligned} u &= cx + sy \qquad & x &= cu - sv \qquad & U_M &= cX_M + sY_M \\ v &= -sx + cy \qquad & y &= su + cv \qquad & V_M &= -sX_M + cY_M \end{aligned} \tag{2.93}$$

Substituting for u and v in (2.92), the transformed values of u and v in the xy system are:

$$\frac{(cx + sy)^2}{a^2} + \frac{(-sx + cy)^2}{b^2} = 1 \tag{2.94}$$

Multiplying out and collecting terms gives:

$$\left(\frac{c^2}{a^2} + \frac{s^2}{b^2}\right)x^2 + 2\left(\frac{cs}{a^2} - \frac{cs}{b^2}\right)xy + \left(\frac{s^2}{a^2} + \frac{c^2}{b^2}\right)y^2 = 1 \tag{2.95}$$

which may be written as:

$$Ax^2 + 2Bxy + Cy^2 = 1 \tag{2.96}$$

Applying shifts from the xy to XY system, $x = (X - X_M)$ and $y = (Y - Y_M)$:

$$\begin{aligned} &AX^2 + 2BXY + CY^2 - 2(AX_M + BY_M)X - 2(BX_M + CY_M)Y \\ &+ (AX_M^2 + 2BX_M Y_M + CY_M^2 - 1) = 0 \end{aligned} \tag{2.97}$$

which may be written as:

$$AX^2 + 2BXY + CY^2 + 2DX + 2EY + F = 0 \tag{2.98}$$

Eqn. (2.98) is identical to the original generalised eqn. 2.91. Comparing (2.95) and (2.96) it can be seen that:

$$A = \left(\frac{c^2}{a^2} + \frac{s^2}{b^2}\right) \qquad C = \left(\frac{s^2}{a^2} + \frac{c^2}{b^2}\right) \qquad 2B = 2cs\left(\frac{1}{a^2} - \frac{1}{b^2}\right) \tag{2.99}$$

Using the standard trigonometrical identities:

$$\begin{aligned} &2cs = 2\cos\alpha \sin\alpha = \sin 2\alpha \\ &c^2 - s^2 = \cos^2\alpha - \sin^2\alpha = \cos 2\alpha \end{aligned} \tag{2.100}$$

it can further be seen that:

$$\frac{2B}{A - C} = 2\frac{cs}{c^2 - s^2} = \frac{\sin 2\alpha}{\cos 2\alpha} = \tan 2\alpha \tag{2.101}$$

From a comparison of (2.98) and (2.97):

$$\begin{aligned} AX_M + BY_M &= -D \\ BX_M + CY_M &= -E \end{aligned} \tag{2.102}$$

which by standard algebraic manipulation gives:

$$X_M = \frac{-DC - EB}{AC - B^2} \qquad Y_M = \frac{-BD + AE}{B^2 - AC} \tag{2.103}$$

As can be seen from the analysis, the 6 parameters A, B, C, D, E, F of the generalised eqn. 2.98 are themselves based on only 5 parameters a, b, α, X_M, Y_M. In general, the parameters A and C are positive and one may be set to the value 1 by dividing through to obtain, for example, the following linear solution equation for measured point (X_i, Y_i):

$$X_i^2 + 2B'X_iY_i + C'Y_i^2 + 2D'X_i + 2E'Y_i + F' = 0 \tag{2.104}$$

As explained, 5 measured points on the ellipse will generate 5 linear observation equations which can be solved directly by standard matrix algebra. Expressing (2.101) and (2.103) in terms of the actual solution parameters *B'*, *C'*, *D'*, *E'*, *F'*:

$$X_M = \frac{-D'C' - E'B'}{C' - B'^2} \qquad Y_M = \frac{-B'D' + E'}{B'^2 - C'} \qquad \tan 2\alpha = \frac{2B'}{1 - C'} \tag{2.105}$$

If required, the axial parameters of the ellipse, *a* and *b*, can be determined as follows. The ellipse equation in the UV system is found by applying the following rotational transformation from the XY system.

$$X = cU - sV \qquad Y = sU + cV \tag{2.106}$$

Substituting for X and Y in eqn. 2.104 and collecting terms results in:

$$\begin{aligned}&(c^2 + 2B'cs + C's^2)U^2 + [2B'(c^2 - s^2) - 2cs(1 - C')]UV \\ &+ (C'c^2 - 2B'cs + s^2)V^2 + (2D'c + 2E's)U + (2E'c - 2D's)V + F' = 0\end{aligned} \tag{2.107}$$

This can be written as follows as follows:

$$\overline{A}U^2 + 2\overline{B}UV + \overline{C}V^2 + 2\overline{D}U + 2\overline{E}V + \overline{F} = 0 \tag{2.108}$$

In the UV system it is simple to show $\overline{B} = 0$, which leads to the same result for tan 2α as expressed in eqn. 2.105. It is also possible to show that the semi axes are then given by:

$$\begin{aligned}a &= \sqrt{\frac{\overline{C}\overline{D}^2 + \overline{A}\overline{C}^2 - \overline{A}\overline{C}\overline{F}}{\overline{A}^2\overline{C}}} \\ b &= \sqrt{\frac{\overline{C}\overline{D}^2 + \overline{A}\overline{C}^2 - \overline{A}\overline{C}\overline{F}}{\overline{A}\overline{C}^2}}\end{aligned} \tag{2.109}$$

2.3.1.4 Curves

Consider the requirement that a polynomial with *k*+1 points $P_i(x_i, y_i)$, $i = 0 \ldots k$, be described by a closed curve. If the curve should pass through the vertices of a polygon, the process is referred to as *interpolation*. If the curve should be an optimal fit to the polygon, it is referred to as *approximation*. Curves in general are usually defined by polynomials whose order and curvature properties can be varied with respect to the application.

Polynomials

A polynomial of degree *n* is a function of the form:

$$Q(x) = a_n x^n + a_{n-1} x^{n-1} + \ldots a_1 x^1 + a_0 \quad (2.110)$$

that is defined by $n+1$ coefficients.

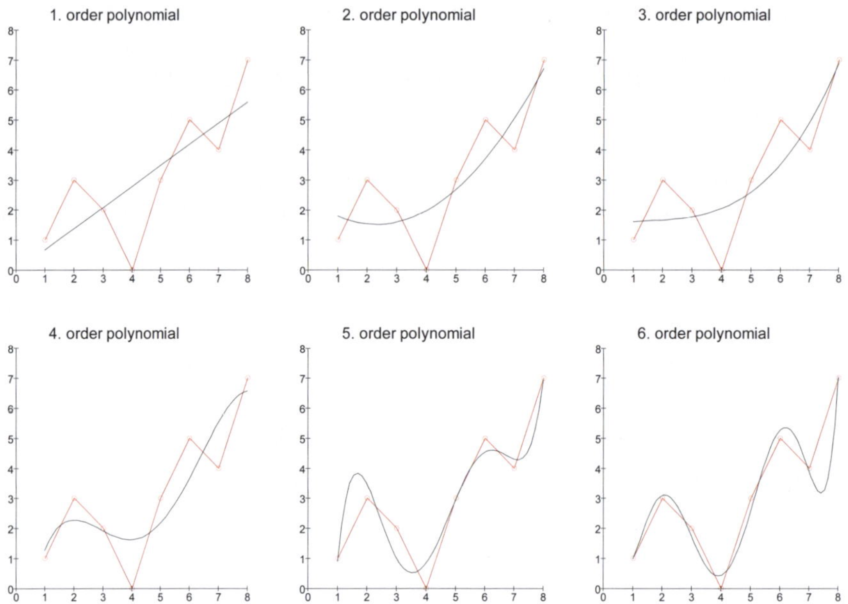

Fig. 2.23: Polygon with 8 data points and polynomial interpolations of different order

All points in a data set are used to determine the polynomial coefficients, if necessary by least-squares adjustment. For over-determined solutions, the polynomial does not normally pass through the vertices of the polygon defined by the points. In particular it does not intersect the end points. Polynomials of higher degree quickly tend to oscillate between the points (Fig. 2.23).

A more natural curve shape is obtained if the polygon is approximated by piecewise polynomials. A piecewise polynomial $Q(x)$ is a set of k polynomials $q_i(t)$, each of order n, and $k+1$ nodes[1] $x_0, \ldots, x_k$, with:

$$Q(x) = \{q_i(t)\} \quad (2.111)$$
$$\text{for } x_i \le t \le x_{i+1} \text{ and } i = 0, \ldots, k-1$$

Using additional constraints it is possible to generate an approximation curve that is both continuous and smooth. All methods which follow generate a curve that passes through the end points of the polygon and which can be differentiated n–1 times at all points. Approximations based on cubic splines ($n = 3$) are of major importance for they provide a practical level of smoothness with a minimum polynomial degree.

[1] Nodes are the given points defining a curve, i.e. the vertices of a polygon.

Splines

Splines are used to interpolate between the points of a polygon, i.e. the curve passes through all the points. For this purpose basic B-spline functions of degree n and order $m = n+1$ are suitable. They are recursively defined for a set of nodes $x_0, x_1, \ldots, x_{k-1}$:

$$B_{i,0}(t) = \begin{cases} 1 & \text{for } x_i \leq t \leq x_{i+1} \\ 0 & \text{otherwise} \end{cases}$$

$$B_{i,n}(t) = \frac{t - x_i}{x_{i+n} - x_i} B_{i,n-1}(t) + \frac{x_{i+n+1} - t}{x_{i+n+1} - x_{i+1}} B_{i+1,n-1}(t) \tag{2.112}$$

for $x_i \leq t \leq x_{i+n+1}$.

Optimal smoothness at the data point points is required for spline interpolation, i.e. continuous derivatives up to order $n-1$ should exist. This criterion is fulfilled by the following linear combination of $k+1$ nodes:

$$S_n(t) = \sum_{i=0}^{n+k-1} a_i B_{i,n}(t) \tag{2.113}$$

For the frequently used cubic spline function ($n = 3$)

$$S_3(t) = \sum_{i=0}^{k+2} a_i B_{i,3}(t) \tag{2.114}$$

a number $k+3$ of coefficients a_i have to be determined by a corresponding number of equations. Here $k+1$ equations are provided by the data points and the remaining two equations defined by additional constraints. For example, for *natural splines* these are:

$$\begin{aligned} S_3''(x_0) &= 0 \\ S_3''(x_n) &= 0 \end{aligned} \tag{2.115}$$

Fig. 2.24a shows a polygon approximated by a cubic spline. The resulting curve continuously passes through the vertices (nodes). Splines are therefore most effective when the vertices are free of position errors, i.e. no smoothing is desired. However, the shape of the entire curve is affected if only one point changes.

B-Splines

For many technical applications it is more feasible to approximate a given polygon by a curve with the following properties:

- analytical function is simple to formulate
- can be easily extended to higher dimensions, especially for the surface approximations
- smoothness at vertices is easy to control
- variation of a node has only a local effect on the shape of the curve

The requirements are met by B-spline approximations which are a combination of base functions (2.112) for each point to be interpolated P(t):

$$\mathrm{P}(t) = \begin{cases} x(t) = \sum_{i=0}^{k} x_i B_{i,n}(t) \\ y(t) = \sum_{i=0}^{k} y_i B_{i,n}(t) \end{cases} \qquad 0 \le t \le k - n + 1 \tag{2.116}$$

It is obvious that the spline base functions are directly "weighted" by the coordinates of the vertices instead of the computed coefficients. The smoothness of the curve is controlled by the order $m = n+1$, whereby the curve becomes smoother with increasing order. Fig. 2.24c and d show B-spline approximations of order $m = 3$ and $m = 5$. In addition, the computed curve always lies inside the envelope of the polygon, in contrast to normal spline or polynomial interpolation. Moreover, the approach can be extended directly to three-dimensional polygons (surface elements).

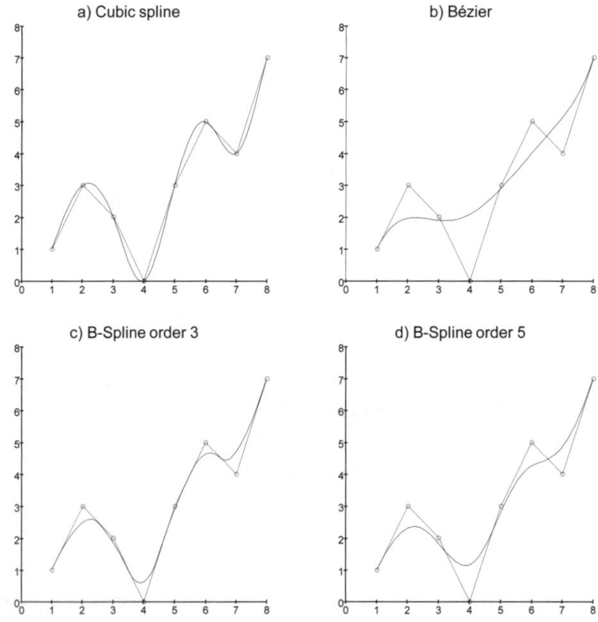

Fig. 2.24: Spline interpolation, and Bézier and B-spline approximation

Bézier approximation

The Bézier approximation has been developed by the car industry. Here a given polygon is approximated by a curve that has optimal smoothness but does not pass through the vertices. The approximation

$$P(t) = \begin{cases} x(t) = \sum_{i=0}^{k} x_i BE_{i,k}(t) \\ y(t) = \sum_{i=0}^{k} y_i BE_{i,k}(t) \end{cases} \qquad 0 \le t \le 1 \tag{2.117}$$

is similar to the B-spline approximation but is based on the Bernstein polynomials

$$BE_{i,k}(t) = \frac{k!}{i!(k-i)!} t^i (1-t)^{k-i} \qquad 0 \le t \le 1 \tag{2.118}$$

All points in the polygon data set are used for the computation of the curve. The approach can be extended directly to three-dimensional polygons (see section 2.3.3.2).

Fig. 2.24b shows the curve which results from Bézier approximation of a polygon. The continuous curve does not pass through the vertices, but shows an "averaged" shape. Bézier curves are therefore very suitable for applications where the data points are not free of error and smoothing is required.

2.3.2 Analytical geometry in 3D space

2.3.2.1 Straight line

The form of a straight line in 3D space can be derived directly from the straight line in 2D space. Thus a straight line between two points $P_1(x_1,y_1,z_1)$ and $P_2(x_2,y_2,z_2)$ is given by the proportional relationships:

$$\frac{x - x_1}{x_2 - x_1} = \frac{y - y_1}{y_2 - y_1} = \frac{z - z_1}{z_2 - z_1} \tag{2.119}$$

and in parametric form:

$$\mathbf{x} = \mathbf{x}_1 + t \cdot (\mathbf{x}_2 - \mathbf{x}_1)$$
$$\begin{bmatrix} x \\ y \\ z \end{bmatrix} = \begin{bmatrix} x_1 \\ y_1 \\ z_1 \end{bmatrix} + t \cdot \begin{bmatrix} x_2 - x_1 \\ y_2 - y_1 \\ z_2 - z_1 \end{bmatrix} = \begin{bmatrix} x_0 \\ y_0 \\ z_0 \end{bmatrix} + t \cdot \begin{bmatrix} a \\ b \\ c \end{bmatrix} \tag{2.120}$$

Here $P_0(x_0,y_0,z_0)$ is any point on the line. The direction cosines are defined by

$$\cos\alpha = \frac{x_2 - x_1}{d} = a$$
$$\cos\beta = \frac{y_2 - y_1}{d} = b \qquad \text{where } d = \sqrt{(x_2 - x_1)^2 + (y_2 - y_1)^2 + (z_2 - z_1)^2} \tag{2.121}$$
$$\cos\gamma = \frac{z_2 - z_1}{d} = c$$

At first glance it looks as though there are 6 independent parameters for a straight line in 3D space. However, taking into account the condition

$$a^2 + b^2 + c^2 = 1$$

there are only two direction parameters that are linearly independent. In addition, the coordinate z_0 of a point on the straight line can be derived from the corresponding x_0 and y_0 coordinates. Hence, 4 independent parameters remain in order to describe a straight line in space:

1. direction vector: $(a,b,1)$
2. point on the line: $z_0 = -a \cdot x_0 - b \cdot y_0$ (2.122)

For numerical reasons these two criteria are only valid for straight lines which are approximately vertical (parallel to the z axis). Arbitrarily oriented straight lines must therefore first be transformed into a vertical direction.

Intersection of two straight lines

The intersection point of two straight lines in space only exists if both lines lie in a common plane, otherwise the lines are skew. In this case the shortest distance e between them is defined along a direction which is perpendicular to both. For two lines g_i, $i = 1 \ldots 2$, each defined by a point $P_i(x_i, y_i, z_i)$ and direction cosine a_i, b_i, c_i the shortest distance e between them is given by:

$$e = \frac{\pm \begin{vmatrix} x_1 - x_2 & y_1 - y_2 & z_1 - z_2 \\ a_1 & b_1 & c_1 \\ a_2 & b_2 & c_2 \end{vmatrix}}{\sqrt{a^2 + b^2 + c^2}} \tag{2.123}$$

where

$$a = \begin{vmatrix} a_1 & b_1 \\ a_2 & b_2 \end{vmatrix} \qquad b = \begin{vmatrix} b_1 & c_1 \\ b_2 & c_2 \end{vmatrix} \qquad c = \begin{vmatrix} c_1 & a_1 \\ c_2 & a_2 \end{vmatrix}$$

For consistency, the point of intersection S is then defined at half this distance, $e/2$, between both lines (Fig. 2.25). Using the factors

$$\lambda = -\frac{\begin{vmatrix} x_1 - x_2 & y_1 - y_2 & z_1 - z_2 \\ a & b & c \\ a_2 & b_2 & c_2 \end{vmatrix}}{\begin{vmatrix} a_1 & b_1 & c_1 \\ a_2 & b_2 & c_2 \\ a & b & c \end{vmatrix}} \qquad \mu = -\frac{\begin{vmatrix} x_1 - x_2 & y_1 - y_2 & z_1 - z_2 \\ a & b & c \\ a_1 & b_1 & c_1 \end{vmatrix}}{\begin{vmatrix} a_1 & b_1 & c_1 \\ a_2 & b_2 & c_2 \\ a & b & c \end{vmatrix}}$$

the spatial coordinates of points S_1 and S_2 at the ends of the perpendicular reduce to

$$\begin{aligned} x_{S1} &= x_1 + \lambda \cdot a_1 & x_{S2} &= x_2 + \mu \cdot a_2 \\ y_{S1} &= y_1 + \lambda \cdot b_1 & y_{S2} &= y_2 + \mu \cdot b_2 \\ z_{S1} &= z_1 + \lambda \cdot c_1 & z_{S2} &= z_2 + \mu \cdot c_2 \end{aligned}$$

and hence the point of intersection S:

$$x_S = \frac{x_{S1} + x_{S2}}{2} \qquad y_S = \frac{y_{S1} + y_{S2}}{2} \qquad z_S = \frac{z_{S1} + z_{S2}}{2} \tag{2.124}$$

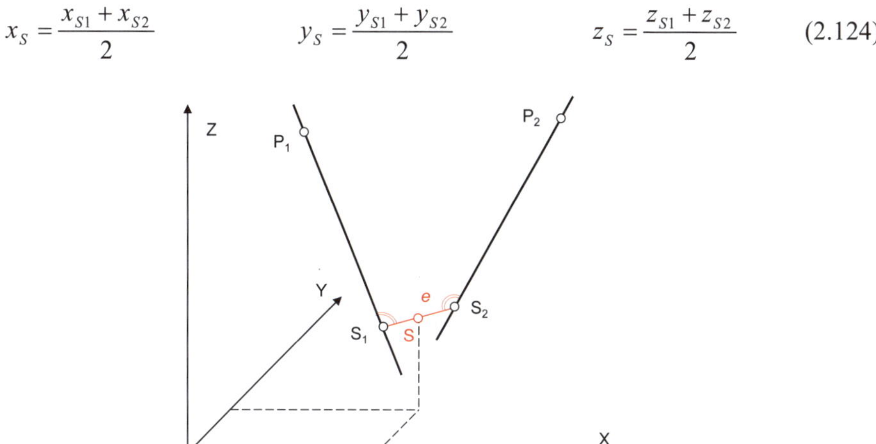

Fig. 2.25: Intersection of two straight lines in 3D space

The intersection angle φ between both lines is given by:

$$\cos\varphi = a_1 \cdot a_2 + b_1 \cdot b_2 + c_1 \cdot c_2 \tag{2.125}$$

The intersection of two straight lines in space is used for spatial intersection in stereo photogrammetry (see section 4.3.6.2). Here the distance *e* provides a quality measure for the intersection.

Regression line in space

The calculation of a best-fit straight line in space can be derived directly from the algorithm presented in section 2.3.1.1. The distance of a point $P_i(x_i,y_i,z_i)$ from the straight line defined by the point $P_0(x_0,y_0,z_0)$ and the direction cosine a,b,c is given by:

$$d_i = \sqrt{u_i^2 + v_i^2 + w_i^2} \tag{2.126}$$

where

$$u_i = c(y_i - y_0) - b(z_i - z_0)$$
$$v_i = a(z_i - z_0) - c(x_i - x_0)$$
$$w_i = b(x_i - x_0) - a(y_i - y_0)$$

The fitted line passes through P_0, the centroid of all points on the line. As in the two-dimensional case, the spatial direction of the line is defined by the eigenvector which corresponds to the largest eigenvalue of the matrix **B**:

$$\mathbf{B} = \mathbf{A}^T \cdot \mathbf{A}$$

$$\text{where} \quad \mathbf{A} = \begin{bmatrix} x_1 - x_0 & y_1 - y_0 & z_1 - z_0 \\ \vdots & \vdots & \\ x_n - x_0 & y_n - y_0 & z_n - z_0 \end{bmatrix} \tag{2.127}$$

2.3.2.2 Plane

Parameters

A plane in space is defined by $n \geq 3$ points which must not lie on a common straight line. The analytical form of a plane is given by:

$$A \cdot x + B \cdot y + C \cdot z + D = 0 \tag{2.128}$$

A plane in 3D space is therefore analogous to a straight line in 2D (see eqn. 2.72 for comparison). The vector $\mathbf{n}(A,B,C)$ is defined as vector normal to the plane with direction cosines

$$\cos\alpha = \frac{A}{\sqrt{A^2 + B^2 + C^2}} = a$$
$$\cos\beta = \frac{B}{\sqrt{A^2 + B^2 + C^2}} = b \tag{2.129}$$
$$\cos\gamma = \frac{C}{\sqrt{A^2 + B^2 + C^2}} = c$$

Given a point $P_0(x_0,y_0,z_0)$ on the plane with normal vector having direction cosines (a,b,c), then all points $P(x,y,z)$ on the plane are defined by the following equation:

$$a(x-x_0)+b(y-y_0)+c(z-z_0)=0 \tag{2.130}$$

Given a plane that is formed by 3 points P_1,P_2,P_3, any other point P on the plane meets the condition (Fig. 2.26):

$$\begin{vmatrix} x-x_1 & y-y_1 & z-z_1 \\ x_2-x_1 & y_2-y_1 & z_2-z_1 \\ x_3-x_1 & y_3-y_1 & z_3-z_1 \end{vmatrix}=0 \tag{2.131}$$

This determinant corresponds to the volume of a parallelepiped defined by its three vectors. It can be taken as a definition of the coplanarity condition used in relative orientation (see section 4.3.3.1).

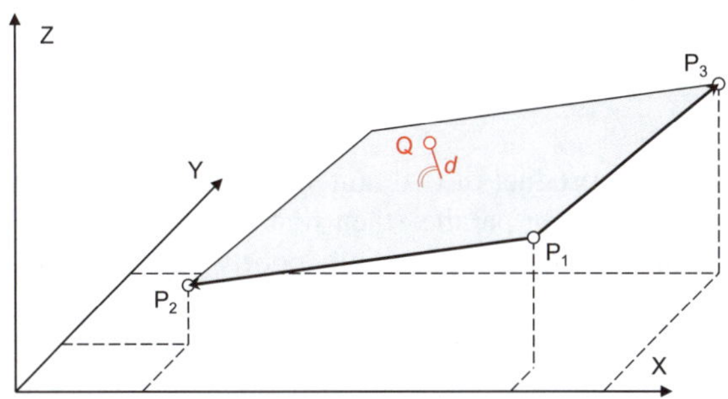

Fig. 2.26: Definition of a plane in space

The distance of a point $Q(x,y,z)$ from the plane is given by (see eqn. 2.130 for comparison):

$$d=a(x-x_0)+b(y-y_0)+c(z-z_0) \tag{2.132}$$

Intersection of line and plane

Given a straight line defined by point $P(x_G,y_G,z_G)$ and direction cosines (a_G,b_G,c_G)

$$\begin{bmatrix} x \\ y \\ z \end{bmatrix}=\begin{bmatrix} x_G \\ y_G \\ z_G \end{bmatrix}+t\cdot\begin{bmatrix} a_G \\ b_G \\ c_G \end{bmatrix} \tag{2.133}$$

and a plane defined by point $P(x_E,y_E,z_E)$ and direction cosines (a_E,b_E,c_E):

$$a_E(x-x_E)+b_E(y-y_E)+c_E(z-z_E)=0 \tag{2.134}$$

Substituting in (2.134) for the variable point from (2.133), the solution for line parameter *t* is:

$$t = \frac{a_E(x_E - x_G) + b_E(y_E - y_G) + c_E(z_E - z_G)}{a_E a_G + b_E b_G + c_E c_G} \tag{2.135}$$

The denominator becomes zero if the line is parallel to the plane. The coordinates of the point of intersection are obtained if the solution for *t* is substituted in (2.133).

As an example, the intersection of line and plane is used in photogrammetry for single image analysis in conjunction with object planes (see section 4.2.7.1).

Intersection of two planes

The intersection line of two non-parallel planes has a direction vector $\mathbf{a}(a,b,c)$ which is perpendicular to the unit vectors $\mathbf{n}_1$ and $\mathbf{n}_2$ normal to the planes and can be calculated directly as the vector product of $\mathbf{n}_1$ and $\mathbf{n}_2$:

$$\mathbf{a} = \mathbf{n}_1 \times \mathbf{n}_2 \tag{2.136}$$

The magnitude of the vector product of two unit vectors is the sine of the angle θ between them. If the planes are identical, or parallel, then $\sin\theta = 0$ and the intersection line does not exist. A small value of $\sin\theta$ indicates a potentially poorly defined intersection line.

A point $\mathbf{x}_0$ on the line of intersection is defined by where it intersects a principal coordinate plane, e.g. the xy plane. The intersection line is then given in parametric form:

$$\mathbf{x} = \mathbf{x}_0 + t \cdot \mathbf{a}$$

Best-fit plane

In analogy with best-fitting lines, the best-fit plane is calculated by minimising the distances d_i in eqn. 2.132. The adjusted plane that fits n points is defined by the centroid P_0 of the points and the direction cosines of the normal vector. The matrix **A**, used for the computation of eigenvalues, is identical to the matrix given in (2.127). However, here the direction cosines correspond to the eigenvector with the minimum eigenvalue.

2.3.2.3 Rotationally symmetric shapes

The measurement of rotationally symmetric shapes is of major importance, especially in industrial metrology. They have the common property that they can be described by a single reference axis (straight line in space) and one or more shape parameters (see Table 2.1).

These shapes (3D circle, sphere and cylinder) are often used in practical applications of close-range photogrammetry and are explained below in more detail. For the analysis of other rotationally symmetric shapes (paraboloid, ellipsoid, cone etc.) the reader is directed to further references.

Table 2.1: 3D rotationally symmetric shapes (selection)

Shape	Parameters	Number of points
sphere	centre point x_0,y_0,z_0 radius r	≥ 3
3D circle	centre point x_0,y_0,z_0 normal vector l,m,n radius r	≥ 3
cylinder	axis point x_0,y_0,z_0 direction vector l,m,n radius r	≥ 5

Sphere

A sphere (Fig. 2.27) is defined by

- the centre (x_0,y_0,z_0),
- the radius r.

Here there are 4 independent parameters which require 4 observation equations for a solution. Therefore, a minimum of 4 points must be measured on the surface of the sphere to generate these. The 4 points must not all lie on the same circle. (Any 3 will lie on a circle and the fourth point must lie off the plane of this circle.)

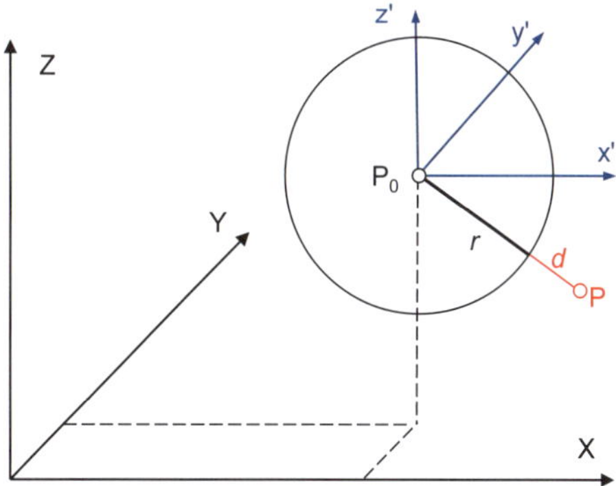

Fig. 2.27: Definition of a sphere

The equation of a sphere with centre (x_0,y_0,z_0) and radius r is given by:

$$(x-x_0)^2+(y-y_0)^2+(z-z_0)^2=r^2 \tag{2.137}$$

Alternatively, a general equation for the circle is as follows:

$$x^2 + y^2 + z^2 + 2ux + 2vy + 2wz + d = 0 \tag{2.138}$$

which can be re-arranged as:

$$(x+u)^2 + (y+v)^2 + (z+w)^2 = u^2 + v^2 + w^2 - d \tag{2.139}$$

(2.139) has the same general form as (2.137) with the centre at $(-u,-v,-w)$ and radius r given by:

$$r = \sqrt{u^2 + v^2 + w^2 - d} \tag{2.140}$$

Expressed in the form of (2.138), the equation is linear in the parameters u, v, w, d and can be used to compute their values directly by substituting coordinates of 4 well-chosen points to create four independent equations. From these, initial values of sphere centre and radius can be derived as indicated above.

For the over-determined case, the distance of a point from the surface of the sphere is given by:

$$d_i = r_i - r \tag{2.141}$$

$$\text{where } r_i = \sqrt{(x_i - x_0)^2 + (y_i - y_0)^2 + (z_i - z_0)^2}$$

The derivatives for use in the design matrix are then given by:

$$\frac{\partial d_i}{\partial x_0} = \frac{x_0 - x_i}{r_i} \qquad \frac{\partial d_i}{\partial y_0} = \frac{y_0 - y_i}{r_i} \qquad \frac{\partial d_i}{\partial z_0} = \frac{z_0 - z_i}{r_i} \qquad \frac{\partial d_i}{\partial r} = -1 \tag{2.142}$$

Circle in 3D space

A 3D circle is a circle located in an arbitrarily oriented plane in space (Fig. 2.28). It is defined by

- the centre (x_0,y_0,z_0),
- the direction cosines (a,b,c) of the normal vector to the plane
- the radius r.

With regard to (2.122), a 3D circle in an approximately horizontal (xy) plane is defined by 6 independent parameters. A minimum number of 6 observations is therefore required to compute the parameters. These can be provided by 3 points on the circumference.

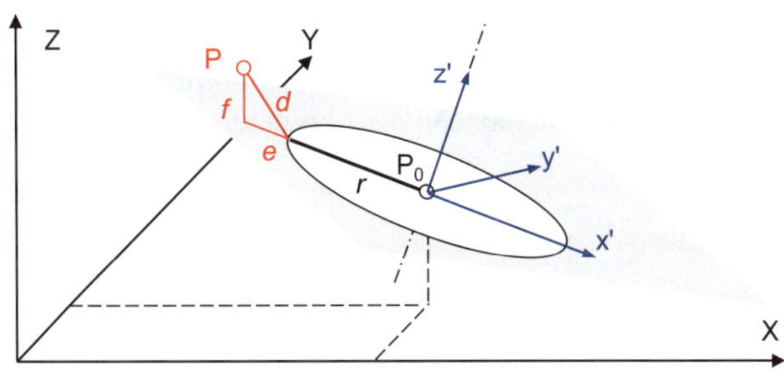

Fig. 2.28: Definition of a 3D circle

Analogously to the best-fit cylinder, a distance can be defined from a fitted point in space to the circle circumference. Here the point not only has a radial distance to the circle but also a perpendicular distance to the plane. The spatial distance is given by

$$d_i^2 = e_i^2 + f_i^2 \tag{2.143}$$

Here e_i is the radial distance analogous to the definition in (2.89) and f_i is the distance from the plane according to (2.132). In order to calculate a best-fit circle, both components must be minimised.

$$\sum_{i=1}^{n} d_i^2 = \sum_{i=1}^{n} e_i^2 + \sum_{i=1}^{n} f_i^2 \tag{2.144}$$

Defining $c = 1$ as in (2.122), there are 6 remaining parameters x_0, y_0, z_0, a, b and r. For the special case $x_0 = y_0 = z_0 = a = b = 0$ the derivatives forming the elements of the design matrix are:

$$\begin{aligned}
\frac{\partial e_i}{\partial x_0} &= -\frac{x_i}{r_i} & \qquad \frac{\partial e_i}{\partial a} &= -\frac{x_i z_i}{r_i} \\
\frac{\partial e_i}{\partial y_0} &= -\frac{y_i}{r_i} & \qquad \frac{\partial e_i}{\partial b} &= -\frac{y_i z_i}{r_i} \\
\frac{\partial e_i}{\partial z_0} &= -\frac{z_i}{r_i} & \qquad \frac{\partial e_i}{\partial r} &= -1
\end{aligned}$$

$$\begin{aligned}
\frac{\partial f_i}{\partial x_0} &= 0 & \qquad \frac{\partial f_i}{\partial a} &= x_i \\
\frac{\partial f_i}{\partial y_0} &= 0 & \qquad \frac{\partial f_i}{\partial b} &= y_i \\
\frac{\partial f_i}{\partial z_0} &= -1 & \qquad \frac{\partial f_i}{\partial r} &= 0
\end{aligned} \tag{2.145}$$

The parameters are determined iteratively in the way described for a best-fit cylinder.

As in the case of the cylinder, discrete points on the circumference of the 3D circle are not necessarily required and its parameters can also be determined by direct measurement of the circle edges (see section 4.4.7.2).

Cylinder

A cylinder (Fig. 2.29) is defined by

- one point (x_0,y_0,z_0) on the axis of the cylinder
- the direction vector (a,b,c) of the axis
- the radius r.

With reference to (2.122), the position and direction of an approximately vertical axis is defined by 4 parameters. Including the radius, an approximately vertical cylinder therefore requires 5 independent parameters. A minimum of 5 observations are therefore required to compute these parameters, i.e. a minimum of 5 points on the cylinder surface are necessary.

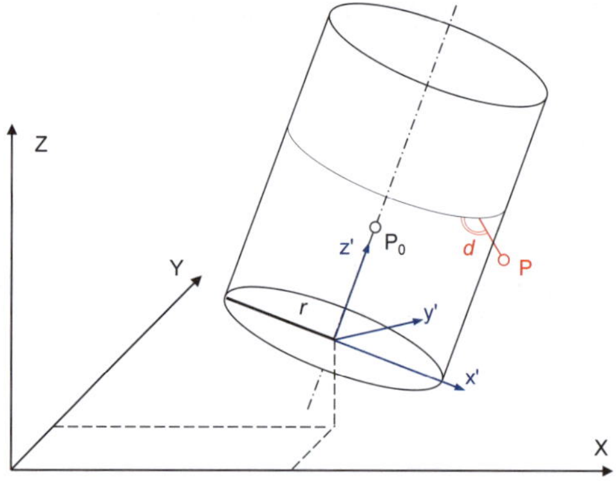

Fig. 2.29: Definition of a cylinder

The distance of a point from the cylinder surface is given by:

$$d_i = r_i - r$$

where

$$r_i = \frac{\sqrt{u_i^2 + v_i^2 + w_i^2}}{\sqrt{a^2 + b^2 + c^2}} \tag{2.146}$$

and

$$u_i = c(y_i - y_0) - b(z_i - z_0)$$
$$v_i = a(z_i - z_0) - c(x_i - x_0)$$
$$w_i = b(x_i - x_0) - a(y_i - y_0)$$

In the special case where $x_0 = y_0 = a = b = 0$ the above relations simplify to

$$r_i = \sqrt{x_i^2 + y_i^2}$$

The derivatives required to set up the design matrix **A** are given by:

$$\frac{\partial d_i}{\partial x_0} = -\frac{x_i}{r_i} \qquad \frac{\partial d_i}{\partial y_0} = -\frac{y_i}{r_i} \qquad \frac{\partial d_i}{\partial r} = -1 \tag{2.147}$$
$$\frac{\partial d_i}{\partial a} = -\frac{x_i \cdot z_i}{r_i} \qquad \frac{\partial d_i}{\partial b} = -\frac{y_i \cdot z_i}{r_i}$$

The following procedure provides one possible algorithm for the computation of a best-fit cylinder with given initial values of (x_0,y_0,z_0), (a,b,c) and r. These initial values can be found, for example, by first measuring 3 points which are well distributed near a circle of the cylinder. A circumscribing circle to these points then provides the necessary initial parameter values.

1. Translate the data points P_i onto a local origin near the axis
2. Rotate the axis into an approximately vertical (z) direction with rotation matrix $\mathbf{R}(a,b,c)$:

   Steps 1 and 2 temporarily transform an arbitrarily oriented cylinder into a local system[1] x'y'z' where the transformed cylinder is oriented vertically and the axis passes through the origin (see Fig. 2.29)
3. Set up and solve the normal system of equations using (2.147)
4. Correct the unknowns and reverse the transformation back into the original coordinate system

Steps 1–4 are repeated until the unknowns do not change appreciably.

An example of the use of cylinders in close-range photogrammetry is in process plant (pipeline) modelling. Note that discrete 3D points on the cylinder surface are not necessarily required and a cylinder can also be determined by photogrammetric measurement of its edges in an image (see section 4.4.7.2).

2.3.3 Surfaces

Objects with surfaces which cannot be described by the above geometric elements are, in the first instance, usually represented by a dense distribution of 3D surface points. From these 3D point clouds, triangular mesh generation can create digital surface models of suitable detail. Analytical functions can also be used in a similar way to polynomials (see section 2.3.1.4) in order to approximate the shape of the surface.

[1] Exceptionally the notation x'y'z' is used. This is normally reserved for image coordinates.

Fig. 2.30: Example of a 2½D surface **Fig. 2.31:** Example of a 3D surface

Surfaces which can be defined as a function $Z = f(X,Y)$ are known as *2½D surfaces*. Here every point on a horizontal XY plane is related to exactly one unique height value Z. Terrain models and simple component surfaces are examples of 2½D surfaces (Fig. 2.30). In contrast, objects with holes and occlusions have true *3D surfaces* where a point on the surface is defined by a function $f(X,Y,Z) = 0$. A sculpture or cup with a handle are examples for such 3D objects (Fig. 2.31).

2.3.3.1 Digital surface model

A 3D point cloud represents a *digital surface model* (DSM) if its point density (grid spacing) is sufficient for describing changes in surface shape. The point distribution can have a regular structure (e.g. $\Delta X = \Delta Y = \text{const.}$) or an irregular spacing. Object edges (breaklines) can be represented by special point codes or by additional vector-based data such as polygons.

Triangle meshing

The simplest way to generate a closed surface from the point cloud is by triangle meshing (Fig. 2.32), where every three adjacent 3D points combine to form a triangular surface element. Delaunay triangle meshing offers an appropriate method of creating such a triangular mesh. This identifies groups of three neighbouring points whose maximum inscribed circle does not include any other surface point.

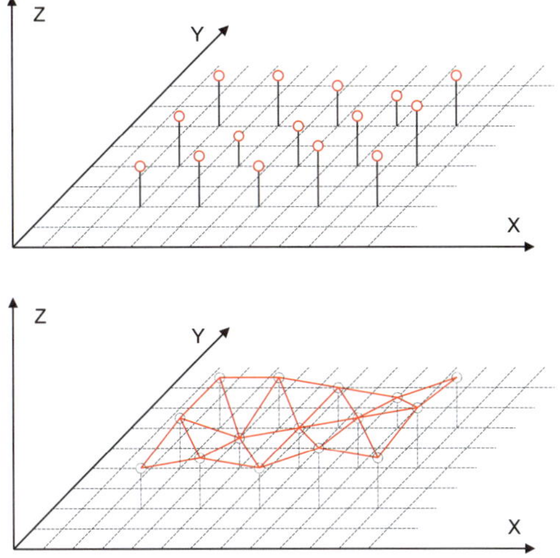

Fig. 2.32: Triangle mesh from a 3D point cloud

Each triangle can be defined as a plane in space using eqn. 2.130 and the result is a polyhedron representation or wire-frame model of the surface.

There is normally no topological relation between the 3D points. Differential area elements must be established between adjacent points in order to generate a topologically closed surface which enables further processing as a surface description. The approximation of a surface by small planar surface elements has the advantage that it is easy to perform further calculations of, say, normal vectors or intersections with straight lines. These are required, for example, in visualisation using *ray tracing* techniques (see section 5.3.3.1). If triangular elements rather than polygons are used for surface descriptions then the planarity of surface elements is guaranteed.

Most commonly a triangular mesh is stored as a collection of triplets, where each triplet represents the three corner points of the triangle. Each corner point is represented again as a triplet of its X, Y and Z coordinates. One example of a popular file format based on this representation is the STL (stereo lithography) format used for rapid prototyping and 3D printing. It creates a block for each triangle consisting of the normal vector (to discern the inside from the outside of the object) and the vertex triplet. The content of a STL file is shown in Fig. 2.33 on the left. One problem with this format can be seen when comparing the first vertex of the two triangles stored. Obviously the format duplicates vertices which are shared by neighbouring triangles. This is not an efficient use of memory. An alternative way to represent a triangle mesh is to keep a separate list of unique vertices. The list of triangles then stores the corner points as a triplet of indices to the list of vertices. A popular file format using this scheme is the PLY format (polygon file format). Directly following the header is a list of coordinate triplets for the vertices. This is followed by a list of polygons. Each polygon starts with the number of vertices and then contains one index for each vertex. Again an example is provided in Fig. 2.33 on the right. From the list of vertex indices (2, 0, 1 and 2, 3, 0) it can be seen that the two triangles share two vertices (indexed 0 and 2). Thus they actually share an edge. Such formats are commonly referred to as indexed triangle meshes. A complementary form of representation centred on edges is the half-edge structure, which will not be detailed here.

STL ASCII file:

```
solid example
  facet normal -0.282 0.312 0.991
    outer loop
      vertex  -70.312 347.656 -736.759
      vertex  -70.313 345.269 -735.94
      vertex  -67.665 347.656 -735.93
    endloop
  endfacet
  facet normal -0.849 0.172 0.500
    outer loop
      vertex  -70.312 347.656 -736.759
      vertex  -72.130 347.656 -739.843
      vertex  -70.313 345.269 -735.938
    endloop
  endfacet
```

PLY ASCII file:

```
ply
format ascii 1.0
element vertex 4
property float x
property float y
property float z
element face 2
property list uchar int vertex_indices
end_header
-70.313 345.269 -735.938
-67.665 347.656 -735.938
-70.313 347.656 -736.759
-72.130 347.656 -739.844
3 2 0 1
3 2 3 0
```

Fig. 2.33: Two file formats storing the same triangle mesh (four vertices and two triangles)

Interpolation

Additional points can easily be interpolated within a given triangular element. For a tilted plane defined in a local coordinate system x'y'z', with origin located in one of the vertices of the triangle (Fig. 2.34), then the equation for the plane is given by

$$z = a_0 + a_1 x' + a_2 y' \tag{2.148}$$

The coefficients can be calculated as follows:

$$\begin{bmatrix} a_0 \\ a_1 \\ a_2 \end{bmatrix} = \frac{1}{x'_2 y'_3 - x'_3 y'_2} \begin{bmatrix} x'_2 y'_3 - x'_3 y'_2 & 0 & 0 \\ y'_2 - y'_3 & y'_3 & -y'_2 \\ x'_3 - x'_2 & -x'_3 & x'_2 \end{bmatrix} \cdot \begin{bmatrix} z_1 \\ z_2 \\ z_3 \end{bmatrix} \tag{2.149}$$

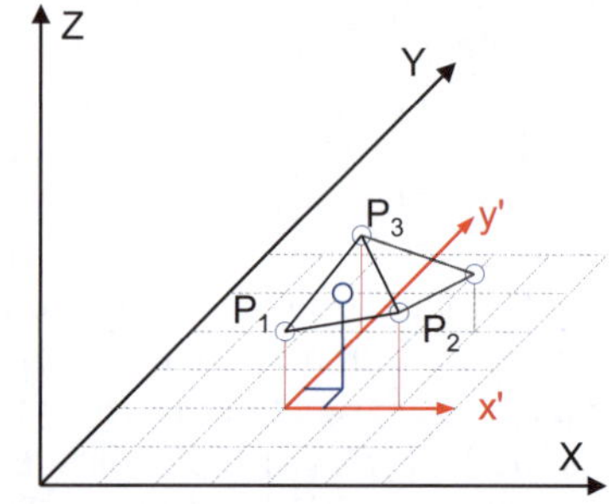

Fig. 2.34: Interpolation within a triangular mesh

For meshes defined by four points, additional points can be calculated by bilinear interpolation according to eqn. 2.15.

2.3.3.2 B-spline and Bézier surfaces

Three-dimensional surfaces can be represented directly by a general form of the B-spline used for curves in a plane (see section 2.3.1.4). Given a three-dimensional network of m+1 x n+1 nodes (surface model, see Fig. 2.32), a B-spline surface approximation gives:

$$Q(s,t) = \begin{cases} x(s,t) = \sum_{i=0}^{m} \sum_{j=0}^{n} x_{ij} B_{i,\alpha}(s) B_{j,\beta}(t) \\ y(s,t) = \sum_{i=0}^{m} \sum_{j=0}^{n} y_{ij} B_{i,\alpha}(s) B_{j,\beta}(t) \\ z(s,t) = \sum_{i=0}^{m} \sum_{j=0}^{n} z_{ij} B_{i,\alpha}(s) B_{j,\beta}(t) \end{cases} \tag{2.150}$$

where

$0 \leq s \leq m - \alpha + 1$ and $0 \leq t \leq n - \beta + 1$

The result is a quadratic approximation when $\alpha = \beta = 2$ and a cubic spline approximation when $\alpha = \beta = 3$. The determination of the basic functions B is equivalent to the two-dimensional case.

In an analogous way, Bézier approximations can be generated for 3D elements. They are mainly used for the construction of industrial free-form surfaces, for example in the automotive industry for the representation of car body surfaces.

2.3.4 Compliance with design

In industrial dimensional metrology, the measurement of geometric elements is normally related to a component's functionality, i.e. the measurement result is used to analyse how well the measured object meets its design function. The elements discussed above, which are a least-squares best-fit (L2 approach, see section 2.4.2), may be used only in cases when the measured point coordinates have random, normally distributed errors. Non-systematic errors in the production of components might not be detected due to the smearing effect of the L2 approach. Depending on the function of a component, additional geometric descriptions are required (see Table 2.2).

The *minimum zone element* meets the condition that the maximum deviation between component surface and calculated element is a minimum. The *minimum circumscribed element* is the calculated element that includes all measured points within a minimum area. The *maximum inscribed element* is the maximum size of the element that excludes all measured points.

Table 2.2: Selected approximation criteria for standard form elements (according to Weckenmann 1993)

Element	Best-fit circle	Minimum zone	Minimum circumscribed	Maximum inscribed
point	●		●	
line	●	●		
circle	●	●	●	●
plane	●	●		
sphere	●	●	●	●
cylinder	●	●	●	●
cone	●	●	●	●

Fig. 2.35 shows different estimations of a circle applied to a measured bore. If the bore is to be matched to a cylindrical shaft, the circular shape of the bore must fulfil the minimum circumscribed condition whilst the maximum inscribed condition must be met by the shaft.

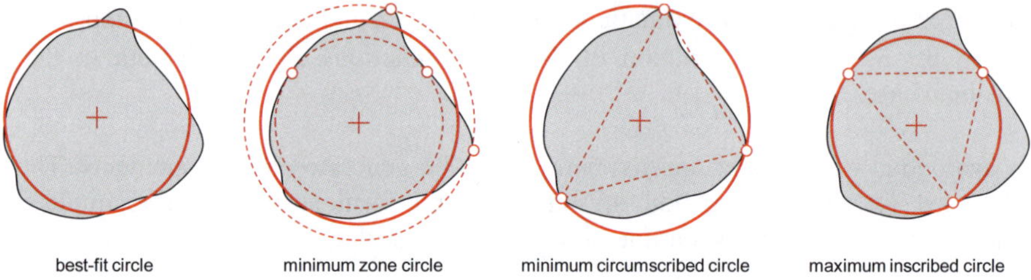

Fig. 2.35: Different estimates of a circle calculated for a bore (according to Weckenmann 1993)

2.4 Adjustment techniques

2.4.1 The problem

This section provides a summary of some important techniques for the computation of over-determined, non-linear systems of equations by adjustment methods. These are essential for the understanding of numerous photogrammetric calculations. In general the task is to determine a number of unknown parameters from a number of observed (measured) values which have a functional relationship to each other. If more observations are available than required for the determination of the unknowns, there is normally no unique solution and the unknown parameters are estimated according to functional and stochastic models. See specialist literature for a more detailed explanation of adjustment methods and applications.

2.4.1.1 Functional model

A number of observations n (measured values) form an observation vector **L**:

$$\mathbf{L} = (L_1, L_2, ..., L_n)^T \qquad \text{: observation vector} \qquad (2.151)$$

Since the elements of the observation vector are measured data they are regarded as having small random error effects but are free of systematic defects.

A number u of unknown parameters must be determined. These form the *vector of unknowns* **X**, also called the parameter vector.

$$\mathbf{X} = (X_1, X_2, ..., X_u)^T \qquad \text{: vector of unknowns} \qquad (2.152)$$

The number of observations is assumed to be greater than the number of unknowns.

$$n > u$$

The *functional model* describes the relation between the "true" observation values $\tilde{\mathbf{L}}$ and the "true" values of the unknowns $\tilde{\mathbf{X}}$. This relationship is expressed by the vector of functions $\boldsymbol{\varphi}$ of the unknowns:

$$\tilde{\mathbf{L}} = \boldsymbol{\varphi}(\tilde{\mathbf{X}}) = \begin{bmatrix} \varphi_1(\tilde{\mathbf{X}}) \\ \varphi_2(\tilde{\mathbf{X}}) \\ \vdots \\ \varphi_n(\tilde{\mathbf{X}}) \end{bmatrix} \qquad \text{: functional model} \qquad (2.153)$$

Since the true values are normally not known, the observation vector $\tilde{\mathbf{L}}$ is replaced by the sum of the measured observations **L** and corresponding vector of residuals **v**. Similarly the vector of unknowns is replaced by the estimated (adjusted) unknowns $\hat{\mathbf{X}}$. As a result, the following non-linear correction equations are obtained:

$$\hat{\mathbf{L}} = \mathbf{L} + \mathbf{v} = \boldsymbol{\varphi}(\hat{\mathbf{X}}) \qquad (2.154)$$

If approximate values $\mathbf{X}^0$ of the unknowns are available, the vector of unknowns can be expressed as the following sum:

$$\hat{\mathbf{X}} = \mathbf{X}^0 + \hat{\mathbf{x}} \qquad (2.155)$$

i.e. only the small unknown values $\hat{\mathbf{x}}$ must be determined.

From the values in $\mathbf{X}^0$, approximate values of the observations can then be calculated using the functional model:

$$\mathbf{L}^0 = \boldsymbol{\varphi}(\mathbf{X}^0) \qquad (2.156)$$

In this way *reduced observations* (observed minus computed) are obtained:

$$\mathbf{l} = \mathbf{L} - \mathbf{L}^0 \tag{2.157}$$

For sufficiently small values of $\hat{\mathbf{x}}$, the correction equations can be expanded into a Taylor series around the approximate values $\mathbf{X}^0$, ignoring terms after the first:

$$\begin{aligned}\mathbf{L} + \mathbf{v} &= \varphi(\mathbf{X}^0) + \left(\frac{\partial \varphi(\mathbf{X})}{\partial \mathbf{X}}\right)_0 \cdot (\hat{\mathbf{X}} - \mathbf{X}^0) \\ &= \mathbf{L}^0 + \left(\frac{\partial \varphi(\mathbf{X})}{\partial \mathbf{X}}\right)_0 \cdot \hat{\mathbf{x}}\end{aligned} \tag{2.158}$$

After introduction of the *Jacobian matrix* **A**, also known as the design, model or coefficient matrix:

$$\underset{n,u}{\mathbf{A}} = \left(\frac{\partial \varphi(\mathbf{X})}{\partial \mathbf{X}}\right)_0 = \begin{bmatrix} \left(\frac{\partial \varphi_1(\mathbf{X})}{\partial X_1}\right)_0 & \left(\frac{\partial \varphi_1(\mathbf{X})}{\partial X_2}\right)_0 & \cdots & \left(\frac{\partial \varphi_1(\mathbf{X})}{\partial X_u}\right)_0 \\ \left(\frac{\partial \varphi_2(\mathbf{X})}{\partial X_1}\right)_0 & \left(\frac{\partial \varphi_2(\mathbf{X})}{\partial X_2}\right)_0 & \cdots & \left(\frac{\partial \varphi_2(\mathbf{X})}{\partial X_u}\right)_0 \\ \vdots & \vdots & \ddots & \vdots \\ \left(\frac{\partial \varphi_n(\mathbf{X})}{\partial X_1}\right)_0 & \left(\frac{\partial \varphi_n(\mathbf{X})}{\partial X_2}\right)_0 & \cdots & \left(\frac{\partial \varphi_n(\mathbf{X})}{\partial X_u}\right)_0 \end{bmatrix} \tag{2.159}$$

the linearised correction equations are obtained:

$$\underset{n,1}{\hat{\mathbf{l}}} = \underset{n,1}{\mathbf{l}} + \underset{n,1}{\mathbf{v}} = \underset{n,u}{\mathbf{A}} \cdot \underset{u,1}{\hat{\mathbf{x}}} \tag{2.160}$$

The Jacobian matrix **A** consists of derivatives which describe the functional relation between the parameters and which are calculated from approximate values. The vector of unknowns $\hat{\mathbf{x}}$ contains the estimated parameters and **l** is the vector of reduced observations. A computation scheme is given in section 2.4.2.2

2.4.1.2 Stochastic model

The stochastic properties of the unknowns **L** are defined by the *covariance matrix* $\mathbf{C}_{ll}$:

$$\underset{n,n}{\mathbf{C}_{ll}} = \begin{bmatrix} \sigma_1^2 & \rho_{12}\sigma_1\sigma_2 & \cdots & \rho_{1n}\sigma_1\sigma_n \\ \rho_{21}\sigma_2\sigma_1 & \sigma_2^2 & \cdots & \rho_{2n}\sigma_2\sigma_n \\ \vdots & \vdots & \ddots & \vdots \\ \rho_{n1}\sigma_n\sigma_1 & & \cdots & \sigma_n^2 \end{bmatrix} \tag{2.161}$$

where σ_i : standard deviation of observation L_i, $i = 1..n$
ρ_{ij}: correlation coefficient between L_i and L_j, $i \neq j$

Introducing the multiplication factor σ_0^2, the *cofactor matrix* $\mathbf{Q}_{ll}$ of observations is obtained:

$$\mathbf{Q}_{ll} = \frac{1}{\sigma_0^2}\mathbf{C}_{ll} = \mathbf{P}^{-1} \tag{2.162}$$

where $\mathbf{P}_{ll}$ is the weight matrix

The covariance matrix is the only component containing information about the accuracy of the functional model in the adjustment process. It is therefore called the *stochastic model* (see section 2.4.3.1) In the case of independent observations, the correlation coefficients become zero and the covariance matrix is reduced to a diagonal matrix. This is the standard case for many adjustment problems where either independent observations are given, or no significant knowledge about correlations between observations is available.

The weight matrix **P** then becomes:

$$\underset{n,n}{\mathbf{P}} = \begin{bmatrix} \frac{\sigma_0^2}{\sigma_1^2} & & & \\ & \frac{\sigma_0^2}{\sigma_2^2} & & \\ & & \ddots & \\ & & & \frac{\sigma_0^2}{\sigma_n^2} \end{bmatrix} = \begin{bmatrix} p_1 & & & \\ & p_2 & & \\ & & \ddots & \\ & & & p_n \end{bmatrix} \tag{2.163}$$

In this case an observation L_i with standard deviation $\sigma_i = \sigma_0$ has weight

$$p_i = \frac{\sigma_0^2}{\sigma_i^2} = 1$$

and **P** becomes the identity matrix **I**. σ_0 is the true value of the *standard deviation of unit weight* (standard deviation of an observation with weight = 1). It can be regarded as a multiplication constant. Refer to sections 2.4.2.1 and 2.4.3.1 for a definition of this parameter.

Usually the true standard deviation σ is not known in practical applications and the empirical standard deviation s is used instead. Here s denotes the a priori standard deviation, while $\hat{s}$ represents the a posteriori standard deviation (adjusted standard deviation). The empirical standard deviation is only meaningful in cases of significant redundancy.

2.4.2 Least-squares method (Gauss-Markov linear model)

The Gauss-Markov adjustment model is based on the idea that the unknown parameters are estimated with maximum probability. Assuming a data set with an infinite number of measured values and normally distributed errors (non-centrality parameter $\Delta = 0$, i.e. no systematic errors), the following condition for the residuals results:

$$\mathbf{v}^T \cdot \mathbf{P} \cdot \mathbf{v} \rightarrow \min \tag{2.164}$$

For independent observations it reduces to

$$\sum_{i=1}^{n} p_i \cdot v_i^2 \rightarrow \min \tag{2.165}$$

It is known as a *least-squares adjustment* or *minimisation using the L2 norm*. The Gauss-Markov model ensures that estimations of the unknown parameters are unbiased and have minimum variance.

2.4.2.1 Adjustment of direct observations

Consider a number of direct measurements of a single unknown value, e.g. from repeated measurements of the distance between two points by laser range measurement. The functional model is then reduced to the extent that the required quantity is simply the mean of the observations.

In measurements where observations are considered to be equally accurate, the weights p_i are simplified to $p_i = 1$.

For observations of varying accuracy, the corresponding weights are estimated from the a priori standard deviations of the original observations and the observation of unit weight (s_i and s_0 respectively):

$$p_i = \frac{s_0^2}{s_i^2} \quad \text{: weight of observation } i \tag{2.166}$$

Alternatively, where measurements are considered to be equally accurate, and an improved value for a particular quantity is obtained by averaging a number of repeated measurements, then this improved average can be given a weight which corresponds to the number of measurements in the set. (A single measurement has weight 1, an average based on 6 repetitions has weight 6, etc.).

The estimated unknown is obtained by the geometric (weighted) average:

$$\hat{x} = \frac{p_1 l_1 + p_2 l_2 + \ldots + p_n l_n}{p_1 + p_2 + \ldots + p_n} = \frac{\sum_{i=1}^{n} p_i l_i}{\sum_{i=1}^{n} p_i} \tag{2.167}$$

The residual of an observation i gives:

$$v_i = \hat{x} - l_i \tag{2.168}$$

After adjustment the a posteriori standard deviation of unit weight is given by

$$\hat{s}_0 = \sqrt{\frac{\sum p \cdot v^2}{n-1}} \tag{2.169}$$

The a posteriori standard deviation of the original observation i is given by

$$\hat{s}_i = \frac{\hat{s}_0}{\sqrt{p_i}} \tag{2.170}$$

The standard deviation of the average value is, in this case, equal to the standard deviation of the adjusted observations:

$$\hat{s}_{\hat{x}} = \frac{\hat{s}_0}{\sqrt{[p]}} \tag{2.171}$$

2.4.2.2 General least squares adjustment

Normally values of interest must be measured indirectly. For example, photogrammetric triangulation by the intersection of measured directions produces, indirectly, the 3D coordinates of required target points. This section describes a generally applicable adjustment process.

Let the following linearised functions define an adjustment problem:

$$\underset{n,1}{\hat{\mathbf{l}}} = \underset{n,1}{\mathbf{l}} + \underset{n,1}{\mathbf{v}} = \underset{n,u}{\mathbf{A}} \cdot \underset{u,1}{\hat{\mathbf{x}}} \qquad \text{: functional model}$$

$$\mathbf{Q}_{ll} = \frac{1}{s_0^2}\mathbf{C}_{ll} = \mathbf{P}^{-1} \qquad \text{: stochastic model}$$

with n observations and u unknowns, $n>u$. To set up the weight matrix **P**, the a priori standard deviations of observations s_i, and the a priori standard deviation of unit weight s_0, are required. They could, for example, be derived from the empirically known accuracy of a measuring device:

$$p_i = \frac{s_0^2}{s_i^2} \qquad \text{: weight of observation } i$$

After generation of initial values, setting up of the Jacobian matrix **A**, and calculation of reduced observations **l**, the following computation scheme may be used in order to calculate the vector of unknowns $\hat{\mathbf{x}}$:

1) $$\underset{n,n}{\mathbf{P}} = \underset{n,n}{\mathbf{Q}_{ll}^{-1}} \qquad \text{: weight matrix}$$

2) $$\underset{u,u}{\mathbf{N}} \cdot \underset{u,1}{\hat{\mathbf{x}}} - \underset{u,1}{\mathbf{n}} = \underset{u,1}{\mathbf{0}} \qquad \text{: normal equations} \tag{2.172}$$

where

$$\underset{u,u}{\mathbf{N}} = \underset{u,n}{\mathbf{A}^T} \cdot \underset{n,n}{\mathbf{P}} \cdot \underset{n,u}{\mathbf{A}} \qquad \text{: matrix of normal equations}$$

$$\underset{u,1}{\mathbf{n}} = \underset{u,n}{\mathbf{A}^T} \cdot \underset{n,n}{\mathbf{P}} \cdot \underset{n,1}{\mathbf{l}} \qquad \text{: absolute term}$$

3) $$\underset{u,u}{\mathbf{Q}} = \underset{u,u}{\mathbf{N}^{-1}} \qquad \text{: solving the normal equations} \tag{2.173}$$

$$\underset{u,1}{\hat{\mathbf{x}}} = \underset{u,u}{\mathbf{Q}} \cdot \underset{u,1}{\mathbf{n}} = (\underset{u,n}{\mathbf{A}^T} \cdot \underset{n,n}{\mathbf{P}} \cdot \underset{n,u}{\mathbf{A}})^{-1} \cdot \underset{u,n}{\mathbf{A}^T} \cdot \underset{n,n}{\mathbf{P}} \cdot \underset{n,1}{\mathbf{l}}$$

: where **Q**: cofactor matrix of unknowns

4) $$\underset{n,1}{\mathbf{v}} = \underset{n,u}{\mathbf{A}} \cdot \underset{u,1}{\hat{\mathbf{x}}} - \underset{n,1}{\mathbf{l}} \qquad \text{: residuals} \tag{2.174}$$

5) $$\underset{n,1}{\hat{\mathbf{l}}} = \underset{n,1}{\mathbf{l}} + \underset{n,1}{\mathbf{v}} \qquad \text{: adjusted observations}$$

$$\underset{n,1}{\hat{\mathbf{L}}} = \underset{n,1}{\mathbf{L}} + \underset{n,1}{\mathbf{v}}$$

6) $$\underset{u,1}{\hat{\mathbf{X}}} = \underset{u,1}{\mathbf{X}^0} + \underset{u,1}{\hat{\mathbf{x}}} \qquad \text{: vector of unknowns}$$

7) $$\hat{s}_0 = \sqrt{\frac{\mathbf{v}^T \cdot \mathbf{P} \cdot \mathbf{v}}{n-u}} \qquad \text{: standard deviation a posteriori}$$

8) $$\underset{u,u}{\mathbf{C}} = \hat{s}_0^2 \cdot \underset{u,u}{\mathbf{Q}} \qquad \text{: variance-covariance matrix}$$

9) $\underset{n,1}{\hat{\mathbf{L}}} \overset{!}{=} \varphi\,(\underset{n,1}{\hat{\mathbf{X}}})$: final computing test

For most non-linear problems, initial values are only approximate and multiple iterations are required to reach an accurate solution (e.g. bundle adjustment, see section 4.4). In this case the corrected approximate values in iteration k of step (6) are used as new starting values for the linearised functional model of next iteration k+1, until the sum of added corrections for the unknowns is less than a given threshold.

$$\mathbf{X}^0_{k+1} = \mathbf{X}^0_k + \hat{\mathbf{x}}_k$$

In order to solve the normal system of equations (2) in step (3), the Jacobian matrix **A** has to be of full column rank.

$$r = rank(\underset{n,u}{\mathbf{A}}) = u$$

This requirement means that the included observations allow a unique solution for the vector of unknowns and that the inverse of the normal equation matrix **N** exists. For adjustment problems where some observations are missing for a unique solution, a rank defect d is detected:

$$d = u - r \quad \text{: rank defect} \tag{2.175}$$

This problem occurs, for example, in the adjustment of points in coordinate systems which are not uniquely defined by known reference points, or other suitable observations (*datum defect*).

The resulting singular system of normal equations can be solved with the help of the Moore-Penrose inverse (see section 4.4.3.3) or by including suitable constraints.

2.4.2.3 Levenberg-Marquardt algorithm

In computer vision, a bundle adjustment is often solved with the Levenberg-Marquardt algorithm (LMA). This will not be explained in detail here but its essential operational difference will be presented.

Both procedures are least-squares adjustments, but LMA offers a refinement in the form of a damping or regularisation term which essentially prevents a subsequent iteration from having worse starting values than its preceding iteration.

In the bundle adjustment, a correction (solution) vector is calculated in the following form (compare with eqn. 2.173):

$$\hat{\mathbf{x}} = (\mathbf{A}^T \cdot \mathbf{P} \cdot \mathbf{A})^{-1} \cdot \mathbf{A}^T \cdot \mathbf{P} \cdot \mathbf{l} \tag{2.176}$$

Using the Levenberg-Marquardt algorithm, the formulation is as follows:

$$\hat{\mathbf{x}} = (\mathbf{A}^T \cdot \mathbf{P} \cdot \mathbf{A} + \lambda \cdot diag(\mathbf{A}^T \cdot \mathbf{P} \cdot \mathbf{A}))^{-1} \cdot \mathbf{A}^T \cdot \mathbf{P} \cdot \mathbf{l} \tag{2.177}$$

Here the parameter λ regularises the iterations. If the solution improves from one iteration to the next, then λ is reduced (often by a factor of 10) and the LM formulation becomes closely similar to the conventional formulation because the term in λ gradually disappears.

On the other hand, if the solution degrades between iterations then λ is increased, again typically by a factor of 10.

A damped iterative solution can result which may be more robust than the conventional approach.

2.4.2.4 Conditional least squares adjustment

The above method of general least squares adjustment is based on a set of observation equations that model the measured observations as a function of the unknowns. An extended adjustment model results when additional constraints are incorporated between the unknowns. This method may be called the *conditional least squares adjustment*. The following cases are examples of such constraints between unknowns (see section 4.4.2.3):

- Coordinates of a number of adjusted object points must be located on a common geometric element, e.g. a straight line, plane or cylinder.
- Two adjusted object points must have a fixed separation resulting, for example, from a high accuracy distance measurement between them.

The correction equations derived earlier are then extended by a number, r', of non-linear constraints:

$$\psi(\tilde{\mathbf{X}}) = \begin{bmatrix} \psi_1(\tilde{\mathbf{X}}) \\ \psi_2(\tilde{\mathbf{X}}) \\ \vdots \\ \psi_{r'}(\tilde{\mathbf{X}}) \end{bmatrix} = \mathbf{0} \qquad \text{: constraints} \tag{2.178}$$

Using approximate values, these constraint equations are linearised in an analogous way to the observation equations:

$$\underset{r',u}{\mathbf{B}} = \left(\frac{\partial \psi(\mathbf{X})}{\partial \mathbf{X}}\right)_0 \qquad \text{: linearised constraint equations} \tag{2.179}$$

Inconsistencies $\mathbf{w}$ result from the use of approximate values instead of expected values for the unknowns:

$$\mathbf{B} \cdot \hat{\mathbf{x}} = -\mathbf{w} \qquad \text{: vector of inconsistencies} \tag{2.180}$$

The linearised functional model reduces to:

$$\begin{aligned} \mathbf{A} \cdot \hat{\mathbf{x}} - \mathbf{l} &= \mathbf{v} \\ \mathbf{B} \cdot \hat{\mathbf{x}} + \mathbf{w} &= \mathbf{0} \end{aligned} \tag{2.181}$$

The Gauss-Markov model (2.164) must be extended as follows:

$$\mathbf{v}^T \cdot \mathbf{P} \cdot \mathbf{v} + 2\mathbf{k} \cdot (\mathbf{B} \cdot \hat{\mathbf{x}} + \mathbf{w}) \rightarrow \min \tag{2.182}$$

which leads further to the following extended normal system of equations:

$$\begin{bmatrix} \mathbf{A}^T \cdot \mathbf{P} \cdot \mathbf{A} & \mathbf{B}^T \\ \mathbf{B} & \mathbf{0} \end{bmatrix} \cdot \begin{bmatrix} \hat{\mathbf{x}} \\ \mathbf{k} \end{bmatrix} + \begin{bmatrix} -\mathbf{A}^T \cdot \mathbf{P} \cdot \mathbf{l} \\ \mathbf{w} \end{bmatrix} = \mathbf{0} \qquad \text{: normal equations} \tag{2.183}$$

$$\overline{\mathbf{N}} \cdot \overline{\mathbf{x}} + \overline{\mathbf{n}} = \mathbf{0}$$

Here **k** is the vector of Lagrangian multipliers. The numerical values of **k** are not normally of interest, although the condition that $\mathbf{A}^T\mathbf{P}\mathbf{v} + \mathbf{B}^T\mathbf{k} \equiv 0$ can be tested for validity. Only the first u elements of the solution vector $\overline{\mathbf{x}}$ are therefore important.

The a posteriori standard deviation is then given by:

$$\hat{s}_0 = \sqrt{\frac{[pvv]}{n-u+r'}} = \sqrt{\frac{\sum p \cdot v^2}{n-u+r'}} \tag{2.184}$$

The redundancy f (degrees of freedom defined by the number of excess observations) changes to $f = n - u + r'$. Additional constraints can therefore increase redundancy or they can effectively compensate for missing observations which lead to a rank defect (see also free net adjustment, section 4.4.3.3).

2.4.3 Quality measures

Fig. 2.36 illustrates the relationship between the true value $\tilde{X}$, the expected value μ_x, the mean or adjusted value $\hat{x}$ and the single observation x_i. True value and expected value can differ due to systematic errors Δ_x. The true deviation η_i is the sum of a systematic component Δ_x and a random component ε_i.

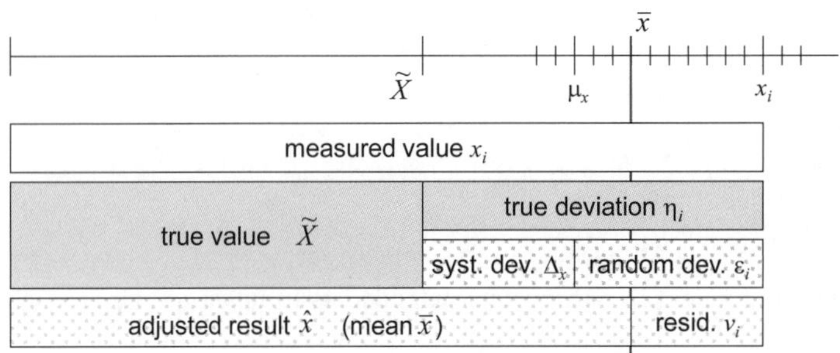

Fig. 2.36: True, stochastic and systematic deviation and residual (after Möser et al. 2000)

Since true value and expected value are unknown with a finite number of measurements, quality assessment of measured values is based on their residuals v_i. The quality values discussed below are based on statistical measures. Depending on application, the quality of a measured value (e.g. fit between cylinder and bore) must potentially be assessed by taking into account relevant associated conditions (see section 2.3.4 and section 7.2.2).

2.4.3.1 Precision and accuracy

The accuracy of observations and adjusted unknowns are of prime interest when analysing quality in an adjustment procedure. The calculated stochastic values provide information about the quality of the functional model with respect to the input data. This criterion is referred to as *precision* since it describes an internal quality of the adjustment process. In contrast, the term *accuracy* should only be used if a comparison to reference data of higher accuracy is performed. However, in practice accuracy is widely used as a general term for quality.

Standard deviation

Using the cofactor matrix **Q** or the covariance matrix **C** (see section 2.4.1.2), the standard deviations of unknowns can be obtained:

$$\mathbf{Q}_{\hat{x}\hat{x}} = \mathbf{Q} = \begin{bmatrix} q_{11} & q_{12} & \cdots & q_{1u} \\ q_{21} & q_{22} & \cdots & q_{2u} \\ \vdots & \vdots & \ddots & \vdots \\ q_{u1} & q_{u2} & \cdots & q_{uu} \end{bmatrix} \quad \text{: cofactor matrix of unknowns} \qquad (2.185)$$

The cofactor matrix of adjusted observations is derived from **Q** and the design matrix **A** as follows:

$$\mathbf{Q}_{\hat{l}\hat{l}} = \mathbf{A} \cdot \mathbf{Q} \cdot \mathbf{A}^T \quad \text{: cofactor matrix of adjusted observations} \qquad (2.186)$$

The a posteriori (empirical) *standard deviation of unit* weight is given by:

$$\hat{s}_0 = \sqrt{\frac{\mathbf{v}^T \cdot \mathbf{P} \cdot \mathbf{v}}{n-u}} \tag{2.187}$$

with redundancy: $r = n-u$

If the a posteriori standard deviation $\hat{s}_0$ diverges from the a priori standard deviation s_0, two possible sources of error are indicated. Firstly, the stochastic model may be set up incorrectly, although it should be noted that s_0 does not affect the numerical values of the adjusted unknowns. Secondly, the functional model may be incomplete. For example, unmodelled systematic errors will affect the values of the unknowns.

According to (2.162) and (2.185) the standard deviation of a single unknown x_j is given by

$$\hat{s}_j = \hat{s}_0 \cdot \sqrt{q_{jj}} \tag{2.188}$$

where q_{jj} are the elements of the principal diagonal of matrix $\mathbf{Q}$.

Root mean square

In many cases, adjustment results are reported as root mean square errors instead of the above defined standard deviation. An RMS value (*root mean square*) is simply the square root of the arithmetic mean of the squares of a set of numbers. Normally the numbers represent a set of differences or changes which are of some particular interest, such as differences between measured and nominal values or positional changes resulting from a deformation.

The root mean square error (RMS error or RMSE) indicates the RMS error of adjusted observations with respect to the mean of adjusted observations:

$$RMSE = \sqrt{\frac{\sum (X_i - \bar{X})^2}{n}} \tag{2.189}$$

For large *n,* the RMSE is equal to the empirical standard deviation. This is because the standard deviation of a simple set of error values would have the same form but use (*n*-1) in place of *n*. As *n* becomes large, the difference between *n* and (*n*-1) becomes negligible.

Span

The span *R* denotes the maximum separation between two observations of a set of measurements.

$$R = X_{max} - X_{min} \tag{2.190}$$

The span is not unbiased as the observations may contain blunders. However, it is important in metrology since, for manufacturing purposes, it may necessary that all measured values lie within particular limits (tolerance, see section 7.2.1.8). Hence, the span implicitly describes a confidence interval of 100% probability (see section 2.4.3.2). The span can also be defined as the difference between the minimum and maximum residuals in a data set.

2.4.3.2 Confidence interval

It is generally assumed that the observations in an adjustment process have a normal (Gaussian) random error distribution. Given a normally distributed random variable l with expected value μ and standard deviation σ, the *probability density* is given by:

$$f(x) = \frac{1}{\sigma\sqrt{2\pi}} \cdot \exp\left(-\frac{1}{2}\frac{(x-\mu)^2}{\sigma^2}\right) \tag{2.191}$$

The error of the random variable is defined by:

$$\varepsilon = l - \mu \qquad \text{: random error} \tag{2.192}$$

This is valid for a normally distributed sample with an infinite number of sample points and an expected value defined as:

$$\mu = E\{x\} = \tilde{x} \qquad \text{: expected value (true value)} \tag{2.193}$$

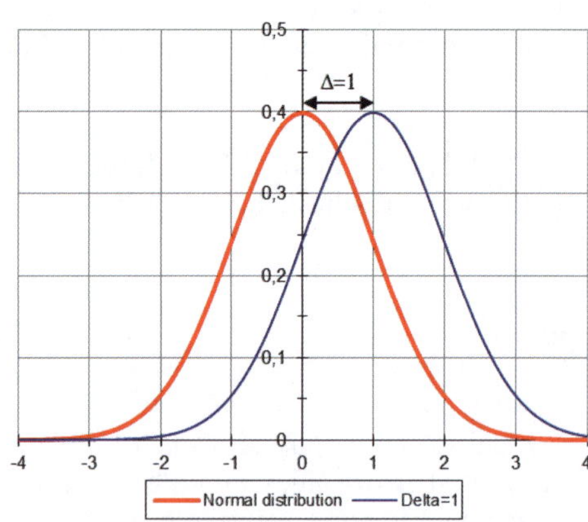

Fig. 2.37: Standardised Gaussian distribution

Fig. 2.37 shows the probability density function of the normalised Gaussian distribution ($\mu = 0$, $\sigma = 1$) and, for comparison, the systematically shifted distribution corresponding to the non-centrality parameter $\Delta = 1$[1]. The area underneath the curve, between specified limits on the horizontal axis, corresponds to the probability that the error of a random

[1] In the following it is assumed that no systematic deviations exist, hence $\Delta = 0$.

variable lies between these limits. The total area under the curve = 1 and the probability limits are usually defined as a symmetrical factor of the standard deviation.

$$P\{-k \cdot \sigma < \varepsilon < k \cdot \sigma\}$$

Table 2.3: Probability of error $|\varepsilon| < k \cdot \sigma$ at different degrees of freedom

Gaussian distribution		Student distribution			
k	*P*	*P*	*P*	*P*	*P*
	$f=\infty$	$f=2$	$f=5$	$f=10$	$f=20$
1	68.3%	57.7%	63.7%	65.9%	67.1%
2	95.4%	81.6%	89.8%	92.7%	94.1%
3	99.7%	90.5%	97.0%	98.7%	99.3%

Table 2.3 shows that, in the case of an infinitely large data set (degrees of freedom $f=\infty$), the probability is 68.3% that all deviations are within a single standard deviation of the true value ($k=1$). The probability rises to 95.4% for 2 standard deviations ($k=1.96$ for $P=$ 95%). Lastly, only 0.3% of all errors lie outside limits defined by 3 standard deviations.

In the case of large but finite data sets, the Gaussian distribution is replaced by the t-distribution (Student distribution). The probability *P* that a deviation is within a factor *k* of the standard deviation, increases with increasing degrees of freedom. For very large degrees of freedom, the t-distribution becomes equivalent to the Gaussian distribution.

For real (finite) data sets, only estimates $\hat{x}$ and $\hat{s}$ of the true values μ and σ can be computed. However, an interval between two limiting values C_u and C_o can be defined, within which $\hat{x}$ is determined with probability *P* (Fig. 2.38). This confidence interval is given by

$$P\{C_u \leq \hat{x} \leq C_o\} = 1-\alpha$$
$$P\{\hat{x} < C_u\} = \frac{\alpha}{2} \tag{2.194}$$

with confidence level $1-\alpha$.

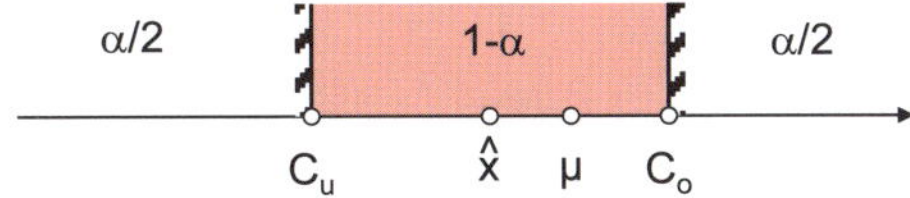

Fig. 2.38: Confidence interval

The confidence limits for empirical estimate $\hat{x}$ with a given empirical standard deviation are defined as:

$$
\begin{aligned}
C_u &= \hat{x} - t_{f,1-\alpha/2} \cdot s_{\hat{x}} \\
C_o &= \hat{x} + t_{f,1-\alpha/2} \cdot s_{\hat{x}}
\end{aligned}
\tag{2.195}
$$

Here t is a quantile of the t-distribution. For example, $t_{5,\,0.975} = 2.57$ corresponds to a confidence level of 95% ($\alpha = 0.05$) and $f = 5$ degrees of freedom. The confidence interval therefore increases with finite number of excess measurements, i.e. the confidence that estimate $\hat{x}$ lies between defined limits is reduced. Fig. 2.39 shows the limiting curves of confidence intervals for different degrees of freedom and different confidence levels.

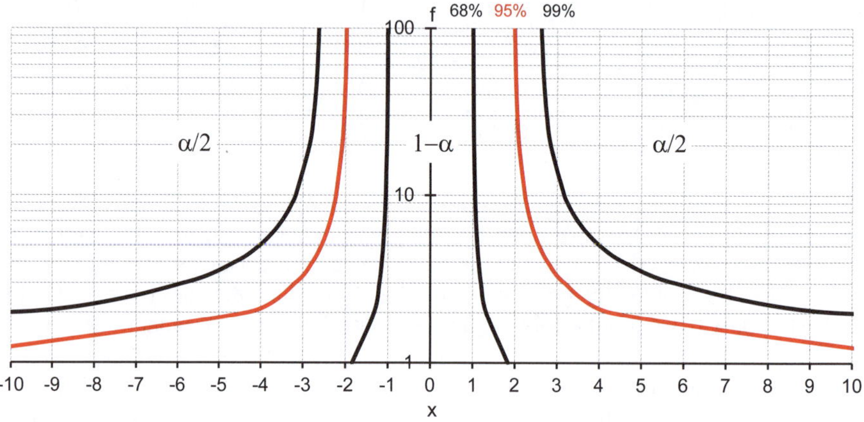

Fig. 2.39: Confidence intervals with different t-distributions

2.4.3.3 Correlations

In addition to standard deviations, dependencies between adjusted parameters can also be investigated in order to assess the quality of an adjustment result. They indicate the extent to which an unknown can be calculated and hence the adequacy of the functional model and geometric configuration of the observations.

According to (2.161) the covariance matrix provides the correlations between single parameters:

$$
\mathbf{C}_{\hat{x}\hat{x}} = \hat{s}_0^2 \cdot \mathbf{Q}_{\hat{x}\hat{x}} = \begin{bmatrix} \hat{s}_1^2 & \rho_{12}\hat{s}_1\hat{s}_2 & \cdots & \rho_{1u}\hat{s}_1\hat{s}_u \\ \rho_{21}\hat{s}_2\hat{s}_1 & \hat{s}_2^2 & \cdots & \rho_{2u}\hat{s}_2\hat{s}_u \\ \vdots & \vdots & \ddots & \vdots \\ \rho_{u1}\hat{s}_i\hat{s}_1 & \rho_{u2}\hat{s}_i\hat{s}_2 & \cdots & \hat{s}_u^2 \end{bmatrix}
\tag{2.196}
$$

The correlation coefficient ρ_{ij} between two unknowns i and j is defined by[1]:

[1] Here the notation ρ is used for the empirical correlation coefficient in order to avoid confusion with the redundancy number.

$$\rho_{ij} = \frac{\hat{s}_{ij}}{\hat{s}_i \cdot \hat{s}_j} \qquad -1 \le \rho_{ij} \le +1 \tag{2.197}$$

Higher correlation coefficients indicate linear dependencies between parameters. They should be avoided particularly because the inversion of the normal equation matrix, and hence the adjustment solution, can then become numerically unstable.

2.4.3.4 Reliability

The *reliability* of an adjustment process indicates the potential to control the consistency of the observations and the adjustment model. It depends on the number of excess observations (total redundancy) and the geometric configuration (configuration of images). Reliability gives a measure of how well gross errors (outliers) can be detected in the set of observations.

Essential information about reliability can be derived from the cofactor matrix of residuals:

$$\mathbf{Q}_{vv} = \mathbf{Q}_{ll} - \mathbf{A} \cdot \mathbf{Q} \cdot \mathbf{A}^T \qquad \text{: cofactor matrix of residuals} \tag{2.198}$$

The total redundancy in an adjustment is given by

$$r = n - u = trace(\mathbf{Q}_{vv} \cdot \mathbf{P}) = \sum r_i \tag{2.199}$$

where r_i are the elements of the principal diagonal of the redundancy matrix $\mathbf{R} = \mathbf{Q}_{vv}\,\mathbf{P}$.

$$\mathbf{R} = \mathbf{Q}_{vv} \cdot \mathbf{P} = \begin{bmatrix} r_{11} & & & \\ & r_{22} & & \\ & & \ddots & \\ & & & r_{nn} \end{bmatrix} \qquad \text{: redundancy matrix} \tag{2.200}$$

r_i is denoted as the *redundancy number* of an observation l_i with respect to the total redundancy r where

$$0 \le r_i \le 1$$

The redundancy number of an observation indicates the relative part of an observation which is significantly used for the estimation of the unknowns $(1-r_i)$, or which is not used (r_i). Small redundancy numbers correspond to weak configurations which are hard to control, whilst high redundancy numbers enable a significant control of observations. If an observation has a redundancy number $r_i = 0$, it cannot be controlled by other observations. Hence, a gross error in this observation cannot be detected but it has a direct influence on the estimation of unknowns. If an observation has a very high redundancy number (0.8 to 1), it is very well controlled by other observations. When optimising an adjustment, such observations can initially be eliminated without a significant effect on the adjustment result.

The relation between residuals and observations is defined by:

$$\mathbf{v} = \mathbf{A} \cdot \hat{\mathbf{x}} - \mathbf{l} = -\mathbf{R} \cdot \mathbf{l} \tag{2.201}$$

Hence, for gross (systematic) observation errors $\Delta\mathbf{l}$:

$$\Delta\mathbf{v} = -\mathbf{R} \cdot \Delta\mathbf{l} \tag{2.202}$$

Eqn. 2.202 does permit the detection of gross errors to be quantified because gross errors do not have correspondingly large residuals when redundancy numbers are small. According to Baarda, a normalised residual is therefore used:

$$w_i = \frac{v_i}{\hat{s}_{v_i}} \tag{2.203}$$

The standard deviation of a residual is derived either from the cofactor matrix or redundancy numbers as follows:

$$\hat{s}_{v_i} = \hat{s}_0 \sqrt{(\mathbf{Q}_{vv})_{ii}} = \hat{s}_{l_i} \sqrt{(\mathbf{Q}_{vv} \cdot \mathbf{P})_{ii}} = \hat{s}_{l_i} \sqrt{r_i} \tag{2.204}$$

Here it is obvious that a redundancy number of $r_i = 0$ leads to an indeterminate value of w_i and no error detection is then possible. The normalised residuals are normally distributed with expectation 0 and standard deviation 1. To detect a gross error they are compared with a threshold k:

$$|w_i| \begin{cases} > k: \text{gross error} \\ \leq k: \text{no gross error} \end{cases} \tag{2.205}$$

In order to compute the threshold value k, a statistical test is used where the value δ_0 (non-centrality parameter) is defined:

$$\delta_0 = \delta_0\,(\alpha, \beta)$$

where α : probability of identifying an *error-free* value as a gross error (significance level)

β : probability of identifying a *defective* value as a gross error (statistical power)

This test establishes a null hypothesis which states that only randomly distributed errors may occur and the expected value of the normalised residuals must therefore be zero.

$$E\{w_0\} = 0 \tag{2.206}$$

The probability of a false decision is equal to α (type 1 error). This is the decision that residuals w_i lie outside the range ±k and are therefore excluded. Here k denotes a quantile of the t-distribution.

If a gross error occurs, the expected value of the corresponding normalised residual is not equal to zero and has standard deviation 1:

$$E\{w_a\} \neq 0 \tag{2.207}$$

Given the alternative hypothesis that only observations where $|w_i| > k$ are identified as gross errors, a possible number of outliers still remain in the data set. The probability of this false decision (type 2 error) is 1-β (Fig. 2.40).

Using (2.202) and (2.203), a lower expected value can be defined for a gross error that can be detected significantly with statistical power, β.

$$E\{\Delta l_a\} = \frac{\delta_0}{\sqrt{r_i}} \cdot \hat{s}_{l_i} = \delta'_0 \cdot \hat{s}_{l_i} \tag{2.208}$$

The following term is normalised with respect to $\hat{s}_{l_i}$:

$$\delta'_{0,i} = \frac{E\{\Delta l_a\}}{\hat{s}_{l_i}} = \frac{\delta_0}{\sqrt{r_i}} \tag{2.209}$$

It serves as a measure of *internal reliability* of an adjustment system. It defines the factor by which a gross observation error Δl_a must be larger than $\hat{s}_{l_i}$ in order to be detected with probability β.

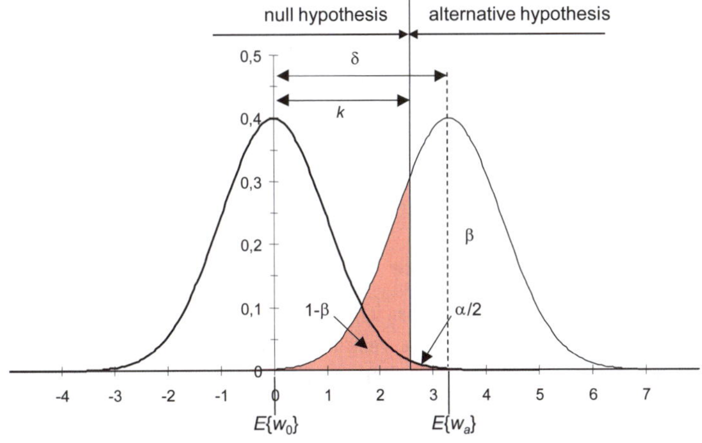

Fig. 2.40: Probability densities for null and alternative hypothesis

Table 2.4: Test statistics and internal reliability for different significance numbers and statistical power

	α=5%	α=1%	α=0.1%
	β=75%	β=93%	β=80%
k	1.96	2.56	3.29
δ'_0	3.9	4.0	4.1

Table 2.4 shows some test statistics resulting from a chosen significance level α and statistical power β. The selected probabilities lead to similar measures of the *internal reliability*. It becomes clear that with increasing significance level (i.e. decreasing confidence level) the statistical power is reduced and is therefore less effective.

In general a value of $\delta'_0 = 4$ is appropriate for photogrammetric bundle adjustment (see section 4.4). It is also an appropriate value for the decision threshold k in inequality (2.205). In order to use this standard value for bundle adjustments, a high redundancy is required. Eqn. 2.209 clearly shows that the test value depends on the a posteriori standard deviation of a single observation and hence is a function of the standard deviation of unit weight $\hat{s}_0$. Smaller gross errors can therefore only be detected when $\hat{s}_0$ becomes sufficiently small after iteration that its value is of the order of the precision of the observations (measurement precision).

When planning a measuring project, the internal reliability can be calculated prior to knowledge of actual observations (measurement values) because the necessary information can be obtained from the Jacobian matrix **A** and the (assumed) a priori accuracy values for the observations (see section 7.3.2).

During or after an iterative adjustment, the internal reliability is used as criterion for the automated elimination of gross data errors.

The *external reliability* indicates the influence of defective observations on the estimated unknowns. For this purpose the vector of internal reliability values, defined as in (2.208), is used in the system of equations, defined in (2.173). For each unknown it is possible to compute a total number of n values of external reliability, each dependent on an observation.

2.4.4 Error detection in practice

It is difficult to avoid gross data errors (outliers) in real projects. In photogrammetric applications they typically occur as a result of faulty measurements, errors in point identification, or mistakes in image numbering. Gross errors must be eliminated from the data set because they affect all estimated unknowns and standard deviations, leading to a significantly distorted adjustment result.

The residuals calculated in the adjustment should not be used directly for the detection of outliers. Residuals not only result from errors in the set of observations but also from errors in the geometric model, i.e. a functional model of the adjustment which is incomplete. Model and data errors can both be present and their effects may overlap.

Most approaches for detection and elimination of gross errors are based on the assumption that only very few outliers exist, and in extreme cases only one. The method of least-squares adjustment described above optimally disperses the observation errors over all observations in the data set, with larger deviations affecting the unknowns more than smaller. Where gross errors are present this results in a smearing effect. The ability to recognise a gross error is therefore limited, especially if several such outliers occur at the same time. Where there is an unfavourable geometric configuration of unknowns and number of outliers, even error-free observations may be identified as having gross errors. It is therefore critically important to eliminate only those observations that can be identified without doubt as errors. The elimination of outliers should always be associated with an analysis of the entire measurement task.

2.4.4.1 Data snooping

Baarda's *data snooping* is a method of error detection based on the value of internal reliability derived in section 2.4.3.4. It is based on the assumption that only *one* gross error can be identified in the data set at any time. The process iteratively searches for and eliminates gross errors. After each iteration of the adjustment that observation is eliminated which, on the basis of the decision function (2.205), corresponds to the largest normalised residual w_i. The complete adjustment procedure is set up again and the computation repeated until no gross errors remain in the set of observations.

In cases where several large residuals w_i exist, and where their geometric configuration ensures they are independent of each other, it is possible to detect more than one outlier simultaneously. However, one should still check carefully those observations which are suspected as gross errors.

2.4.4.2 Variance component estimation

The internal reliability value used in data snooping is a function of the standard deviation of the adjusted observations. These are derived from the covariance matrix by multiplication with the standard deviation of unit weight $\hat{s}_0$. Since $\hat{s}_0$ is a global value influenced by all residuals, it is really only useful for observations of equal accuracy. Data sets with different types of observations, or different levels of accuracy, should therefore be divided into separate groups with homogeneous accuracy.

In order to set up the weight matrix **P**, each separate observation group g is assigned its own a priori variance:

$s_{0,g}^2$: a priori variance of unit weight

This variance, for example, can be derived from the existing known accuracy of a measuring device used for that specific observation group.

After computing the adjustment, the a posteriori variance can be determined:

$$\hat{s}_{0,g}^2 = \frac{\mathbf{v}_g^T \mathbf{P}_g \mathbf{v}_g}{r_g} \quad \text{: a posteriori variance of unit weight}$$

$$\text{where } r_g = \sum (r_i)_g \tag{2.210}$$

Using the a posteriori variance of unit weight, it is possible to adjust the a priori weights in succeeding adjustments until the following condition is achieved:

$$Q_g = \frac{\hat{s}_{0,g}}{s_{0,g}} = 1 \tag{2.211}$$

Subsequently, the normal data snooping method can be used.

Taking eqn. 2.162 into account, the variance of unit weight can be used to calculate the variance of a complete observation group.

2.4.4.3 Robust estimation with weighting functions

The comments above indicate that the residuals resulting from an adjustment process are not directly suitable for the detection of one or more gross errors. Different approaches have therefore been developed for defining the weights p_i as a function of the residuals in successive iterations. If the weighting function is designed such that the influence of a gross error is reduced as the error becomes larger, then it is referred to as *robust estimation* (robust adjustment). One possible approach is given by the following function:

$$p'_i = p_i \cdot \frac{1}{1 + \left(a \cdot |v_i|\right)^b} \tag{2.212}$$

For $v_i = 0$ it reduces to $p'_i = 1$ and for $v_i = \infty$ it reduces to $p'_i = 0$. The parameters a and b form the curve of a bell-shaped weighting function. With

$$a_i = \frac{\sqrt{p_i}}{\sqrt{r_i} \cdot \hat{s}_0 \cdot k} = \frac{1}{\hat{s}_{v_i} \cdot k} \tag{2.213}$$

the parameter a is controlled by the redundancy number of an observation. Definition of parameters b and k is done empirically. With a correct choice of parameters, the quotient Q (see preceding section) converges to 1.

Kruck proposes a weighting function that is also based on redundancy numbers:

$$p'_i = p_i \cdot \frac{\tan\left(\frac{1-r_m}{c}\right)}{\tan\left(\frac{1-r_i}{c}\right)} \tag{2.214}$$

The constant c is defined empirically, r_m is referred to as the *average redundancy* or *constraint density*:

$$r_m = \frac{n-u}{n} \tag{2.215}$$

The weights become constant values when the stability $r_i = r_m$ is reached (*balanced observations*).

Procedures for robust estimation are primarily designed to reduce the effect of *leverage points*. Leverage points, in the sense of adjustment, are those observations which have a significant geometric meaning but only small redundancy numbers. Gross errors at leverage points affect the complete result but are difficult to detect. Using balanced weights, leverage points and observations with gross errors are temporarily assigned the same redundancy numbers as every other observation. As a result they can be detected more reliably. After elimination of all defective observations, a final least-squares adjustment is calculated using the original weights.

2.4.4.4 Robust estimation according to L1 norm

In recent times more attention has been paid to the principle of adjustment according to the L1 norm, especially for gross error detection in weak geometric configurations. For example, it is used to calculate approximate orientation values using data sets containing gross errors.

The L1 approach is based on the minimisation of the absolute values of the residuals, whilst the L2 approach (least squares, see section 2.4.2) minimises the sum of squares of the residuals:

$$\text{L1 norm:} \quad \sum |v| \quad \rightarrow \min \tag{2.216}$$

$$\text{L2 norm:} \quad \sum v^2 \quad \rightarrow \min$$

The solution of the system of equations using the L1 approach is a task in linear optimisation. It is much more difficult to handle than the L2 approach in terms of mathematics, error theory and computation algorithms. One solution is given by the Simplex algorithm known from linear programming.

The L1 approach is also suitable for balancing weights according to eqn. 2.215. In theory it is possible with the L1 norm to process data sets with up to 50% gross errors. The reason is that the L1 solution uses the median value whereas the L2 solution is based on the arithmetic mean which has a smearing effect.

After error elimination based on the L1 norm, the final adjustment should be calculated according to the least-squares approach.

2.4.4.5 RANSAC

RANSAC (random sample consensus) describes an adjustment algorithm for any functional model which is based on a voting scheme and is particularly robust in the presence of outliers, e.g. even as high as 80%. The idea is based on the repeated calculation of a target function by using the minimum number of observations. These are randomly selected from the full set of observations. Subsequently, all other observations are tested against this particular solution. All observations which are consistent with the calculated solution within a certain tolerance are regarded as valid measurements and form a consensus set. The solution with the maximum number of valid observations (largest consensus set) is taken as the best result. Outliers are rejected and an optional final least-squares adjustment is calculated.

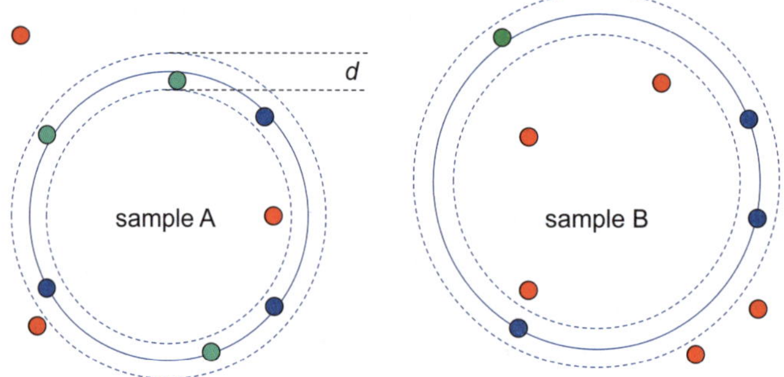

Fig. 2.41: RANSAC approach for a best-fit circle
blue: selected RANSAC points; red: detected outliers; green: valid points

Fig. 2.41 shows the RANSAC principle applied to the calculation of a best-fit circle with a set of valid observations and outliers. For each of the two samples A and B, a circle is calculated from three randomly selected points. The other observations are tested against the circle tolerance *d*. Out of all samples the solution with the maximum number of valid observations is finally selected (sample A in this case). Remaining outliers are rejected from the data set.

The success of the RANSAC algorithm depends mainly on the selection of tolerance parameter *d* and the termination criteria, e.g. the number of iterations or the minimum size of the consensus set.

RANSAC is widely used in photogrammetry and computer vision for solving tasks such as relative orientation (section 4.3.3), feature detection in point clouds (section 5.6.2) or general shape fitting of geometric primitives.

2.4.5 Computational aspects

2.4.5.1 Linearisation

In order to linearise the functional model at initial values, two methods are available:

- exact calculation of the first derivative
- numerical differentiation

Exact calculation of the derivatives

$$\left(\frac{\partial \varphi(X)}{\partial X} \right)_0$$

may require considerably more manual programming effort for complex functions such as those expressed by rotation matrices.

In contrast, numerical differentiation is based on small changes to the initial values of unknowns in order to calculate their effects on the observations:

$$\begin{aligned} L^0_{+\Delta x} &= \varphi(X^0 + \Delta X) \\ L^0_{-\Delta x} &= \varphi(X^0 - \Delta X) \end{aligned} \tag{2.217}$$

The difference quotients are then:

$$\frac{\Delta\varphi(X)}{\Delta X} = \frac{L^0_{+\Delta x} - L^0_{-\Delta x}}{2\Delta X} \approx \frac{\partial \varphi(X)}{\partial X} \tag{2.218}$$

In a computer program, the function φ can be entered directly, for example in a separate routine. The set-up of the Jacobian matrix **A** and subsequent adjustment procedure can then be programmed independently of the functional model. Only the increment ΔX need be adjusted if necessary.

Compared with numerical differentiation, the exact calculation of derivatives leads to faster convergence. If suitable initial values are available, then after a number of iterations both adjustment results are, for practical purposes, identical.

2.4.5.2 Normal systems of equations

In order to solve solely for the solution vector $\hat{\mathbf{x}}$, efficient decomposition algorithms can be used which do not require the inverse of the normal equation matrix **N**, for example the Gaussian algorithm. However, for many photogrammetric and geodetic calculations a quality analysis based on the covariance matrix is required and so the inverse of **N** must be computed. The dimension of the normal system of equations based on (2.172) is u x u elements. For photogrammetric bundle adjustments, the number of unknowns u can easily range from a few hundred up to a few thousand. Often a direct inversion of the normal equation matrix is not possible or else consumes too much computation time.

Usually the matrix **N** is factorised according to Cholesky. Using the triangular rearrangement

$$\mathbf{C}^T \cdot \mathbf{C} = \mathbf{N} \tag{2.219}$$

inserted into (2.172)

$$\mathbf{C}^T \cdot \mathbf{C} \cdot \hat{\mathbf{x}} = \mathbf{n} \tag{2.220}$$

and the substitution

$$\mathbf{C} \cdot \hat{\mathbf{x}} = \mathbf{g} \tag{2.221}$$

the equation becomes

$$\mathbf{C}^T \cdot \mathbf{g} = \mathbf{n} \tag{2.222}$$

For the computation of **g,** and subsequently $\hat{\mathbf{x}}$, only a diagonal matrix must be inverted.

2.4.5.3 Sparse matrix techniques and optimisation

The computational effort to solve the normal system of equations is mainly a function of the dimension of matrix **C**. Since matrix **N** can consist of numerous zero elements, relating to unknowns which are not connected by an observation, then these elements are also present in **C**.

Sparse techniques provide efficient use of RAM (Random-Access Memory). Instead of a full matrix, a profile is stored which, for each column of a matrix, only records elements from the first non-zero value up to the principal diagonal, together with a corresponding index value.

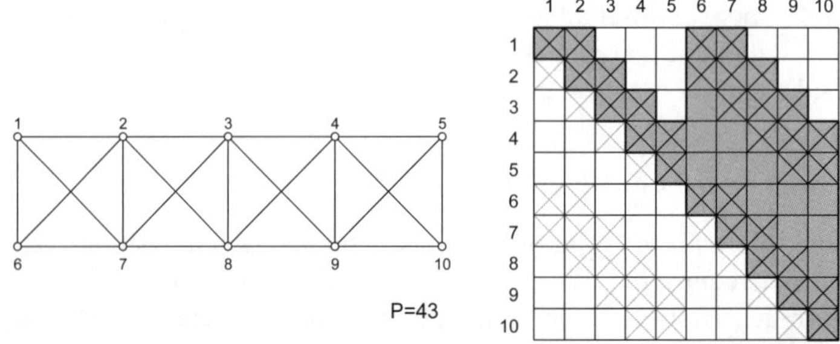

Fig. 2.42: Order of point numbers without optimisation (after Kruck 1983)

Fig. 2.42 shows an example of a network of observations and the corresponding structure of the normal equation matrix. The crosses indicate connections between unknowns while the blank fields have zero values. For example, point 2 is connected to points 1, 3, 6, 7 and 8.

In this example, the size of the profile to be stored, i.e. the number of stored matrix elements, amounts to P = 43. In order to reduce the profile size without modifying the functional relationships, the point order can be sorted (Banker's algorithm), leading to the result of Fig. 2.43. The profile size in this case has been reduced to P = 31.

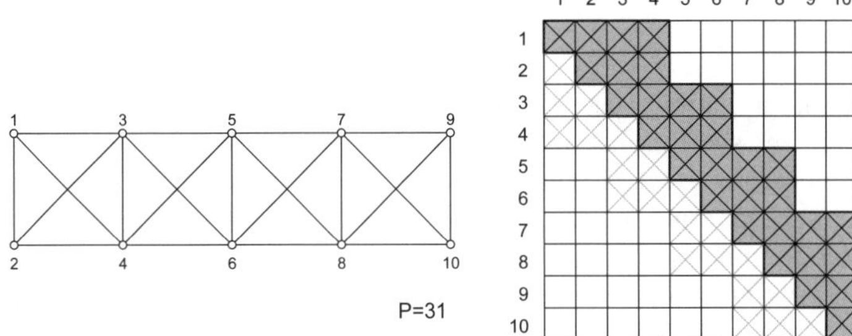

Fig. 2.43: Order of point numbers after optimisation (after Kruck 1983)

Since the resulting computational effort for matrix inversion is a quadratic function of the profile size, optimisation is of major importance for solving large systems of equations.

3 Imaging technology

Photogrammetric imaging technologies for close-range measurement purposes impact upon all elements in the measurement process, from the preparation of the measuring object prior to imaging, through image acquisition, to subsequent analysis of the image content. Following an introduction to the physics behind optical imaging, common photogrammetric imaging concepts are briefly presented. These are followed by the geometric analysis defining the camera as a measuring tool. Current components and technologies for 2D image acquisition are then reviewed in the next two sections. From there, discussion moves to signalisation (targeting) and its illumination, which are critical in achieving photogrammetric accuracy.

Historically, photogrammetry uses 2D imaging to deliver 3D object data. Since the mid 1990s, measurement systems have been available which are designed to deliver directly very dense, 3D object surface data in a process generally known as 3D imaging. Many of these systems, such as area surface scanners, make use of photogrammetric principles. However, photogrammetry can also embrace devices such as terrestrial laser scanners (TLS) which do not operate like cameras. A common element in 3D imaging is the use of projected light in the data acquisition process, which is summarised in section 3.7.

3.1 Physics of image formation

3.1.1 Wave optics

3.1.1.1 Electro-magnetic spectrum

In photogrammetry the usable part of the electromagnetic spectrum is principally restricted to wavelengths in the visible and infra-red regions. This is due to the spectral sensitivity of the imaging sensors normally employed, such as photographic emulsions, CCD and CMOS sensors which respond to wavelengths in the range 200 nm to 1100 nm (visible to near infrared). In special applications, X-rays (X-ray photogrammetry), ultraviolet and longer wavelength thermal radiation (thermography) are also used. However, microwaves (radar) are only used in remote sensing from aircraft and satellite platforms. Table 3.1 summarises the principal spectral regions, with associated sensors and applications, which are relevant to photogrammetry.

The relationship between wavelength λ, frequency ν und speed of propagation c is given by:

$$\lambda = \frac{c}{\nu} \tag{3.1}$$

The propagation of electromagnetic radiation is described using either a wave propagation model or a photon stream. Both models have relevance in photogrammetry. The wave properties of light are employed in the description of optical imaging and its aberrations, as well as refraction and diffraction. The particle properties of light are useful in understanding the transformation of light energy into electrical energy in image sensors (CCD, CMOS), see also section 3.4.1.

Table 3.1: Regions of the electromagnetic spectrum with example application areas

Radiation	Wavelength	Sensor	Application area
X-ray	10 pm – 1 nm	X-ray machine	medical diagnosis, non-destructive testing
ultraviolet	300 nm – 380 nm	CCD, CMOS, silicon (Si) detector	remote sensing, UV reflectography
visible light	380 nm – 720 nm	CCD, CMOS, film	photography, photogrammetry, remote sensing
near infrared	720 nm – 3 µm	CCD, CMOS, film	IR reflectography, photogrammetry, remote sensing
mid infrared	3 µm – 5 µm	germanium (Ge) lead selenide (PbSe) and lead sulphide (PbS) dectors	remote sensing
thermal infrared	8 µm – 14 µm	micro-bolometer, quantum detector	thermography, material testing, building energy efficiency
terahertz	30 µm – 3 mm	Golay cell	spectroscopy, body scanning
microwave	1 cm – 30 cm	radar antenna	radar remote sensing
radio wave	30 cm – 10 m	coil	magnetic resonance imaging (MRI)

3.1.1.2 Radiometry

According to quantum theory, all radiation is composed of quanta of energy (photons in the case of light). The radiant energy is a whole multiple of the energy in a single quantum of radiation which is related to the reciprocal of the photon's associated wavelength according to the following equation:

$$E = h \cdot \nu = h \cdot \frac{c}{\lambda} \tag{3.2}$$

where

h: Planck's constant $6.62 \cdot 10^{-34}$ Js

The spectral emission of a black body at absolute temperature T is defined by Planck's law:

$$M_{\lambda} = \frac{c_1}{\lambda^5}\left[\exp\left(\frac{c_2}{\lambda \cdot T}\right) - 1\right]^{-1} \quad \text{: Planck's law} \tag{3.3}$$

where

$c_1 = 3.7418 \cdot 10^{-16}$ W m^2

$c_2 = 1.4388 \cdot 10^{-2}$ K m

This states that the radiant power is dependent only on wavelength and temperature. Fig. 3.1 shows this relationship for typical black-body temperatures. Radient power per unit area of the emitting source and per unit solid angle in the direction of emission is defined as radience. The area under the curves represents the total energy in Watts per m². The example of the sun at a temperature of 5800 K clearly shows the maximum radiant power at a wavelength around 580 nm, which is in the yellow part of the visible spectrum. In contrast, a body at room temperature (20° C) radiates with a maximum at a wavelength around 10 μm and a very much smaller power.

The radiant power maxima are shifted towards shorter wavelengths at higher temperatures according to Wien's displacement law (see straight line in Fig. 3.1).

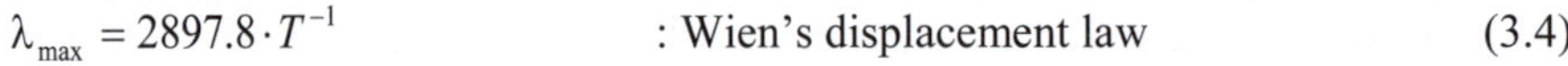

$$\lambda_{max} = 2897.8 \cdot T^{-1} \qquad \text{: Wien's displacement law} \qquad (3.4)$$

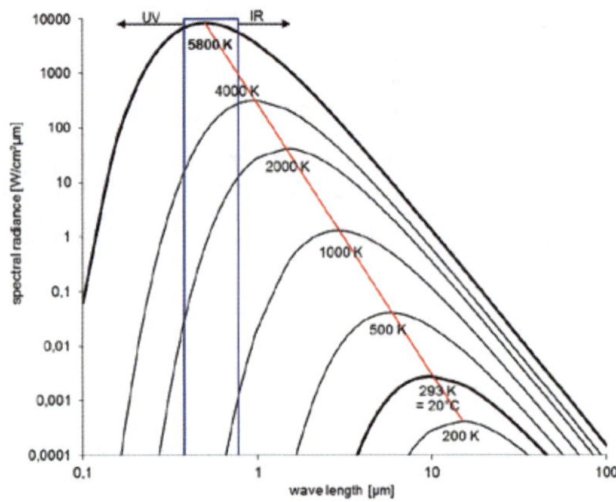

Fig. 3.1: Spectrally dependent radiant power at different black-body temperatures

3.1.1.3 Refraction and reflection

The *refractive index n* is defined as the ratio of the velocities of light propagation through two different media (frequency is invariant):

$$n = \frac{c_1}{c_2} \qquad \text{: refractive index} \qquad (3.5)$$

In order to define refractive indices for different materials, c_1 is assigned to the velocity of propagation in a vacuum or in air:

$$n = \frac{c_{air}}{c} \qquad (3.6)$$

The refractive index of pure water has been determined to be $n = 1.33$ whilst for glass the value varies between 1.45 and 1.95 depending on the material constituents of the glass. In general, homogeneous and isotropic media are assumed. A ray of light passing from a low density media to a more dense media is refracted towards the normal at the media interface T, as denoted by Snell's law (Fig. 3.2, Fig. 3.3).

$$n_1 \sin \varepsilon_1 = n_2 \sin \varepsilon_2 \qquad \text{: Snell's law of refraction} \tag{3.7}$$

The law can also be expressed as a function of the tangent:

$$\frac{\tan \varepsilon_1}{\tan \varepsilon_2} = \sqrt{n^2 + (n^2 - 1)\tan^2 \varepsilon_1} \tag{3.8}$$

where $n = \frac{n_2}{n_1}$ and $n_2 > n_1$

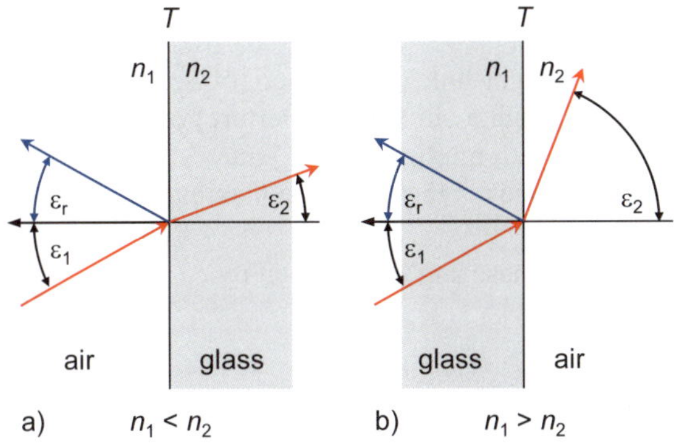

Fig. 3.2: Refraction and reflection

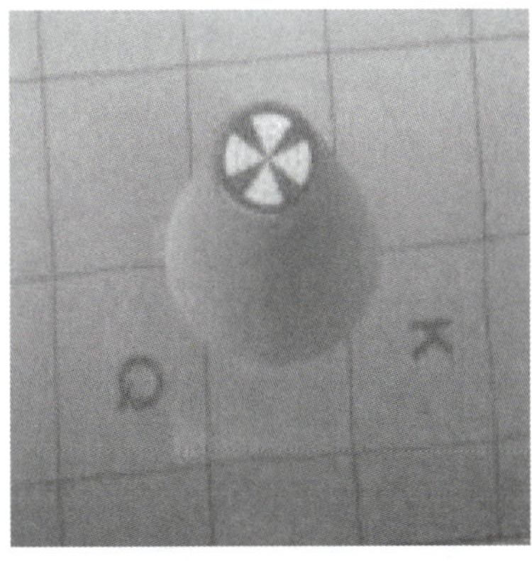

a) Still water surface

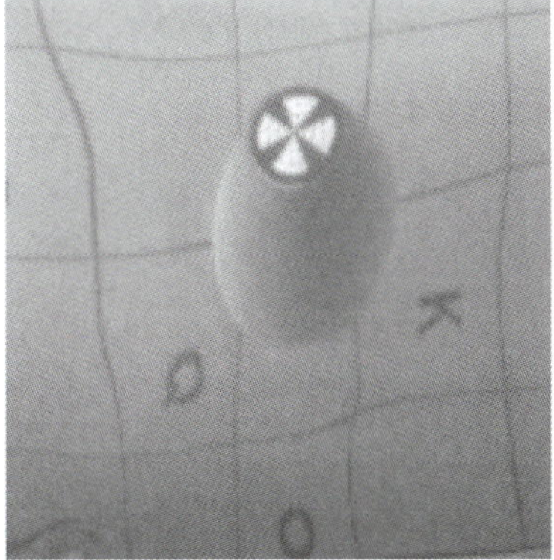

b) Disturbed water surface

Fig. 3.3: Refraction effects caused by taking an image through a water surface

For the case of *reflection* it holds that $n_2 = -n_1$ and the law of reflection follows:

$$\varepsilon_r = -\varepsilon_1 \qquad \text{: law of reflection} \tag{3.9}$$

The velocity of propagation of light depends on the wave length. The resulting change of refraction is denoted as *dispersion* (Fig. 3.4). In an optical imaging system this means that different wavelengths from the object are refracted at slightly different angles which lead to chromatic aberration in the image (see section 3.1.3.2).

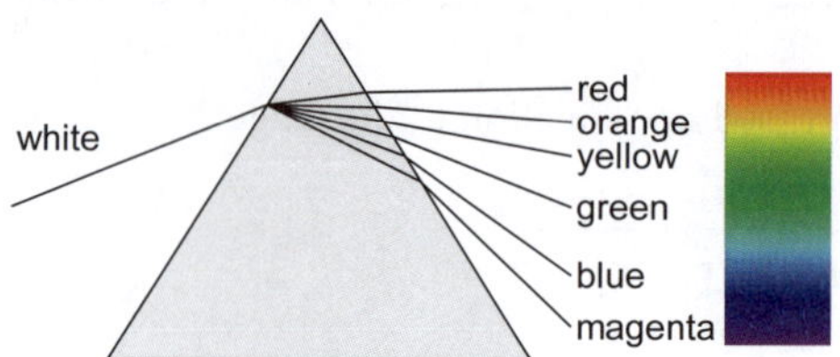

Fig. 3.4: Dispersion by a glass prism

3.1.1.4 Diffraction

Diffraction occurs if the straight-line propagation of spherical wave fronts of light is disturbed, e.g. by passing through a slit (linear aperture) or circular aperture. The edges of the aperture can be considered multiple point sources of light which also propagate spherically and interfere with one another to create maxima and minima (Fig. 3.5, Fig. 3.6).

The intensity I observed for a phase angle φ is given by:

$$I(\varphi) = \frac{\sin x}{x} = \mathrm{sinc}(x) \tag{3.10}$$

where $x = \frac{\pi d' \sin\varphi}{\lambda}$

d': slit width
λ: wavelength
φ: phase angle

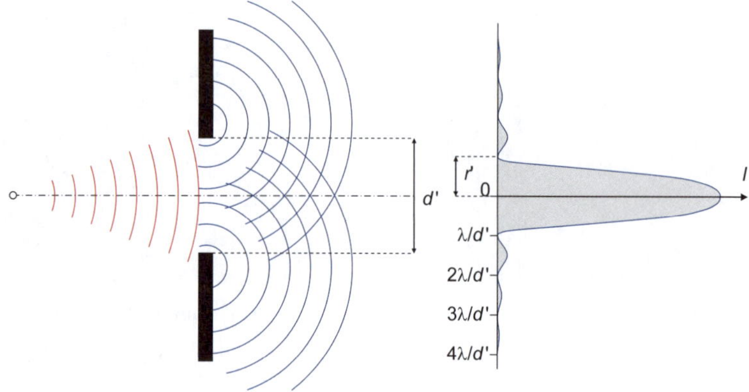

Fig. 3.5: Diffraction caused by a slit

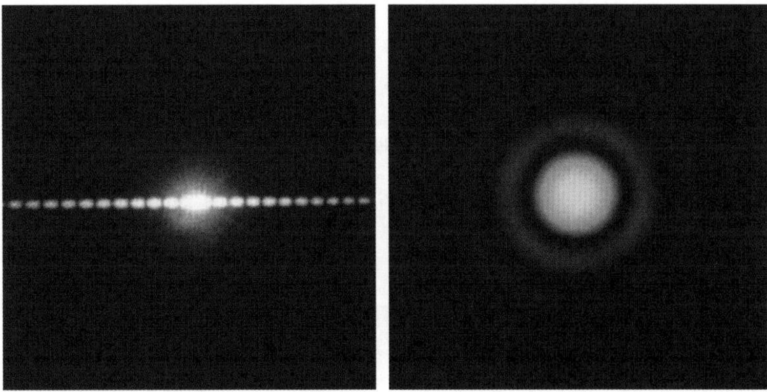

Fig. 3.6: Diffraction at a slit (left) and circular aperture (right)

The intensity becomes a minimum for the values

$$\sin \varphi_n = n\frac{\lambda}{d'} \qquad n = 1, 2, \ldots \qquad (3.11)$$

Diffraction at a circular aperture results in a diffraction pattern, known as an Airy disc, with concentric lines of interference. Bessel functions are used to calculate the phase angle of the maximum intensity of the diffraction disc which leads to:

$$\sin \varphi_n = b_n\frac{2\lambda}{d'} \qquad b_1 = 0.61;\ b_2 = 1.12;\ b_3 = 1.62 \ldots \qquad (3.12)$$

The radius of a diffraction disc at distance f' and $n = 1$ is given by:

$$d = 2 \cdot 1.22\frac{\lambda}{d'}f' \qquad (3.13)$$

where

f' : focal length

$k = \frac{f'}{d'}$: f/number

Diffraction not only occurs at limiting circular edges such as those defined by apertures or lens mountings but also at straight edges and (regular) grid structures such as the arrangement of sensor elements on imaging sensors (see section 3.4.1). Diffraction-limited resolution in optical systems is discussed further in section 3.1.4.1.

In conjunction with deviations of the lens equation 3.14, (defocusing) diffraction yields the *point spread function* (PSF). This effect is dependent on wavelength and described by the contrast or modulation transfer function (see sections 3.1.4.3 and 3.1.5).

3.1.2 Optical imaging

3.1.2.1 Geometric optics

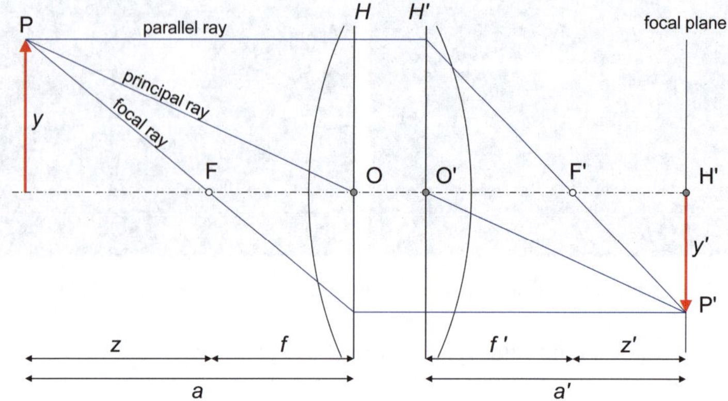

Fig. 3.7: Geometrical construction for a typical thin lens system

The optical imaging model for thin lenses is illustrated in Fig. 3.7. The well-known *thin lens equations* can be derived as follows:

$$\frac{1}{a'}+\frac{1}{a}=\frac{1}{f'} \tag{3.14}$$

$$z \cdot z' = -f'^2 \tag{3.15}$$

where

a:	object distance
a':	image distance ≈ principal distance c
f, f':	external and internal focal length
z, z':	object and image distances relative to principle foci, F and F'
H_1, H_2:	object-side and image-side principal planes
O, O':	object-side and image-side principal points

If the transmission media are different in object and image space, eqn. 3.14 is extended as follows:

$$\frac{n}{a}+\frac{n'}{a'}=\frac{n'}{f'} \tag{3.16}$$

where

n:	refractive index in object space
n':	refractive index in image space

The *optical axis* is defined by the line joining the principal points O and O' which are the centres of the principal planes *H* and *H'*. The object and image side *nodal points* (see Fig.

3.7) are those points on the axis where an imaging ray makes the same angle with the axis in object and image space. If the refractive index is the same on the object and image sides, then the nodal points are identical with the principal points. For a centred lens system, the principal planes are parallel and the axis is perpendicular to them. The optical axis intersects the image plane at the *autocollimation point,* H'. With centred lenses and principal planes orthogonal to the axis, the autocollimation point is the point of symmetry for lens distortion and corresponds to the *principal point* in photogrammetry (see section 3.3.2).

In addition the imaging scale is given in analogy to (3.43)[1]:

$$\beta' = \frac{y'}{y} = \frac{a'}{a} = -\frac{z'}{f'} = 1 : m \tag{3.17}$$

or in the case of different transmission media:

$$\beta' = \frac{n}{n'}\frac{a'}{a} \tag{3.18}$$

According to eqn. 3.14, an object point P is focused at distance a' from the image side principal plane H'. Points at other object distances are not sharply focused (see section 3.1.2.3). For an object point at infinity, $a' = f'$. To a good approximation, image distance a' corresponds to the principal distance c in photogrammetry. m denotes image scale (magnification), as commonly used in photogrammetry.

3.1.2.2 Apertures and stops

The elements of an optical system which limit the angular sizes of the transmitted bundles of light rays (historically denoted as light pencils) can be described as *stops*. These are primarily the lens rims or mountings and the aperture and iris diaphragms themselves. Stops and diaphragms limit the extent of the incident bundles and, amongst other things, contribute to amount of transmitted light and the depth of field.

The most limiting stop is the *aperture stop,* which defines the *aperture* of a lens or imaging system. The object-side image of the aperture is called the *entrance pupil EP* and the corresponding image-side image is known as the *exit pupil E'P.* The f/number is defined as the ratio of the image-side focal length f' to the diameter of the entrance pupil d':

$$\text{f/number} = \frac{f'}{d'} \tag{3.19}$$

In symmetrically constructed *compound lenses*, the diameters of the pupils *E'P* and *EP* are equal. In this case *E'P* and *EP* are located in the principal planes (Fig. 3.8). In that case only, the incident and emerging rays make the same angle $\tau = \tau'$ with the optical axis. Asymmetrical lenses are produced when the component lens do not provide a symmetrical

[1] In optics β' is used instead of $M = 1/m$

structure. For example in wide-angle or telephoto lens designs where the asymmetrical design can place one or more of the nodes outside of the physical glass boundaries or the aperture is asymmetrically positioned between the principal planes (Fig. 3.9). Different angles of incidence and emergence give rise to radially symmetric lens distortion $\Delta r'$ (see also sections 3.1.3.1 and 3.3.2.2).

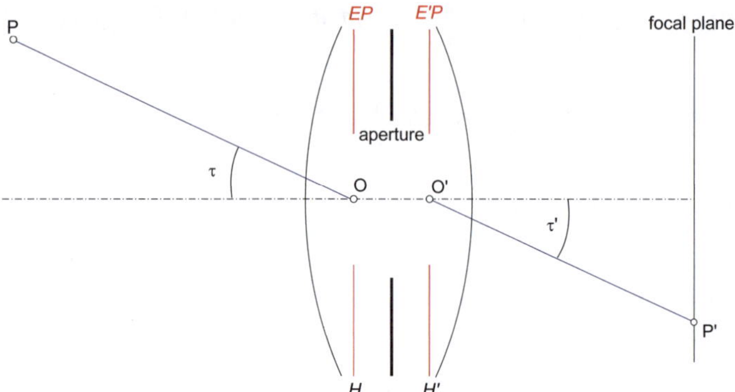

Fig. 3.8: Symmetrical compound lens

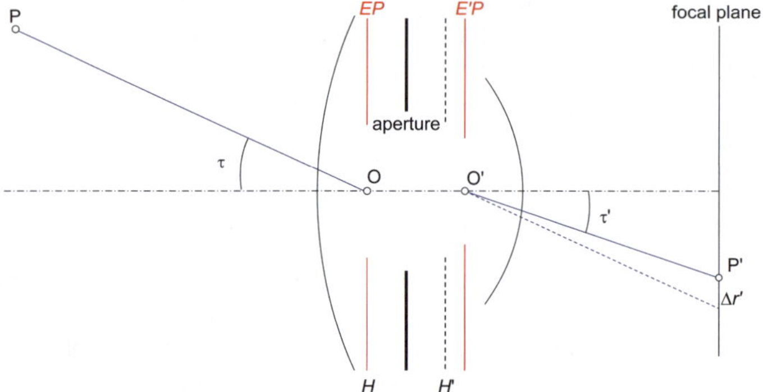

Fig. 3.9: Asymmetrical compound lens

3.1.2.3 Focussing

In practical optical imaging a point source or light is not focused to a point image but to a spot known as the *circle of confusion* or *blur circle*. An object point is observed by the human eye as sharply imaged if the diameter of the circle of confusion u' is under a resolution limit. For film cameras u' is normally taken as around 20–30 µm and for digital cameras as around 1–3 pixels. It is also common to make the definition on the basis of the eye's visual resolution limit of $\Delta\alpha = 0.03°$. Transferring this idea to a normal-angle camera lens of focal length f' ($\approx$ image diagonal d'), a blur circle of the following diameter u' will be perceived as a sharp image:

$$u' = f' \cdot \Delta\alpha \approx \frac{1}{2000} d' \tag{3.20}$$

Example 3.1:

It is required to find the permissible diameter of the blur circle for a small-format film camera, as well as for two digital cameras with different image formats.

	Film camera	Digital camera 1	Digital camera 2
Pixels:	n/a - analogue	2560 x 1920	4368 x 2912
Pixel size:		3.5 µm	8.2 µm
Image diagonal:	43.3 mm	11.2 mm	43.0 mm
Blur circle:	22 µm	5.6 µm = 1.6 pixel	21 µm = 2.6 pixel

The blur which can be tolerated therefore becomes smaller with smaller image format. At the same time the demands on lens quality increase.

Based on the blur which can be tolerated, not only does an object point P at distance a appear to be in sharp focus but also all points between P_v and P_h (Fig. 3.10).

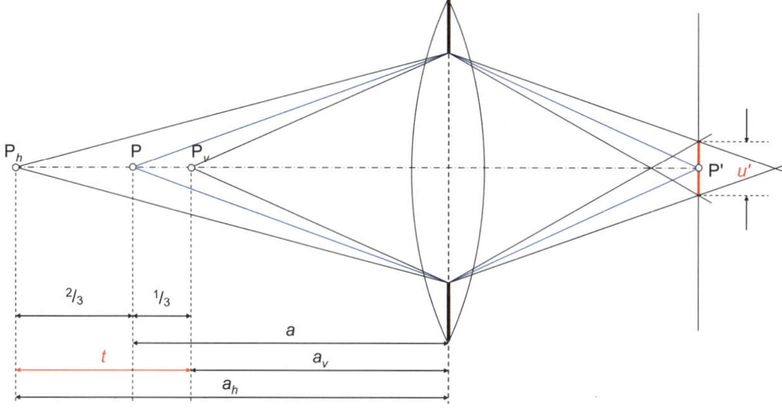

Fig. 3.10: Focusing and depth of field

The distance to the nearest and the furthest sharply defined object point can be calculated as follows:

$$a_v = \frac{a}{1+K} \qquad a_h = \frac{a}{1-K} \tag{3.21}$$

where $K = \frac{k(a-f)u'}{f^2}$

and

k: f/number
f: focal length
a: focused object distance

By re-arrangement of (3.21) the diameter of the blur circle u' can be calculated:

$$u' = \frac{a_h - a_v}{a_h + a_v} \cdot \frac{f^2}{k(a-f)} \tag{3.22}$$

The *depth of field* is defined by

$$t = a_h - a_v = \frac{2u'k(1+\beta')}{\beta'^2 - \left(\frac{u'k}{f}\right)^2} \tag{3.23}$$

Hence, for a given circle of confusion diameter, depth of field depends on the f/number of the lens k and the imaging scale β'. The depth of field will increase if the aperture is reduced, the object distance is increased, or if the focal length is reduced through the use of a wider angle lens. Fig. 3.11 shows the non-linear curve of the resulting depth of field at different scales and apertures. Fig. 3.13a,b shows an actual example.

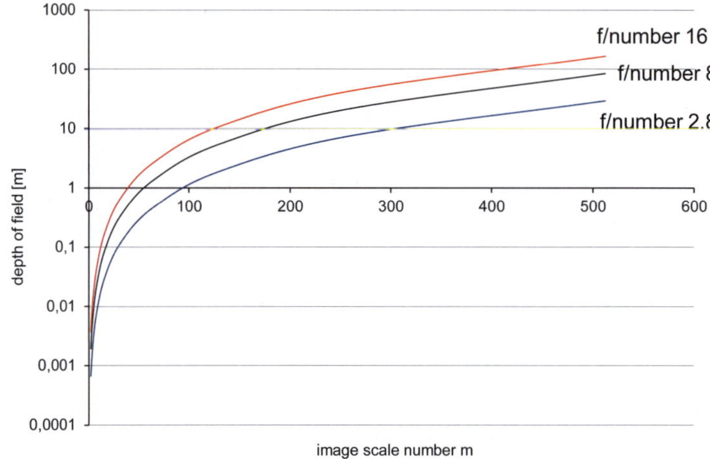

Fig. 3.11: Depth of field as a function of image scale number (u' = 20 µm)

For large object distances (3.23) can be simplified to

$$t = \frac{2u'k}{\beta'^2} \tag{3.24}$$

Example 3.2:

Given a photograph of an object at distance a =5 m and image scale number m = 125, i.e. β' = 0.008. The depth of field for each aperture k = 2.8 and k = 11 is required.

Solution:

1. For k = 2.8 $\quad t = \frac{2 \cdot u' \cdot k}{\beta'^2} = \frac{2 \cdot 2 \cdot 10^{-5} \cdot 2.8}{0.008^2} = 1.8\,\text{m}$

2. For k = 11 $\quad t = 6.9\text{m}$

When imaging objects at infinity, depth of field can be optimised if the lens is not focused at infinity but to the *hyperfocal distance b*. Then the depth of field ranges from the nearest acceptably sharp point a_v to ∞.

$$b = \frac{f^2}{u'k} \quad \text{: hyperfocal distance} \tag{3.25}$$

Sufficient depth of field has to be considered especially carefully for convergent imagery, where there is variation in scale across the image, and when imaging objects with large depth variations. Depth of field can become extremely small for large image scales (small *m*), for example when taking images at very close ranges. Sharp focusing of obliquely imaged object plane can be achieved using the Scheimpflug condition (section 3.1.2.4).

It is worth noting that under many circumstances a slight defocusing of the image can be tolerated if radially symmetric targets are used for image measurement. On the one hand the human eye can centre a symmetrical measuring mark over a blurred circle and on the other hand the optical centroid remains unchanged for digital image measurement.

3.1.2.4 Scheimpflug condition

For oblique imaging of a flat object, the lens eqn. 3.14 can be applied to all object points provided that the object plane, image plane and lens plane (through the lens perspective centre and orthogonal to the optical axis) are arranged such that they intersect in a common line (Fig. 3.12). This configuration is known as the Scheimpflug condition and can be realised, for example, using a *view camera* (Fig. 3.91) or special *tilt-shift lenses*, also called *perspective control lenses* (Fig. 3.100).

Fig. 3.12 shows the imaging configuration which defines the Scheimpflug principle. The relative tilts of the three planes leads to the condition:

$$\frac{1}{a'_1} + \frac{1}{a_1} = \frac{1}{a'_2} + \frac{1}{a_2} \tag{3.26}$$

Compare this with eqn. 3.27 and it is clear that all points on the tilted object plane are sharply imaged. Fig. 3.13c shows an actual example of the effect.

The opposing tilts of the lens and image planes by angle α results in a separation of the principal point H' (foot of the perpendicular from perspective centre to image plane) and the point of symmetry of radial distortion S' (point of intersection of optical axis and image plane).

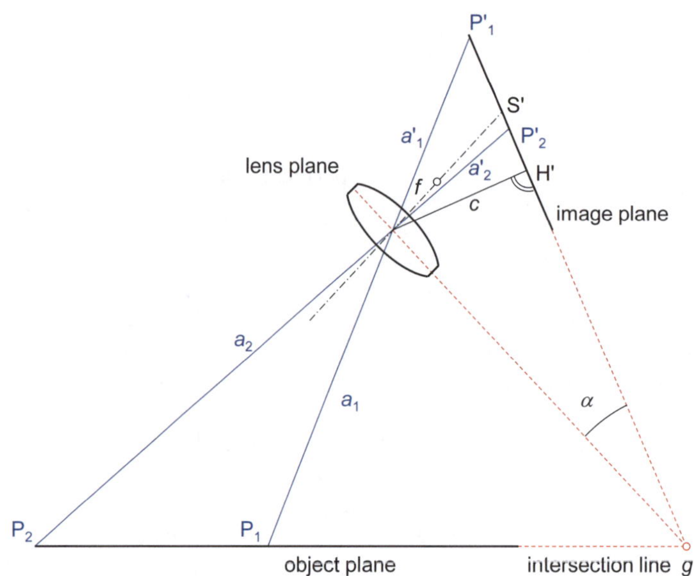

Fig. 3.12: Arrangement of object, lens and image planes according to the Scheimpflug principle

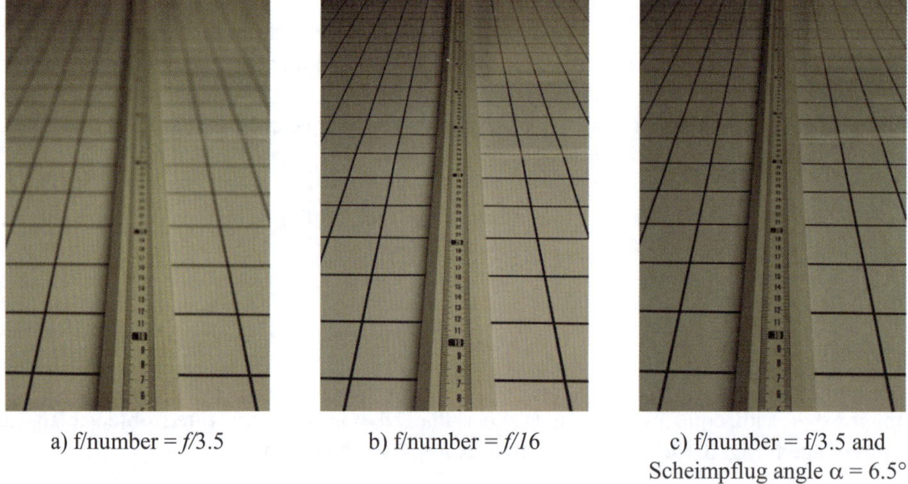

a) f/number = *f*/3.5 b) f/number = *f*/16 c) f/number = f/3.5 and Scheimpflug angle α = 6.5°

Fig. 3.13: Depth of field at different f/numbers and with the Scheimpflug condition active

3.1.3 Aberrations

In optics, *aberrations* are deviations from an ideal imaging model, and aberrations which arise in monochromatic light are known as *geometric distortions*. Aberrations which arise through dispersion are known as *chromatic aberrations*. In photogrammetry, individual optical aberrations, with the exception of radial distortion, are not normally handled according to their cause or appearance as derived from optical principles. Instead, only their effect in the image is described. However, a physically-based analysis is appropriate when several aberrations combine to create complex image errors.

3.1.3.1 Distortion

In the ideal case of Fig. 3.8 the angle of incidence τ is equal to the angle of emergence τ'. As the position of entrance pupil and exit pupil do not usually coincide with the principal planes, an incident ray enters according to angle τ, and exits according to a different angle τ'. This has a distorting effect $\Delta r'$ in the image plane which is radially symmetric with a point of symmetry S'. In the ideal case S' is identical with the autocollimation point H'. The sign, i.e. direction, of $\Delta r'$ depends on the construction of the lens, in particular the position of the aperture.

Fig. 3.14 shows how *barrel distortion* arises when the aperture is moved towards the object, and *pincushion distortion* arises when it moves towards the image. When the image is free of distortion it may be called *orthoscopic*.

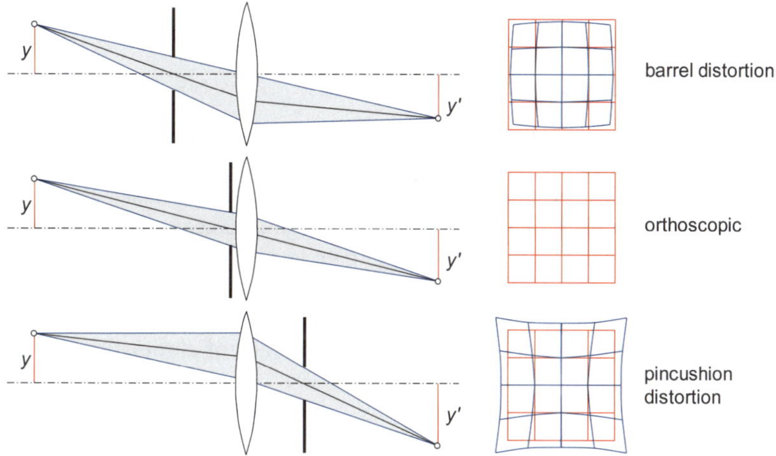

Fig. 3.14: Distortion as a function of aperture position

Fig. 3.15: Pincushion (left) and barrel distortion (right) in an actual image

Radial distortion can be interpreted as a radially dependent scale change in the image. The relative distortion is defined as:

$$\frac{\Delta r'}{r'} = \frac{\Delta x'}{x'} = \frac{\Delta y'}{y'} \tag{3.27}$$

From the definition of image scale in (3.17) and dividing numerator and denominator in (3.27) by y, it can be seen that the relative distortion is also equivalent to the relative scale change as follows:

$$\frac{\Delta \beta'}{\beta'} = \frac{\Delta y'}{y'} = f(\tau) \tag{3.28}$$

For $\Delta y'/y' < 0$ the image is too small, i.e. the image point is shifted towards the optical axis and the result is seen as barrel distortion. Correspondingly, when $\Delta y'/y' > 0$ the result is pincushion distortion. Fig. 3.15 shows the effect in an actual image. The analytical correction of distortion is discussed in section 3.3.2.2.

3.1.3.2 Chromatic aberration

Chromatic aberration in a lens is caused by dispersion which depends on wavelength, also interpreted as colour, hence the name. *Longitudinal chromatic aberration*, also called *axial chromatic aberration*, has the consequence that every wavelength has its own focus. A white object point is therefore focused at different image distances so that an optimal focus position is not possible (see Fig. 3.16). Depending on lens quality, the effect can be reduced by using different component lenses and coatings in its construction. If the image plane is positioned for sharp imaging at mid-range wavelengths, e.g. green, then image errors will appear in the red and blue regions. In practice, the presence of chromatic aberration limits the sharpness of an image.

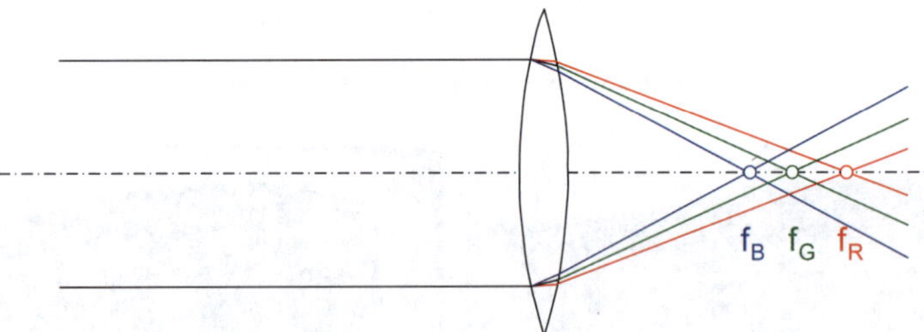

Fig. 3.16: Longitudinal chromatic aberration

Lateral chromatic aberration, also called *transverse chromatic aberration*, has the consequence that an object is imaged at different scales depending on radial position in the image. For monochromatic light the effect is equivalent to radial distortion. For polychromatic light, the aberration causes a variable radial shift of colour.

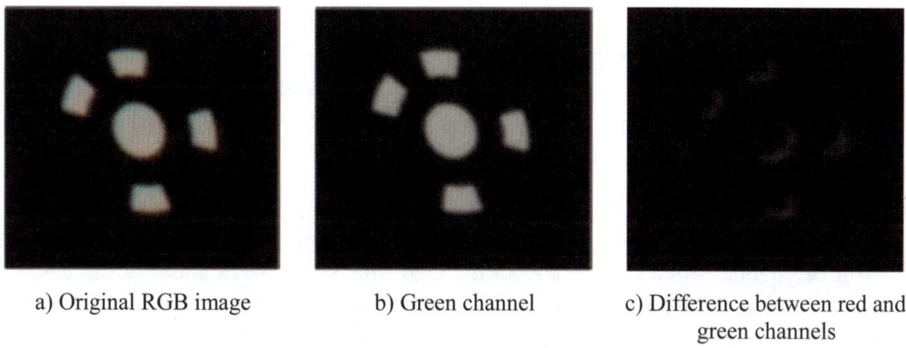

a) Original RGB image b) Green channel c) Difference between red and green channels

Fig. 3.17: Colour shift at a black/white edge

The effect can easily be seen in digital colour images. Fig. 3.17a shows white targets with poor image quality in a colour image. The colour errors are clearly visible at the target edges, as they would occur at the edge of any other imaged object. Fig. 3.17b shows the green channel only. This has an image quality sufficient for photogrammetric point measurement. The difference image in Fig. 3.17c makes clear the difference between the red and green channels.

In black-and-white cameras chromatic aberration causes blurred edges in the image. In colour cameras, multi-coloured outlines are visible, particularly at edges with black/white transitions. In these cases, colour quality also depends on the method of colour separation (e.g. Bayer filter, see section 3.4.1.4) and the image pre-processing common in digital cameras, e.g. in the form of colour or focus correction.

3.1.3.3 Spherical aberration

Spherical aberration causes rays to focus at different positions in the z' direction depending on their displacement from the optical axis. The effect is greater for off-axis object points (with bundles of rays angled to the axis).

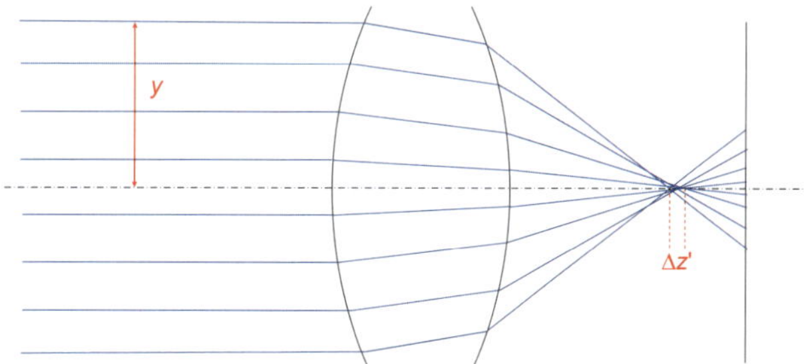

Fig. 3.18: Spherical aberration

The offset $\Delta z'$ in the longitudinal direction varies quadratically with the displacement y (Fig. 3.18). The effect can be considerably reduced by stopping down (reducing the aperture size). In the image the effect causes a reduction in contrast and a softening of focus. Photogrammetrically it can be concluded that spherical aberration results in a shift of perspective centre at different field angles (off-axis locations). Good quality lenses, as normally used in photogrammetry, are mostly free of spherical aberration so that this source of error need not be taken into account in photogrammetric camera calibration.

3.1.3.4 Astigmatism and curvature of field

Astigmatism and *curvature of field* apply to off-axis object points. They are incident at different angles to the refractive surfaces of the component lenses and have different image effects in orthogonal directions. These orthogonal directions are defined by the *meridian* and *sagittal* planes, M_1M_2 and S_1S_2 (Fig. 3.19). The meridian plane is defined by the optical axis and the *principal ray* in the imaging pencil of light, this being the ray through the centre of the entrance pupil. The two planes no longer meet at a single image point but in two image lines (Fig. 3.20). An object point therefore appears in the image as a blurred oval shape or an elliptical spot.

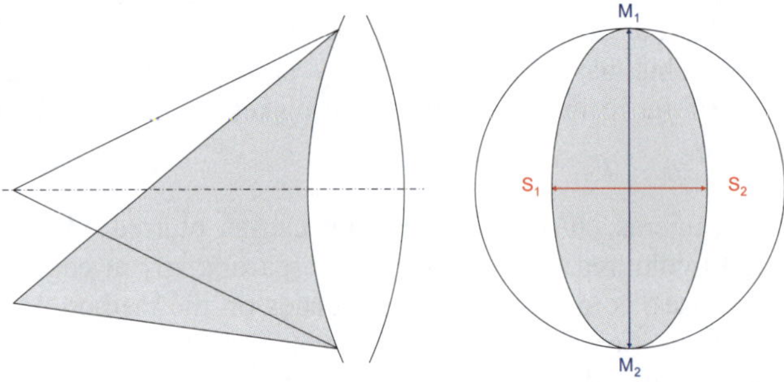

Fig. 3.19: Elliptical masking of off-axis ray bundles (after Marchesi 1985)

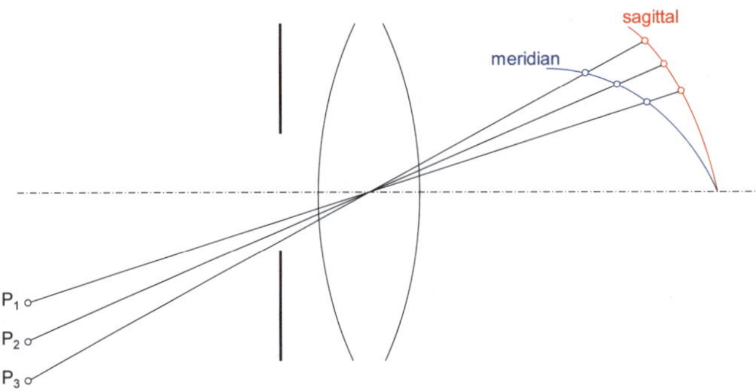

Fig. 3.20: Astigmatism (after Schröder 1990)

The curved image surfaces, which are produced as shown in Fig. 3.20, are known as *curvature of field*. If the meridian and sagittal surfaces are separated then the effect is true astigmatism with differing curvatures through the image space. Where significant curvature is present sensor placement within the field becomes critical in maintaining image quality across the image format.

The effects on the image of astigmatism and curvature of field are, like spherical aberration, dependant on the off-axis position of the object point (incoming ray pencils are at an angle to the axis). The imaging error can be reduced with smaller apertures and appropriately designed curvatures and combinations of component lenses. No effort is made to correct this in photogrammetry as it is assumed that lens design reduces the error below the level of measurement sensitivity. Camera manufacturer Contax has historically produced camera with a curved image plane in order to optimise image sharpness in the presence of field curvature.

3.1.3.5 Coma

In addition to spherical aberration and astigmatism, asymmetric errors occur for wide and off-axis light pencils which are known as *coma*. They have the effect that point objects are imaged as comet-shaped figures with an uneven distribution of intensity.

3.1.3.6 Light fall-off and vignetting

For conventional lenses, the luminous intensity I effective at the imaging plane is reduced with increasing field angle τ according to the $\cos^4$ law:

$$I' = I \cos^4 \tau \tag{3.29}$$

Hence the image gets darker towards its periphery (Fig. 3.21, Fig. 3.22a). The effect is particularly observable for super-wide-angle lenses where it may be necessary to use a concentrically graduated neutral density filter in the optical system to reduce the image intensity at the centre of the field of view. Fisheye lenses avoid the $\cos^4$ law through the use of different projections which reduce the fall-off in illumination at the expense of image distortions (section 3.3.6). The $\cos^4$ reduction in image intensity can be amplified if vignetting, caused by physical obstructions due to mounting parts of the lens, is taken into account (Fig. 3.22b). The fall-off in light can be eliminated by analytical correction of the colour value, but at the expense of increased image noise.

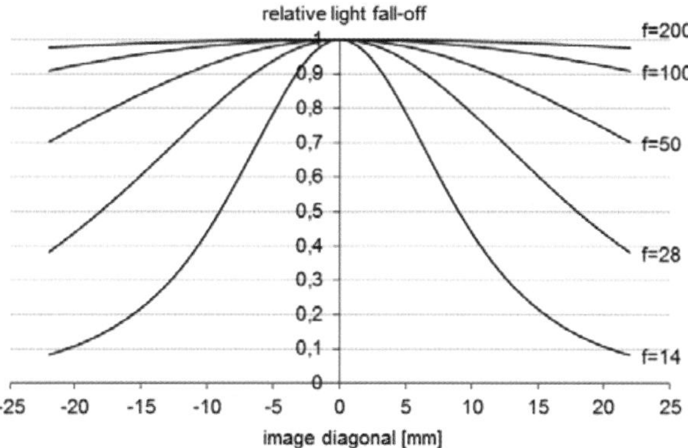

Fig. 3.21: Relative fall-off in light for small format images at different focal lengths (in mm)

a) Light fall-off

b) Vignetting

Fig. 3.22: Images showing loss of intensity towards the edges

3.1.4 Resolution

3.1.4.1 Resolving power of a lens

The resolving power of a lens is limited by diffraction and aperture. For a refractive index of $n = 1$, the diameter of the central Airy diffraction disc is obtained from (3.13) und (3.19) as follows:

$$d = 2.44 \cdot \lambda k \tag{3.30}$$

which corresponds to an angular resolution of

$$\delta = 1.22 \cdot \frac{\lambda}{d'} \tag{3.31}$$

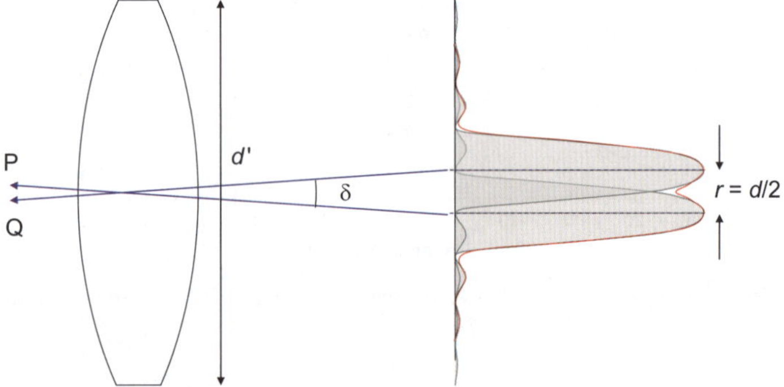

Fig. 3.23: Resolving power defined by the separation of two Airy discs (compare with Fig. 3.5)

Two neighbouring point objects can only be seen as separate images when their image separation is greater than $r = d/2$ (Fig. 3.23). At that limiting separation the maximum of one diffraction disc lies on the minimum of the other. At greater apertures (smaller f/numbers) the resolving power increases.

Example 3.3:

The following diffraction discs with diameter d are generated by a camera for a mid-range wavelength of 550 nm (yellow light):

Aperture $f/2$: $d = 2.7$ µm; Aperture $f/11$: $d = 15$ µm

A human eye with a pupil diameter of $d' = 2$ mm and a focal length of $f' = 24$ mm has an f/number of $f/12$. At a wavelength $\lambda = 550$ nm, and a refractive index in the eyeball of $n = 1.33$, a diffraction disc of diameter $d = 12$ mm is obtained. The average separation of the rod and cone cells which sense the image is 6 µm, which exactly matches the diffraction-limited resolving power.

3.1.4.2 Geometric resolving power

The geometric *resolving power* of a film or an imaging system defines its capability to distinguish between a number of black and white *line pairs* with equal spacing, width and contrast in the resulting image. Therefore it is a measure of the information content of an image.

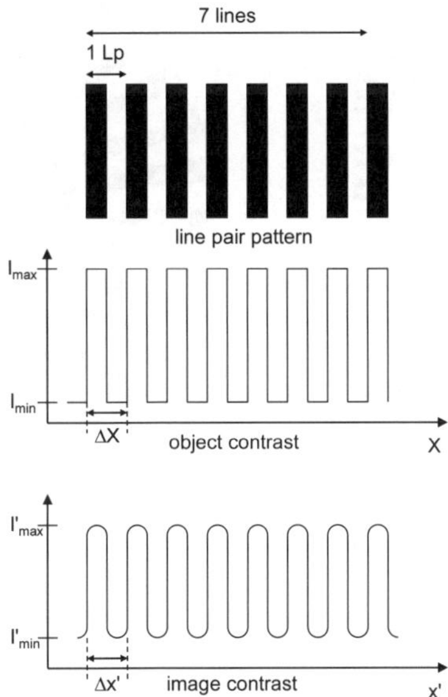

Fig. 3.24: Line pair pattern and contrast transfer

The resolving power *RP* is measured visually as the number of *line pairs per millimetre* (Lp/mm). Alternatively the terms *lines per millimetre* (L/mm) or *dots per inch* (dpi) may be used[1]. Such terms describe the ability of the imaging system to distinguish imaged details, with the interest usually being in the maximum distinguishable spatial frequency attainable (Fig. 3.24, see also section 3.1.5.1).

$$F = \frac{1}{\Delta X} \qquad f = \frac{1}{\Delta x'} \tag{3.32}$$

The spatial frequencies F in object space with respect to f in image space are the reciprocals of the corresponding line spacings ΔX in object space with respect to $\Delta x'$ in image space.

Resolving power can be measured by imaging a test pattern whose different spatial frequencies are known (Fig. 3.25). For the example, the *Siemens star* consisting of 72 sectors (36 sector pairs) allows the maximum resolving power of the imaging system to be determined in Lp/mm by relating the number of sectors to the circumference (in mm) of the inner circle where the sectors are no longer distinguishable.

$$RP = \frac{36}{\pi d} \tag{3.33}$$

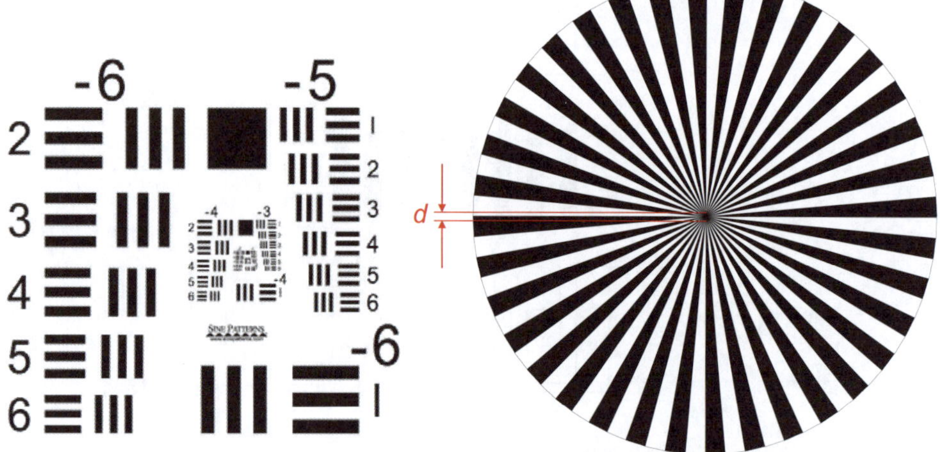

Fig. 3.25: Test chart and Siemens star for the measurement of resolving power

The minimum resolved structure size in object space (structure resolution) ΔX is calculated from image scale and resolving power as follows:

$$\Delta X = m \cdot \Delta x' = m \cdot \frac{1}{AV} \tag{3.34}$$

[1] With unit L/mm only black lines are counted, with Lp/mm black and white lines (pairs) are counted, i.e. both notions are comparable since a black line is only visible if bordered by white lines.

Example 3.4:

In the Siemens star printed above, the diameter of the non-resolved circle is about 1.5 mm (observable with a magnifying glass). Thus the print resolution of this page can be computed as follows:

1. Resolving power: $RP = \frac{36}{\pi \cdot 1.5} \approx 8 \text{ L/mm}$

2. Line size: $\Delta x' = 1 / RP = 0.13 \text{ mm}$

3. Converted to dpi: $RP = \frac{25.4}{\Delta x'} = 194 \text{ dpi} \approx 200 \text{ dpi}$

The applicability of resolving power to opto-electronic sensors is discussed in section 3.4.1.5.

3.1.4.3 Contrast and modulation transfer function

Original high contrast object pattern

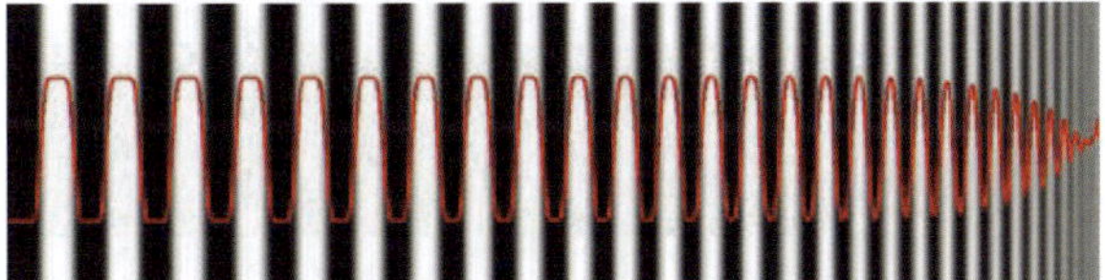

Corresponding image of high contrast pattern

Sinusoidal contrast pattern in object

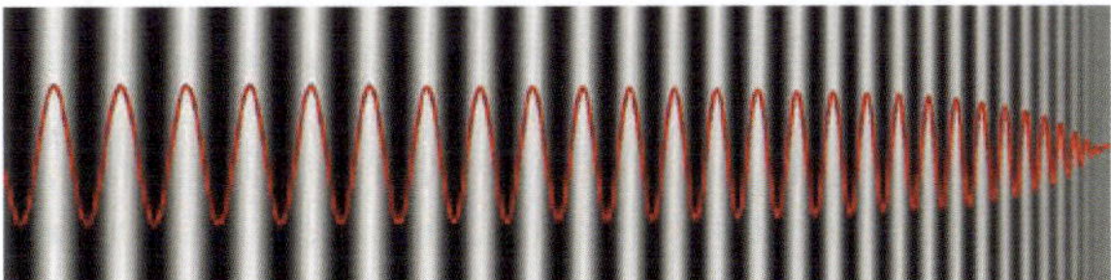

Corresponding image of sinusoidal contrast pattern

Fig. 3.26: Contrast and modulation transfer

The actual resolving power of a film or a complete imaging system depends on the contrast of the original object, i.e. for decreasing contrast signal transfer performance is reduced, particularly at higher spatial frequencies. A contrast-independent formulation of the resolving power is given by the *contrast transfer function* (CTF).

The object contrast K and the imaged contrast K' are functions of the minimum and maximum intensities I of the fringe pattern (Fig. 3.24, Fig. 3.26):

$$K(f) = \frac{I_{max} - I_{min}}{I_{max} + I_{min}} \qquad K'(f) = \frac{I'_{max} - I'_{min}}{I'_{max} + I'_{min}} \tag{3.35}$$

Hence the contrast transfer CT of a spatial frequency f follows:

$$CT(f) = \frac{K'(f)}{K(f)} \tag{3.36}$$

For most imaging systems contrast transfer varies between 0 and 1. The contrast transfer function (CTF) defines the transfer characteristic as a function of the spatial frequency f (Fig. 3.27). Here the resolving power RP can be defined by the spatial frequency that is related to a given minimum value of the CTF (e.g. 30% or 50%). Alternatively RP can be determined as the intersection point of an application-dependent threshold function of a receiver or observer that cannot resolve higher spatial frequencies. The threshold function is usually a perception-limiting function that describes the contrast-dependent resolving power of the eye with optical magnification.

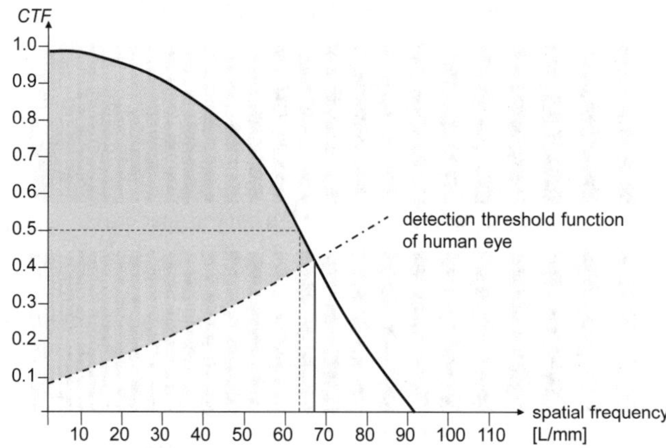

Fig. 3.27: Contrast transfer function (CTF)

In an analogous way, if the rectangular function of Fig. 3.24 is replaced by a sinusoidal function, the contrast transfer function is known as the *modulation transfer function* (MTF).

For an optical system an individual MTF can be defined for each system component (atmosphere, lens, developing, scanning etc.). The total system MTF is given by multiplying the individual MTFs (Fig. 3.28):

$$MTF_{total} = MTF_{imageblur} \cdot MTF_{lens} \cdot MTF_{sensor} \cdot \ldots \cdot MTF_{n} \tag{3.37}$$

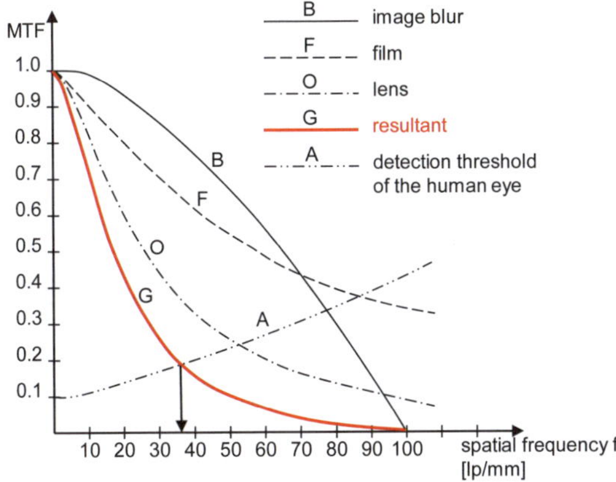

Fig. 3.28: Resulting total MTF for an analogue aerial imaging system (after Kocecny & Lehmann 1984)

3.1.5 Fundamentals of sampling theory

3.1.5.1 Sampling theorem

A continuous analogue signal is converted into a discrete signal by sampling. The amplitude of the sampled signal can then be transferred into digital values by a process known as quantisation (Fig. 3.29).

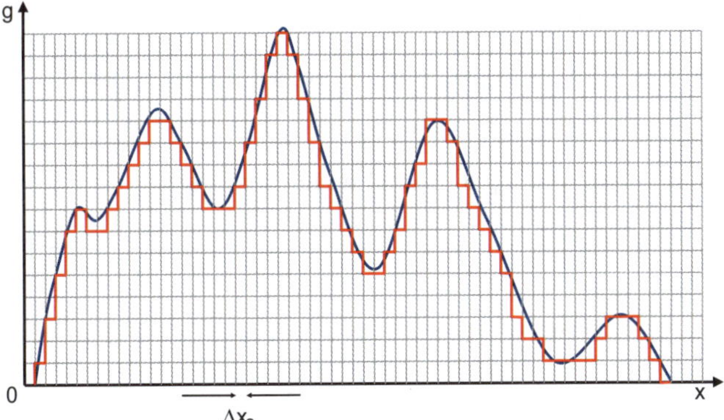

Fig. 3.29: Sampling and quantisation

If *sampling* is performed using a regular array of detector or sensor elements of spacing Δx_S, then the sampling frequency f_A can be expressed as:

$$f_A = \frac{1}{\Delta x_S} \tag{3.38}$$

According to *Shannon's sampling theorem*, the *Nyquist frequency* f_N defines the highest spatial frequency that can be reconstructed by f_A without loss of information:

$$f_N = \frac{1}{2} f_A = \frac{1}{2\Delta x_S} \qquad \text{: Nyquist frequency} \tag{3.39}$$

Spatial frequencies f higher than the Nyquist frequency are *undersampled,* and they are displayed as lower frequencies, an effect known as *aliasing* (see Fig. 3.30).

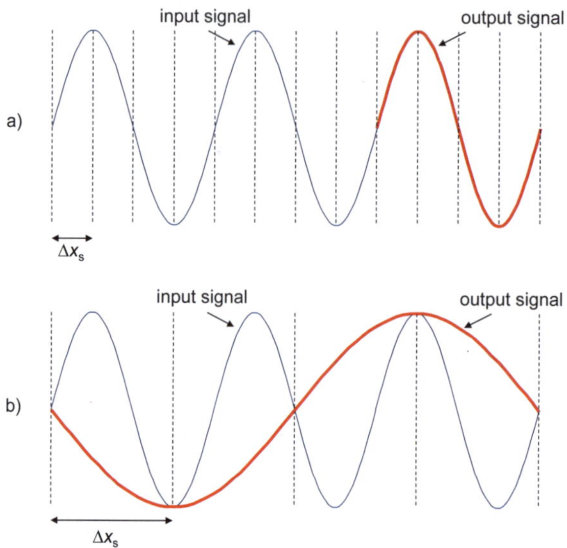

Fig. 3.30: (a) Nyquist sampling; (b) undersampling/aliasing

To avoid sampling errors, as well as to provide a good visual reproduction of the digitised image, it is advisable to apply a higher sampling rate as follows:

$$f_A \approx 2.8 \cdot f \tag{3.40}$$

The transfer characteristic of the sampling system can be described with the modulation transfer function (MTF). With respect to a normalised frequency f/f_N the MTF falls off significantly above 1 (= Nyquist frequency). If the aliasing effects (Fig. 3.30) are to be avoided, the system must consist of a band-pass filter (anti-aliasing filter) that, in the ideal case, cuts off all frequencies above the Nyquist frequency (Fig. 3.31). As an optical low-pass filter, it is possible to use a lens with a resolving power (section 3.1.4.1) somewhat lower than the pixel pitch of the imaging sensor.

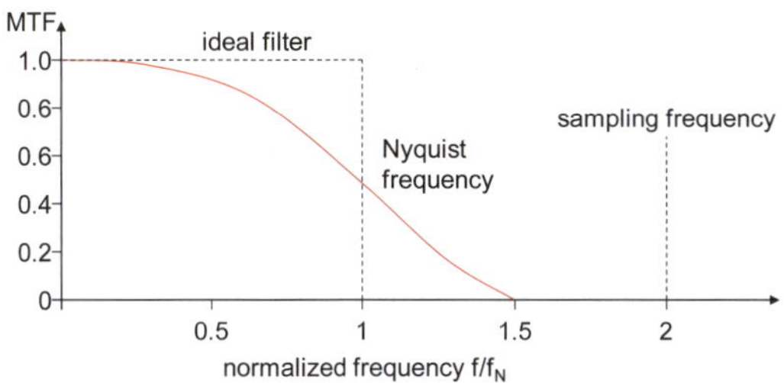

Fig. 3.31: MTF as a function of the normalised sampling frequency

3.1.5.2 Detector characteristics

The sampling device consists of one or several detectors (sensor elements) of limited size with constant spacing with respect to each other. It is important to realise that, because of the need to place a variety of electronic devices in the sensing plane, not all of the area of each detector element is likely to be light sensitive. Sampling and transfer characteristics are therefore a function of both the size of the light-sensitive detector area (aperture size Δx_D) and of the detector spacing (pixel spacing Δx_S). In contrast to the sampling scheme of Fig. 3.30 real sampling integrates over the detector area.

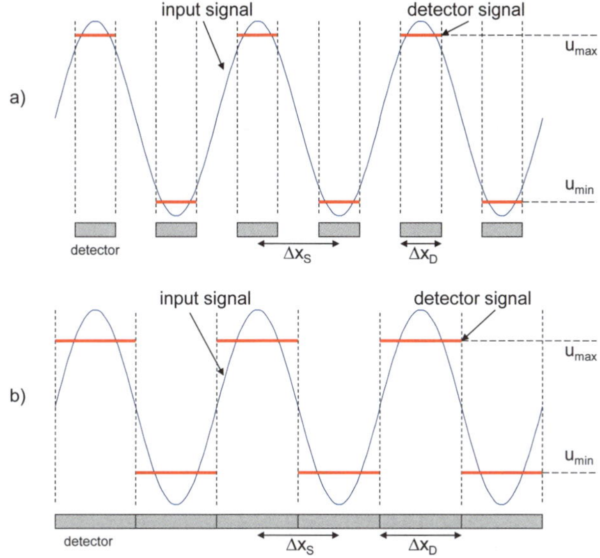

Fig. 3.32: Detector signals
a) light sensitive size equal to half detector spacing; b) light sensitive size equal to detector spacing

Fig. 3.32a displays the detector output when scanning with a sensor whose light-sensitive elements are of size Δx_D and have gaps between them. This gives a detector spacing of $\Delta x_S = 2\ \Delta x_D$ (e.g. interline-transfer sensor, Fig. 3.77). In contrast, Fig. 3.32b shows the sampling result with light sensitive regions without gaps $\Delta x_S = \Delta x_D$ (e.g. frame-transfer sensor, Fig. 3.75). For the latter case the detector signal is higher (greater light sensitivity), however, dynamic range ($u_{max}-u_{min}$) and hence modulation are reduced.

The MTF of a sampling system consisting of sensor elements is given by:

$$MTF_{Detector} = \frac{\sin(\pi \cdot \Delta x_D \cdot f)}{\pi \cdot \Delta x_D \cdot f} = \text{sinc}(\pi \cdot \Delta x_D \cdot f) \tag{3.41}$$

The sinc function was previously introduced for diffraction at aperture slits, see section 3.1.1.4. The function shows that a point signal (Dirac pulse) also generates an output at adjacent detector elements. In theory this is true even for elements at an infinite distance from the pulse. Together with possible defocusing this gives rise to the point spread function (PSD) which, for example, causes sharp edges to have somewhat blurred grey values. Consequently both MTF and PSF can be reconstructed from an analysis of edge profiles.

The MTF becomes zero for $f=k / \Delta x_D$ where $k = 1,2..n$. The first zero-crossing ($k = 1$) is given by the frequency:

$$f_0 = \frac{1}{\Delta x_D} \tag{3.42}$$

The first zero crossing point can be regarded as a natural resolution limit. Fig. 3.33 shows a typical MTF of a detector system. Negative values correspond to reverse contrast, i.e. periodic black fringes are imaged as white patterns, and vice versa. Usually the MTF is shown up to the first zero-crossing only.

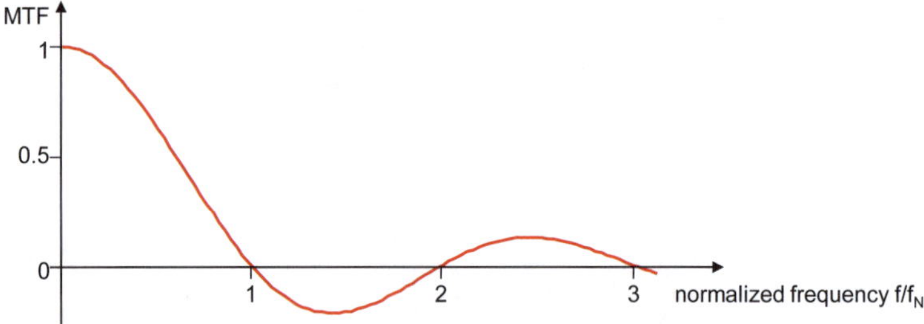

Fig. 3.33: MTF of a detector system

3.2 Photogrammetric imaging concepts

3.2.1 Offline and online systems

With the availability of digital imaging systems it is now possible to implement a seamless data flow from image acquisition to analysis of the photogrammetric results (compare with Fig. 1.9). When image acquisition and evaluation take place in different places or at different times, this is known as *offline photogrammetry*. In contrast, when the acquired images are immediately processed and the relevant data used in a connected process, then this is known as *online photogrammetry* (Fig. 3.34).

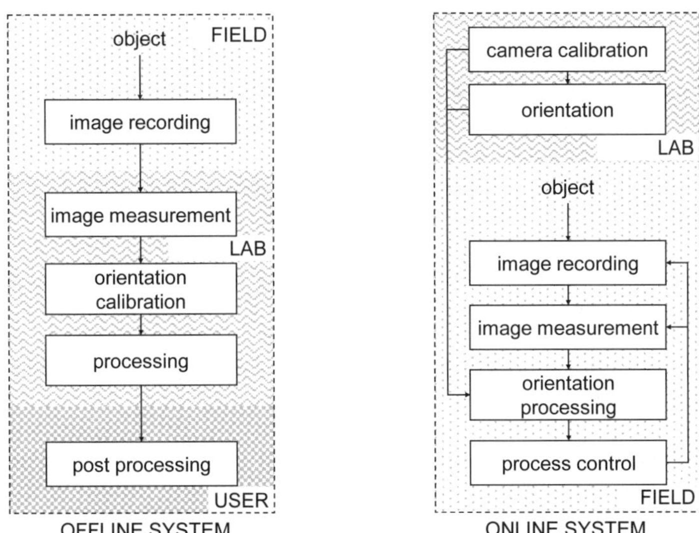

Fig. 3.34: Chain of processing in offline and online systems

Both concepts have a direct impact on the relevant imaging technology. In many cases, cameras used off-line can subsequently be calibrated from the recorded sequence of images (see section 7.3.1). Both analogue and digital cameras may be used here. In contrast, pre-calibrated digital cameras, which are assumed to be geometrically stable over longer periods of time, are typically used in online applications. By appropriate use of reference information in the object space such as control points, the orientation of online systems can be checked during measurement and, if necessary, corrected.

3.2.1.1 Offline photogrammetry

In offline applications it is typical to acquire what, in principle, can be an unlimited number of images which can then be evaluated at a later time and in a different place, possibly also by different users. If the imaging configuration is suitable (see section 7.3.2), the recording cameras can be simultaneously calibrated during the photogrammetric object reconstruction by employing a bundle adjustment (section 4.4). In this way it is possible also to use cameras with a lower level of mechanical stability as the photogrammetric recording system. Because object measurement and camera calibration take place simultaneously and with a high level of redundancy, the highest measurement accuracies can be achieved with offline systems. Depending on application, or technical and economic restrictions, it may also be sensible to make use of pre-calibrated cameras or metric cameras.

Examples of the use of offline photogrammetry are conventional aerial photography, the production of plans and 3D models in architecture, archaeology or facilities management, accident recording, the measurement of industrial equipment and components, as well as image acquisition from unmanned aerial vehicles (UAVs) or mobile platforms.

3.2.1.2 Online photogrammetry

Online photogrammetric systems have a limited number of cameras. There are single, stereo and multi-camera systems which at given time intervals deliver three-dimensional object information. Systems with image sensors integrated at fixed relative positions are generally pre-calibrated (known interior orientation) and oriented (exterior orientation). Depending on stability, this geometry remains constant over longer periods of time. Examples of mobile and stationary online systems with a fixed configuration of cameras are shown in Figs. 6.21 and 6.24. Online systems with variable camera configurations offer the option of on-site orientation which is normally achieved with the aid of reference objects. An example is shown in Fig. 6.19.

Online systems are commonly connected to further operating processes, i.e. the acquired 3D data are delivered in real time in order to control the operation of a second system. Examples here include stereo navigation systems for computer-controlled surgery, positioning systems in car manufacture, image-guided robot positioning or production-line measurement of pipes and tubes.

3.2.2 Imaging configurations

In photogrammetry, imaging configuration is the arrangement of camera stations and viewing directions for object measurement. Usually in photogrammetry the following imaging configurations are distinguished:

- single image acquisition
- stereo image acquisition
- multi-image acquisition

3.2.2.1 Single image acquisition

Three-dimensional reconstruction of an object from a single image is only possible if additional geometric information about the object is available. Single image processing is typically applied for rectification (of plane object surfaces, see section 4.2.6) and orthophotos (involving a digital surface model, see section 4.2.7), and plane object measurements (after prior definition of an object plane, see section 4.2.7.1).

The achievable accuracy of object measurement depends primarily on the image scale (see section 3.3.1, Fig. 3.37 to Fig. 3.39) and the ability to distinguish those features which are to be measured within the image. In the case of oblique imaging, the image scale is defined by the minimum and maximum object distances.

3.2.2.2 Stereo image acquisition

Stereo imagery represents the minimum configuration for acquiring three-dimensional object information. It is typically employed where a visual or automated stereoscopic evaluation process is to be used. Visual processing requires near parallel camera axes similar to the normal case (Fig. 3.35a), as the human visual system can only process images comfortably within a limited angle of convergence. For digital stereo image processing (*stereo image matching*, see section 5.5), the prerequisites of human vision can be ignored with the result that more convergent image pairs can be used (Fig. 3.35b). Staggered camera positions (Fig. 3.35c) provide parallel viewing directions but a different scale in each image. Provided points of detail to be measured are visible in both images, such differences in image scale can be accommodated through differential zooming in the viewing optics, or mathematically through the use of an appropriate scale factor.

In the simplest case, three-dimensional object reconstruction using image pairs is based on the measurement of image parallax $px' = x' - x''$ (Fig. 3.35a) that can be transformed directly into a distance measure h in the viewing direction (see section 4.3.6.2). More generally, image coordinates ($x'y'$, $x''y''$) of homologous (corresponding) image points can be measured in order to calculate 3D coordinates by spatial intersection (see section 4.4.7.1). The accuracy of the computed object coordinates in the viewing direction will generally differ from those parallel to the image plane. Differences in accuracy are a function of the intersection angle between homologous image rays, as defined by the *height-to-base ratio* h/b.

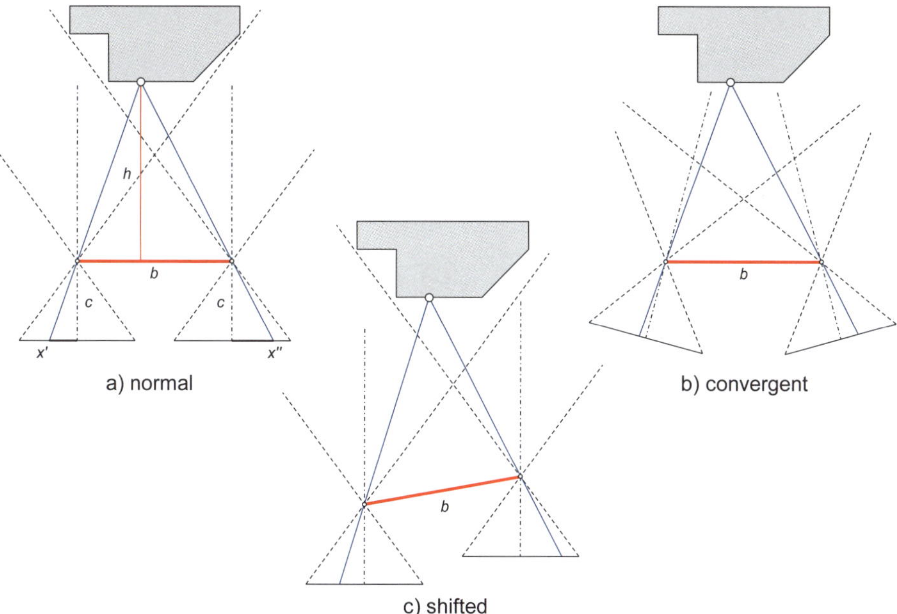

Fig. 3.35: Stereo image configurations

Stereo imagery is most important for the measurement of non-signalised (targetless) object surfaces that can be registered by the visual setting of an optical floating mark (see section

4.3.6.3). Special stereometric cameras have been developed for stereo photogrammetry (see section 3.5.4). An example application in industry is the measurement of free-form surfaces where one camera can be replaced by an oriented pattern projector (see section 3.7.2).

3.2.2.3 Multi-image acquisition

Multi-image configurations (Fig. 3.36) are not restricted with respect to the selection of camera stations and viewing directions. In principle, the object is acquired by an unlimited number of images from locations chosen to enable sufficient intersecting angles of bundles of rays in object space. At least two images from different locations must record every object point to be coordinated in 3D.

Object coordinates are determined by multi-image triangulation (bundle adjustment, see section 4.4, or spatial intersection, see section 4.4.7.1). If a sufficient number and configuration of image rays (at least 3–4 images per object point) are provided, uniform object accuracies in all coordinates can be obtained (see section 3.3.1.2).

In close-range photogrammetry multi-image configurations are the most common case. They are required in all situations where a larger number of different viewing locations are necessary due to the object structure (e.g. occlusions, measurement of both interior and the exterior surfaces) or to maintain specified accuracy requirements. Images can be arranged in strips or blocks (Fig. 4.52) or as all-around configurations (see Fig. 4.53) but, in principle, without any restrictions (examples in Fig. 1.6, Fig. 1.8, Fig. 8.2).

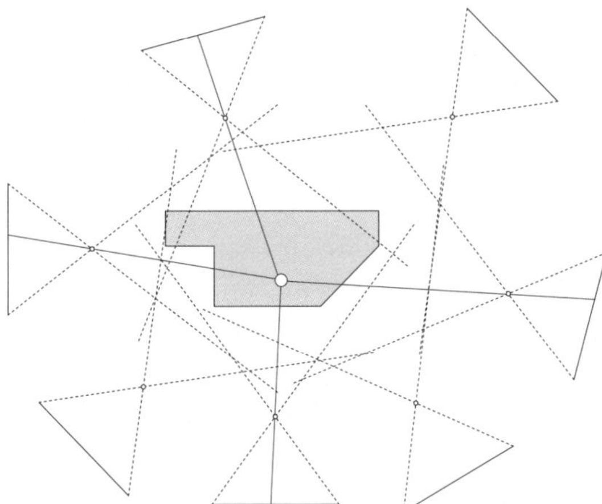

Fig. 3.36: Multi-image acquisition (all-around configuration)

Where the configuration provides a suitable geometry, multi-image configurations enable the simultaneous calibration of the camera(s) by self-calibrating bundle adjustment procedures (see section 4.4.2.4).

3.3 Geometry of the camera as a measuring device

3.3.1 Image scale and accuracy

Image scale and achievable accuracy are the basic criteria of photogrammetric imaging and will dominate the choice of camera system and imaging configuration.

3.3.1.1 Image scale

The image scale number or magnification *m* is defined by the ratio of object distance *h* to the principal distance *c* (lens focal length plus additional shift to achieve sharp focus). It may also be given as the ratio of a distance in object space *X* to the corresponding distance in image space *x*', assuming X is parallel to x' (compare with section 3.1.2):

$$m = \frac{h}{c} = \frac{X}{x'} = \frac{1}{M} \tag{3.43}$$

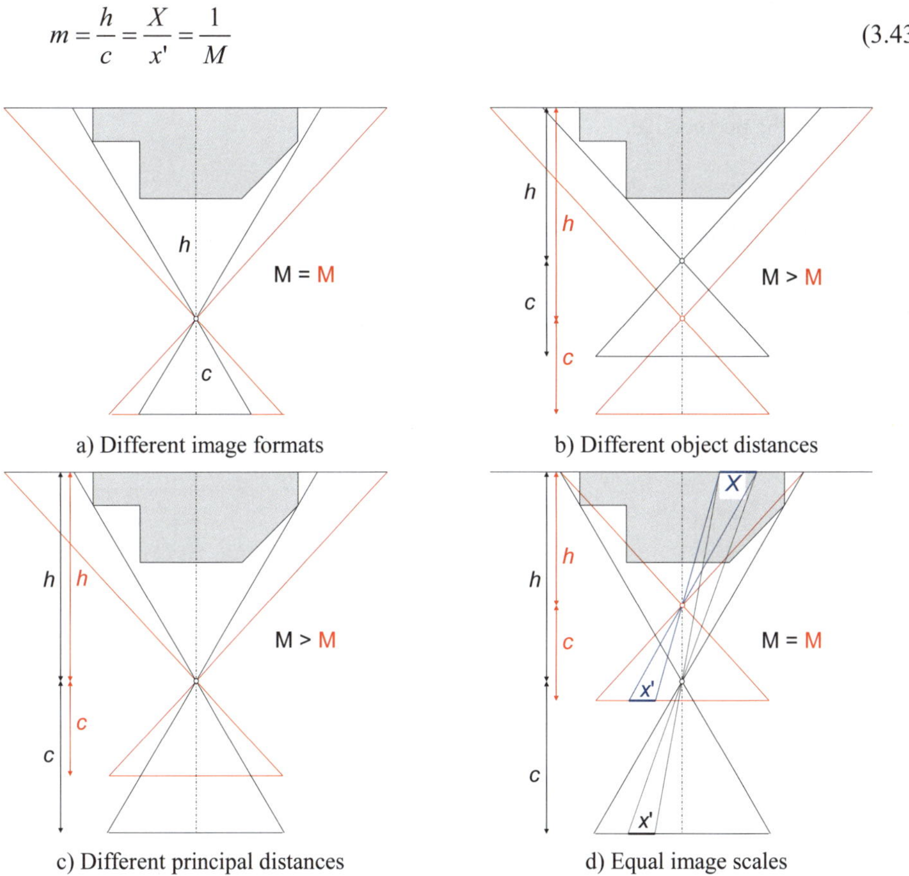

Fig. 3.37: Dependency of image scale on image format, focal length and object distance

In order to achieve a sufficient accuracy and detect fine detail in the scene, the selected image scale *m* must take into account the chosen imaging system and the surrounding environmental conditions. Fig. 3.37 illustrates the relationship between object distance, principal distance and image format on the resulting image scale.

Using a camera with a smaller image format and the same image scale (and principle distance) at the same location, the imaged object area is reduced (Fig. 3.37a). In contrast, a larger image scale can be achieved if the object distance is reduced for the same image format (Fig. 3.37b).

For a shorter object distance (Fig. 3.37b), or a longer principal distance (Fig. 3.37c), a larger image scale will result in a correspondingly reduced imaged object area, i.e. the number of images necessary for complete coverage of the object will increase

Fig. 3.37d shows that equal image scales can be obtained by different imaging configurations. With respect to image scale it can be concluded that the selection of imaging system and camera stations is often a compromise between contrary requirements. Note however that any change in the position of the camera with respect to the object will result in a different perspective view of the object. Conversely, changing the focal length of the lens, or altering the camera format dimensions, whilst maintaining the camera position, will not alter the perspective.

Example 3.5:

Given a camera with image format $s' = 60$ mm and a principal distance of $c = 40$ mm (wide angle). Compute the object distance h, where an object size of 7.5 m is imaged over the full format.

Solution:

1. Image scale number: $m = \frac{X}{x'} = \frac{7500}{60} = 125$ $(M = 1{:}125)$

2. Object distance: $h = m \cdot c = 125 \cdot 40 = 5000mm = 5m$

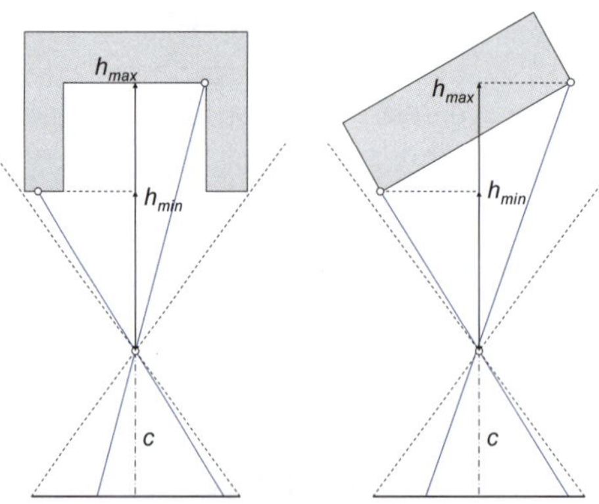

Fig. 3.38: Different image scales

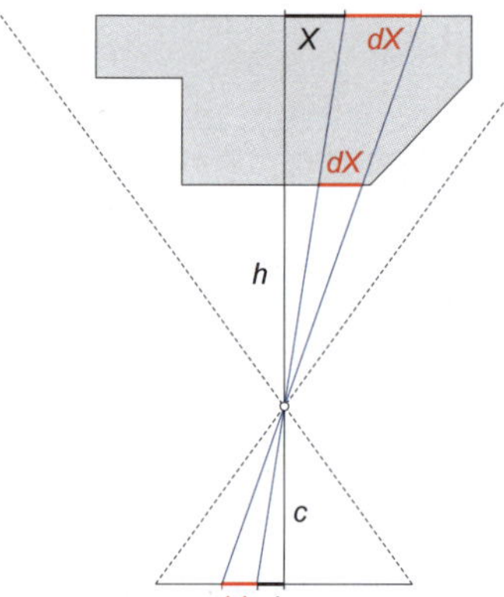

Fig. 3.39: Single image acquisition

An image has a uniform scale only in the case of a plane object which is viewed normally (perpendicular camera axis). For small deviations from the normal an average image scale number related to an average object distance can be used for further estimations. In practical imaging configurations large deviations in image scale result mainly from

- large spatial depth of the object and/or
- extremely oblique images of a plane object

For these cases (see Fig. 3.38) minimum and maximum image scales must be used for project planning and accuracy estimations.

3.3.1.2 Accuracy estimation

The achievable accuracy[1] in object space of a photogrammetric measurement requires assessment against an independent external standard, but the precision of the derived coordinates can be estimated approximately according to Fig. 3.39.

Differentiation of eqn. 3.43 shows that the uncertainty of an image measurement *dx*' can be transferred into object space by the image scale number *m*:

$$dX = m \cdot dx' \tag{3.44}$$

Applying the law of error propagation the standard deviation (mean error) gives:

$$s_X = m \cdot s_{x'} \tag{3.45}$$

In many cases a relative precision, rather than an absolute value, that is related to the maximum object dimension *S*, or the maximum image format *s*' is calculated:

$$\frac{s_X}{S} = \frac{s_{x'}}{s'} \tag{3.46}$$

Eqn. 3.46 shows that a larger image format results in better measurement precision.

Example 3.6:

Given an object with object size $S = 7.5$ m photographed at an image scale of $m = 125$. What is the necessary image measuring accuracy $s_{x'}$, if an object precision of $s_X = 0.5$ mm is to be achieved?

Solution:

1. According to (3.45): $s_{x'} = \frac{s_X}{m} = \frac{0.5}{125} = 0.004\text{mm} = 4\mu\text{m}$

2. Relative accuracy: $\frac{s_X}{S} = 1:15000 = 0.007\%$

---

[1] Here the term "accuracy" is used as a general quality criteria. See section 2.4.3.1 for a definition of accuracy, precision and reliability.

Firstly, the achievable object precision according to (3.45) indicates the relationship between scale and resulting photogrammetric precision. Furthermore, it is a function of the imaging geometry (number of images, ray intersection angles in space) and the extent to which measured features can be identified. A statement of relative precision is only then meaningful if the measured object, processing methods and accuracy verification are effectively described.

Eqn. 3.45 must be extended by a *design factor q* that provides an appropriate weighting of the imaging configuration:

$$s_X = q \cdot m \cdot s_{x'} = \frac{q_D}{\sqrt{k}} m \cdot s_{x'} \tag{3.47}$$

The design parameter q_D is related to the intersection geometry of the imaging configuration, and k defines the mean number of images per camera location. For a normal case in practice $k = 1$ and therefore $q_D = q$.

Practical values for the design factor q vary between 0.4–0.8 for excellent imaging configurations, (e.g. all-around configuration, see Fig. 3.36), and up to 1.5–3.0 or worse for weak stereo configurations (see Fig. 3.35).

If the object is targeted (marked, or signalised, e.g. by circular high-contrast retro-targets) and imaged by an all-around configuration, (3.47) provides a useful approximation for all three coordinate axes, such that $q = 0.5$ can be achieved. In cases where the object cannot be recorded from all sides, accuracies along the viewing direction can differ significantly from those in a transverse direction. As an example, the achievable precision in the viewing direction Z for a normal stereo pair (see section 4.3.6.2) can be estimated by:

$$s_Z = \frac{h^2}{b \cdot c} s_{px'} = m \frac{h}{b} s_{px'} \tag{3.48}$$

Here b defines the distance between both camera stations (stereo base) and $s_{px'}$ the measurement precision of the x-parallax; base b and principal distance c are assumed to be free of error. Measurement precision in the viewing direction depends on the image scale (h/c) and on the intersection geometry, as defined by the height-to-base ratio (h/b).

Example 3.7:

Given a stereo image pair with an image scale of $M = 1{:}125$, an object distance of $h = 5$ m and a base length of $b = 1.2$ m. Compute the achievable precision in the XY-direction (parallel to the image plane) and in the Z-direction (viewing direction) respectively, given a parallax measurement precision of $s_{px'} = 4$ µm (assume $s_{px'} = s_{x'}\sqrt{2}$).

1. Precision in Z: $$s_Z = m \cdot \frac{h}{b} \cdot s_{px'} = 125 \cdot \frac{5}{1.2} \cdot 0.004 = 2.1\text{mm}$$

2. Precision in X,Y where $q_{XY} = 1$:
$$s_X = s_Y = m \cdot s_{x'} = 125 \cdot 0.004 = 0.5\text{mm}$$

3. Design factor from (3.47): $$q_Z = \frac{s_Z}{m \cdot s_{x'}} = 4.2$$

The example shows that the precision in the viewing direction is reduced by a factor of 4.

3.3.2 Interior orientation of a camera

The interior orientation of a camera comprises all instrumental and mathematical elements which completely describe the imaging model within the camera. By taking proper account of the interior orientation, the real camera conforms to the pinhole camera model. It is a requirement of the model that there exists a reproducible image coordinate system so that geometric image values (e.g. measured image coordinates) can be transformed into the physical-mathematical imaging model.

a) Original image b) Compression c) Extract

Fig. 3.40: Original and processed images

Fig. 3.40 shows an arbitrarily selected, original digital image that has been stored in exactly this form and therefore has a reproducible relationship to the camera. The applied processing (compression and extraction) results in a loss of reference to the image coordinate system fixed in the camera, unless the geometric manipulation can be reconstructed in some way.

3.3.2.1 Physical definition of the image coordinate system

The image coordinate system must not only be defined physically with respect to the camera body, but must also be reconstructable within the image. Three common methods are found in photogrammetric imaging systems (Fig. 3.41):

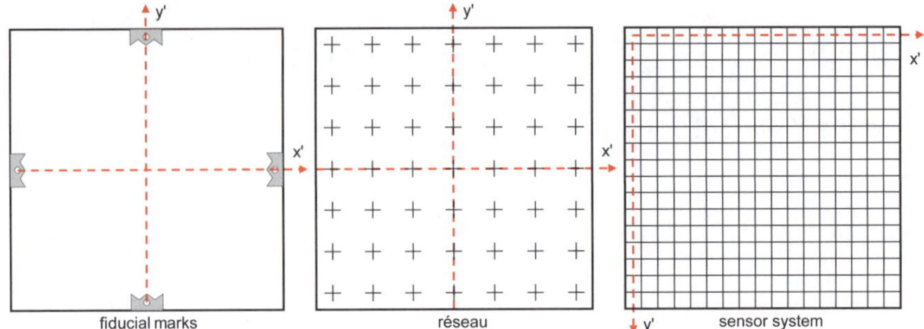

Fig. 3.41: Image coordinate systems

1. Fiducial marks:

The image coordinate system is defined by at least four unique fiducial marks each of which has calibrated nominal coordinates. During exposure the fiducial marks are either imaged in contact with, or projected onto the image. Fiducial marks are typically found in metric film cameras (analogue cameras) and are used in association with a mechanically flattened image plane (examples in Fig. 3.42).

Fig. 3.42: Variations of fiducial marks

2. Réseau:

A réseau consists of a thin plane glass plate with a grid of calibrated reference points (réseau crosses) etched onto it. The plate is mounted directly in front of, and in contact with the film (sensor), and projected onto the image at exposure (examples in Fig. 3.43). Deviations of planarity or film deformations can later be compensated numerically by the imaged réseau. In variant form the réseau can be implemented by a grid of back-projected reference points. Réseaus were used in analogue semi-metric cameras (Fig. 1.33) and as the réseau-scanning method in some scanning systems (Fig. 1.36, section 3.5.5.2).

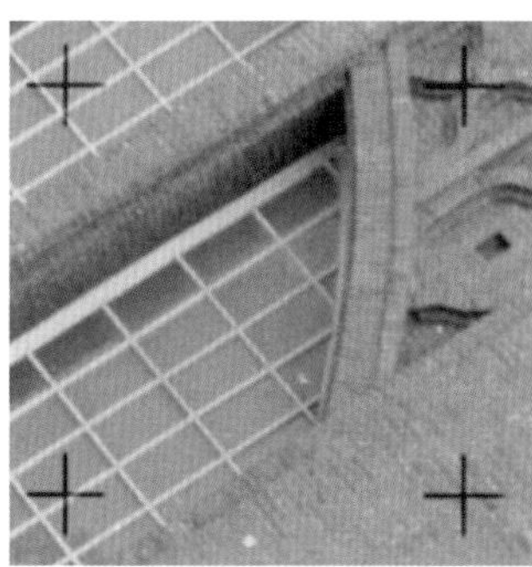

Fig. 3.43: Réseau crosses projected onto a film

3. Sensor coordinate system:

Artificial reference points for the definition of the image coordinate system are not necessary for digital cameras if there is a unique relationship between digital image and opto-electronic image sensor. This is true for fixed-mounted matrix sensors (CCD and CMOS arrays) with direct digital read-out of the image information. Usually the origin of these systems is defined in the upper left corner, hence a left-handed coordinate system results. If the digital image is geometrically altered, e.g. resized or cropped, then the reference to the camera is lost and, in general, the image coordinate system cannot be reconstructed (examples in Fig. 3.40).

Conversion of pixel coordinates u,v to image coordinates x',y' is done using pixel separations Δx_S, Δy_S and the sensor dimensions s'_x, s'_y as follows:

$$x' = -\frac{s'_x}{2} + \Delta x_S \cdot u \qquad\qquad y' = \frac{s'_y}{2} - \Delta y_S \cdot v \qquad (3.49)$$

3.3.2.2 Perspective centre and distortion

Mathematically, the perspective centre is defined by the point through which all straight-line image rays pass. For images created with a compound lens (multiple lens components) both an external and an internal perspective centre can be defined. Each is defined by the intersection point of the optical axis with the entrance pupil *EP* and the exit pupil *E'P*, respectively (see section 3.1.2). The position and size of the entrance and exit pupils are defined by the lens design and its limiting aperture (Fig. 3.8, Fig. 3.44). Hence, the position of the perspective centre depends on the chosen aperture and, due to the influence of dispersion, is additionally dependent on wavelength.

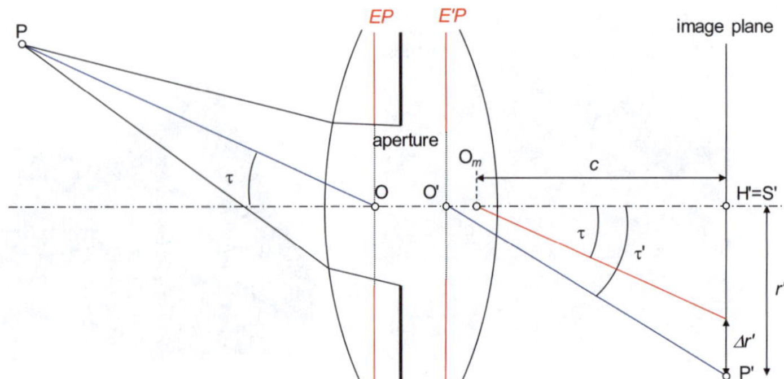

Fig. 3.44: Perspective centres O, O' and principal distance *c* (after Kraus 1994)

In the ideal case of Fig. 3.8 the angle of incidence τ is equal to the exit angle τ', and the principal distance *c* is equal to the image distance *a*' (between principal plane and image plane). As the position of entrance pupil and exit pupil do not usually coincide with the principal planes, an incident ray enters at angle τ, and exits at a different angle τ'. This effect is radially symmetric with a point of symmetry S'. Compared with the ideal case, an image point P' is shifted by an amount Δ*r*' that is known as *radial distortion* (see also Fig. 3.9 and Fig. 3.14).

$$\Delta r' = r' - c \cdot \tan\tau \tag{3.50}$$

In this formulation the distortion is therefore linearly dependent on the principle distance (see also Fig. 3.55).

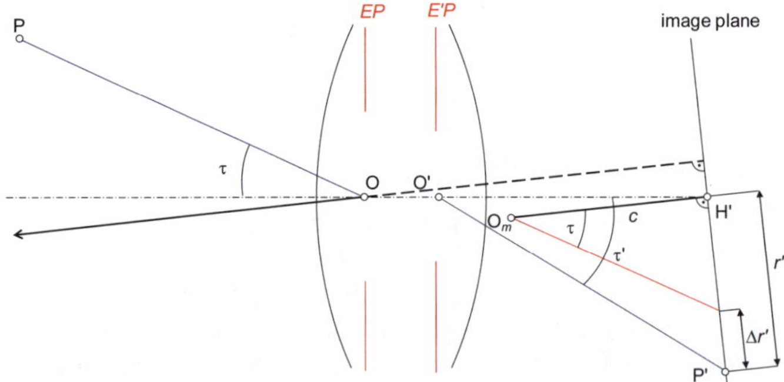

Fig. 3.45: Definition of principal point for a tilted image plane (after Kraus 2000)

The *point of autocollimation* H' is defined as the intersection point of the optical axis of the lens OO' and the image plane. The mathematical perspective centre O_m, used for photogrammetric calculations, is chosen to maximise the symmetry in the distortion in the image plane. The principal distance *c* is normally chosen such that the sum of the distortion components across the whole image format is minimised. If the optical axis is not normal to the image plane, O_m is not positioned on the optical axis (Fig. 3.45). H' is also known as the *principal point*. In real cameras, principal point, point of symmetry and the centre of the image can all be separate. In principle, the image coordinate system can be arbitrarily defined (Fig. 3.46).

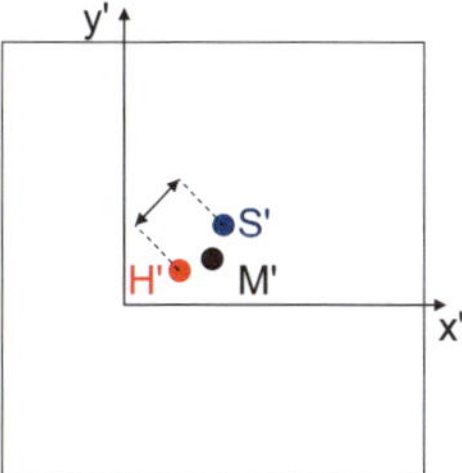

Fig. 3.46: Possible locations of principal point, point of symmetry, centre of image and origin of image coordinate system

Hence the photogrammetric reference axis is defined by the straight line O_mH'. In object space it is given by the parallel ray passing through O[1]. The image radius r' of an image point, and the residual of radial distortion $\Delta r'$, are defined with respect to the principal point H' (see section 3.3.2.3).

In practice, every lens generates distortion. The radial distortion described above can be reduced to a level of $\Delta r' < 4$ µm for symmetrically designed high-quality lenses (Fig. 3.47). In contrast, asymmetric lens designs produce significantly larger distortion values, especially towards the corners of the image. Fig. 3.48 illustrates the complex light paths inside an asymmetric lens which in this case using two custom aspheric optical surfaces and a floating element group to counter lens aberrations.

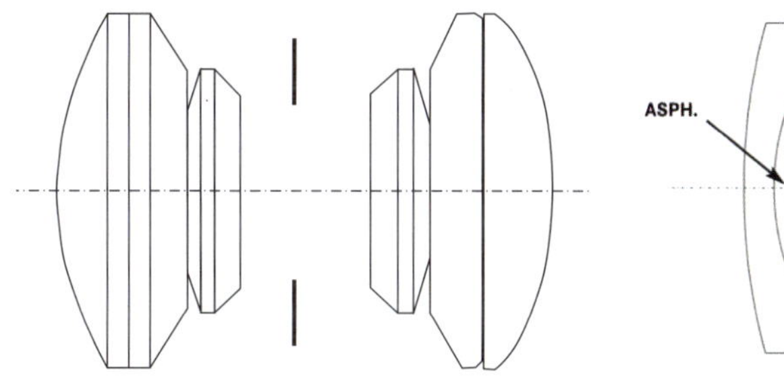

Fig. 3.47: Example of a symmetric lens design (Lametar 8/f=200 for Zeiss UMK, $\Delta r'_{max} = \pm 4$ µm at $r' < 90$ mm)

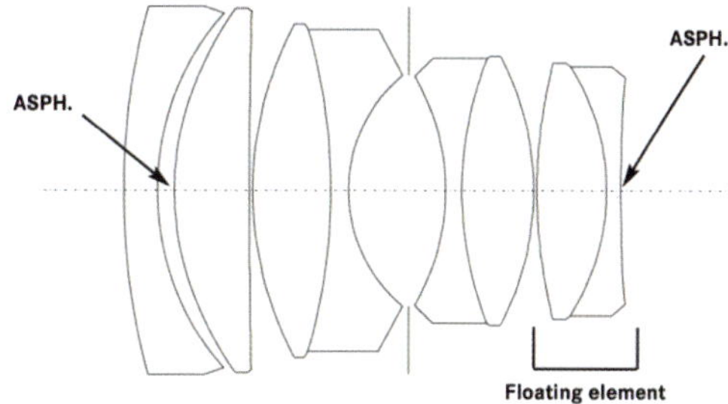

Fig. 3.48: Example of a state of the art low light asymmetric lens design (Leica 50/0.95 Noctilux-M, $\Delta r' = \pm 200$ µm at $r' = 20$ mm, $\Delta r' = \pm 60$ µm at $r' = 10$ mm) (Leica)

In contrast, tangential distortions are attributable to decentring and tilt of individual lens elements within the compound lens (Fig. 3.58 shows the effect in image space). For good quality lenses, these distortions are usually 10 times smaller than radial distortion and thus can be neglected for many photogrammetric purposes. However, the simple low-cost lenses which are increasingly used have been shown to exhibit significantly larger tangential and

[1] Usually the notations O or O' are used, even when O_m is meant.

asymmetric radial distortion values. Distortions in the range of more than 30 µm are possible and are attributable to the low cost of these lenses, combined with the small size of their individual elements.

3.3.2.3 Parameters of interior orientation

A camera can be modelled as a spatial system that consists of a planar imaging area (film or electronic sensor) and the lens with its perspective centre. The parameters of interior orientation of a camera define the spatial position of the perspective centre, the principal distance and the location of the principal point with respect to the image coordinate system defined in the camera. They also encompass deviations from the principle of central perspective to include radial and tangential distortion and often affinity and orthogonality errors in the image.

Fig. 3.49 illustrates the schematic imaging process of a photogrammetric camera. Position and offset of the perspective centre, as well as deviations from the central perspective model, are described with respect to the image coordinate system as defined by reference or fiducial points (film based system) or the pixel array (electronic system). The origin of the image coordinate system is located in the image plane. For the following analytical calculations the origin of the image coordinate system is shifted to coincide with the perspective centre according to Fig. 2.1.

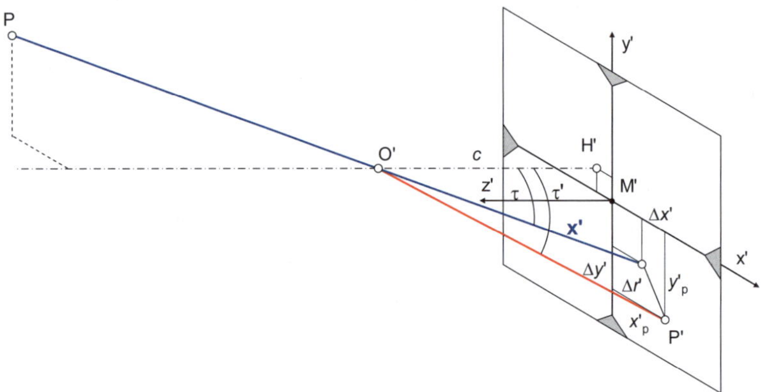

Fig. 3.49: Interior orientation

Hence, the parameters of interior orientation are (see section 3.3.2.1):

- Principal point H':

  Foot of perpendicular from perspective centre to image plane, with image coordinates (x'_0, y'_0). For commonly used cameras approximately equal to the centre of the image: $H' \approx M'$.

- Principal distance *c*:

  Perpendicular distance to the perspective centre from the image plane in the negative z' direction. When focused at infinity, c is approximately equal to the focal length of the lens ($c \approx f'$).

- Parameters of functions describing imaging errors:

  Functions or parameters that describe deviations from the central perspective model are dominated by the effect of symmetric radial distortion $\Delta r'$.

If these parameters are given, the (error-free) imaging vector **x'** can be defined with respect to the perspective centre, and hence also the principal point:

$$\mathbf{x'} = \begin{bmatrix} x' \\ y' \\ z' \end{bmatrix} = \begin{bmatrix} x'_p - x'_0 - \Delta x' \\ y'_p - y'_0 - \Delta y' \\ -c \end{bmatrix} \tag{3.51}$$

where
x'_p, y'_p: measured coordinates of image point P'
x'_0, y'_0: coordinates of the principal point H'
$\Delta x', \Delta y'$: correction values for errors in the image plane

The parameters of interior orientation are determined by camera calibration (see section 3.3.2.5).

3.3.2.4 Metric and semi-metric cameras

The expression *metric camera* is used for photogrammetric cameras with a stable optical and mechanical design. For these cameras the parameters of interior orientation can be calibrated in the factory (laboratory) and are assumed to be constant over a long period of time. Usually metric cameras consist of a rigidly mounted fixed-focus lens with minimal distortion. In addition, they have a flat image plane. A *semi-metric camera* meets the above metric camera requirements only with respect to a plane image surface and its corresponding plane image coordinate system. These specifications are fulfilled by a réseau for analogue film cameras, and the physical surface of the imaging sensor for digital cameras.

For a semi-metric camera, the spatial position of the principal point is only given approximately with respect to the image coordinate system. It is not assumed to be constant over a longer period of time. Movements between perspective centre (lens) and image plane (sensor) can, for example, result from the use of variable focus lenses, or an unstable mounting of lens or imaging sensor. Fig. 3.50 illustrates this effect for a series of 90 images acquired with a Fuji FinePix S1 Pro digital camera held in different orientations. The observed variations can only be handled by a calibration which varies with the image (image-variant camera calibration).

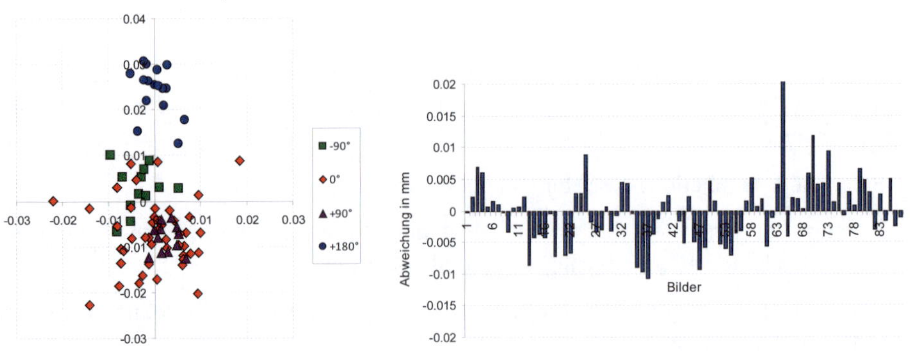

a) Variation of principal point (in mm) b) Variation of principal distance (in mm)

Fig. 3.50: Variation in perspective centre position across a series of images (Fuji S1 Pro)

Cameras, such as analogue photographic cameras, without a suitable photogrammetric reference system and/or without a planar image surface, are known as *amateur* or *non-metric cameras*. In special cases, such as accident investigation or the reconstruction of old buildings, e.g from old postcards, it is necessary to use amateur cameras for photogrammetric analysis. Normally unique reference points, related to the principal point, do not exist and even original image corners often cannot be reconstructed. In such cases, a suitable analytical method for image orientation is the direct linear transformation (DLT, see section 4.2.4.1). This method does not require an image coordinate system. Fig. 3.51 shows an example of an image where neither fiducial marks nor image corners in the camera body are visible.

Fig. 3.51: Example of an image taken by a non-metric camera (historical picture of Worms cathedral)

3.3.2.5 Determination of interior orientation (calibration)

In photogrammetry, the determination of the parameters of interior orientation is usually referred to as *calibration*. State-of-the-art close-range techniques employ analytical calibration methods to derive the parameters of the chosen camera model indirectly from photogrammetric image coordinate observations. For this purpose the imaging function is extended by the inclusion of additional parameters that model the position of the perspective centre and image distortion effects.

Usually calibration parameters are estimated by bundle adjustment (simultaneous calibration, section 4.4.2.4). Depending on available object information (reference points, distances, constraints) suitable imaging configurations must be chosen (see section 7.3).

The necessity for the periodic calibration of a camera depends on the accuracy specifications, the mechanical construction of the camera and environmental conditions on site at the time of measurement. Consequently, the time and form of the most appropriate calibration may vary:

- One-time factory calibration:

  Imaging system: metric camera
  Method: factory or laboratory calibration (see section 7.3.1.1)
  Reference: calibrated test instruments (e.g. goniometer or comparators)
  Assumption: camera parameters are valid for the life of the camera

- Long-term (e.g. annual) checks:

  Imaging system: metric camera
  Method: laboratory or test-field calibration
  Reference: calibrated test instruments, reference points, scale bars, plumb lines
  Assumption: camera parameters are valid for a long period of time

- Calibration immediately before object measurement:

  Imaging system: semi-metric camera, metric camera
  Method: test field calibration, self-calibration (see section 7.3.1.2)
  Reference: reference points, reference lengths within the test field, straight lines
  Assumption: camera parameters do not alter until the time of object measurement

- Calibration integrated into object reconstruction:

  Imaging system: semi-metric camera, metric camera
  Method: self-calibration, on-the-job calibration (see section 7.3.1.4)
  Reference: reference points, distances on object, straight lines
  Assumption: constant interior orientation during image network acquisition

- Calibration of each individual image:

  Imaging system: semi-metric camera
  Method: self-calibration with variable interior orientation (see section 4.4.2.4)
  Reference: reference points, distances on object, straight lines
  Assumption: only limited requirements regarding camera stability, e.g. constant distortion values

In the default case, the parameters of interior orientation are assumed to be constant for the duration of image acquisition. However, for close-range applications in it may be the case that lenses are changed or re-focused during an image sequence, and/or mechanical or thermal changes occur. Every change in camera geometry will result in a change of interior orientation which must be taken into account in the subsequent evaluation by use of an individual set of parameters for each different camera state.

From the standpoint of camera calibration the difference between metric and semi-metric cameras becomes largely irrelevant as the stability of interior orientation depends on the required accuracy. Consequently, even metric cameras are calibrated on-the-job if required by the measuring task. On the other hand, semi-metric digital cameras can be calibrated in advance if they are components of multi-camera online systems.

3.3.3 Standardised correction functions

Deviations from the ideal central perspective model, attributable to imaging errors, are expressed in the form of correction functions $\Delta x'$, $\Delta y'$ to the measured image coordinates. Techniques to establish these functions have been largely standardised and they capture the effects of radial, tangential and asymmetric distortion, as well as affine errors in the image coordinate system. Extended models for special lenses and imaging systems are described in section 3.3.4.

3.3.3.1 Symmetric radial distortion

According to section 3.1.3.1, distortion is related to the principal point, i.e. the measured image coordinates x'_P, y'_P must first be corrected by a shift of origin to the principal point at x'_0, y'_0:

$$\begin{aligned} x^\circ &= x'_P - x'_0 \\ y^\circ &= y'_P - y'_0 \end{aligned} \quad \text{: image coordinates relative to principal point}$$

where

$$r' = \sqrt{x^{\circ 2} + y^{\circ 2}} \quad \text{: image radius, distance from the principal point} \tag{3.52}$$

The correction of the image coordinates x°, y° for distortion is then given by:

$$\begin{aligned} x' &= x^\circ - \Delta x' \\ y' &= y^\circ - \Delta y' \end{aligned} \quad \text{: corrected image coordinates} \tag{3.53}$$

The distortion corrections $\Delta x'$, $\Delta y'$ must be calculated using the final image coordinates x', y' but must be initialised using the approximately corrected coordinates x°, y°. Consequently, correction values must be applied iteratively (see section 3.3.5).

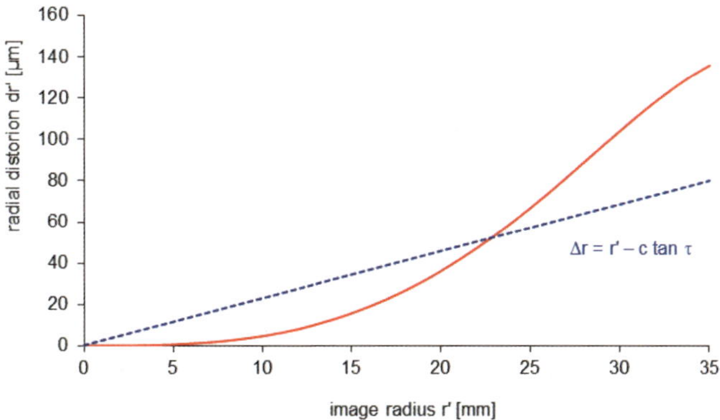

Fig. 3.52: Symmetric radial distortion

Symmetric radial distortion, commonly known as radial distortion, constitutes the major imaging error for most camera systems. It is attributable to variations in refraction at each individual component lens within the camera's compound lens. It depends also on the wavelength, aperture setting, focal setting and the object distance at constant focus.

Fig. 3.52 shows the effect of radial distortion as a function of the radius of an imaged point. In the example, the distortion increases with distance from the principal point. For standard lenses it can reach more than 100 µm in the image corners. The *distortion curve* is usually modelled with a polynomial series (Seidel series) with distortion parameters K_1 to K_n:

$$\Delta r'_{rad} = K_1 \cdot r'^3 + K_2 \cdot r'^5 + K_3 \cdot r'^7 + \ldots \tag{3.54}$$

For most lens types the series can be truncated after the second or third term without any significant loss of accuracy. Then the image coordinates are corrected proportionally:

$$\Delta x'_{rad} = x' \frac{\Delta r'_{rad}}{r'} \qquad \Delta y'_{rad} = y' \frac{\Delta r'_{rad}}{r'}$$

The distortion parameters defined in (3.54) are numerically correlated with image scale or principal distance. In order to avoid these correlations, a linear part of the distortion function is removed. This is equivalent to a rotation of the distortion curve into the direction of the r' axis, thus resulting in a second zero-crossing. This generates a distortion function in which only the differences from the straight line $\Delta r' = r' - c \cdot \tan\tau$ (see eqn. 3.50) in Fig. 3.52 must be modelled.

$$\Delta r'_{rad} = K_0 r' + K_1 r'^3 + K_2 r'^5 + K_3 r'^7 \tag{3.55}$$

Alternatively a polynomial of the following type is used:

$$\Delta r'_{rad} = A_1 r'(r'^2 - r_0^2) + A_2 r'(r'^4 - r_0^4) + A_3 r'(r'^6 - r_0^6) \tag{3.56}$$

By simple rearrangement of (3.56) it can be shown that this has the same effect as eqn. 3.55:

$$\Delta r'_{rad} = A_1 r'^3 + A_2 r'^5 + A_3 r'^7 - r'(A_1 r_0^2 + A_2 r_0^4 + A_3 r_0^6) \tag{3.57}$$

Here the term in brackets is a constant analogous to K_0. However r_0 cannot be chosen arbitrarily as there is a dependency on parameters A_1, A_2, A_3. In practice r_0 should be chosen such that minimal and maximal distortion values are more or less equal with respect to the complete image format (*balanced radial distortion*). Usually r_0 is set to approximately 2/3 of the maximum image radius. Fig. 3.53 shows a typical radial distortion curve according to (3.56), Fig. 3.54 displays the corresponding two-dimensional effect with respect to the image format. Balanced radial lens distortion is only necessary when the distortion correction is applied using analogue methods, for example by specially shaped cams within the mechanical space rod assembly. Note that r_0 is not commonly used in the calibration of digital systems since in digital data processing the correction is numerical rather than mechanical.

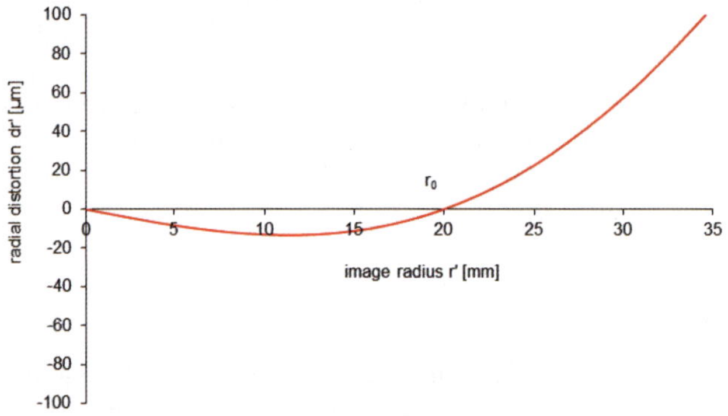

Fig. 3.53: Typical balanced lens distortion curve (Rollei Sonnar 4/150)

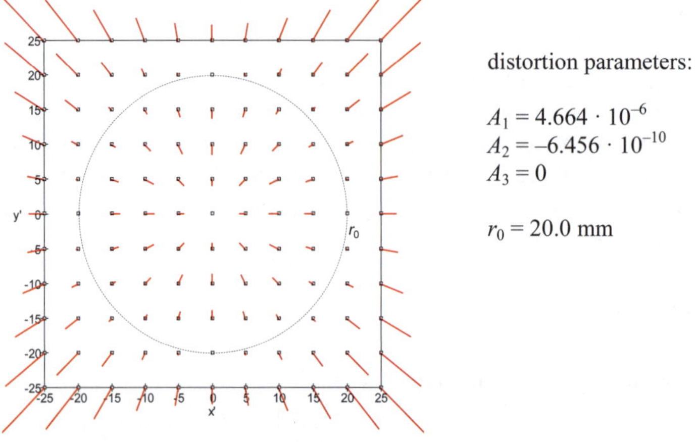

distortion parameters:

$A_1 = 4.664 \cdot 10^{-6}$
$A_2 = -6.456 \cdot 10^{-10}$
$A_3 = 0$

$r_0 = 20.0$ mm

Fig. 3.54: Effect of radial distortion (data from Table 3.2)

Eqn. 3.50 and Fig. 3.55 show that the effect of introducing the linear term K_0 or r_0 can also be achieved by a change in principal distance.

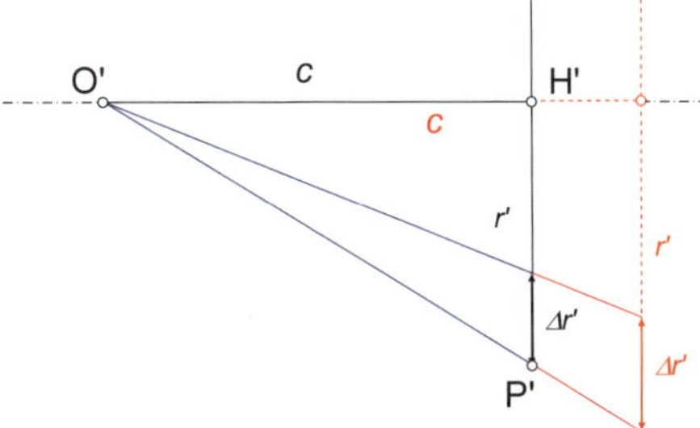

Fig. 3.55: Linear relationship between c and $\Delta r'$

Table 3.2: Correction table for distortion (see Fig. 3.53, Fig. 3.54, all values in mm)

r'	Δr'	r'	Δr'	r'	Δr'	r'	Δr'
0	0.0000	9	-0.0125	18	-0.0057	27	0.0350
1	-0.0018	10	-0.0130	19	-0.0031	28	0.0419
2	-0.0035	11	-0.0133	20	0.0000	29	0.0494
3	-0.0052	12	-0.0132	21	0.0035	30	0.0574
4	-0.0068	13	-0.0129	22	0.0076	31	0.0658
5	-0.0082	14	-0.0122	23	0.0121	32	0.0748
6	-0.0096	15	-0.0112	24	0.0170	33	0.0842
7	-0.0107	16	-0.0098	25	0.0225	34	0.0941
8	-0.0117	17	-0.0080	26	0.0285	35	0.1044

A table of correction values for radial distortion is often derived from a camera calibration (Table 3.2). It can then be used by a real-time processing system to provide correction values for image coordinates by linear interpolation.

The meaning of the distortion parameters

The sign of parameter A_1 determines the form of the distortion as either barrel ($A_1 < 0$) or pincushion ($A_1 > 0$), see Fig. 3.56. A_2 and A_3 model deviations of the distortion curve from the cubic parabolic form. A_2 normally has its primary effect towards the edges of the image (Fig. 3.57). The introduction of the term A_3 makes it possible to model lenses with large distortion values at the image edges (Fig. 3.57).

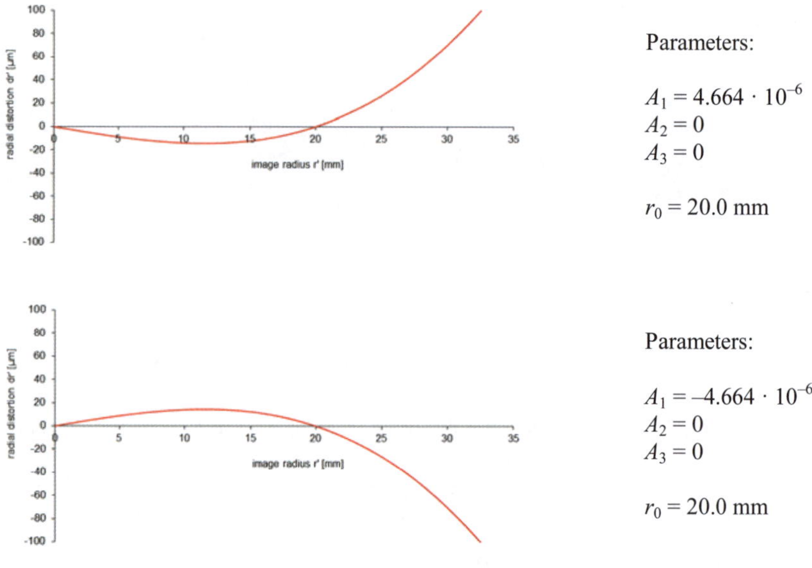

Fig. 3.56: Effect of changing the sign of A_1

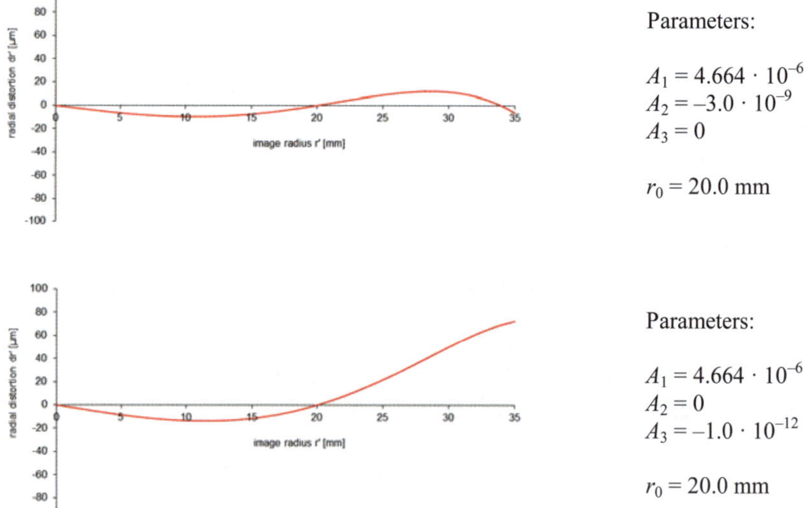

Fig. 3.57: Effects of parameters A_2 and A_3

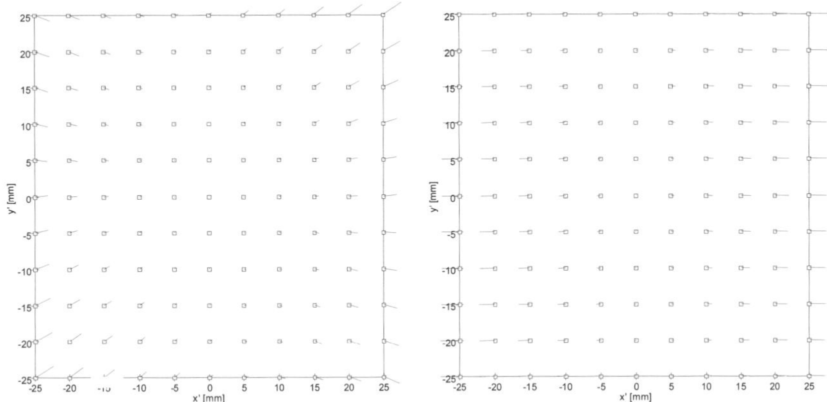

Fig. 3.58: Effect of tangential distortion

Fig. 3.59: Effect of affinity

3.3.3.2 Tangential distortion

Tangential or decentring distortion (Fig. 3.58) is mainly caused by decentring and misalignment of individual lens elements within the camera's compound lens. It can be described by the following functions:

$$\begin{aligned} \Delta x'_{tan} &= B_1(r'^2 + 2x'^2) + 2B_2x'y' \\ \Delta y'_{tan} &= B_2(r'^2 + 2y'^2) + 2B_1x'y' \end{aligned} \tag{3.58}$$

For most quality lenses the effects are significantly smaller than for radial distortion and are often only determined where accuracy demands are high. If low-cost lenses are used, as is often the case in surveillance camera systems, significant tangential distortion can be present.

3.3.3.3 Affinity and shear

Affinity and shear are used to describe deviations of the image coordinate system with respect to orthogonality and uniform scale of the coordinate axes (Fig. 3.59). For analogue cameras these effects can be compensated by means of an affine transformation based upon measurements of fiducial marks or a réseau. Digital imaging systems can produce these characteristics if the sensor has light-sensitive elements not on a regular grid, due for example to line scanning, or which are rectangular rather than square. Affinity and shear also exist for images that are transferred in analogue form before being digitised by a frame grabber. The following function can be used to provide an appropriate correction:

$$\begin{aligned} \Delta x'_{aff} &= C_1x' + C_2y' \\ \Delta y'_{aff} &= 0 \end{aligned} \tag{3.59}$$

Parameter C_1 here acts as a scale factor along the x-axis, in a similar way to factor s' in eqn. 3.62.

3.3.3.4 Total correction

The individual terms used to model the imaging errors of most typical photogrammetric imaging systems can be summarised as follows:

$$\begin{aligned}\Delta x' &= \Delta x'_{rad} + \Delta x'_{tan} + \Delta x'_{aff} \\ \Delta y' &= \Delta y'_{rad} + \Delta y'_{tan} + \Delta y'_{aff}\end{aligned} \tag{3.60}$$

Example 3.8:

A calibration of a Canon EOS1000 digital camera with lens f = 18 mm gives the following set of correction parameters:

$A_1 = -4.387 \cdot 10^{-4}$	$A_2 = 1.214 \cdot 10^{-6}$	$A_3 = -8.200 \cdot 10^{-10}$
$B_1 = 5.572 \cdot 10^{-6}$	$B_2 = -2.893 \cdot 10^{-6}$	
$C_1 = -1.841 \cdot 10^{-5}$	$C_2 = 4.655 \cdot 10^{-5}$	$r_0 = 8.325$ mm

Compute the effects of individual distortion terms for two image points, the first located in the centre of the image and the second in one corner:

	x'_1	y'_1	x'_2	y'_2	
x',y'	1.500	1.500	11.100	7.400	mm
A_1, A_2, A_3	34.4	34.4	–215.4	–143.6	µm
B_1, B_2	0.0	0.0	1.9	0.2	µm
C_1, C_2	0.0	0.0	0.1	0.0	µm
total	34.4	34.4	–213.4	–143.4	µm

This example indicates that the effect of radial distortion predominates. However, if the accuracy potential of this camera of about 0.2 µm (1/50th pixel) is to be reached, the other sources of image errors must be taken into account.

3.3.4 Alternative correction formulations

3.3.4.1 Simplified models

In Tsai's approach, only one parameter k is used in addition to principal distance, principal point and two scaling parameters. The distorted image coordinates, with origin at the principal point, are given by:

$$\begin{aligned}x^\circ &= \frac{2x'}{1+\sqrt{1-4kr'^2}} \\ y^\circ &= \frac{2y'}{1+\sqrt{1-4kr'^2}}\end{aligned} \tag{3.61}$$

This widely used model does not include any high-order distortion effects and is not therefore equivalent to a Seidel polynomial with two or three parameters (A_1, A_2, A_3). Typical lenses with large distortion components are not fully modelled by this approach and the disadvantage increases towards the image edges. In addition, asymmetric and tangential distortions are not taken into account at all.

In computer vision the parameters of interior orientation are typically expressed by a *calibration matrix* **K** which consists of five degrees of freedom (principal distance c, principal point x'_0, y'_0, shear s and differential scale m' between the axes).

$$\mathbf{K} = \begin{bmatrix} 1 & s' & x'_0 \\ 0 & 1+m' & y'_0 \\ 0 & 0 & 1 \end{bmatrix} \cdot \begin{bmatrix} c & 0 & 0 \\ 0 & c & 0 \\ 0 & 0 & 1 \end{bmatrix} = \begin{bmatrix} c & cs' & x'_0 \\ 0 & c(1+m') & y'_0 \\ 0 & 0 & 1 \end{bmatrix} \tag{3.62}$$

K is part of a general 3x4 transformation matrix from object space into image space by homogeneous coordinates (see section 2.2.2.3). Lens distortion cannot directly be integrated in this method and must be modelled by a position-dependent correction matrix $\mathbf{dK}(x', y')$.

3.3.4.2 Additional parameters

An approach by Brown (1971) was developed specifically for large format analogue aerial cameras, but can also be applied to digital cameras:

$$\begin{aligned} \Delta x'_{Brown} &= D_1 x' + D_2 y' + D_3 x' y' + D_4 y'^2 + D_5 x'^2 y' + D_6 x' y'^2 + D_7 x'^2 y'^2 + \\ &\quad [E_1(x'^2 - y'^2) + E_2 x'^2 y'^2 + E_3(x'^4 - y'^4)]x'/c + \\ &\quad [E_4(x'^2 + y'^2) + E_5(x'^2 + y'^2)^2 + E_6(x'^2 + y'^2)^3]x' + E_7 + E_9\, x'/c \\ \Delta y'_{Brown} &= D_8 x' y' + D_9 x'^2 + D_{10} x'^2 y' + D_{11} x' y'^2 + D_{12} x'^2 y'^2 + \\ &\quad [E_1(x'^2 - y'^2) + E_2 x'^2 y'^2 + E_3(x'^4 - y'^4)]y'/c + \\ &\quad [E_4(x'^2 + y'^2) + E_5(x'^2 + y'^2)^2 + E_6(x'^2 + y'^2)^3]y' + E_8 + E_9\, y'/c \end{aligned} \tag{3.63}$$

In addition to parameters for modelling deformations in the image plane (D_1 to D_{12}) terms for the compensation of lack of flatness in film or sensor are included (E_1 to E_6). These are formulated either as a function of the radial distance or of the tangent of the imaging angle (x'/c or y'/c). However, the approach is more difficult to interpret geometrically, and it can easily be over-parameterised, leading to a dependency or correlation between individual parameters. For a self-calibrating bundle adjustment, individual parameters should be tested for their significance and correlation with respect to each other. Any parameters which fail such tests should be eliminated, starting with the weakest first.

An example of an extension to the parameter set to accommodate digital cameras is given below:

$$\begin{aligned}\Delta x'_{Beyer} &= \Delta x'_0 - \Delta c\, x'/c + K_1 x' r'^2 + K_2 x' r'^4 + K_3 x' r'^6 \\ &\quad + P_1(r'^2 + 2x'^2) + 2P_2 x' y' - C_1 x' + C_2 y' \\ \Delta y'_{Beyer} &= \Delta y'_0 - \Delta c\, y'/c + K_1 y' r'^2 + K_2 y' r'^4 + K_3 y' r'^6 \\ &\quad + 2P_1 x' y' + P_2(r'^2 + 2y'^2) + C_2 x'\end{aligned} \tag{3.64}$$

Here the parameters K describe radial distortion, P describes tangential (decentring) distortion and C models affinity and shear. The parameters $\Delta x'_0$, $\Delta y'_0$ and Δc are used for small corrections to the spatial position of the perspective centre. Together with the factors x'/c and y'/c they have an effect similar to r_0 in eqn. 3.56.

The correction of radial distortion according to (3.54) can be formulated as a function of image angle rather than image radius:

$$\Delta r = W_1\theta^2 + W_2\theta^3 + W_3\theta^4 + \ldots\ldots + W_i\theta^{i+1} \tag{3.65}$$

where

$\theta = \arctan\left(\frac{1}{c}\sqrt{x'^2 + y'^2}\right)$: angle between optical axis and object point

x', y': corrected image coordinates, projection of the object point in the image

c: principal distance

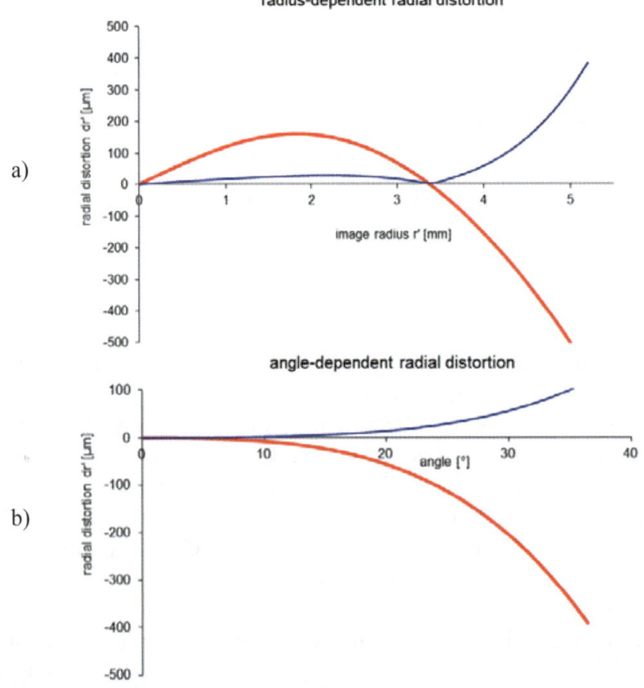

Fig. 3.60: Radial distortion (red) as a function of radius (a) and image angle (b) for a lens with large distortion, with the uncertainty of the distortion curve for each case shown in blue

A distortion model based on angle is particularly suitable for central perspective optics with large symmetric radial distortion. In this case the distortion curve has a lower slope into the corners (Fig. 3.60). The increasing uncertainty is in general due to a reduced density of imaged points with increasing image radius.

3.3.4.3 Correction of distortion as a function of object distance

Strictly speaking, the above approaches for the correction of lens distortion are valid only for points on an object plane that is parallel to the image plane, and focused, according to the lens eqn. 3.14, at a distance *a'* or with respect to the interior perspective centre. Imaging rays of points outside this object plane pass through the lens along a different optical path and hence are subject to different distortion effects. As an example, Fig. 3.61 shows the distortion curves of object points with different image scales, i.e. at different object distances.

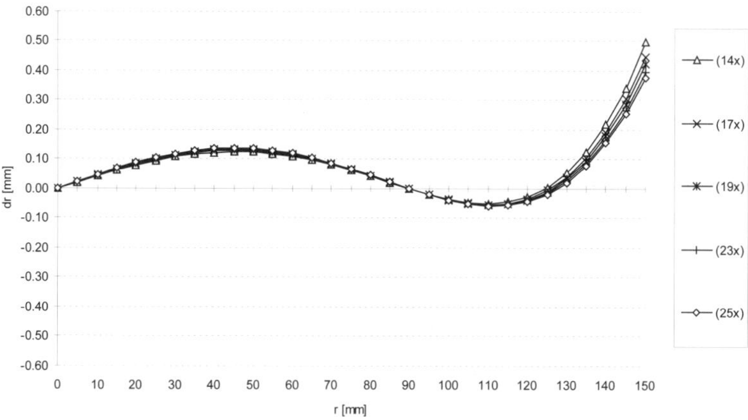

Fig. 3.61: Lens distortion curves for different image scales (Dold 1997)

This effect can be taken into account by a correction dependent on distance. This requires introduction of a scaling factor:

$$\gamma_{SS'} = \frac{c_{S'}}{c_S} = \frac{S'}{S} \cdot \frac{(S-c)}{(S'-c)} \tag{3.66}$$

where c_S: principal distance (image distance) of object distance S
c_S: principal distance (image distance) of object distance S'

For a given set of distortion parameters $K_{1S'}$, $K_{2S'}$, $K_{3S'}$ applying to an object plane at distance S', according to (3.55), the correction of radial distortion for object points at a focused distance S can be calculated as follows:

$$\Delta r'_{SS'} = \gamma_{SS'}^2 K_{1S'} r'^3 + \gamma_{SS'}^4 K_{2S'} r'^5 + \gamma_{SS'}^6 K_{3S'} r'^7 \tag{3.67}$$

This model is suitable for high-precision measurements made at large scales ($m < 30$) and lenses with relatively steep distortion curves. As the effect of distortion dependent on distance increases with image scale (decreasing object distance), an empirically estimated correction factor $g_{SS'}$ can be introduced that can be determined for each individual lens:

$$\Delta r'_{SS'} = \Delta r'_S + g_{SS'}(\Delta r'_{S'} - \Delta r'_S) \tag{3.68}$$

Tangential distortion can also be formulated as a function of the focused distance S:

$$\begin{aligned} \Delta x'_S &= \left(1-\frac{c}{S}\right)\left[P_1(r'^2+2x') + 2P_2x'y'\right] \\ \Delta y'_S &= \left(1-\frac{c}{S}\right)\left[P_2(r'^2+2y') + 2P_1x'y'\right] \end{aligned} \tag{3.69}$$

In contrast to the above approaches, the following set of parameters for the correction of distance-dependent distortion can be estimated completely within a self-calibrating bundle adjustment. However it should be noted that incorporation of such corrections within a self-calibration require very strong networks that contain many images taken at each distance setting in order to provide a robust and reliable parameter set.

$$\Delta r'_{dist} = \frac{1}{Z^*}\left[D_1r'(r'^2-r_0^2) + D_2r'(r'^4-r_0^4) + D_3r'(r'^6-r_0^6)\right] \tag{3.70}$$

where

Z^*: denominator of collinearity equations (4.9) $\approx S$ (object distance)

Extension of (3.60) leads to the following total correction of imaging errors:

$$\begin{aligned} \Delta x' &= \Delta x'_{rad} + \Delta x'_{tan} + \Delta x'_{aff} + \Delta x'_{dist} \\ \Delta y' &= \Delta y'_{rad} + \Delta y'_{tan} + \Delta y'_{aff} + \Delta y'_{dist} \end{aligned} \tag{3.71}$$

Usually, distance-dependant distortion does not exceed more than 1µm at the edge of the image. Hence, it must only be considered for high-accuracy measurement tasks where sub-micron image measuring accuracies are required. This is relevant to large-scale industrial applications with analogue large-format cameras, but is particularly relevant to high-resolution digital cameras that provide an accuracy potential of better than 0.5 µm. The need will be most prevalent where there is a large range of depth over the object, or objects, to be recorded.

3.3.4.4 Image-variant calibration

Cameras and lenses, which have so little mechanical stability that the geometry of the imaging system can vary within a sequence of images, can be modelled by an image-variant process. Here individual parameters defining the perspective centre (principal distance and principal point) are determined individually for every image j. In contrast, distortion parameters are normally assumed to be constant for the entire image sequence. Image coordinate adjustment is then done with correction terms which are calculated for every image as a function of the perspective centre's location.

$$\begin{aligned}\Delta x'_{var} &= \Delta x'_{\{\Delta c, \Delta x'_0, \Delta y'_0\}_j} \\ \Delta y'_{var} &= \Delta y'_{\{\Delta c, \Delta x'_0, \Delta y'_0\}_j}\end{aligned} \tag{3.72}$$

Fig. 3.50 shows an example for a digital camera, where principal distance and principal point vary due to mechanical handling and the effects of gravity. The calculation of image-variant camera parameters is done using an extended bundle adjustment.

3.3.4.5 Correction of local image deformation

Local image deformations are imaging errors which appear only in specific areas of the image format and are not therefore covered by the global correction functions described above.

Local deformations are due, for example, to local lack of flatness in the image plane. The lack of flatness often seen in analogue film cameras can be corrected by application of the réseau method, i.e. by local geometric transformation of the image onto a reference grid (Fig. 3.62).

Fig. 3.62: Example of simulated film deformation on an image with a réseau

A similar analytical formulation can be adopted to describe the lack of flatness of digital image sensors. A finite element approach, based on localised correction nodes, is used. This requires a grid of two-dimensional correction vectors at the intersection points of the grid lines. Corrections within the grid are computed using an interpolation process, typically following the linear process in the following equation (Fig. 3.63):

$$\begin{aligned}x_{corr} &= (1 - x_l - y_l + x_l \cdot y_l) \cdot k_{x[i,j]} & y_{corr} &= (1 - x_l - y_l + x_l \cdot y_l) \cdot k_{y[i,j]} \\ &+ (x_l - x_l \cdot y_l) \cdot k_{x[i+1,j]} & &+ (x_l - x_l \cdot y_l) \cdot k_{y[i+1,j]} \\ &+ (y_l - x_l \cdot y_l) \cdot k_{x[i,j+1]} & &+ (y_l - x_l \cdot y_l) \cdot k_{y[i,j+1]} \\ &+ x_l \cdot y_l \cdot k_{x[i+1,j+1]} & &+ x_l \cdot y_l \cdot k_{y[i+1,j+1]}\end{aligned} \tag{3.73}$$

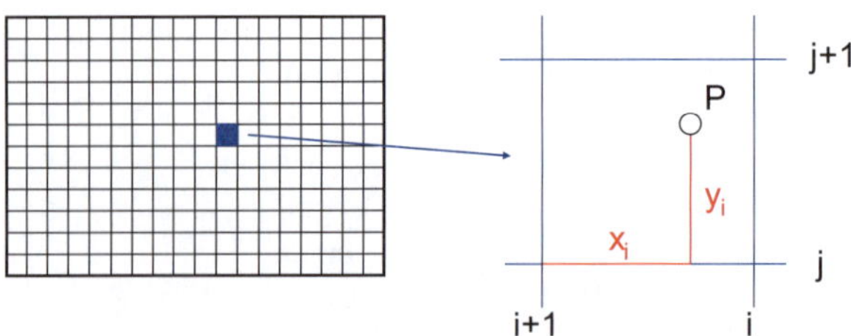

Fig. 3.63: Interpolation within the correction grid

Here x_{corr} is the corrected value of the image coordinate x, coordinates x_l,y_l represent the local position of the image point within the grid element and $k_{x[i,j]}$, $k_{x[i+1,j]}$, $k_{x[i,j+1]}$, $k_{x[i+1,j+1]}$ are the correction vector components at the corresponding grid intersection points. An analogous correction applies to image coordinate y. The collinearity equations (4.9) are extended by the above formulation so that the grid parameters can be estimated by a bundle adjustment (see section 4.4).

In order to separate out the noise component, i.e. the random measurement error in the image point, from the sensor deformation error and other lens imaging errors not otherwise taken into account, deformation conditions at the nodes are introduced as pseudo-equations:

$$\begin{aligned} 0 &= (k_{x[i,j-1]} - k_{x[i,j]}) - (k_{x[i,j]} - k_{x[i,j+1]}) \\ 0 &= (k_{x[i-1,j]} - k_{x[i,j]}) - (k_{x[i,j]} - k_{x[i+1,j]}) \\ 0 &= (k_{y[i,j-1]} - k_{y[i,j]}) - (k_{y[i,j]} - k_{y[i,j+1]}) \\ 0 &= (k_{y[i-1,j]} - k_{y[i,j]}) - (k_{y[i,j]} - k_{y[i+1,j]}) \end{aligned} \tag{3.74}$$

These nodal conditions are applied in both x and y directions in the image plane. Within the system of equations, this leads to a new group of observations. The accuracy with which the equations are introduced depends on the expected “roughness” of the unflatness parameters in the correction grid, as well as the number of measured points in the imaging network which appear on each grid element. In addition, the equations protect against potential singularities within the complete set of adjustment equations when there are grid elements which contain no image points.

As an example, Fig. 3.64 shows a calculated correction grid for a digital camera (Canon EOS1000) fitted with a zoom lens. The grids show similar trends despite setting the lens at two different zoom settings. In addition to the simultaneously determined distortion parameters, effects are modelled in image space which have similar characteristics in both cases. However, these are not necessary pure sensor deformations such as lack of flatness. The finite element grid compensates for all remaining residual errors in the image sequence.

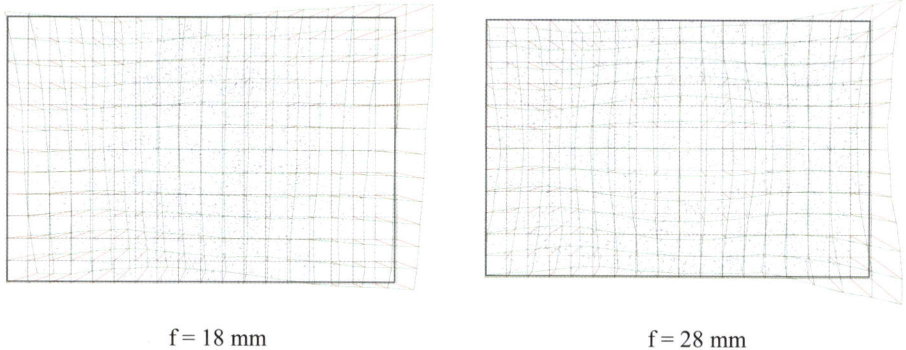

Fig. 3.64: Computed correction grids for the same camera body with a zoom lens at different settings

The finite element method of calibration is also suitable for modelling ray paths in complex optics, for example when using a stereo-mirror attachment (see Fig. 3.103).

A technique known as vision-ray calibration models the imaging ray in object space for every pixel on an imaging sensor, without reference to the properties of the lens or other optical elements in the camera. The basic idea is to associate a ray with every pixel which defines the direction to the corresponding object point via two correction vectors in the imaging plane.

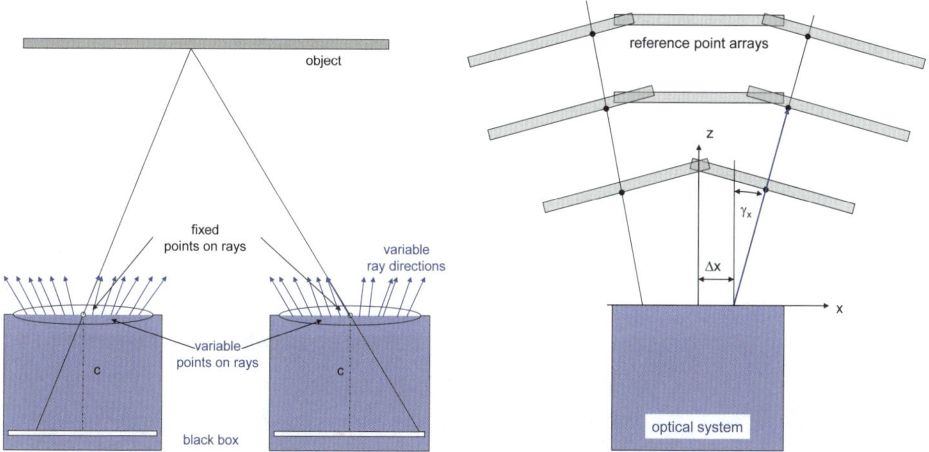

Fig. 3.65: Geometric imaging model for vision-ray calibration (after Schulte et al. 2006)

Here an optical system is no longer regarded as a central perspective, pinhole camera model with functions describing the image errors but as a black-box system which has a ray starting point and direction for every pixel (Fig. 3.65). In principle, every point on the virtual image plane can have its own object point vector which can correct for all optical and electronic imaging errors.

Determining all the ray directions is not without problem and demands a very dense spread of reference points in object space. For example, this can be created by imaging a reference point array at different distances from the imaging system. If this results in a three-dimensional array of reference points which fill the measurement volume, then the system can be fully calibrated. Four parameters are generated for every vision ray, i.e. and offset Δx and Δy as well as two angles γ_x and γ_y (Fig. 3.65).

3.3.5 Iterative correction of imaging errors

Measured image points can be corrected a priori if the parameters of interior orientation are known. Example cases are the calculation of object coordinates by space intersection, or the resampling of distortion-free images. However, there are often misunderstandings about the sequence and sign of corrections to image errors. For clarity it is necessary to know the derivation of the correction values and details about the software implementation, which is often not available in practice.

The correction model described in (3.60) is based on the assumption that the error corrections are calculated by self-calibrating bundle adjustment. This incorporates the collinearity model (4.9), which enables image points to be calculated for points defined in object space. Any possible distortion values are then added to these image coordinates.

$$\begin{aligned} x' &= f(X,Y,Z,X_0,Y_0,Z_0,\omega,\varphi,\kappa,c,x'_0) + \Delta x'(x',y') \\ y' &= f(X,Y,Z,X_0,Y_0,Z_0,\omega,\varphi,\kappa,c,y'_0) + \Delta y'(x',y') \end{aligned} \tag{3.75}$$

After adjustment, the correction parameters relate to adjusted object coordinates and orientation parameters or, expressed differently, to "error-free" image coordinates. Consequently, these correction values are directly applicable to image coordinates calculated by applying the collinearity equations from object space to image space. However, since the corrections $\Delta x'$, $\Delta y'$ depend on the current values of image coordinates, in the case where image coordinates, measured in a distorted image, must be corrected for previously established distortion parameters, then corrections must be applied iteratively. In this process, the currently corrected image positions are the starting point for the subsequent calculation of corrections. The process continues until the computed corrections are insignificant.

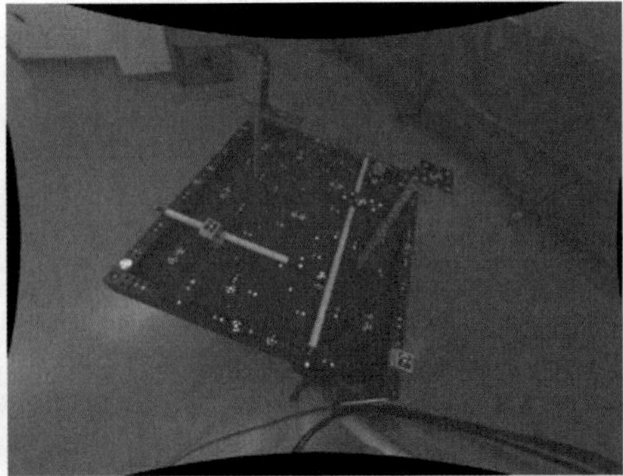

Fig. 3.66: Example of a significantly distorted image (left) and its corrected version (right)

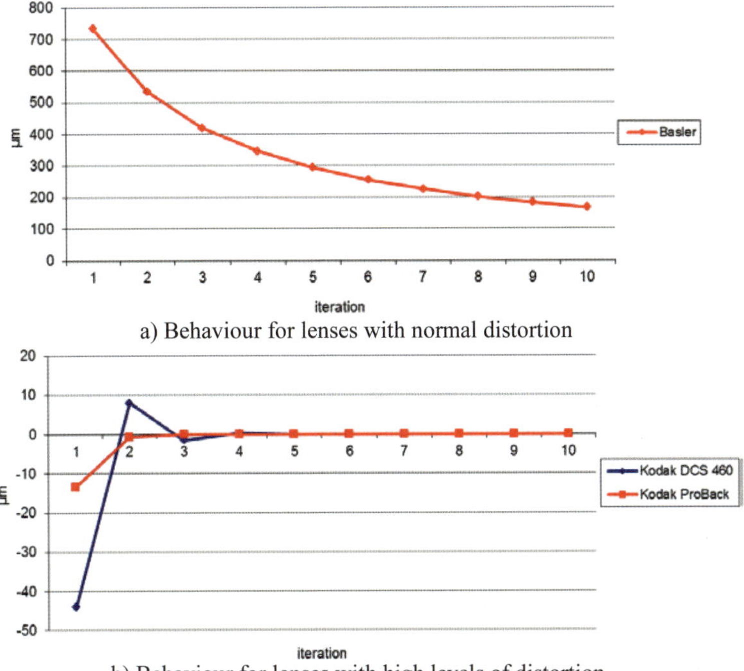

a) Behaviour for lenses with normal distortion

b) Behaviour for lenses with high levels of distortion

Fig. 3.67: Iterative correction of distortion

Table 3.3 summarises the effect of iterative correction of image coordinates for typical camera systems. It becomes obvious that significant errors can occur for cameras with larger image formats and higher distortion values (DCS 645M: 13.4 µm, DCS 460: 30 µm). In contrast, for the other examples a non-iterative correction can be used (Rollei 6008: 0.8 µm, CCD camera JAI: 0.2 µm). In practice it is always better to default to iterative correction since the additional computation time is generally insignificant. Typically 4 iterations are required.

The iterative correction of image coordinates is of particular practical importance when lenses with high levels of distortion are used. Table 3.3 shows the effect on image coordinates for typical camera systems when no iterative correction is made (row label "Correction due to a single iteration"). Fig. 3.67a shows that cameras with wide angles of view (short focal lengths) and high distortion parameters, have distortion errors which cannot be ignored without iterative correction (row label "Additional corrections due to further iterations": DCS 645M: 13.5 µm, DCS 460: 42.3 µm). Cameras with low-distortion lenses (Rollei 6008 with Sonnar 4/150) or small angles of view (JAI analogue camera with 8mm lens) can be corrected without iteration (Rollei 6008: 0.8 µm, JAI: 0.2 µm). If a high-distortion wide angle lens is evaluated (e.g. the Basler camera with a 4.8mm Pentax lens, see sample images in Fig. 3.66), then correction values are, firstly, high (in this case 238 µm) and, secondly, generated in a slow convergence process (Fig. 3.67b) which may even diverge in the corners.

Table 3.3: Effects of iterative distortion correction for typical camera systems

Camera	Rollei 6008	Kodak DCS 645M	Kodak DCS 460	JAI	Basler
Lens	f = 150 mm	f = 35 mm	f = 15 mm	f = 8 mm	f = 4.8 mm
c	–150.5566	–35.6657	–14.4988	–8.2364	–4.3100
x'_0 y'_0	0.023 0.014	–0.0870 0.4020	0.1618 –0.1199	0.2477 0.0839	–0.0405 –0.0011
A_1 A_2 A_3	4.664E-06 –6.456E-10 0	–9.0439E-05 6.3340E-08 –1.5294E-11	–4.5020E-04 1.7260E-06 –4.0890E-10	–1.9016E-03 3.1632E-05 0	–1.4700E-02 3.0396E-04 3.4731E-06
B_1 B_2	0 0	2.5979E-06 –7.2654E-07	–1.4670E-06 –6.6260E-06	–2.6175E-05 –7.3739E-05	1.1358E-05 –4.5268E-05
C_1 C_2	0 0	1.0836E-04 1.0964E-05	–2.5300E-04 –1.6130E-04	9.8413E-03 –7.2607E-06	1.9833E-05 8.9119E-05
x' y'	25.000 25.000	18.000 18.000	13.000 9.000	3.200 2.400	4.400 3.300
imaging angle	13.21°	35.28°	47.36°	24.49°	52.08°
correction due to a single iteration (µm)	76.7	–264.1	222.9	–21.1	–567.4
corrections due to further iterations (µm)	–0.8	–13.4	–42.3	–0.3	238.5

3.3.6 Fisheye projections

As the angle of view of an imaging system increases beyond about 110°, optical performance decreases rapidly. Degradation in image quality is seen in the capability to correctly image straight lines in the object as straight lines in the image and in reduced illumination towards the extremes of the image format following the $\cos^4$ law (section 3.1.3.6). As an example a 110° angle of view would be given by a 15 mm focal length lens on a full frame, FX format digital SLR sensor (Fig. 3.70), such lenses are available, but extremely expensive.

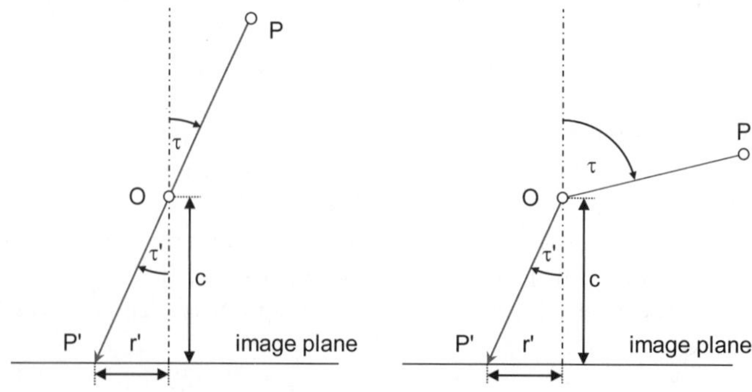

Fig. 3.68: Central perspective projection (left) and fisheye projection (right) (after Schneider 2008)

A solution is to change the imaging geometry from the central perspective projection where incident and exit angles τ and τ' are equal, to one where the incident angle τ from a point P in object space is greater than the exit angle τ' in image space (Fig. 3.68). Fisheye designs allow the projection of a half hemisphere onto the image plane with the optical axis coinciding with the centre of the resultant circular image. If the image format is larger than the resultant image circle, the camera is termed a fisheye system. Conversely, if the format is smaller than the circle, such that the image diagonal is 180°, a quasi-fisheye system is produced.

Three fisheye projections are in optical usage: stereographic, equidistant and orthographic. They are defined by the following equations.

$$r' = c \cdot \tan \tau \qquad \text{: central perspective} \tag{3.76}$$

$$r' = c \cdot \tan \tau / 2 \qquad \text{: stereographic} \tag{3.77}$$

$$r' = c \cdot \tau \qquad \text{: equidistant} \tag{3.78}$$

$$r' = c \cdot \sin \tau \qquad \text{: orthographic} \tag{3.79}$$

When modelling the distortions in a fisheye lens, conventional radial lens distortion corrections (eqn. 3.55) are mathematically unstable beyond the central region of the image format where the gradient of the distortion curve describing the departure from the central perspective case is low. For sensors that capture a significant area of the fisheye image circle, for example a DX sensor with a 10.5 mm fisheye lens or an FX sensor with a 16 mm fisheye lens, it is necessary to apply the appropriate fisheye lens model before using a radial distortion model such as eqn. 3.55, to account for any remaining radial departures from the fisheye projection exhibited by the specific lens in use.

3.4 System components

Electronic imaging systems use opto-electronic sensors for image acquisition instead of photographic emulsions. They directly provide an electronic image that can be digitised by suitable electronic components and transferred to a local processor or host computer for measurement and analysis. Hence the term *electronic imaging system* summarises all system components involved in the generation of a digital image (Fig. 3.69).

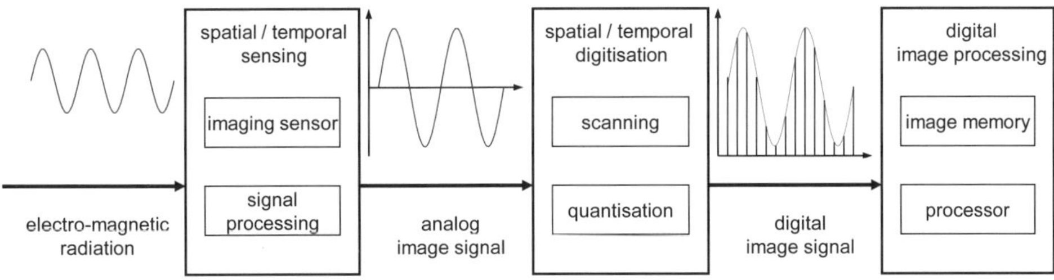

Fig. 3.69: Electronic imaging system

The electro-magnetic radiation (light) emitted or reflected by the object is imaged by a sensor as a function of time (exposure time, integration time) and space (linear or area sensing). After signal enhancement and processing an analogue image signal, in the form of an electric voltage proportional to the amount of light falling on the sensor, is produced. In a second stage, this signal is sampled by means of an analogue to digital converter in order to produce a digital image consisting of a series of discrete numerical values for each light sensitive cell or pixel in the image. This digital image can then be used for further processing such as discrete point measurement, or edge detection.

As far as photogrammetry is concerned, where geometrically quantifiable images are required, the development of digital imaging technology is closely related to the technology of CCD image sensors (*charge coupled devices*). They were invented at the beginning of the 1970s, and can today be found in many imaging applications. More recently CMOS technologies for opto-electronic imaging sensors are gaining in importance (see section 3.4.1.3). Both current generation digital camera systems and legacy analogue imaging systems (see section 3.5) can use either CCD or CMOS sensor types.

Developments over the last two decades indicate continued growth in numbers of sensor elements and image formats for CCD and CMOS sensors. Fig. 3.70 shows typical formats of imaging sensors (valid 2013), where larger formats confer greater pixel area for the same pixel count to deliver better electronic signal to noise at the expense of physically larger and generally more expensive cameras. For example the larger FX and DX sensors represent standard sizes for consumer digital SLR camera systems and their associated lens designs (see section 3.4.2.1), whilst smaller sensors are popular in mobile phones and webcams. Intermediate sensors are becoming popular for cameras typified by mirror-less, SLR-like designs, for example the 18 mm x 13.5 mm "four thirds" system developed by Olympus and Kodak.

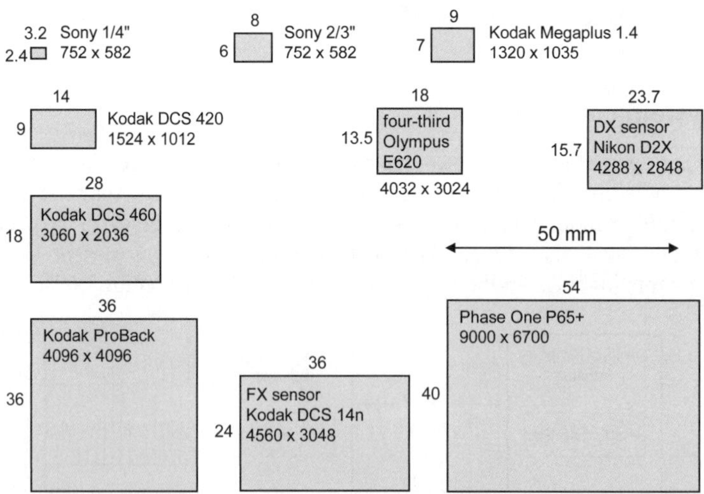

Fig. 3.70: Image formats of some common digital cameras and matrix sensors

3.4.1 Opto-electronic imaging sensors

3.4.1.1 Principle of CCD sensor

Solid-state imaging sensors are exclusively used in digital photogrammetric systems. Solid state imaging sensors consist of a large number of light-sensitive detector elements that are arranged as lines or arrays on semi-conductor modules (linear or area sensor). Each detector element (sensor element) generates an electric charge that is proportional to the amount of incident illumination falling on it. The sensor is designed such that the charge at each individual element can be read out, processed and digitised.

Fig. 3.71 illustrates the principle of a single sensor element. Incident light, in the form of photons, is absorbed in a semi-conducting layer where it generates pairs of electron holes (charged particles). The ability of a sensor element to create a number n_E of charged particles from a number n_P of incident photons is expressed by the external quantum efficiency η_{EXT}. The quantum efficiency depends on the sensor material and wavelength of the incident light.

$$\eta_{EXT} = \frac{n_E}{n_P} \tag{3.80}$$

The negative charged particles are attracted by a positive electrode. Charges are accumulated in proportion to the amount of incident light until saturation or overflow of charge is achieved. The positive electric field of the electrode is generated by a potential well that collects the negative charge. In CCD sensors the detector elements are formed from MOS capacitors (*metal-oxide semiconductor*).

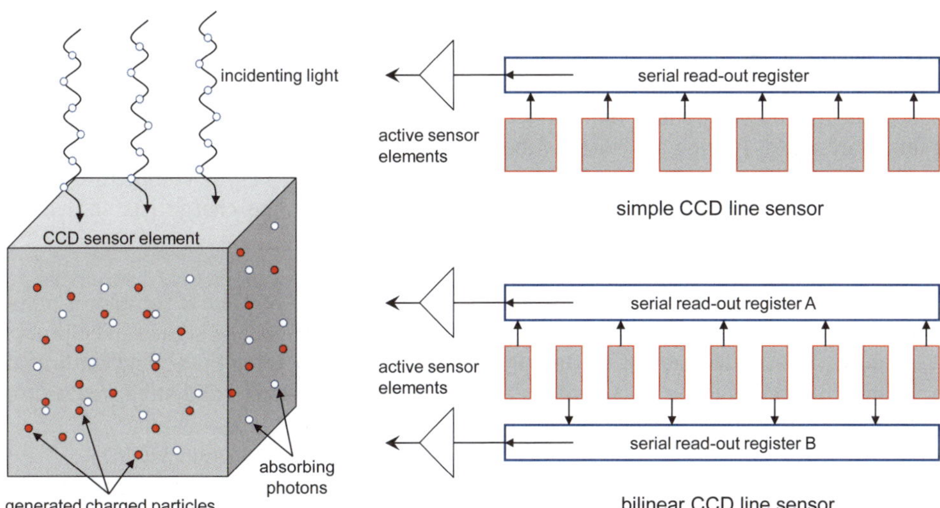

Fig. 3.71: Conversion of photons into charged particles

Fig. 3.72: Principle of simple and bilinear CCD line sensors

Sensor elements can be arranged in lines or two-dimensional arrays. Fig. 3.72 shows the simplified layout of a CCD line sensor. Each active sensor element is directly connected to a serial read-out register that is used to output the generated charge. In contrast, bilinear CCD lines can be resampling into what is effectively a single line to provide increased resolution if the sensor elements are coupled in an alternating manner with two read-out registers.

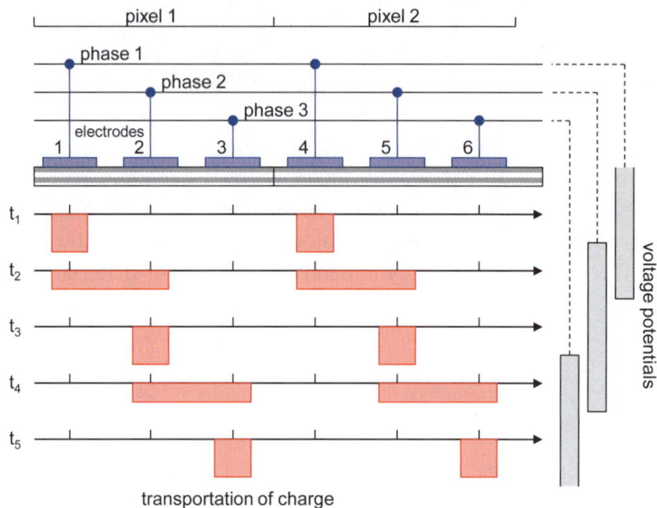

Fig. 3.73: Principle of CCD charge transportation (red-rectangles = bucket-brigade device)

The core problem for such sensor arrangements is the transportation of the charge stored in the sensor element to an output. Fig. 3.73 illustrates a typical solution for a linear arrangement of sensor elements. Here, the individual sensor elements have three electrodes, each connected to a different voltage phase. (In practice, most sensors use a 4-phase technique.) The cumulated charge at electrode 1 cannot discharge at time t_1, since the voltage is high on electrode 1 and low on electrode 2. At time t_2 the voltages on electrodes 1 and 2 are equal, forcing a portion of the charge under electrode 1 to flow to electrode 2. At time t_3 the voltages of electrode 1 and 3 have a low value, i.e. the complete charge is shifted under electrode 2. The result of the sequence is that the charge has shifted one electrode width to the right.

This process is continued until the charge reaches the read-out register at the end of a line. There the charges are read out and transformed into electrical voltage signals. The process is usually denoted as the CCD principle (*charge coupled device*), or bucket-brigade principle. In addition to the CCD principle, the CMOS principle for solid-state area sensors has also become well established (see section 3.4.1.3).

CCD line sensors can consist of more than 12 000 sensor elements. Given a sensor spacing of ca. 4 µm to 20 µm, the length of line sensors can be more than 100 mm. Line sensors are used in a wide variety of devices such as line cameras, fax machines, photo scanners or digital copiers.

3.4.1.2 CCD area sensors

Area sensors, which have their sensor elements arranged in a two-dimensional matrix, are almost exclusively used for photogrammetric image acquisition. In comparison with line sensors, the construction of matrix sensors is more complicated since the read-out process must be accomplished in two dimensions. Examples are shown in Fig. 3.74.

a) Frame-transfer sensor with imaging zone and storage zone (Dalsa)

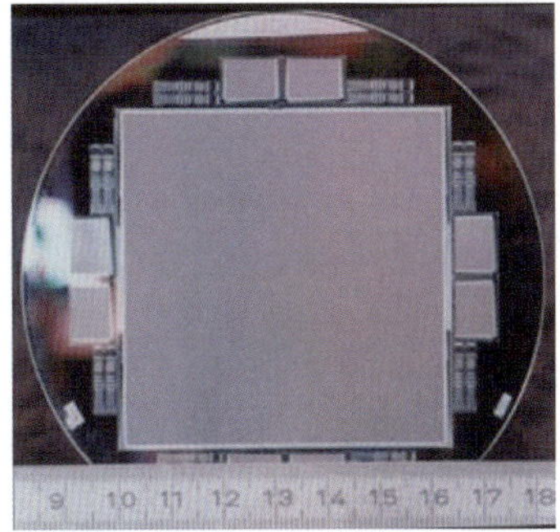

b) A full-frame transfer sensor with 4096 x 4096 elements on a single silicon wafer

Fig. 3.74: CCD matrix sensors

There are three different arrangements of CCD matrix sensors that differ in layout and read-out process:

- frame transfer
- full-frame transfer
- interline transfer

Frame-transfer sensors (FT) consist of a light-sensitive, image-recording zone and an equally sized, opaque storage zone (Fig. 3.74a). Each contains a parallel array of linear CCD sensors. After exposure, charge is moved along the arrays from the imaging zone into the storage zone. From there they are rapidly shifted line by line into the read-out register (Fig. 3.75). Charge transfer from the imaging to the storage zone can be carried out very rapidly, allowing high frame rates to be achieved since the imaging zone can be exposed again whilst the previous image is written out of the camera from the storage zone. Because imaging and storage zones are completely separate areas, the elements in the imaging zone can be manufactured with almost no gaps between them.

A simpler variation is given by the full-frame transfer sensor (FFT, Fig. 3.74b). It consists only of an imaging zone from where charges are directly transferred into the read-out register (Fig. 3.76). During read-out the sensor may not be exposed. In contrast to FT sensors, FFT sensors tend to show greater linear smearing effects since longer transfer times are required. The simpler layout enables the construction of very large sensor areas[1]

[1] The production of very large CCD sensors is limited mainly by economic restrictions (production numbers, quality) rather than technological restrictions.

with very small sensor elements (6–9 µm size). Such layouts are used for high-resolution digital cameras with typically more than 1000 x 1000 sensor elements (manufacturers: e.g. Thomson, Kodak, Fairchild, and Dalsa). Note that the number of FFT sensor elements is often based on integer powers of 2 (512 x 512, 1024 x 1024, 4096 x 4096).

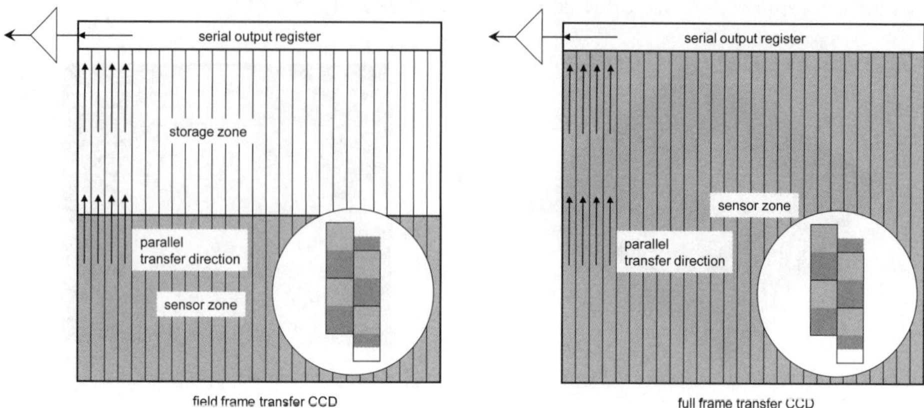

Fig. 3.75: Principle of a frame-transfer sensor **Fig. 3.76:** Principle of a full-frame transfer sensor

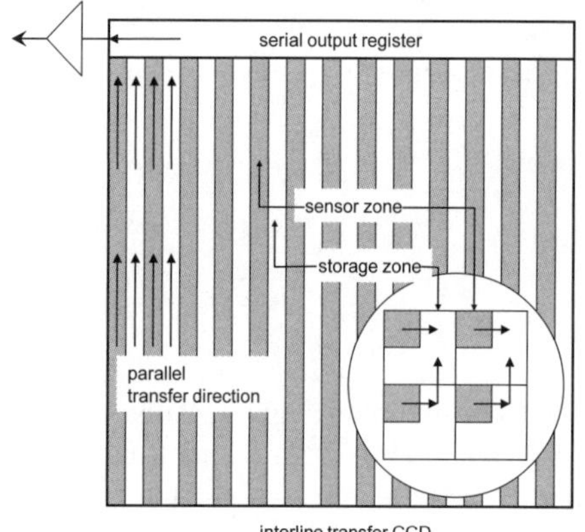

Fig. 3.77: Principle of an interline-transfer sensor

In contrast, interline-transfer sensors (IL) have a completely different layout. Here linear CCD arrays, which are exposed to light, alternate with linear CCD arrays which are opaque to light. Following exposure, charges are first shifted sideways into the opaque arrays which act as transfer columns. Then they are shifted along the columns to the read-out registers (Fig. 3.77). The light sensitive area of the detector covers only about 25% of the total sensor area, compared with 90 to 100% for FT sensors, i.e. IL sensors are less light-sensitive. IL sensors with standard pixel numbers of about 780 x 580 pixels are mainly used for CCD video and TV cameras (especially colour cameras). High-resolution IL sensors have up to 1900 x 1000 pixels.

Current sensor designs typically employ microlens arrays in order to increase the fill factor of each pixel. A microlens array consists of a series of lens elements, each of which is designed to collect the light falling on a region approximating to the area of a single pixel and to direct that light to the smaller light-sensitive region of the actual sensor element (Fig. 3.78, Fig. 3.79). Whilst microlenses significantly enhance pixel fill factor, they typically have the limitation of only being able to receive light over a ±30 degree range of angles. This performance can limit the use of such arrays for extreme wide angle recording unless special optics are used.

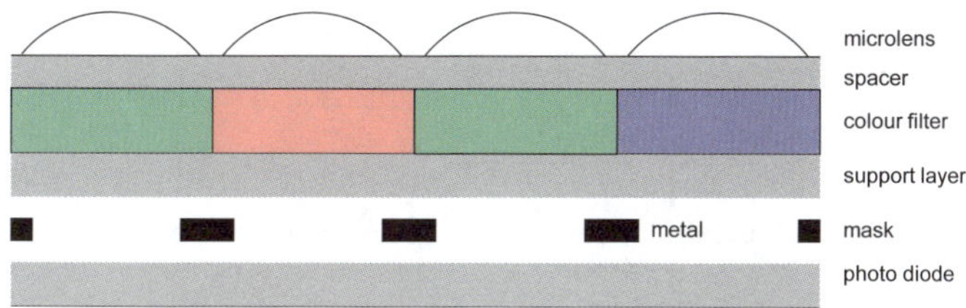

Fig. 3.78: Example of a microlens structure

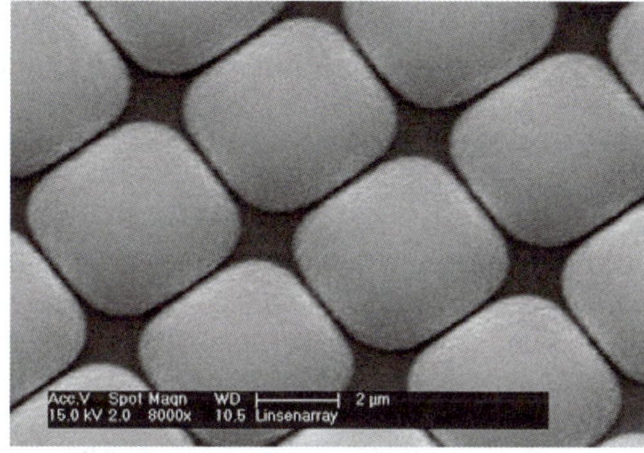

Fig. 3.79: Microlens array (raster electron microscope, 8000x magnification)

3.4.1.3 CMOS matrix sensors

CMOS technology (*complementary metal oxide semi-conductor*) is a widely used technique for the design of computer processors and memory chips. It is increasingly used in the manufacture of opto-electronic imaging sensors, since it has significant advantages over CCD technology:

- only 1/10 to 1/3 power consumption
- lower manufacturing costs
- directly addressable sensor elements; acquisition of arbitrary image windows
- frame rates of more than 2000 frames per second
- can be provided with on-chip processing, e.g. for sensor control or image processing
- high dynamic range and low image noise

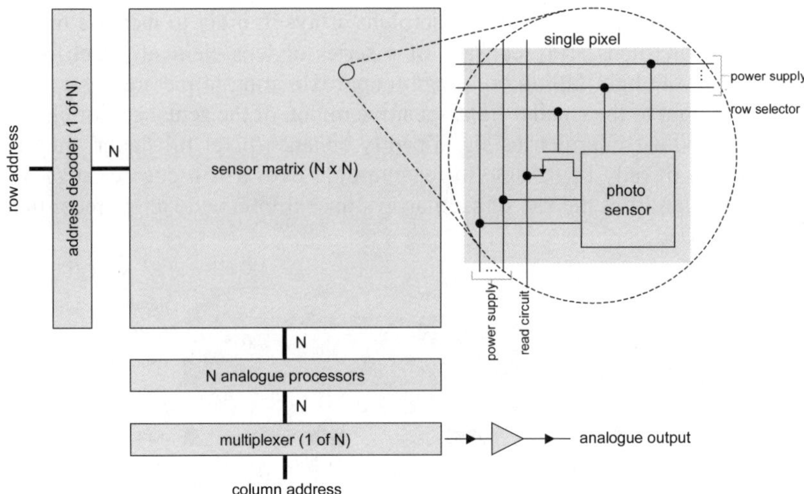

Fig. 3.80: Architecture of a simple 2D CMOS sensor (after Hauschild 1999)

CMOS imaging sensors are already available with 6000 x 4000 sensor elements. They can be designed with multiple sensor layers, each responsive to a selected spectral band (see section 3.4.1.4).

In contrast to CCD sensor elements (MOS capacitors) CMOS detectors are based on photo diodes or transistor elements. The charge generated by the incident light is directly processed by an integrated amplifier and digitiser unit attached to the pixel element. There is, therefore, no sequential charge transfer. As a consequence, individual sensor elements can be operated or processed, and there is a lower sensitivity to blooming and transfer loss. Fig. 3.80 illustrates the basic architecture of a CMOS matrix sensor.

Due to the presence of additional electronic components, the fill factor of CMOS sensors is smaller than for FFT CCD sensors. CMOS sensors are therefore generally provided with microlenses (section 3.4.1.2).

Table 3.4: Features of colour separating methods

	3-chip camera	RGB filter	Colour (mosaic) mask	True colour sensor
number of sensors	3	1	1	1
number of images	1	3	1	1
number of video signals	3	1	1	1
dynamic scenes	yes	no	yes	yes
resolution	full	full	half	full
colour convergency	adjustment	yes	interpolation	yes

3.4.1.4 Colour cameras

In order to create true colour images, incident light must be separated into three spectral bands, typically red, green and blue. Separation can be performed by four common methods (summarised in Table 3.4):

- Parallel or 3-chip method:

  A prism system is used to project incident light simultaneously onto three CCD sensors of the same design. Each sensor is located behind a different colour filter so that each registers the intensity of only one colour channel (Fig. 3.81). Full sensor resolution is retained but exact alignment is required in order to avoid colour shifts. The camera delivers three separated analogue image signals that must be digitised in parallel. The principle is used for professional colour cameras, most of which are used either in TV studios or by mobile film crews.

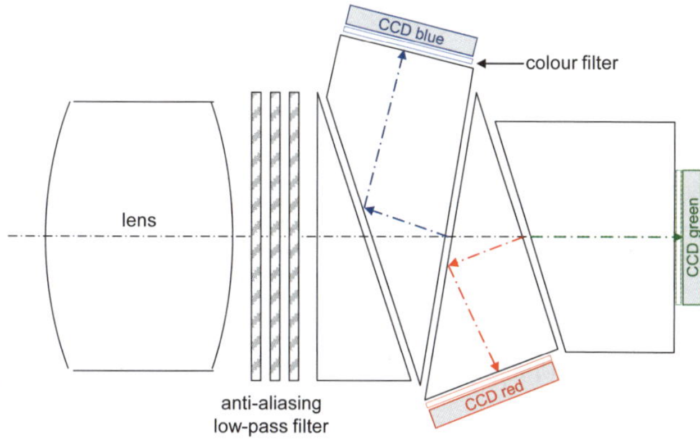

Fig. 3.81: Schematic optical diagram of a 3-CCD or 3-chip camera

- Time-multiplex or RGB filter method:

  Colour separation is performed by the sequential recording of *one* sensor whereby for each image a primary colour filter is introduced into the path of light. Full sensor resolution is preserved, but dynamic scenes cannot be imaged. The method can be applied to both matrix sensor cameras and scanning cameras. The camera delivers a single image signal which must be filtered temporally in order to generate a digital RGB combination from the single colour bands.

- Space-multiplex or colour-mask methods:

  A filter mask is mounted in front of the CCD matrix so that individual sensor elements react to only one colour. Strip or mosaic masks are used, although the Bayer pattern mosaic mask is the most common. In comparison with the previous methods, geometric resolution will decrease since the output of three or more (typically four) sensor elements are combined to form each colour pixel (Fig. 3.82). The principle enables the recording of moving objects and its cost effectiveness means that it used for practically all consumer digital camera systems.

Fig. 3.82: Strip and mosaic masks

When using a Bayer pattern, RGB colour values are calculated by interpolating neighbouring grey values. Because there are twice as many green as red or blue pixels, different calculation methods are used as follows:

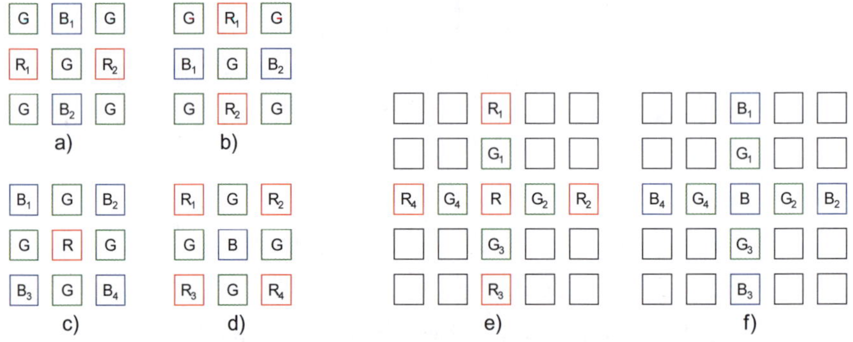

Fig. 3.83: Different combinations used in Bayer colour interpolation

Calculation of red and blue components for a green pixel (Fig. 3.83a und b):

$$R_{green} = \frac{R_1 + R_2}{2} \qquad B_{green} = \frac{B_1 + B_2}{2} \tag{3.81}$$

Calculation of the blue component for a red pixel and the red component for a blue pixel (Fig. 3.83c und d):

$$B_{red} = \frac{B_1 + B_2 + B_3 + B_4}{4} \qquad R_{red} = \frac{R_1 + R_2 + R_3 + R_4}{4} \tag{3.82}$$

Calculation of the green component for red and blue pixels (Fig. 3.83e and f):

$$G_{red} = \begin{cases} (G_1 + G_3)/2 & \text{for } |R_1 - R_3| < |R_2 - R_4| \\ (G_2 + G_4)/2 & \text{for } |R_1 - R_3| > |R_2 - R_4| \\ (G_1 + G_2 + G_3 + G_4)/2 & \text{for } |R_1 - R_3| = |R_2 - R_4| \end{cases}$$

$$G_{red} = \begin{cases} (G_1 + G_3)/2 & \text{for } |B_1 - B_3| < |B_2 - B_4| \\ (G_2 + G_4)/2 & \text{for } |B_1 - B_3| > |B_2 - B_4| \\ (G_1 + G_2 + G_3 + G_4)/2 & \text{for } |B_1 - B_3| = |B_2 - B_4| \end{cases} \quad (3.83)$$

There are additional interpolation functions which differ in their operation with regard to edge sharpness, colour fringes and speed of operation.

- True colour sensor:

  Foveon manufactures a CMOS-based, high-resolution, single-chip, true colour sensor consisting of three layers that are each sensitive to one primary colour (Fig. 3.84). It utilises the property of silicon that light of different wavelengths penetrates to different depths. Hence, this sensor provides the full resolution of a usual CMOS sensor with true-colour registration capability.

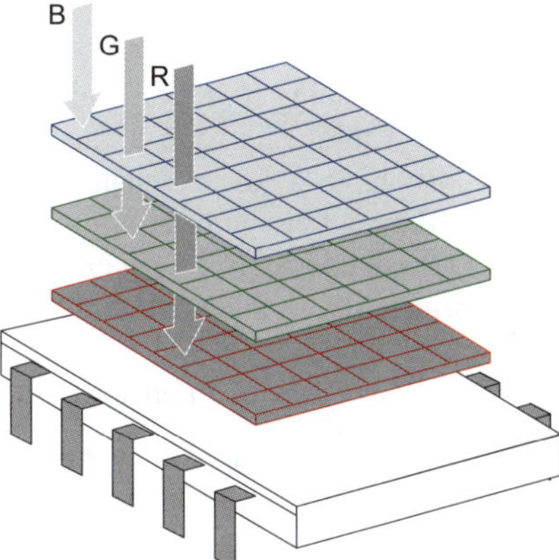

Fig. 3.84: Simplified structure of the Foveon X3 RGB sensor

3.4.1.5 Geometric properties

Resolving power

The theoretical resolving power of monochrome CCD sensors is limited by two factors:

- detector spacing Δx_s (distance between sensor elements) and scanning theorem (Nyquist frequency f_N)
- detector size Δx_d (aperture size) and MTF (limiting frequency f_0)

According to section 3.1.5.2, there are different theoretical resolution limits for FT, FFT and IL sensors due to their different arrangements of sensor elements. Furthermore, for all types, image quality can differ in both x and y directions where rectangular, rather than square, light sensitive regions have been used (Table 3.5).

For FFT sensors, or progressive-scan sensors (no interlaced mode), with square detector elements, resolving power can be expected to be equal in both directions.

IL sensors have approximately the same detector spacing as FT sensors. However, each pixel is split into a light sensitive detector and a shift register. Hence, the resulting theoretical resolving power is four times higher than the Nyquist frequency, and about two times higher than for FT and FFT sensors.

In practice, theoretical resolving power cannot be achieved unless the sampling interval is small enough to push the Nyquist frequency up beyond the effective resolution limit (cut-off) for that MTF. Normally the best that can be achieved is a full-fill system, for example the Kodak FFT in Table 3.5, where the Nyquist frequency is about half the theoretical resolution. In practical systems, frequencies higher than the Nyquist frequency are filtered out in order to avoid aliasing and micro lenses are used in order to provide pixel fill factors close to unity. Micro scanning systems (section 3.5.5.1) are able to achieve closer to theoretical resolving power since they subsample by moving the detector in fractions of a pixel between images.

New sensors with very small detector spacing (e.g. the Kodak KAC-05020 used in mobile phone cameras, see last column in Table 3.5) have a theoretically very high resolving power. However, these sensors also have a lower sensitivity to light and higher image noise. In order to match the pixel dimensions with an appropriate optical resolution, very high quality lenses are required (see section 3.1.4.1 for comparison). The advantage of the small detector size therefore lies principally in the small dimensions of the imaging sensor.

Table 3.5: Resolving power of different CCD sensors

		FT Valvo NXA	FFT Kodak	IL Sony	CMOS Kodak
detector spacing in x	Δx_{sx} [µm]	10.0	9.0	11.0	1.4
detector spacing in y	Δx_{sy} [µm]	7.8	9.0	11.0	1.4
detector size in x	Δx_{dx} [µm]	10.0	9.0	5.5	1.4
detector size in y	Δx_{dy} [µm]	15.6	9.0	5.5	1.4
Nyquist frequency in x	f_{Nx} [lp/mm]	50	55	45	357
Nyquist- frequency in y	f_{Ny} [lp/mm]	64	55	45	357
theoretical resolution in x	f_{0x} [lp/mm]	100	111	180	714
theoretical resolution in y	f_{0y} [lp/mm]	64	111	180	714

In comparison to photographic emulsions, recent opto-electronic sensors have equal or even better resolutions but, at the same time, usually much smaller image formats. A comparable resolution is achieved with sensor element sizes of about 7 µm or less.

In the digital photographic industry, alternative image quality measures are currently in use. The value *MTF50* defines the spatial frequency in lp/mm where the MTF is equal to 50%. With *line widths per picture height* (LW/PH) digital cameras are classified as a function of line width instead of line pairs. LW/PH is equal to 2 x lp/mm x (picture height in mm). The term *cycles or line pairs per pixel* (c/p or lp/p) is used to give an indicator of the performance of a pixel.

Geometric accuracy

The geometric accuracy of matrix sensors is mainly influenced by the precision of the position of sensor elements. Due to the lithographic process used to manufacture semi-conductors, CCD matrix sensors have regular detector positions of better than 0.1–0.2 µm, corresponding to 1/60 to 1/100 of the size of a sensor element. This does not mean that the resulting image can be evaluated to this accuracy. Several electronic processing steps are performed between image acquisition and digital storing that may degrade image geometry and contrast.

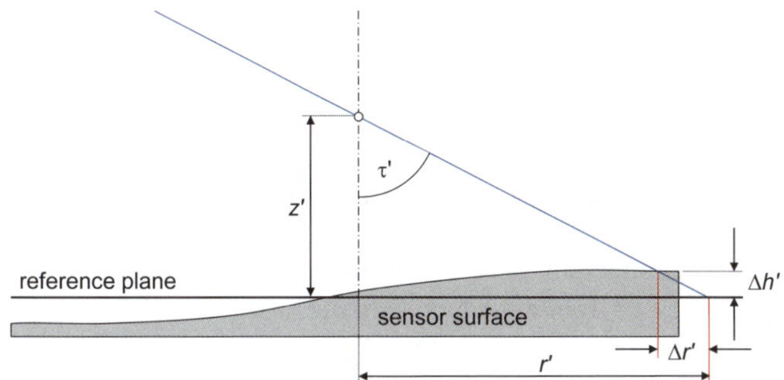

Fig. 3.85: Lateral displacement for an uneven sensor surface

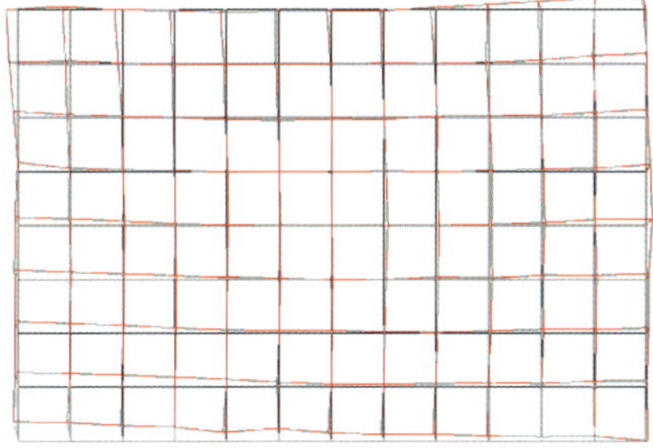

Fig. 3.86: Sensor deformations (red) after finite element calibration (Kodak DCS460)

An additional effect is given by the possible lack of flatness of the sensor surface. For sensor areas of 1500 x 1000 pixels departures of up to 10 µm from a plane surface have been demonstrated. Depending on the viewing angle, a perpendicular displacement of a sensor element causes a corresponding lateral shift in the image (Fig. 3.85). If there is a non-systematic lack of flatness in sensor surface, the usual approaches to distortion correction fail (additional parameters). A suitable correction model based on finite elements has been presented in section 3.3.4.4. Fig. 3.86 shows deformations of an imaging sensors computed by this approach.

$$\Delta r' = \Delta h' \frac{r'}{z'} = \Delta h' \tan \tau' \tag{3.84}$$

According to (3.84), image displacement is greater at shorter focal lengths z' (wide angle lenses), increasing distance r' from the optical axis and greater lack of flatness $\Delta h'$.

3.4.1.6 Radiometric properties

Light falling onto a CCD sensor is either reflected on the sensor surface, absorbed within the semi-conducting layer, or transmitted if high-energy photons are present. Absorption happens if the wave length of light is shorter than a threshold wave length λ_g which is a property of the radiated material and defined according to:

$$\lambda_g = \frac{h \cdot c_0}{E_g} \tag{3.85}$$

where c_0: velocity of propagation = $3 \cdot 10^8$ m/s
h: Planck's constant = $6.62 \cdot 10^{-34}$ Js
E_g: energy between conduction and valence band

Silicon used in the manufacture of CCD sensors has an E_g value of 1.12 eV, thus a threshold wavelength of λ_g = 1097nm which is in the near infrared. In comparison with film and the human eye, CCD sensors show a significantly wider spectral sensitivity (Fig. 3.87). Optionally, infrared absorbing filters can be attached to the sensor in order to restrict incident radiation to visible wavelengths. The spectral sensitivity of CCD sensors with Bayer colour masks (see section 3.4.1.4) varies between the individual colour channels. Typically, the blue channel has the lowest sensitivity (Fig. 3.88).

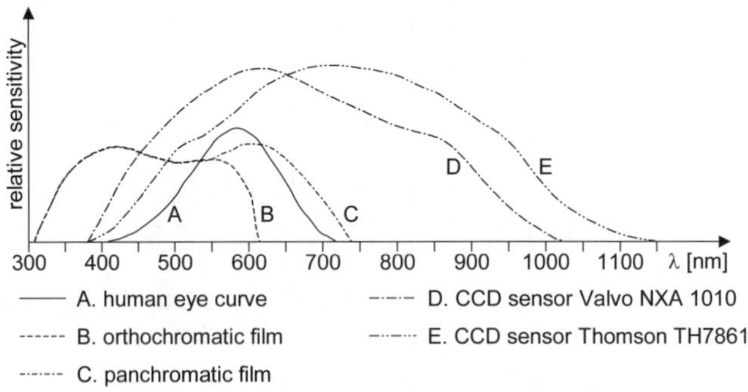

Fig. 3.87: Spectral sensitivity of different sensors

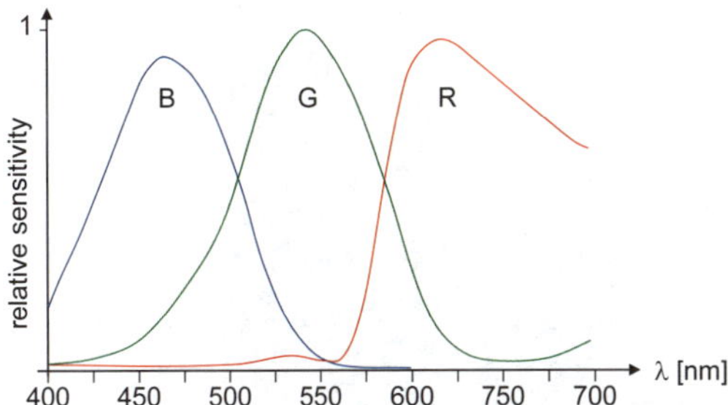

Fig. 3.88: Spectral sensitivity of a colour CCD sensor (Sony ICX 098BQ)

Independent of the image forming incident light, thermal effects in the semi-conducting layer can generate small charges which appear as background noise in the sensor signal. This background noise is known as dark current since it occurs independently of any image illumination and can be observed in total darkness.

Artificial cooling of the sensor can reduce background noise, i.e. the radiometric dynamic range is improved. Cooling by 5–10° C reduces noise by a factor of 2. Artificially cooled sensors are typically used in applications with low light intensity and long integration times (e.g. imaging sensors for astronomy).

The radiometric dynamic range is defined by the signal-to-noise ratio (SNR):

$$SNR = \frac{S}{\sigma_S} = 20 \log \frac{S}{\sigma_S} [\text{dB}] \tag{3.86}$$

S is the maximum signal amplitude and σ_S is denotes the system noise, caused by a combination of photon shot noise, dark current and circuit noise. For CCD matrix sensors typical SNR lies somewhere between 1000:1 (approx. 60 dB) and 5000:1 (approx. 74 dB). If an additional noise for the subsequent A/D conversion is assumed, the CCD sensor signal should be digitised with at least 10–12 bits per pixel. In low-light conditions or where exposure times are very short (e.g. high frequency cameras, section 3.5.3), the SNR has a significant role. In static situations the dynamic range can be increased by multiple exposures with different apertures and exposure settings, as well as analytical combination of individual images (High Dynamic Range Imaging, HDR photography).

For very bright areas in the scene (hot spots) the sensor may be subject to saturation or overflow effects where the imaged signal flows into adjacent pixels. Such effects are caused by movement of charged particles into neighbouring sensor elements (*blooming*), and by continuous charge integration during the read-out process (*smear*). In photogrammetric applications these effects can be observed for brightly illuminated retro-reflective targets resulting in an incorrect determination of the target centre. Blooming and smear can also occur within images of natural scenes where large variations in lighting and contrast are present (example in Fig. 3.89).

Fig. 3.89: Blooming effect for an architectural image (camera: Kodak DCS 420)

3.4.2 Camera technology

3.4.2.1 Camera types

Three generic camera designs can be identified among the wide variety of camera systems that are available to the photographer:

- Viewfinder or compact cameras:

  Historically a viewfinder camera uses a viewing lens that is separated from the actual image taking lens. Such camera designs are generally very light in weight as the optical systems and image formats are compact. However, direct observation of the image is not possible for focusing, depth of field control or lens interchange. More seriously for close-range work, the difference in content between the viewfinder image and the recorded image give rise to parallax errors. An overlay in the viewfinder can indicate the area imaged by the camera, but it is not possible to fully correct the difference in perspective. The viewfinder displays an upright image as would be seen by the eye.

  Digital camera technology has replaced the optical viewfinder with a live display of the camera image on a LCD screen integrated into the camera. The screen may be located on the back of the camera or on hinge mechanism to allow comfortable viewing for high and low vantage points. Many consumer compact digital cameras and mobile phone cameras are of this type (see Fig. 3.111 and Fig. 3.116).

- Single-lens reflex camera:

  In a single-lens reflex (SLR) camera, viewing is done directly through the camera lens by means of a plane mirror which deflects the path of rays into a viewfinder (Fig. 3.90). Before exposure, the mirror is flipped out of the optical path. Single lens reflex cameras with a simple box viewfinder (often equipped with a magnifier) consist of a ground-glass focusing screen where the image can be viewed at waist level from above. The focus screen image appears upside down and reversed (mirrored horizontally). This feature was used in many mid-format analogue cameras. The addition of a pentagonal prism or penta mirror to the viewfinder light path provides an upright unreversed view of the scene. Cameras employing this principle with a digital sensor are termed Digital SLRs or DSLRs (see Fig. 3.112 to Fig. 3.115).

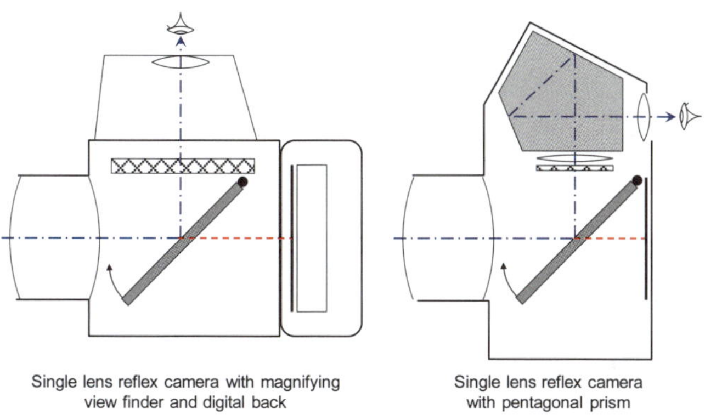

Fig. 3.90: Single lens reflex cameras (after Marchesi 1985)

Fig. 3.112 to Fig. 3.115 show examples of digital single-lens reflex cameras. Here it is also normally possible to view the live camera image on the LCD display.

As with viewfinder cameras, the incorporation of direct display to an LCD panel can add a direct view of the captured digital image. A camera design, known as a *bridge camera*, removes the mirror from the system and replaces the ground glass screen with a digital display. Most modern DSLR cameras maintain the mirror and ground glass viewing system, but support direct live viewing on an LCD screen at the back of the camera in a similar manner to the consumer compact camera. Often termed "live view", this method requires the mirror to be in the up position so that light can reach the CCD or more commonly CMOS sensor.

- Studio camera:

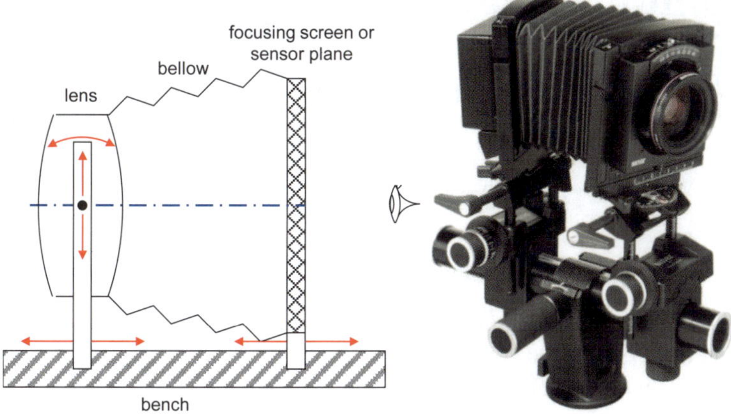

Fig. 3.91: Studio camera (Sinar)

Studio cameras (Fig. 3.91) allow for the individual translation and rotation of lens and image planes. By tilting the lens, special focus settings can be enabled, e.g. Scheimpflug's condition (section 3.1.2.4) which maximises depth of field in a particular plane. When the lens is shifted the perspective imaging properties are changed, e.g. to avoid convergent lines for vertical architectural pictures. Studio cameras represent the ultimate in photographic quality by virtue of their large image sizes (5"x7" and 10"x8" being common). Due to their bulky nature and relatively cumbersome deployment they are typically used in professional studio applications or for landscape and architectural work.

3.4.2.2 Shutter

The shutter is used to open the optical path for the duration of time necessary for correct exposure. For conventional camera systems two basic types are used, the focal-plane shutter and the inter-lens or leaf shutter.

Fig. 3.92: Principle of focal plane shutter

The majority of 35mm single lens reflex cameras use a focal-plane shutter that is mounted directly in front of the image plane and, during exposure, moves a small slit across it (Fig. 3.92). Variation in the size of the slit allows a variety of short duration shutter settings. Focal plane shutters are easy to design and provide shutter times of less than 1/4000 s. If the camera moves parallel to the shutter movement (e.g. photos from a moving platform), imaging positions are displaced, i.e. at the different slit positions the image has a different exterior orientation.

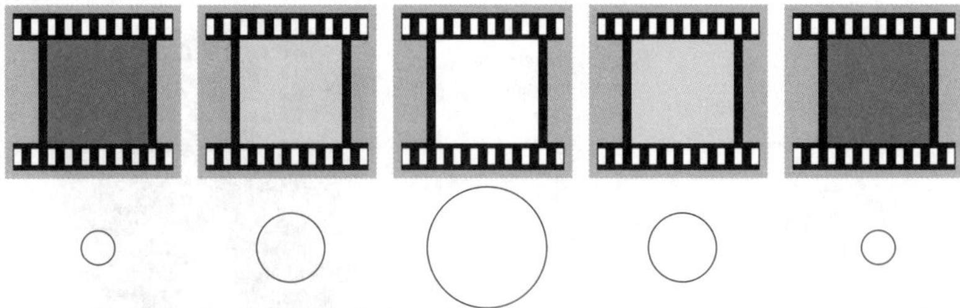

Fig. 3.93: Principle of the inter-lens shutter

Inter-lens shutters are typically mounted between the lens elements and close to the aperture stop. Because of this, each lens must have its own shutter. Mechanically sprung blades are used to open the shutter radially (Fig. 3.93). In the figure, the circles represent the diameter of the shutter as it opens and closes and the corresponding tone in the image shows the level of light reaching the image plane. In practice the opening and closing of the shutter can be regarded as instantaneous, which means that the complete image is exposed simultaneously with the same projection even if the camera platform moves. Inter-lens shutters are mechanically more complex than focal-plane shutters since the individual elements must be sprung open and then closed. Consequently, shortest shutter times are restricted to about

1/1000 s. In photogrammetry they are usually encountered in low-cost viewfinder cameras and in professional medium and large format cameras.

Both CMOS and CCD sensors can employ an electronic shutter either to replace a mechanical shutter or to supplement it. Several technologies are in use today:

- Interline odd/even line:

  All even lines in the image are captured followed by all odd lines. The two sets of lines are then presented as a pair of half images to form a full frame consecutively. The initial purpose of this standard was to provide a high frame rate (60 Hz) in early TV transmission with limited given bandwidth.

- Progressive scan:

  Lines are progressively read out and exposed line by line as an electronic shutter is moved across the array (see also section 3.5.2). This process is analogous to the higher speed focal plane SLR shutters which sweep two blinds with a variable gap across the sensor.

- Global shutter:

  All locations start integration at the same time and are then read out either into a covered region of the sensor. If light continues to reach the sensor this risks some image smear. An additional mechanical shutter can be closed to block further light reaching the sensor.

- Global reset:

  This is similar to a progressive scan system, but the whole sensor area starts acquiring light at the same point in time. The data is read out line by line so that the first lines receive less integration time than the later lines. This is often used in cameras with a trigger output so that an electronic flash can be used to freeze motion and background low light does not have a significant effect on pixels read out at a later time.

3.4.2.3 Image stabilisation

Modern lenses and digital cameras can be equipped with electronic image stabilisation which ensures that blurring caused by hand-held operation is reduced. The vibrations, for which this compensates, lie approximately in the range 1 to 10 Hz and are corrected either in the camera body or within the lens.

Image stabilisers in the camera body have a movement sensor which detects the accelerations acting on the camera and immediately applies a counteracting movement of the imaging sensor (Fig. 3.94a). The compensating movement of the imaging sensor is implemented with piezoelectric or electromagnetic elements which can generate shifts of up to three millimetres. Cameras with built-in image stabilisers can utilise this option with any lens.

Image stabilisation at the lens (Fig. 3.94b) is implemented with a correction lens (or lenses) moved by piezoelectric elements. This lens group alters the imaging path in a way which compensates for the blurring movement. To make use of this feature the lens requires a digital interface which receives a signal from the camera when exposure takes place.

Both image stabilisation techniques directly influence the imaging geometry and thereby continuously alter the interior orientation of the camera. This function should therefore be switched off for photogrammetric recording. Geometric changes due to image stabilisation can only be accommodated by use of image-variant camera calibration (see section 3.3.4.4).

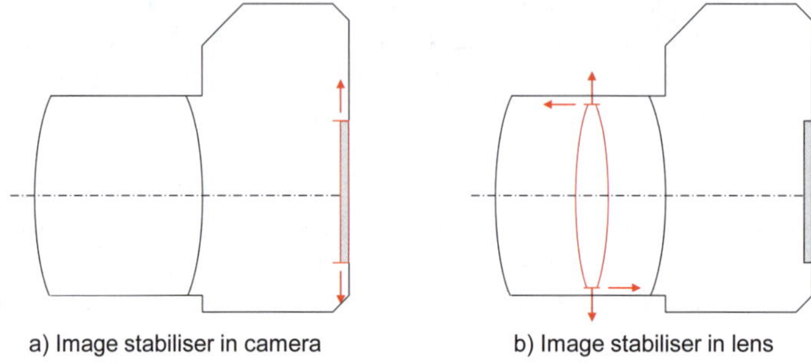

Fig. 3.94: Techniques for automatic image stabilisation

3.4.3 Lenses

3.4.3.1 Relative aperture and f/number

The light gathering capacity of a lens is measured by the relative aperture which is the ratio of the iris diameter d' of the entrance pupil to the focal length f (see Fig. 3.95):

$$\text{Relative aperture} = \frac{d'}{f} \tag{3.87}$$

For a lens with an entrance pupil of $d' = 20$ mm and a focal length of $f = 80$ mm, the relative aperture is 1:4.

The f/number is given by the reciprocal of relative aperture:

$$\text{f/number} = \frac{f}{d'} \tag{3.88}$$

A higher f/number corresponds to a smaller relative aperture, i.e. the light gathering capacity is reduced. Changes in f/number follow a progression whereby the area of aperture (hence the amount of gathered light) changes by a factor of 2 from step to step (Table 3.6). The f/number also alters the depth of field (see section 3.1.2.3).

Table 3.6: Standard f/number sequence

1	1.4	2	2.8	4	5.6	8	11	16	22	...

3.4.3.2 Field of view

The field of view (format angle) $2\alpha'$ of a lens is defined by its focal length and the diameter of the entrance pupil *EP* (Fig. 3.95):

$$\tan\alpha' = \frac{d'}{2f} \tag{3.89}$$

In contrast, the format angle (field angle) 2Ω is given by the maximum usable image angle with respect to the diagonal of a given image format s' and the principal distance c:

$$\tan\Omega = \frac{s'}{2c} \tag{3.90}$$

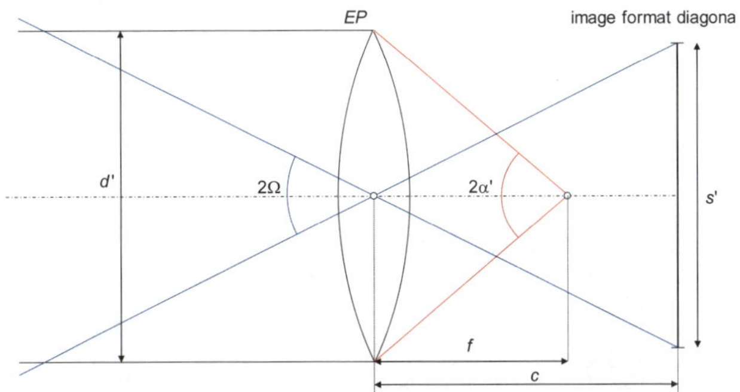

Fig. 3.95: Field of view and format angle

Table 3.7: Resulting focal lengths for typical lens types and image formats

Lens type Image diagonal	Format-angle 2Ω	Video [8 × 6 mm²] 10 mm	Small format [36 × 24 mm²] 43 mm	Medium format [60 × 60 mm²] 84 mm	Large format [230 × 230 mm²] 325 mm
telephoto (small angle)	15 – 25°	> 22 mm	> 90 mm	> 190 mm	610 mm
normal (default)	40 – 70°	7 – 13 mm (8 mm)	40 – 60 mm (50 mm)	70 – 115 mm (80 mm)	280 – 440 mm (300 mm)
wide angle	70 – 100°	4 – 7 mm	18 – 35 mm	35 – 60 mm	135 – 230 mm
fish-eye	> 100°	<4 mm	< 18 mm	< 35 mm	< 135 mm

Format angle is a convenient method of distinguishing between different basic lens types (see Table 3.7). As a rule of thumb, the focal length of a normal lens is approximately equal to the diagonal of the image format. Small image formats (found in video cameras, mobile phones and low-cost digital cameras) require short focal lengths in order to produce wide angles of view.

3.4.3.3 Super wide-angle and fisheye lenses

In close-range photogrammetry, short focal length lenses with wide fields of view are often selected because they allow for shorter object distances, greater object coverage and, consequently, more favourable ray intersection angles. Wide-angle lenses have fields of view typically in the range 60–75°. Lenses with larger fields of view designed to maintain the central perspective projection are known as super wide-angle lenses (approx. 80–120°). Whilst they have increased image aberrations and illumination fall-off towards the extremes of the image format (section 3.1.3.6), their use in photogrammetry is common as it supports working in cluttered environments with short taking distances. Fisheye lenses utilise a different optical design that departs from the central perspective imaging model to produce image circles of up to 180° (section 3.3.6).

In general, the effect of radial distortion increases with larger fields of view. Fig. 3.96 shows examples of images taken with lenses of different focal lengths and fields of view. For some lenses the distortion is clearly visible and here the 15 mm quasi-fisheye lens shows the greatest distortion, but the illumination fall-off identifiable in the 14mm super wide angle lens image is noticeably absent. Fisheye lenses must be modelled using the appropriate fish eye projection model (see section 3.3.6) as the polynomials described in section 3.3.3.1 will be unstable due to the very high distortion gradients.

a) f = 14 mm

b) f = 15 mm

c) f = 20 mm

Fig. 3.96: Imaging with a) super-wide-angle; b) fisheye; c) wide angle lenses

3.4.3.4 Zoom lenses

Zoom or variofocal lenses enable a varying focal length to be produced from a single lens system. Designs may also permit constant focusing and maintenance of the same relative aperture as focal length is changed. Fig. 3.97 illustrates the principle of a zoom lens where moving a central lens group gives a change in focal length whilst motion of a second group of lenses provides focus compensation.

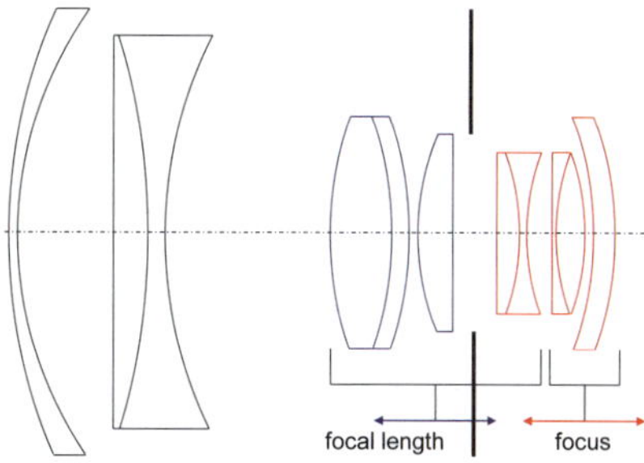

Fig. 3.97: Principle of a zoom lens (after Marchesi 1985)

According to section 3.3.2.2, a change in focal length results in a new interior orientation. Due to the zoom lens construction, not only the spatial position of the perspective centre will change (Fig. 3.98) but parameters of radial and tangential distortion will also change (Fig. 3.99). Whilst zoom lenses can be calibrated photogrammetrically, off-the-shelf designs cannot be assumed to provide high mechanical stability. Thus, whilst they provide great flexibility, they are seldom used in practice for accurate work.

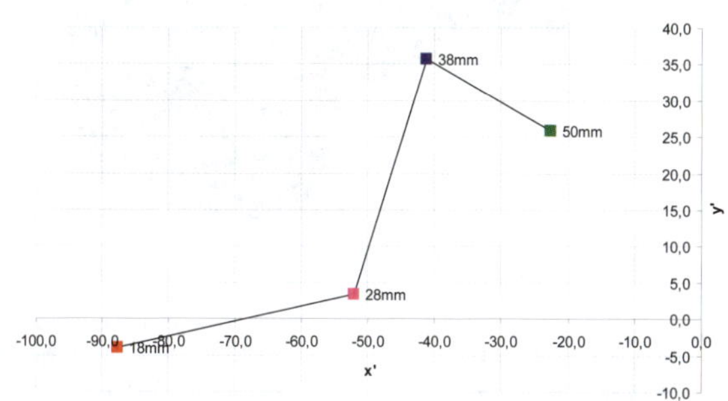

Fig. 3.98: Variation of the principal point in µm for the different focal lengths of a zoom lens (Canon EOS 1000D with Sigma DC 18–50 mm)

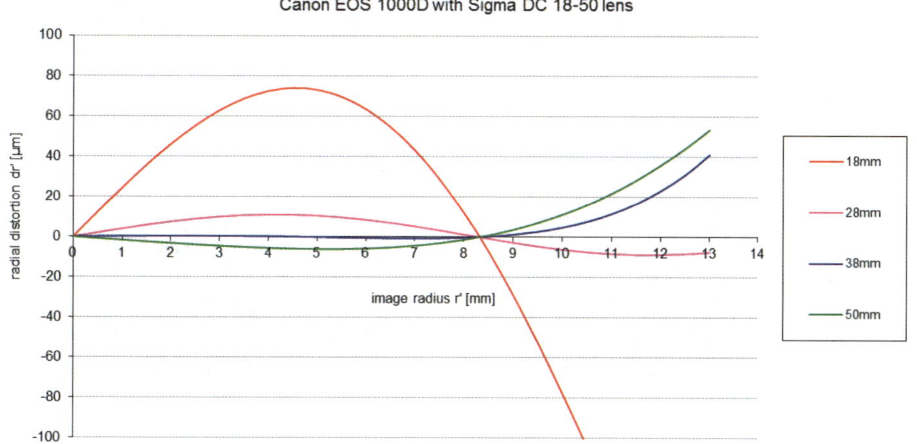

Fig. 3.99: Variation in radial distortion for different focal settings of a zoom lens

3.4.3.5 Tilt-shift lenses

Tilt-shift lenses are special lenses which permit a lateral shift and tilt of the lens relative to the image sensor (Fig. 3.100). Using the shift function the optical axis can be displaced in order to eliminate converging lines in oblique views. As an example, Fig. 3.101 shows the image of a building façade in which the converging effect of perspective has been eliminated by shifting the lens in the vertical direction. In both cases the camera is set to point horizontally.

The tilt function permits the use of the Scheimpflug condition (section 3.1.2.4) which enables sharp imaging of an obliquely imaged object plane.

Fig. 3.100: A digital SLR camera fitted with tilt-shift lens

a) Conventional image

b) Image using shift function

Fig. 3.101: Correction of converging lines by the use of a tilt-shift lens

3.4.3.6 Telecentric lenses

Telecentric lenses are designed such that all object points are imaged at equal image scales regardless of their distance. A two-stage telecentric lens consists of two lens groups where the object-side principal plane of the second (right) system coincides with the focal plane of the first (left) system (Fig. 3.102). The aperture stop is also positioned at this location. An object point P located within one focal length of the first system is virtually projected into P'. The second system projects P' sharply into the image plane at P". Since all points in object space lie on a parallel axis (hence distance-independent), light rays are projected through identical principal rays and are therefore imaged at the same position in the focal plane.

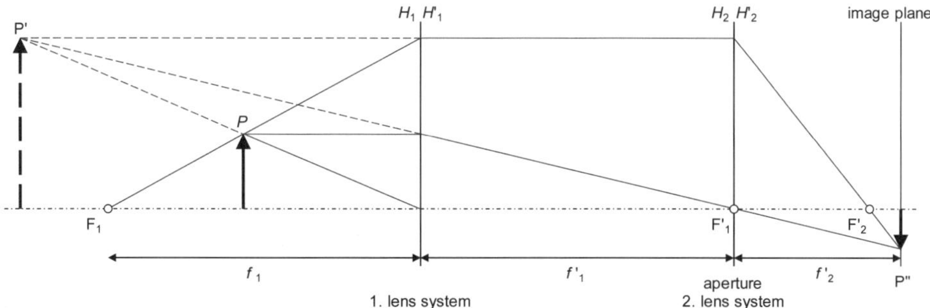

Fig. 3.102: Two-stage telecentric lens (after Schmidt 1993)

Limits to the bundles of rays place limits on the size of the object which can be seen in the image plane. The limiting factor will be either the maximum diameter of the aperture stop or the lens. Telecentric lenses are mainly used for imaging small objects (∅<100 mm) in the field of two-dimensional optical metrology. The imaging model does not correspond to a central projection but can be modelled as a parallel projection.

3.4.3.7 Stereo image splitting

With the aid of beam splitters and mirrors it is possible, using only a single camera, to make stereo and multiple image recordings with only one exposure. This possibility is of particular interest for recording dynamic processes where synchronisation of camera imaging is mandatory. A single camera, split image system is intrinsically synchronised and is therefore suitable for recording fast-changing events.

Fig. 3.103 shows the principle of a stereo mirror attachment with a central lens. With this arrangement, it is possible to use an existing camera/lens combination or the camera in combination with other lenses. Assuming that the mirror and beam splitting surfaces are planar, the arrangement generates the equivalent of two virtual cameras with perspective centres O' and O", each of which provides a central perspective image onto half of the image sensor area.

The stereo base *b* is the separation between the points O' and O". The photogrammetric exterior orientation is defined at these points, i.e. a change in the tilt or position of the mirror attachment causes a change in the imaging geometry analogous to a conventional measurement with separate cameras. The reference point for the interior orientation of the camera is the perspective centre O located within the lens. The field of view τ for the half image is derived from half of the horizontal format $s/2$ and the principal distance c. The inclination angle of the inner mirror must meet the requirement $\beta_1 > \tau$ so that the outer ray at the edge of the image can still be reflected by the mirror. The outer mirror has an inclination angle β_2 and, relative to the inner mirror, a rotation angle of α. Angle β_2 must be smaller than β_1 so that optical axes h_1 and h_2 converge, thus ensuring stereoscopic coverage of the object space. At the same time, β_2 should be set with respect to τ such that the edge rays r_1 and r_2 diverge in order to capture an object space wider than the base b.

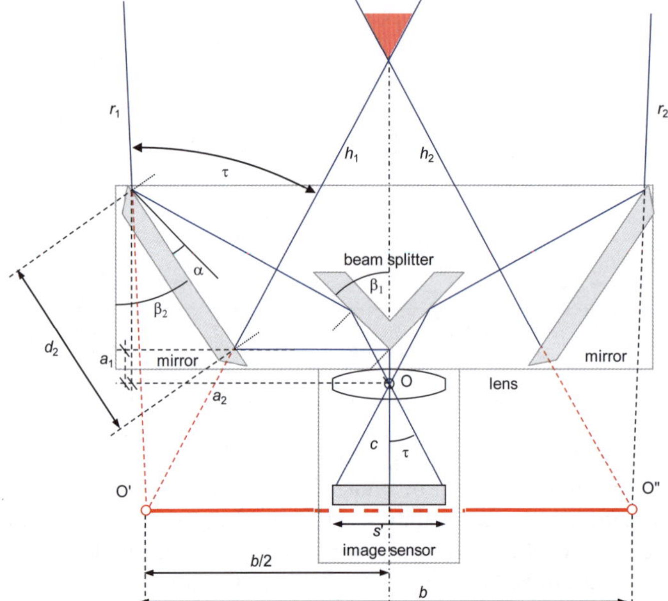

Fig. 3.103: Principle of the stereo-mirror attachment

The size of the outer mirror is primarily a function of the field of view of the lens, as well as its rotation angle. Mirror size increases linearly with decreasing focal length and increasing mirror offset. The resulting horizontal (stereoscopic) measurement range is limited by the inner and outer imaging rays. The vertical measurement range continues to be defined by the field of vie and vertical image format.

High demands are made of the planarity of the mirror surfaces. Departures from planarity result in non-linear image deformations which can only be removed by a significant calibration effort.

3.4.4 Filters

Various filters are employed in analogue and digital imaging procedures. These absorb or transmit different parts of the optical spectrum. The following types of filter are typically used. Fig. 3.104 illustrates their transmission properties.

- Ultraviolet blocking filter:
  UV blocking filters are mostly used where the light conditions have a strong UV component, e.g. in snow. They absorb all radiation under about 380 nm.
- Infrared blocking filter:
  IR blocking filters are generally employed to suppress the natural sensitivity of digital image sensors to light in the infrared part of the spectrum (see section 3.4.1.6). They work from about 720 nm and are an integrated filter layer in many digital cameras.
- Band-pass filter:
  Band-pass filters are designed to transmit a limited range of wavelengths. All wavelengths outside the band defined by the central wavelength and the half-power width are suppressed. Band-pass filters are used in optical systems where a monochromatic or spectrally narrow illumination is used and extraneous light outside this illumination band must be excluded from reaching the imaging sensor.

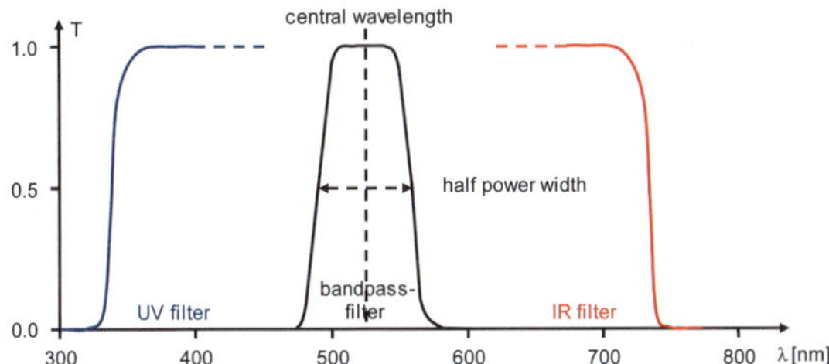

Fig. 3.104: Transmission properties of UV and IR blocking filters

3.5 Imaging systems

Imaging systems closely follow the consumer market with significant advances being made in design and capability. Photogrammetric system development parallels these advances, from making use of the very first analogue CCD camera systems to building on the lessons learned to quickly develop systems based on low-cost webcams and mass market DSLR systems for example.

Historically, analogue cameras have been used for photogrammetry. These are typically small-format cameras delivering an analogue signal in real-time (25–30 frames per second). The analogue signal must be transmitted, usually through a cable, to a dedicated analogue to digital converter, which usually takes the form of a frame grabber board within a host computer. Typical images are between 640 x 480 and 780 x 580 pixels. Imaging systems based on analogue cameras are diminishing in importance, becoming increasingly historical. Section 3.5.1 describes the operation of the *analogue video camera* in detail since such cameras are still found in many legacy photogrammetric systems.

The majority of analogue video cameras output standardised video signals based on the specifications for analogue image transfer for video and TV technology (*analogue video*). Established standard analogue video formats deliver image signals with an *aspect ratio* of 4:3. CCIR (Europe) and RS-170 (America, Asia) are the two standards in use world-wide, whose specifications only differ slightly (see Table 3.8).

Digital cameras have largely superseded analogue cameras. They incorporate the analogue to digital conversion electronics, and in many cases local image storage, within the camera unit. The design results in significantly better output image quality at reduced cost and complexity compared to analogue cameras as interference and timing errors introduced through the external cabling systems and specialist frame grabbers of analogue cameras are avoided.

Whilst there are no standard dimensions for single digital images, the requirement for mass market television has led to the emergence of new *digital video* standards (ITU Rec 709). These define HDTV (*high definition television*) formats with a 16:9 aspect ratio. Based on 1920 pixels per active line, these include 1080i (with 1080 interlaced lines) and 1080p (1080 progressively scanned lines). Frame rates of 24, 25, 30, 50 and 60 frames per second are defined. Broadcast of HDTV typically requires these images aim to fill a 1900 x 1080 pixel display. Bandwidth limitations necessitate the image stream to be compressed, with popular compression schemes being MPEG-2 and MPEG-4 AVC. The effect of compression on local image quality and geometry is image content specific and its effect needs to be carefully tested prior to photogrammetric use.

A convenient digital camera classification considers those systems needing a connecting cable to capture images and those which do not. Note that analogue cameras are of the cabled type:

- Cabled digital cameras:

  These follow a similar design concept to analogue cameras in that a cable providing power and transfer of the digital image from the camera unit to a host computer is required. Such systems are available in a wide range of different sizes and formats. This type of camera is typified by small-format cameras delivering digital image signals in real-time (25–60 frames per second). Typical image sizes are between 780 x 580 pixels and upwards. For example cameras with 1900 x 1100 pixels, utilising low cost CMOS sensors, are increasingly common in line with recently established, high-definition TV standards (see the later part of section 3.5.1.1). Webcams and GigE Ethernet cameras are typical of those falling into this category, both types being able to capture user selectable single frames or digital image sequences and transfer them to a host computer.

- Self-contained digital cameras:

  This type of digital camera comprises a plethora of consumer and professional level imaging systems where local analogue to digital version, image storage and battery power are used to allow the camera to be used as an independent imaging device. Many devices have basic image processing and viewing functions included. Typical examples include mobile phone cameras, consumer compact cameras and professional digital SLR cameras. Trends in these systems include: increasing pixel count for both single images and image sequences; onboard processing for example to increase apparent image sharpness, stitch images together, combine images to increase dynamic range and the capability to correct image distortions; wireless connectivity to enable direct upload to the internet or output to a printer. Self-contained systems generally deliver high-resolution images, usually between 1000 x 1000 and 6700 x 9000 pixels (valid 2013).

Image capture rates of up to 60 frames per second to several seconds per image can be achieved.

A series of specialist sub-types of camera can be conveniently defined linked to specific purposes. Examples of specialist sub-types and their use in photogrammetry include: *high speed cameras* (see section 3.5.3); stereo and multi-camera systems (see section 3.5.4); micro and macro-scanning cameras (see section 3.5.5); panoramic cameras (see section 3.5.6) and; thermal imaging cameras (see section 3.5.7). As with more general purpose systems, the trend is for digital systems replacing analogue systems.

3.5.1 Analogue cameras

3.5.1.1 Analogue video cameras

In the case of the analogue video CCIR format, a video signal is based on 25 frames per second generated in *interlaced scanning* mode (50 field images per second). Each full frame consists of a sequential pair of one odd and one even field. A full CCIR standard frame has 625 lines (Table 3.8). Each individual line is led by a line synchronisation signal (*horizontal sync*), each image (*frame*) commences with an image synchronisation signal (*vertical sync*). Interlaced imaging was designed to overcome bandwidth limitations in broadcast television by sending sequential odd and even fields. The time period of a single CCIR frame is 1/25s = 40 ms, i.e. a single line requires 40ms / 625 = 64 µs. Some of the image lines contain no image information, such that 576 active image lines remain. Subtracting the time for synchronisation signals, each active image line has a time period of 52 µs. Assuming a sampling of 744 image points per line, the time required for one image pixel is 69 ns.

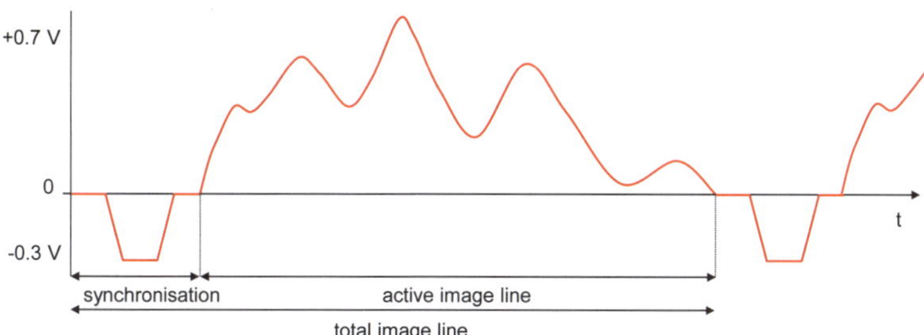

Fig. 3.105: An analogue video signal

Analogue video cameras are equipped with an area sensor that can be read out within a video cycle (25–30 images per second). They can differ concerning usable image format and number of sensor elements. With respect to physical sensor size, definitions originally developed for vidicon camera tubes are still in use today with sensor format dimensions classified as equivalent vidicon tube diameters in the sequence 1/4", 1/3", 1/2", 2/3" and 1" (inch).

Table 3.8: CCIR and RS-170 analogue video specifications

	CCIR	RS-170	Remarks
full frame frequency	25 Hz	30 Hz	
field frequency	50 Hz	60 Hz	
number of image lines	625	525	
active image lines	576	480	
duration of an image line	64 µs	63.5 µs	
frequency of an image line	15.6 kHz	15.7 kHz	
duration of an active line	52 µs	52.5 µs	only image information
image points per line	744	752	no standard, depends on sampling frequency used
duration of an image point	69 ns	70 ns	
frequency of an image point	14.5 MHz	14.5 MHz	≈ pixel-clock frequency

Table 3.9: Example image formats and outputs of some small format cabled camera units

Format (typical) [mm]	Diagonal of corresponding vidicon tube [inch]	Sensor diagonal [mm]	Pixel count and size [µm]	Example	Output
3.2 x 2.4	1/4"	4.0	752 x 582 4.2 x 4.2	Sony DXC-LS1P	analogue S-VHS from dedicated control unit
4.8 x 3.6	1/3"	6.0	752 x 582 7.4 x 7.4	Pulnix TMC-63M	analogue CCIR signal
6.4 x 4.8	1/2"	8.0	768 x 494 8.4 x 9.8	Pulnix TM-7	analogue RS170 signal
8.8 x 6.6	2/3"	11.0	768 x 593 11.0 x 11.0	Sony XC-77	analogue CCIR
9.1 x 6.9	2/3"	11.4	1300 x 1030 6.7 x 6.7	Pulnix TM-1300	10-bit per pixel digital output and SVGA analogue signal
9.1 x 9.1	1"	12.7	1024 x 1024 9.0 x 9.0	Pulnix TM-1001	8 bit per pixel digital output and analogue CCIR signal
12.3 x 12.3	1"	16.0	1024 x 1024 12.0 x 12.0	Dalsa CA-D4/T	12 bit per pixel digital output
14.0 x 8.0	1"	16.1	1920 x 1035 7.3 x 7.6	Sony XCH-1125	8 bit per pixel digital output

The number of sensor lines in an analogue video camera are arranged to comply with the video standard. Usually between 480 and 580 lines are used whereby each line consists of approximately 600 to 780 sensor elements. According to the image formats, displayed in

Table 3.9, the size of a single sensor element (pixel) varies between 4 µm (1/4" cameras) and 12 µm (1" cameras). Cameras with square pixels and cameras with rectangular pixels are possible. The trend for sensors is towards higher resolution and larger areas with CMOS sensors dominating the market by virtue of low cost and power requirement at the expense of the higher image quality of CCD based devices.

3.5.1.2 Analogue camera technology

In general analogue cameras are not designed for high-precision geometric measurement tasks. The opto-mechanical arrangement of the imaging sensor, camera body, filters and lens will often not meet the dimensional stability requirements demanded by photogrammetric measurement needs. In addition, in many cases the image signal is processed internally, resulting in a loss of the precise geometric relationship between imaging sensor and recorded image.

Fig. 3.106 displays the schematic layout of an analogue camera. The imaging sensor is embedded into a ceramic housing and covered by a thin glass plate in order to make it mechanically robust. Often a diffuser is mounted in front as a low-pass filter to compensate for aliasing effects. In some cases, a matrix of semi-cylindrical lenses (microlenses) are used instead, one lens per pixel. An optional infra-red cut-off filter can be included to absorb wavelengths longer than 700 nm.

Usually a C-mount or CS-mount lens adapter with screw thread is used. For off-the-shelf cameras it cannot be assumed that the individual optical components making up the lens are sufficiently well aligned to the optical axis, nor have a homogeneity that meets photogrammetric requirements. In addition, many video camera lenses have large radial and tangential distortions that must be modelled (optical distortions of tens of pixels being common). Provided that the opto-mechanical properties of the camera are physically stable, modelling is a routine process.

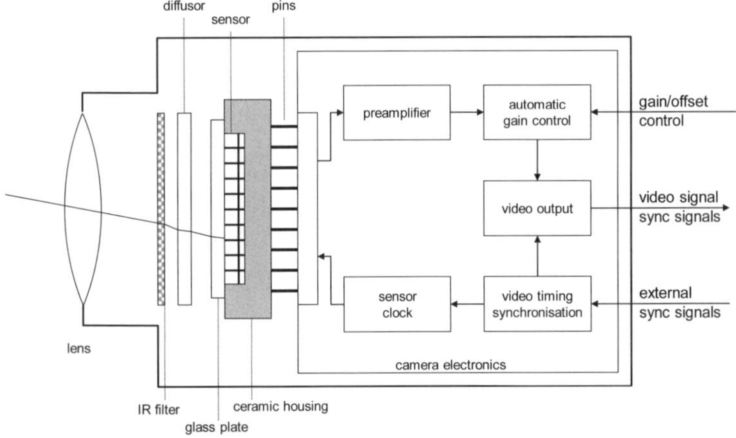

Fig. 3.106: Schematic layout of an analogue camera (after Shortis & Beyer 1996)

After sensor read-out the image signal is amplified and pre-processed. Subsequently its contrast and brightness may be automatically adjusted through the use of an *automatic gain control* whereby the analogue signal is amplified to the maximum amplitude. The level of amplification is controlled by the brightest and the darkest locations in the image. Many cameras allow gain control to be controlled externally (example in Fig. 3.107). Since automatic gain control is a post-processing step, loss of geometric quality and undesired brightness changes cannot be excluded. In addition, and most noticeably under poor lighting conditions, gain control can increase image noise. Automatic gain control should be switched off under conditions of sufficient and constant illumination.

Fig. 3.107: Digital image without (left) and with (right) automatic gain control at constant illumination

Both image sensor and video signal require high-frequency synchronisation signals that are delivered by a pulse generator (*sensor clock*). Synchronisation can also be controlled by an external signal, e.g. for the synchronised image acquisition of several cameras, or for the synchronisation of a video frame grabber for the precise digitisation of a video signal (see section 3.5.1.3).

3.5.1.3 Digitisation of analogue video signals

In order to provide an output stream of digital images, the analogue video signal must be digitised by an analogue-to-digital converter (ADC) which is synchronised to the video frame rate of the sensor (e.g. 25 frames per second). *Frame grabbers* are additional, computer-integrated hardware devices containing an ADC and associated timing electronics, and connected via a data transfer bus. The digitised video image can be displayed in real-time either on the graphical user-interface of the computer, or transferred to computer memory or a digital recording device such as a digital tape.

Fig. 3.108 shows the simplified layout of a frame grabber for digitising a monochrome video signal. Using a video multiplexer the system can switch between different video sources. By means of PLL synchronisation (*phase locked loop*) horizontal synchronisation pulses are extracted from the video signal in order to correctly drive the analogue-to-digital converter (ADC). After the detection of the beginning of a line, the ADC scans the analogue signal at a predefined scanning frequency to deliver, typically, 8-bit grey values. The grey values are stored line by line in image memory after they have passed an input lookup table (LUT) where they are optionally transformed, e.g. converted into a negative image (see section 5.2.1.2). For image display, the image memory content passes through an output LUT before being converted into an analogue video image. Digital access to image data, lookup tables and other data processing functions of the frame grabber are enabled through a computer interface.

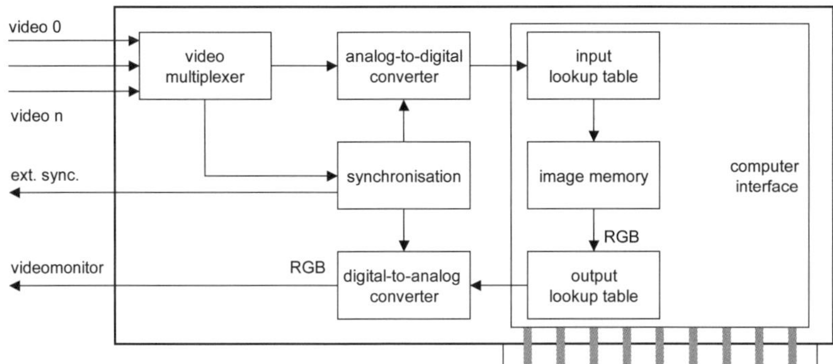

Fig. 3.108: Schematic layout of a frame grabber

The geometry of the digitised video image is affected by the scanning process. The start of an image line is detected by an electronic synchronisation process called a Pixel Lock Loop (PLL) which contributes a potential uncertainty of about 0.1 pixels. This uncertainty is known as *line jitter*. Variations of the scanning frequency lead to deviations within an image line called *pixel jitter*. Usually the scanning frequency is generated internally in the frame grabber rather than using the signal driving the sensor, hence it may not be identical to the pixel frequency of the CCD sensor. According to Table 3.8 the pixel frequency is about 14.5 MHz depending on the imaging sensor. If the scanning frequency is different, the resulting pixel size along a line (x) will alter:

$$s'_x = s_x \frac{n_X}{n_C} = \frac{s_x n_X}{f_A t_X} \tag{3.91}$$

where

s'_x: scanned horizontal pixel size
s_x: horizontal size of a CCD sensor element
n_X: number of CCD elements per line
n_C: number of scanned pixels per line
f_A: scanning frequency
t_X: active time of an image line (52 μs for CCIR)

As a result, the digitised pixel size can differ in both x and y directions. For photogrammetric applications this effect must be corrected by introducing an affine parameter into the model of interior orientation (see section 3.3.3.3). Photogrammetric video-based measuring systems must therefore be calibrated together with the frame grabber.

Example 3.9:

Given an analogue CCD camera with 604 x 576 sensor elements and an image format of 6.0 mm x 4.5 mm (size of sensor elements: 10.0 µm x 7.8 µm). The connected frame grabber enables a scanning rate of 10 MHz (size of image memory 512 x 512 pixels). What is the resulting size of a digitised pixel and the required scanning frequency to generate square pixels?

Solution:

1. Scanned pixels per line: $n_C = f_A \cdot t_X = 10\text{MHz} \cdot 52\mu\text{s} = 520$

With an image memory size of 512 pixels per line, 8 pixels are erased from each line.

2. Scanned pixel size: $s'_x = \frac{s_x \cdot n_X}{n_C} = \frac{10\mu\text{m} \cdot 604}{520} = 11.6\mu\text{m}$

Hence the scanned pixel size differs from the original size of the sensor element.

3. Pixel aspect ratio: $m = \frac{s'_x}{s'_Y} = \frac{11.6\mu\text{m}}{7.8\mu\text{m}} = 1.4 \approx 3:2$

4. Scanning frequency: $f_A = \frac{s_x \cdot n_X}{s'_x \cdot t_X} = \frac{10\mu\text{m} \cdot 604}{7.8\mu\text{m} \cdot 52\mu\text{s}} = 14.89\text{MHz}$

For a scanning frequency of 14.89 MHz, the scanned horizontal pixel size is equal to the vertical size of a sensor element, which remains unchanged with scanning.

Due to limits in available space for image storage, there may be some loss of image, i.e. it is possible that neither all pixels in a row, nor all rows in an image, are placed in the storage area.

If an analogue camera is synchronised by a trigger from the frame grabber (or vice versa), accuracies of about 0.05 pixels (ca. 0.5 µm) in the image plane can be achieved under practical conditions. An additional accuracy improvement can be obtained if the video scanning is controlled by a pixel-synchronous pulse from the camera (*pixel clock*). Today there are video cameras with internal digital output where analogue-to-digital conversion is performed inside the camera. As a result, timing signals at the sensor can be identical to those at the ADC.

Analogue cameras require a warm-up period of up to 2 hours. During warm up, image coordinate displacements of several tenths of a pixel have been shown to occur. Fig. 3.109 shows an example of drift measured under controlled laboratory conditions. The figure shows coordinate differences extracted from five images of a warm-up time series. In the x direction, small shifts only can be observed which, after 10 minutes, are below the measuring uncertainty of around ±0.02 pixel In contrast, in the y direction a drift of about 0.15 pixel can be observed within the first 20 minutes. Part of this effect is caused by temperature increases on mechanical components, but electronic devices inside the camera also contribute as they warm up.

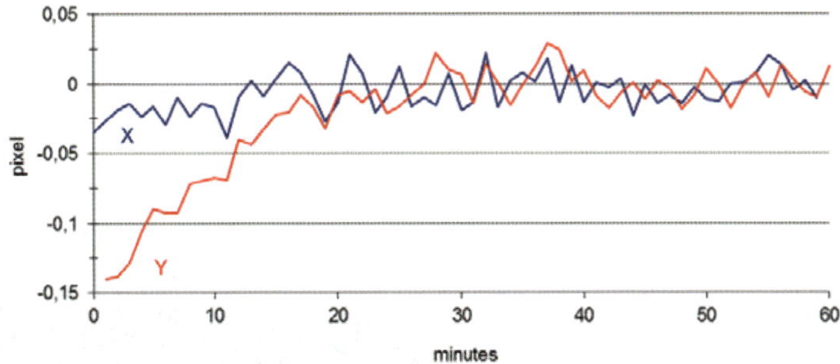

Fig. 3.109: Drift of measured sensor coordinates during warm-up

In photogrammetry, CCD video cameras are generally used in specialised real-time applications with relative accuracies of the order of 1:5000 to 1:15 000.

3.5.2 Digital cameras

Early cabled digital cameras were designed to be compatible with video standards by reading pixel data out of the sensor in interlaced scanning mode (see section 3.4.1.2). However the capability to depart from established interlaced standards and deliver images in a single, full-frame block was of great benefit to the photogrammetric community. The technology, *progressive scan* (also called *slow scan*) read out in full-frame mode, i.e. the complete image is exposed at one and the same time. Given a fast shutter speed, relative movements (also vibrations) between camera and object no longer lead to blurring effects for the resulting video image (Fig. 3.110). In addition to providing a standard video output, these cameras require a separate host computer digital interface in order to access the progressive scan signal or, where the camera includes onboard memory and digitisation electronics (see section 3.5.1.3), to directly access the digital images produced.

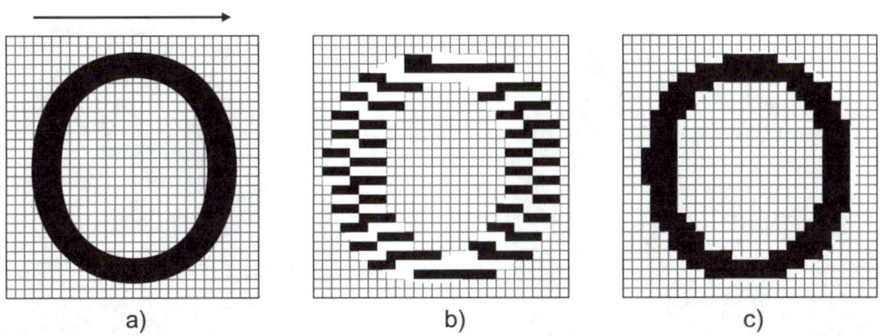

Fig. 3.110: An input moving image (a) and its resulting capture by an interlaced (b) and a progressive-scan system (c)

Note that due to the numerous applications of digital cabled cameras, especially for the commercial video market and general surveillance tasks, the choice is immense. Home-video and TV cameras, especially camcorders, are often based on mass-produced IL sensors. FT sensors are often used in technical applications since, by virtue of larger pixel fill factors, they provide higher light sensitivity in comparison with IL sensors.

Following the first developments at the end of the 1980s (e.g. the Kodak Megaplus, 1320 x 1035 pixel FT sensor), a growing number of high-resolution imaging sensors have become available at economic cost. In this field, consumer-orientated developments in digital photography are the driving force in that they offer a wide range of imaging systems for both amateurs (low-resolution examples in Fig. 3.111), and professional photographers (high-resolution monochrome and colour camera examples in Fig. 3.112 and Fig. 3.113).

a) Casio Exilim EX Z-1000

b) Sony Ericsson DSC W190

Fig. 3.111: Examples of digital compact cameras

High-resolution digital cameras with up to 60 Mpixel are becoming readily available for photogrammetric practice, examples including still-video cameras, scanning cameras, digital camera backs or specialised metric cameras (examples in Fig. 3.114, Fig. 3.115). In combination with established photographic camera technologies, they provide powerful and user-friendly systems. Data transfer is either performed offline by means of an internal storage device (e.g. SmartMedia, CompactFlash, MicroDrive, SD Card) or online with a connected computer. In addition, there are imaging systems with integrated processors where a variety of image processing functions are performed directly inside the camera.

Fig. 3.112: Nikon D3

Fig. 3.113: Leica S2

Table 3.10: A selection of digital imaging systems

Supplier	Type	Pixels	Pixel size [µm]	Format [mm]	Sensor type	Image size [MPixel]
Compact cameras						
Sony	DSC W190	4040 x 3032	1.4	5.6 x 4.2	CMOS	12.2
Casio	Exilim EX-Z1000	3678 x 2736	1.9	6.9 x 5.2	CCD	10.1
Leica	M8	3916 x 2634	6.8	26.7 x 17.9	CCD	10.3
SLR cameras						
Sigma	SD 14	2652 x 1768	7.8	20.7 x 13.8	CMOS RGB	4.5
Canon	EOS 5D	4368 x 2912	8.2	35.8 x 23.9	CMOS	12.8
Leica	S2	7500 x 5000	6.0	45.0 x 30.0	CCD	37.5
Nikon	D3X	6048 x 4032	5.9	35.9 x 24.0	CMOS	25.7
Nikon	D3	4256 x 2832	8.5	36.0 x 24.0	CMOS	12.9
Digital camera backs / sensors						
Mamiya	ZD	5328 x 4000	9.0	48.0 x 36.0	CCD	21.7
Hasselblad	H4D-60	8956 x 6708	6.0	53.7 x 40.2	CCD	60.1
Alpa	12 WA / Leaf	6666 x 4992	7.2	48.0 x 36.0	CCD	33.3
Rollei	6008digital metric	7228 x 5428	6.8	48.9 x 36.9	CCD	39.0
Leaf	Aptus 75	6666 x 4992	7.2	48.0 x 36.0	CCD	33.3
Phase One	P65+	8984 x 6732	6.0	53.9 x 40.4	CCD	60.5
Digital metric cameras						
AXIOS 3D	SingleCam	776 x 582	8.3	6.4 x 4.8	CCD	0.5
Rollei	AIC P65	8924 x 6732	6.0	53.9 x 40.4	CCD	60
GSI	INCA 3	3500 x 2350	4.8	16.7 x 11.2	CCD	8.2

Table 3.10 shows a selection of high-resolution digital cameras. The table lists digital compact cameras, digital SLR cameras, cameras with digital backs and digital photogrammetric cameras. Due to rapid technological progress in this area, the table can only provide a recent view (valid: 2013) without claiming completeness or implying long-term availability in the market.

There are several trends in the current development of digital cameras:

- smaller pixel and sensor sizes for compact and mobile phone cameras
- larger image resolutions at economic cost in SLR image formats
- combination camera systems able to selectively provide still and video images from the same sensor

- formats up to 54 mm x 40 mm in digital camera backs or medium format cameras
- smaller choice of cameras designed specially for optical measurement technology

In close-range photogrammetry, digital compact cameras are only used for measurements with lower accuracy requirements, for example applications such as accident recording or texturing of 3D city models. Digital SLRs are used in many photogrammetric applications due to their favourable price/performance ratio. Specialised SLRs with robust metal housings and high quality lenses can achieve high measurement accuracies with appropriate calibration.

In principle, digital cameras with small image format (examples in Fig. 3.112 and Fig. 3.113) permit the use of standard lenses for this type of camera so that a wide choice of lenses is available, particularly for wide-angle use. However, cameras with very small pixel sizes have greater noise in the image and are less sensitive to light. In addition, the corresponding lenses are subject to high demands with respect to resolution which are often not fulfilled, resulting in a lower image quality than available from cameras with a lower number of pixels.

Medium-format digital cameras (Fig. 3.114, Fig. 3.115) are mainly used in high-end applications requiring a maximum number of pixels and large image format. Due to their high cost, these medium-format cameras do not have a significant presence in close-range photogrammetry.

Fig. 3.114: Mamiya ZD

Fig. 3.115: Hasselblad H4D-60

Purpose-built digital metric cameras, specially designed for measuring purposes, guarantee high geometric stability due to their opto-mechanical design. In particular, lens and image sensor are rigidly connected in the camera housing. An integrated ring flash is used to illuminate retro-reflecting targets. The INCA3 metric camera (Fig. 3.116) from Geodetic Systems Inc. (GSI) has a resolution of 3500 x 2350 pixel and built-in image processing. The processor can compress images as well as automatically recognise and measure targets so that for standard applications using retro-reflective targets only the measured image coordinates need be transferred. This camera is for demanding industrial tasks.

Fig. 3.116: GSI INCA 3

Fig. 3.117: Rollei AIC

The Rollei AIC digital metric camera (Fig. 3.117) utilises CCD sensors with 22 to 60 Mpixel from the company PhaseOne. Although it has terrestrial applications it is primarily used for aerial photography. Here it appears in various arrangements with two or four cameras which are rigidly mounted in a common housing.

The SingleCam metric camera from AXIOS 3D (Fig. 3.162) is a mechanically stable video camera with an optimised assembly of lens, sensor and housing. It has an integrated, monochromatic, LED ring flash for recording retro-reflecting targets. The camera is shock-resistant to 50 times the acceleration of gravity without any change in interior orientation.

3.5.3 High-speed cameras

High-speed cameras allow the recording of fast-changing scenes with image frequencies much higher than for standard video cameras. Today, high-speed cameras exist that can record more than 1200 images per second at resolutions of the order of 2000 x 2000 pixel. They are mainly used for monitoring dynamic production processes, but are also used for analysing object movement in industrial, medical and research applications. High-speed cameras are offered by different suppliers (e.g. PCO, Microtron). As an illustrative example, Fig. 3.118 shows a subset of an image sequence recorded by the pco.dimax (1920 x 1080 pixels, 1100 frames per second). The duration of the recording sequence is limited by the internal or external storage. This camera type can be of the cabled or self-contained variety.

In general high speed imaging systems can be characterised as follows:

- Progressive-scan CCD sensor:

  Progressive-scan sensors have the ability to read out image data on a line-by-line basis. Whilst removed from television standards, this method minimises image displacements due to subject movements which can occur during interline sampling (Fig. 3.110). This type of camera can also allow the number of lines in the output image to be reduced in order to increase image data rates.

- CMOS sensors:

  As an alternative to progressive-scan sensors, CMOS imaging sensors are available that provide exposure times of a few microseconds using an electronic shutter which is integrated in the sensor. These sensors facilitate direct addressing of individual pixels, but have the disadvantage of increased image noise when compared with CCD sensors.

- Electronic shutter:

  Electronic shutters control the integration time of the imaging sensor with exposure times of down to 50 µs, corresponding to 1/20 000 s.

- Solid-state image memory:

  Solid-state image memory modules permit immediate storage of the image sequence either in the camera or nearby. The maximum number of images, and the maximum recording period, depend on the image memory capacity and the (selectable) sensor resolution.

- Analogue or digital image output:

  The image sequence stored in memory can be read out either as a standard analogue video signal, or as a digital image sequence.

- External trigger signals:

  External trigger signals serve to control the image acquisition process by allowing the camera to be synchronised with an external event, e.g. a particular point on a machine production cycle or the simultaneous acquisition of image sequence from multiple cameras.

- Recording of additional information:

  Storing of additional information (image number, data rate etc.) allows for subsequent image sequence analysis.

- Onboard processing:

  CMOS sensors combined with specialised Field Programmable Gate Array (FPGA) processors are able to process incoming images in real-time. For example the AICON TraceCam F is a specialised photogrammetric high-speed camera (Fig. 3.119). It has a 1.3 MPixel CMOS image sensor and an FPGA capable of automatically measuring up to 10 000 bright targets at an image frequency of up to 500 Hz with full sensor resolution. Only the measured image coordinates are transferred to the externally connected computer and the current image is then immediately overwritten. In this way, recording sequences can be of any duration. The camera is used, for example, to measure wheel movements on a moving vehicle.

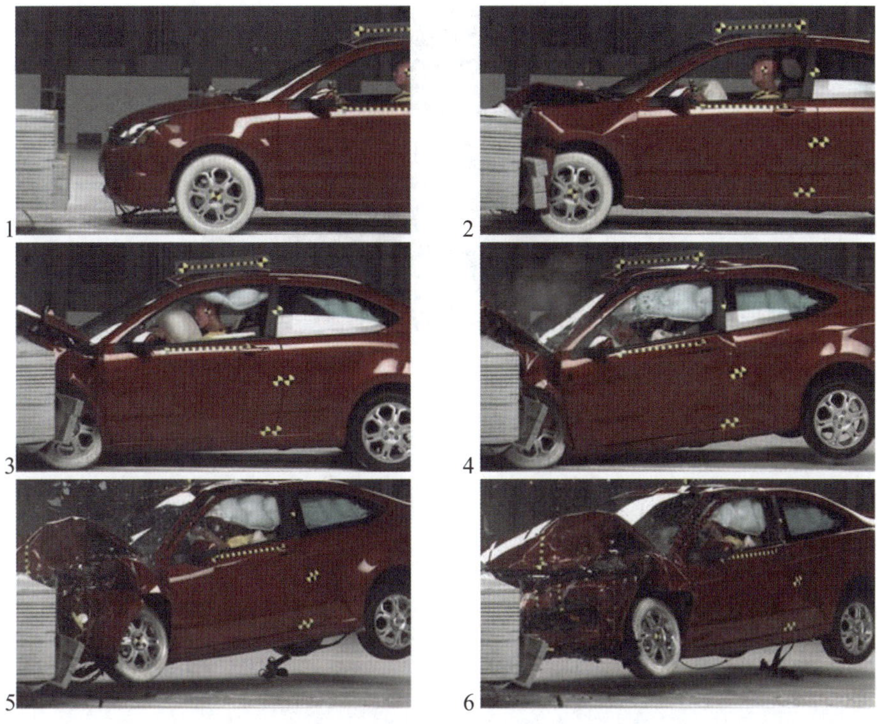

Fig. 3.118: Subset of an image sequence (PCO)

a) AICON TraceCam F

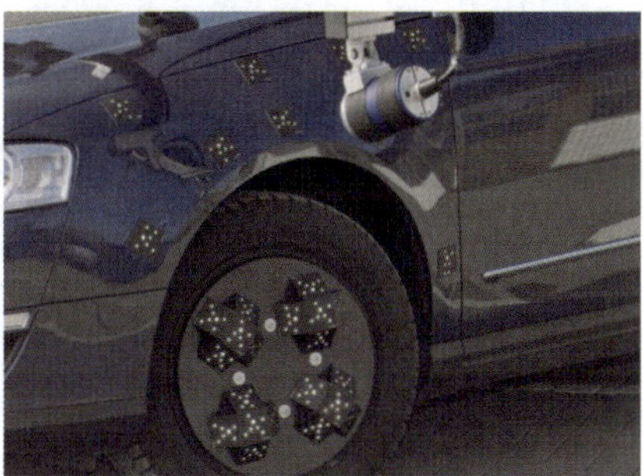

b) TraceCam in use for wheel movements

Fig. 3.119: High-speed camera with integrated point measurement (AICON)

Synchronised high-speed stereo images can be recorded using a stereo mirror attachment. (see section 3.4.3.7). Fig. 3.120 shows a high-speed camera with the stereo mirror attached. There are demanding requirements in respect of planarity and alignment of the mirrors, and significant asymmetric distortion effects if these are not met (Fig. 3.121).

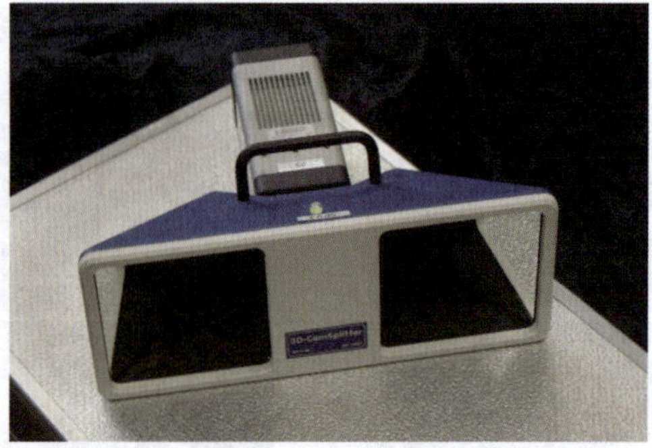

Fig. 3.120: High-speed camera with stereo mirror attachment (HS Vision)

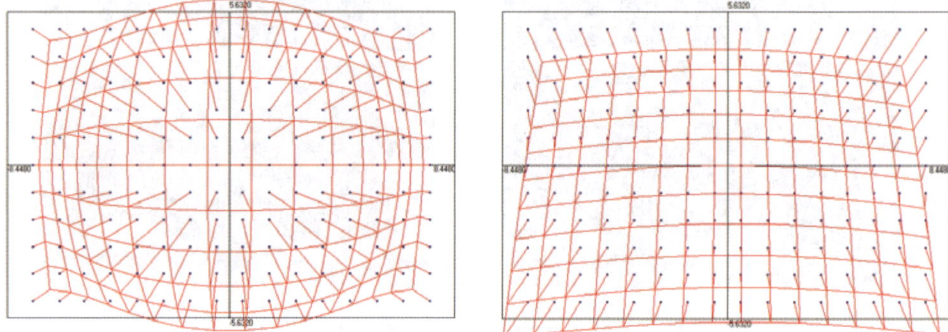

Fig. 3.121: Radial and tangential distortion in a camera with a stereo mirror attachment (Weinberger Visario, focal length 12.5 mm)

Fig. 3.122: High-speed stereo images taken with the stereo mirror attachment (Volkswagen)

Fig. 3.122 shows two images from a stereo recording sequence. Only half the sensor format is available for the left and right hand parts of the image. Between each half image there is a narrow strip with no useful image data.

3.5.4 Stereo and multi-camera systems

Stereo and multi-camera systems, in which the cameras are rigidly mounted in a single housing, are suitable for 3D applications requiring the synchronous recording of at least two images (Fig. 3.123). These systems permit the imaging of object points with simultaneous calculation of their 3D coordinates by the technique of intersection (see section 4.4.7.1). The cameras have largely constant orientation parameters and are normally calibrated in the factory. For example, the CamBar stereo camera system from AXIOS 3D (Fig. 3.123a) is shock-resistant to 75 g and specified to operate in the range 15-30° C, without any change in interior orientation of an individual camera or the relative orientation between cameras in the housing.

a) Stereo camera for use in medical navigation applications (AXIOS 3D)

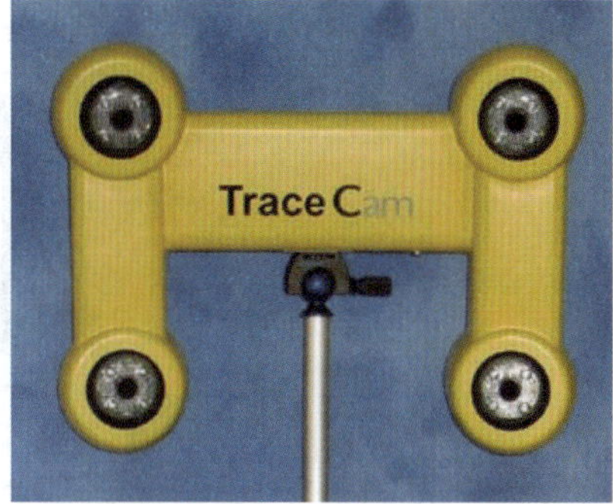

b) 4-camera system for monitoring movements of crash dummies (AICON)

Fig. 3.123: Stereo and multi-camera systems

Another form of multi-camera system is represented by three-line systems which have three linear imaging elements. Each element has a cylindrical lens which creates a line image of a LED target point, with the image being recorded by a linear CCD array set in the image plane at right angles to the image line (Fig. 3.124(a)). This effectively defines a plane in space on which the target point lies. By orienting the central element at right angles to the outer elements, the target point can be located by the intersection of three planes in space (Fig. 3.125). Linear array technology permits a measurement frequency of up to 3000 Hz.

Stereo and multi-camera systems are often used for navigation tasks in which objects must be absolutely tracked or positioned in space, or located relative to other objects. One of the most common application areas is in medical navigation. Here the systems are employed during operations to track the movements of surgical instruments. Typical industrial applications include the positioning of parts and robots (see also section 6.4).

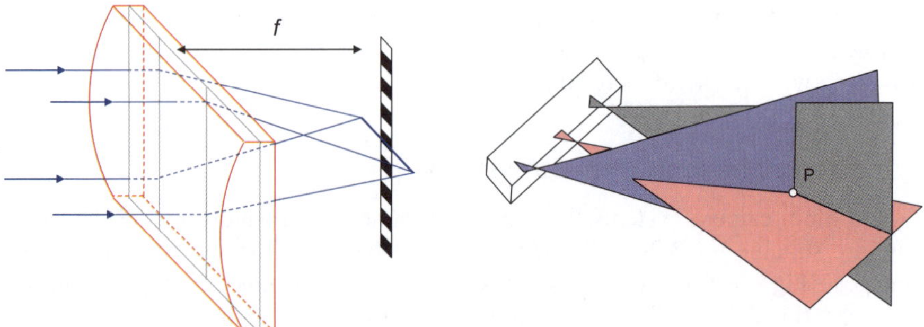

a) A line sensor and a cylindrical lens producing a line location for an illuminated point target

b) Three-line measurement principle

Fig. 3.124: Line target imaging and three-line measurement

Fig. 3.125: Three-line camera (Nikon Metrology)

3.5.5 Micro and macro-scanning cameras

Since the resolution of digital cameras was a limiting factor some years ago, scanning cameras were developed in order to increase pixel resolution by sequential scanning of an object or scene (see also Table 3.10). Two basic scanning principles can be distinguished: *micro scanning* where small sub-pixel steps of the sensor are made sequentially within the same image format area to increase spatial resolution and *macro scanning* where the sensor is sequentially stepped by a significant portion of its total dimension to expand the image format. Such systems can only deliver high quality images if there is no relative movement between camera and object. Depending on implementation, image sizes of the order of 3000 x 2300 pixels to 20 000 x 20 000 pixels can be obtained.

3.5.5.1 Micro scanning

In the case of micro-scanning cameras, piezo elements are used to translate an interline-transfer sensor horizontally and vertically, in fractions of the sensor element size (micro-scan factor) (Fig. 3.126). A high resolution output image is created by alternating the storage of single images. The final image has a geometric resolution that is increased by the micro-scan factor in both directions. From a photogrammetric standpoint, the micro-scanning principle results in a reduction of effective pixel size whilst maintaining the usable image format of the camera.

Cameras based on this principle are the RJM JenScan (4608 x 3072 pixel), Kontron ProgRes 3012 (4608 x 3480 pixel) and Jenoptik Eyelike (6144 x 6144 pixel, Fig. 3.128). Even higher resolutions can be achieved if micro scanning is combined with the principle of macro scanning (see below).

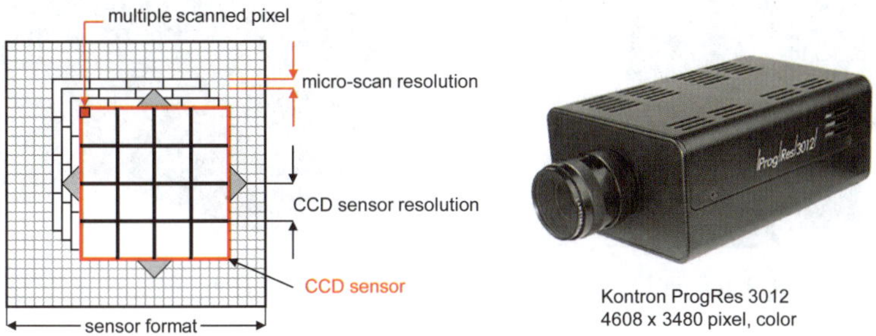

Fig. 3.126: Principle and example of a micro-scanning camera

3.5.5.2 Macro scanning

Macro-scanning systems shift a linear or area CCD sensor in large steps over a large image format. The position of the separate partial images is determined either mechanically or by an optical location technique. This enables the images to be combined into one complete, large-format image.

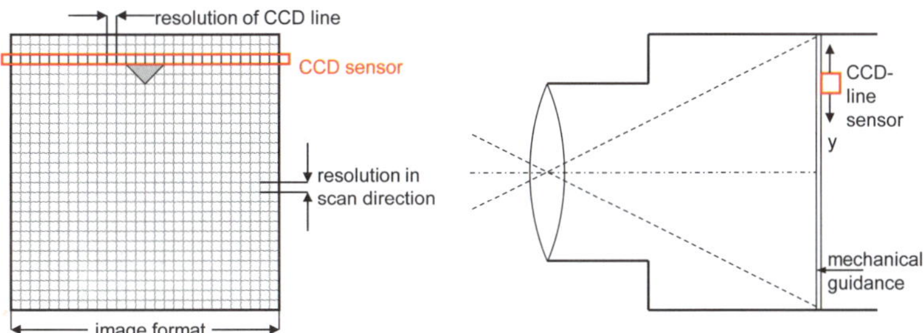

Fig. 3.127: Principle of a line-scanning camera

Fig. 3.127 illustrates the principle of a line-scanning camera. A single linear CCD array, set parallel to one axis of the image, is mechanically moved along the other image axis in order to scan the entire image area. Image resolution is therefore given in one direction by the resolution of the array and in the other direction by the step resolution of the mechanical guide.

An example of a high-resolution, line-scan camera is the PentaconScan 6000 shown in Fig. 3.129. This camera has a linear RBG CCD sensor which is scanned across an image format of 40 mm x 40 mm and, at its highest resolution, can deliver images with 10 000 x 10 000 pixel for each colour channel. Due to limitations in the precision which is achievable in the mechanical guidance mechanism, these systems are principally designed for professional still photography and not for photogrammetric applications.

Fig. 3.128: Jenoptik Eyelike

Fig. 3.129: Pentacon Scan 6000

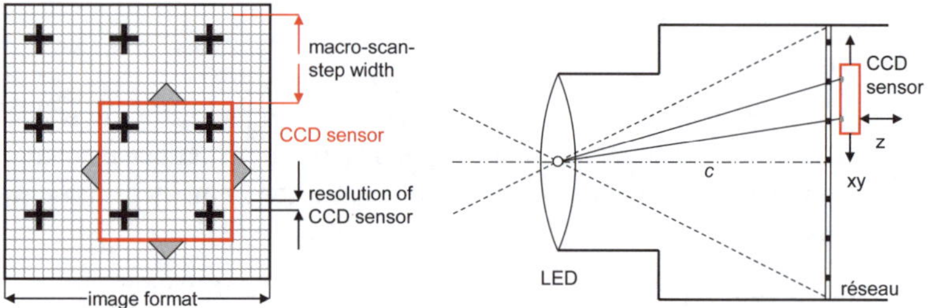

Fig. 3.130: Principle of the réseau-scanning camera

The macro-scanning principle can also be employed with area sensors. The two-dimensional mechanical positioning technique which is required in this case is considerably more complex and therefore more costly to produce. The réseau-scanning principle offers an alternative solution. Here the individual sensor locations are determined by measuring the images of réseau crosses in the sub-image delivered by the sensor. This technique does not require an accurate mechanical positioning mechanism (Fig. 3.130). In addition to movement in the xy direction, the sensor can also be shifted in the z direction, parallel to the optical axis. This makes it possible to focus without altering the parameters of interior orientation. An example of a réseau-scanning camera is the Rollei RSC which is no longer in production.

3.5.6 Panoramic cameras

3.5.6.1 Line scanners

Digital panoramic cameras with a scanning line sensor form a special case of the macro-scan technique. A vertically mounted sensor line is rotated about a vertical axis, thereby imaging the surrounding object area (rotating line scanner). The mathematical projection model is a central perspective projection in the vertical direction and mapped by rotation angle to a cylindrical projection surface in the horizontal direction. Some systems internally convert the camera output to a spherical projection. Providing the particular panoramic imaging geometry is taken into account, photogrammetric methods such as bundle adjustment and spatial intersection can be applied in an analogous way to conventional images. Scanning panoramic cameras based on CCD line sensors can achieve image sizes of the order of 50 000 x 10 000 pixels, often known as gigapixel images.

An eccentricity vector **e** between perspective centre and rotation axis must be determined by camera calibration (Fig. 3.131). For photogrammetric processing of panoramic images the mathematical model is extended for the x' direction whereby the image coordinate x' (column position in image) is defined as a function of the rotation angle α and the eccentricity vector **e**. A number of panoramic cameras are currently available on the market, e.g. the SpheroCam HDR from Spheron (50 Mpixels), Panoscan Mark III (max. 9000 x 6000 pixels) or the KST EyeScan M3 metric (Fig. 3.132, max. 54 000 x 10 200 pixels). Most cameras provide 360° images and can also produce smaller, user-definable sections.

Panoramic images can also be generated from single frame images if the individual images have a sufficient overlap. A geometrically exact stitching of such panoramas can only be performed if the single images are acquired around a common rotation axis, and if the interior orientation parameters are known.

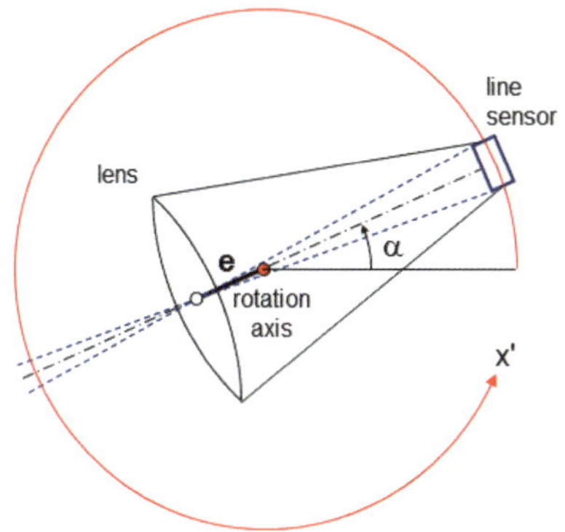

Fig. 3.131: Principle of a panoramic scanning camera

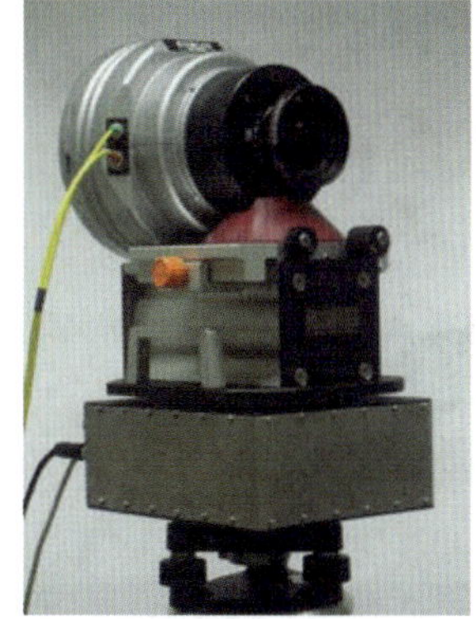

Fig. 3.132: Digital panorama camera KST EyeScan M3

3.5.6.2 Panorama stitching

A panoramic image can be generated from a number of central-perspective, single images which overlap horizontally or vertically (see example in Fig. 3.135). The individual images can be merged together in various ways:

- Manual merging:

  The overlapping images are interactively positioned relative to one another until there is a good match in the areas of overlap.

- Automatic merging:

  Procedures for automatically creating the panoramic image are based on detecting corresponding regions in the overlaps using feature extraction or correlation methods (see section 5.5.2). Once these common locations are found, a simple geometric transformation is used to make a best fit between them in the overlaps. The technique makes no use of a global geometric model, such as a cylindrical projection, and this gives rise to residual errors where neighbouring images are connected.

- Photogrammetric merging:

  The photogrammetric construction of a panorama takes into account the outer and inner orientation of the individual images. In an analogous way to orthophoto creation (section 4.2.7.3), the resultant panorama is generated on a cylindrical surface with the colour values calculated by back-projection into the original images.

Fig. 3.133: Digital camera on a panoramic adapter

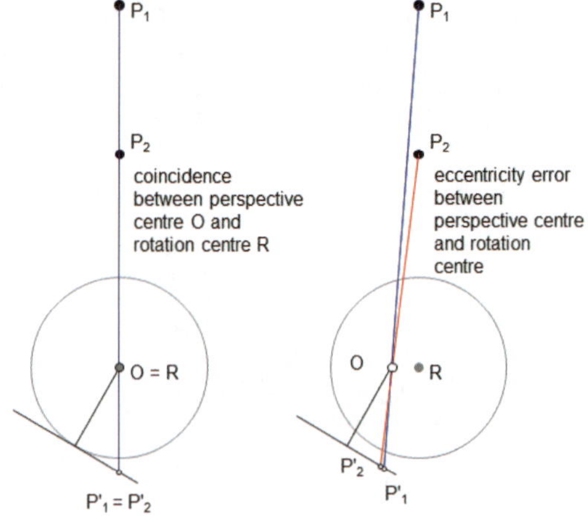

Fig. 3.134: Alignment of perspective centre with rotation centre by observing collinear points

The geometrically correct generation of a panoramic image requires that the rotation axis of the camera passes through the perspective centre. Only then do all imaged rays pass through the same point, as would be the case in a perfect panoramic image. To achieve this condition it is advantageous to connect the camera to an adjustable panoramic adapter, one which permits a shift along the camera axis (Fig. 3.133). This adjustment can be achieved by simple means, e.g. the imaging of two points in line with the rotation axis (Fig. 3.134). When correctly adjusted, both object points overlap in the same image point. When the ideal configuration is achieved and the rotation axis passes through the perspective centre, then the resulting panoramic image has the same central-perspective properties as a normal image, although in this case the horizontal field of view is 360°. If there is a residual displacement error then the camera's perspective centre moves on a circle around the rotation axis. Note that in panoramic photography the perspective centre is also known as the nodal point.

As an alternative to the use of a camera, a panoramic image can also be generated by a video theodolite, or imaging total station with integrated camera (Fig. 3.140).

Fig. 3.135: A digital cylindrical panorama created by stitching together overlapping camera images

3.5.6.3 Panoramas from fisheye lenses

360° panoramas can also be created by combining two images taken with fisheye lenses (see section 3.3.6) which are pointing in opposite directions. Fig. 3.136 shows two original fisheye images and Fig. 3.137 the resultant panorama. The source of the images is a system designed for mobile mapping (see section 6.7.1).

a) Camera view to front

b) Camera view to rear

Fig. 3.136: Images from two cameras with fisheye lenses which are pointing in opposite directions (Cyclomedia)

Fig. 3.137: Digital panorama created from the images in Fig. 3.136 (Cyclomedia)

3.5.6.4 Video theodolites and total stations

Video-theodolites are based on a conventional electronic theodolite with a digital video camera integrated into the optical path. While electronic angular read-out is used for acquiring horizontal and vertical directions, the video image can be digitised and processed to enable automatic target detection and point measurement. These instruments are similar in principle to the photo-theodolite (see Fig. 1.26).

Fig. 3.138 illustrates the design of the Leica TM 3000 V video-theodolite. The telescope image is projected onto a digital sensor by means of a beam splitter. The use of a beam splitter means that a visual observation is still possible. An optical coupling device serves for the switching between telescope image and a wide-angle overview image. Target coordinates are measured inside the digitised camera image. These measurements can be corrected for horizontal and vertical angles if a suitable camera calibration is available. Angular accuracies of about 0.5 arc sec can be achieved if well-defined targets are used (see section 3.6.1).

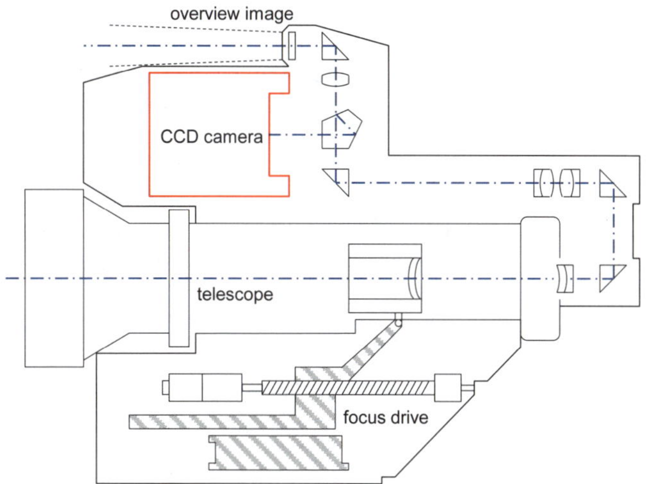

Fig. 3.138: Schematic diagram of the Leica TM 3000 V video-theodolite

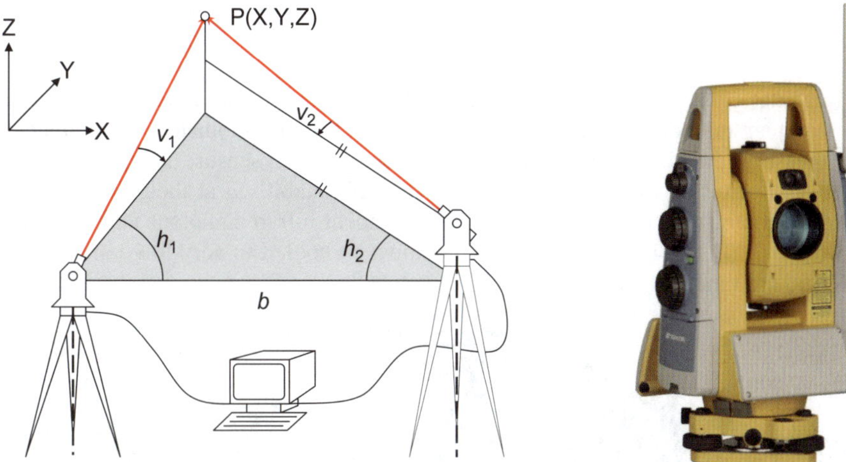

Fig. 3.139: Theodolite measuring system

Fig. 3.140: IS3 Imaging total station (Topcon)

If the video-theodolite is equipped with servo-motors for driving its vertical axis, horizontal axis and focus, as shown for the TM 3000 V, target points can be imaged and measured automatically. This enables a system to be configured using a pair of video-theodolites which automatically locate three-dimensional object points by spatial intersection of measured angles, i.e. the principle of triangulation (Fig. 3.139). This type of system was developed for industrial use in the 1990s, examples being the Kern SPACE and Wild ATMS, but they are no longer in production and have been superseded by systems based on total stations and laser trackers which can directly measure 3D coordinates as the next comments indicate.

By integrating an opto-electronic distance meter, the video-theodolite becomes an imaging total station. With this single instrument, 3D (spherical) coordinates can be acquired by the principle of polar measurement, i.e. two angles and range to target. The imaging total station makes possible automatic target acquisition. In addition, images from the integrated camera can be stored with their orientation data and, for example, merged together into a panoramic image.

As an example, Fig. 3.140 shows the IS3 imaging total station from Topcon. This has a wide angle camera and a 30x camera coaxial with the telescope's optics.

3.5.7 Thermal imaging cameras

Thermal imaging cameras (thermographic cameras) are used for analytical tasks in building research and materials science. They deliver a thermal image using wavelengths in the range 2.5 to 14 µm (medium to near infrared part of the spectrum, see Fig. 3.1). Typical thermal sensitivity of the cameras lies approximately in the range –40° C to +2500° C. Until now, thermal imaging cameras have been rarely used for geometric and photogrammetric analyses because they are costly and, with sensor resolutions in the range 320 x 240 pixel to 640 x 480 pixel, have a relatively poor resolution compared with modern CCD and CMOS sensors.

Thermal imaging cameras (example shown in Fig. 3.141) work either with thermal detectors or quantum detectors. Thermal detectors (micro-bolometers, pyro-electric detectors) measure the temperature directly incident on the sensor, independently of wavelength. The cameras do not need to be cooled, only stabilised at their own constant temperature. In contrast, quantum detectors absorb incident infrared photons and operate in a similar way to CCD and CMOS sensors. They must be cooled to very low temperatures (60-140 K) and are functionally dependent on wavelength.

Fig. 3.141: Thermal imaging camera (InfraTec)

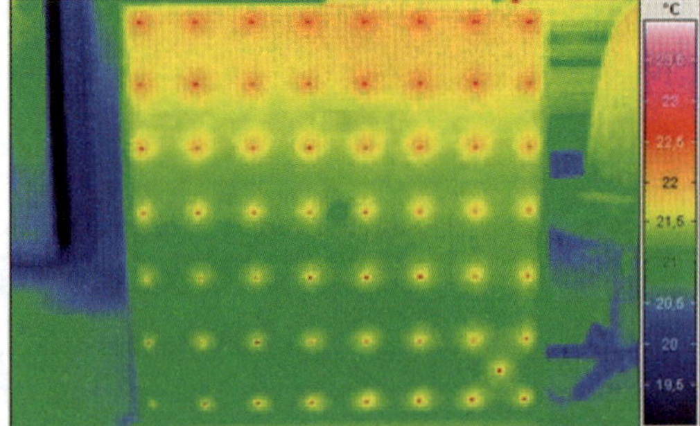

Fig. 3.142: Thermographic test-field image

Typical sizes of detector elements (pixel sizes) lie in the range 35 µm to 50 µm. According to eqn. 3.13, the diffraction-limited resolution of thermal imaging cameras is around 12 µm for a wavelength of 5 µm and an f/number of 1. Because of the requirements imposed by the refractive indices of thermal radiation, the lenses which can be used are not constructed from normal glass but from germanium. They are expensive to produce and therefore represent the principal component cost of a thermal imaging camera.

Like conventional cameras, thermal imaging cameras can be geometrically calibrated. Test fields for this purpose have target points which radiate in the thermal region, for example active infrared LEDs, heated or cooled metal discs, or retro-reflecting targets exposed to a suitable infrared illumination source. Fig. 3.142 shows a thermal image of such a photogrammetric test field.

3.6 Targeting and illumination

3.6.1 Object targeting

In many applications, locations to be measured on an object need to be marked with artificial targets, e.g.

- to identify natural object feature points which cannot otherwise be identified accurately,
- as uniquely defined points for comparative measurements,
- as control points for geodetic measurement,
- for automatic point identification and measurement,
- for accuracy improvement.

The physical size and type of target to be used depends on the chosen imaging configuration (camera stations and viewing directions, image scale, resolution) and illumination (light source, lighting direction). The manufacture of very small and the logistics of very large targets can be prohibitive.

3.6.1.1 Targeting material

3.6.1.1.1 Retro-reflective targets

Retro-reflective targets (retro-targets) are widely used in practice, particularly in industrial applications. They consist of a retro-reflective material that is either covered by a black pattern mask, or is stamped into the target shape. Usually retro-targets are circular in shape, but they can be manufactured in any size and shape (examples in Fig. 3.143). In addition, there are spherical retro-reflective targets (retro-spheres) which can be viewed over a wider range of directions (see section 3.6.1.3). The retro-reflective material consists either of a dense arrangement of small reflective balls ($\varnothing \approx 80$ µm), or an array of micro-prisms.

a) Circular target with point number

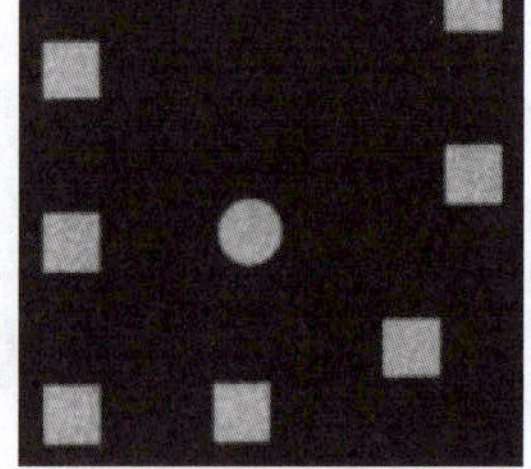

b) Circular target with area code

c) Circular target with ring code

Fig. 3.143: Examples of retro-reflective targets

In order to achieve high contrast target images, retro-reflective targets must be illuminated from a position close to the camera axis (e.g. by a ring flash, see Fig. 3.161). The resulting images can be simply and fully automatically measured since, for all practical purposes, only the high-contrast measurement locations, without any background information, are imaged (examples in Fig. 7.22).

The microscopic balls or prisms of a retro-target are attached to a base material. Incident light is reflected internally within each ball and returns parallel to its incident path. Optionally, the material may have a protective plastic coating to allow the surface to function under wet conditions, but at the expense of a reduced light return. For masked targets, balls can partly be occluded at the edge of the target, so that the measured centre is laterally displaced. The shift depends on the viewing direction and gives rise to a 3D target location which is above the physical centre of the target if a triangulation method is applied (Fig. 3.144). The opposite effect occurs for stamped targets and leads to a triangulated 3D point below physical target level. Both effects can result in a shift of about 50 µm and should be corrected by the processing software.

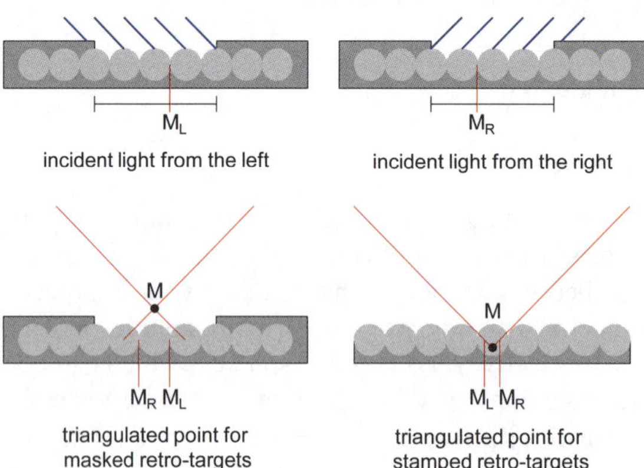

Fig. 3.144: Position of virtual centre of circular retro-targets (after Dold 1997)

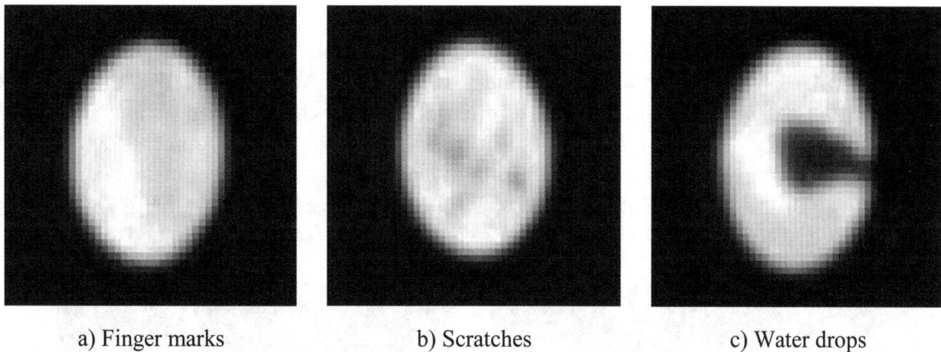

a) Finger marks b) Scratches c) Water drops

Fig. 3.145: Degraded retro-targets

Retro-reflective target materials are sensitive to surface marking caused by fingerprints, dust, liquids and mechanical abrasion, as well as humidity and aging. In these cases reflectivity and target outline are affected, which may lead to degradation in 3D measurement quality. This particularly affects measurement techniques which determine the target centre from the grey value distribution across the target image (e.g. centroid methods and template matching, see section 5.4.2). Fig. 3.145 shows examples of degraded retro-targets. In such cases the plastic-coated versions may be of more practical use, since the ability to wipe these clean can more than offset their reduced light return and viewing angle.

Further drawbacks of retro-targets are caused by:

- relatively high manufacturing costs, particularly for large or coded targets (section 3.6.1.5),
- finite target dimensions which restrict the target density which can be applied to the object surface,
- self-adhesive targets can generally only be used once, therefore requiring a new set of targets for each object to be measured,
- method of illumination produces images which show only the targets and no other object information,
- limitations when using particularly large targets (∅ > 10 mm).

Other target materials

The drawbacks of retro-reflective targets mentioned above demands the use of alternative materials or targeting techniques for a number of applications.

Circular plastic targets with a central target point are suitable for long-term targeting of outdoor objects (e.g. buildings, bridges) and are also useful for geodetic measurements. They can be produced in almost any size. Fig. 3.146 shows examples of circular plastic targets.

Targets printed onto paper or adhesive film are simple to manufacture, for example with a laser printer. Self-adhesive targets can be used for temporary targeting of locations where no long-term reproducibility in measurement is required. At minimal cost, this type of targeting offers an almost unlimited range of target designs with sharp edges and additional information. Since printed targets are not retro-reflective, image contrast and target detectability strongly depend on high quality photography and sophisticated feature extraction algorithms during post-processing. The adhesive must also be selected with regard to ease of removal, potential damage to the object surface and resistance to heat.

Target with retro-reflective insert
(∅ = 75 mm)

Target with a bright plastic centre
(∅ = 50 mm)

Fig. 3.146: Examples of artificial circular targets

Self-luminous (active) targets are more complex. They are used in those applications where no artificial illumination is possible or where a recording or measurement process is controlled by switching active targets on or off. Self-luminous targets can, for example, be designed with an LED (*light emitting diode*) that is mounted behind a semi-transparent plastic cover. These targets provide optimal contrast and sharp edges (Fig. 3.147). As examples, LEDs are incorporated into manual probes used in online measuring systems (Fig. 3.156).

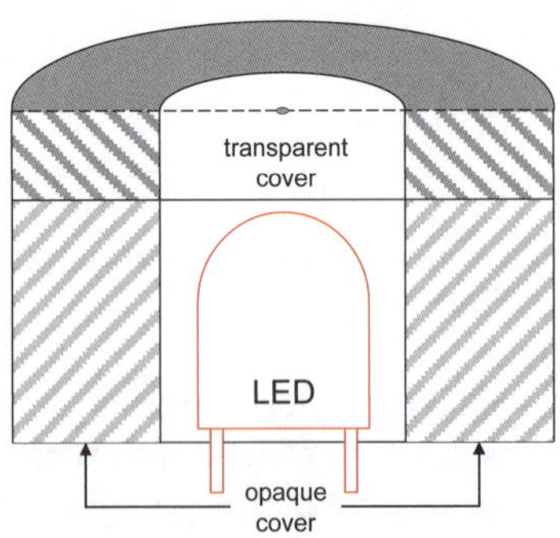

Fig. 3.147: Conceptual design for a luminous target

3.6.1.2 Circular targets

Because of their radially symmetric form, circular targets are very suitable for representing a target point (the centre) by a surface. Determination of the target centre is invariant to rotation and, over a wide range, also invariant to scale. They are suitable for both manual, interactive image measurement as well as automated digital techniques for point determination.

The centre of circular targets can be found in analogue images by manually centring a measuring mark, which is a circle or dot, over the target image. In a digitised image, the target centre can be computed by a centre-of-gravity calculation, correlation with a reference pattern or numerical calculation of the centre of the target's outline circle or ellipse (section 5.4.2).

In order for accurate measurements to be made, the diameter of the circular target images must be matched to the target detection and measurement process. For analogue measurement systems, diameters of the order of 100–150% of the floating mark diameter are appropriate. For digital processing systems, it is generally accepted that target images should be at least 5 pixels in diameter. In addition to practical considerations, such as the maximum target size for attaching to an object, the maximum target diameter is also a function of the maximum allowable eccentricity between the true image position of the circle centre and the centre of the target's elliptical image.

Fig. 3.148 displays the effect of perspective eccentricity of an imaged circle. The circle centre C is imaged as C' whilst the ellipse centre E' is displaced by the eccentricity *e*'. Only in the case where circle and image planes are parallel, both points are identical: C' = E'.

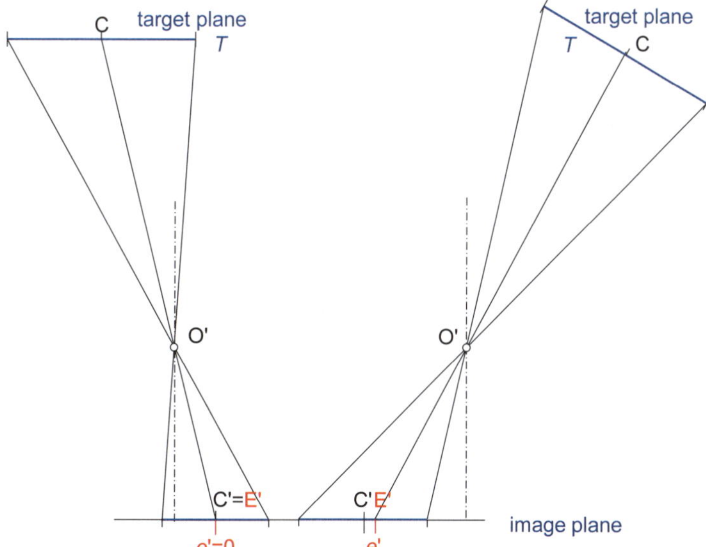

Fig. 3.148: Image eccentricity for a circular target with parallel and inclined target planes

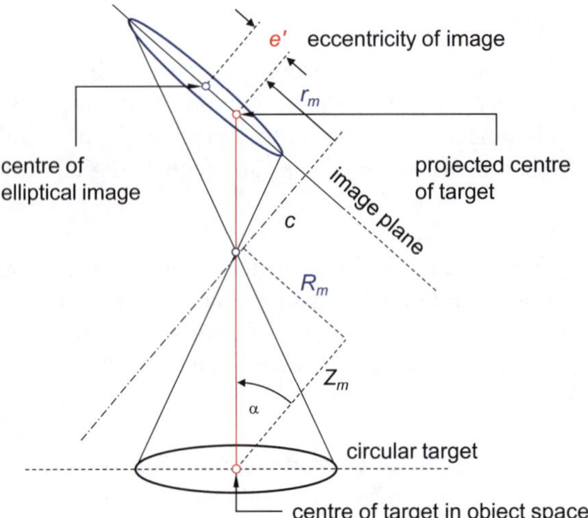

Fig. 3.149: General case eccentricity in imaging a circular target

Fig. 3.149 demonstrates the eccentricity *e*' between the calculated ellipse centre and the actual target centre. The degree of eccentricity depends on the size of the target, viewing direction, lateral offset from the optical axis and image scale. It can be estimated as follows:

$$e' = r_m - \frac{c}{2} \cdot \left(\frac{R_m + \frac{d}{2} \cdot \sin(90-\alpha)}{Z_m - \frac{d}{2} \cdot \cos(90-\alpha)} + \frac{R_m - \frac{d}{2} \cdot \sin(90-\alpha)}{Z_m + \frac{d}{2} \cdot \cos(90-\alpha)} \right) \tag{3.92}$$

where e': eccentricity of projection
d: target diameter in object space
r_m: radial offset of target in image
α: viewing direction = angle between image plane and target plane
R_m: lateral offset of target from optical axis
Z_m: distance to target
c: principal distance

Essentially, targets which are smaller and/or placed at greater distances result in eccentricities which are negligible (smaller than 0.5 µm). For greater image scales, larger image formats (e.g. aerial photo format), bigger targets, strongly convergent camera directions and high accuracy requirements, then eccentricity can be significant and must be taken into account during processing.

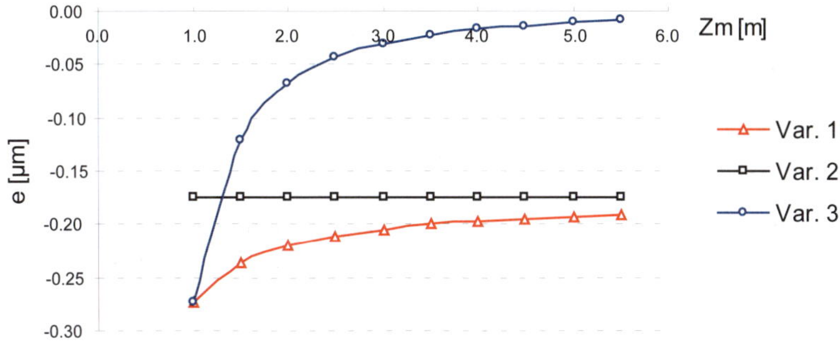

Fig. 3.150: Eccentricity of projection with respect to object distance
where r_m = 35 mm (medium format)
c = 40 mm (wide angle)
α = 60° (maximum effect)

Var. 1: variable target diameter in object space with
$d' = d / m$ = 170–200 µm (17–20 pixel at pixel size of 10 µm)

Var. 2: variable target diameter in object space = constant
target diameter in image space with d' = 160 µm (16 pixel)

Var. 3: constant target diameter in object space = variable
target diameter in image space with d' = 36–200 µm (4–20 pixel)

The effect of eccentricity is extremely complex for the multi-photo convergent image configurations that characterise most high-accuracy photogrammetric measurement. It is often assumed that the effect is compensated by the parameters of interior and exterior orientation if they are estimated using self-calibration techniques. However, for high-precision applications (<0.5 µm in image space) it is recommended that small targets are used. In using small targets, the target diameter, or level of ring flash illumination for retro-targets, must be capable of generating target image diameters of the order of 5–10 pixels if demanding measuring accuracies are to be achieved.

Eccentricity does not matter if the target centre is calculated as the centre of a directly defined circle in 3D space, as implemented by the contour algorithm (section 4.4.7.2) or bundle adjustment with geometric 3D elements.

Example 3.10:

Fig. 3.150 shows three variant calculations of eccentricity. The example is based on a medium-format camera (image radius r_m = 35 mm) with a wide-angle lens (c = 40 mm) and distances to the target vary between 1 m and 6 m. The angle between image plane and target plane is 60°, giving maximum eccentricity. The three variants differ with respect to the selected target diameter.

For the first variant, the diameter increases in proportion to distance from the camera (object distance). This gives image diameters between 170 µm and 200 µm. For a pixel size of 10 µm, this is equivalent to diameters of 17–20 pixels. In this variant the maximum eccentricity is 0.27 µm, decreasing with increasing object distance.

For the second variant, the diameter is chosen so that the image diameter remains constant at 160 µm, independent of object distance. The eccentricity is constant at 0.17 µm.

For the third variant, the diameter is constant (d = 5 mm). This results in image diameters between 36 µm and 200 µm (4–20 pixels). The maximum eccentricity is 0.27 µm.

3.6.1.3 Spherical targets

Spherical targets have the following advantages over flat targets:

- spherical targets are always imaged as an ellipse with an eccentricity that is radial about the optical axis;
- from Fig. 3.151 it can be seen that the ellipse is the result of a conic projection. Note that the base of the projective conic is smaller than the sphere diameter;
- spherical targets can be viewed consistently over a much wider angular range;
- spheres can be used as touch probes.

However, there are disadvantages with regard to mechanical construction and optical properties:

- the brightness of the image falls off rapidly towards the edges of the sphere;
- the edge of the sphere appears blurred;
- the mounting of the sphere can disrupt the isolation of the target edge in the image;
- sphere diameters for practical use lie between 5 mm and 20 mm;
- retro-reflecting spheres are expensive to manufacture.

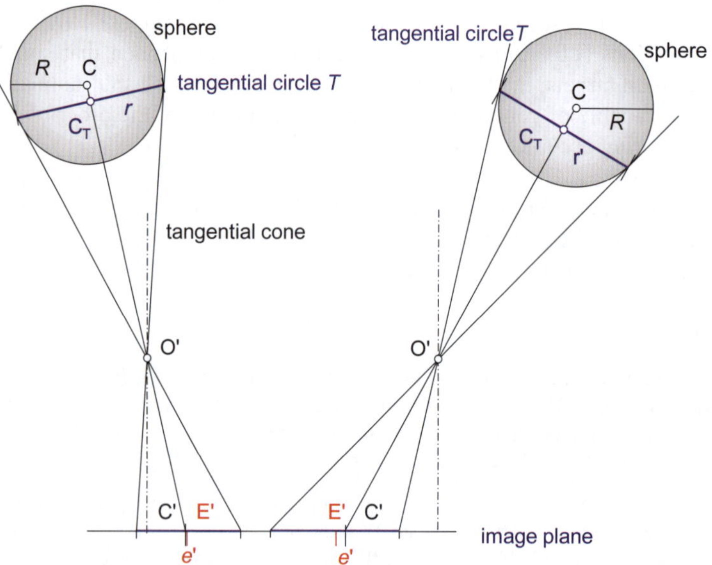

Fig. 3.151: Eccentricity of projection for a spherical target

The most important advantage of spherical targets is the fact that they can be viewed from almost any direction. Whereas flat, retro-reflective targets can only be viewed over an angle of ±45°, retro-reflecting spheres can be viewed over a range of 240° (Fig. 3.152). They are used, for example, in hand-held probes for photogrammetric navigation systems or as reference points on engineering tools and components (Fig. 3.153).

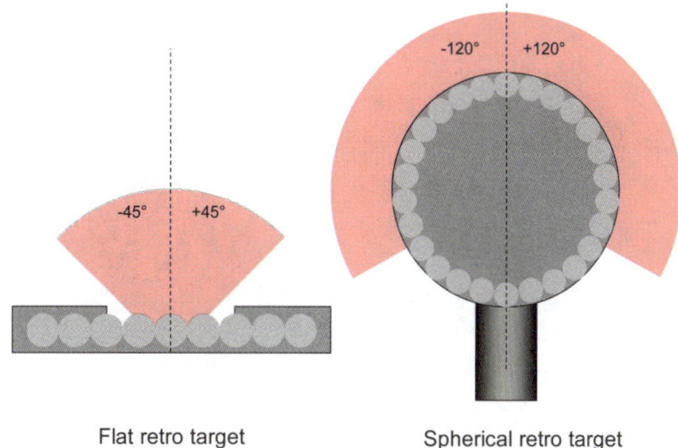

Fig. 3.152: Example viewing angles of flat and spherical retro-targets

a) Retro-reflecting spheres (above: IZI, below: Atesos)

b) Hand-held probe for optical navigation system (AXIOS 3D)

Fig. 3.153: Application of spherical targets

3.6.1.4 Patterned targets

Some applications make use of targets where the target point is defined by the intersection of two lines. Cross-shaped, checkerboard and sectored targets are examples here. Their advantage is the direct definition of a centre point using well defined edges and their good separation from the background. A disadvantage is the greater effort in locating the centre using digital methods and, in comparison with circular targets, the greater dependency on rotation angle. Fig. 3.154 shows examples of targets whose centre id defined by the intersection of two lines in object space or two curves in image space.

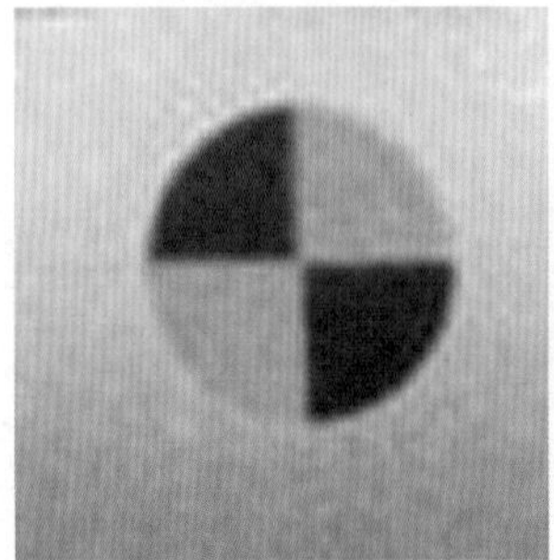 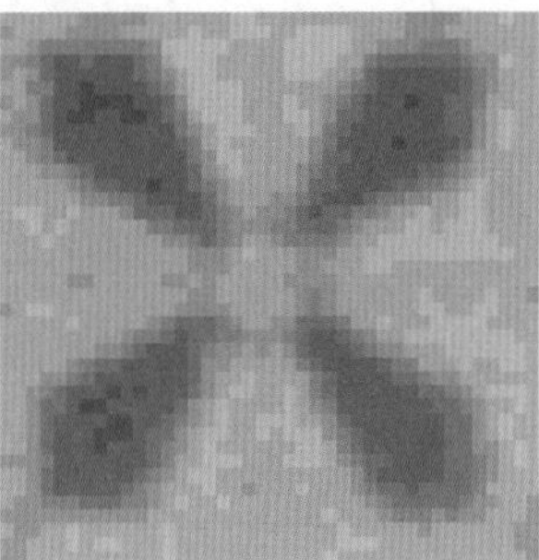

Fig. 3.154: Examples of line-pattern targets

3.6.1.5 Coded targets

Targets with an additional pattern which encodes an individual point identification number, can be used to automate point identification. The codes, like product bar codes, are arranged in lines, rings or regions around the central target point (Fig. 3.143b and c, Fig. 3.155). Patterns can be designed which encode more than several hundred point identification numbers.

Coded targets should meet the following requirements:

- invariance with respect to position, rotation and size
- invariance with respect to perspective or affine distortion
- robust decoding with error detection (even with partial occlusions)
- precisely defined and identifiable centre
- sufficient number of different point identification numbers
- pattern detectable in any image
- fast processing times for pattern recognition
- minimum pattern size
- low production costs

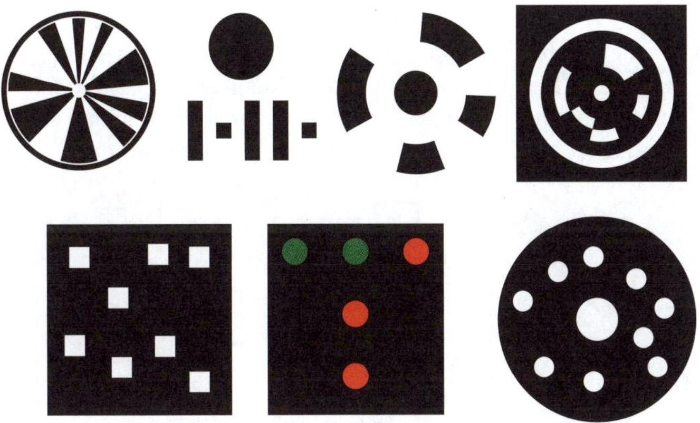

Fig. 3.155: Selection of coded targets
upper row: barcode patterns, lower row: shape and colour patterns

The point identification number is decoded by image analysis of the number and arrangement of the elements defining the code. The patterns displayed in Fig. 3.155 (upper row) are based on barcode techniques where the code can be reconstructed from a series of black and white marks (bit series) spread around a circle or along a line. The number of coded characters is limited by the number of bits in the barcode. The patterns displayed in Fig. 3.155 (lower row) consist of clusters of targets grouped in a local coordinate system where the target identification number is a function of the local point distribution. For decoding, image coordinates of all the sub-targets are computed. The planar image pattern can then be perspectively transformed onto reference points in the pattern so that target identification can be deduced from the map of transformed coordinate positions. Alternatively, targets in the group can be given different colours so that the point identification number can be deduced from the locations of these differently coloured points.

3.6.1.6 Probes and hidden-point devices

Targeted probing devices have been developed for measuring "hidden" areas on an object which are not directly visible to the camera system, although it is also convenient to use them for general-purpose measurement. Fig. 3.156 shows probes with an offset measuring sharp point P which is determined indirectly by the 3D coordinates of a number of auxiliary points Q_i attached to the probe in a known local configuration. These points can either be arranged on a line (2–3 points) or spatially distributed (≥3 points) with known 3D reference coordinates in the probe's local coordinate system. The points Q can be either passive retro-reflective targets or active, self-luminous LEDs. Alternatively, point P can be a ruby sphere such as those found on CMM touch probes. In such a case, a correction must be made in software for the offset caused by the significant radius of the sphere. Hand-held probes are used in many industrial applications where they are tracked by at least two cameras simultaneously, with a new image pair being taken for each point measured by the probe. Such a system is often termed an online system.

If the probe is provided with an additional measuring point R (Fig. 3.156), which can be shifted relative to the reference points Q, then this can act as a switch, for example to trigger a measurement. If R is on a lever which is depressed by the operator, it is shifted slightly away from its nominal position. The photogrammetric tracking system can register this movement and trigger a recording of the measuring point position.

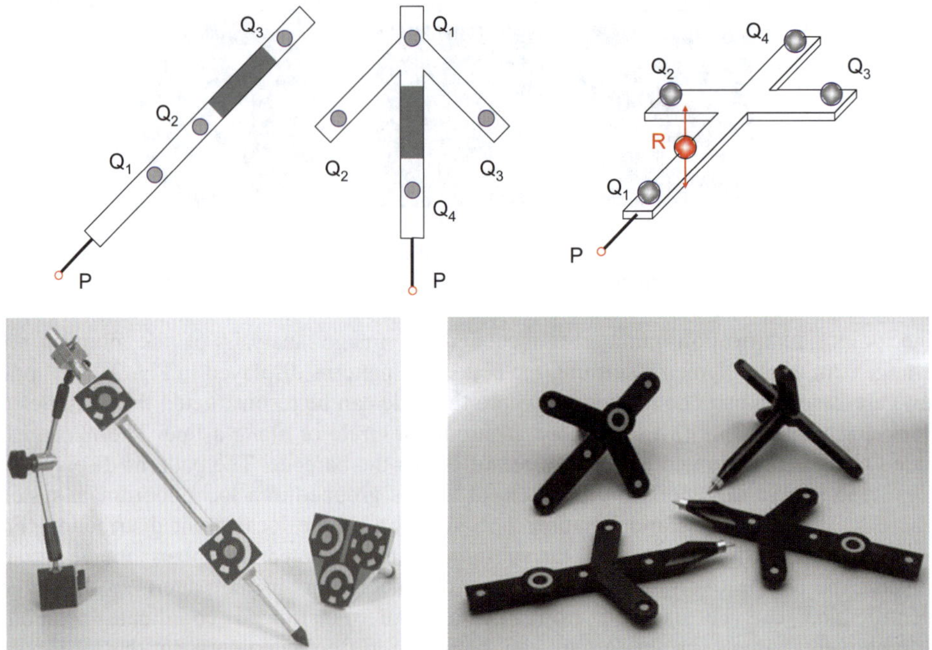

Fig. 3.156: Principle and examples of targeted measuring probes

In order to measure the centre of drilled holes or edges of components, specially designed target adapters provide a fixed relationship between measurable targets and the actual point or edge to be measured. Manufacturing tolerances in the adapter directly contribute to the overall accuracy of the indirectly measured object point.

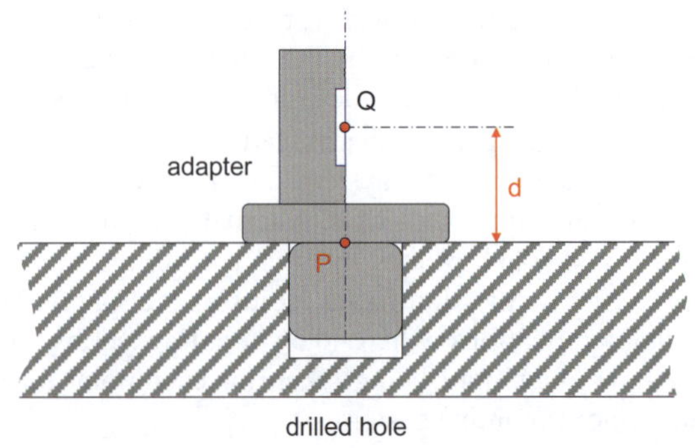

Fig. 3.157: Principle of adapter for drilled hole measurement

Fig. 3.157 shows the principle of an adapter for indirect measurement of the centre of a drilled hole. The visible target is mounted such that the measured point Q is located on the drilling axis at a known offset *d* to the hole centre P.

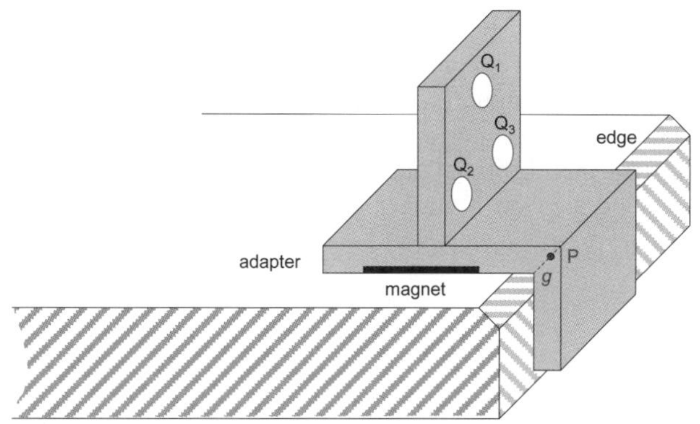

Fig. 3.158: Principle of an edge adapter

The edge adapter displayed in Fig. 3.158 has three target points Q_i, with a known local relationship to the straight line *g* which, on attachment of the adapter, represents an edge point P. Such adapters can be equipped with magnets for rapid mounting on metallic surfaces. If the points Q define a coded target, the photogrammetric system can identify the adapter type directly and calculate P automatically. However, multiple-point adapters with coded targets require significantly more surface area, so adapter size increases.

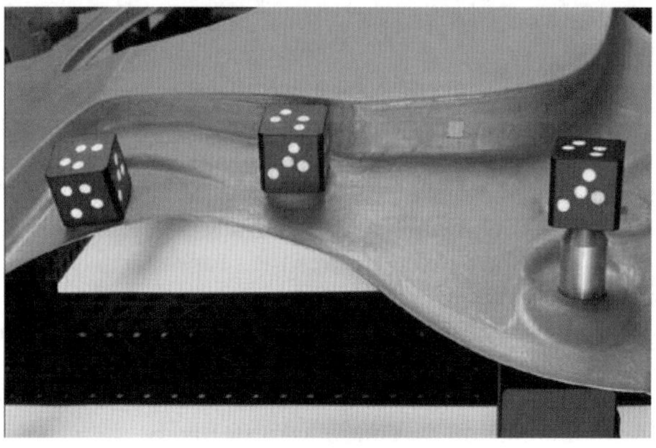

Fig. 3.159: AICON adapter cubes

The adapter cubes developed by AICON (Fig. 3.159) have a coded target on each of 5 faces, as well as a measurement point or edge on the sixth side. The points in the coded targets, and the measurement point, are manufactured to have a relative reference coordinate accuracy of around 10 µm. By measuring the points automatically and analysing the spatial distances between them, a particular cube can be identified and its measurement point located in 3D.

3.6.2 Illumination and projection techniques

3.6.2.1 Electronic flash

Electronic flash systems can be attached and synchronised to almost all modern photographic and many digital camera systems. Whilst the electronic flash output may occur in 1/70 000 s or less, camera shutters use synchronisation times ranging between 1/30 s to 1/250 s. Electronic flash performance is characterised by the *guide number Z*. The guide number indicates the ability of the flash to illuminate an object at a given object distance a in metres and a given f/number k (Fig. 3.160). Higher guide numbers indicate greater light output.

$$a = \frac{Z}{k} \tag{3.93}$$

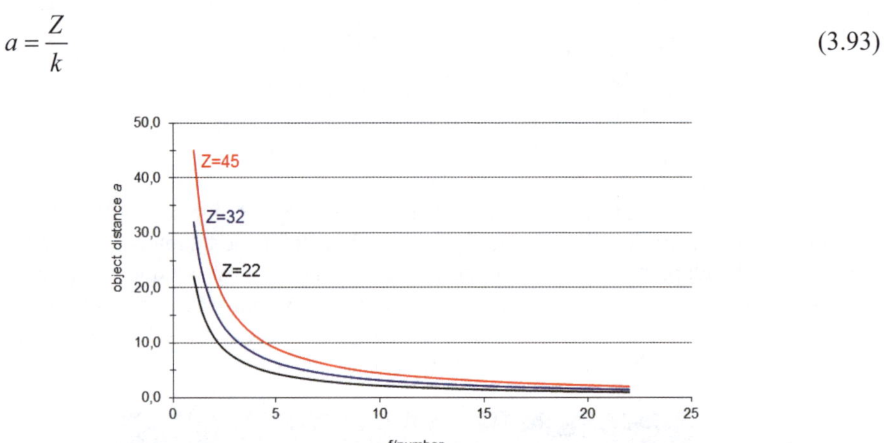

Fig. 3.160: Usable object distances for different guide numbers and f/numbers

Ring flashes, which are circular flash tubes mounted concentrically around the lens, are of special importance for photogrammetry (see Fig. 3.161). They are mainly used for illuminating objects to which retro-reflective targets are attached. Their light output, which is concentric and close to the lens axis, ensures the retro-reflective target images are well separated from the background. It is important to note that electronic flash units with automatic exposure options utilise light measurement systems that are calibrated for general-purpose photography. Correctly illuminated retro-targets do not require as much light output and it is generally necessary to use manual controls in order to obtain consistent image quality. When illumination is controlled effectively, it is possible to make the background disappear almost completely in the image such that only reflecting target images remain (example in Fig. 7.22). Such a situation is ideal for rapid, automated target detection and measurement.

Modern LED technology makes it possible to construct almost any form of illumination geometry, for example on a ring or flat field. LED illumination is available in the colours blue, green, yellow, red and white, as well as infrared, which permits the generation of almost any colour mix. They are extremely robust, durable and have low energy requirements. If LEDs are used for high-voltage flash illumination, very high illumination levels can be achieved. Fig. 3.162 shows a digital metric camera with an integrated LED ring flash operating in the infrared.

Fig. 3.161: Digital SLR camera with ring flash

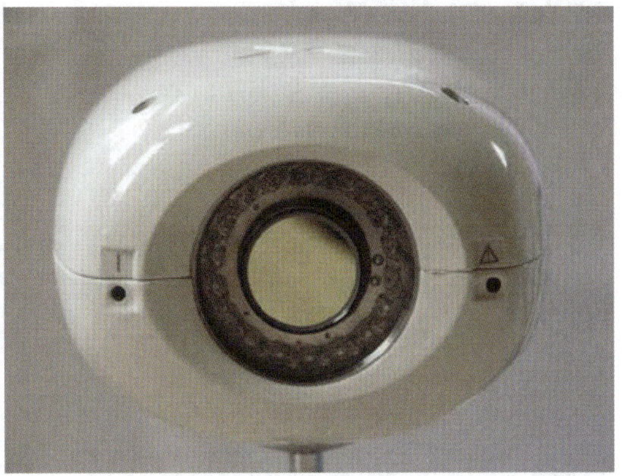

Fig. 3.162: Digital metric camera with integrated LED ring flash (AXIOS 3D)

3.6.2.2 Pattern projection

Arbitrary light patterns can be projected onto object surfaces by means of analogue slide projectors or computer-controlled digital projectors. These devices, are typically used if the object does not provide sufficient natural texture to enable a photogrammetric surface reconstruction. They are also known as structured light projectors, particularly if the projected image is designed to form part of the surface reconstruction algorithm. Fig. 3.163 shows examples of projectors used for industrial photogrammetry. The GSI Pro-Spot system can project up to 22 000 circular dots onto an object. These can be measured like conventional targets to produce a dense object surface measurement.

LCD (liquid crystal display) and LCOS (liquid crystal on silicon) fringe projectors have increased in importance since they can project line patterns for surface measurement by structured light methods (see section 6.5.1). A reflective condensor optics illuminates a CCD array where the programmed line pattern is projected through a lens onto the object surface (Fig. 3.164). LCD video projectors are also available which can project video and computer images with arbitrary patterns and in colour. Current projectors have resolutions up to 1600 x 1400 lines.

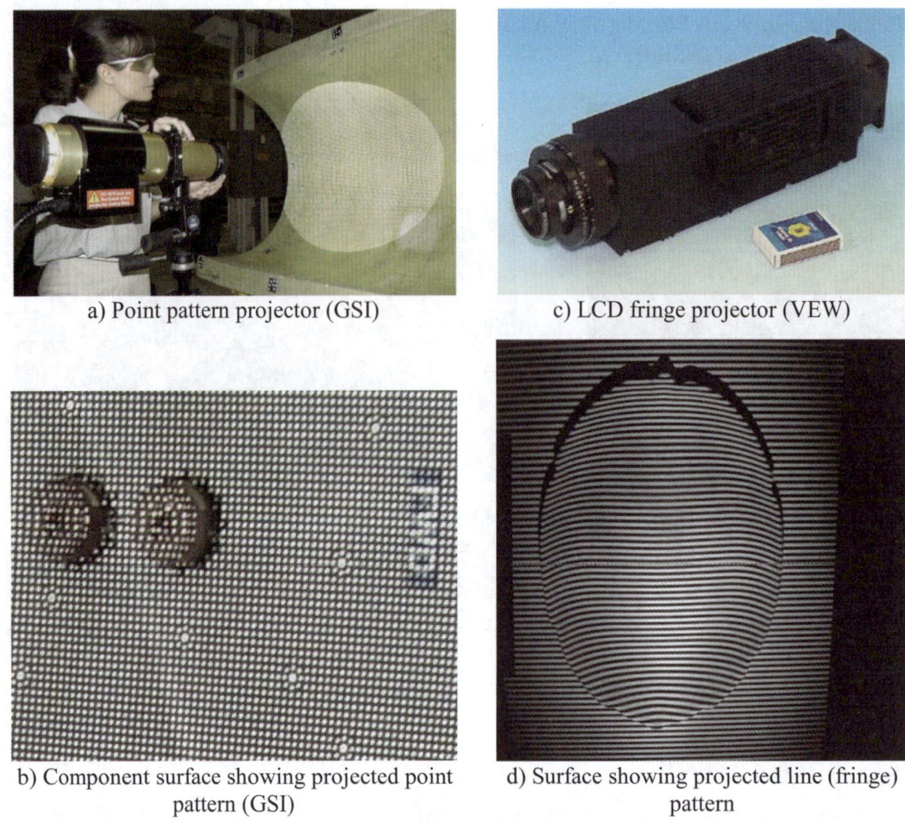

a) Point pattern projector (GSI)

c) LCD fringe projector (VEW)

b) Component surface showing projected point pattern (GSI)

d) Surface showing projected line (fringe) pattern

Fig. 3.163: Surface pattern projection

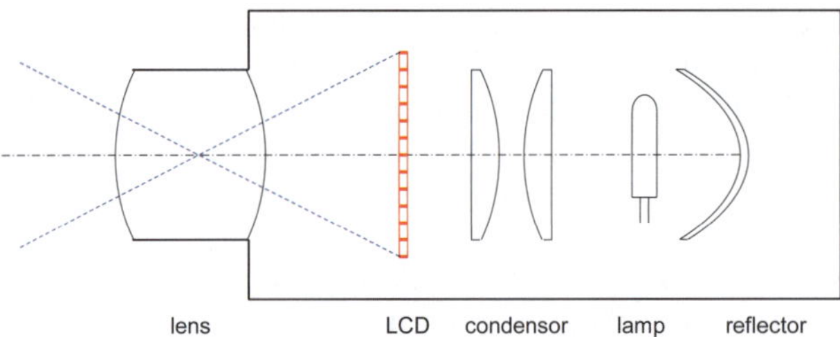

Fig. 3.164: Schematic design of a LCD-fringe projector

An alternative solution uses a digital micromirror device (DMD). Movable micromirrors of about 16 µm size are attached to a semiconductor. They can be controlled digitally in order to reflect or absorb incident light. Micromirror chips are available with 1280 x 1024 points or more and switching times of 20 ms. They can be used as both projectors and monitors.

If a projector is calibrated and oriented it can be treated as a camera in the photogrammetric process. In such a situation the position of a projection point within the slide or digital projection plane is analogous to an image coordinate measurement and the collinearity equations can be used to compute its direction. This approach is used for projectors integrated into structured light systems for optical 3D metrology.

3.6.2.3 Laser projectors

Laser projectors can be used to create a structured pattern on an object surface, e.g. with points or lines. In contrast to other light sources, high illumination powers can be achieved even for eye-safe laser classes (up to Class 2).

Laser projectors can be classified into three groups:

- Point projection:

  A single laser spot is projected onto the object, for example with a laser pointer. The resulting pattern is not a perfect circle and oblique projection angles make the spot elliptical. Interference between the coherent laser light and the roughness of the incident surface gives rise to speckle effects and generates an inhomogeneous intensity distribution within the laser spot which depends on viewpoint and any relative motion between laser, surface and viewpoint. As a result the optical centroid does not correspond to the geometric centre (Fig. 3.165). For these reasons, laser point projection is seldom used for high accuracy photogrammetric targeting.

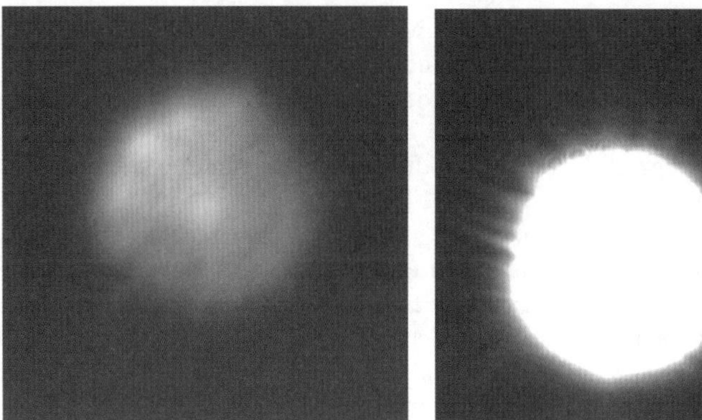

Fig. 3.165: Magnified image of laser points

- 1D line projection:

  Laser lines can be projected by means of a cylindrical lens (see Fig. 3.124) mounted in front of the laser source. Projected lines are used for triangulation with light-section methods (Fig. 3.169).

- 2D pattern projection:

  Two-dimensional laser patterns can be created by special lenses and diffractive elements mounted in front of the laser source (Fig. 3.166). This is used, for example, in the Kinect sensor (see section 3.7.3).

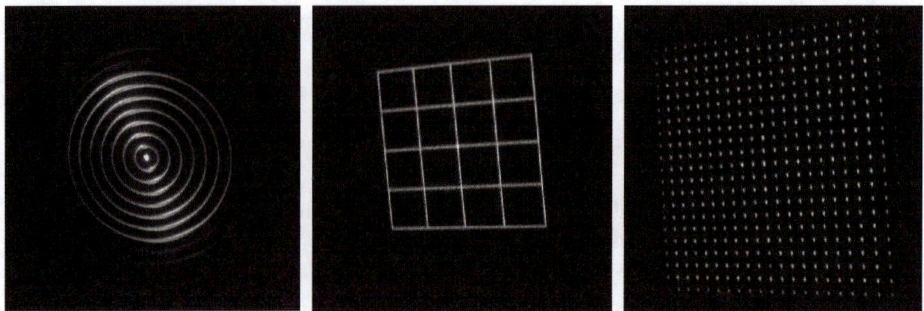

Fig. 3.166: Laser projection of two-dimensional patterns by front lenses

Arbitrary patterns can be generated by fast 2D laser scanners. A laser beam is reflected by two galvanometer mirrors which rotate at high frequencies (up to 15 MHz) (principle as in Fig. 3.170 but without range detection). If the projection frequency is higher than the integration time of the camera (or the human eye), a continuous two-dimensional pattern is visible in the image. Such systems can be used in industrial measurement to set out information on a surface.

3.6.2.4 Directional lighting

Directional lighting techniques are of major importance since the measurement of particular object areas, for example edges, can be enhanced by the controlled generation of shadows. This applies, for example, to the illumination of object edges which, without further targeting, are to be located by image edge detection (see section 4.4.7.2). Under such situations the physical edge of the object and the imaged edge must be identical.

Depending on the relative position of object surface and camera, directional lighting must be chosen such that one surface at the object edge reflects light towards the camera, while the other surface is shadowed, or does not reflect into the camera (Fig. 3.167). In the image, contrast characteristics and edge structure depend on the reflective properties of the two surfaces, and on the sharpness of the actual object edge. For example, a right angled edge can be expected to give better results than a curved or bevelled edge.

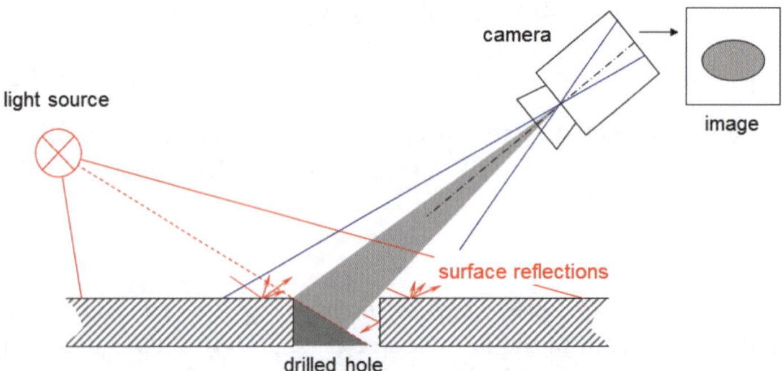

Fig. 3.167: Directional lighting for edge measurement

Fig. 3.168 shows the influence of different types of illumination on the images of object edges on a metal workpiece. Illumination from the viewing direction of the camera (a) results, as expected, in a poor image of the edge. If light is incident from the opposite direction (b), the edge of the drilled hole is correctly located in the object surface. Conditions permitting, a diffuse illumination source located below the drilled hole (c) will also provide a suitable, as required for example in the optical measurement of pipes and tubes (section 6.4.3.1).

a) Directly from the camera

b) According to Fig. 3.167

c) Diffuse from below

Fig. 3.168: Illumination of a drill hole in a metal component

3.7 3D cameras and range systems

Range cameras deliver dense range maps without post processing combining a regular sampling of the scene with depth information (see section 5.6). Systems are reliant on either triangulation, time of flight or phase comparison to produce the range component of the image. Characteristic systems described below include close-range triangulation where a profile is collected, laser spot based systems where single points are sampled consecutively and full field cameras where the complete image is captured at a given instant.

3.7.1 Laser-based systems

3.7.1.1 Laser triangulation

Fig. 3.169 shows the principle of the laser triangulation (light-section) method. A laser beam is projected through a cylindrical lens to generate a light plane (see Fig. 3.124). An array imaging sensor is arranged with an offset to the laser diode in order to form a triangle with known (calibrated) geometric parameters. The reflected laser plane is deformed as a function of the distance to the object.

A full 3D measurement of an object surface can only be provided by a scanning procedure, e.g. if the triangulation sensor is mounted on a CMM arm (Fig. 6.35b) or positioned by a laser tracker (Fig. 3.178). Laser triangulation sensors have typical accuracies for distance measurement in the order of 0.1mm in a distance of up to 500mm. They allow for measurement rates of up to 1000 Hz.

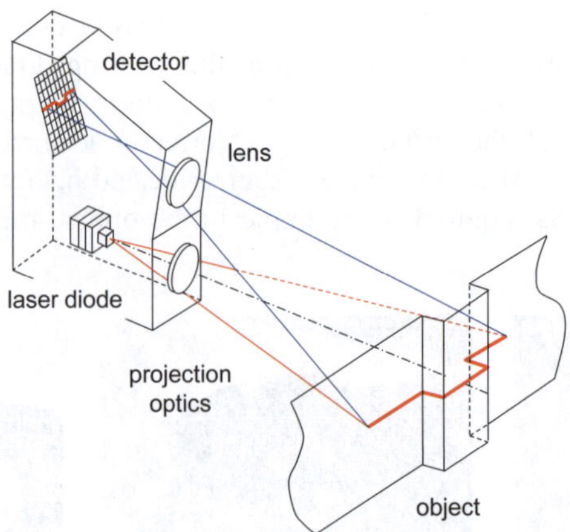

Fig. 3.169: Principle of light-section method (after Schwarte 1997)

3.7.1.2 Laser scanners

3D laser scanning describes the three-dimensional measurement of the surface of an object through the analysis of the reflected light from a laser beam which is scanned over the object surface. Polar coordinates, with an origin at the scanning system are defined by angle and range measurements. If the reflected intensity is also registered, an object image is recorded where each measured 3D point is associated with an intensity value dependent on the reflectance of the object surface to the scanning laser wavelength. Laser scanners were initially developed for airborne applications in which a laser scan of the ground from an aircraft produces a digital terrain model. They are increasingly used in close-range applications, commencing as weld checking tools and expanding out to range from the recording of building interiors to the measurement of complex structures (terrestrial laser scanning, TLS).

Commercially available systems differ mainly in terms of physical principle, measuring frequency, measuring accuracy, range of operation, beam diameter and costs. Whilst some scanner designs have a restricted horizontal and vertical view (less than a hemisphere), panoramic scanners provide all-around 360° measurement from a single location.

Fig. 3.170 shows the schematic layout of a laser scanner with two mirror galvanometers and phase-difference measurement of two modulated signals. Phase-difference measurement enables high measuring frequencies ($> 2.5 \cdot 10^9$ points per second).

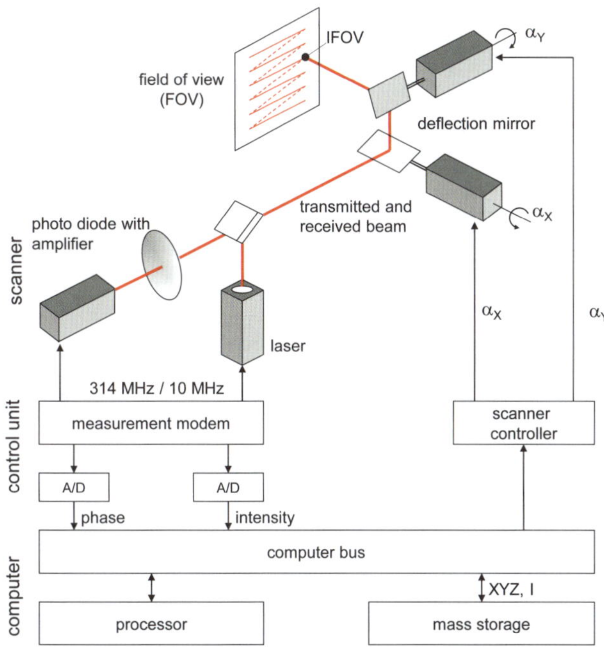

Fig. 3.170: Principle of laser scanning (after Wehr 1998)

Most commercially available laser scanners implement range detection by time-of-flight measurement. Time-of-flight, or pulse measurement, is less sensitive with respect to object surface variations and is therefore generally preferred for reflectorless surface measurements. Measurement rates are about one order of magnitude slower than for phase-difference measurement. Fig. 3.171 to Fig. 3.173 show three typical instruments. Table 3.11 summarises their technical specification.

Fig. 3.171: Zoller & Fröhlich Imager 5010c

Fig. 3.172: Leica Scanstation C10

Fig. 3.173: Riegl VZ 400

Table 3.11: Examples of laser scanning systems with technical data drawn from company product information

	Zoller + Fröhlich Imager 5010c	Leica Scanstation C10	Riegl VZ 400
instrument type	panorama View	camera View	panorama View
distance measurement	phase difference	time of flight	time of flight
laser wavelength power beam diameter	 1500 nm (class 1) 3.5 mm at 0.1 m < 0.3mrad	 532 nm 1 mW (class 3R) 6 mm at 50 m	 near infrared (class 1) 7 mm / 0.35 mrad
resolution	0.6 mm at 10 m	0.1 mm at 10 m	1 mm
range distance horizontal vertical	 0.3 to 187.3 m 360° 320°	 1.5 to 300 m 40° 40°	 1.5 to 600 m 360° 100° (–40° ... ±60°)
points per sec	< 1 016 000	< 50 000	< 122 000
accuracy distance angle coodinate	 < 2.2 mm at 50 m [a] ± 0.007° no information	 ± 4 mm at 50 m [b] ± 12" ± 6 mm at 50 m [b]	 ± 5 mm ± 1.8" [c] ± 5 mm/100 m
weight	11 kg	15 kg	9.6 kg
other features	dual-axis tilt sensor, laser plummet, integrated HDR camera	dual-axis tilt sensor, laser plummet	optional camera; multiple target capability; inclination sensor; GPS receiver; laser plummet

[a] 1σ noise, unfiltered raw data at 127 000 points/sec data rate; [b] 1σ noise; [c] resolution

If the recorded data (*X,Y,Z,intensity*) are formatted as a regular array, two data sets can be derived

1. A range image where the value (grey or colour) assigned to each pixel is proportional to the range of the object surface from the scanner (Fig. 3.174a,b)

2. An intensity image where the value assigned to each pixel is a function of the strength of the return signal from the surface of the object (Fig. 3.174c).

The intensity image appears similar to a photographic image. However it only records reflection values for the wavelength of the scanner's laser.

The principal feature of laser scanning is the fast three-dimensional measurement of a (large) object surface or scene with high point density. A typical scanning system has a 3D coordinate accuracy of the order of 3-10 mm at a distance of 10 m. In addition to angle and range measurement, the accuracy is also dependent on the stability of the instrument station (e.g. a tripod).

a) Grey-value range image

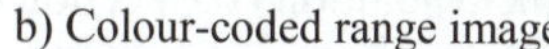

b) Colour-coded range image

c) Intensity image

d) RGB image overlay

Fig. 3.174: 3D laser scanning of a sculpture

Terrestrial laser scanners can be combined with a digital camera. If the camera has been calibrated, and oriented with respect to the scanner's internal coordinate system, the generation of a colour image, (Fig. 3.174d), can be achieved by mapping pixel data from a camera onto the point cloud in one of three ways.

1. A digital camera within the scanner viewing through the mirror system, or mounted onto the rear of the mirror, collects imagery either during scanning or in separate imaging passes of the scanner. This method has the advantage of minimizing parallax between the image and point cloud, but the disadvantage of reduced image quality due

to utilisation of the scanning optics. Image quality limitations can be partially solved by mounting a small camera on the reverse of the mirror.

2. An external camera mounted on the scanner either at a known offset or slid into place for a second imaging pass of the scanner. Cameras used in this case are typically DSLR systems capable of excellent image quality. Limitations are in accurate determination of the camera-to-scanner geometry for pixel-to-point-cloud mapping and, more fundamentally, to differences in parallax caused by the disparate projection centres.

3. An external camera used separately from the scanner. In this case very high quality images can be obtained, either from a system placed on a tripod in nominally the same location as the scanner or by mapping a network of photogrammetric images into the scanning geometry and selecting appropriate images to colour the point cloud based on local image magnification to each scanned surface, angle of view to the surface and image content. This last solution has the advantage of being able to exploit special purpose high dynamic range imagery for example, but requires a combined network adjustment solution (see section 4.4.2.3).

Combined 3D point-cloud and image data can be used to improve point-cloud registration to external information, for example a CAD model of the environment or to generate additional products through the use of image analysis by monoplotting (section 4.2.7) for example. True orthophotos can also be derived from oriented images and 3D surface models (see also section 4.2.7.3).

Since measurements are made by the reflection of a laser beam from a surface, 3D laser scanners can record large numbers of points without any need for targeting. Like photogrammetry, laser scanning requires lines of sight so complex objects must be recorded from several instrument stations whose individual point clouds must be transformed into a common coordinate system by means of common points or matching procedures. As with other area-based scanning methods, unstructured data acquisition is a major disadvantage. Usually point clouds must be manually or automatically thinned and structured in order to prepare the data for further processing. Image data, from one or more of the acquisition methods mentioned above, can also be useful for registration of point clouds from multiple stations.

3.7.1.3 Laser trackers

Laser trackers use a directed laser beam to measure the 3D coordinates of the centre of a retro-reflecting corner cube. This is typically mounted in a spherical housing (spherically mounted retro-reflector - SMR) and used as a manual touch probe to measure object points and surfaces (Fig. 3.176). Angle encoders on the two deflecting axes provide horizontal and vertical angle values, and range to the reflector is determined either by interferometry or, now more commonly, by an absolute optical range-measuring technique. Servo drives on the rotation axes, and return-beam sensing, keep the beam on track as the reflector is moved, enabling a continuous, dynamic path measurement. Laser trackers are used for the measurement of individual object points, scanning large free-form surfaces, as large scale deflection gauges, and for the dynamic tracking of moving objects.

High-precision angle and range measurement results in accuracies of better than 0.1 mm, at object distances of up to 40 m, for measured 3D coordinates.

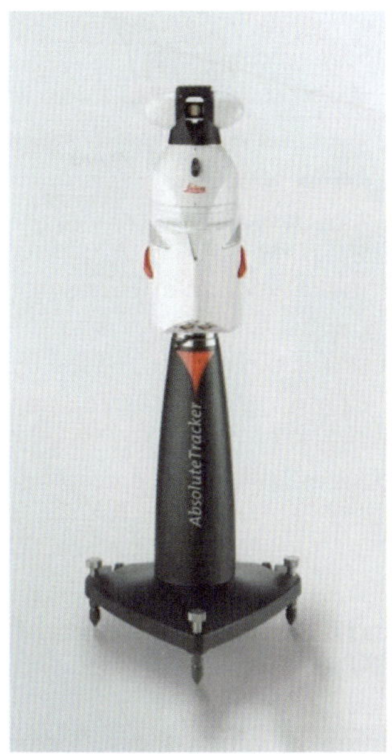

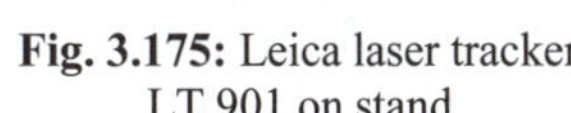
Fig. 3.175: Leica laser tracker LT 901 on stand

Fig. 3.176: Reflector ball (above) and object measurement (below)

Fig. 3.177: Faro laser tracker

There are three manufacturers of 3D laser trackers: Leica Geosystems (part of Hexagon Metrology), Faro Technologies Inc., and Automated Precision Inc. (API). Fig. 3.175 and Fig. 3.177 show typical examples of these systems.

Leica Geosystems and API both offer an enhanced system which tracks a target probe in all six degrees of freedom (6DOF). The Leica system is illustrated in (Fig. 3.178). Here the probing device (T-Probe) has an embedded retro-reflector which is tracked in 3D by the laser beam as normal. LED targets, known with the reflector in a local probe coordinate system, surround the reflector and are imaged by a digital zoom camera (T-Cam) mounted above the tracker head. By means of a standard photogrammetric space resection (section 4.2.3) the angular orientation of the probe can be calculated. With known 3D location and angular orientation, an offset touch point, used for measuring object points, can be located in 3D. In addition to a hand-held probe there is also a hand-held surface scanner (T-Scan) and a special probe for robot and machine control (T-Mac).

The operational concept of the API probe has not been published but is believed to use tilt sensing for roll angle measurement and a pinhole camera view of the laser beam behind the retro-reflecting prism which detects the probe's pitch and yaw angles.

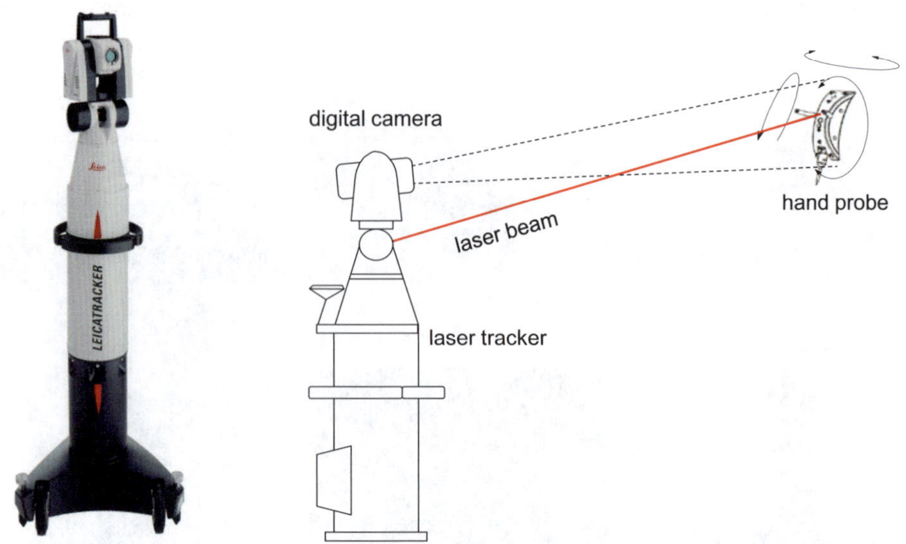

Fig. 3.178: Laser tracker Leica LT 800 combined with a digital camera (T-Cam) for measurement of position and orientation of a hand-held probe (T-Probe)

The LaserTRACER from Etalon offers a laser tracking solution in which the reflector position is found by multiple interferometric distance measurement only, i.e. no angle measurement is made, although angle tracking is required for its operation. The technique is similar to trilateration in geodesy, where a point is located by distance measurement from three known locations. Measuring techniques can utilise either multiple instruments or one instrument moved into multiple locations. The application area is principally the calibration of coordinate measuring machines (CMMs), milling machines and similar.

3.7.2 Fringe projection systems

3.7.2.1 Stationary fringe projection

Methods of stationary fringe projection are based on a fixed grid of fringes generated by a projector and observed by one camera. The grid has a cyclical structure, normally with a square or sine-wave intensity distribution with constant wave length λ.

Fig. 3.179 shows the principle of phase measurement for parallel fringe projection and parallel (telecentric) image acquisition at an angle α. The wave length λ corresponds to a height difference ΔZ_0:

$$\Delta Z_0 = \frac{\lambda}{\sin \alpha} \tag{3.94}$$

The height difference ΔZ with respect to a reference plane corresponds to the phase difference $\Delta\varphi$:

$$\Delta Z = \frac{\Delta\varphi}{\tan\alpha} \tag{3.95}$$

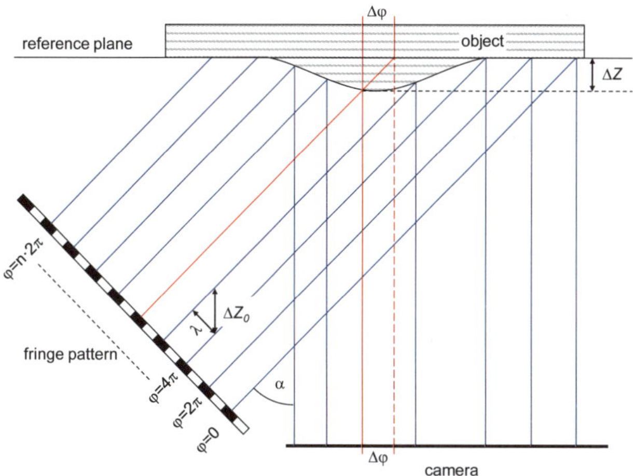

Fig. 3.179: Fringe projection with phase measurement by telecentric imaging

Phase difference is only unique in the range $-\pi...+\pi$ and so only provides a differential height value within this range. Absolute height change also requires counting the fringe number to determine the additional whole number of height units. The measuring method is therefore only suitable for continuous surfaces which enable a unique matching of fringes. A resolution in height measurement of about $\lambda/20$ can be obtained.

The telecentric configuration illustrated in Fig. 3.179 limits the size of measured objects to the diameter of the telecentric lens. If larger objects are to be measured the method has to be extended to perspective imaging. Stationary fringe projection can also be applied to dynamic tasks, for example the measurement of moving objects.

3.7.2.2 Dynamic fringe projection (phase-shift method)

Phase measurement can be performed directly using intensity values in the image. For this purpose the projected fringes are regarded as an interferogram. For the intensities of an interferogram at a fringe position *n,* the following equation applies:

$$I_n(x,y) = I_0\left(1+\gamma(x,y)\cos(\delta(x,y)+\varphi_m)\right) \tag{3.96}$$

where

I_0: constant or background intensity

$\gamma(x, y)$: fringe modulation

$\delta(x, y)$: phase

φ_m: phase difference

The equation above contains the three unknowns I_0, $\gamma(x,y)$ and $\delta(x,y)$. Hence at least three equations of this type are required for a solution. They can be obtained by m sequential shifts of the fringes by the difference φ_n (Fig. 3.181a).

$$\varphi_m = (n-1)\varphi_0 \tag{3.97}$$

where

m: number of shifts

$n = 1 \ldots m,$ where $m \geq 3$

$\varphi_0 = 2\pi / n$

The measuring principle is known as the *phase-shift method*. When $m = 4$ samples taken at equally spaced intervals of $\pi/2$ (Fig. 3.181a), the phase of interest $\delta(x, y)$ reduces to:

$$\delta = \arctan\frac{I_2 - I_4}{I_3 - I_1} \qquad \delta = \delta(x,y); \quad I_n = I_n(x,y) \tag{3.98}$$

Finally, the height profile is given by

$$Z(x,y) = \frac{\lambda}{2 \cdot 2\pi}\delta(x,y) \tag{3.99}$$

As with stationary fringe projection, the result is unique only in the interval $-\pi\ldots+\pi$, so that integer multiples of 2π must be added for a complete determination of profile. This process is known as *demodulation* or *unwrapping* (Fig. 3.180). Discontinuities in the object surface lead to problems in the unique identification of the fringe number. A unique solution is possible by using fringes of varying wavelength (Fig. 3.181b) or Gray-coded fringes (section 3.7.2.3).

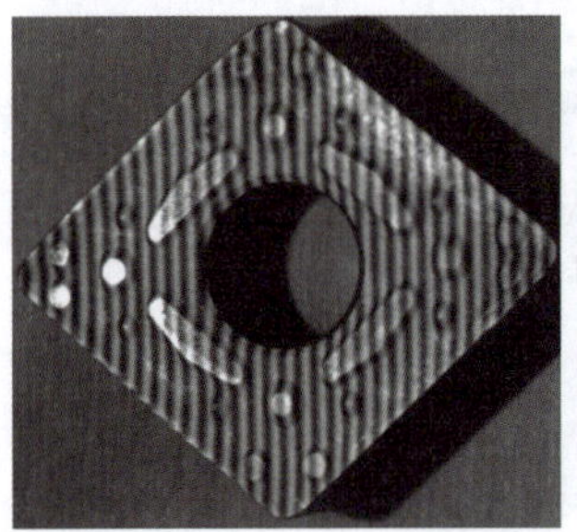

a) Object with projected fringes

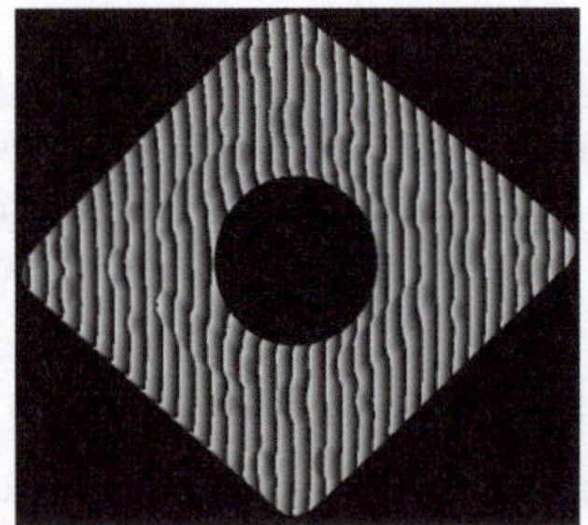

b) Phase image with phase discontinuities

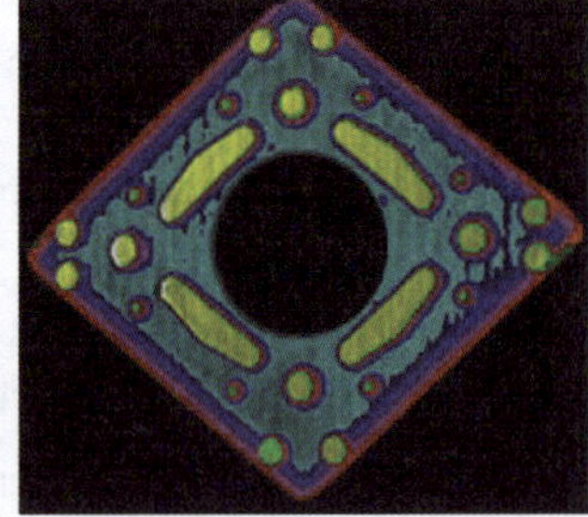

c) Demodulated height model

Fig. 3.180: Phase-shift method (AICON/Breuckmann)

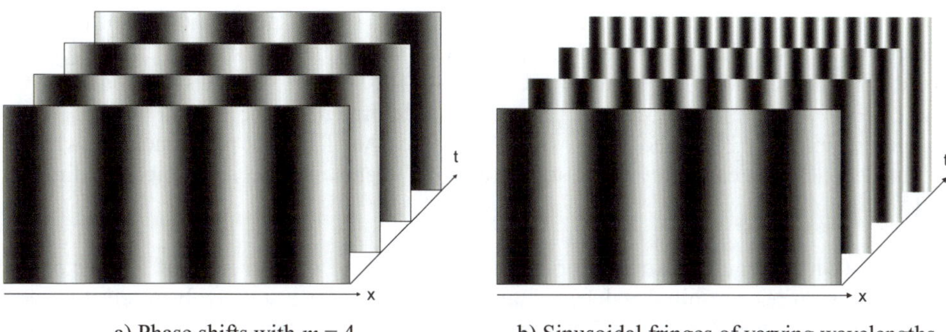

a) Phase shifts with $m = 4$ b) Sinusoidal fringes of varying wavelengths

Fig. 3.181: Sinusoidal patterns used in the phase-shift method

The accuracy of height measurement is about $\lambda/100$. The interior and exterior orientation of projector and camera must be found by calibration. Each pixel (x, y) is processed according to eqn. 3.98. The computations can be solved using fast look-up table operations, with the result that height measurements can be processed for all pixels in an image (for example 780 x 570 pixel) in less than one second. Examples of measuring systems are presented in sections 3.7.2.4 and 3.7.2.5.

3.7.2.3 Coded light (Gray code)

Solving ambiguities is a major problem for fringe projection methods, especially for discontinuous surfaces. In contrast, the *coded-light* or *Gray code* technique provides an absolute method of surface measurement by fringe projection.

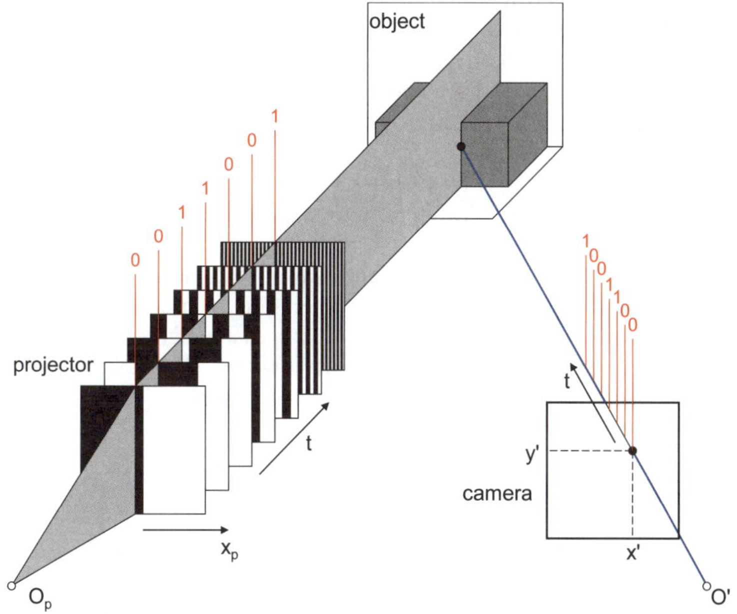

Fig. 3.182: Coded-light approach (after Stahs & Wahl 1990)

The projector generates m coded fringes sequentially, so that perpendicular to the fringe direction x_p a total of 2^m different projection directions can be identified by an m-digit code word (Fig. 3.182 shows an example bit order 0011001 for $m = 7$). A synchronised CCD camera observes the fringe pattern reflected and deformed by the object surface. The m images acquired are binarised and stored as bit values 0 or 1 in a bit plane memory which is m bits deep. Hence, each grey value at position (x',y') denotes a specific projection direction x_p from O_p.

This procedure requires known orientation parameters of camera and projector but otherwise requires no initial values related to the object. It is relatively insensitive with respect to changing illumination and the reflection properties of the object. Continuity of the surface is not a requirement. The accuracy of the method is limited to about 1:500. It is therefore mostly used for fast surface measurements of lower accuracy, for example as a preliminary to subsequent measurement by phase-shift.

3.7.2.4 Single-camera fringe-projection systems

Measuring systems which operate according to the phase-shift method consist of a fringe projector and a camera with a base separation appropriate for the measurement volume (larger base for bigger volume, but with accuracy decreasing with measurement volume). Camera and projector must be calibrated and spatially oriented to one another. Because of the ambiguities of phase measurement inherent in eqn. 3.98, the systems are frequently combined with a coded light technique (section 3.7.2.3) which provides approximate values for the object surface. Alternatively, fringes of varying wavelength can be projected.

The projector is normally designed using a liquid crystal display (LCD), liquid crystal on silicon (LCOS) or micromirror array (DMD). As a rule, the camera is a digital video camera with up to 2000 x 2000 pixels. The performance of this method is determined mainly by the reflective properties of the surface. Usually, homogeneous, diffusely reflecting surfaces are required. Specular reflections and hot spots must be avoided by preparation of the surface (for example by dusting with white powder) and provision of suitable ambient lighting conditions.

Dynamic fringe projection is applicable only to static objects. Depending on camera and projector used, as well as the required measured point density, measurement frequency lies between 0.2 Hz and 50 Hz. The basic configuration of just one camera typically results in occluded areas of the object which are not therefore measured. The most important advantage of the phase-shift method is the fast measurement of a few hundreds of thousands of surface points. At the same time, however, the problem of thinning and structuring of the measured point cloud arises, as is necessary for further processing of the 3D data, for example for the control of production tools.

A measuring volume between $0.1 \times 0.1 \times 0.1\ m^3$ and $1 \times 1 \times 1\ m^3$ is typical for fringe projection systems. In order to measure larger objects, mobile fringe projection systems are used in combination with photogrammetric methods or mechanical positioning systems for spatial orientation. Relative accuracies of 1:8000 can be achieved (see section 3.7.2.5). Single camera fringe projection systems are offered, amongst others, by AICON/Breuckmann, Steinbichler or Vialux.

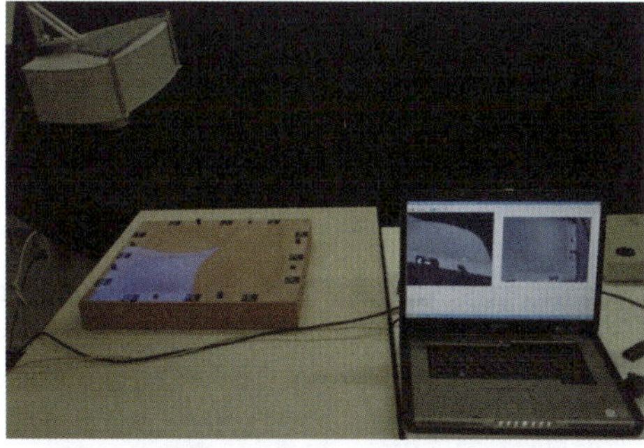

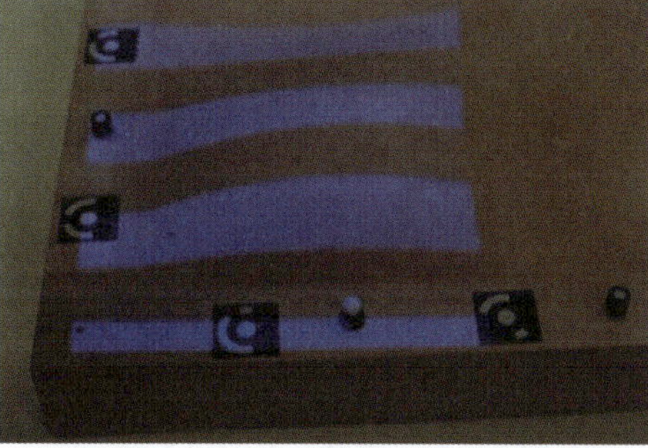

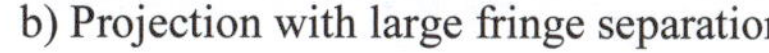

a) Vialux fringe projection system

b) Projection with large fringe separation

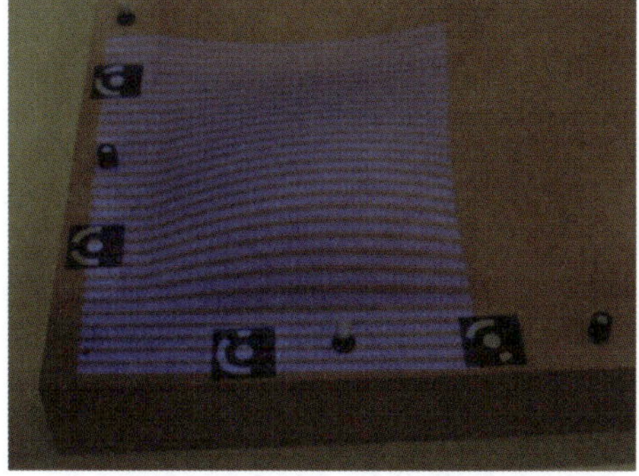

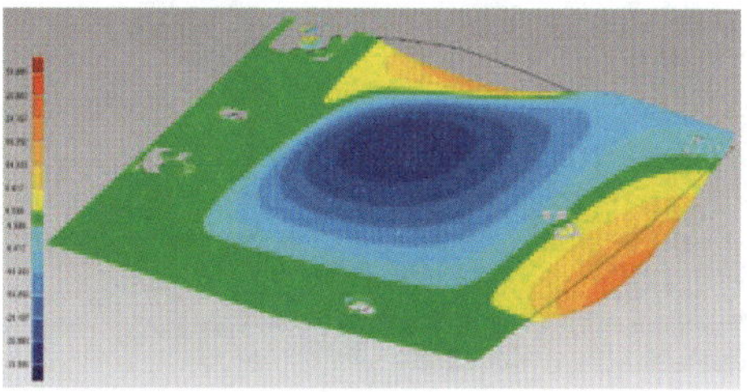

c) Projection with small fringe separation

d) 3D measurement result

Fig. 3.183: Scanning an object with a single-camera, fringe-projection system (Vialux)

Fig. 3.183 shows fringe projection at different wavelengths and the resulting 3D point cloud. Due to the limited measurement volume, the object must be scanned in several parts which are combined into a full model with the aid of common reference points (see section 5.6.2) or through accurate robotic placement (see section 6.5.3.3).

3.7.2.5 Multi-camera fringe-projection systems

In their basic configuration, the fringe projection methods mentioned above use one projector and one CCD camera. Because the projector serves as a component of the measurement system, uncertainties in its geometry adversely affect the results. In addition, for objects with large surface variations there are frequently areas of occlusion and shadow which cannot be observed by a single camera. Furthermore, smooth surfaces often give rise to over illumination or highlights in the image.

Consequently, a number of advantages are offered by multi-camera systems with active fringe projection:

- reduced measuring uncertainty as a result of greater redundancy (number of cameras);
- fewer areas of occlusion or highlights;

- no requirement to calibrate the projector;
- possibility of measuring moving objects using synchronised multi-imaging;
- greater flexibility in measurement by variation of the projected pattern and the relative arrangement of projector and cameras.

Calibration and orientation of multi-camera systems with active illumination follow the principles of test-field calibration (section 7.3.1.2). If required, the orientation parameters of the projector can also be determined in this process since the projector can be regarded as an inverse camera.

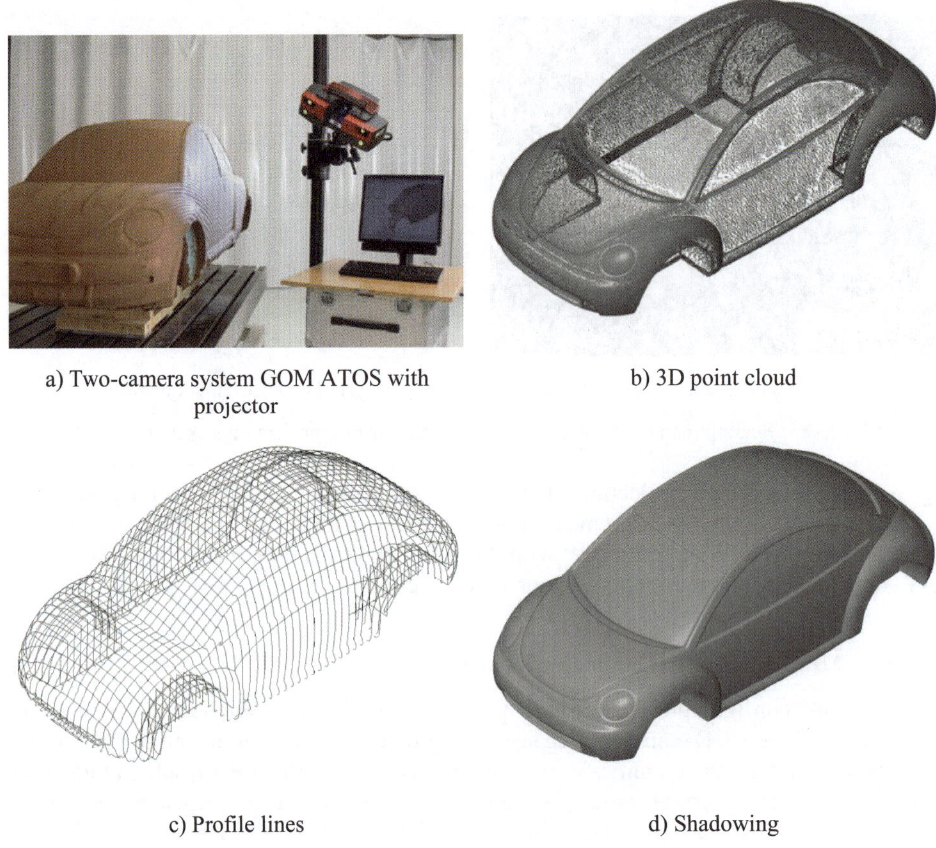

a) Two-camera system GOM ATOS with projector

b) 3D point cloud

c) Profile lines

d) Shadowing

Fig. 3.184: Measurement of a design model by a photogrammetric fringe projection system (GOM)

Fig. 3.184 shows a mobile two-camera system with a pattern projector. Two convergent CCD video cameras are arranged on a fixed base and an pattern projector is mounted between them. The measuring volume of a single recording lies between 12 x 7 x 8 mm^3 and 750 x 500 x 500 mm^3 depending on the system configuration. A relative accuracy between 1:2000 and 1:10 000 can be achieved. This type of measurement system is offered, amongst others, by GOM and AICON/Breuckmann.

Using several spatially distributed camera stations, even more complex and larger objects can be measured. For this purpose the object is prepared with suitable targets that can be measured photogrammetrically to high accuracy and then act as reference points for connecting part models. Then the 3D point clouds generated by each single view can be transformed to a complete object model.

3.7.3 Low-cost consumer-grade range 3D cameras

Recent demand from the gaming and entertainment industry has boosted the development of low-cost consumer-grade range cameras. Whilst the underlying technology has been available for several years, interest in these sensors initially remained low. Interest grew with Microsoft's release of the Kinect, a low-cost motion-sensing input device for their video-game console. As with consumer digital cameras, the mass market effect of this sensor (over 10 million units in circulation), has had a significant impact on the photogrammetric community where 3D sensing systems typically number in the 1000s. The affordability and versatility of the Kinect and similar devices, along with a growing programming community, have led to its use in a wide range of sensing applications such as 3D scene reconstruction and even engineering tasks. However the short (1-6 m) effective maximum working range of this class of sensor, coupled with its limited accuracy (5 to 20 mm) limit, its suitability for many engineering measurement requirements.

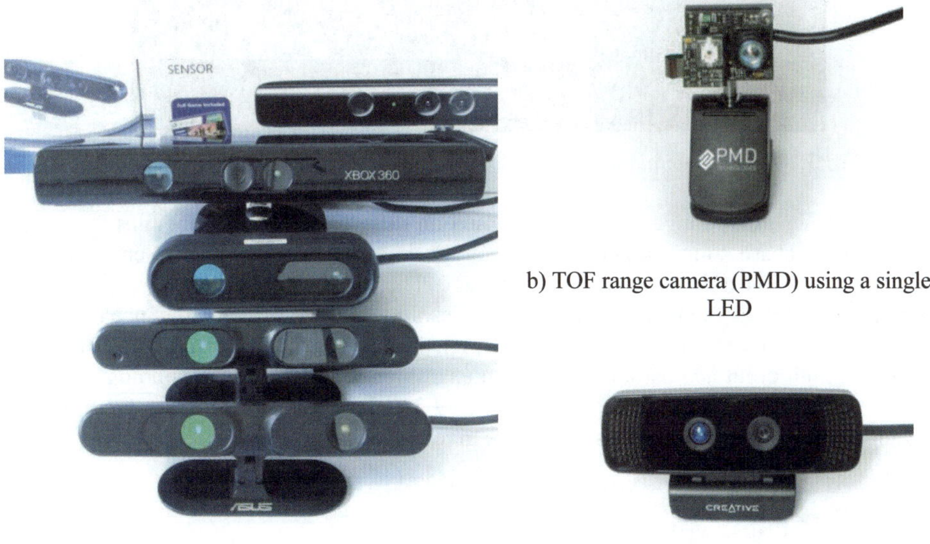

b) TOF range camera (PMD) using a single LED

a) Structured-light range cameras using the PrimeSense chipset, including the Microsoft Kinect

c) TOF range camera with separate RGB camera (Intel)

Fig. 3.185: Selection of low-cost consumer-grade 3D cameras

Consumer grade range cameras fall into two categories: range cameras based on triangulation and range cameras based on the time-of-flight (TOF) principle. In Fig. 3.185a various sensors based on triangulation are shown. They all use the same chipset by

PrimeSense for the depth computation. The technology can be described as structured light using a fixed pattern projection. The pattern is a speckle dot pattern generated by a diffractive optical element and a near-infrared laser diode. The projected pattern is shown in Fig. 3.186. Sensors using TOF are shown in Fig. 3.185b and c. Fig. 3.185b is the PMD CamBoard nano reference design to demonstrate the capabilities of PMD's photonic mixer technology. The camera uses a single LED to emit sinusoidal modulated infrared light. A synchronised imaging chip takes four samples of one period of the sinus signal. This allows the computation of the phase-shift and subsequently depth for each pixel.

Fig. 3.186: Dot pattern projected for structured light using the PrimeSense chipset

While consumer grade range cameras are based on different physical principles to acquire range, they do share some properties. As they are intended for gaming and entertainment they are not optimised for measurement accuracy. These sensors typically have a relatively low resolution (pixel count), for example VGA resolution or less. However, since they were designed to acquire motion they provide a high frame rate, typically up to 60 Hz. This results in a high point acquisition rate. As an example the PrimeSense technology delivers 640x480 pixel at 30 Hz – a point rate of over 9.2 million points a second. Examples of range images acquired with these systems are given in Fig. 5.93.

4 Analytical methods

4.1 Overview

This chapter deals with the analytical methods which are essential for the calculation of image orientation parameters and object information (coordinates, geometric elements). The methods are based on measured image coordinates derived from both analogue image measurement and digital image processing.

Due to their differing importance in practice practical use, the analytical methods of calculation are classified according to the number of images involved. It is common to all methods that the relationship between image information and object geometry is established by the parameters of interior and exterior orientation. Procedures usually have the following stages:

- provision of object information (reference points, distances, geometric elements)
- measurement of image points for orientation (image coordinates)
- calculation of orientation parameters (interior and exterior orientation)
- object reconstruction from oriented images (new points, geometric elements)

Depending on method and application, these stages are performed either sequentially, simultaneously or iteratively in a repeated number of processing steps.

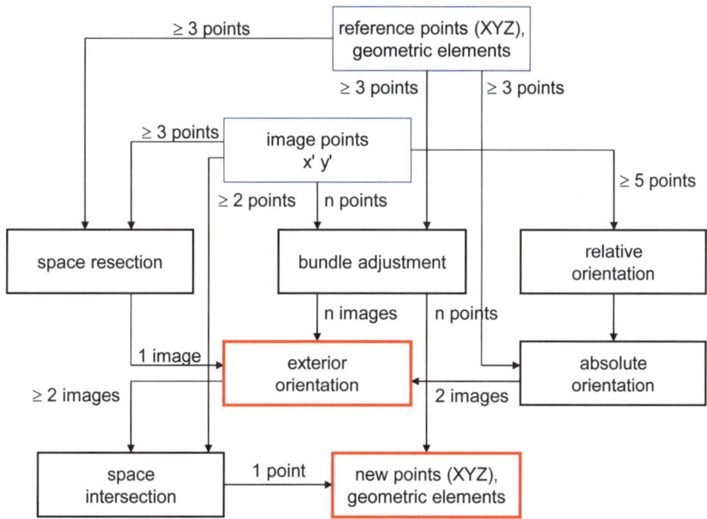

Fig. 4.1: Methods and data flow for orientation and point determination

Fig. 4.1 is a simplified illustration of typical methods for the determination of orientation parameters (exterior only) and for the reconstruction of 3D object geometry (new points, geometric elements). Both procedures are based on measured image coordinates and known object information. It is obvious that 3D point determination for object reconstruction cannot be performed without the parameters of exterior orientation, i.e. position and orientation of an image in space.

If orientation data and coordinates of reference points or measured object points are available, further object reconstruction can be performed according to one or more of the following methods:

- Numerical generation of points and geometric elements:

  Using images of known orientation for the determination of additional object coordinates and geometric elements by, for example, spatial intersection

- Graphical object reconstruction:

  Extraction of graphical and geometric information to create maps, drawings or CAD models

- Rectification or orthophoto production:

  Using various projections, the transformation of the measurement imagery into image-based products such as photo maps, image mosaics and 3D animations

With regard to the number of images involved, the following methods can be identified:

- Single image analysis (see section 4.2):

  Analysis of single images which takes into account additional geometric information and constraints in object space (straight lines, planes, surface models etc.). Here a distinction is made between the calculation of object coordinates and the production of rectified images and orthophotos (Fig. 4.2).

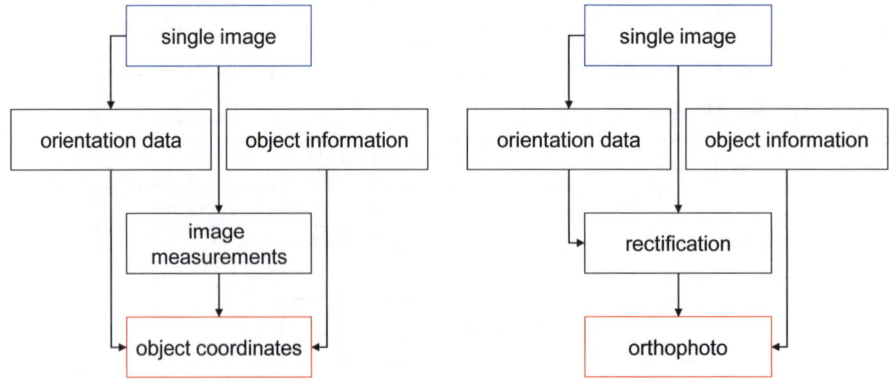

Fig. 4.2: Methods of single image processing

- Stereo image processing (see section 4.3):

  Visual or digital processing of image pairs based on the principles of stereoscopic image viewing and analysis, in particular for the measurement of natural features (non-targeted points) and the capture of free-form surfaces (Fig. 4.3).

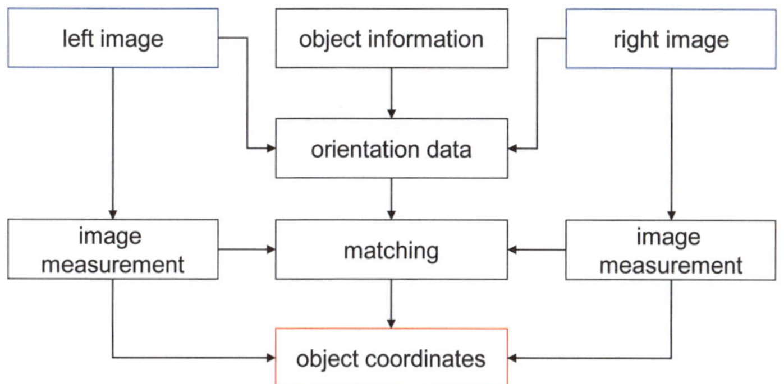

Fig. 4.3: Methods of stereo image processing

- Multi-image processing (see sections 4.4):

  Simultaneous evaluation of an unlimited number of images of an object, e.g. for camera calibration or the reconstruction of complex object geometries.

If the geometric imaging model is appropriately modified, the techniques of orientation and 3D object reconstruction can be extended to panoramic applications (section 4.5) and multimedia photogrammetry (section 4.6). In addition to traditional methods using discrete points, analytical processing of geometric elements (straight line, plane, cylinder etc.) is being increasingly employed.

4.2 Processing of single images

4.2.1 Exterior orientation

4.2.1.1 Standard case

The *exterior orientation* consists of six parameters which describe the spatial position and orientation of the camera coordinate system with respect to the global object coordinate system (Fig. 4.4). The standard case in aerial photography of a horizontal image plane is also used as the basic model in close-range photogrammetry. Terrestrial photogrammetry is covered in the next section as a special case.

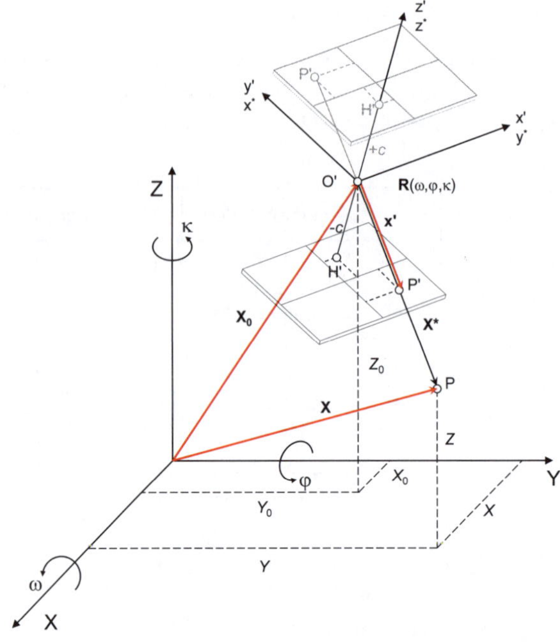

Fig. 4.4: Exterior orientation and projective imaging

The camera coordinate system has its origin at the perspective centre of the image (see section 2.1.1). It is further defined by reference features fixed in the camera (fiducial marks, réseau, sensor system). It can therefore be reconstructed from the image and related to an image measuring device (comparator). In photogrammetry this procedure is often known as *reconstruction of interior orientation*[1].

The spatial position of the image coordinate system is defined by the vector $\mathbf{X}_0$ from the origin to the perspective centre O'. The orthogonal rotation matrix $\mathbf{R}$ defines the angular orientation in space. It is the combination of three independent rotations ω,φ,κ about the coordinate axes X,Y,Z (see section 2.2.2.1).

$$\mathbf{X}_0 = \begin{bmatrix} X_0 \\ Y_0 \\ Z_0 \end{bmatrix} \qquad \text{: position of perspective centre} \qquad (4.1)$$

$$\mathbf{R} = \mathbf{R}_\omega \cdot \mathbf{R}_\varphi \cdot \mathbf{R}_\kappa$$

$$\mathbf{R} = \begin{bmatrix} r_{11} & r_{12} & r_{13} \\ r_{21} & r_{22} & r_{23} \\ r_{31} & r_{32} & r_{33} \end{bmatrix} \qquad \text{: rotation matrix} \qquad (4.2)$$

[1] Commonly abbreviated to *interior orientation*.

The elements of the rotation matrix r_{ij} can be defined either as trigonometric functions of the three rotation angles or as functions of four algebraic variables (see section 2.2.2.1).

With given parameters of exterior orientation, the direction from the perspective centre O' to the image point P' (image vector **x'**) can be transformed into an absolutely oriented spatial ray from the perspective centre to the object point P.

The exterior orientation further describes the spatial transformation (rotation and shift) from camera coordinates x^*, y^*, z^* into object coordinates X,Y,Z (see Fig. 4.4):

$$\begin{aligned}
\mathbf{X} &= \mathbf{X}_0 + \mathbf{R} \cdot \mathbf{x}^* = \mathbf{X}_0 + \mathbf{X}^* \\
X &= X_0 + r_{11}x^* + r_{12}y^* + r_{13}z^* \\
Y &= Y_0 + r_{21}x^* + r_{22}y^* + r_{23}z^* \\
Z &= Z_0 + r_{31}x^* + r_{32}y^* + r_{33}z^*
\end{aligned} \tag{4.3}$$

4.2.1.2 Special case of terrestrial photogrammetry

For the special case of conventional terrestrial photogrammetry, the camera axis is approximately horizontal. In order to avoid singularities in trigonometric functions, the rotation sequence must either be re-ordered (see section 2.2.2.1) or the image coordinate system must be defined by axes x' and z' (instead of x' and y', see section 2.1.1). In this case, image acquisition systems which provide angle measurements, e.g. video theodolites, can use rotation angles ω (tilt about horizontal axis), κ (roll around optical axis) and φ or α (azimuth) instead of the standard sequence ω,φ,κ (Fig. 4.5). It must be remembered here that geodetic angles are positive clockwise.

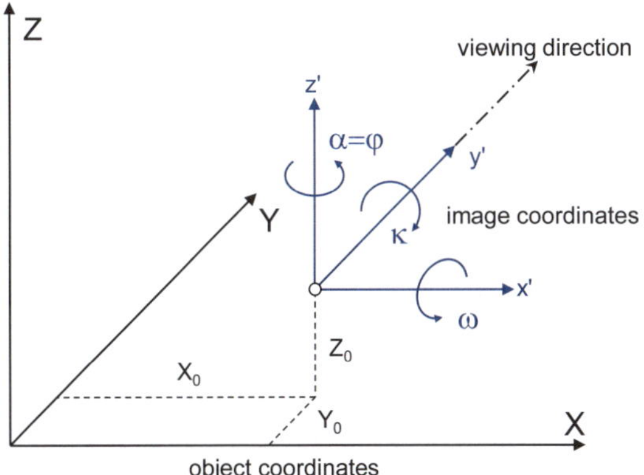

Fig. 4.5: Exterior orientation for terrestrial photogrammetry

The modified rotation order φ,ω,κ leads to the following rotation matrix (compare with eqn. 2.32):

$$\mathbf{R}_{terr.} = \begin{bmatrix} r_{11} & r_{12} & r_{13} \\ r_{21} & r_{22} & r_{23} \\ r_{31} & r_{32} & r_{33} \end{bmatrix}$$
$$= \begin{bmatrix} \cos\varphi\cos\kappa - \sin\varphi\sin\omega\sin\kappa & -\sin\varphi\cos\omega & \cos\varphi\sin\kappa + \sin\varphi\sin\omega\cos\kappa \\ \sin\varphi\cos\kappa + \cos\varphi\sin\omega\sin\kappa & \cos\varphi\cos\omega & \sin\varphi\sin\kappa - \cos\varphi\sin\omega\cos\kappa \\ -\cos\omega\sin\kappa & \sin\omega & \cos\omega\cos\kappa \end{bmatrix} \tag{4.4}$$

4.2.2 Collinearity equations

The central projection in space is at the heart of many photogrammetric calculations. Thus the coordinates of an object point P can be derived from the position vector to the perspective centre $\mathbf{X}_0$ and the vector from the perspective centre to the object point $\mathbf{X}^*$ (Fig. 4.4):

$$\mathbf{X} = \mathbf{X}_0 + \mathbf{X}^* \tag{4.5}$$

The vector $\mathbf{X}^*$ is given in the object coordinate system. The image vector $\mathbf{x}'$ may be transformed into object space by rotation matrix $\mathbf{R}$ and a scaling factor m. Then, since it is in the same direction as $\mathbf{X}^*$:

$$\mathbf{X}^* = m \cdot \mathbf{R} \cdot \mathbf{x}' \tag{4.6}$$

Hence, the projection of an image point into a corresponding object point is given by: [1]

$$\mathbf{X} = \mathbf{X}_0 + m \cdot \mathbf{R} \cdot \mathbf{x}'$$
$$\begin{bmatrix} X \\ Y \\ Z \end{bmatrix} = \begin{bmatrix} X_0 \\ Y_0 \\ Z_0 \end{bmatrix} + m \cdot \begin{bmatrix} r_{11} & r_{12} & r_{13} \\ r_{21} & r_{22} & r_{23} \\ r_{31} & r_{32} & r_{33} \end{bmatrix} \cdot \begin{bmatrix} x' \\ y' \\ z' \end{bmatrix} \tag{4.7}$$

The scale factor m is an unknown value which varies for each object point. If only one image is available then only the direction to an object point P can be determined but not its absolute position in space. The 3D coordinates of P can only be computed if this spatial direction intersects another geometrically known element (e.g. intersection with a second ray from another image or intersection with a given surface in space).

By inverting equation (4.7), adding the principal point H'(x'_0,y'_0) and introducing correction terms $\mathbf{\Delta x'}$ (image distortion parameters), the image coordinates are given by (see also eqn. 3.51):

[1] In equations (4.7), z' appears where, on the basis of Fig. 4.4, one would expect to see –c. For reasons of generality, and in order to extend the algebra to certain types of imaging systems such as panoramic cameras, the value z' appears in equations throughout.

$$\mathbf{x'}-\mathbf{x'}_0-\mathbf{\Delta x'}=\frac{1}{m}\cdot\mathbf{R}^{-1}\cdot(\mathbf{X}-\mathbf{X}_0)$$

$$\begin{bmatrix} x'-x'_0-\Delta x \\ y'-y'_0-\Delta y \\ z' \end{bmatrix}=\frac{1}{m}\cdot\begin{bmatrix} r_{11} & r_{21} & r_{31} \\ r_{12} & r_{22} & r_{32} \\ r_{13} & r_{23} & r_{33} \end{bmatrix}\cdot\begin{bmatrix} X-X_0 \\ Y-Y_0 \\ Z-Z_0 \end{bmatrix} \tag{4.8}$$

Note that the inverse rotation matrix is equal to its transpose. By dividing the first and second equations by the third equation, the unknown scaling factor m is eliminated and the *collinearity equations* follow:

$$\begin{aligned} x' &= x'_0+z'\cdot\frac{r_{11}\cdot(X-X_0)+r_{21}\cdot(Y-Y_0)+r_{31}\cdot(Z-Z_0)}{r_{13}\cdot(X-X_0)+r_{23}\cdot(Y-Y_0)+r_{33}\cdot(Z-Z_0)}+\Delta x' \\ y' &= y'_0+z'\cdot\frac{r_{12}\cdot(X-X_0)+r_{22}\cdot(Y-Y_0)+r_{32}\cdot(Z-Z_0)}{r_{13}\cdot(X-X_0)+r_{23}\cdot(Y-Y_0)+r_{33}\cdot(Z-Z_0)}+\Delta y' \end{aligned} \tag{4.9}$$

These equations describe the transformation of object coordinates (X,Y,Z) into corresponding image coordinates (x',y') as functions of the interior orientation parameters $(x'_0,y'_0,c,\Delta x',\Delta y')$ and exterior orientation parameters $(X_0,Y_0,Z_0,\omega,\varphi,\kappa)$ of *one* image.

An alternative form is given if the object coordinate system is transformed by shifting to the perspective centre and orienting parallel to the image coordinate system. Within the local coordinate system which results, object coordinates are denoted by x^*,y^*,z^* (see Fig. 4.4):

$$\begin{bmatrix} x^* \\ y^* \\ z^* \end{bmatrix}=\mathbf{R}^{-1}\begin{bmatrix} X-X_0 \\ Y-Y_0 \\ Z-Z_0 \end{bmatrix} \quad \text{: spatial translation and rotation} \tag{4.10}$$

Multiplying out and substitution results in the following transformation equation:

$$\begin{aligned} \mathbf{x}^* &= \mathbf{R}^{-1}\cdot(\mathbf{X}-\mathbf{X}_0) \\ &= \mathbf{R}^{-1}\cdot\mathbf{X}-\mathbf{R}^{-1}\cdot\mathbf{X}_0 \\ &= \mathbf{X}_0^*+\mathbf{R}^*\cdot\mathbf{X} \end{aligned} \quad \text{: spatial translation and rotation} \tag{4.11}$$

where $\mathbf{R}^*=-\mathbf{R}^{-1}$.

By introducing the image scale $1/m$ and corrections for shift of principal point and image errors, the collinearity equations for image coordinates are again obtained:

$$\begin{bmatrix} x' \\ y' \end{bmatrix} = \frac{z'}{z^*}\begin{bmatrix} x^* \\ y^* \end{bmatrix} + \begin{bmatrix} x'_0 \\ y'_0 \end{bmatrix} + \begin{bmatrix} \Delta x' \\ \Delta y' \end{bmatrix} \quad \text{: projection into image plane} \tag{4.12}$$

$$m = \frac{z^*}{z'}$$

In the following, principal point and image error corrections are ignored. By multiplying out eqn. 4.12 and multiplying through with $1/Z^*_0$, the following spatial projection equations are obtained:

$$x' = \frac{\frac{z' r_{11}}{Z_0^*} X + \frac{z' r_{21}}{Z_0^*} Y + \frac{z' r_{31}}{Z_0^*} Z + \frac{z'}{Z_0^*} X_0^*}{\frac{r_{13}}{Z_0^*} X + \frac{r_{23}}{Z_0^*} Y + \frac{r_{33}}{Z_0^*} Z + \frac{Z_0^*}{Z_0^*}}$$

$$y' = \frac{\frac{z' r_{12}}{Z_0^*} X + \frac{z' r_{22}}{Z_0^*} Y + \frac{z' r_{32}}{Z_0^*} Z + \frac{z'}{Z_0^*} Y_0^*}{\frac{r_{13}}{Z_0^*} X + \frac{r_{23}}{Z_0^*} Y + \frac{r_{33}}{Z_0^*} Z + \frac{Z_0^*}{Z_0^*}} \tag{4.13}$$

If the fractional terms are replaced by coefficients a',b',c', the three dimensional image equations are obtained as follows (see for comparison eqn. 2.20 and eqn. 4.22):

$$x' = \frac{a'_0 + a'_1 X + a'_2 Y + a'_3 Z}{1 + c'_1 X + c'_2 Y + c'_3 Z}$$

$$y' = \frac{b'_0 + b'_1 X + b'_2 Y + b'_3 Z}{1 + c'_1 X + c'_2 Y + c'_3 Z} \tag{4.14}$$

The collinearity equations demonstrate clearly that each object point is projected into a unique image point, if it is not occluded by other object points. The equations effectively describe image generation inside a camera by the geometry of a central projection.

The equations (4.9) form the fundamental equations of analytical photogrammetry. It is important to note that, since the observed measurements stand alone on the left-hand side, these equations are suitable for direct use as observation equations in an over-determined least-squares adjustment (see section 2.4.2). For example, the collinearity equations are used to set up the equation system for spatial intersection (section 4.4.7.1), space resection (section 4.2.3) and bundle triangulation (section 4.4). Additionally, they offer the mathematical basis for the generation of orthophotos and the principle of stereo plotting systems (see section 6.3.1).

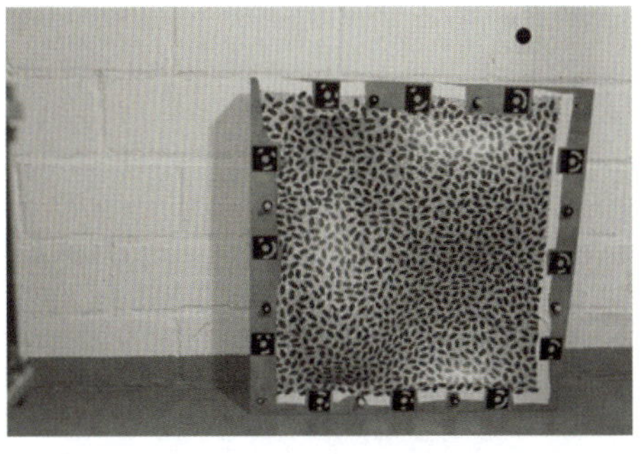

a) Original image (taken by Nikon D2X, f = 24 mm)

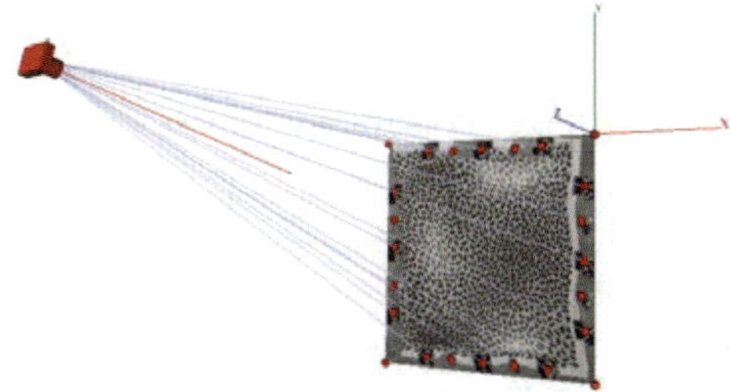

b) Position of the camera (and image) in the object coordinate system

Fig. 4.6: Exterior orientation of an image

Example 4.1:

The following data are available for the image illustrated in Fig. 4.6:

Interior orientation:	$c = -24.2236$ mm	$x'_0 = 0.0494$ mm	$y'_0 = -0.2215$ mm
Exterior orientation:	$X_0 = -471.89$ mm	$Y_0 = 11.03$ mm	$Z_0 = 931.07$ mm
	$\omega = -13.059°$	$\varphi = -4.440°$	$\kappa = 0.778°$
Object coordinates:	$X_1 = -390.93$ mm	$Y_1 = -477.52$ mm	$Z_1 = 0.07$ mm
	$X_2 = -101.54$ mm	$Y_2 = -479.19$ mm	$Z_2 = 0.10$ mm

Image coordinates of the object points are required in accordance with equations (2.28) and (4.9).

Image coordinates:	$x'_1 = 0.0104$ mm	$y'_1 = -6.5248$ mm
	$x'_2 = 6.7086$ mm	$y'_2 = -6.5162$ mm

From the object and image separations of both points, an approximate image scale is calculated as $m \approx 42$.

4.2.3 Space resection

Orientation of single images is taken to mean, in the first instance, the process of calculating the parameters of exterior orientation. Since direct determination, for example by angle or distance measurement, is not usually possible, methods of indirect orientation are employed. These make use of XYZ reference points whose image coordinates may be measured in the image. Common calculation procedures can be divided into two groups:

1. Calculation of exterior orientation based on collinearity equations:

   The method of space resection provides a non-linear solution that requires a minimum of three XYZ reference points in object space and approximate values for the unknown orientation parameters (see also section 4.2.3.1).

2. Calculation of exterior orientation based on projective relations:

   The most popular method in this group is the Direct Linear Transformation (DLT). It requires a minimum of five XYZ reference points, but provides a direct solution without the need for approximate values (see also section 4.2.4). Linear methods in projective geometry function in a similar way.

4.2.3.1 Space resection with known interior orientation

Space resection is used to compute the exterior orientation of a single image. The procedure requires known *XYZ* coordinates of at least three object points P_i which do not lie on a common straight line. The bundle of rays through the perspective centre from the reference points can fit the corresponding points in image plane P'_i in only one, unique, position and orientation of the image (Fig. 4.7).[1]

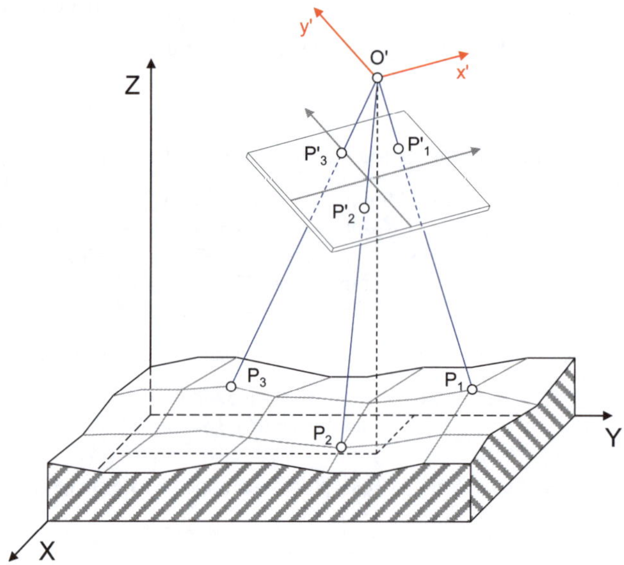

Fig. 4.7: Space resection

Using the measured image coordinates of the reference points, and with known parameters of interior orientation, the following system of correction equations can be derived from the collinearity equations (4.9):

$$\begin{aligned} x'+vx' &= F(\underline{X_0, Y_0, Z_0, \omega, \varphi, \kappa}, x'_0, c, \Delta x', X, Y, Z) \\ y'+vy' &= F(\underline{X_0, Y_0, Z_0, \omega, \varphi, \kappa}, y'_0, c, \Delta y', X, Y, Z) \end{aligned} \tag{4.15}$$

Function F is a representation for equations (4.9) in which the underlined values are introduced as unknowns. This system can be linearised at approximate values by Taylor-series expansion and solved by least-squares adjustment.

---

[1] Strictly speaking, multiple solutions exist with only three reference points but a single unique solution is possible if at least one more reference point is added (further information in section 4.2.3.4).

Each of the measured image points provides two linearised correction equations:

$$
\begin{aligned}
vx'_i &= \left(\frac{\partial x'}{\partial X_0}\right)^0 dX_0 + \left(\frac{\partial x'}{\partial Y_0}\right)^0 dY_0 + \left(\frac{\partial x'}{\partial Z_0}\right)^0 dZ_0 \\
&+ \left(\frac{\partial x'}{\partial \omega}\right)^0 d\omega + \left(\frac{\partial x'}{\partial \varphi}\right)^0 d\varphi + \left(\frac{\partial x'}{\partial \kappa}\right)^0 d\kappa - (x'_i - x'^0_i) \\
vy'_i &= \left(\frac{\partial y'}{\partial X_0}\right)^0 dX_0 + \left(\frac{\partial y'}{\partial Y_0}\right)^0 dY_0 + \left(\frac{\partial y'}{\partial Z_0}\right)^0 dZ_0 \\
&+ \left(\frac{\partial y'}{\partial \omega}\right)^0 d\omega + \left(\frac{\partial y'}{\partial \varphi}\right)^0 d\varphi + \left(\frac{\partial y'}{\partial \kappa}\right)^0 d\kappa - (y'_i - y'^0_i)
\end{aligned}
\tag{4.16}
$$

Here x_i' and y_i' are the measured image coordinates, x'^0_i and y'^0_i are the image coordinates which correspond to the approximate orientation parameters. If quaternions are used to define rotations (see eqn. 2.33) then the partial derivatives with respect to ω,φ,κ, must be replaced by derivatives with respect to the algebraic parameters *a, b, c, d*.

Simplification of the collinearity equations (4.9), by substituting k_x and k_y for the numerators and N for the denominator, leads to

$$
x' = x'_0 + z'\frac{k_X}{N} + \Delta x' \qquad\qquad y' = y'_0 + z'\frac{k_Y}{N} + \Delta y' \tag{4.17}
$$

from which the derivatives of (4.16) are given by:

$$
\begin{aligned}
\frac{\partial x'}{\partial X_0} &= \frac{z'}{N^2} \cdot (r_{13} k_X - r_{11} N) \\
\frac{\partial x'}{\partial Y_0} &= \frac{z'}{N^2} \cdot (r_{23} k_X - r_{21} N) \\
\frac{\partial x'}{\partial Z_0} &= \frac{z'}{N^2} \cdot (r_{33} k_X - r_{31} N) \\
\frac{\partial x'}{\partial \omega} &= \frac{z'}{N} \cdot \left\{ \frac{k_X}{N} \cdot \left[r_{33}(Y - Y_0) - r_{23}(Z - Z_0) \right] - r_{31}(Y - Y_0) + r_{21}(Z - Z_0) \right\} \\
\frac{\partial x'}{\partial \varphi} &= \frac{z'}{N} \cdot \left\{ \frac{k_X}{N} \cdot \left[k_Y \cdot \sin\kappa - k_X \cdot \cos\kappa \right] - N \cdot \cos\kappa \right\} \\
\frac{\partial x'}{\partial \kappa} &= \frac{z'}{N} \cdot k_Y
\end{aligned}
\tag{4.18a}
$$

$$\frac{\partial y'}{\partial X_0} = \frac{z'}{N^2} \cdot (r_{13}k_Y - r_{12}N) \tag{4.18b}$$

$$\frac{\partial y'}{\partial Y_0} = \frac{z'}{N^2} \cdot (r_{23}k_Y - r_{22}N)$$

$$\frac{\partial y'}{\partial Z_0} = \frac{z'}{N^2} \cdot (r_{33}k_Y - r_{32}N)$$

$$\frac{\partial y'}{\partial \omega} = \frac{z'}{N} \cdot \left\{ \frac{k_Y}{N} \cdot [r_{33}(Y - Y_0) - r_{23}(Z - Z_0)] - r_{32}(Y - Y_0) + r_{22}(Z - Z_0) \right\}$$

$$\frac{\partial y'}{\partial \varphi} = \frac{z'}{N} \left\{ \frac{k_Y}{N} [k_Y \sin\kappa - k_X \cos\kappa] + N \sin\kappa \right\}$$

$$\frac{\partial y'}{\partial \kappa} = -\frac{z'}{N} \cdot k_X$$

The coefficients r_{ij} can be derived from the rotation matrix **R** according to (2.31).

Example 4.2:

The following data are available for the space resection of the image in Fig. 4.6:

Interior orientation:
$c = -24.2236$ mm $x'_0 = 0.0494$ mm $y'_0 = -0.2215$ mm
$A_1 = -1.697 \cdot 10^{-4}$ $A_2 = 2.944 \cdot 10^{-7}$ $A_3 = -1.258 \cdot 10^{-10}$ $r_0 = 8.8$mm
$B_1 = 1.631 \cdot 10^{-5}$ $B_2 = -8.073 \cdot 10^{-6}$
$C_1 = -1.131 \cdot 10^{-4}$ $C_2 = 2.717 \cdot 10^{-5}$

Reference points:
$X_1 = -390.93$ mm $Y_1 = -477.52$ mm $Z_1 = 0.07$ mm
$X_2 = -101.5$ mm $Y_2 = -479.19$ mm $Z_2 = 0.11$ mm
$X_3 = -23.22$ mm $Y_3 = -256.85$ mm $Z_3 = -0.09$ mm
$X_4 = -116.88$ mm $Y_4 = -21.03$ mm $Z_4 = 0.07$ mm
$X_5 = -392.17$ mm $Y_5 = -21.46$ mm $Z_5 = 0.11$ mm
$X_6 = -477.54$ mm $Y_6 = -237.65$ mm $Z_6 = -0.16$ mm

Corrected image coordinates:
$x'_1 = -0.0395$ mm $y'_1 = -6.3033$ mm
$x'_2 = 6.6590$ mm $y'_2 = -6.2948$ mm
$x'_3 = 9.0086$ mm $y'_3 = -1.3473$ mm
$x'_4 = 7.3672$ mm $y'_4 = 4.5216$ mm
$x'_5 = 0.2936$ mm $y'_5 = 4.7133$ mm
$x'_6 = -2.0348$ mm $y'_6 = -0.7755$ mm

The parameters of exterior orientation are required:

Exterior orientation:
$X_0 = -471.89$ mm $Y_0 = 11.00$ mm $Z_0 = 931.11$ mm

$$\mathbf{R} = \begin{bmatrix} 0.99691 & -0.01353 & -0.07741 \\ 0.03071 & 0.97382 & 0.22524 \\ 0.07234 & -0.22692 & 0.97122 \end{bmatrix}$$

$\omega = -13.057°$ $\varphi = -4.440°$ $\kappa = 0.778°$

4.2.3.2 Space resection with unknown interior orientation

For images from cameras with unknown parameters of interior orientation (e.g. amateur cameras) the number of unknown parameters, ignoring distortion in the first instance, increases by 3 (c, x'_0, y'_0) to a total of 9. Two more additional reference points, providing 4 additional image observation equations, are required for the solution (a minimum of 5 reference points in total).

If all reference points lie approximately on a plane, then the normal system of equations for the resection is singular, since the problem can be solved by an 8 parameter projective transformation between image and object planes (see equations 2.17 and 4.42). However, if one of the unknown parameters, such as the principal distance, is fixed to an arbitrary value, a unique solution can be computed.

If a suitable spatial distribution of object points is available, the space resection approach can be used to calibrate the parameters of interior orientation c, x'_0, y'_0, $\Delta x'$, $\Delta y'$ from only one image. The number of elements to be determined increases to a total of 11 if the parameters A_1 and A_2 for radial distortion are introduced (6 for exterior and 5 for interior orientation). In this case a minimum of 6 spatially distributed XYZ reference points are required.

4.2.3.3 Approximate values for resection

In some situations, approximate values for the unknown parameters of exterior orientation can be readily determined by one of the following methods:

- Approximate values by direct measurement:

  Approximate orientation values can possibly be measured on site, for example by geodetic methods or the use of GPS and INS sensors. The camera position can often be determined by simple means such as estimation off a site plan. Even rotation angles can be estimated sufficiently well in simple configurations, for example without oblique or rolled views.

- Small image rotations:

  If the coordinate axes of the image system are approximately parallel to the object coordinate system, initial rotation angles can be approximated by the value zero. The parameters for translation X_0 and Y_0 can be estimated from the centroid of the reference points; the object distance (Z–Z_0) can be determined from the principal distance and approximately known image scale.

- General solution:

  A general solution for approximate values of the six parameters of outer orientation can be derived with the aid of three reference points as the next method explains.

4.2.3.4 Resection with minimum object information

In the general case of arbitrary position and orientation of an image, approximate values can no longer be easily estimated but must be computed. This can be performed according to the following scheme which may also be called space resection with minimum object information. It is noted that other solutions, including algebraic solutions, exist but are not discussed here.

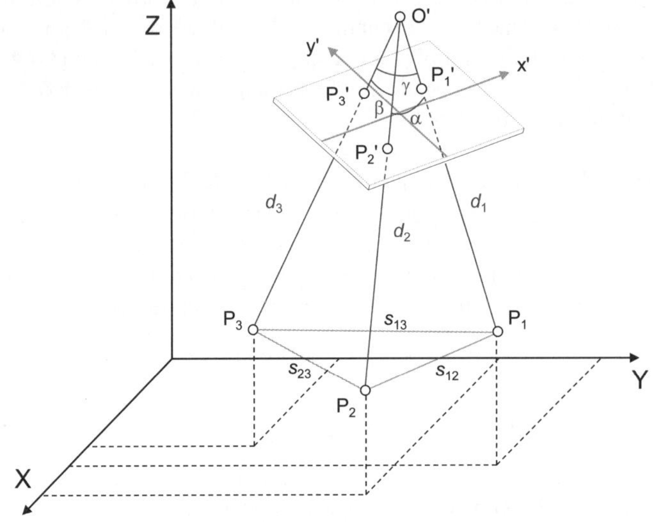

Fig. 4.8: Tetrahedron for space resection

Given a minimum of three XYZ reference points, the position of the perspective centre O' is first determined. As shown in Fig. 4.8 a tetrahedron can be formed by the perspective centre and the three object points. From the simple properties of triangles, the angles α,β,γ of the tetrahedrons $O'P'_1P'_2P'_3$ and $O'P_1P_2P_3$ can be calculated from the measured image coordinates of P'_1, P'_2 and P'_3 and the principal distance c. However, the side lengths of the tetrahedron, d_i, $i = 1 \ldots 3$, necessary to complete the location of O', remain unknown at this stage.

If each of the tetrahedron sides is rotated into the plane of the three object points, the configuration of Fig. 4.9 is obtained. Taking as an example the triangle formed by object points P_1, P_2 and the required perspective centre O', it is obvious that a circle K_1 can be constructed from known distance s_{12} and angle α calculated previously, and that O' lies on this circle. Similarly, circle K_2 can be constructed from s_{23} and circle K_3 from s_{13}. Three spindle tori shapes are formed when the three circles are rotated about their corresponding chords P_1P_2, P_2P_3, and P_3P_1 and their intersection points provide possible solutions for the spatial position of O'.

However, rather than attempting to compute the intersection of these relatively complex shapes, the following iterative search strategy estimates the unknown side lengths of the tetrahedron and from these the perspective centre can be more simply calculated as described below. The search is made as follows. Starting near P_1 in circle K_1, a test point R is stepped around the circle and distances d_{R1} and d_{R2} calculated at test positions. Distance d_{R1} is an estimate of side length d_1 and d_{R2} an estimate of side length d_2. At each test position, distance d_{R2} is transferred into circle K_2 where a corresponding value d_{R3} can be calculated from angle β and the known distance s_{23}. Finally d_{R3} is transferred into circle K_3 where, in a similar way, distance d_1 is again estimated as value d'_{R1}. At the end of the loop a difference $\Delta d_1 = d_{R1} - d'_{R1}$ results. If R is stepped further around the circle until it approaches P_2, the sign of the difference at some position will change. At this position sufficiently good approximate values for the side lengths d_i are available.

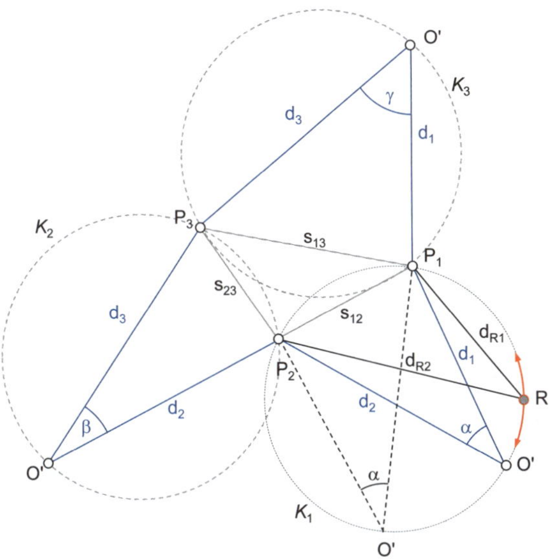

Fig. 4.9: Approximate values for space resection

When transferred into circle K_2, d_{R2} in general creates two possible positions for O' and therefore two possible values of distance d_{R3}. When transferred into circle K_3 each value of d_{R3} generates two possible positions for O' which therefore leads to a total number of four possible solutions for the position of the perspective centre. To ensure that all possible solutions are investigated, and correct solutions with $\Delta d_1 = 0$ are found, the circles must be searched in order of increasing size, starting with the smallest.

Now the rotation and translation of the image coordinate system with respect to the object coordinate system can be determined. Using the side lengths estimated above for one of the solutions, the coordinates of reference points P_i, $i = 1\ldots3$, are calculated in the camera coordinate system xyz, with origin O', which coincides with the image coordinate system x'y'z' (Fig. 4.10):

$$\mathbf{x}_i = \frac{d_i}{d_i{}'} \cdot \mathbf{x'}_i \qquad \text{where } d' = \sqrt{x'^2 + y'^2 + z'^2} \tag{4.19}$$

$$x_i = d_i \frac{x'_i}{d'_i} \qquad y_i = d_i \frac{y'_i}{d'_i} \qquad z_i = d_i \frac{z'_i}{d'_i}$$

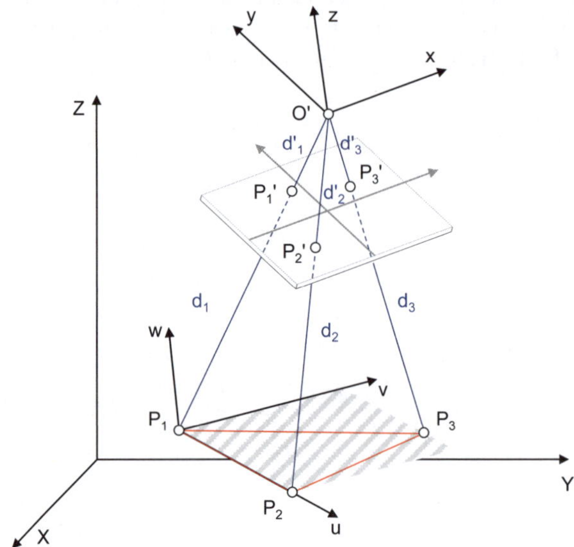

Fig. 4.10: Similarity transformation for space resection

Approximate values for the rotation parameters can then be derived by a spatial similarity transformation (see section 2.2.2.2), since the coordinates of P_1, P_2, P_3 are known in both the xyz and the XYZ systems (Fig. 4.10).

$$\mathbf{X}_i = \mathbf{X}_0 + m \cdot \mathbf{R} \cdot \mathbf{x}_i \quad \text{for } i = 1,2,3 \tag{4.20}$$

Since the distances, P_iP_{i+1} are the same in both systems, the scale factor, m, will be unity. The rotation matrix **R** rotates the xyz coordinate system parallel to the XYZ system. Finally the required translation, $\mathbf{X}_0$, may be found using the xyz coordinates $\mathbf{x}_i$ of a reference point, P_i, and rotated by **R**. For example, using P_1:

$$\mathbf{X}_0 = \mathbf{X}_1 - \mathbf{R} \cdot \mathbf{x}_1 \tag{4.21}$$

For the reasons given above, the spatial position of O', and the corresponding rotation matrix, are not unique. The ambiguity can be resolved by use of a fourth reference point. Its transformed coordinates, found from (4.9), may be compared with the reference coordinates. A good match indicates the correct solution.

The approximate values found in this way can then initiate a linearized solution to generate optimized values. The normal system of equations is set up as described elsewhere and corrective additions added to the solutions in a sequence of iterative calculations until these additions lie below a threshold value.

$$
\begin{aligned}
X_0^{k+1} &= X_0^k + dX_0^k \qquad & \omega^{k+1} &= \omega^k + d\omega^k \\
Y_0^{k+1} &= Y_0^k + dY_0^k \qquad & \varphi^{k+1} &= \varphi^k + d\varphi^k \\
Z_0^{k+1} &= Z_0^k + dZ_0^k \qquad & \kappa^{k+1} &= \kappa^k + d\kappa^k
\end{aligned}
$$

Amongst other applications, space resection is used to compute initial orientation values from approximate object point coordinates (see section 4.4.4.1).

4.2.3.5 Quality measures

In addition to the accuracy of image coordinate measurement, the quality of the resection depends on the number and distribution of reference points. Measured image points should ideally fill the image format. If all the reference points are located on or close to a common straight line, the normal system of equations becomes singular or numerically weak. Similarly there is no solution if object points and perspective centre are all located on the same *danger surface* such as a cylinder (see also section 4.3.3.6).

As in other adjustment problems, the a posteriori standard deviation of unit weight s_0 can be used as a quality criterion for the resection. It represents the internal accuracy of the observations, in this case the measured image coordinates. In addition the standard deviations of the estimated orientation parameters can be analysed. They can be derived from the covariance matrix and hence depend on s_0.

4.2.4 Linear orientation methods

4.2.4.1 Direct linear transformation (DLT)

Using the *Direct Linear Transformation* (DLT) it is possible, by solving a linear system of equations, to determine the orientation of an image without the need for approximate initial values. The method is based on the collinearity equations, extended by an affine transformation of the image coordinates. No image coordinate system fixed in the camera is required.

The transformation equation of the DLT is given by:

$$
\begin{aligned}
x &= \frac{L_1 X + L_2 Y + L_3 Z + L_4}{L_9 X + L_{10} Y + L_{11} Z + 1} \\
y &= \frac{L_5 X + L_6 Y + L_7 Z + L_8}{L_9 X + L_{10} Y + L_{11} Z + 1}
\end{aligned}
\tag{4.22}
$$

Here x and y are the measured comparator or image coordinates and X,Y,Z are the 3D coordinates of the reference points. The coefficients L_1 to L_{11} are the DLT parameters to be estimated and from these the parameters of interior orientation (3) and exterior orientation (6) can be derived. The two remaining elements describe shearing and scaling of the affine transformation. By re-arrangement of eqn. 4.22 the following linear system is obtained:

$$\begin{aligned} &L_1X + L_2Y + L_3Z + L_4 - xL_9X - xL_{10}Y - xL_{11}Z - x = 0 \\ &L_5X + L_6Y + L_7Z + L_8 - yL_9X - yL_{10}Y - yL_{11}Z - y = 0 \end{aligned} \tag{4.23}$$

In order to solve this system of equations for n reference points ($n \geq 6$) according to the usual model

$$\mathbf{v} = \mathbf{A} \cdot \hat{\mathbf{x}} - \mathbf{l}$$

the design matrix **A** is set up as follows:

$$\underset{n,u}{\mathbf{A}} = \begin{bmatrix} X_1 & Y_1 & Z_1 & 1 & 0 & 0 & 0 & 0 & -x_1X_1 & -x_1Y_1 & -x_1Z_1 \\ 0 & 0 & 0 & 0 & X_1 & Y_1 & Z_1 & 1 & -y_1X_1 & -y_1Y_1 & -y_1Z_1 \\ X_2 & Y_2 & Z_2 & 1 & 0 & 0 & 0 & 0 & -x_2X_2 & -x_2Y_2 & -x_2Z_2 \\ 0 & 0 & 0 & 0 & X_2 & Y_2 & Z_2 & 1 & -y_2X_2 & -y_2Y_2 & -y_2Z_2 \\ \vdots & \vdots & \vdots & \vdots & \vdots & \vdots & \vdots & \vdots & \vdots & \vdots & \vdots \\ X_n & Y_n & Z_n & 1 & 0 & 0 & 0 & 0 & -x_nX_n & -x_nY_n & -x_nZ_n \\ 0 & 0 & 0 & 0 & X_n & Y_n & Z_n & 1 & -y_nX_n & -y_nY_n & -y_nZ_n \end{bmatrix} \tag{4.24}$$

Determination of the 11 DLT parameters requires a minimum of 6 reference points. Since equations (4.24) are linear in the unknowns, L_i, no approximate values of the unknown parameters are required. Because of the affine transformation applied to the measured image coordinates, there is no need for an image coordinate system defined by reference points fixed in the camera, such as fiducial marks. Instead it is possible to make direct use of measured comparator coordinates and, in general, coordinates from an arbitrary image measuring device with non-orthogonal axes and different axial scale factors, e.g. pixel coordinates. Images from non-metric cameras, which have no image coordinate system or have an unknown interior orientation, can therefore be evaluated by this method.

Because of its robust linear form, the DLT is also used for the calculation of approximate exterior orientation parameters prior to a bundle adjustment. The more familiar orientation parameters can be derived from the DLT parameters as follows:

With

$$L = \frac{-1}{\sqrt{L_9^2 + L_{10}^2 + L_{11}^2}}$$

the parameters of interior orientation are obtained as follows:

$$x'_0 = -L^2 \cdot (L_1 \cdot L_9 + L_2 \cdot L_{10} + L_3 \cdot L_{11})$$
$$y'_0 = -L^2 \cdot (L_5 \cdot L_9 + L_6 \cdot L_{10} + L_7 \cdot L_{11})$$
: coordinates of principal point (4.25)

$$c_x = -\sqrt{L^2 \cdot (L_1^2 + L_2^2 + L_3^2) - x'^2_0}$$
$$c_y = -\sqrt{L^2 \cdot (L_5^2 + L_6^2 + L_7^2) - y'^2_0}$$
: principal distance (different scales in x and y)

The parameters of exterior orientation, as defined by the elements of rotation matrix **R,** are given by

$$r_{11} = \frac{L \cdot (x'_0 \cdot L_9 - L_1)}{c_x} \qquad r_{12} = \frac{L \cdot (y'_0 \cdot L_9 - L_5)}{c_y} \qquad r_{13} = L \cdot L_9$$
$$r_{21} = \frac{L \cdot (x'_0 \cdot L_{10} - L_2)}{c_x} \qquad r_{22} = \frac{L \cdot (y'_0 \cdot L_{10} - L_6)}{c_y} \qquad r_{23} = L \cdot L_{10} \qquad (4.26)$$
$$r_{31} = \frac{L \cdot (x'_0 \cdot L_{11} - L_3)}{c_x} \qquad r_{32} = \frac{L \cdot (y'_0 \cdot L_{11} - L_7)}{c_y} \qquad r_{33} = L \cdot L_{11}$$

It is sometimes possible in this process that the determinant of the rotation matrix **R** is negative. This must be checked and, if necessary, the matrix must be multiplied by –1 to make its determinant positive.

The position of the perspective centre is given by

$$\begin{bmatrix} X_0 \\ Y_0 \\ Z_0 \end{bmatrix} = -\begin{bmatrix} L_1 & L_2 & L_3 \\ L_5 & L_6 & L_7 \\ L_9 & L_{10} & L_{11} \end{bmatrix}^{-1} \cdot \begin{bmatrix} L_4 \\ L_8 \\ 1 \end{bmatrix} \qquad (4.27)$$

To avoid possible numerical uncertainties, the elements of the rotation matrix must be normalised to an orthonormal matrix (see section 2.2.2.1). The individual rotation angles can be further derived according to (2.31), with due regard to the ambiguities indicated. The DLT model can be extended by correction terms for radial distortion.

Together with the benefits of the DLT mentioned above, there are some drawbacks. If the parameters of interior orientation are known, the DLT has an excess of parameters. In addition, singular or weakly conditioned systems of equations arise if all reference points are located on a common plane, or if the denominator in equations (4.22) is close to zero. Measurement errors in the image coordinates, and errors in reference point coordinates, cannot be detected by the DLT and this results in false parameters. Finally the minimum number of 6 reference points cannot always be provided in real applications.

Example 4.3:

Using the six image and reference points from example 4.2, at DLT is calculated:

DLT parameters:
$L_1 = 0.025598$ $L_2 = 0.000789$ $L_3 = 0.001132$
$L_4 = 10.34162$ $L_5 = -0.000347$ $L_6 = 0.025004$
$L_7 = -0.005910$ $L_8 = 4.986401$ $L_9 = 0.000082$
$L_{10} = -0.000238$ $L_{11} = -0.001003$

According to eqn. (4.25ff), the following orientation parameters are then derived:

Interior orientation:
$c_x = -24.7718$ mm $c_y = -24.8403$ mm
$x'_0 = -0.7259$ mm $y'_0 = 0.0598$ mm

Exterior orientation:
$X_0 = -446.89$ mm $Y_0 = 20.20$ mm $Z_0 = 955.42$ mm

$$\mathbf{R} = \begin{bmatrix} 0.99665 & -0.01331 & -0.07939 \\ 0.03458 & 0.97254 & 0.23060 \\ 0.07262 & -0.23233 & 0.96981 \end{bmatrix}$$

$\omega = -13.375°$ $\varphi = -4.552°$ $\kappa = 0.765°$

Compared with the result of the space resection calculation, the DLT shows significant departures from this but still provides sufficiently good approximate values to initialize an optimized space resection.

4.2.4.2 Perspective projection matrix

The creation of an image point from an object point can also be formulated using the methods of projective geometry. Here object and image coordinates are expressed as homogeneous vectors[1]. Object coordinates $\boldsymbol{X}$ are transformed into image space with image coordinates $\boldsymbol{x'}$ using a 3x4 projection matrix **P.**

$$\boldsymbol{x'} = \mathbf{P} \cdot \boldsymbol{X} \tag{4.28}$$

where

$$\mathbf{P} = \boldsymbol{K} \cdot \mathbf{R} \cdot [\boldsymbol{I} \mid -\boldsymbol{X_0}] \tag{4.29}$$

P is based on the parameters of exterior orientation **R** and $\boldsymbol{X_0}$, as well as a calibration matrix $\boldsymbol{K}$ which contains five parameters of interior orientation.

$$\boldsymbol{K} = \begin{bmatrix} 1 & s' & x'_0 \\ 0 & 1+m' & y'_0 \\ 0 & 0 & 1 \end{bmatrix} \cdot \begin{bmatrix} c & 0 & 0 \\ 0 & c & 0 \\ 0 & 0 & 1 \end{bmatrix} = \begin{bmatrix} c & cs' & x'_0 \\ 0 & c(1+m') & y'_0 \\ 0 & 0 & 1 \end{bmatrix} \tag{4.30}$$

In addition to the coordinates of the perspective centre (x'_0,y'_0,c), s' and m' describe the shear and differential scales of the image coordinate axes.

With the six parameters of exterior orientation, **P** therefore has 11 independent transformation parameters. They can be solved in a linear system of equations and transformed into the 11 DLT parameters.

[1] Homogeneous matrices and vectors are written here in bold, italic script.

Written in full, (4.29) give the following system of equations:

$$\mathbf{P} = \begin{bmatrix} c & cs' & x'_0 \\ 0 & c(1+m') & y'_0 \\ 0 & 0 & 1 \end{bmatrix} \cdot \begin{bmatrix} r_{11} & r_{12} & r_{13} \\ r_{21} & r_{22} & r_{23} \\ r_{31} & r_{32} & r_{33} \end{bmatrix} \cdot \left[\begin{array}{ccc|c} 1 & 0 & 0 & -X_0 \\ 0 & 1 & 0 & -Y_0 \\ 0 & 0 & 1 & -Z_0 \end{array}\right] \tag{4.31}$$

(4.28) can therefore be written as follows:

$$\boldsymbol{x}' = \begin{bmatrix} x' \\ y' \\ 1 \end{bmatrix} = \begin{bmatrix} c & cs' & x'_0 \\ 0 & c(1+m') & y'_0 \\ 0 & 0 & 1 \end{bmatrix} \cdot \begin{bmatrix} r_{11} & r_{12} & r_{13} \\ r_{21} & r_{22} & r_{23} \\ r_{31} & r_{32} & r_{33} \end{bmatrix} \cdot \left[\begin{array}{ccc|c} 1 & 0 & 0 & -X_0 \\ 0 & 1 & 0 & -Y_0 \\ 0 & 0 & 1 & -Z_0 \end{array}\right] \cdot \begin{bmatrix} X \\ Y \\ Z \\ 1 \end{bmatrix} \tag{4.32}$$

From the right-hand side, (4.32) expresses a translation of the perspective centre, followed by a rotation **R**, and finally a perspective scaling and correction to the principal point, in image space.

Distortion effects are taken into account in the above model by the addition of correction values $\Delta x', \Delta y'$:

$$\boldsymbol{K}' = \begin{bmatrix} c & cs' & x'_0 + \Delta x' \\ 0 & c(1+m') & y'_0 + \Delta y' \\ 0 & 0 & 1 \end{bmatrix} = \begin{bmatrix} 1 & 0 & \Delta x' \\ 0 & 1 & \Delta y' \\ 0 & 0 & 1 \end{bmatrix} \cdot \begin{bmatrix} c & cs' & x'_0 \\ 0 & c(1+m') & y'_0 \\ 0 & 0 & 1 \end{bmatrix} \tag{4.33}$$

4.2.5 Object position and orientation by inverse resection

4.2.5.1 Position and orientation of an object with respect to a camera

An inverse space resection enables the determination of the spatial position and orientation of an object with respect to the 3D camera coordinate system. The procedure is sometimes known as the 6 DOF (degrees of freedom) calculation for the target object.

Based on the parameters of exterior orientation described in equations 4.1 and 4.2, the relationship between coordinates of a point, **X** in the object system and $\mathbf{x}^*$ in the camera system (see Fig. 4.4), is given by:

$$\mathbf{X} = \mathbf{X}_0 + \mathbf{R} \cdot \mathbf{x}^* \tag{4.34}$$

Rearranging (4.34) gives:

$$\mathbf{x}^* = \mathbf{R}^{-1} \cdot (\mathbf{X} - \mathbf{X}_0) \tag{4.35}$$

where $\mathbf{x}^*$ gives the coordinates with respect to the camera system of a point on the object.

When expanded, eqn. 4.35 gives:

$$\begin{aligned}\mathbf{x}^* &= -(\mathbf{R}^{-1} \cdot \mathbf{X}_0) + \mathbf{R}^{-1} \cdot \mathbf{X} \\ &= \mathbf{X'}_0 + \mathbf{R'} \cdot \mathbf{X}\end{aligned} \tag{4.36}$$

This has the same form as eqn. 4.34 and can be interpreted as an inverse operation, i.e. object coordinates in the camera system rather than vice versa. If $\mathbf{X_0}$ and $\mathbf{R}$ are obtained from a space resection, then $\mathbf{X'_0}$ and $\mathbf{R'}$ are the inverse resection values. Typically this is applied to the monitoring of a known object. If the camera remains fixed during an imaging sequence, the spatial motion of a known object can be fully determined by repeated inverse space resections.

4.2.5.2 Position and orientation of one object relative to another

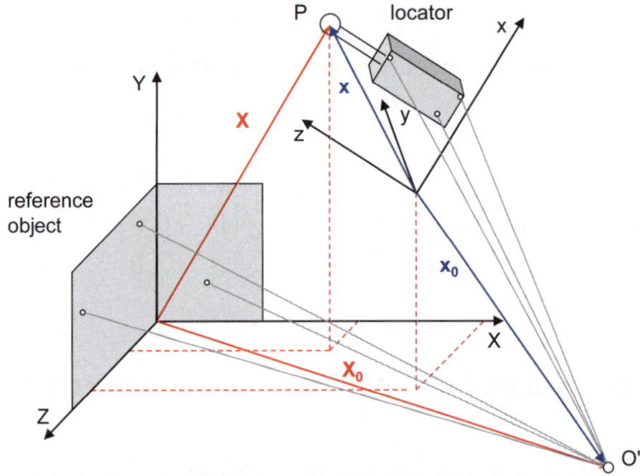

Fig. 4.11: 6 DOF relation between two objects and a camera

The spatial relationship between two objects can also be calculated by space resection, provided that both objects appear in the same image. In this case a reference object is defined with its own local coordinate system, XYZ, in which the position and orientation of a second object is to be determined. This second object, here called the probe or *locator,* has a separate coordinate system, xyz (Fig. 4.11). Two space resections are calculated using control points given in each of these two coordinate systems (see also eqn. 4.3):

$$\mathbf{X} = \mathbf{X}_0 + m \cdot \mathbf{R}_R \cdot \mathbf{x}^* \quad \text{: resection on reference points} \tag{4.37}$$

$$\mathbf{x} = \mathbf{x}_0 + m \cdot \mathbf{R}_L \cdot \mathbf{x}^* \quad \text{: resection on locator points} \tag{4.38}$$

Rearranging (4.38) gives:

$$\mathbf{x}^* = \frac{1}{m} \cdot \mathbf{R}_L^{-1}(\mathbf{x} - \mathbf{x}_0) \tag{4.39}$$

Substituting for $\mathbf{x}^*$ in (4.39) from (4.37) eliminates the scale factor m to give:

$$\begin{aligned} \mathbf{X} &= \mathbf{X}_0 + \mathbf{R}_R \cdot \mathbf{R}_L^{-1} \cdot (\mathbf{x} - \mathbf{x}_0) \\ \mathbf{X} &= \mathbf{X}_0 + \mathbf{R} \cdot (\mathbf{x} - \mathbf{x}_0) \end{aligned} \tag{4.40}$$

in which

x: position of a locator point P (e.g. a probing tip, see Fig. 4.11) within its xyz system

X: position of the same point P in the XYZ reference system (which is the result required)

$\mathbf{x}_0$, $\mathbf{X}_0$: coordinates of the projection centre within locator xyz and reference XYZ systems respectively

$\mathbf{R}_L$, $\mathbf{R}_R$: rotation matrices of camera axes with respect to locator xyz and reference XYZ systems respectively

R: rotation matrix of the locator xyz axes with respect to the XYZ axes

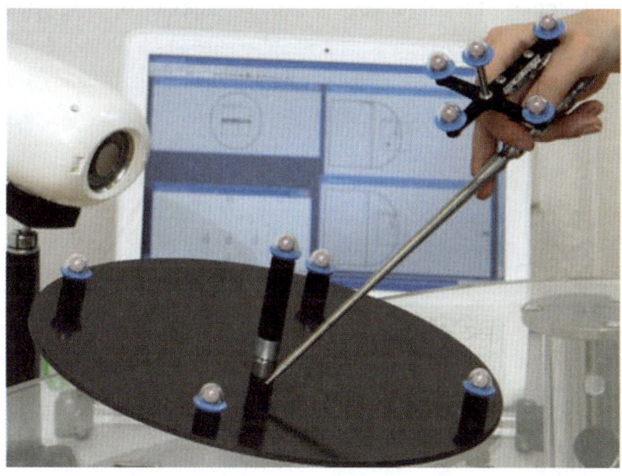

Fig. 4.12: Example of a 6DOF application with reference object and locator (AXIOS 3D)

Fig. 4.12 illustrates an application in which a hand-held measurement probe is monitored by a single camera. The 6DOF calculation is made in real time at a rate of 25 Hz. Since the reference object is also simultaneously located, any movement of the reference object or camera has no effect on the calculation of relative 6DOF between reference and probe. Fig. 3.119 shows another industrial example of the same principle, used to measure the spatial movements of a wheel on a moving car.

The accuracy of the calculated 6DOF values depends on a number of parameters, including:

- focal length of camera,
- physical extent of the observed targets,
- distance of targets from camera,
- accuracy of the reference target locations and the image measurement.

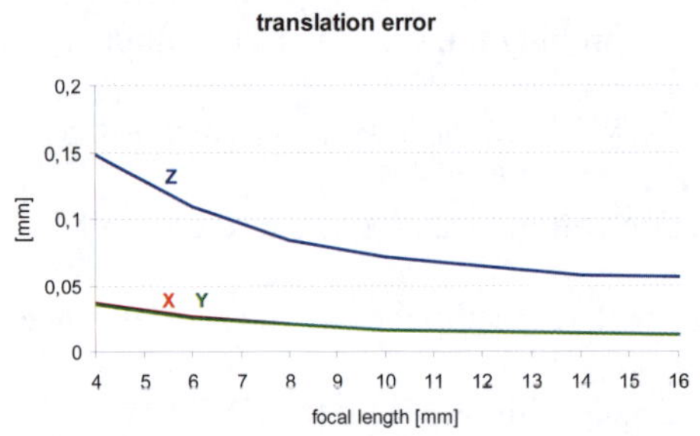

a) Position error as a function of focal length

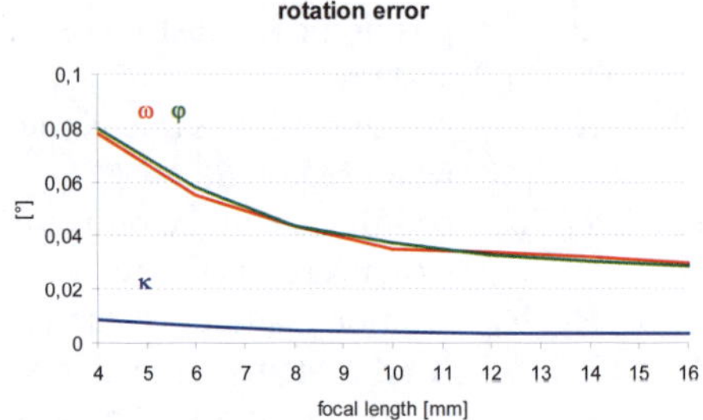

b) Rotation error as a function of focal length

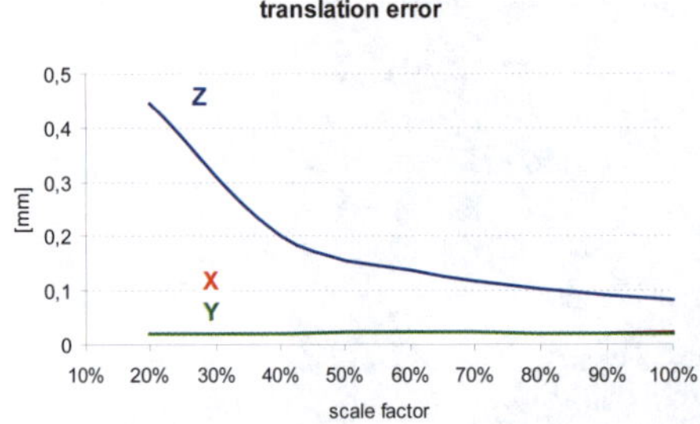

c) Position error as a function of locator size

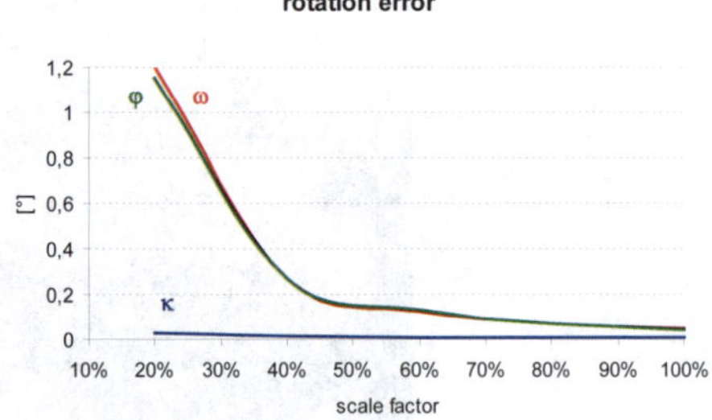

d) Rotation error as a function of locator size

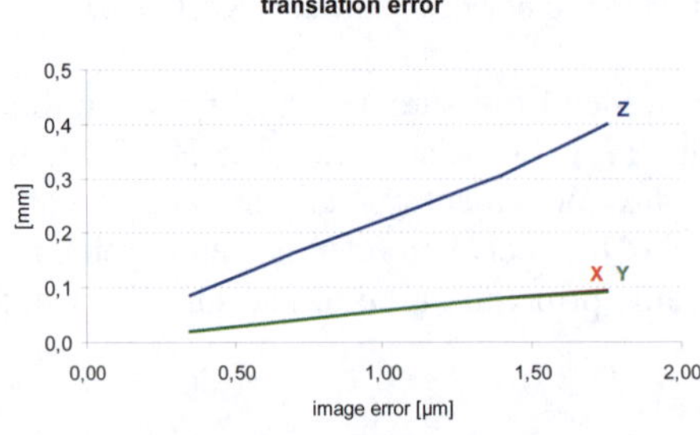

e) Position error as a function of image measurement accuracy

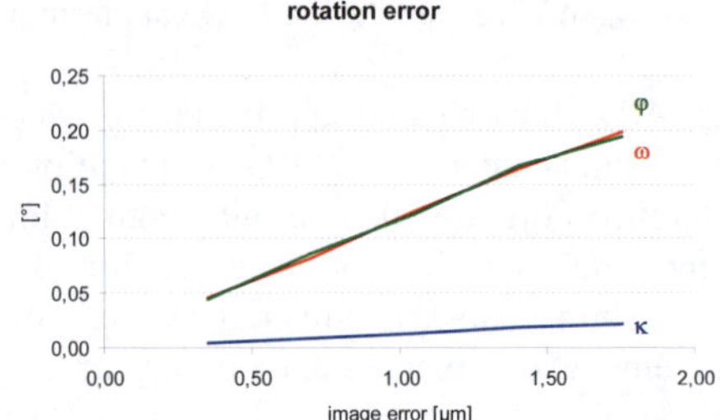

f) Rotation error as a function of image measurement accuracy

Fig. 4.13: Simulated errors in 6DOF calculation

In general, rotations of either body around an axis perpendicular to the image plane (roll angle), as well as lateral shifts (parallel to the image plane), can be determined very accurately. Determinations of movement in the camera's viewing direction (Z), and around the two remaining axes of rotation, is less accurate and significantly more sensitive to measurement and errors in camera orientation and calibration. Fig. 4.13 shows examples of position and rotation errors in a 6DOF calculation relating to the scenario illustrated by the application in Fig. 4.12.

4.2.6 Projective transformation of a plane

4.2.6.1 Mathematical model

A special case is the reconstruction of plane object surfaces. The central perspective projection of an object plane onto the image plane is described by the projective transformation (see section 2.2.1.5).

By setting $Z = 0$ in eqn. 4.14, the transformation equations for the central projection of a plane are obtained:

$$
\begin{aligned}
x' &= \frac{a'_0 + a'_1 X + a'_2 Y}{1 + c'_1 X + c'_2 Y} \\
y' &= \frac{b'_0 + b'_1 X + b'_2 Y}{1 + c'_1 X + c'_2 Y}
\end{aligned}
\tag{4.41}
$$

or alternatively in the inverse form:

$$
\begin{aligned}
X &= \frac{a_0 + a_1 x' + a_2 y'}{c_1 x' + c_2 y' + 1} \\
Y &= \frac{b_0 + b_1 x' + b_2 y'}{c_1 x' + c_2 y' + 1}
\end{aligned}
\tag{4.42}
$$

In order to determine the eight parameters of eqn. 4.42, at least four reference points are required on the plane, no three of which may lie on a common straight line.

The (over-determined) system of equations is solved by an iterative least-squares adjustment. Using the approximate initial values:

$$
\begin{aligned}
&a_1 = b_2 = 1 \\
&a_0 = a_2 = b_0 = b_1 = c_1 = c_2 = 0
\end{aligned}
$$

and the substitution

$$
N_i = c_1 x'_i + c_2 y'_i + 1
$$

the following observation equations are derived:

$$\begin{aligned} v_{Xi} &= \frac{1}{N_i} da_0 + \frac{x'_i}{N_i} da_1 + \frac{y'_i}{N_i} da_2 + \frac{x'_i X_i}{N_i} dc_1 + \frac{y'_i X_i}{N_i} dc_2 - \frac{lx_i}{N_i} \\ v_{Yi} &= \frac{1}{N_i} db_0 + \frac{x'_i}{N_i} db_1 + \frac{y'_i}{N_i} db_2 + \frac{x'_i Y_i}{N_i} dc_1 + \frac{y'_i Y_i}{N_i} dc_2 - \frac{ly_i}{N_i} \end{aligned} \tag{4.43}$$

where

$$-lx_i = a_0 + a_1 x'_i + a_2 y'_i - c_1 x'_i X_i - c_2 y' X_i - X_i$$
$$-ly_i = b_0 + b_1 x'_i + b_2 y'_i - c_1 x'_i Y_i - c_2 y' Y_i - Y_i$$

The approximate solution values at iteration k are adjusted by the computed corrections to the unknowns and the process repeated until the changes are no longer significant, i.e.

$$a_0^{k+1} = a_0^k + da_0^k$$

and so on

Equations 4.42 are non-linear in the coefficients a,b,c. A direct, non-iterative calculation of the unknown parameters is possible if linear equations (2.18) are used. This results in the following equations:

$$\begin{aligned} a_0 + a_1 x_i + a_2 y_i - c_1 x_i X_i - c_2 y_i X_i &= X_i \\ b_0 + b_1 x_i + b_2 y_i - c_1 x_i Y_i - c_2 y_i Y_i &= Y_i \end{aligned} \tag{4.44}$$

which can be solved directly according to the scheme

$$\mathbf{A} \cdot \hat{\mathbf{x}} = \mathbf{l} \tag{4.45}$$

Using the transformation parameters of (4.42) further image coordinates can be transformed into object coordinates.

Fig. 4.14 shows the position of five plane object points and the corresponding image coordinates with their related transformation parameters. For the special case where $c_1 = 0$ and $c_2 = 0$, the projective transformation (4.42) reduces to an affine transformation (eqn. 2.8). For the further case of parallel object and image planes (Fig. 4.14 top right), eqn. 4.42 can be replaced by the plane similarity transformation (2.2).

By the use of homogeneous coordinates, the projective transformation of a plane can be formulated as a homogeneous transformation.

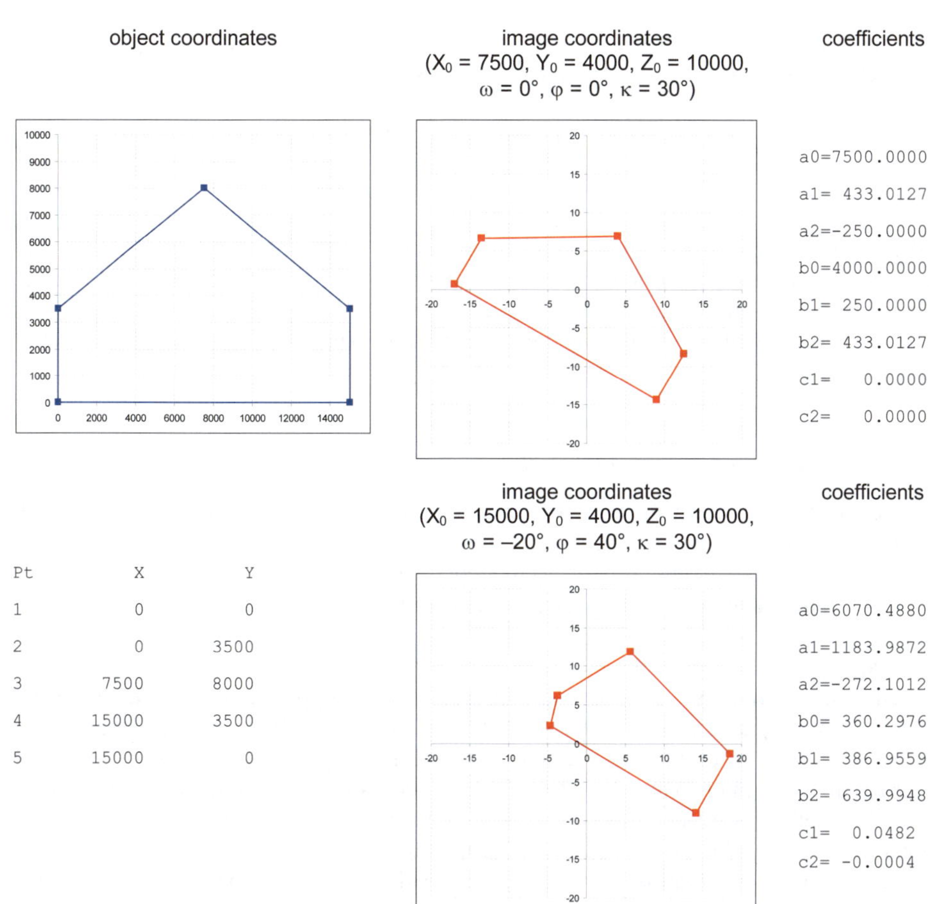

Pt	X	Y
1	0	0
2	0	3500
3	7500	8000
4	15000	3500
5	15000	0

Fig. 4.14: Projective transformation of a plane pentangle

$$\boldsymbol{X} = \boldsymbol{H} \cdot \boldsymbol{x}$$

$$\begin{bmatrix} w \cdot X \\ w \cdot Y \\ w \end{bmatrix} = \begin{bmatrix} h_{11} & h_{12} & h_{13} \\ h_{21} & h_{22} & h_{23} \\ h_{31} & h_{32} & h_{33} \end{bmatrix} \cdot \begin{bmatrix} x' \\ y' \\ 1 \end{bmatrix} \tag{4.46}$$

This formulation is known as *homography*. Since the matrix $\boldsymbol{H}$ can be arbitrarily scaled without altering its projective properties, this shows that there are eight degrees of freedom in the transformation, as there are in eqn. 4.42. Thus the matrix could also be normalized to $h_{33}=1$. From this a set of linear equations is obtained which can be solved directly.

$$\begin{aligned} w \cdot X &= h_{11} \cdot x' + h_{12} \cdot y' + h_{13} \\ w \cdot Y &= h_{21} \cdot x' + h_{22} \cdot y' + h_{23} \\ w &= h_{31} \cdot x' + h_{32} \cdot y' + h_{33} \end{aligned} \tag{4.47}$$

If the last equation is plugged into the first two and all terms containing lower case coordinates are moved to the right of the equation we obtain:

$$\begin{aligned} h_{33} \cdot X &= h_{11} \cdot x' + h_{12} \cdot y' + h_{13} - h_{31} \cdot x' \cdot X - h_{32} \cdot y' \cdot X \\ h_{33} \cdot Y &= h_{21} \cdot x' + h_{22} \cdot y' + h_{23} - h_{31} \cdot x' \cdot Y - h_{32} \cdot y' \cdot Y \end{aligned} \tag{4.48}$$

If the normalization to $h_{33} = 1$ is applied, the same equations as (4.44) are obtained. Alternatively the term to the left of the equal sign can be moved to the right side and the equation set to zero. In this case an eigenvalue decomposition will provide the solution. The eigenvector corresponding to the smallest eigenvalue contains the desired transformation parameters.

4.2.6.2 Influence of interior orientation

The model of central projection described above assumes straight line rays through the perspective centre. Although the spatial position of the perspective centre (principal distance c, principal point x'_0, y'_0) is modelled by the parameters of the projective transformation, it is not possible to compensate here for the effects of lens distortion.

The image coordinates must therefore be corrected for lens distortion before applying the projective transformation. For optimal accuracy, when applying distortion parameters derived from a separate process, care should be taken in case they are correlated with the interior and exterior orientation parameters. Further usage of these parameters might involve a different mathematical model where, strictly speaking, the distortion parameters should be applied with their full variance-covariance matrix from the bundle adjustment. However, this transfer of correlation information is rarely done. A determination of distortion parameters which is numerically independent of the orientation parameters can, for example, be done using the plumb-line method (see section 7.3.1.3).

4.2.6.3 Influence of non-coplanar object points

Object points which do not lie in the plane of the reference points have a positional error in image space after projective transformation. This shift depends on the height above the reference plane and on the position of the point in the image. The image plane is here assumed to be parallel to the reference plane. Since Fig. 4.15 represents a vertical plane containing the projection centre, the shift $\Delta r'$ is radial with respect to the principal point:

$$\Delta r' = \frac{r'}{h} \Delta h \tag{4.49}$$

Multiplying by the image scale gives the corresponding shift in the reference plane:

$$\Delta r = \frac{h}{c} \Delta r' = m \cdot \Delta r' \tag{4.50}$$

The equations above can also be used to determine the maximum height of a point above the reference plane which will not exceed a specified shift in image space.

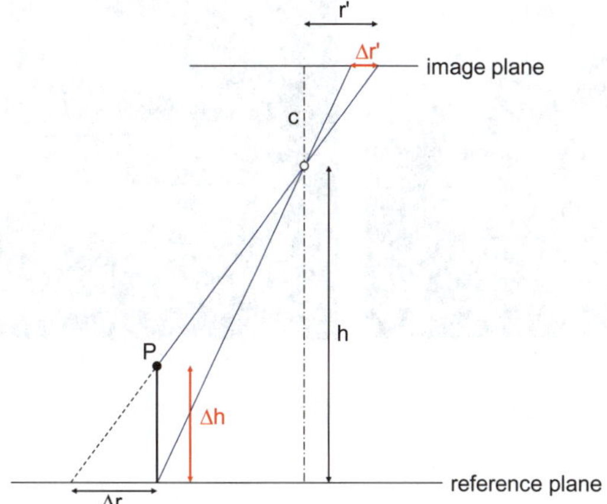

Fig. 4.15: Shift in image space caused by height differences

Example 4.4:

For the measurement of a flat plate, a digital video camera is used with following specifications: $c = 8$ mm, $h = 2.5$ m, $r'_{max} = 5$ mm. The maximum offset above the object plane must be calculated which ensures that the resulting shift in object space Δr is less than 1mm.

1. Image scale: $$m = \frac{h}{c} = \frac{2.5\,\text{m}}{0.008\,\text{m}} = 312$$

2. From (4.49) and (4.50) $$\Delta h = \Delta r \frac{c}{r'_{\max}} = 1.0\,\text{mm} \cdot \frac{8\,\text{mm}}{5\,\text{mm}} = 1.6\,\text{mm}$$

4.2.6.4 Plane rectification

In addition to coordinate determination, the projective transformation is also used as the mathematical basis for optical or digital image rectification. The aim is the transformation of an analogue or digital image into a new geometric (graphical) projection, in most cases a parallel projection.

For rectification of a plane, the complete image format, or a defined area of interest, is transformed into the reference system (target system) by the parameters of a single projective transformation, i.e. the same transformation coefficients are applied to every point in the source image area. In contrast, for non-planar objects every image point must be rectified as a function of its corresponding XYZ coordinates (differential rectification or orthophoto production, see section 4.2.7.3).

In close-range photogrammetry, digital rectification has gained in importance, e.g. for the production of rectified image mosaics of building façades, or for the superposition of natural textures onto CAD elements (see section 5.3). Fig. 4.16 shows an example of the plane rectification of a facade. Object areas outside the reference plane are distorted.

a) Original image

b) Rectification onto facade plane

Fig. 4.16: Example of plane rectification

4.2.6.5 Measurement of flat objects

If the projective transformation between any object plane and the image is known, measured values such as coordinates, lengths, areas or angles can be determined directly in the original image. Fig. 4.17 illustrates the measurement of flat panels on a roof. The parameters of the projective transformation are determined beforehand with the aid of a reference cross visible in the middle of the image. Every pixel or coordinate in the image can therefore be directly transformed into object coordinates with respect to the reference plane (and vice versa).

Fig. 4.17: Image showing overlay results from in-plane measurement

Measurement accuracy depends strongly on the perspective of the image. As usual, it reduces with increasing object distance, which results in measurements of very variable accuracy in the image. Fig. 4.18 shows simulated error vectors which would be expected in object points measured by oblique imaging of a flat roof. In this case the control points (blue) are in the lower right corner. The systematic and, in object space asymmetric, error distribution is obvious.

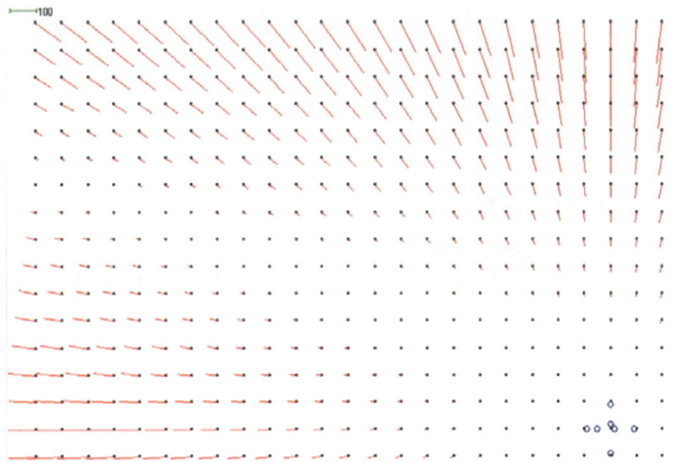

Fig. 4.18: Simulated errors of object points measured using an oblique image of a flat roof

4.2.7 Single image evaluation of three-dimensional object models

The spatial ray defined by an image point can be intersected with the object surface, if interior and exterior orientation of the image are known and a geometric model of the surface exists. The object model can be defined by a known mathematical element (e.g. straight line, plane, cylinder), or by a dense grid of points (digital surface or elevation model). The intersection point of the ray and the object model is the desired 3D object coordinate, as the next two sections explain in more detail.

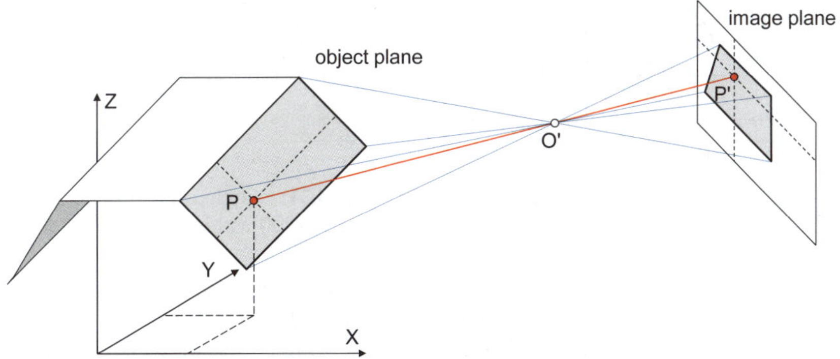

Fig. 4.19: Single image evaluation within an object plane

4.2.7.1 Object planes

The principle of photogrammetric point determination within an object plane is illustrated in Fig. 4.19. The 3D coordinates of the object point P result from the intersection of the object surface and the ray defined by the image point P' and the perspective centre O'. The object plane can be defined, for example, by prior photogrammetric measurement of three points. The parameters of the plane can be calculated from a least-squares best-fit adjustment (see section 2.3.2.2).

4.2.7.2 Digital surface models

An arbitrary or free-form object surface can be approximated by a suitably dense grid of 3D points (digital surface model, DSM, see section 2.3.3.1). The point grid can be regular (e.g. $\Delta X = \Delta Y = \text{const.}$) or non-regular. Object edges (breaklines) can be modelled by special point codes or by additional vector data (polygons). Inside the object model further points can be interpolated.

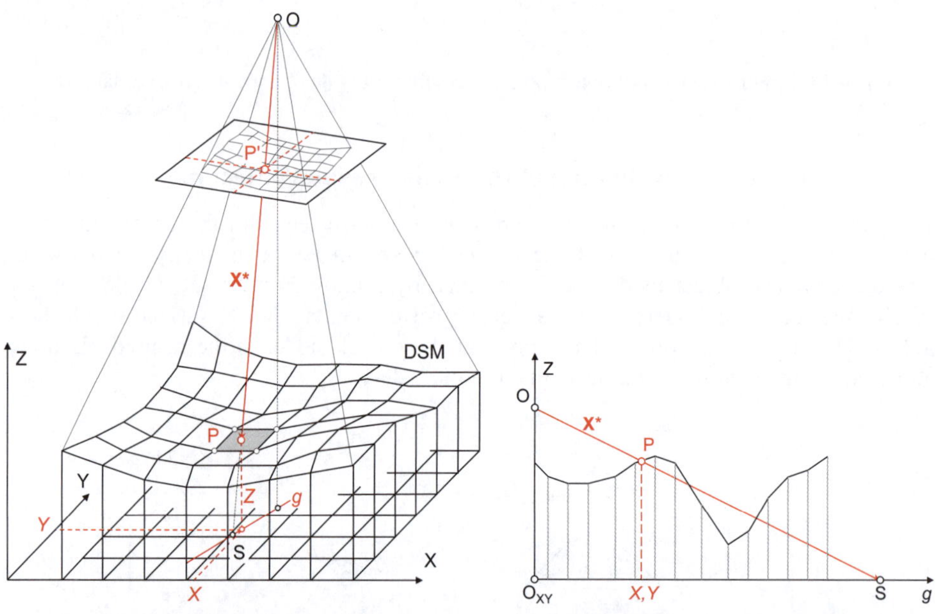

Fig. 4.20: Point determination in a digital surface model (DSM)

Fig. 4.21: Search and interpolation of object point P(*XYZ*) within the height profile above *g*

Fig. 4.20 shows the principle of spatial point determination from a single image using a digital surface model (DSM). The spatial direction defined by the measured image coordinates x', y' and c in the image coordinate system (image vector $\mathbf{x'}$) is transformed into the spatial vector $\mathbf{X}^*$ using the exterior orientation parameters (similarity transform with arbitrary scale, e.g. $m = 1$). This ray intersects the DSM at point P using a local surface plane defined by the four adjacent points.

In order to calculate the point of intersection, the straight line g is constructed between the intersection point S of $\mathbf{X}^*$ and the XY plane, and the foot of the perpendicular O_{XY} from the perspective centre to the XY plane (Fig. 4.21). A search for point P starts along this line at O_{XY} until its interpolated height Z lies within two Z values of adjacent profile points.

This procedure is known as *monoplotting*. It is not very popular in close-range photogrammetry, but is gaining importance in the field of CAD technology. As an example Fig. 4.22 shows the determination of the water line of a moving ship model using one oriented image. The object surface model is built up from a dense series of CAD profiles that are intersected with the water line derived from the image.

Fig. 4.22: Monoplotting for the determination of the a water line of a ship by intersection of the imaged water surface with the CAD surface of the ship model

a) Measurement image overlaid with simulated cylinder in object space

b) Image detail with overlaid 3D point cloud

Fig. 4.23: Monoplotting applied to the measurement of pipework by intersecting image rays with the 3D point cloud from a terrestrial laser scanner (Riegl, Phocad)

Fig. 4.23 shows an example of monoplotting using a 3D point cloud from a terrestrial laser scanner which is equipped with a high-resolution camera (see section 3.7.1.2). Here geometrical primitives (e.g. cylinders) are extracted from the measurement data. The evaluation is semi-automatic, requiring visual interpretation of the image. The enlargement of the image extract in Fig. 4.23b makes clear that a fully automatic evaluation of the complex point cloud is not practicable without additional image information.

4.2.7.3 Differential rectification

For the production of an orthophoto by differential rectification, each point (infinitesimally small object patch) is projected individually according to its XYZ coordinates in order to create a new image which is a parallel projection of the object surface. A digital orthophoto displaying a rectangular area of the ground plane XY is calculated in the following steps (Fig. 4.24):

1. Definition of the rectangular area of interest on the object:

   lower left corner: X_1, Y_1
   upper right corner: X_2, Y_2

2. Definition of output scale and print resolution of the orthophoto:

   The output scale (map scale) m_o of the orthophoto and the required resolution Δx_k, Δy_k for printing (output) define the grid width (point distance) in object space.

3. Definition of grid width $\Delta X,\Delta Y$, used to "scan" the object:

   $$\Delta X = m_o \cdot \Delta x_k$$
   $$\Delta Y = m_o \cdot \Delta y_k$$

4. For each grid point $(X,Y)_i$ the corresponding Z_i value is interpolated in the given surface model:

   $$X_i = X_1 + j \cdot \Delta X \qquad Z_i = Z(X_i, Y_i)$$
   $$Y_i = Y_1 + k \cdot \Delta Y$$

5. Using the collinearity equations (4.9), and the given parameters of interior and exterior orientation, the image coordinate $(x',y')_i$ corresponding to $(X,Y,Z)_i$ is calculated:

   $$x'_i = F(X_0, Y_0, Z_0, \omega, \varphi, \kappa, x'_0, c, \Delta x', X_i, Y_i, Z_i)$$
   $$y'_i = F(X_0, Y_0, Z_0, \omega, \varphi, \kappa, y'_0, c, \Delta y', X_i, Y_i, Z_i)$$

6. At position $(x',y')_i$ the stored colour value is extracted from the digital image. The colour value is usually interpolated, since $(x',y')_i$ are floating point numbers which do not match the integer pixel raster of the image (see section 5.3.2):

   $$g'_i = g(x',y')_i$$

7. The interpolated colour value g'_i is transferred into the output image at position $(x,y)_i$:

$$x_i = x_1 + j \cdot \Delta x_k$$
$$y_i = y_1 + k \cdot \Delta y_k$$

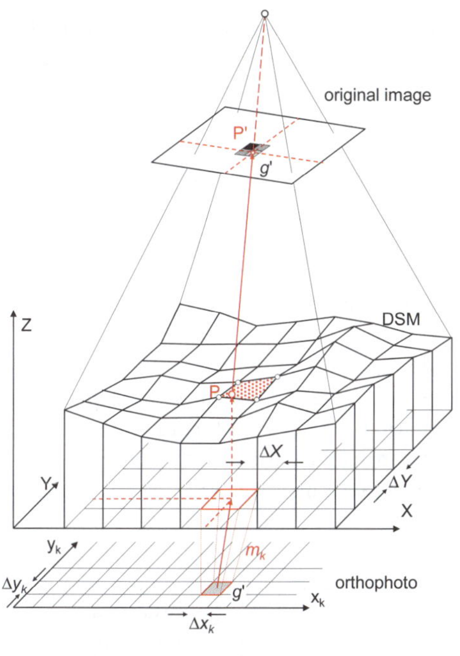

a) Principle of orthophoto calculation

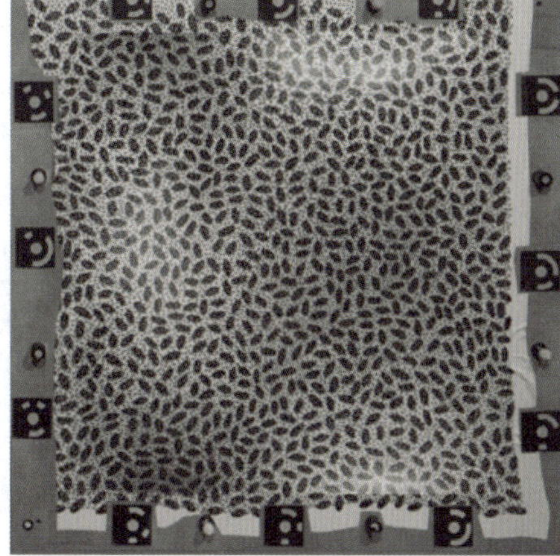

b) Orthophoto calculated using surface model

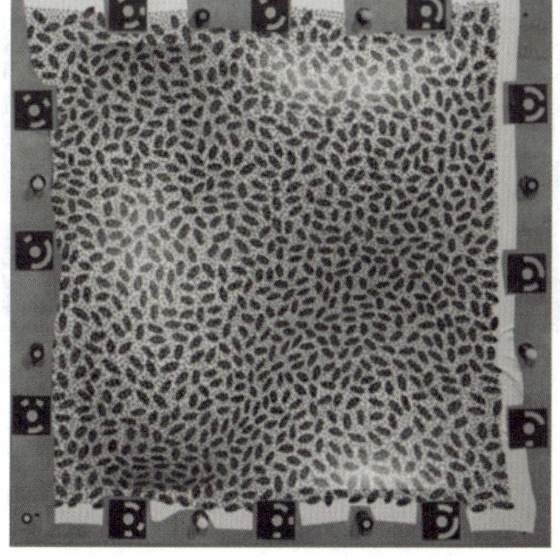

c) Rectification by plane transformation

d) Difference image between b) and c)

Fig. 4.24: Creation of orthophotos

The method described above transforms each point of the orthophoto into the original image. In order to reduce the computational effort, the *anchor-point method* can be used. This transforms only a coarse grid of points from which local colour values can be linearly interpolated in the image. Fig. 4.24b shows an orthophoto of the image from Fig. 4.6 created using the corresponding surface model. In comparison, Fig. 4.24c shows the rectification calculated by plane projective transformation. At first glance, this looks identical to the orthophoto. However, the difference between the images, shown in Fig. 4.24d, demonstrates clear differences between the methods at all points where there are significant height differences from the reference plane (black: no difference, white: large difference).

Object areas which are not described by the surface model are therefore shifted and distorted in image space. Occluded object areas lead to empty (black) spaces in the orthophoto, and can only be projected by additional images from different stations. The latter method however requires either extensive *ray tracing* or simultaneous multi-image processing, for example using object-based multi-image matching (see section 5.5.6).

Differential rectification is a general approach which, in principle, can be applied to any projections and object surfaces. In such cases it is necessary only to replace the transformations of steps 3 and 4 with functions relating the 2D coordinates of the original image to the 3D object coordinates. This method can handle, for example, cylindrical or spherical projections as well as the plane projective transformation described above (example in Fig. 4.25).

An orthophoto can also be generated by combining a number of partial views or original images into a common image mosaic. This procedure, used in the production of an orthophoto mosaic, can be employed in close-range photogrammetry for the rectification of facades, or for texture mapping of CAD models (Fig. 4.26, section 5.3.3).

Fig. 4.25: Original image and cylindrical projection (west apse of the monastery church at Gernrode (Fokus Leipzig)

Fig. 4.26: 3D model with rectified image areas and superimposed textures (church at Bad Zwischenahn)

4.3 Processing of stereo images

4.3.1 Stereoscopic principle

4.3.1.1 Stereoscopic matching

The photogrammetric processing of image pairs has, in many ways, a particular significance. For a start, two images of an object, taken from different positions, represent the minimum condition necessary for a 3D object measurement. Then there is the fact that human vision is highly developed and enables the stereoscopic viewing and analysis of image pairs.

The process of evaluating stereo images is typically divided into three steps:

1. Determination of homologous image features (e.g. corresponding points in both images).
2. Orientation of the image pair (potentially also with calibration of the camera(s))
3. 3D object measurement (e.g. measurement of free-form surfaces).

Depending on application, the steps can be executed in the sequence given, or in combination, or in a different order.

The essential task in stereo and multi-image processing is the solution of the correspondence problem, i.e. the matching of identical (homologous) image points. This depends first of all on the visible object texture which influences the accuracy and uniqueness of the association. Fig. 4.27 shows an example of varying textures, patterns and targets (extracts from the image pair in Fig. 4.33), which make clear that unique or ambiguous matches are possible, depending on pattern.

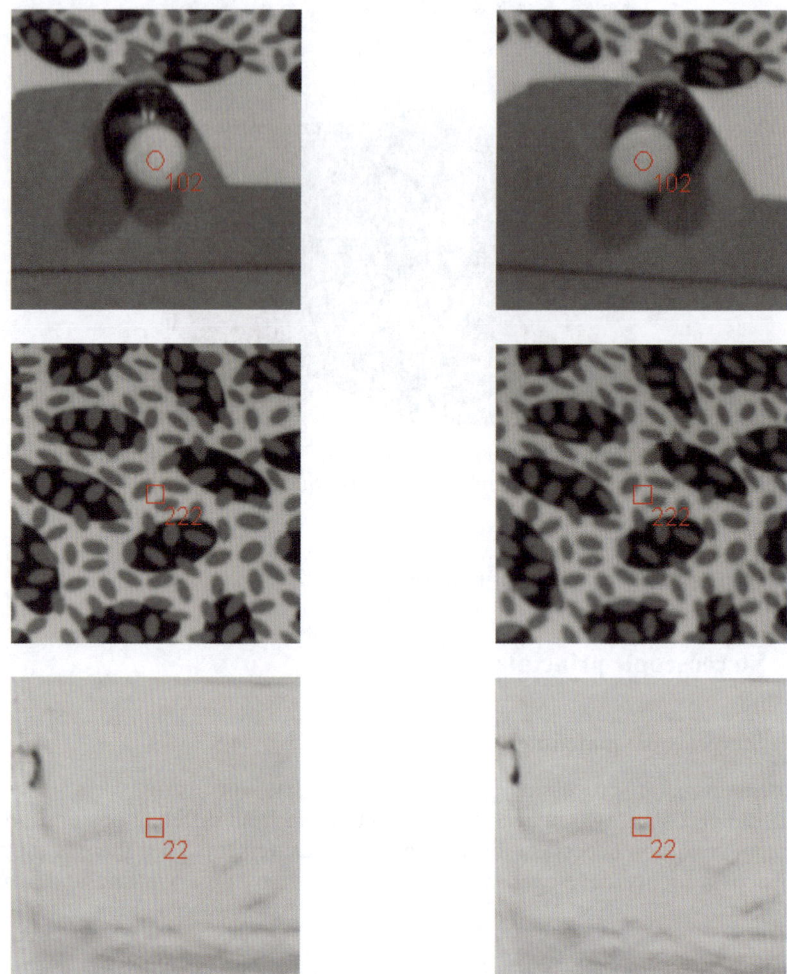

Fig. 4.27: Correspondence problem resulting from different object surface patterns in a stereo image

4.3.1.2 Tie points

Tie points are measured, homologous points in the images, i.e. they represent the same object point. They assist in the geometric connection between two or more images and need not be reference points. More specifically, they must be selected to cover a sufficient area in image and object space in order to provide a robust connection between the images.

Homologous points can be identified visually, either by stereoscopic viewing or by monoscopic measurement of single images. Non-targeted object points can be identified more reliably by stereoscopic viewing. Correspondence between homologous points can also be performed by digital stereo image matching (see section 5.5.3.1). Here similarities in grey level patterns are compared in order to match corresponding points (image correlation).

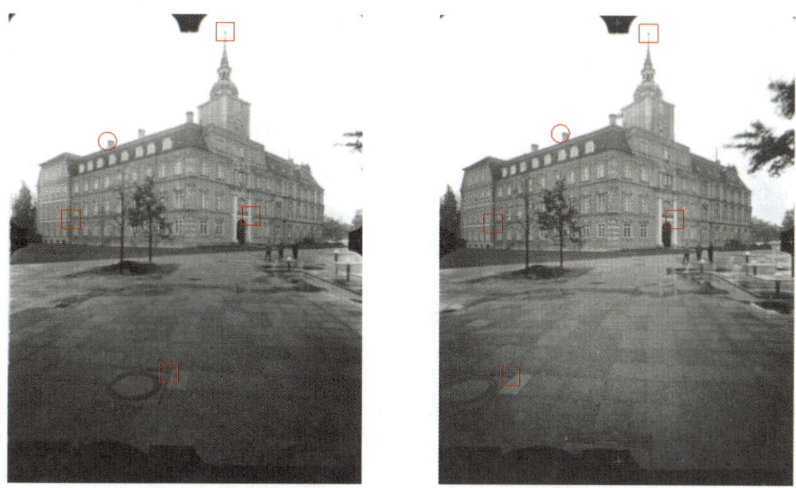

Fig. 4.28: Tie points in a stereo pair
□: correctly matched points; ○: wrongly matched points

Normally there is no orientation information available during the tie point measurement stage and so there are few controls to prevent false measurements. In larger photogrammetric projects, gross errors (blunders) are therefore almost always present in the observations due to errors in measurement or identification. Fig. 4.28 shows an example of four correct and one incorrect tie points.

4.3.1.3 Orientation of stereo image pairs

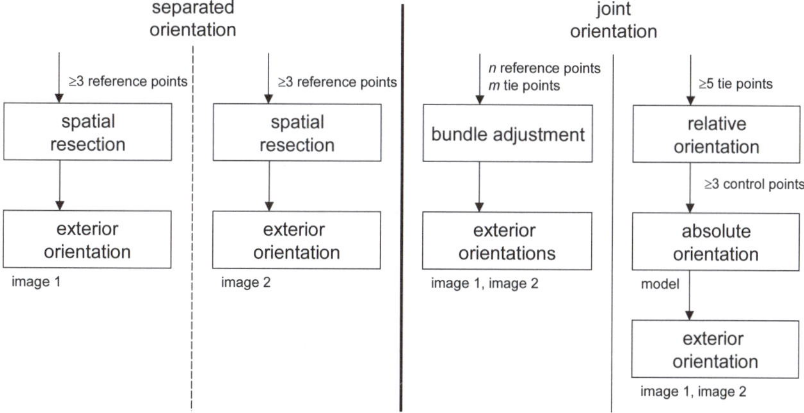

Fig. 4.29: Orientation methods for stereo images

The orientation of a stereo pair provides exterior orientation parameters of both images (Fig. 4.29). In principle, this task can be solved separately for each image by space resection (see section 4.2.3) but three-dimensional reference points (full reference points) are then required for each photo (see Fig. 4.7). The reference points can be identical or different for each image. In this procedure the geometric relationship between the two images in a stereo model is not used.

Fig. 4.28 shows a typical stereo pair where both images cover the object with an overlap of at least 50 % (a standard overlap is 60 %, see also Fig. 3.35a). This feature can be employed in the orientation of both images.

The one-step solution employs the principle of bundle triangulation (section 4.4, Fig. 4.50) for the special case of two images. Here the orientation elements of both images are determined simultaneously in one step using the image coordinates of the reference points and additional tie points.

The traditional two-step solution of this problem works as follows. In the first step the correspondence between the images, and the coordinates of model points, are determined in a local coordinate system (relative orientation, see section 4.3.3). In the second step the transformation into the global object coordinate system is performed using reference points (absolute orientation, see section 4.3.5).

4.3.1.4 Normal case of stereo photogrammetry

The normal case of stereo photogrammetry is, in fact, the special case in which two identical ideal cameras have parallel axes pointing in the same direction at right angles to the stereo base. With respect to an XYZ coordinate system located in the left perspective centre, object coordinates can be derived from the ratios indicated in Fig. 4.30. Using real cameras, the normal case can only be achieved with low accuracy requirements (example in Fig. 1.29). Its use is mainly in the calculation of approximate coordinates and the estimation of the achievable measurement accuracy.

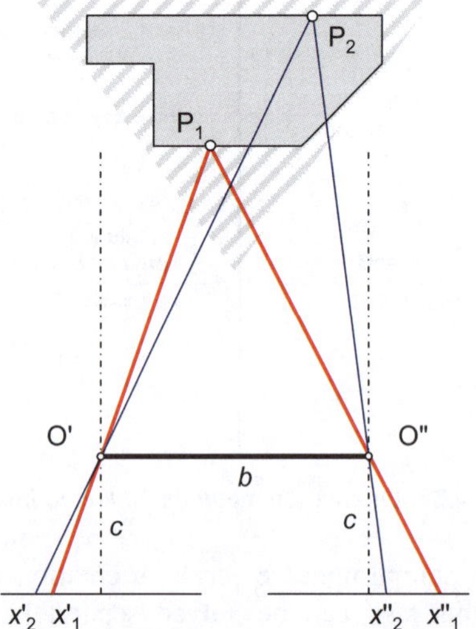

Fig. 4.30: Normal case of stereo photogrammetry

Stereo images, which approximate to the normal case, can easily be evaluated by visual inspection. Ideally, objects are only observed with a distance-dependent, horizontal shift between the images, and a vertical shift, which is detrimental to stereoscopic viewing, will not exist. This horizontal shift is known as x-parallax or disparity. X-parallax increases with shorter object distances and is zero for objects at infinity. In the example of Fig. 4.28, the x-parallax for point P_1, with $px_1 = x'_1 - x''_1$, is greater than for point P_2 with $px_2 = x'_2 - x''_2$.

Real stereo image pairs can be digitally transformed into the normal stereo case (normal-case images, section 4.3.3.5) if their relative orientation is known (relative orientation, section 4.3.3).

4.3.2 Epipolar geometry

Fig. 4.31 shows the geometry of a stereo pair imaging any object point P. The base **b**, and the projected rays **r'** and **r"** from each perspective centre to the object point, define an *epipolar plane*, sometimes called a *basal plane*. This plane intersects the image planes along lines *k*' and *k*", which are known as *epipolar lines*. In the case of convergent images the epipolar lines are convergent. In the special case of normal stereo photogrammetry, the epipolar lines are parallel to the x' direction (Fig. 4.32, see also section 4.3.6.2).

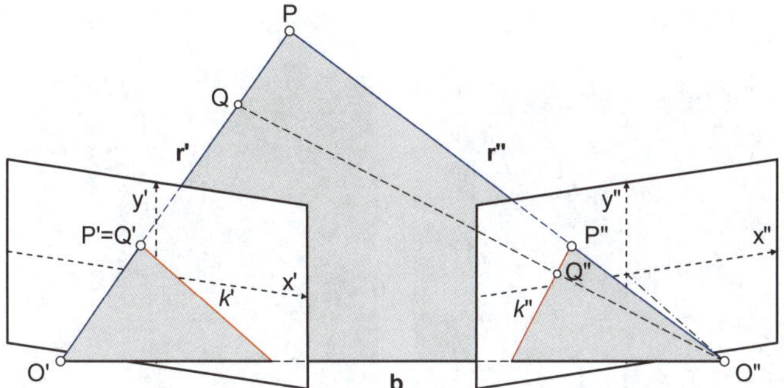

Fig. 4.31: Epipolar plane for convergent images

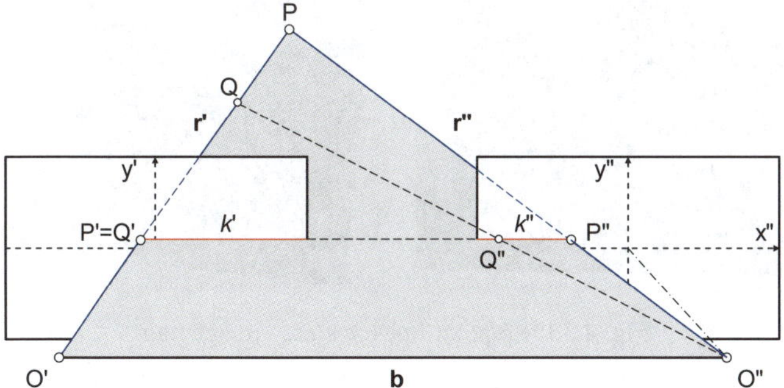

Fig. 4.32: Epipolar plane for normal case of stereo photogrammetry

The importance of epipolar geometry lies in the fact that, assuming an error-free ray intersection, an image point P" in the right image, corresponding to P' in the left image, must lie on the epipolar plane and hence on the epipolar line *k*". Thus the search space for matching corresponding points can be significantly reduced. Assuming an additional object point Q lying on ray O'P, it is obvious that the difference in distance (depth) between Q and P results in a parallax along the epipolar line *k*". In the normal case of stereo photogrammetry the parallax is purely in the x' direction (x-parallax or horizontal parallax).

If the orientation parameters are known, the position of the epipolar line *k*" corresponding to P' (or vice versa) can be calculated. Given an arbitrary image point P', and with projection equations 4.7, two points P and Q on the ray **r'** can be calculated for two different arbitrary values of scaling factor *m*. The XYZ coordinates of points P and Q can subsequently be projected into the right image using the collinearity equations (4.9). The epipolar line *k*" is then defined by the straight line containing image points P" and Q".

Fig. 4.33 shows a convergent stereo image pair, and the images derived from them which correspond to the normal case of stereo photogrammetry, see section 4.3.3.5. For a point measured in the left and right images, the corresponding epipolar lines are shown. In the convergent case they run at an angle through the images and in the normal case they are run parallel on identical y image-coordinate values.

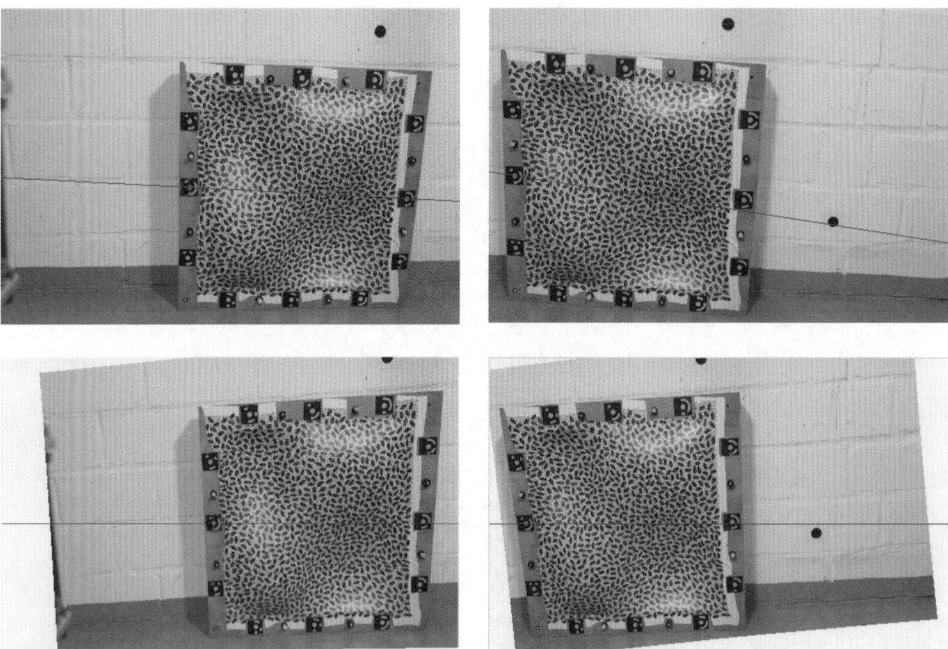

Fig. 4.33: Epipolar lines in stereo image pair

The epipolar lines can also be calculated from the parameters of relative orientation which are derived below (see section 4.3.3.4).

4.3.3 Relative orientation

Relative orientation describes the translation and rotation of one image with respect to its stereo partner in a common local model coordinate system. It is the first stage in the two-step orientation of a stereo image pair (see Fig. 4.29).

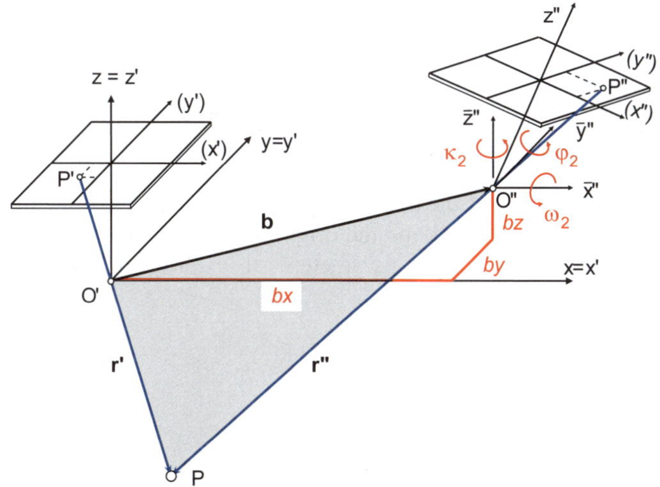

Fig. 4.34: Model coordinate system and relative orientation (left image fixed)

The numerical method of relative orientation can be easily developed for the following case. A local three-dimensional model coordinate system xyz is located in the perspective centre of the first (left) image and oriented parallel to its image coordinate system Fig. 4.34). The parameters of *exterior* orientation of the left image with respect to the model coordinate system are therefore already given:

$$\begin{aligned} x_{01} &= 0 \qquad & \omega_1 &= 0 \\ y_{01} &= 0 & \varphi_1 &= 0 \\ z_{01} &= 0 & \kappa_1 &= 0 \end{aligned} \tag{4.51}$$

Now the second (right) image is oriented in the model coordinate system by 3 translations and 3 rotations:

$$\begin{aligned} x_{02} &= bx \qquad & \omega_2 \\ y_{02} & & \varphi_2 \\ z_{02} & & \kappa_2 \end{aligned} \tag{4.52}$$

The base space vector **b** between the perspective centres O' and O" is defined by the base components *bx*, *by* and *bz*. It is stated in section 4.3.3.1 that the condition for correct relative orientation is that all pairs of homologous rays must be coplanar with the base. Suppose that the right-hand perspective centre is displaced along the base line towards O' and that the image is not rotated. It is clear that the homologous rays of Fig. 4.34 will still be coplanar with the base and that they will intersect in a point lying on the line between O' and P'. Consideration of similar triangles shows that the scale of the model will be directly proportional to the length of the base. That is to say, the model coordinate system can be scaled by an arbitrary factor depending on our choice of base length. One of the base components is therefore set to a constant value, commonly

$$bx = 1$$

Five independent elements *by*, *bz* and ω_2, φ_2, κ_2 therefore remain for the definition of the relative orientation.

For an alternative formulation of the relative orientation, the x axis of the model coordinate system is defined by the stereo base and the origin of the system is located in the left hand perspective centre (Fig. 4.35). The parameters of exterior orientation in the model coordinate system are then given by:

$$\begin{aligned} x_{01} &= 0 & \quad & \omega_1 \\ y_{01} &= 0 & & \varphi_1 \\ z_{01} &= 0 & & \kappa_1 \end{aligned}$$

$$\begin{aligned} x_{02} &= bx & \quad & \omega_2 = 0 \\ y_{02} &= 0 & & \varphi_2 \\ z_{02} &= 0 & & \kappa_2 \end{aligned}$$

The five elements to be determined are here expressed by five independent rotation angles ω_1, φ_1, κ_1 and φ_2, κ_2. Instead of ω_1 (rotation about x axis), ω_2 may be used as an alternative. The scale is again set to an arbitrary value, normally with $bx = 1$.

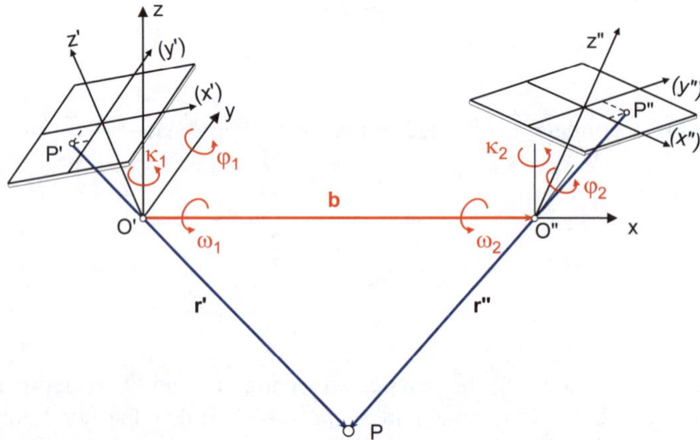

Fig. 4.35: Model coordinate system with base defining the x axis

4.3.3.1 Coplanarity constraint

The computational solution of relative orientation utilizes the condition that an object point P and the two perspective centres O' and O" must lie in a plane (*coplanarity constraint*). This is the epipolar plane defined by vectors **b**, **r'** and **r"**, which also contains the image points P' and P".

The coplanarity constraint is only fulfilled if rays **r'** and **r"** strictly intersect in object point P, i.e. if the positions of image points P' and P", as well as the orientation parameters, are free of error. For each pair of homologous image points, one coplanarity constraint equation can be derived. Consequently, in order to calculate the five unknown orientation parameters, a minimum of five homologous points (tie points) with measured image coordinates is required. The constraint is equivalent to the minimization of y-parallaxes at all observed points P. The term y-parallax is defined by eqn. 4.60.

The coplanarity constraint can be expressed using the scalar triple product of the three vectors. They lie in a plane if the volume of the parallelepiped they define is zero:

$$(\mathbf{b} \times \mathbf{r'}) \cdot \mathbf{r''} = 0 \qquad (4.53)$$

Alternatively, equation (4.53) can be expressed by the determinant of the following matrix. The base vector **b** is replaced by the three base components, the image vector **r'** is replaced by the image coordinates in the left image and the image vector **r"** is given by the image coordinates of the right image, transformed by the relative rotation parameters.

$$\Delta = \begin{vmatrix} 1 & x' & \bar{x}'' \\ by & y' & \bar{y}'' \\ bz & z' & \bar{z}'' \end{vmatrix} = 0 \qquad (4.54)$$

$$\quad \mathbf{b} \quad \mathbf{r'} \quad \mathbf{r''}$$

where $$\begin{bmatrix} \bar{x}'' \\ \bar{y}'' \\ \bar{z}'' \end{bmatrix} = \begin{bmatrix} a_{11} & a_{12} & a_{13} \\ a_{21} & a_{22} & a_{23} \\ a_{31} & a_{32} & a_{33} \end{bmatrix} \cdot \begin{bmatrix} x'' \\ y'' \\ z'' \end{bmatrix}$$

$$\mathbf{r''} = \mathbf{A} \cdot \mathbf{x''}$$

Here **A** is the rotation matrix of the right-hand image, so that the coefficients a_{11}, a_{12} ... are functions of the rotation angles $\omega_2, \varphi_2, \kappa_2$. Different principal distances for both images may be introduced as $z' = -c_1$ and $z'' = -c_2$. For each measured tie point P_i one observation equation can be established using (4.54).

If the calculation of relative orientation is solved with the aid of the collinearity equations (4.9), then the 5 homologous point pairs give rise to 5 x 2 x 2 = 20 observation equations. Opposite these are five unknowns of relative orientation as well as 5 x 3 = 15 unknown model coordinates, so that again 5 tie points provide a minimum solution. The determination of relative orientation using the fundamental and essential matrices is discussed in section 4.3.4.

4.3.3.2 Calculation

The calculation of the five elements of relative orientation follows the principle of least-squares adjustment (see section 2.4.2.2). Based on the coplanarity condition, the following correction equation can be set up for each tie point:

$$v_\Delta = \frac{\partial\Delta}{\partial by}dby + \frac{\partial\Delta}{\partial bz}dbz + \frac{\partial\Delta}{\partial\omega_2}d\omega_2 + \frac{\partial\Delta}{\partial\varphi_2}d\varphi_2 + \frac{\partial\Delta}{\partial\kappa_2}d\kappa_2 + \Delta^0 \tag{4.55}$$

In the case of approximately parallel viewing directions, the initial values required for linearisation are as follows:

$$by^0 = bz^0 = \omega_2{}^0 = \varphi_2{}^0 = \kappa_2{}^0 = 0 \tag{4.56}$$

Δ^0 is the volume of the parallelepiped calculated from the initial values. The differentials can again easily be computed using the following determinants:

$$\frac{\partial\Delta}{\partial by} = \begin{vmatrix} 0 & x' & \bar{x}'' \\ 1 & y' & \bar{y}'' \\ 0 & z' & \bar{z}'' \end{vmatrix} \qquad \frac{\partial\Delta}{\partial bz} = \begin{vmatrix} 0 & x' & \bar{x}'' \\ 0 & y' & \bar{y}'' \\ 1 & z' & \bar{z}'' \end{vmatrix}$$

$$\frac{\partial\Delta}{\partial\omega_2} = \begin{vmatrix} 1 & x' & 0 \\ by & y' & -\bar{z}'' \\ bz & z' & \bar{y}'' \end{vmatrix} \qquad \frac{\partial\Delta}{\partial\varphi_2} = \begin{vmatrix} 1 & x' & -\bar{y}''\sin\omega_2 + \bar{z}''\cos\omega_2 \\ by & y' & \bar{x}''\sin\omega_2 \\ bz & z' & -\bar{x}''\cos\omega_2 \end{vmatrix} \tag{4.57}$$

$$\frac{\partial\Delta}{\partial\kappa_2} = \begin{vmatrix} 1 & x' & -\bar{y}''\cos\omega_2\cos\varphi_2 - \bar{z}''\sin\omega_2\cos\varphi_2 \\ by & y' & \bar{x}''\cos\omega_2\cos\varphi_2 - \bar{z}''\sin\varphi_2 \\ bz & z' & \bar{x}''\sin\omega_2\cos\varphi_2 + \bar{y}''\sin\varphi_2 \end{vmatrix}$$

The approximate values are iteratively improved by the adjusted corrections until there is no significant change.

Here the standard deviation of unit weight s_0 provides little information about achieved accuracy because the volumes of the parallelepipeds are used as observations instead of the measured image coordinates. Residuals of the estimated orientation elements result in skew intersection of the rays **r'** and **r''**, thus generating y-parallaxes in model space. It is therefore advantageous to analyse the quality of relative orientation using the calculated model coordinates.

Example 4.5:

The following data are available to calculate the relative orientation of the image pair in Fig. 4.33.

	Image 1 (left):		Image 2 (right)	
Interior orientation:	$c = -24.2236$ mm		$c = -24.2236$ mm	
	$x'_0 = 0.0494$ mm		$x'_0 = 0.0494$ mm	
	$y'_0 = -0.2215$ mm		$y'_0 = -0.2215$ mm	
Image coordinates:				
P_1	$x'_1 = -0.0395$	$y'_1 = -6.3033$	$x''_1 = -8.1592$	$y''_1 = -6.1394$
P_2	$x'_2 = 6.6590$	$y'_2 = -6.2948$	$x''_2 = -1.0905$	$y''_2 = -6.5887$
P_3	$x'_3 = 9.0086$	$y'_3 = -1.3473$	$x''_3 = 1.1945$	$y''_3 = -1.4564$
P_4	$x'_4 = 7.3672$	$y'_4 = 4.5216$	$x''_4 = -0.8836$	$y''_4 = 4.8255$
P_5	$x'_5 = 0.2936$	$y'_5 = 4.7133$	$x''_5 = -8.3009$	$y''_5 = 5.2626$
P_6	$x'_6 = -2.0348$	$y'_6 = -0.7755$	$x''_6 = -10.4401$	$y''_6 = -0.2882$

Parameters of relative orientation are required (left image fixed).

Base components: $bx = 1$ $by = -0.0642$ $bz = -0.1278$

Rotation angles: $\omega = 1.4514°$ $\varphi = 4.1282°$ $\kappa = 2.5186°$

It can be seen that the base components in the y and z directions are significantly smaller than in the x direction and that convergence in the configuration is most apparent in the φ rotation angle about the y axis.

4.3.3.3 Model coordinates

The relationship between image and model coordinates can be expressed by the following ratios:

$$\frac{x}{x'} = \frac{y}{y'} = \frac{z}{z'} = \lambda$$: scale factor for a particular point in left image

$$\frac{x - bx}{\overline{x}''} = \frac{y_2 - by}{\overline{y}''} = \frac{z - bz}{\overline{z}''} = \mu$$: scale factor for the same point in right image (4.58)

Elimination of model coordinates gives the scale factors as:

$$\lambda = \frac{bx \cdot \overline{z}'' - bz \cdot \overline{x}''}{x' \cdot \overline{z}'' - z' \cdot \overline{x}''} \qquad \mu = \frac{bx \cdot z' - bz \cdot x'}{x' \cdot \overline{z}'' - z' \cdot \overline{x}''} \tag{4.59}$$

and hence the model coordinates

$$x = \lambda \cdot x' \qquad z = \lambda \cdot z' \tag{4.60}$$

$$y_1 = \lambda \cdot y' \qquad y_2 = by + \mu \cdot \overline{y}''$$

$$y = \frac{y_1 + y_2}{2} \qquad\qquad py = y_2 - y_1$$

Due to uncertainties in measurement there are two solutions for the model coordinates in the y direction, i.e. corresponding rays are skew and do not exactly intersect, which results in y-parallax *py*.

Additional arbitrary homologous image points can be measured in the relatively oriented model, and transformed into model coordinates *xyz* using equations (4.60). They describe a three-dimensional object surface, correctly shaped, but at an arbitrarily defined scale resulting from our arbitrary choice, $bx = 1$. The transformation of model coordinates into a global object coordinate system at true scale is performed by absolute orientation (see section 4.3.5). The set of equations 4.60 describe a special case of spatial intersection (see also sections 4.3.6.2, 4.4.7.1).

Example 4.6:

The homologous points in example 4.5 have the following model coordinates:

Model coordinates	x	y	z
P_1	–0.0044	–0.6795	–2.6113
P_2	0.7338	–0.6937	–2.6693
P_3	0.9510	–0.1422	–2.5572
P_4	0.7307	0.4485	–2.4026
P_5	0.0285	0.4569	–2.3484
P_6	–0.2064	–0.0787	–2.4567

The z coordinates are approximately 2.5 times the base length 1. This gives rise to an average height-to-base ratio of 2.5:1.

4.3.3.4 Calculation of epipolar lines

The equation of the epipolar line *k*" in the right-hand image is given in parametric form as:

$$k": \quad \mathbf{k}" = \mathbf{p}" + t(\mathbf{q}" - \mathbf{p}") \tag{4.61}$$

Here **k"** is the locus of points on the straight line through image points **p"** and **q"**, which correspond to the arbitrary model points P and Q lying on the ray **r'** (Fig. 4.31). If the parameters of relative orientation (exterior orientation of both images in the model coordinate system) are inserted into the collinearity equations (4.9), the image coordinates in the right-hand image are obtained (with $z" = -c_2$):

$$\begin{aligned} x"_i &= z" \frac{r_{11}(x_i - bx) + r_{21}(y_i - by) + r_{31}(z_i - bz)}{r_{13}(x_i - bx) + r_{23}(y_i - by) + r_{33}(z_i - bz)} \\ y"_i &= z" \frac{r_{12}(x_i - bx) + r_{22}(y_i - by) + r_{32}(z_i - bz)}{r_{13}(x_i - bx) + r_{23}(y_i - by) + r_{33}(z_i - bz)} \end{aligned} \tag{4.62}$$

Here r_{ik} are the elements of the rotation matrix of $\omega_2, \varphi_2, \kappa_2$. The perspective centre O' can be used in place of point P:

$$
\mathrm{P}: \quad \begin{aligned} x_P &= 0 \\ y_p &= 0 \\ z_p &= 0 \end{aligned}
$$

Point Q is given by multiplication of the image vector **x'** by an arbitrary scaling factor λ:

$$
\mathrm{Q}: \quad \begin{aligned} x_Q &= -\lambda x' \\ y_Q &= -\lambda y' \\ z_Q &= -\lambda z' \end{aligned}
$$

By inserting the model coordinates into (4.62), the image coordinates of points **p"** and **q"** are obtained and hence the straight line equation of the epipolar line. Due to unavoidable measurement errors, the search for point P", the homologous point to P', should not be done along straight line *k*" but within a narrow band either side of this line.

4.3.3.5 Calculation of normal-case images

Digitised convergent stereo images can be rectified by epipolar resampling in order to correspond to the normal case of stereo photogrammetry. After rectification they are suitable for ocular stereo viewing. In addition, the epipolar lines are parallel to the x' direction, enabling simplified algorithms for stereo image matching to be applied (see section 5.5.3.1).

Fig. 4.36 illustrates the spatial position of normal-case images with respect to the stereo model. With given exterior orientations for both images (e.g. in model coordinate system xyz), three-dimensional image coordinates $x'_n, y'_n, -c_n$ in the normal-case images can be transformed using (4.9) into the image coordinates x', y' of the original image (and analogously for the second image). Rectification is performed when, for all points in the images, the colour level of the original image $g'(x', y')$ is copied to position x'_n, y'_n in the normal-case image (see section 5.3).

Fig. 4.33 shows a strongly convergent stereo image pair and the normal-case stereo images derived from them. Homologous points then lie on the same y coordinate values. The principal points of the normal-case images are exactly at the image centres. The large areas with no image information are the result of the convergence of the original images and could be removed.

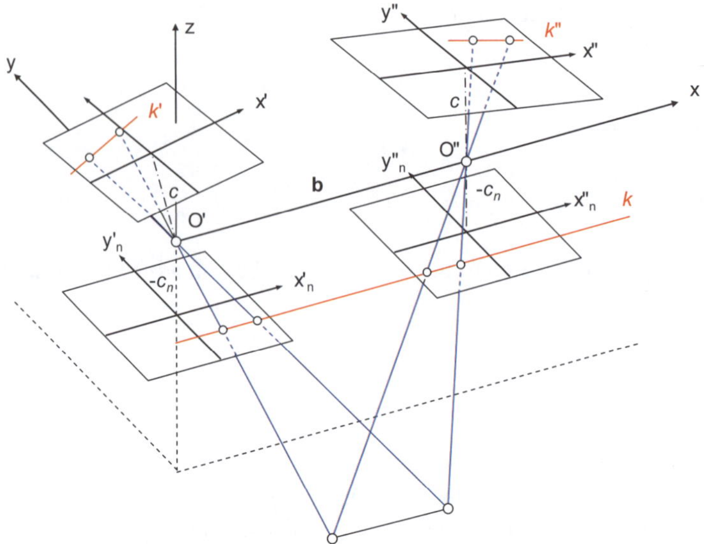

Fig. 4.36: The geometry of normal-case stereo images

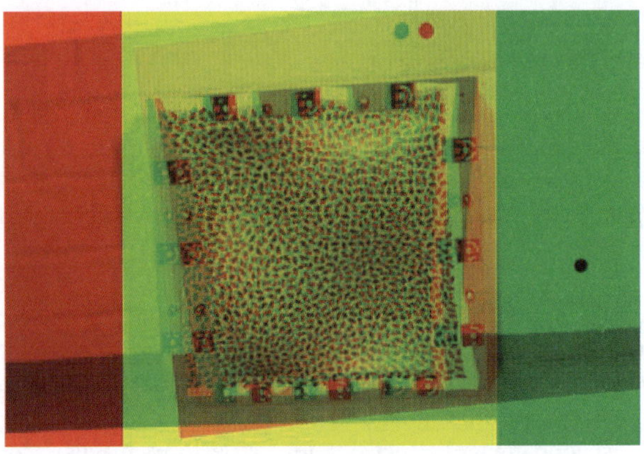

Fig. 4.37: Anaglyph image derived from normal-case stereo images

Fig. 4.37 shows an anaglyph stereo image which has been created from the normal-case image pair of Fig. 4.33. Note that the x-parallaxes of homologous points vary with object distance.

4.3.3.6 Quality of relative orientation

The existence of y-parallax py (defined in 4.60) at a point in the model indicates failure of homologous rays to intersect at that point. The y-parallaxes, considered over the whole model, may be used as a measure of the quality of relative orientation; y-parallax at photo scale gives a normalised figure.

If the y-parallax at a point, i, in the model is py_i, then the y-parallax at photo scale may be taken as

$$py'_i = \frac{z'}{z_i} py_i \tag{4.63}$$

Assuming that by and bz are small compared to bx, the following expression, based on a number of tie points, n, gives a measure of the quality of the relative orientation:

$$s_{py'} = \frac{1}{n}\sqrt{\sum_{i=1}^{n} py'^2_i} \tag{4.64}$$

The intersection angle α of homologous image rays is the angle between the two spatial vectors **r'** and **r''** where:

$$\cos\alpha = \frac{\mathbf{r}'^T \cdot \mathbf{r}''}{|\mathbf{r}'| \cdot |\mathbf{r}''|} \tag{4.65}$$

Taking all n tie points into account, the mean intersection angle can be calculated:

$$\bar{\alpha} = \frac{1}{n}\sum_{n} \alpha \tag{4.66}$$

The mean intersection angle also approximately describes the ratio of the stereo base b to the mean object distance h as given by (Fig. 4.38):

$$\tan\frac{\bar{\alpha}}{2} \approx \frac{b}{2h} \tag{4.67}$$

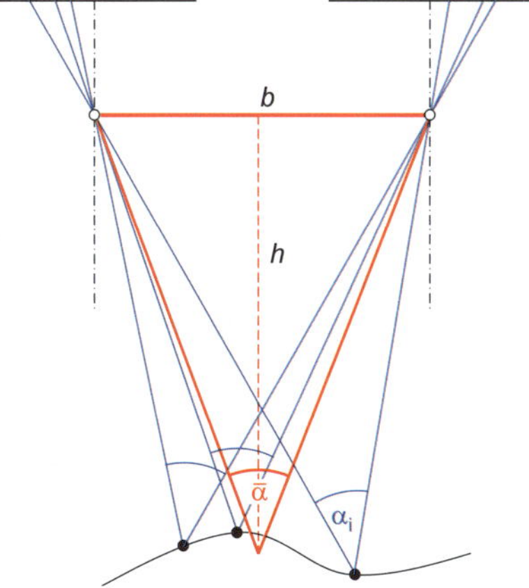

Fig. 4.38: Mean intersection angle

Accuracy of relative orientation and point determination will be optimized when the mean intersection angle is close to a right angle.

The quality of relative orientation depends on the following criteria:

- Accuracy of image coordinates:

  The accuracy of image coordinates depends partly on the measuring accuracy of the instrument and partly on the ability to identify matching points in both images. Image patches with poor structure can be matched less accurately than areas with a significant grey level structure, regardless of the image processing method (visual interpretation or digital image matching).

- Number and distribution of tie points in model space:

  Tie points should be chosen in model space to ensure a robust geometric link between both images. A point distribution as recommended by von Gruber is particularly suitable. This has a tie point in each corner of the model space and one in the middle of each long side. This distribution is strictly possible only in the normal case (Fig. 4.39).

  If object structure, for example containing large homogeneous areas, does not allow an optimum distribution of homologous points, model errors, which cannot be controlled, may occur in the regions not covered. If all tie points lie on a common straight line, the resulting normal system of equations becomes singular.

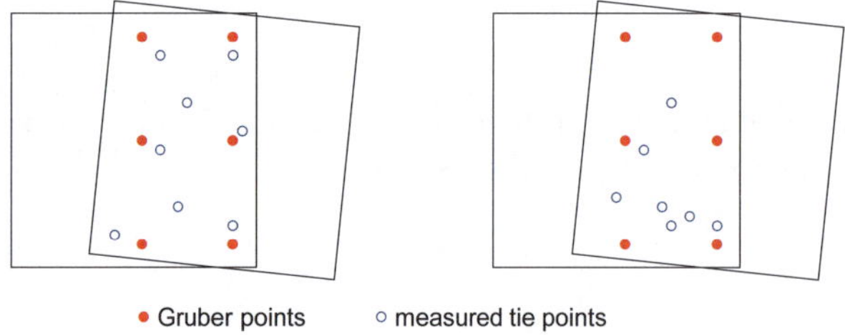

Fig. 4.39: Good and bad distribution of tie points in model space

  To properly control the relative orientation, at least 8-10 well distributed tie points should be measured.

- Height-to-base ratio:

  If the base of the stereo model is small relative to object distance (height) then ray intersection angles are poor. The parameters of relative orientation are then determined with greater uncertainty. As mentioned above, an optimal configuration is achieved with intersection angles of around 90 degrees.

- Distribution of tie points in object space:

  There are a few exceptional cases where singular or weakly conditioned normal equations occur, even though there is a good point distribution in model space. Amongst other cases this applies to the *danger cylinder*, where the object points used as tie points and the perspective centres of both images lie on a common cylindrical surface (Fig. 4.40). This effect can also occur where object surfaces have small curvatures and the imaging lens has a long focal length. The same problem exists for the space resection, if the image to be oriented is also located on a danger surface.

The result of this configuration is that the cameras do not have unique positions on the cylindrical surface because at different camera positions the tie points subtend the same angles and so have the same image positions (a fundamental property of circles).

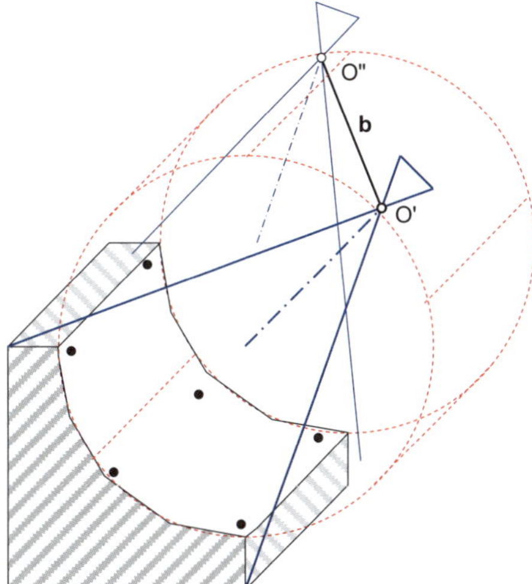

Fig. 4.40: Danger cylinder above a curved surface

4.3.3.7 Special cases of relative orientation

The method of relative orientation is widely used for traditional stereo image analysis on analytical stereo instruments. These applications typically have parallel imaging directions which permit stereoscopic viewing (as shown in Fig. 4.32). The orientation elements are then relatively small so that initial values for the iterative adjustment can be zero.

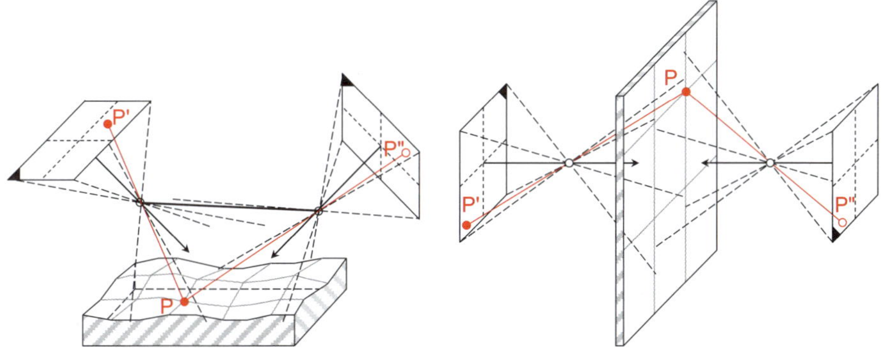

Fig. 4.41: Convergent image pair configurations

Close range photogrammetry, in contrast, often involves arbitrary, convergent, multi-image configurations. In this case relative orientation is not used as the actual orientation method but only as one step in the calculation of approximate values for the subsequent bundle adjustment (see section 4.4.4.1). Here the image pairs may have orientation values that differ significantly from those of the normal case (examples in Fig. 4.41).

Fig. 4.42 shows the distribution in image space of 12 homologous points in an image pair, in which the right-hand image has significant tilts with respect to the left-hand image.

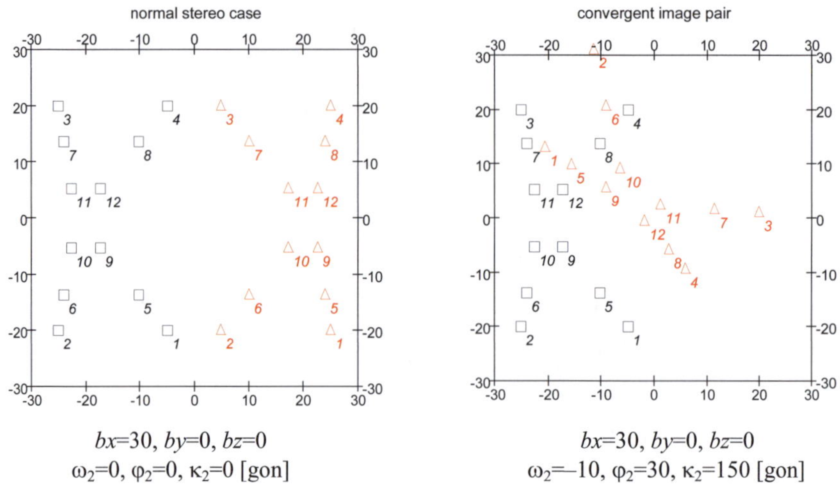

bx=30, *by*=0, *bz*=0 $\omega_2=0$, $\varphi_2=0$, $\kappa_2=0$ [gon]

bx=30, *by*=0, *bz*=0 $\omega_2=-10$, $\varphi_2=30$, $\kappa_2=150$ [gon]

Fig. 4.42: Image point distribution for different orientations
□ left image, Δ right image

It is not possible to define approximate values for the relative orientation of arbitrarily oriented images according to eqn. 4.56. Instead the methods of spatial similarity transformation (see section 2.2.2.2) or space resection (see section 4.2.3) can be applied. For these methods, tie points require 3D object coordinates which may be calculated by transforming image coordinates in both images by an approximate scale factor.

In multi-image applications it is possible that two or more images are exposed at the same point but with different orientations. It is also possible that images are located behind one another on the same viewing axis. In these and similar cases, both images cover a common model space which does not provide distinct intersections at object points (Fig. 4.43). The calculation of relative orientation then leads to poor results or fails completely. Such images can, of course, be included with others in a multi-image bundle adjustment.

A further special case occurs for relative orientation using images of strictly planar surfaces. This happens often in close-range applications, for example in the measurement of flat façades or building interiors. In this case only 4 tie points are required because both bundles of rays can be related to each other by an affine transformation with 8 parameters. In order to solve the adjustment problem, the planarity of the object surface can be handled by an additional constraint equation. This constraint can replace one of the required coplanarity conditions so that only four tie points are necessary.

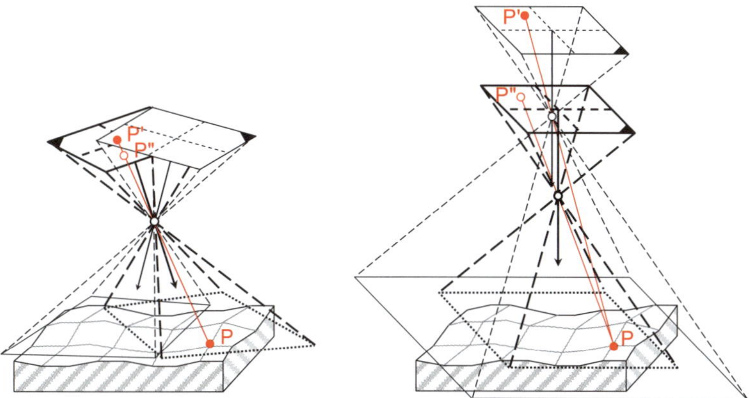

Fig. 4.43: Overlapping image pairs with insufficient spatial ray intersections

4.3.4 Fundamental matrix and essential matrix

The relationship between two images can also be derived with the aid of projective geometry. According to eqn. 4.29, a projection matrix can be defined for each image:

$$\mathbf{P}_1 = \boldsymbol{K}_1 \cdot \mathbf{R}_1 \cdot [\boldsymbol{I} \mid \boldsymbol{0}] \qquad \mathbf{P}_2 = \boldsymbol{K}_2 \cdot \mathbf{R}_2 \cdot [\boldsymbol{I} \mid -\boldsymbol{b}] \tag{4.68}$$

where

$\mathbf{R}_1, \mathbf{R}_2$: rotation matrices

$\boldsymbol{K}_1, \boldsymbol{K}_2$: calibration matrices

$\boldsymbol{b}$: base vector

For relative orientation (fixed left-hand image) $\mathbf{R}_1 = \mathbf{I}$. The image coordinates of homologous points in both images can be transformed into a local camera coordinate system which initially has an arbitrary principal distance $c_1 = c_2 = 1$.

$$\boldsymbol{x}_1 = \mathbf{R}_1^{-1} \cdot \boldsymbol{K}_1^{-1} \cdot \boldsymbol{x}' \qquad \boldsymbol{x}_2 = \mathbf{R}_2^{-1} \cdot \boldsymbol{K}_2^{-1} \cdot \boldsymbol{x}'' \tag{4.69}$$

The transformed image coordinates now lie in the same coordinate system as the base vector $\boldsymbol{b}$ and can be used in the following coplanarity condition (compare with eqn. 4.53):

$$\boldsymbol{x}_1 \cdot (\boldsymbol{b} \times \boldsymbol{x}_2) = \boldsymbol{x}_1^T \cdot \mathbf{S}_b \cdot \boldsymbol{x}_2 = 0 \tag{4.70}$$

$$\text{where } \mathbf{S}_b = \begin{bmatrix} 0 & -bz & by \\ bz & 0 & -bx \\ -by & bx & 0 \end{bmatrix}$$

The vector product of vectors $\boldsymbol{b}$ and $\boldsymbol{x}_2$ can therefore be expressed using the skew symmetric matrix $\mathbf{S}_b$. Insertion of (4.69) in (4.70) leads to the linearized condition:

$$\boldsymbol{x}_1^T \cdot \mathbf{S}_b \cdot \boldsymbol{x}_2 = \boldsymbol{x}'^T \cdot (\boldsymbol{K}_1^{-1})^T \cdot \mathbf{S}_b \cdot \mathbf{R}_2^{-1} \cdot \boldsymbol{K}_2^{-1} \cdot \boldsymbol{x}'' = 0 \tag{4.71}$$

Using

$$F = (K_1^{-1})^T \cdot \mathbf{S}_b \cdot \mathbf{R}_2^{-1} \cdot K_2^{-1} \qquad \text{: fundamental matrix} \tag{4.72}$$

the coplanarity condition for homologous image points can be expressed in the simple form:

$$x'^T \cdot F \cdot x'' = 0 \qquad \text{: coplanarity condition} \tag{4.73}$$

The fundamental matrix F is a homogeneous 3x3 matrix, i.e. multiplication by a scalar does not alter the projection. As a consequence, it can be described by 9–1 = 8 degrees of freedom. F contains all necessary relative orientation data, including the parameters of interior orientation. Using the fundamental matrix, the epipolar line in the partner image can be calculated:

$$x'^T \cdot F = k'' = \begin{bmatrix} A_{k''} \\ B_{k''} \\ C_{k''} \end{bmatrix} \qquad \text{: epipolar line in right image} \tag{4.74}$$

Here $A_{k''}, B_{k''}, C_{k''}$ correspond to the parameters of a straight line defined according to eqn. 2.72.

If the interior orientation of both images is known (known calibration matrix K), the fundamental matrix reduces to the *essential matrix* E:

$$x'^T_k \cdot \mathbf{S}_b \cdot \mathbf{R}_2^{-1} \cdot x''^T_k = x'^T_k \cdot E \cdot x''^T_k = 0 \tag{4.75}$$

where

$$x'_k = K_1^{-1} \cdot x'$$

$$x''_k = K_2^{-1} \cdot x''$$

$$E = \mathbf{S}_b \cdot \mathbf{R}_2^{-1}$$

At least eight homologous points are required to calculate the fundamental matrix F. In comparison only five are required to solve for the essential matrix E. The linear system of equations in each case can be solved, for example, using the *single value decomposition.*

4.3.5 Absolute orientation

4.3.5.1 Mathematical model

Absolute orientation describes the transformation of the local model coordinate system xyz, resulting from a relative orientation with arbitrary position, rotation and scale, into the object coordinate system XYZ via reference points. Reference points are object points measured in the model coordinate system which have one or more known coordinate components in object space (e.g. XYZ, XY only or Z only). The reference points can be identical to the tie points already used for relative orientation, or they can be measured subsequently as model points in the relatively oriented model.

Absolute orientation consists of a spatial similarity transformation with three translations, three rotations and one scaling factor as described in equation 2.40 (see section 2.2.2.2). In order to solve the system of equations, a minimum of seven suitable point elements are required, for example taken from three spatially distributed XYZ reference points.

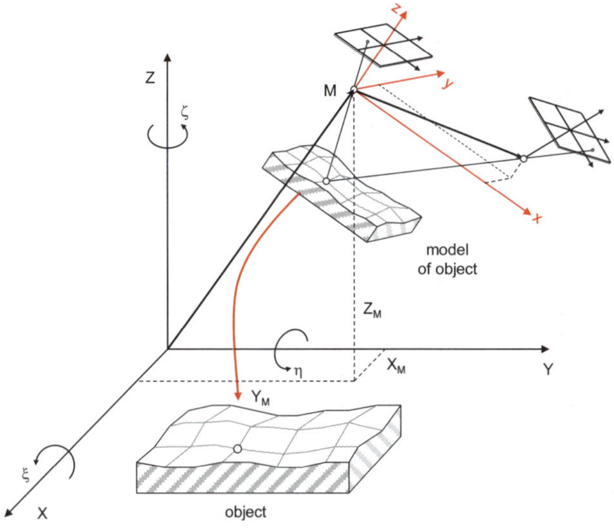

Fig. 4.44: Absolute orientation

Fig. 4.44 illustrates the transformation of the model coordinate system, xyz with origin at M, into the object coordinate system, XYZ. The coordinates of M in the XYZ system are $\mathbf{X_M}$. The rotation matrix **R** is a function of the three rotation angles ξ,η,ζ about the axes XYZ. The transformation for a model point with coordinates *xyz* (vector **x**) is given by:

$$\begin{aligned}\mathbf{X} &= F(X_M, Y_M, Z_M, m, \xi, \eta, \zeta, x, y, z) \\ &= \mathbf{X_M} + m \cdot \mathbf{R} \cdot \mathbf{x}\end{aligned} \tag{4.76}$$

or

$$\begin{bmatrix} X \\ Y \\ Z \end{bmatrix} = \begin{bmatrix} X_M \\ Y_M \\ Z_M \end{bmatrix} + m \cdot \begin{bmatrix} r_{11} & r_{12} & r_{13} \\ r_{21} & r_{22} & r_{23} \\ r_{31} & r_{32} & r_{33} \end{bmatrix} \cdot \begin{bmatrix} x \\ y \\ z \end{bmatrix}$$

where *m* is the scale factor between model and object coordinates.

Equations 4.76 are non-linear and are solved in the usual way; if there is redundancy the solution will be based on a least-squares adjustment (see section 2.4.2.2) in which each coordinate component of a reference point provides one linearised correction equation:

$$
\begin{aligned}
X + v_X &= dX_M + \frac{\partial F}{\partial m} dm + \frac{\partial F}{\partial \xi} d\xi + \frac{\partial F}{\partial \eta} d\eta + \frac{\partial F}{\partial \zeta} d\zeta + X^0 \\
Y + v_Y &= dY_M + \frac{\partial F}{\partial m} dm + \frac{\partial F}{\partial \xi} d\xi + \frac{\partial F}{\partial \eta} d\eta + \frac{\partial F}{\partial \zeta} d\zeta + Y^0 \\
Z + v_Z &= dZ_M + \frac{\partial F}{\partial m} dm + \frac{\partial F}{\partial \xi} d\xi + \frac{\partial F}{\partial \eta} d\eta + \frac{\partial F}{\partial \zeta} d\zeta + Z^0
\end{aligned}
\tag{4.77}
$$

4.3.5.2 Definition of the datum

In many close-range applications, 3D reference points (full reference points) are available. Each reference point therefore provides three correction equations. In aerial photogrammetry it is possible that some reference points only have known plan position (XY) and others only height (Z), resulting in a reduced set of correction equations. In order to solve the absolute orientation, at least 2 X coordinates, 2 Y coordinates and 3 Z coordinates must be available (see also section 4.4.3.2).

The reference points should be well distributed over the object space to be transformed. If all reference points lie on a common straight line, a singular or weak system of equations results. In over-determined configurations of control points, inconsistencies between coordinates (network strain) can distort the transformation parameters and give rise to higher standard deviations in the transformation.

4.3.5.3 Calculation of exterior orientations

From the parameters of relative and absolute orientation for an image pair, the exterior orientation parameters of each image can be calculated.

The position of the perspective centre, $\mathbf{X}_{0i}$, of an image i is derived from the origin of the model coordinate system $\mathbf{X_M}$ and the transformed components of the base $\mathbf{b}$:

$$\mathbf{X}_{0i} = \mathbf{X_M} + m \cdot \mathbf{R}_{\xi\eta\zeta} \cdot \mathbf{b}_i \tag{4.78}$$

For the left-hand image ($i = 1$) the base components are zero, hence $\mathbf{X}_{0i} = \mathbf{X_M}$.

In order to calculate the rotation matrix of image i, $\mathbf{R}_i(\omega\varphi\kappa)$, with respect to the object system, the rotation matrix $\mathbf{A}_i(\omega\varphi\kappa)$ of the relative orientation is pre-multiplied by the rotation matrix $\mathbf{R}(\xi\eta\zeta)$ of the absolute orientation:

$$\mathbf{R}_i(\omega\varphi\kappa) = \mathbf{R}(\xi\eta\zeta) \cdot \mathbf{A}_i(\omega\varphi\kappa) \tag{4.79}$$

After absolute orientation, object coordinates are available for the model points. As an alternative, therefore, the parameters of exterior orientation can also be determined by space resection using the transformed model coordinates in object space.

4.3.5.4 Calculation of relative orientation from exterior orientations

If the parameters of exterior orientation, in the object coordinate system, are available for both images of the stereo pair, then the parameters of relative orientation can be derived.

For relative orientation with origin in the left-hand image, rotation matrices of relative orientation $\mathbf{A}_i$ are obtained by multiplication of the exterior orientation matrices $\mathbf{R}_i$ by the inverse rotation of the left image:

$$\mathbf{A}_1 = \mathbf{R}_1^{-1} \cdot \mathbf{R}_1 = \mathbf{I} \qquad \mathbf{A}_2 = \mathbf{R}_1^{-1} \cdot \mathbf{R}_2 \tag{4.80}$$

The required base components in the model coordinate system are obtained from the vector between the two perspective centres, transformed by the inverse rotation of the left-hand image. The resulting vector $\mathbf{B}$ is then scaled to create a base vector $\mathbf{b}$ where $bx = 1$.

$$\mathbf{X}_{0i} = \mathbf{0} \qquad \begin{aligned} \mathbf{B} &= \mathbf{R}_1^{-1} \cdot (\mathbf{X}_2 - \mathbf{X}_1) = \begin{bmatrix} Bx & By & Bz \end{bmatrix}^T \\ \mathbf{b} &= \frac{1}{Bx} \cdot \mathbf{B} = \begin{bmatrix} 1 & by & bz \end{bmatrix}^T \end{aligned} \tag{4.81}$$

Example 4.7:

The exterior orientation parameters of the stereo pair in Fig. 4.33 are given as follows:

	Image 1 (left):	Image 2 (right)
Translation:	$X_{01} = -471.887$ mm	$X_{02} = -78.425$ mm
	$Y_{01} = 11.033$ mm	$Y_{02} = -12.630$ mm
	$Z_{01} = 931.071$ mm	$Z_{02} = 916.554$ mm
Rotation:	$\omega_1 = -13.0592°$	$\omega_2 = -11.6501°$
	$\varphi_1 = -4.4402°$	$\varphi_2 = -0.3134°$
	$\kappa_1 = 0.7778°$	$\kappa_2 = 3.4051°$

Solution for relative orientation:

Rotations are calculated using eqn. 4.80

$$\mathbf{A}_1 = \begin{bmatrix} 1 & 0 & 0 \\ 0 & 1 & 0 \\ 0 & 0 & 1 \end{bmatrix} \qquad \mathbf{A}_2 = \begin{bmatrix} 0.99647 & -0.04378 & 0.07160 \\ 0.04572 & 0.99862 & -0.02556 \\ -0.07038 & 0.02874 & 0.99710 \end{bmatrix}$$

Rotation angles: $\omega = 1.4687°$ $\varphi = 4.1059°$ $\kappa = 2.5162°$

Base components are calculated as:

$$\mathbf{B} = \begin{bmatrix} 390.471 \\ -25.071 \\ -49.890 \end{bmatrix} \qquad \mathbf{b} = \begin{bmatrix} 1 \\ -0.0642 \\ -0.1277 \end{bmatrix}$$

The result is largely identical to the relative orientation calculated in example 4.5. Any small differences are due to the fact that the parameters of exterior orientation (used here) originate in a bundle adjustment composed of 16 images.

4.3.6 Stereoscopic processing

4.3.6.1 Principle of stereo image processing

Stereo processing covers all visual or computational methods for the processing of a stereo image pair. Traditionally, it has greatest application in aerial photogrammetry.

In close-range work, stereo photogrammetry is used in the following example applications:

- Visual processing of natural features:

  The reconstructed object is measured in a stereo plotting instrument using binocular, stereoscopic optical viewing systems. The operator observes an optically or digitally generated "floating mark", the apparent spatial position of which is under his or her control. A measurement is taken when the floating mark appears to lie on the virtual surface of the object; the point measured corresponds to the pair of homologous image points simultaneously viewed stereoscopically. The movement of the floating mark on to the surface is controlled interactively by the operator. Single points may be measured or continuous lines may also be measured as the operator moves the floating mark over the virtual surface.

- Visual or digital reconstruction of free-form surfaces:

  Object surfaces of arbitrary shape can be evaluated by stereo photogrammetry if the surface structure permits the identification (matching) of homologous points. Surfaces with insufficient visual pattern or structure must therefore be prepared with a suitable texture, e.g. by pattern projection or other method. Image processing is performed either by the visual method above, or by image processing algorithms which implement stereo image matching of corresponding points (see section 5.5). The final goal is the complete 3D reconstruction of the free-form surface, for example as a digital surface model.

- Image acquisition with stereometric cameras:

  Stereometric cameras (see section 3.5.4) are usually configured to correspond to the normal case of stereo photogrammetry. They provide a simple method of imaging and of photogrammetric object reconstruction which, to a large extent, avoids complicated orientation procedures.

- Point-by-point (tactile) object measurement with online dual camera systems:

  Online photogrammetric systems comprising two digital metric cameras can be treated as stereo systems, although they can easily be extended to incorporate more than two cameras. The object is measured by spatial intersection of targeted points (targets, probes) which can be detected and identified automatically. If the exposure of both cameras is synchronised, the object can be measured by hand-held contact probes (see section 6.4.2.2).

- Control of vision-based machines (e.g. autonomous robots):

  There are a number of applications in computer vision (stereo vision, shape from stereo) where a scene is analysed by stereo-based algorithms which reflect the mechanisms of natural human vision. Examples are the control of autonomous robots in unknown environments (avoidance of collisions) and the control of production tools (see example in Fig. 6.10).

The principle of stereo processing is based on the correspondence of homologous points lying in an epipolar plane. The epipolar plane intersects the image planes in epipolar lines (see section 4.3.2). For the normal case of stereo photogrammetry (Fig. 4.30) the epipolar lines are parallel and depth information can be determined by measuring the x-parallax px'.

4.3.6.2 Point determination using image coordinates

Coordinate calculation in normal case

The normal case of stereo photogrammetry is, in fact, the special case in which two identical cameras have parallel axes pointing in the same direction at right angles to the stereo base. With respect to an XYZ coordinate system located in the left perspective centre, object coordinates can be derived from the ratios indicated in Fig. 4.45:

Parallel to the image plane:

$$X = \frac{h}{c} \cdot x' = m \cdot x' \qquad Y = \frac{h}{c} \cdot y' = m \cdot y' \tag{4.82}$$

In the viewing direction:

$$\frac{h}{c} = \frac{b}{x' - x''} = m$$

and it follows that:

$$Z = h = \frac{b \cdot c}{x' - x''} = \frac{b \cdot c}{px'} \tag{4.83}$$

(4.82) and (4.83) can also be derived from the collinearity equations (4.9). The rotation angles of both images are zero. The right image is shifted in the X direction by the base length b with respect to the left image. Hence it follows that:

$$x'_{01} = y'_{01} = x'_{02} = y'_{02} = 0$$

$$X_{01} = Y_{01} = Z_{01} = Y_{02} = Z_{02} = 0 \qquad X_{02} = b$$

$$\omega_1 = \varphi_1 = \kappa_1 = \omega_2 = \varphi_2 = \kappa_2 = 0 \qquad \mathbf{R}_1 = \mathbf{R}_2 = \mathbf{I}$$

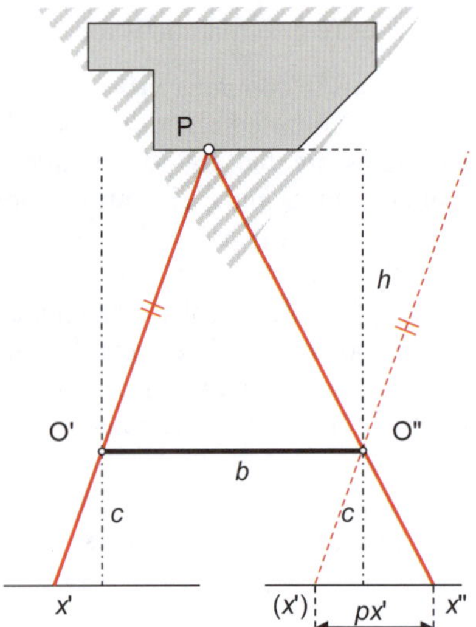

Fig. 4.45: Normal case of stereo photogrammetry

The x-parallax *px'* is measured either by visual examination and coordinate measurement in a stereo plotter (stereo comparator), or by methods of digital image matching. As an example, Fig. 4.46 shows the measurement of two object points. Point P_1 on the manhole cover closest to the cameras has a much larger x-parallax than the more distant point P_2 at the top of the tower. The *y'* image coordinates are almost equal, i.e. y-parallax does not exist.

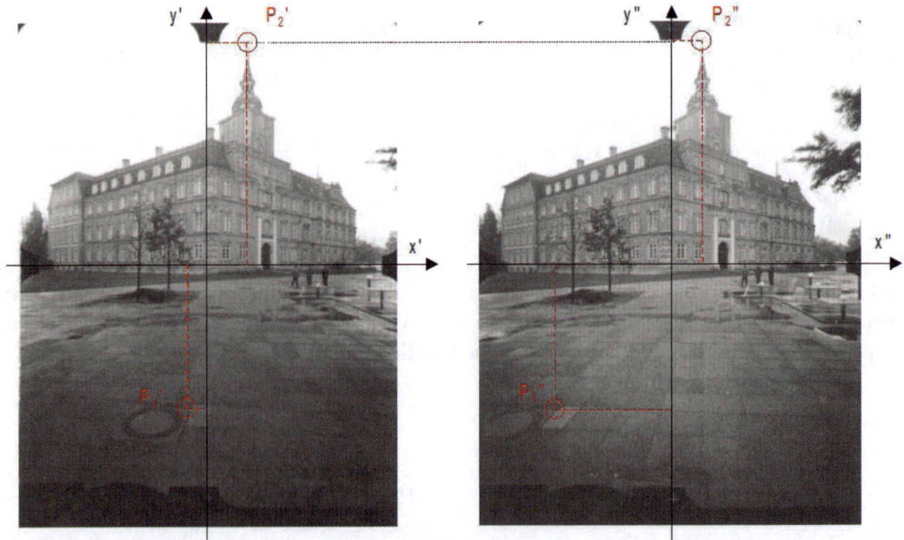

Fig. 4.46: Measurement of two object points in a stereo image pair

Example 4.8:

The following image coordinates have been measured for the example image pair above (SMK 120, $b = 1.2$ m, $c = 64$ mm). The object coordinates of the two points are to be calculated.

Solution:

1. Point P_1	$x' = -3.924$ mm	$x'' = -23.704$ mm
	$y' = -29.586$ mm	$y'' = -29.590$ mm
	$px' = x' - x'' = 19.780$ mm	
2. Point P_2	$x' = 7.955$ mm	$x'' = 6.642$ mm
	$y' = 45.782$ mm	$y'' = 45.780$ mm
	$px' = x' - x'' = 1.313$ mm	
3. Z coordinate (distance)	$Z = \frac{b \cdot c}{px'}$	$Z_1 = 3.88\text{m}$ $Z_2 = 58.49\text{m}$
4. XY coordinates	$X_1 = -0.24\text{m}$ $Y_1 = -1.79\text{m}$ $m = 61$	$X_2 = 7.27\text{m}$ $Y_2 = 41.84\text{m}$ $m = 914$

The result shows that point P_1 lies beneath the left-hand camera at a distance of 3.88 m. Point P_2 is located to the right and above the left camera at a distance of 58.49 m and with a smaller image scale.

Accuracy

Differentiation of eqn. 4.83, and application of error propagation, gives the following accuracy estimation of the object coordinate in viewing direction Z (c and b are assumed to be free of error).

$$s_Z = \frac{Z^2}{b \cdot c} s_{px'} = \frac{h}{b} \cdot \frac{h}{c} s_{px'} = q \cdot m \cdot s_{px'} \tag{4.84}$$

The equation shows that the accuracy in the viewing direction is a function of the accuracies of parallax measurement, image scale $m = h/c$ and height-to-base ratio h/b, which corresponds to the design factor q introduced in section 3.3.1.2. The equation also shows that, since b and c are constant for any particular case, the accuracy falls off in proportion to the square of the distance, Z. The height-to-base ratio, or more correctly distance-to-base ratio, describes the intersection geometry. If the base is small compared with the distance, ray intersection is weak and accuracy in the viewing direction is poor.

In general, parallax measurement accuracy can be estimated as

$$s_{px'} = \frac{s_{x'}}{\sqrt{2}} \tag{4.85}$$

i.e. it is slightly more accurate than a single measured image coordinate. In monoscopic measurement, error propagation for $px' = x' - x''$ gives:

$$s_{px'} = s_{x'}\sqrt{2} \tag{4.86}$$

The accuracy of the X and Y coordinates can be similarly derived from eqn. 4.82:

$$\begin{aligned} s_X &= \sqrt{\left(\frac{x'}{c}s_Z\right)^2 + \left(\frac{Z}{c}s_{x'}\right)^2} = \sqrt{\left(\frac{x'}{c}q\cdot m\cdot s_{px'}\right)^2 + (m\cdot s_{x'})^2} \\ s_Y &= \sqrt{\left(\frac{y'}{c}s_Z\right)^2 + \left(\frac{Z}{c}s_{y'}\right)^2} = \sqrt{\left(\frac{y'}{c}q\cdot m\cdot s_{px'}\right)^2 + (m\cdot s_{y'})^2} \end{aligned} \tag{4.87}$$

In (4.87) the dominant term is the second summand of the square root. Object accuracy can therefore usually be estimated as

$$s_X = s_Y = m\cdot s_{x'y'} \tag{4.88}$$

The following object accuracies result in example 4.8 above:

Example 4.9:

Assume image coordinates are measured to an accuracy of 10 µm, resulting in a parallax accuracy of 7 µm. Principal distance and base length are assumed free of error.

Object accuracies:

1. Point P_1: $s_X = 0.6$ mm $s_Y = 0.6$ mm $s_Z = 1.4$ mm $q = h/b = 3.2$
2. Point P_2: $s_X = 9.1$ mm $s_Y = 9.1$ mm $s_Z = 315$ mm $q = h/b = 48$

It is clear that the accuracy of point P_2 is significantly decreased in Z and coordinate determination of this point is not practicable.

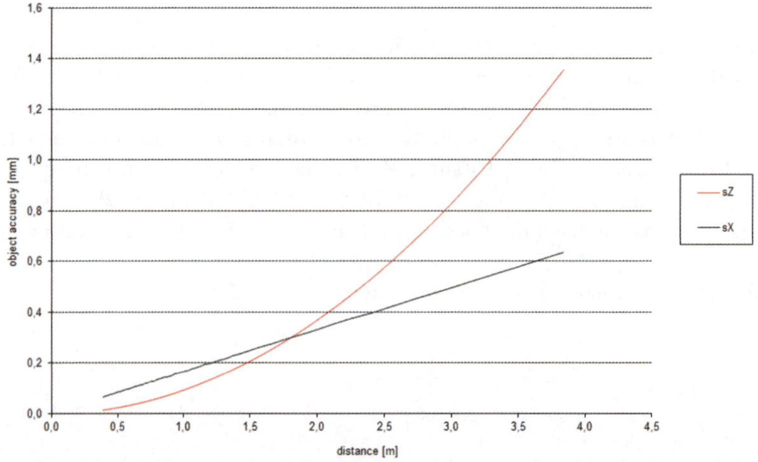

Fig. 4.47: Object accuracy for the normal case of stereo photogrammetry (from example 4.8)

Fig. 4.47 shows the object accuracies for additional point distances taken from example 4.8. For object distances less than about 1.7 m ($q = 1.4 = 1{:}0.7$), the accuracy s_Z in the viewing direction is higher than in the other directions. At longer distances the increase in uncertainty s_Z is quadratic whilst the increase in s_X remains linear.

Intersection in the general stereo case

A stereo pair which is not configured according to the strictly normal case has orientation parameter values which are not equal to zero. In addition the images can have arbitrary parameters of interior orientation.

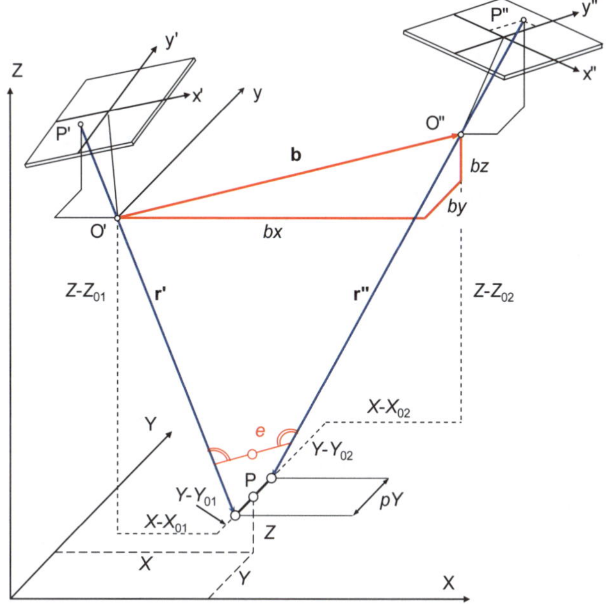

Fig. 4.48: Spatial intersection for the general stereo case

Object coordinates *XYZ* can be calculated by spatial intersection of the rays **r'** and **r''** if the parameters of interior and exterior orientation of both images are known (Fig. 4.48). Both spatial rays are defined by the measured image coordinates, transformed by the orientation parameters. For the special case of a stereo pair, the spatial intersection can be calculated as follows (see also the calculation of model coordinates, section 4.3.3.3):

1. Transformation of image coordinates:

$$\begin{bmatrix} X' \\ Y' \\ Z' \end{bmatrix} = \begin{bmatrix} X_{01} \\ Y_{01} \\ Z_{01} \end{bmatrix} + \mathbf{R}_1 \cdot \begin{bmatrix} x' \\ y' \\ z' \end{bmatrix} \qquad \begin{bmatrix} X'' \\ Y'' \\ Z'' \end{bmatrix} = \begin{bmatrix} X_{02} \\ Y_{02} \\ Z_{02} \end{bmatrix} + \mathbf{R}_2 \cdot \begin{bmatrix} x'' \\ y'' \\ z'' \end{bmatrix} \tag{4.89}$$

2. Stereo base components:

$$bx = X_{02} - X_{01}$$
$$by = Y_{02} - Y_{01} \qquad (4.90)$$
$$bz = Z_{02} - Z_{01}$$

For the simple version of the intersection, the skew rays intersect the XY plane at elevation Z of object point P, giving rise to two possible solutions (Fig. 4.48), i.e.

$$X = X_1 = X_2$$
$$Z = Z_1 = Z_2 \qquad (4.91)$$
$$Y = \frac{Y_1 + Y_2}{2}$$

3. Scale factors:

The scale factors for the transformation of image coordinates are:

$$\lambda = \frac{bx \cdot (Z''-Z_{02}) - bz \cdot (X''-X_{02})}{(X'-X_{01}) \cdot (Z''-Z_{02}) - (X''-X_{02}) \cdot (Z'-Z_{01})}$$
$$\mu = \frac{bx \cdot (Z'-Z_{01}) - bz \cdot (X'-X_{01})}{(X'-X_{01}) \cdot (Z''-Z_{02}) - (X''-X_{02}) \cdot (Z'-Z_{01})} \qquad (4.92)$$

4. Object coordinates:

$$X = X_{01} + \lambda \cdot (X'-X_{01}) \qquad Y_1 = Y_{01} + \lambda \cdot (Y'-Y_{01})$$
$$Z = Z_{01} + \lambda \cdot (Z'-Z_{01}) \qquad Y_2 = Y_{02} + \mu \cdot (Y''-Y_{02}) \qquad (4.93)$$
$$Y = \frac{Y_1 + Y_2}{2} \qquad pY = Y_2 - Y_1$$

Here the Y-parallax in object space pY is a quality measure for coordinate determination. It is zero when the two rays exactly intersect. However, pY may be zero if image measuring errors occur in the direction of epipolar lines.

The solution is not completely rigorous but works in most cases where the base is approximately aligned with the X direction and where bx is large in comparison with by and bz.

In the general case of two images with arbitrary orientations, point P is calculated as the mid point of the shortest distance e between both rays (Fig. 4.48; for calculation see section 2.3.2.1). The spatial intersection can also be expressed as an over-determined adjustment problem based on the collinearity equations. In this form it can be extended to more than two images (see section 4.4.7.1).

Example 4.10:

For the stereo pair in Fig. 4.33 there are image coordinates of homologous points from example 4.5, as well as parameters for interior and exterior orientation in the object coordinate system of example 4.7. Object point coordinates computed from a spatial intersection are required.

Solution according to eqn. 4.93:

Object coordinates:	X [mm]	Y [mm]	Z [mm]	pY [mm]
P_1	−390.9432	−477.5426	0.0168	0.0043
P_2	−101.5422	−479.1967	0.1027	0.0069
P_3	−23.2276	−256.8409	−0.0839	−0.0065
P_4	−116.8842	−21.0439	0.0190	0.0138
P_5	−392.1723	−21.4735	0.0974	0.0140
P_6	−477.5422	−237.6566	−0.1844	−0.0055

Separate evaluation of the stereo pair shows measurement noise in the object space of the order of 10 µm to 20 µm. Taking the object-space y-parallaxes pY as a measure of the quality of the spatial intersection, it can be seen that these lie within the measurement noise.

4.3.6.3 Point determination with floating mark

Setting a floating mark onto the surface

The term *floating mark* is used here for a digitally generated stereoscopic mark that can be moved through the virtual 3D space of the stereo model. The floating mark is set onto the object surface in order to measure a surface point. Although the floating mark primarily serves the interactive and visual analysis of the stereo image, its principle is also useful for automatic, digital stereo measurement.

The numerical reconstruction of homologous rays is performed in analytical or digital stereoplotters (see section 6.3.1). Using separate optical paths, the operator observes two floating marks which fuse into one common mark if set correctly onto the object surface. When the floating mark appears to touch the surface the corresponding XYZ coordinates are recorded.

The XYZ coordinates which correspond to a spatially controlled floating mark can be transformed into image coordinates using equations (4.9). Fig. 4.49 shows how the transformed marks only identify corresponding image patches (homologous points) if the XYZ coordinates represent a point P on the object surface. A measuring position below P at point R, as well a position above P at point Q, result in image points which do not correspond, namely the imaging positions corresponding to the intersections of the non-homologous rays with the object surface. The correct position of the floating mark is controlled either by a visual check or by a digital correspondence algorithm applied to the two calculated image positions. The mechanical effort is reduced to a separate real-time shift of both image planes. Starting from an approximate position, the XYZ coordinates of the floating mark are iteratively corrected until correspondence of both images is achieved. It may be noted that this approach enables the measurement of Z above regular XY grid positions.

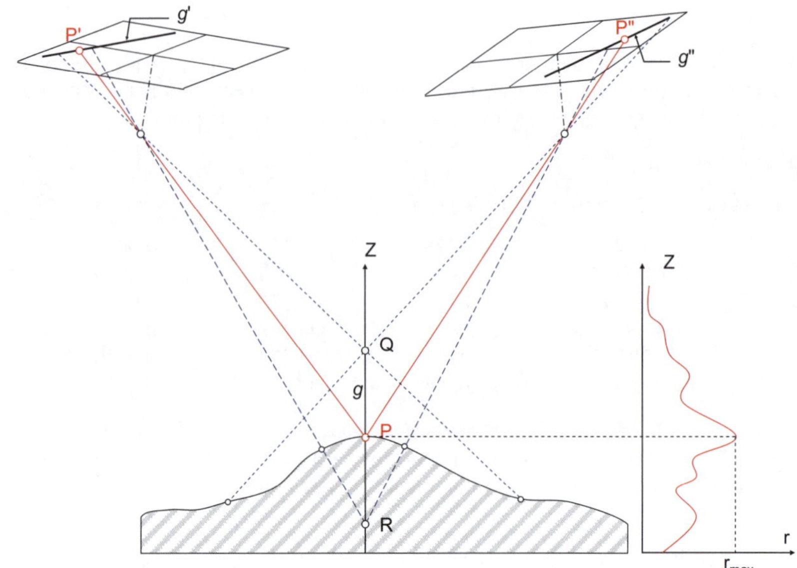

Fig. 4.49: Imaging positions and correlation coefficient *r* of a vertically shifted floating mark

Vertical line locus

The image lines *g*' and *g*'' correspond to the projection of the vertical line *g* which passes through P (Fig. 4.49). These straight lines are epipolar lines only if *g* is located in the epipolar plane of P. The image lines *g*' and *g*'' are known as vertical line loci (VLL), in allusion to aerial photogrammetry (Z ≈ viewing direction). With given orientation parameters they can be easily calculated by a variation of the *Z* coordinate of P.

In order to measure a surface point, it is possible to calculate all points P_i at small intervals ΔZ between two points (e.g. Q and R), re-project them into the images and search for the best correspondence on the straight lines *g*' and *g*''.

For arbitrarily oriented object surfaces, *g* should lie in the direction of the normal vector to the surface at the target point. The method is not restricted to the stereo case but can be extended to an unlimited number of images per point.

4.4 Multi-image processing and bundle adjustment

4.4.1 General remarks

4.4.1.1 Objectives

Bundle adjustment (bundle triangulation, bundle block adjustment, multi-image triangulation, multi-image orientation) is a method for the simultaneous numerical fit of an unlimited number of spatially distributed images (bundles of rays). It makes use of photogrammetric observations (measured image points), survey observations and an object coordinate system (Fig. 4.50). Using tie points, single images are merged into a global model in which the object surface can be reconstructed in three dimensions. The connection to a global object coordinate system can be provided by a minimum number of reference points so that larger areas without reference points can be bridged by multi-image sub-sets. The most important geometric constraint is that all corresponding (homologous) image rays should intersect in their corresponding object point with minimum inconsistency.

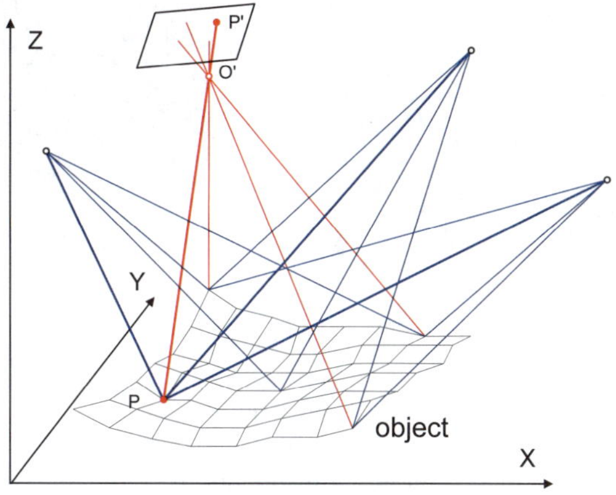

Fig. 4.50: Multi-image triangulation

In an over-determined system of equations, an adjustment technique estimates 3D object coordinates, image orientation parameters and any additional model parameters, together with related statistical information about accuracy and reliability. All observed (measured) values, and all unknown parameters of a photogrammetric project, are taken into account within one simultaneous calculation which ensures that homologous rays optimally intersect. In this way, the ray bundles provide strong geometry for a dense, high-accuracy measurement network (example in Fig. 4.51). The bundle triangulation therefore represents the most powerful and accurate method of image orientation and point determination in photogrammetry.

Bundle adjustment is a very general technique which combines elements of geodetic and photogrammetric triangulation, space resection and camera calibration. These are individually well understood and so the practical problems in the implementation of bundle adjustment do not lie in the mathematical formulations but in the following areas:

- solution of large systems of normal equations (up to a few thousand unknowns)
- generation of approximate values for the unknowns
- detection and elimination of gross data errors

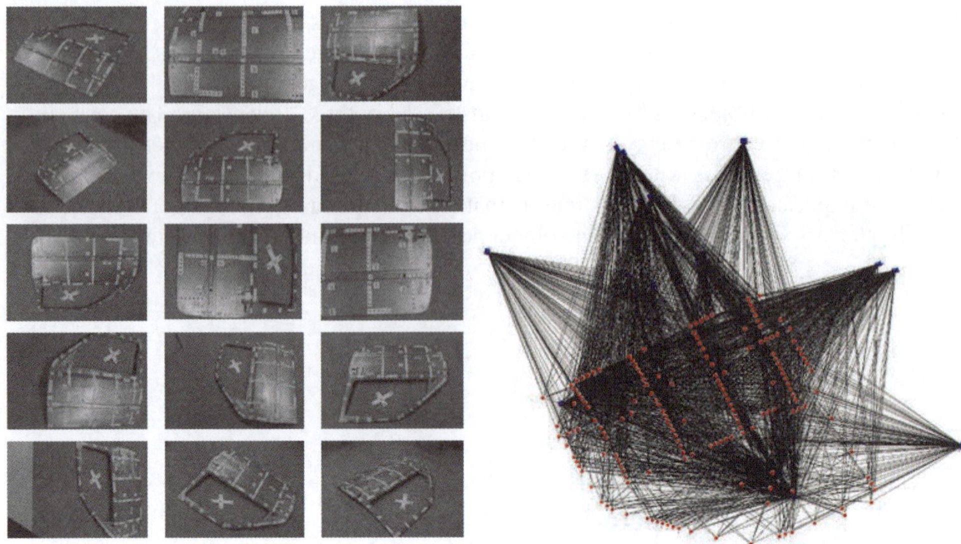

a) Images in a multi-image triangulation

b) Corresponding 3D network formed by the ray bundles

Fig. 4.51: Multi-image network for measuring a car door

The development of practical bundle adjustments is closely related to increases in computing power. In this respect it is worth noting that similar programs for aerial photogrammetry have largely been developed independently of those for close-range applications.

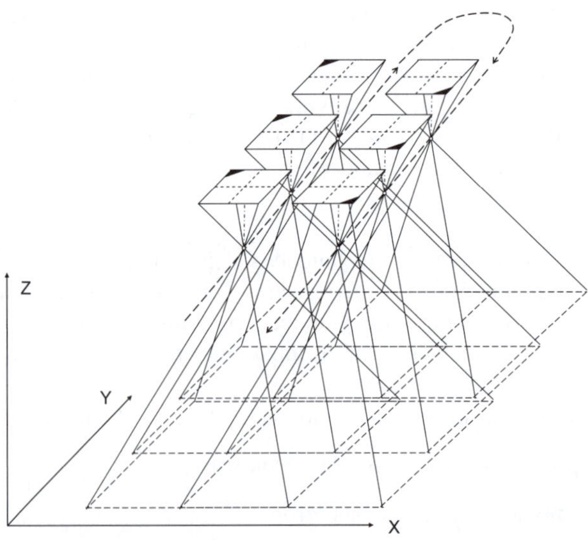

Fig. 4.52: Image configuration for aerial photogrammetry

The triangulation of aerial images is characterised mainly by

- predominant use of regular strip arrangements of images (Fig. 4.52) and hence
- advantageous structure of normal system of equations and
- easier generation of approximate values (e.g. rotations approximately zero);
- large numbers of images and object points and
- use of only *one* calibrated metric camera.

In contrast, the standard case in close-range photogrammetry is characterised by:

- irregularly arranged, arbitrary and often unfavourable image configurations (Fig. 4.53)
- more complex structure of normal system of equations
- arbitrarily oriented object coordinate systems
- more demanding solutions for generating approximate values
- combined adjustment of survey observations and conditions
- (several) imaging systems to be calibrated simultaneously

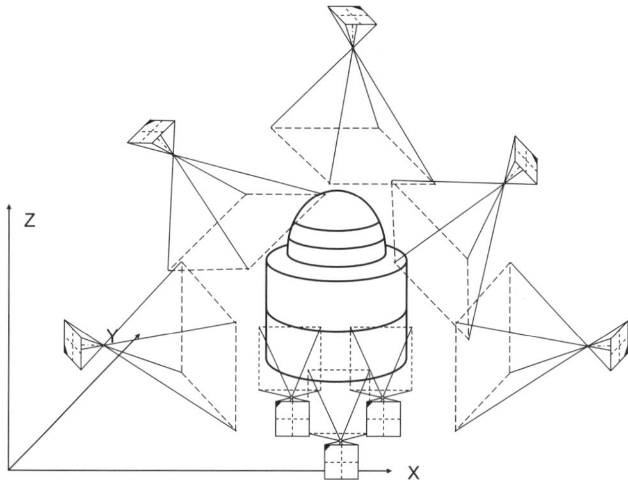

Fig. 4.53: Arbitrary close-range image configuration

Since the early 1980s the bundle adjustment has been accepted in all areas of photogrammetry. As a result of diverse requirements and applications there are many different bundle adjustment packages on the market.

Since its introduction for close-range use, the method of bundle adjustment has considerably widened the application spectrum as a result of its ability to handle almost arbitrary image configurations with few restrictions on the image acquisition systems.

4.4.1.2 Data flow

Fig. 4.54 shows the principle data flow for a bundle adjustment process. Input data for bundle adjustments are typically photogrammetric image coordinates generated by manual or automatic (digital) image measuring systems. Each measured image point is stored together with a unique point identifier and the corresponding image number (Fig. 4.55). This is sufficient to reconstruct the three-dimensional shape of the object surface, as represented by the measured object points.

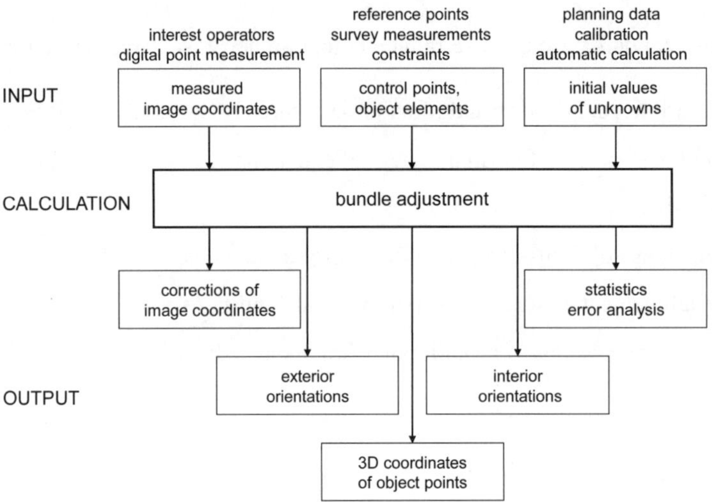

Fig. 4.54: Data flow for bundle adjustment process

```
71   2     0.1160    16.9870  0.0017  0.0017   1   1
71   3   -13.1730    16.9660  0.0017  0.0017   1   1
71   5   -14.9590   -12.9070  0.0017  0.0017   1   1
71   6     0.0480   -12.9780  0.0017  0.0017   1   1
71   7    15.0780   -12.9560  0.0017  0.0017   1   1
71   8    14.1720     2.9660  0.0017  0.0017   1   1
71   9     0.7820     3.8060  0.0017  0.0017   1   1
71  10   -19.2630    -6.3410  0.0017  0.0017   1   1
72   2    10.9920    12.6640  0.0017  0.0017   1   1
72   3     1.6690    21.3820  0.0017  0.0017   1   1
72   5   -20.9080     2.6880  0.0017  0.0017   1   1
72   6   -11.8780    -7.6560  0.0017  0.0017   1   1
72   7    -1.9050   -18.9510  0.0017  0.0017   1   1
72   8    10.1100    -7.3660  0.0017  0.0017   1   1
72   9     0.9680     3.1290  0.0017  0.0017   1   1
72  10   -19.0800     9.7620  0.0017  0.0017   1   1
98   1    -0.0790     2.8960  0.0017  0.0017   1   1
98   2    -8.9560     0.6350  0.0017  0.0017   1   1
98   3   -20.5460    -2.3040  0.0017  0.0017   1   1
98   4   -12.3520    -7.4400  0.0017  0.0017   1   1
98   5     0.4360   -15.4010  0.0017  0.0017   1   1
98   6    13.2960    -8.6670  0.0017  0.0017   1   1
98   7    21.9810    -4.0580  0.0017  0.0017   1   1
98   8     9.2680    -0.0460  0.0017  0.0017   1   1
98   9     0.2590    -1.9800  0.0017  0.0017   1   1
98  10   -10.3580   -12.7920  0.0017  0.0017   1   1
```

Fig. 4.55: Example of an image coordinate file
with image number, point number, x', y', sx', sy', code1, code2

Additional information in the object space (e.g. measured distances, angles, points, straight lines, planes) can also be taken into account. They provide the definition of an absolute scale and the position and orientation of the object coordinate system (datum definition). This information is entered into the system as, for example, reference point files or additional observations (e.g. constraints between object points, see section 4.4.2.3).

In order to linearise the functional model, approximate values must be generated. For simpler image configurations they can be extracted from planning data or project sketches. The generation of approximate values for more complex configurations (larger number of images, arbitrary orientations) is performed by iterative calculation methods (see section 4.4.4).

The principal results of bundle adjustment are the estimated 3D coordinates of the object points. They are given in an object coordinate system defined by reference points or free net adjustment (see section 4.4.3).

In addition, the exterior orientation parameters of all images are estimated. These can be further used, for example, in analytical plotters or for subsequent spatial intersections computed outside the bundle adjustment. The interior orientation parameters are estimated if the cameras are calibrated simultaneously within the adjustment.

In order to analyse the quality of the bundle adjustment, it is possible to calculate image coordinate residuals (corrections), standard deviations of object points and orientation data, correlations between parameters and reliability numbers for the detection of gross errors.

4.4.2 Mathematical model

4.4.2.1 Adjustment model

The mathematical model of the bundle adjustment is based on the collinearity equations (see section 4.2.2).

$$
\begin{aligned}
x' &= x'_0 + z' \cdot \frac{r_{11}\cdot(X-X_0)+r_{21}\cdot(Y-Y_0)+r_{31}\cdot(Z-Z_0)}{r_{13}\cdot(X-X_0)+r_{23}\cdot(Y-Y_0)+r_{33}\cdot(Z-Z_0)} + \Delta x' \\
y' &= y'_0 + z' \cdot \frac{r_{12}\cdot(X-X_0)+r_{22}\cdot(Y-Y_0)+r_{32}\cdot(Z-Z_0)}{r_{13}\cdot(X-X_0)+r_{23}\cdot(Y-Y_0)+r_{33}\cdot(Z-Z_0)} + \Delta y'
\end{aligned}
\tag{4.94}
$$

The structure of these equations allows the direct formulation of primary observed values (image coordinates) as functions of all unknown parameters in the photogrammetric imaging process. The collinearity equations, linearised at approximate values, can therefore be used directly as observation equations for a least-squares adjustment according to the Gauss-Markov model.

It is principally the image coordinates of homologous points which are used as observations[1]. The following unknowns are iteratively determined as functions of these observations:

- three-dimensional object coordinates for each new point i (total u_P, 3 unknowns each)
- exterior orientation of each image j (total u_I, 6 unknowns each)
- interior orientation of each camera k (total u_C, 0 or ≥ 3 unknowns each)

The bundle adjustment is completely general and can, for example, represent an extended form of the space resection (see section 4.2.3, eqn. 4.15):

$$\begin{aligned} x'_i + vx'_i &= F(\underline{X_{0j}, Y_{0j}, Z_{0j}, \omega_j, \varphi_j, \kappa_j, x'_{0k}, c_k, \Delta x'_k, X_i, Y_i, Z_i}) \\ y'_i + vy'_i &= F(\underline{X_{0j}, Y_{0j}, Z_{0j}, \omega_j, \varphi_j, \kappa_j, y'_{0k}, c_k, \Delta y'_k, X_i, Y_i, Z_i}) \end{aligned} \tag{4.95}$$

where i: point index
j: image index
k: camera index

The non-linear equations 4.94 are linearised using a Taylor series expansion with approximate values for all unknowns inside the brackets in (4.95). Here the derivatives, already determined in equations 4.18, are extended by the derivatives with respect to object coordinates:

$$\begin{aligned} \frac{\partial x'}{\partial X} &= -\frac{z'}{N^2} \cdot (r_{13} k_X - r_{11} N) & \frac{\partial y'}{\partial X} &= -\frac{z'}{N^2} \cdot (r_{13} k_Y - r_{12} N) \\ \frac{\partial x'}{\partial Y} &= -\frac{z'}{N^2} \cdot (r_{23} k_X - r_{21} N) & \frac{\partial y'}{\partial Y} &= -\frac{z'}{N^2} \cdot (r_{23} k_Y - r_{22} N) \\ \frac{\partial x'}{\partial Z} &= -\frac{z'}{N^2} \cdot (r_{33} k_X - r_{31} N) & \frac{\partial y'}{\partial Z} &= -\frac{z'}{N^2} \cdot (r_{33} k_Y - r_{32} N) \end{aligned} \tag{4.96}$$

If the interior orientation parameters are introduced as unknowns, the following derivatives are added:

$$\begin{aligned} \frac{\partial x'}{\partial x'_0} &= 1 & \frac{\partial y'}{\partial y'_0} &= 1 \\ \frac{\partial x'}{\partial c} &= -\frac{k_X}{N} & \frac{\partial y'}{\partial c} &= -\frac{k_Y}{N} \end{aligned} \tag{4.97}$$

Derivatives with respect to additional parameters of distortion are introduced in a similar way (see section 3.3.3). If linearisation is done numerically (see section 2.4.5.1), the

[1] Additional observations such as object point coordinates, distances or directions are introduced in section 4.4.2.3.

projection equations and selected distortion model can be programmed directly into the source code and a rigorous differentiation is not required.

In standard form, the linearised model is given by

$$\underset{n,1}{\mathbf{l}} + \underset{n,1}{\mathbf{v}} = \underset{n,u}{\mathbf{A}} \cdot \underset{u,1}{\hat{\mathbf{x}}}$$

and the corresponding system of normal equations is

$$\underset{u,u}{\mathbf{N}} \cdot \underset{u,1}{\hat{\mathbf{x}}} + \underset{u,1}{\mathbf{n}} = \underset{u,1}{\mathbf{0}} \tag{4.98}$$

where

$$\underset{u,u}{\mathbf{N}} = \underset{u,n}{\mathbf{A}^T} \cdot \underset{n,n}{\mathbf{P}} \cdot \underset{n,u}{\mathbf{A}}$$

$$\underset{u,1}{\mathbf{n}} = \underset{u,n}{\mathbf{A}^T} \cdot \underset{n,n}{\mathbf{P}} \cdot \underset{n,1}{\mathbf{l}}$$

The solution vector and its covariance matrix are estimated in an iterative adjustment:

$$\underset{u,1}{\hat{\mathbf{x}}} = \underset{u,u}{\mathbf{Q}} \cdot \underset{u,1}{\mathbf{n}} = (\underset{u,n}{\mathbf{A}^T} \cdot \underset{n,n}{\mathbf{P}} \cdot \underset{n,u}{\mathbf{A}})^{-1} \cdot \underset{u,n}{\mathbf{A}^T} \cdot \underset{n,n}{\mathbf{P}} \cdot \underset{n,1}{\mathbf{l}} \tag{4.99}$$

where

$$\underset{u,u}{\mathbf{Q}} = \underset{u,u}{\mathbf{N}^{-1}} \qquad \text{: cofactor matrix}$$

$$\underset{u,u}{\mathbf{K}} = \hat{s}_0^2 \cdot \underset{u,u}{\mathbf{Q}} \qquad \text{: variance-covariance matrix}$$

4.4.2.2 Normal equations

The number of unknowns in the adjustment system can be calculated as follows:

$$u = u_I \cdot n_{images} + u_P \cdot n_{points} + u_C \cdot n_{cameras} (+u_{datum})$$

where

$u_I = 6$: parameters of exterior orientation per image

$u_P = 3$: XYZ coordinates of new points

$u_C = 0 \ldots \geq 3$: parameters of interior orientation per camera

In addition to the unknown orientation parameters and point coordinates, up to seven parameters are still required for datum definition. However, these can be eliminated by use of reference points or appropriate condition equations (datum defect, see section 4.4.3.1). Table 4.1 gives examples of the number of observations and unknowns for different image configurations.

Table 4.1: Number of unknowns and observations for different image configurations

	Example 1 aerial set-up (Fig. 4.56)	u	u_{total}	Example 2 closed loop set-up (Fig. 4.58)	u	u_{total}	Example 3 test field calibration (Fig. 7.21)	u	u_{total}
n_{images}	8	6	48	16	6	96	8	6	48
n_{points}	14	3	42	25	3	75	13+6	3	57
$n_{ref.\ pts.}$	6	0	0	0	0	0	3 distances	0	0
$n_{cameras}$	1	0	0	1	5	5	1	7	7
u_{datum}	0		0	7		(7)	6		(6)
u_{total}			90			176			112
n_{obs}			110			384			304
$r=n-u$			20			208			192

Example 1 represents a regular arrangement for aerial photogrammetry or the measurement of plane façades (Fig. 4.56). The number of unknowns is $u = 90$ if 8 images and 14 new points are assumed. The datum is defined by 6 reference points. For image acquisition, a metric camera with known interior orientation is used.[1]

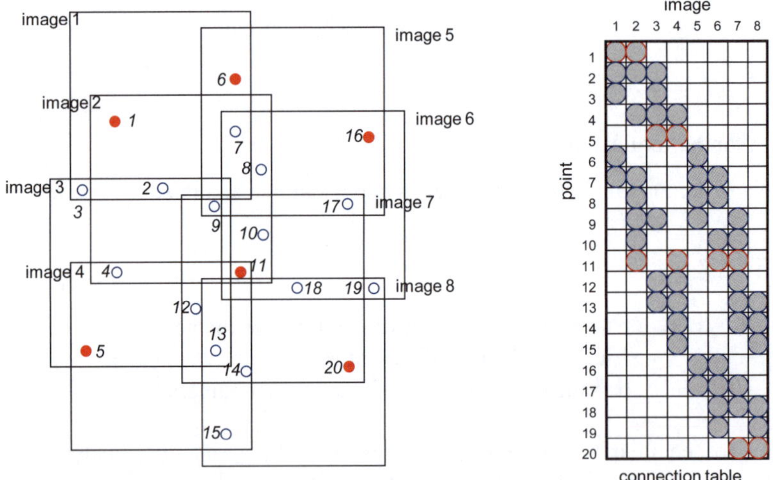

Fig. 4.56: Example 1: aerial or façade arrangement

The connection matrix in Fig. 4.56 shows which point is measured in which image. The numbering of object points by measurement strip produces the typical diagonal structure of connections, which is also seen in the structure of the Jacobian matrix **A** and the resulting

[1] For a clearer illustration, a reduced number of object points is used in this example.

normal equations (Fig. 4.57). In order to solve the normal equations, the order of observations can be further optimised by suitable sorting algorithms (see section 2.4.5.1).

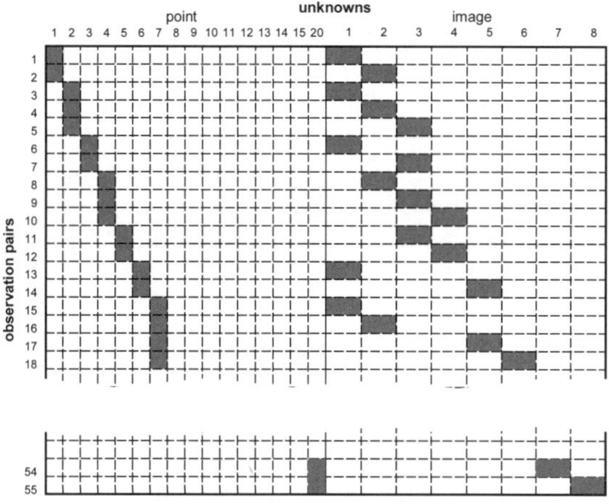

Fig. 4.57: Structure of design matrix **A** according to example 1

On average each object point is measured in 2.8 images. Each measured image point provides 2 observations. Hence, with a total number of $n = 110$ observations and a total redundancy number of $r = n\text{-}u = 110–90 = 20$, redundancy in the adjustment system is relatively weak.

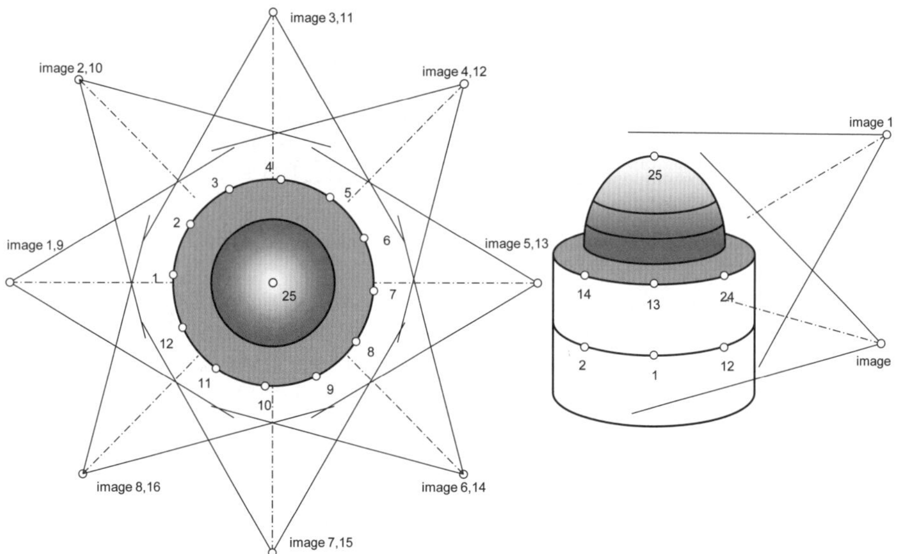

Fig. 4.58: Example 2: closed-loop arrangement

Example 2 shows a typical closed-loop image configuration for an object formed from a cylinder and hemisphere (Fig. 4.58). The object is recorded in two image acquisition sets which have a relative vertical shift (bottom-set images 1–8, top-set images 9–16). The object point on top of the dome appears in all images. As there are no reference points available, the datum defect of 7 is eliminated by a free net adjustment (see section 4.4.3.3). A non-metric camera is used so 5 parameters of interior orientation must be simultaneously calibrated. A total of 176 unknowns must therefore be estimated.

The corresponding connection matrix (Fig. 4.59) shows an average of 7.7 images for each object point. The redundancy is therefore much higher than in example 1 (aerial or façade arrangement). With 384 measured image coordinates, the total redundancy is $r = 384–176 = 208$.

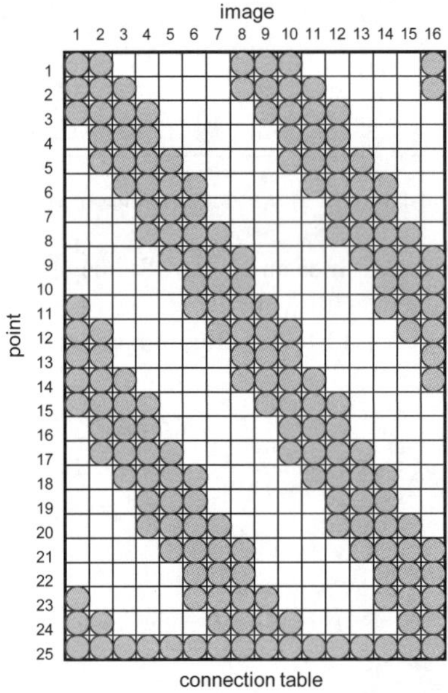

Fig. 4.59: Connection matrix of example 2 (Fig. 4.58)

However, the extent to which the adjustment system can be calculated, and the quality of the results, are less a question of total redundancy than the geometric configuration of the system. Consider the arrangement of example 1 (Fig. 4.56) which allows for the determination of plane coordinates (parallel to the image plane) to an acceptable accuracy whilst a different point accuracy, which varies as a function of the height-to-base ratio, applies along the viewing direction. It is not practicable to perform a camera calibration with this set-up. In contrast, the arrangement of example 2 (Fig. 4.58) represents a very stable geometry which can provide not only self-calibration but also a high and homogenous point accuracy in all three coordinate axes.

Finally example 3 in Table 4.1 refers to an arrangement for test field calibration according to Fig. 7.21 (see section 7.3.1.2). If the test field, in this case with 13 points and 3 distances (6 points), is completely covered by each of the 8 images, a total number of $u = 112$

unknowns and $n = 304$ observations results. Even with this simple set-up a high redundancy and stable geometric configuration are both achieved.

4.4.2.3 Combined adjustment of photogrammetric and survey observations

The system of equations in (4.94) describes the original model of bundle triangulation by defining the image coordinates x',y' (observations) as a function of the unknowns, specifically of the object coordinates X,Y,Z. Additional information about the object or additional non-photogrammetric measurements are not considered in (4.94).

An extended model for the bundle adjustment takes additional (survey) observations into account as, for example, measured distances, directions or angles. Other constraints on the object can also be integrated, such as known points, coordinate differences, straight lines, planes or surfaces having rotational symmetry (Fig. 4.60).

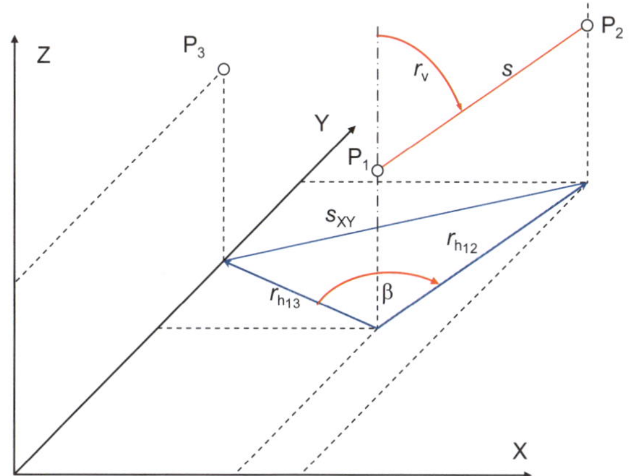

Fig. 4.60: Survey observations

All additional observations can be weighted according to their accuracy or importance and therefore have a rigorous stochastic treatment in the adjustment process.

Coordinates, coordinate differences and distances

It is particularly easy to introduce observed coordinates, coordinate differences or distances. The following observation equations result from the introduction of

object coordinates:

$$
\begin{aligned}
X &= X \\
Y &= Y \\
Z &= Z
\end{aligned}
\tag{4.100}
$$

coordinate differences:

$$\Delta X = X_2 - X_1$$
$$\Delta Y = Y_2 - Y_1 \qquad (4.101)$$
$$\Delta Z = Z_2 - Z_1$$

slope distances:

$$s = \sqrt{(X_2 - X_1)^2 + (Y_2 - Y_1)^2 + (Z_2 - Z_1)^2} \qquad (4.102)$$

distances in XY plane:

$$s_{XY} = \sqrt{(X_2 - X_1)^2 + (Y_2 - Y_1)^2} \qquad (4.103)$$

distances in XZ plane:

$$s_{XZ} = \sqrt{(X_2 - X_1)^2 + (Z_2 - Z_1)^2} \qquad (4.104)$$

distances in YZ plane:

$$s_{YZ} = \sqrt{(Y_2 - Y_1)^2 + (Z_2 - Z_1)^2} \qquad (4.105)$$

Exterior orientations

Known position or orientation data of a camera can be taken into account by the following additional observation equations:

exterior orientation:

$$X_0 = X_0 \qquad \omega = \omega$$
$$Y_0 = Y_0 \qquad \varphi = \varphi \qquad (4.106)$$
$$Z_0 = Z_0 \qquad \kappa = \kappa$$

Normally the exterior orientation cannot be measured directly. Older metric cameras may be combined with surveying instruments; in this case surveyed angles (azimuth, elevation) can be introduced as rotation angles of exterior orientation; the rotation matrix must be properly defined (see section 4.2.1.2). In addition, exterior orientation parameters may be known from previous calculations and so can be processed with a corresponding weight.

Relative orientations

For the orientation and calibration of stereo cameras which are stably mounted relative to each other, it is useful to introduce constraint equations which fix the relative orientation at all imaging locations:

base constraint for fixed relative orientation:

$$\sqrt{(X_{01} - X_{02})^2 + (Y_{01} - Y_{02})^2 + (Z_{01} - Z_{02})^2} = b \qquad (4.107)$$

rotation constraint for fixed relative orientation:

$$\begin{aligned} a_{11}\cdot b_{11}+a_{12}\cdot b_{12}+a_{13}\cdot b_{13}&=\alpha\\ a_{21}\cdot b_{21}+a_{22}\cdot b_{22}+a_{23}\cdot b_{23}&=\beta\\ a_{31}\cdot b_{31}+a_{32}\cdot b_{32}+a_{33}\cdot b_{33}&=\gamma \end{aligned} \tag{4.108}$$

Here a_{ij} and b_{ij} are the elements of the rotation matrices **A** and **B** of the left and right image. Each of the constraint equations forms a scalar product, hence α, β, γ define the rotation angles between the axes of the two cameras.

Directions and angles

Survey directions and angles, for example observed by theodolite, can also be introduced. In a conventional levelled use where the XY plane is horizontal and the theodolite's vertical axis corresponds to the Z axis (Fig. 4.60), the equations are:

horizontal direction:

$$r_h=\arctan\left(\frac{X_2-X_1}{Y_2-Y_1}\right) \tag{4.109}$$

horizontal angle:

$$\beta=\arctan\left(\frac{X_3-X_1}{Y_3-Y_1}\right)-\arctan\left(\frac{X_2-X_1}{Y_2-Y_1}\right) \tag{4.110}$$

vertical angle:

$$r_v=\arctan\left(\frac{Z_2-Z_1}{\sqrt{(X_2-X_1)^2+(Y_2-Y_1)^2+(Z_2-Z_1)^2}}\right) \tag{4.111}$$

This can be extended to complete sets of surveyed directions (several measured horizontal and vertical directions and distances from one station). With modification, the equations can also apply to non-levelled instruments such as laser trackers. With the above observation types, pure 3D survey nets can also be adjusted.

Auxiliary coordinate systems

The introduction of auxiliary coordinate systems is a very elegant way of formulating additional information in object space. An auxiliary coordinate system $\overline{X},\overline{Y},\overline{Z}$ is a 3D coordinate system arbitrarily oriented in space and used to define additional observations or constraints. For example, this can be local reference point configurations with a defined relation to each other or local geometric elements (e.g. rotationally symmetric shapes).

The auxiliary coordinate system can be transformed into the object coordinate system X,Y,Z using a spatial similarity transformation (see section 2.2.2.2):

$$\overline{\mathbf{X}} = \overline{\mathbf{X}}_0 + \overline{m} \cdot \mathbf{R}_{\overline{\omega\varphi\kappa}} \cdot (\mathbf{X} - \mathbf{X}_0) \tag{4.112}$$

or

$$\begin{bmatrix} \overline{X} \\ \overline{Y} \\ \overline{Z} \end{bmatrix} = \begin{bmatrix} \overline{X}_0 \\ \overline{Y}_0 \\ \overline{Z}_0 \end{bmatrix} + \overline{m} \cdot \begin{bmatrix} r_{11} & r_{12} & r_{13} \\ r_{21} & r_{22} & r_{23} \\ r_{31} & r_{32} & r_{33} \end{bmatrix} \cdot \begin{bmatrix} X - X_0 \\ Y - Y_0 \\ Z - Z_0 \end{bmatrix}$$

where

X,Y,Z: spatial coordinates in global system

$\overline{X},\overline{Y},\overline{Z}$: spatial coordinates in local system

X_0,Y_0,Z_0: centroid of point cloud in XYZ system (constant coordinate values in the adjustment process)

$\overline{X}_0,\overline{Y}_0,\overline{Z}_0$: origin of system XYZ with respect to system $\overline{X}\,\overline{Y}\,\overline{Z}$

$\mathbf{R}_{\overline{\omega\varphi\kappa}}$: rotation matrix transforming XYZ parallel to $\overline{X}\,\overline{Y}\,\overline{Z}$

$\overline{m}$: scaling factor between both systems

The seven transformation parameters must be introduced as unknowns in the adjustment; hence approximate values must be provided. It is now possible to define functional relationships between object points within the auxiliary coordinate system. They can be expressed by parameters B_i, which can be transformed into the object coordinate system using (4.112). Parameters B_i can describe simple geometric conditions and also object surfaces of higher order.

$$f(X,Y,Z,\overline{\omega},\overline{\varphi},\overline{\kappa},\overline{X}_0,\overline{Y}_0,\overline{Z}_0,\overline{m},B_i) = g(\overline{X},\overline{Y},\overline{Z},B_i) = C \tag{4.113}$$

C is a constant of the result of constraint equation, expressed as function f in the object coordinate system or as function g in the auxiliary coordinate system. An arbitrary object plane is therefore defined by

$$\overline{X} = C \tag{4.114}$$

A straight line in the auxiliary coordinate system is defined by

$$\overline{X} = C \qquad \overline{Y} = C \tag{4.115}$$

A rotational solid is given by

$$\overline{X}^2 + \overline{Y}^2 + h(\overline{Z},B_i) = C \tag{4.116}$$

The function h defines the shape of the solid in the direction of the rotational axis, e.g. for a circular cylinder with radius r

$$h(\overline{Z},B_i) = 0 \qquad \text{and} \qquad C = r^2 \tag{4.117}$$

or for a sphere with radius r:

$$h(\overline{Z}, B_i) = \overline{Z}^2 \qquad \text{and} \qquad C = r^2 \tag{4.118}$$

Applications of additional observations

With the aid of the additional observations above, the adjustment system can be significantly influenced:

- Weak image configurations can be stabilised by introducing additional object information.
- Surveyed or other measured object data, and photogrammetric observations, can be adjusted in one step, e.g. in order to minimise net strains (inconsistencies between object points) or to handle measurement data from different sources in a balanced way.
- Unlimited numbers of known distances (scales) between new points can be observed and processed according to their accuracy.
- Single object points can be forced onto an object surface by geometric conditions, e.g. points on a cylindrical surface (pipe line).[1]
- Information about exterior orientation provided by instruments such as inertial navigation units (INU), GPS location or gyroscopes, can support the adjustment process. Mechanically defined conditions, such as the movement of a camera along a straight line or circle, can also be added as additional observations. This applies equally to geometric constraints between the orientation parameters of different cameras, e.g. for stereo cameras or camera/projector arrangements.
- Additional information which is introduced with an unrealistic weight can negatively affect the adjustment result. This point is particularly important in practice and demands a careful choice of weights and analysis of results.

4.4.2.4 Adjustment of additional parameters

Self-calibration

Functions for the correction of imaging errors are referred to as *additional parameter functions*. Functional models which describe real optical characteristics of image acquisition systems (parameters of interior orientation) have been summarised in section 3.3. When these parameters are determined within the bundle adjustment, the procedure is known as *self-calibration.*

The linearised model of the adjustment is extended by derivatives for principal distance and principal point, as given in (4.97). Again, approximate values of the additional unknowns must be provided, although the following initial values are usually sufficient:

---

[1] Conditions include both weak constraints defined by additional observations with standard deviations, as well as fixed constraints which force an exact condition.

principal distance: $c \approx f$ (focal length)
principal point: $x'_0 \approx y'_0 \approx 0$
radial distortion: $A_1 \approx A_2 \approx 0$
decentring distortion: $B_1 \approx B_2 \approx 0$
affinity and shear: $C_1 \approx C_2 \approx 0$

The ability to determine individual parameters depends, on one hand, on the modelling of the physical imaging process. Parameter sets based on faulty physical assumptions can lead to weakly conditioned systems of equations, over-parametrized equations or high correlations between parameters. On the other hand, interior orientation parameters can only be reliably calculated if image configuration and distribution of object points are well chosen. Section 7.3.2 summarizes suitable image configurations for self-calibration.

The parameters of interior orientation can additionally be handled as observed values with a corresponding weight. The observation equations in this case are given by:

interior orientation:

$$\begin{aligned} c &= c & A_1 &= A_1 \\ x'_0 &= x'_0\,, & A_2 &= A_2 \\ y'_0 &= y'_0 & &etc. \end{aligned} \tag{4.119}$$

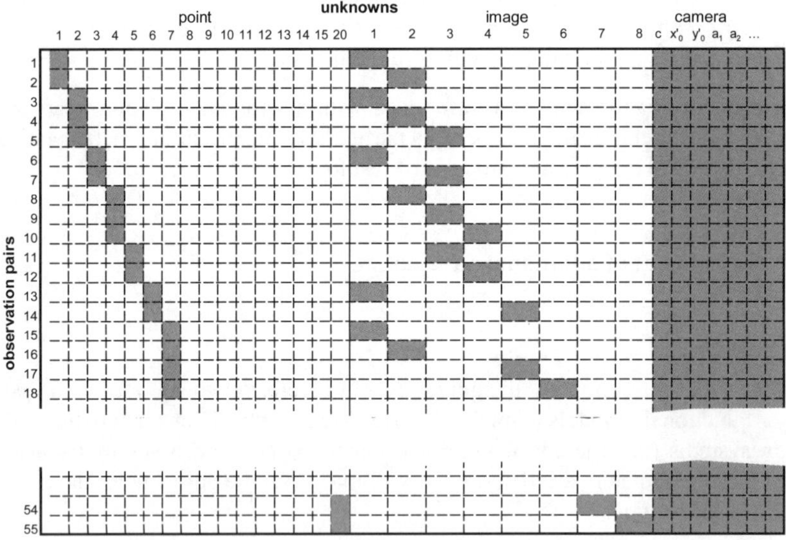

Fig. 4.61: Structure of design matrix **A**, extended by additional unknowns for the interior orientation of a camera (compare Fig. 4.57)

If additional parameters are introduced, the structure of the normal system of equations changes significantly. While the design matrix illustrated in Fig. 4.57 shows a distinctive diagonal structure with a large number of zero elements, the normal equation matrix of adjustment systems with additional parameters contains larger areas with non-zero elements. Each measured image coordinate is functionally connected with the unknowns of interior orientation, hence the right-hand side of the matrix is completely filled (Fig. 4.61).

In order to invert the resulting normal equation matrices, much more storage and computing power is required. Sparse matrix techniques have been successfully applied in practice for the efficient use of computing resources (see section 2.4.5).

Calibration with variable interior orientation

The models normally used for interior orientation assume constant camera parameters during the period of image acquisition. If the camera geometry changes over this period, for example due to refocusing or a change of lens, a "new" camera with its own parameters must be assigned to the corresponding images. The simultaneous determination of more than one group of parameters requires an image configuration appropriate for the calibration of each camera (see section 7.3.2).

If stable camera parameters cannot be guaranteed for longer periods, the interior orientation must be calibrated individually for each image. This approach to image-variant calibration provides corrections for a shift of the perspective centre (camera constant and principal point coordinates) but otherwise assumes stable distortion parameters for all images. For this purpose, the image-variant parameters are introduced as observed unknowns with approximate values and a priori standard deviations which correspond to the expected shift of the perspective centre. The numerical stability of the adjustment is maintained provided there is a suitable number of object points each with an appropriate number of image rays. Depending on image configuration and current state of the camera, applications using digital SLR cameras report an accuracy increase by a factor of 2 to 4 compared with cases in which image-variant parameters are not used (see section 3.3.4.4). The approach can be extended to the simultaneous calculation of a correction grid to allow for sensor deformation (see section 3.3.4.5).

4.4.3 Object coordinate system (definition of datum)

4.4.3.1 Rank and datum defect

A network composed of purely photogrammetric observations leads to a singular system of normal equations because, although the shape of the network can be determined, its absolute position and orientation in space cannot be determined. The resulting system of equations has a *rank defect*

$$d = u - r \tag{4.120}$$

where
u: number of unknowns
r: rank($\mathbf{A}$)

The rank defect is caused by a datum defect in the observed network which, for a three-dimensional network, can be removed by defining 7 additional elements:

3 translations
3 rotations
1 scaling factor

If at least one known distance is observed then the datum defect is reduced by 1. The information is introduced as an additional distance observation according to equations 4.102ff.

Translational datum defects can be eliminated if control or reference points with known object coordinates are observed. Options for this are reviewed in the next section.

Rotational datum defects can also be eliminated by reference points, as well as by directly measured directions or angles.

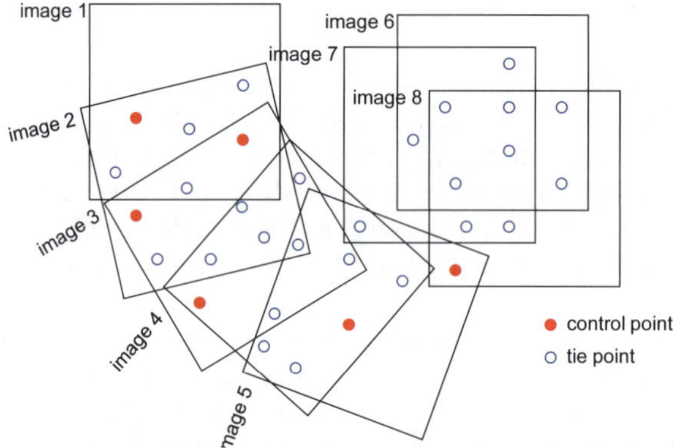

Fig. 4.62: Example of a configuration defect in an arrangement of images. Images 6,7,8 cannot be oriented with respect to the rest of the images

In addition to datum defects, observed networks can contain a configuration defect if, due to missing observations, some portions of the network cannot be determined unambiguously. In photogrammetric networks this problem seldom arises. If some images, for example, contain an insufficient number of tie points, they cannot be oriented with respect to the other images (see example in Fig. 4.62).

4.4.3.2 Reference points

Error-free reference points

Reference points are used for the definition of a global object coordinate system (datum definition). They can be introduced as error-free reference points into the bundle adjustment if their nominal coordinates are known to a high accuracy. Such coordinates could be introduced into the bundle adjustment as constants. Some bundle adjustment programs allow their input as measurements with zero standard deviations. Logically this approach leads to an error-free definition of the datum.

It is assumed, however, that the accuracy of the reference point coordinates is, for example, better by a factor 5–10 than the photogrammetric point determination, and that there are no inconsistencies in reference point coordinates. Errors in reference point coordinates are interpreted as errors in observations and are therefore difficult to detect. The definition of the object coordinate system using error-free reference points gives rise to a fixed datum.

In principle the spatial distribution of reference points should follow the recommendations made for absolute orientation (see section 4.3.5). The stereo model discussed there must here be considered a model defined by all images (bundles), i.e. a minimum of three reference points is required for the definition of the object coordinate system. As a minimum, the following coordinate components must be given by these three reference points (the example components in brackets relate to an image plane parallel to XY and viewing direction parallel to Z):

- a minimum of 2x2 coordinates parallel to the primary object plane (X_1, Y_1, X_2, Y_2)
- a minimum of 3 coordinates perpendicular to the primary object plane (Z_1, Z_2, Z_3)

The minimum configuration can be established, for example, by 2 full reference points (2x *XYZ*) and one additional reference height point (1x *Z*) or 2 plane reference points (2x *XY*) and 3 reference height points (3x *Z*). However, in many applications of close-range photogrammetry there are more than 2 full reference points available.

Reference points should be widely and uniformly distributed over the area covered by the images. Fig. 4.63 (left) shows an image set-up with a suitable distribution of 4 reference points, leading to a stable datum definition and homogeneous accuracies. In Fig. 4.63 (right) the 4 reference points are distributed inefficiently in one corner of the image configuration. As a consequence, the whole system can rotate around this point cloud, which results in discrepancies at more distant points and correspondingly higher standard deviations.

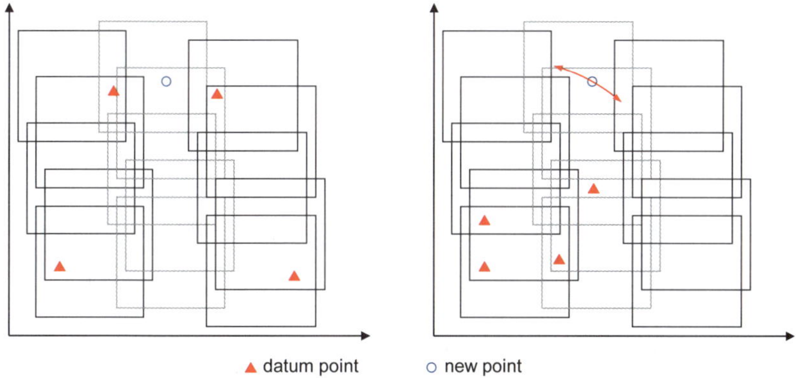

Fig. 4.63: Good (left) and bad (right) distribution of reference points in a multi-image configuration

Reference points used for datum definition must not lie on a common straight line, as the normal system of equations then becomes singular. Unfavourable distributions of reference points which come close to this restriction will result in numerically weak systems of equations.

Coordinates of reference points as observed quantities

Coordinates of reference points can also be introduced as observed quantities with a weight corresponding to their real point accuracy (e.g. depending on the measuring systems used for coordinate determination; see section 4.4.2.3). Within the adjustment system, the coordinates of the reference points are treated as unknowns and receive corrections and accuracy values in the same way as other observations. Standard deviations of weighted reference points can be interpreted as quality measures for the reference points themselves.

Partial compensation for inconsistencies between reference point coordinates can be done by an appropriate variation of weights, provided that these inconsistencies are not directly transferred to the photogrammetric observations. A coordinate system defined in this way is known as a *weighted datum*. Using coordinates of reference points in this way also compensates completely for rank defects in the adjustment system. If the weighted datum also results in a weak definition of the coordinate system, then higher standard deviations for new points are usually obtained.

Unconstrained datum definition using reference points (3-2-1 method)

Using a minimum amount of object information it is possible to define the object coordinate system in order to avoid any possible influence of inconsistencies in the reference points. For this purpose scale is given by a known distance S. The coordinate axes can be then defined according to the following scheme known as the *3-2-1 method* (Fig. 4.64):

1. Fixing the X,Y,Z coordinates of point 1 defines an arbitrary 3D reference point in the object coordinate system which can, for example, represent the origin ($X = Y = Z = 0$).
2. Fixing the Y,Z coordinates of point 2 defines the X axis. At this stage the system can still be rotated about this axis.
3. Fixing the Y coordinate of point 3 defines the XZ plane (alternatively define the XY plane with fixed Z or YZ plane with fixed X). Hence the coordinate system is uniquely defined without any constraints.

In some configurations ambiguities in transformation are possible, e.g. a point could be transformed to the mirror image of itself in a plane. In these cases some very approximate additional data can be used to choose between alternative positions.

Fig. 4.64 shows an example where the following reference coordinates are available:

point	X	Y	Z
P_1	7.0	2.0	2.0
P_2	-	4.5	3.5
P_3	-	2.0	-

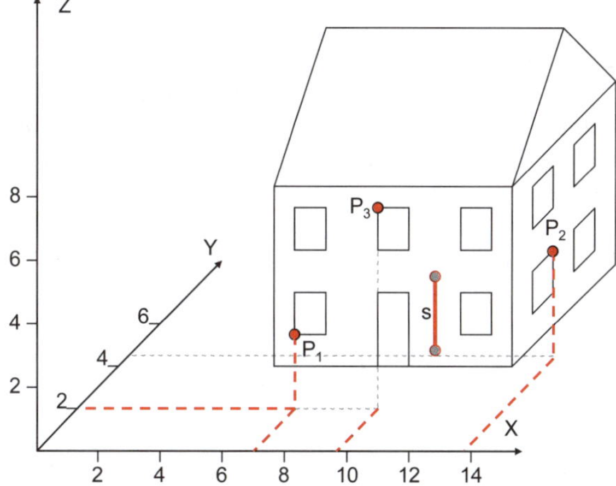

Fig. 4.64: Unconstrained coordinate system definition using reference points (3-2-1 method)

If real reference point coordinates are not available the system can also be defined by fictitious coordinates, e.g.

point	X	Y	Z
P_1	0	0	0
P_2	-	0	0
P_3	-	0	-

Scaling information can, of course, also be derived from the known distances between reference points or simply by the inclusion of a scale bar.

The unconstrained datum definition by reference points does not affect the shape of the photogrammetric network. It is true that the absolute coordinates of object points are related to the arbitrarily selected system origin but the distances between points are independent of the datum. In contrast, the estimated accuracies derived from the covariance matrix are influenced by the datum definition. If datum points are distributed according to Fig. 4.63 (right) then object point accuracy based on datum point accuracy is interpreted too optimistically when they are close to the datum points and too pessimistically in more distant parts of the network.

4.4.3.3 Free net adjustment

If no reference points or equivalent datum definitions are available, the problem of datum definition can be avoided by means of a *free net adjustment* which fits the network onto the initial coordinates of the unknown points. Initial values for new points are required in any case for the linearisation of the correction equations. They can be generated by the procedures described in section 4.4.4.

The initial values of all unknown points (new object points and perspective centres) form a spatial point cloud. This point cloud can be transformed by three translations, three rotations and one scaling factor onto the photogrammetrically determined model of object points, without affecting the shape of the point cloud, i.e. without any geometrical constraint. The photogrammetric observations are not influenced by this transformation.

A rank defect of 7 in the normal equation matrix is avoided if exactly 7 observations can be found which are linearly independent with respect to each other and to the other observations. This requirement is exactly fulfilled if the normal system of equations is extended by a matrix **B** with $d = 7$ rows and u columns, where u is the number of unknowns (see section 2.4.2.4):

$$\underset{d,u}{\mathbf{B}} = \left[\begin{array}{cccccccc|cccccccc} 1 & 0 & 0 & 0 & 0 & 0 & \cdots & 1 & 0 & 0 & \cdots & 1 & 0 & 0 \\ 0 & 1 & 0 & 0 & 0 & 0 & \cdots & 0 & 1 & 0 & \cdots & 0 & 1 & 0 \\ 0 & 0 & 1 & 0 & 0 & 0 & \cdots & 0 & 0 & 1 & \cdots & 0 & 0 & 1 \\ 0 & -Z_{01} & Y_{01} & r_{11} & r_{12} & r_{13} & \cdots & 0 & -Z_1^0 & Y_1^0 & \cdots & 0 & -Z_p^0 & Y_p^0 \\ Z_{01} & 0 & -X_{01} & r_{21} & r_{22} & r_{23} & \cdots & Z_1^0 & 0 & -X_1^0 & \cdots & Z_p^0 & 0 & -X_p^0 \\ -Y_{01} & X_{01} & 0 & r_{31} & r_{32} & r_{33} & \cdots & -Y_1^0 & X_1^0 & 0 & \cdots & -Y_p^0 & X_p^0 & 0 \\ X_{01} & Y_{01} & Z_{01} & 0 & 0 & 0 & \cdots & X_1^0 & Y_1^0 & Z_1^0 & \cdots & X_p^0 & Y_p^0 & Z_p^0 \end{array}\right] \tag{4.121}$$

$$\text{exterior orientations} \qquad\qquad \text{new points } 1..p$$

If scale is defined by a separately measured distance, the potential rank defect decreases to 6 and the last row of matrix **B** can be eliminated.

The first six columns of matrix **B** are related to the unknown parameters of exterior orientation. They are not further discussed as the perspective centres are normally not of interest in the definition of the datum.

The next columns correspond to the unknown object points. Three condition equations are used for the translation of the system, included in the first three rows of matrix **B**:

$$\begin{aligned} dX_1 \quad\quad\quad +dX_2 \quad\quad\quad\quad + \cdots &= 0 \\ dY_1 \quad\quad\quad +dY_2 \quad\quad + \cdots &= 0 \\ dZ_1 \quad\quad\quad +dZ_2 + \cdots &= 0 \end{aligned} \tag{4.122}$$

If the sum of coordinate corrections at all object points becomes zero, the centroid of the initial points is identical to the centroid of the adjusted object points.

The next three rows of matrix **B** contain differential rotations of the coordinate system:

$$\begin{array}{ccccccccc} & -Z_1^0 dY_1 & +Y_1^0 dZ_1 & & -Z_2^0 dY_2 & +Y_2^0 dZ_2 & + & \cdots & = 0 \\ Z_1^0 dX_1 & & -X_1^0 dZ_1 & +Z_2^0 dX_2 & & -X_2^0 dZ_2 & + & \cdots & = 0 \\ -Y_1^0 dX_1 & +X_1^0 dY_1 & & -Y_2^0 dX_2 & +X_2^0 dY_2 & & + & \cdots & = 0 \end{array} \quad (4.123)$$

The final row 7 of matrix **B** defines scale:

$$X_1^0 dX_1 \quad +Y_1^0 dY_1 \quad +Z_1^0 dZ_1 \quad +X_2^0 dX_2 \quad +Y_2^0 dY_2 \quad +Z_2^0 dZ_2 \quad + \quad \cdots \quad = \quad 0 \quad (4.124)$$

The extended system of normal equations then becomes:

$$\begin{bmatrix} \mathbf{A}^T\mathbf{A} & \mathbf{B}^T \\ \mathbf{B} & \mathbf{0} \end{bmatrix} \cdot \begin{bmatrix} \hat{\mathbf{x}} \\ \mathbf{k} \end{bmatrix} = \begin{bmatrix} \mathbf{A}^T\mathbf{l} \\ \mathbf{0} \end{bmatrix} \quad (4.125)$$

Here **k** consists of seven Lagrange multipliers. The solution of the extended system of normal equations is obtained from the pseudo inverse (Moore-Penrose inverse) $\mathbf{Q}^+$

$$\begin{aligned} \begin{bmatrix} \hat{\mathbf{x}} \\ \mathbf{k} \end{bmatrix} &= \begin{bmatrix} \mathbf{A}^T\mathbf{A} & \mathbf{B}^T \\ \mathbf{B} & \mathbf{0} \end{bmatrix}^{-1} \cdot \begin{bmatrix} \mathbf{A}^T\mathbf{l} \\ \mathbf{0} \end{bmatrix} \\ &= \overline{\mathbf{N}}^{-1} \cdot \overline{\mathbf{n}} \\ &= \mathbf{Q}^+ \cdot \overline{\mathbf{n}} \end{aligned} \quad (4.126)$$

The pseudo inverse has the property that the resulting covariance matrix has a minimum trace:

$$trace\{\mathbf{Q}^+\} = \min. \quad (4.127)$$

Hence, the standard deviations of the object points are estimated with minimum quantities. The centroid of object points becomes the origin for the datum which is a fixed point with zero standard deviation. A datum is therefore defined which does not affect the total accuracy of the system.

Full trace minimization

If all unknown new points are used for datum definition, the full trace of the covariance matrix $\mathbf{Q}^+$ is minimised, as explained above. All points therefore contribute to the datum definition but they are not considered free of error and do not have a priori standard deviations.

Fig. 4.65 shows an example of a network for which the datum is defined alternatively by three reference points and one distance (left, 3-2-1 method) and by free net adjustment (right, all points used as datum points). When reference points are used, error ellipses illustrate clearly that standard deviations are smaller for object points close to reference points than for those at the edges of the net. In contrast, error ellipses in the free net adjustment are significantly smaller and more homogenous. Their large semi-axes point towards the centroid of the points. The vertical lines in the ellipses indicate the error in Z, which behaves similarly to those in X and Y.

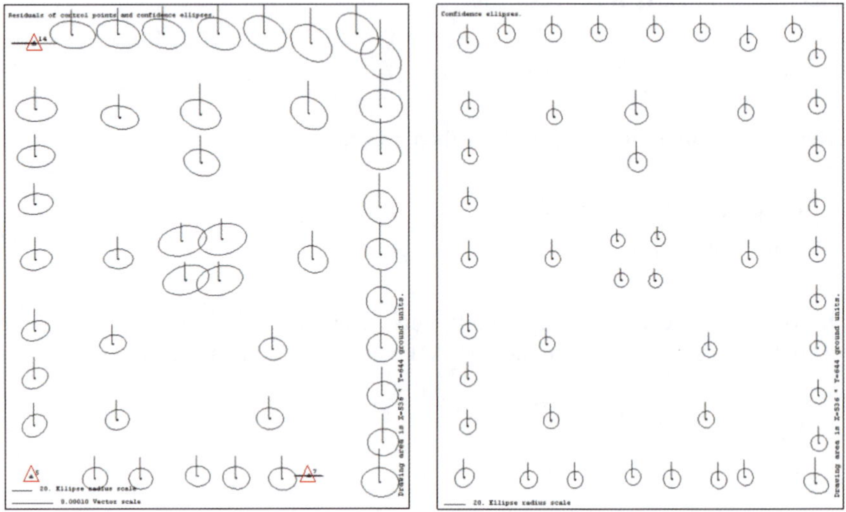

Fig. 4.65: Example network based on reference points (left, red points) and by free net adjustment (right)

Partial trace minimisation

There are some applications where not all the object points should be used for datum definition in a free net adjustment. This can occur, for example, when a subset of new points represents an existing network into which the remaining object points should be optimally fitted. In this case, those columns of matrix **B**, which relate to points not used in datum definition, must be set to zero. However, the rank of **B** must not be smaller than *u–d*.

$$\underset{d,u}{\mathbf{B}} = \begin{bmatrix} 1 & 0 & 0 & 0 & 0 & 0 & \cdots & 1 & 0 & 0 & \cdots & 1 & 0 & 0 \\ 0 & 1 & 0 & 0 & 0 & 0 & \cdots & 0 & 1 & 0 & \cdots & 0 & 1 & 0 \\ 0 & 0 & 1 & 0 & 0 & 0 & \cdots & 0 & 0 & 1 & \cdots & 0 & 0 & 1 \\ 0 & -Z_1^0 & Y_1^0 & 0 & 0 & 0 & \cdots & 0 & -Z_i^0 & Y_i^0 & \cdots & 0 & -Z_p^0 & Y_p^0 \\ Z_1^0 & 0 & -X_1^0 & 0 & 0 & 0 & \cdots & Z_i^0 & 0 & -X_i^0 & \cdots & Z_p^0 & 0 & -X_p^0 \\ -Y_1^0 & X_1^0 & 0 & 0 & 0 & 0 & \cdots & -Y_i^0 & X_i^0 & 0 & \cdots & -Y_p^0 & X_p^0 & 0 \\ X_1^0 & Y_1^0 & Z_1^0 & \underbrace{0 \quad 0 \quad 0} & & & \cdots & X_i^0 & Y_i^0 & Z_i^0 & \cdots & X_p^0 & Y_p^0 & Z_p^0 \end{bmatrix} \tag{4.128}$$

new points non - datum points new points

Eqn. (4.128) shows a modified matrix **B** where all those elements are eliminated which are related to the perspective centres.

Practical aspects of free net adjustment

With free net adjustment, as with the unconstrained datum definition using reference points, the photogrammetric network is not influenced by possible inconsistencies between reference points. The object coordinate residuals are only affected by the photogrammetric observations and the quality of the model. A free net adjustment therefore provides an optimal precision that can be better analysed than standard deviations of unconstrained, or even over-determined, datum definitions.

The free net adjustment is therefore a very flexible tool if

- no reference points are available,
- only the relative positions of object points are of interest, or
- only the quality of the model is to be analysed, for example the model of interior orientation used in self-calibration.

However, the standard deviations of object points are not suitable for a direct assessment of accuracy. They only provide information about the internal quality of a photogrammetric network, i.e. they express how well the observations fit the selected model. Accuracy can only be properly assessed using comparisons with data of higher accuracy (see also section 4.4.5.3).

4.4.4 Generation of approximate values

The generation of approximate values, to be used as starting values or initial values in the iterative solution of a photogrammetric problem, is often a complex task. Approximate values are required for all unknowns to be estimated, i.e. all orientation parameters and all new points. Since arbitrary image configurations in arbitrarily oriented object coordinate systems may well occur in close-range applications, the manual calculation of approximate values is virtually impossible.

Fig. 4.66 depicts (within solid lines) that information which is necessary for the generation of approximate values required by the bundle adjustment, and (in dotted lines) information which is useful but optional. The key component is a module for the automated calculation of approximate values based on measured image coordinates, camera parameters and, if available, coordinates of reference points. This process is also known as *multi-image orientation*, whereby bundle triangulation is expressly not implied.

In many cases, additional information about the image configuration is available, such as surveyed orientation data, parameters derived from free-hand sketches or CAD models. Manual intervention in the procedures for calculating approximate values can sometimes be necessary and such additional information can support this manual process as well as helping to define the coordinate system. Approximate values can, however, also be generated fully automatically.

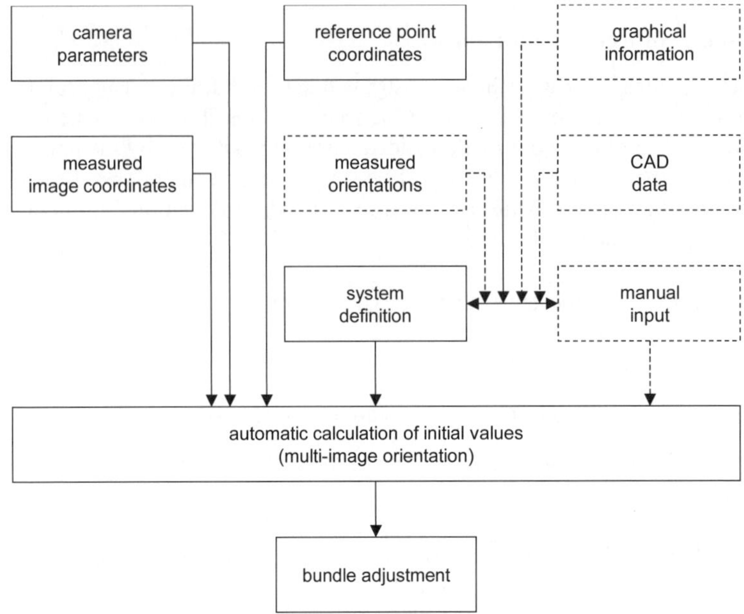

Fig. 4.66: Methods and procedures for the calculation of approximate values

The following principal methods for the generation of approximate values can be identified. They can also be applied in combination:

- Automatic calculation of approximate values:

  In order to generate approximate values automatically, three strategies are feasible for complex photogrammetric configurations:

  - combined intersection and resection
  - successive forming of models
  - transformation of independent models

  All three strategies are based on a step by step determination of the parameters of exterior orientation, as well as the object coordinates, in a process where all images in the project are added sequentially to a chosen initial model.

- Generation of approximate values by automatic point measurement:

  Digital photogrammetric system allow for the automatic identification and measurement of coded targets. Reference and tie points can be matched by this method and, by further applying the above orientation strategies, a fully automatic generation of approximate values is possible.

- Direct measurement of approximate values:

  Approximate values of object points, especially of imaging stations and viewing directions, can be measured directly, for example by survey methods. Fig. 3.140 shows an imaging total station which can do this. It is also possible to use separate measuring equipment for the location of camera stations, such as GPS or inertial navigation units.

4.4.4.1 Strategies for the automatic calculation of approximate values

In order to calculate approximate values automatically, the following information must be provided:

- a file containing camera data (interior orientation data which can be approximate)
- a file containing measured image coordinates (see Fig. 4.55)
- a file containing reference point coordinates (optional, if available) or other information for the definition of the object coordinate system
- other known object information (optional, if available), e.g. approximately known exterior orientations

On the basis of this information, an iterative process is started which attempts to connect all images via approximate orientations. At the same time all measured tie points can be approximately calculated in object space. The bundle adjustment can then be executed.

The following procedures show sample strategies which employ suitable combinations of various methods for orientation and transformation (resection, relative and absolute orientation, similarity transformation, intersection). In all cases, a reasonable starting model formed by two images is defined.

Starting model and order of calculation

From a multi-image configuration, a *starting model* is provided by one image pair for which relative orientation may be computed. The resulting model coordinate system provides an arbitrary 3D coordinate system for including all subsequent images or models. The choice of starting model, and processing sequence of subsequent images, is not arbitrary and is critical to the success of the automatic calculation of approximate values, especially for complex and irregular image configurations.

Theoretically, n images of a multi-image configuration lead to $n(n-1)/2$ possible models. Image pairs with fewer than 5 homologous points would, of course, be eliminated. Quality criteria can be calculated for each possible model and the selected starting model should have the following properties:

- Number of tie points:

  A large number of tie points leads to a more stable relative orientation where possible gross errors (outliers) can eliminated more easily.

starting model: maximum number of tie points

- Accuracy of relative orientation:

  The standard deviation of unit weight of the relative orientation (s_0) should represent the expected image measuring accuracy.

  starting model: minimum s_0

- Mean intersection angle at model points:

  The mean intersection angle of homologous image rays provides information about the quality of a computed relative orientation. Models with small mean intersection angles (e.g. less than 10°) have an unfavourable height-to-base ratio and should not be used as starting models

  starting model: mean intersection angle close to 90°

- Mean residuals of model coordinates:

  The mean intersection offset of homologous image rays which are skew is a quality measure of model coordinate determination.

  starting model: minimum mean intersection offset

- Number of gross errors in relative orientation:

  Models with few or no blunders are preferable for orientation.

  starting model: no blunders

- Image area covered by tie points:

  starting model: maximum image area

By giving appropriate weight to these criteria an optimal starting model, as well as a sorted list of further models in order of calculation, can be selected. In general the model with maximum tie points and best mean intersection angle is a suitable starting model. An unfavourable starting model, chosen without regard to these criteria, can cause the iterative orientation procedure to diverge.

Combination of space intersection and resection

Fig. 4.67 shows the principal steps in generating approximate values by combined space intersections and resections. A starting model is first selected according to the criteria outlined in the preceding section.

This starting model is used to calculate a relative orientation. Subsequent images are oriented to this model by space resection, provided they have at least 3 spatially distributed tie points with model coordinates known from a previous relative orientation. Model coordinates of new unknown object points are then calculated by intersection. When additional points are calculated by intersection, relevant images can be oriented again by resection in order to improve their exterior orientation in the model system. The image configuration is iteratively stabilised in this process since the number of intersected model points continually increases, thus improving the orientation parameters calculated by space resection.

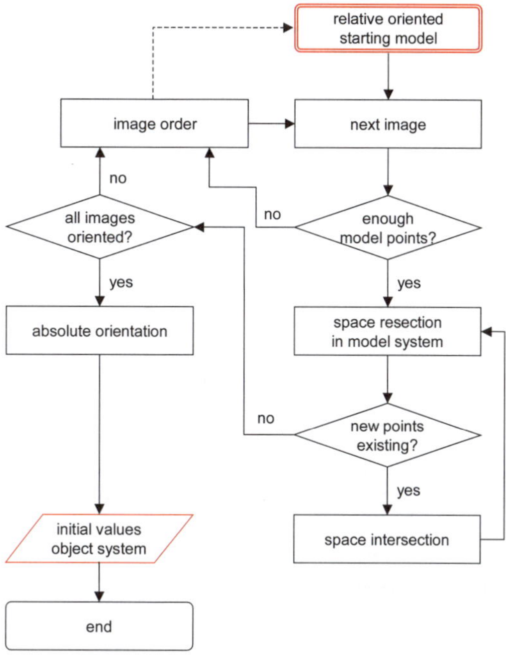

Fig. 4.67: Generation of approximate values with combined space intersection and resection

Once all images have been oriented with respect to the model coordinate system, a final absolute orientation (similarity transformation) can be performed. For this purpose, reference points with known coordinates in both model and object coordinate systems are used. The final result delivers approximate exterior orientation values of all images and approximate coordinates for all 3D object points.

Successive creation of models

Fig. 4.68 illustrates the process of initial value generation by successive creation of models. A relative orientation is first calculated for a suitable starting model. If this model contains enough reference points, an absolute orientation can immediately be performed. If not, overlapping photos can be successively oriented to the initial model if they contain a sufficient number of homologous points. Again, if a set of models contains enough reference points it can be absolutely oriented. In this way approximate values of unknown object points in a model or object coordinate system can be computed. At the end of the process the parameters of exterior orientation can be derived from the parameters of relative and absolute orientation.

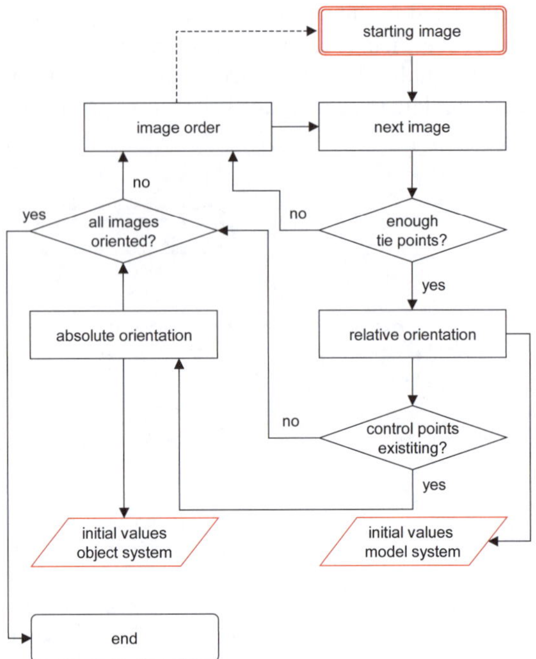

Fig. 4.68: Initial value generation with successive model creation

Transformation of independent models

The process of initial value generation by transformation of independent models is illustrated in Fig. 4.69. A connection matrix of all image pairs (models) is first established. For each possible model a relative orientation is calculated and stored together with corresponding model coordinates and quality estimators.

The starting model is selected according to the following criteria:

a) a maximum number of tie points,

b) appropriate intersection angles and

c) minimum intersection offsets.

The model coordinates of the relatively oriented starting model are used to define a local coordinate system. All other relatively oriented models are subsequently transformed into this local system using their independent model coordinates as input to a 3D similarity transformation.

When all points in all models are calculated with respect to the local system, a final absolute transformation of the complete model into the object coordinate system is calculated using reference points with known object coordinates. As a result, approximate values of object points in the object coordinate system are generated.

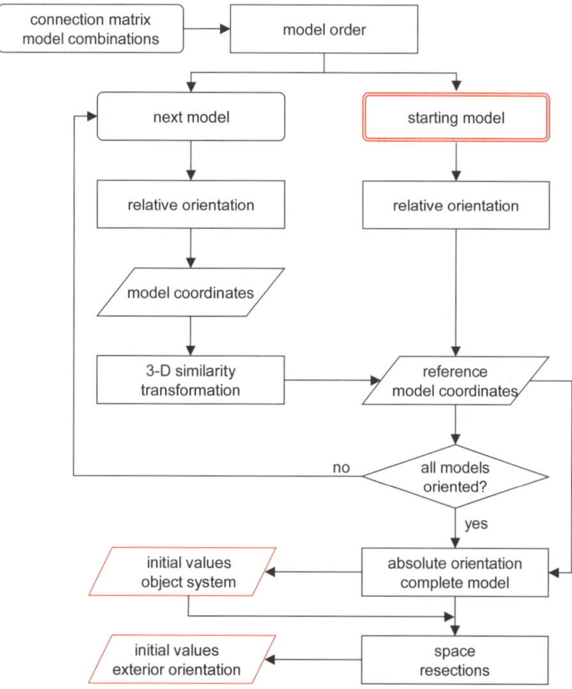

Fig. 4.69: Initial value generation by transformation of independent models

Finally the exterior orientations of all images are calculated by space resection (single image orientation) using the object coordinates computed above. If any object points remain without initial values, these can be provided by space intersection.

4.4.4.2 Initial value generation by automatic point measurement

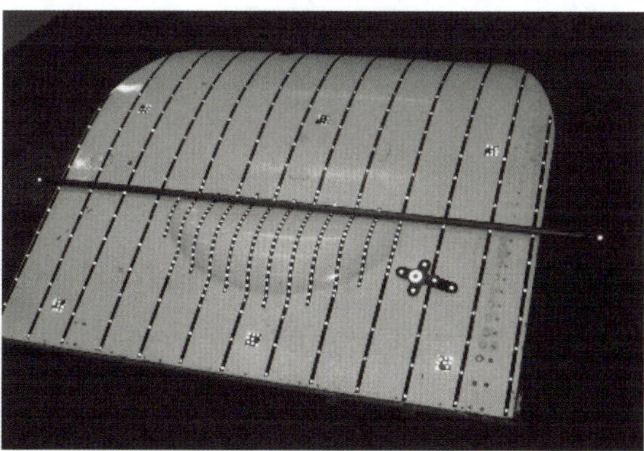

Fig. 4.70: Image with artificial targets and local reference system (GSI)

Fig. 4.70 shows one photo of a multi-image configuration which has been taken to measure a set of targeted points. Targets with coded point numbers have been placed on several points. These can be automatically identified and decoded by the image processing system (see section 3.6.1.5). The majority of points are marked by standard, non-coded targets. In addition a local reference tool (front right) is placed in the object space. It consists of a number of coded targets with calibrated local 3D object coordinates. The reference tool need not be imaged in all photos. The example above also shows a reference scale bar which provides absolute scale but is not relevant to initial value generation.

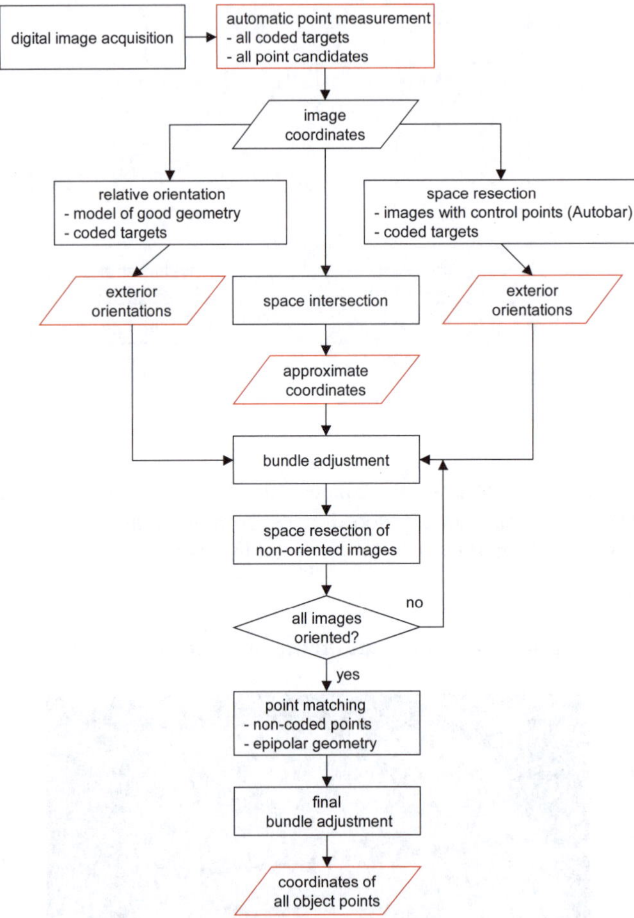

Fig. 4.71: Fully automated generation of initial values and orientation

Approximate values for the freely configured set of images can be generated as follows (Fig. 4.71): A pattern recognition process detects all coded targets and other potential object points. Those photos in which the reference tool is imaged can be individually oriented by space resection into the tool's coordinate system. Remaining images are oriented by relative orientation using the coded targets and the strategies described above for calculation of approximate values. Object coordinates for coded targets are calculated by intersection. A first bundle adjustment generates improved object coordinates and orientation parameters. Remaining photos, which are not at this stage oriented, are iteratively integrated by resection and bundle adjustment until all images are oriented.

A subsequent processing stage identifies and consecutively numbers all non-coded targets using a matching process based on epipolar geometry (see section 5.5.3). A final bundle adjustment provides the coordinates of all object points. This and similar methods form part of digital online and offline measuring systems (see section 6.4.2). Initial value generation and precise coordinate determination are integrated in one common procedure.

4.4.4.3 Practical aspects of the generation of approximate values

The automatic calculation of approximate values (multi-image orientation) is often a time-consuming process for the complex image configurations associated with close-range photogrammetry. The stability of a multi-image project depends mainly on the distribution of object points and on the configuration of bundles of rays. If there are no, or only coarse, approximate values of orientation parameters, then even a very few gross data errors can lead to divergent solutions for initial values. Since the imaging configuration is then poorly defined, the detection of gross errors is more difficult.

As a result, effective systems for generating initial values should incorporate the following features which increase the level of automation and the scope for error detection.

- Use of algebraic rotation matrices:

  Modules for relative orientation, spatial similarity transformation (absolute orientation) and space resection should use rotation matrices based on algebraic functions instead of trigonometric functions. The use of trigonometric functions in the rotation matrices can lead to singularities or ambiguities. In addition, algebraic definitions improve the convergence of solutions (see section 2.2.2.1).

- Robust blunder detection:

  The detection of gross data errors should be sufficiently robust that more than one blunder can be detected and eliminated more or less automatically. Estimation methods based on the L1 norm are particularly suitable for this purpose (see section 2.4.4).

- Detailed reporting of results:

  To analyse the results of initial value generation effectively, statistical data (e.g. standard deviations) and graphical information (position and orientation of object and cameras) should be output.

- Automatic and interactive definition of the order of images:

  With a suitable starting model the order of image orientation can be determined automatically. However, situations occur where the suggested order does not lead to convergence, for example due to images which cannot be controlled by other images. In these cases an interactive definition of image order should be possible.

- Manual activation and deactivation of points and images:

  During the process of initial value generation it can be necessary to deactivate faulty points or images with weak geometry in order to provide an initial image set which can be oriented. Once a sufficient number of images have been oriented, the deactivated images can then be successively added.

- Manual input of approximate values:

  It is efficient to use any additional information about the object or the image configuration which may be available, e.g. from additional or previous measurements. In particular the principal distance should be known in advance, for example approximated by the focal length.

In principle, photogrammetric measuring systems capable of automatic measurement and identification of image points, generate significantly fewer gross errors than interactive systems. With appropriate image configurations these systems therefore provide a fully automatic procedure, from the calculation of initial values through to the final result of the bundle adjustment.

4.4.5 Quality measures and analysis of results

4.4.5.1 Output report

Typical bundle adjustment programs report on the current status of processing including parameters such as number of iterations, corrections to unknowns, error messages. When the program is complete an output report is generated which summarises all results. It should contain the following information:

- list of input files and control parameters, date, project description
- number of iterations and standard deviation of unit weight s_0
- list of observations (image measurements) including corrections, reliability numbers and test values, sorted by images
- mean standard deviations of image coordinates, sorted by image and divided into x' and y' values
- list of blunders detected and eliminated
- list of reference points
- list of adjusted object points (new points) with standard deviations
- mean standard deviations of new points, divided into X, Y and Z values
- maximum corrections with (numeric) identifiers of the corresponding points

- parameters of interior orientation with standard deviations
- correlations between the parameters of interior orientation
- parameters of exterior orientation with standard deviations
- correlations between the parameters of interior and exterior orientation
- list of additional (survey) observations with standard deviations

4.4.5.2 Precision of image coordinates

The precision of image coordinates is calculated from the cofactor matrix:

$$\hat{s}_i = \hat{s}_0 \cdot \sqrt{q_{ii}} \tag{4.129}$$

with q_{ii} the principal diagonal elements of matrix $\mathbf{Q}_{\hat{l}\hat{l}}$ (see equations 2.186ff).

If the bundle adjustment is calculated using only equally weighted photogrammetric image coordinates as observations, $\hat{s}_0$ should represent the accuracy of the instrumental combination of camera and image measuring device.

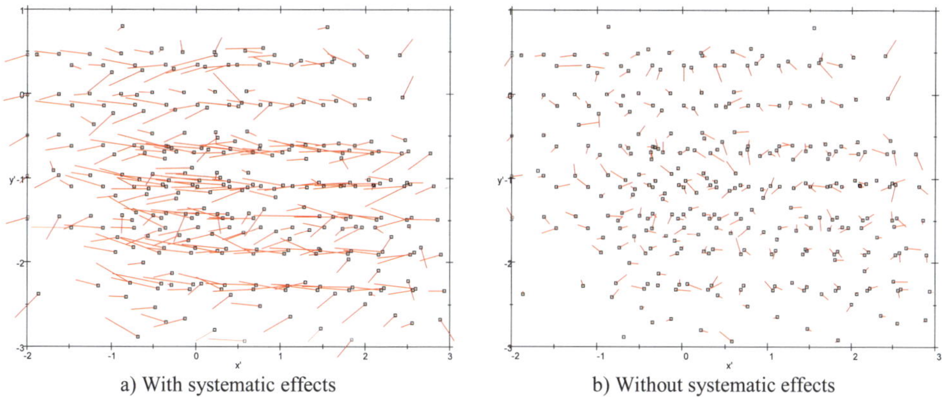

a) With systematic effects   b) Without systematic effects

Fig. 4.72: Residuals of image coordinates

The standard deviations of the measured image coordinates should be similar in both x and y directions. Different values imply that the measuring device or the camera generates systematic errors. This may occur, for example, with digital images generated by an image scanner whose mechanical construction is different in x and y. Bundle adjustment programs with integrated estimation of variance components (see section 2.4.4.2) allow different weighting for separate groups of observations in order to process observations according to their importance or precision.

4.4.5.3 Precision of object coordinates

A graphical analysis of the distribution of residuals should, in the end, show no systematic errors Fig. 4.72). However, a rigorous analysis of point accuracy in object space is not possible with this information.

ADJUSTED OBJECTCOORDINATES					STANDARD DEVIATION		
POINT NO.		X	Y	Z	SIGX	SIGY	SIGZ
NP	1	-550.0000	550.0000	0.0000	0.0000	0.0000	0.0000
NP	2	-548.8025	76.6273	0.0256	0.0185	0.0216	0.0215
NP	3	-550.0000	-401.0688	0.0000	0.0000	0.0262	0.0000
NP	4	-24.2578	-401.1035	0.0791	0.0226	0.0287	0.0237
NP	5	498.7059	-402.0520	0.0756	0.0279	0.0336	0.0272
NP	6	500.3934	72.0957	0.0000	0.0256	0.0306	0.0000
NP	7	502.2132	547.5473	0.2088	0.0281	0.0302	0.0280
NP	8	-23.2051	549.0098	-0.0653	0.0223	0.0244	0.0244
NP	9	-43.1530	97.4204	30.7236	0.0203	0.0245	0.0192
NP	10	294.5810	-544.1801	30.9239	0.0250	0.0284	0.0268
NP	11	-369.2470	716.9463	30.9868	0.0210	0.0185	0.0246
NP	12	-355.2835	-119.2167	499.6979	0.0236	0.0275	0.0267
NP	13	322.5481	256.7477	500.5552	0.0235	0.0340	0.0273
NP	14	-367.8961	-96.4070	500.5341	0.0233	0.0275	0.0251
NP	15	309.9189	279.5831	500.7111	0.0238	0.0331	0.0285
NP	16	107.0291	-218.9378	550.2870	0.0254	0.0313	0.0257
NP	17	-186.2447	333.0639	550.1281	0.0247	0.0287	0.0247
NP	18	130.0717	-206.7312	550.6317	0.0262	0.0323	0.0280
NP	19	-163.1959	345.2691	550.4474	0.0252	0.0292	0.0251
NP	20	-46.6227	95.7521	805.8236	0.0288	0.0354	0.0227
NP	21	-26.8039	106.0982	12.3571	0.0330	0.0356	0.0256
NP	22	-30.1330	104.4621	762.3889	0.0334	0.0381	0.0259
NP	23	-59.3026	88.9372	12.1583	0.0326	0.0348	0.0250
NP	24	-62.7803	87.2304	762.2332	0.0312	0.0365	0.0250
NP	25	296.2902	427.7684	48.8427	0.0257	0.0384	0.0426
NP	26	-507.1044	1.5373	71.8683	0.0219	0.0266	0.0273
NP	27	-258.4435	574.2591	62.1512	0.0347	0.0294	0.0437
NP	28	134.0411	-171.4542	405.3036	0.0362	0.0348	0.0439
				RMS:	0.0258	0.0302	0.0267
X-MAX.:	0.0362	AT POINT NO.:	28				
Y-MAX.:	0.0384	AT POINT NO.:	25				
Z-MAX.:	0.0439	AT POINT NO.:	28				

Fig. 4.73: Report file showing adjusted object coordinates (CAP)

Usually the precision of adjusted object points is of major importance for the quality analysis of a bundle adjustment. The analysis should consider two criteria (example in Fig. 4.73):

- Root mean square error (RMS):

  The root mean square error is a measure of the general precision level of the adjustment. Taking the mean image measuring accuracy $s_{x'y'}$ and the mean image scale m into account, the equation for accuracy estimation can be checked (see section 3.3.1.2):

$$s_{XYZ} = q \cdot m \cdot s_{x'y'} \tag{4.130}$$

  The design factor q reflects the accuracy of the imaging configuration with typical values between 0.7 and 1.5.

- Maximum residuals of single points:

  In contrast to the mean standard deviation, maximum residuals indicate the loss of precision which can be expected at problem points or unfavourable areas within the image configuration. If object coordinates are to be used for further calculation or analysis, then the maximum residuals should stay within specified precision limits.

Both the above quality criteria can be used only to analyse the statistical precision of the photogrammetric procedure. It is necessary here to take into account whether the object coordinate datum was defined without constraints (by minimum number of reference points or by free net adjustment), or if inconsistencies in reference points could be influencing the accuracy of object points.

If image observation accuracy is homogeneous, accuracy differences in object points are mainly caused by:

- different image scales or camera/object distances
- different numbers of image rays per object point
- different intersection angles of image rays
- different image quality of object points
- variations across the image format, e.g. lower quality imaging near the edges where distortion is often less well determined
- inconsistencies in reference points

The true accuracy of a photogrammetric project can be estimated only by comparing photogrammetrically determined points or distances with reference values measured independently to a higher accuracy. However, a rigorous evaluation is possible only with independent reference points which have not already been used as reference points or for reference distances in the bundle adjustment. Only through independent control will all properties of a photogrammetric system become visible and a rigorous accuracy assessment become possible. Suggestions for the verification of photogrammetric systems are further discussed in section 7.2.2.

4.4.5.4 Quality of self-calibration

The adjusted interior orientation parameters, and their correlations, should be carefully examined if the image acquisition system has been calibrated simultaneously in the bundle adjustment.

Fig. 4.74 is a part of a calculation report showing the adjusted interior orientation data, associated standard deviations and a matrix of correlations between the parameters. The following comments can be made about individual parameters:

```
CAMERA NO.          10        R0:    20.0000      FOCAL LENGTH (MM)  S.D. (MM)
                                                  C :     -59.2616     0.0065

PRINCIPAL POINT    S.D. (MM)                      RAD.SYM.DIST.       S.D.
X0:       0.0636     0.0035                       A1:   -0.193D-04  0.476D-06
Y0:       0.1284     0.0040                       A2:    0.556D-08  0.763D-09

CORRELATION BETWEEN PARAMETERS OF INTERIOR ORIENTATION
      C     X0    Y0    A1    A2
  C  1.00
  X0 0.03  1.00
  Y0 0.21  0.03  1.00
  A1 0.04  0.00  0.00  1.00
  A2 0.05  0.00  0.10 -0.97  1.00
```

Fig. 4.74: Result of a self-calibration with correlation between the parameters

- Principal distance c:

  The value of c normally corresponds approximately to the focal length. When plane surfaces are imaged without oblique views, the principal distance cannot be uniquely determined and either the value of c is unreasonable or its corresponding standard deviation is higher than expected. However, this does not mean that the object points are determined to lower accuracy. If the results of the bundle adjustment are intended to calibrate a camera which is subsequently used on other projects, then the principal distance must be calculated accurately and with a standard deviation of order of the image measuring accuracy.

- Principal point x'_0, y'_0:

  The principal point normally lies very close to the foot of the perpendicular from the projection centre to the focal plane. When plane surfaces are imaged without oblique views, the position of the principal point cannot be uniquely determined; in this case, however, this does not mean that the object points are determined to lower accuracy. The importance attached to an accurate knowledge of the principal point will depend on the configuration of the network. If the results of the self-calibration are intended to be subsequently used on other projects, then the position of the principal point must be calculated accurately and with a standard deviation of order of the image measuring accuracy.

  It should be remembered that the definition of radial distortion depends on the location of the principal point. A large shift of the principal point may result in irregular parameters for radial distortion. An iterative pre-correction of measured image coordinates is recommended in those cases.

- Radial (symmetric) distortion A_1, A_2, A_3:

  The parameters of radial distortion are normally the most effective additional parameters. Their related standard deviations should be much smaller than the parameters themselves. Parameter A_2 is often significantly correlated with A_1 (as shown in Fig. 4.74). However, this does not necessarily affect the overall result, especially for the object coordinates. Parameter A_3 can normally be determined to a significant value only in special cases, for example when fisheye lenses are employed.

- Tangential (decentring) distortion B_1, B_2, affinity and shear C_1, C_2:

  The statements concerning A_1 and A_2 can also be applied to the optional additional parameters B_1, B_2, C_1, C_2. For many digital cameras, departures from orthogonality and equality of scale between the axes of the image coordinate system (e.g. due to imaging elements which are not square) are barely detectable. However, cameras utilizing analogue data transfer and a frame grabber should definitely use these parameters.

Correlations between the parameters of interior orientation and exterior orientation should be avoided by an appropriate imaging configuration (see section 7.3.2). They do not need to be considered for point determination if the calculation is performed in a single stage mathematical process. However, if single parameters are extracted for use in further external calculations (e.g. orientation parameters are applied in separate space intersections), correlations, or more exactly the variance-covariance matrix, cannot be further taken into account, which leads to mistakes in the functional model.

4.4.6 Strategies for bundle adjustment

In many practical applications, either the generation of initial values or the complete bundle adjustment may not run successfully at the first attempt, or the final result might not meet the specification. To avoid these problems, a number of practical tips and suggestions for strategies and procedures in bundle adjustments is given below.

4.4.6.1 Simulation

Simulation of the imaging configuration is one method of project planning (see section 7.1). Simulation provides a priori accuracy estimation for optimizing an imaging configuration without real measurements. For this purpose the measuring object must be represented by simulated object points similar to the actual measurement in terms of number and distribution. The required 3D coordinates can be provided by manual input if the object is not too complex. Alternatively the coordinates can be obtained from a CAD model or previous measurements (Fig. 4.75).

The a priori definition of camera stations and viewing directions (exterior orientations) is much more complicated. While orientation data can easily be generated for regular image configuration patterns (example in Fig. 4.52), the measurement of complex structures can often be configured only on site. On the one hand selection of image configurations is more flexible but on the other hand unforeseen problems can often occur in the form of occlusions or restricted camera stations.

Camera data and a priori accuracies must also be defined. Image coordinates can then be calculated using collinearity equations and the simulated object coordinates and orientation data. Using these simulated data as input, a bundle adjustment can be calculated. Here it is only necessary to compute the covariance matrix of the unknowns and the standard deviations of interest, instead of a complete parameter estimation. Now object points, imaging configurations, selection of cameras and accuracies can be varied until a satisfactory adjustment is achieved. By applying the Monte-Carlo method (section 7.1.4), the input data can be altered within specific noise ranges based on normal distribution. The computed output values will vary as a function of the noisy input data.

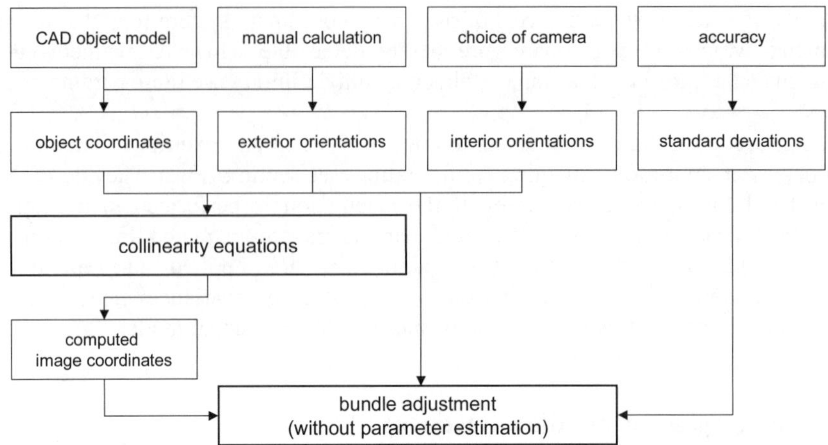

Fig. 4.75: Simulation by bundle adjustment

4.4.6.2 Divergence

Bundle adjustments which do not converge are often a serious problem in practical applications. A major reason for this is that standard statistical methods of error detection only work well if at least one iteration has been successfully calculated. Divergence in adjustments can be caused by:

- Faulty input data:

  Error in data formats, units of measurement, typing errors etc. should be detected by the program but often are not.

- Poor initial values:

  Poor initial values of the unknowns lead to an inadequate linearisation of the functional model.

- Gross errors in the input data:

  Errors in identifying point or image numbers and measuring errors larger than the significance level of the total measurement. They can be detected in part by statistical tests if robust estimation methods are applied (see section 2.4.4).

- Weak imaging geometry:

  Small intersection angles, a small number of rays per object point and/or poor interior orientation data lead to poorly conditioned normal equations and a lower reliability in the adjustment.

The following steps should therefore be applied to handle divergent bundle adjustments:

1) check of input data;
2) controlled program abortion after iteration 0, with checking of differences between initial values and "adjusted" unknowns (see section 2.4.4.1) – high deviations indicate problematic input data;
3) pre-correction of image coordinates by known distortion values;
4) adjustment without camera calibration in the first run; subsequent adding of additional parameters (distortion) when blunders have been eliminated from the input data;
5) check of the geometric configuration of images where gross errors have been detected – the smearing effect of least-squares solutions can lead to misidentification of blunders.

4.4.6.3 Elimination of gross errors

Generally speaking, the bundle adjustment is very sensitive to gross errors (blunders) in the measured data. In complex imaging configurations, gross errors arise easily due to false identification of object points or mistakes in image or point numbering. In contrast, pure measuring errors occur relatively infrequently. The detection of gross errors can fail, especially in geometrically weak configurations where statistical tests based on redundancy numbers are not significant (see section 2.4.3.4). This is of major importance if gross errors in observations occur at *leverage points* which have a strong influence on overall geometry but which cannot be controlled by other (adjacent) observations.

Most bundle adjustment programs permit both manual and automatic elimination of blunders. In both cases only one blunder should be eliminated per program run, usually the one with the largest normalised correction. The adjustment program should allow the corresponding observation to be set as inactive, not deleted, in order that it can be reactivated later in case it is discovered to be correct.

If observations are eliminated, the corresponding object area and image should be analysed. Uncontrolled elimination of observations can lead to weak imaging geometries if the object is recorded by a small number of images, or represented by only a few object points. A system that is based on the few remaining observations cannot be controlled by a rigorous blunder test and may produce a plausible result even if gross errors are still present.

4.4.7 Multi-image processing

This section deals with analytical methods for object reconstruction based on measured image coordinates from an unlimited number of photos. Digital, multi-image methods which additionally process grey values at image points are discussed in chapter 5.

The following methods require known parameters of interior and exterior orientation which usually are calculated by a bundle triangulation (see section 4.4). On this basis object points, surfaces and basic geometric elements can be determined.

4.4.7.1 General space intersection

The general space intersection takes measured image coordinates from multiple images, together with their known orientation parameters, and calculates the spatial point coordinates X,Y,Z. The calculation is based on the collinearity equations (4.9) used as observation equations in a least-squares adjustment:

$$\begin{aligned} x'_i + vx'_i &= F(X_{0j}, Y_{0j}, Z_{0j}, \omega_j, \varphi_j, \kappa_j, x'_{0k}, c_k, \Delta x'_k, \underline{X_i, Y_i, Z_i}) \\ y'_i + vy'_i &= F(X_{0j}, Y_{0j}, Z_{0j}, \omega_j, \varphi_j, \kappa_j, y'_{0k}, c_k, \Delta y'_k, \underline{X_i, Y_i, Z_i}) \end{aligned} \tag{4.131}$$

where i: point index
j: image index
k: camera index

In order to calculate the three unknowns X_i, Y_i, Z_i at least three observations (image coordinates) are required. Two images already provide a redundancy of 1 and with each additional observation the redundancy increases by 2.

To set up the normal equations the derivatives of the unknown object coordinates are calculated according to (4.96).

A global measure of the quality of point determination is given by the shortest distance between the two skew rays (see section 2.3.2.1). If the point accuracy is to be analysed separately for each axis, the covariance matrix must be evaluated. Given the cofactor matrix of unknowns

$$\mathbf{Q}_{\hat{x}\hat{x}} = \begin{bmatrix} q_{XX} & q_{XY} & q_{XZ} \\ q_{YX} & q_{YY} & q_{YZ} \\ q_{ZX} & q_{ZY} & q_{ZZ} \end{bmatrix} \tag{4.132}$$

the standard deviations of adjusted point coordinates are as follows:

$$\begin{aligned} \hat{s}_X &= \hat{s}_0 \cdot \sqrt{q_{XX}} \\ \hat{s}_Y &= \hat{s}_0 \cdot \sqrt{q_{YY}} \\ \hat{s}_Z &= \hat{s}_0 \cdot \sqrt{q_{ZZ}} \end{aligned} \tag{4.133}$$

Hence the mean point error is given by

$$\hat{s}_P = \sqrt{\hat{s}_X^2 + \hat{s}_Y^2 + \hat{s}_Z^2} \tag{4.134}$$

Spectral decomposition of the corresponding variance-covariance matrix leads to eigenvalues λ_i and eigenvectors $\mathbf{s}_i$. It is then possible to calculate the error or confidence ellipsoid which contains the "true" point with a confidence level of 1–α (see section 2.4.3.2)

$$\mathbf{K}_{\hat{x}\hat{x}} = \begin{bmatrix} \mathbf{s}_1 & \mathbf{s}_2 & \mathbf{s}_3 \end{bmatrix} \cdot \begin{bmatrix} \lambda_1 & 0 & 0 \\ 0 & \lambda_2 & 0 \\ 0 & 0 & \lambda_3 \end{bmatrix} \cdot \begin{bmatrix} \mathbf{s}_1^T \\ \mathbf{s}_2^T \\ \mathbf{s}_3^T \end{bmatrix} \quad \text{: spectral decomposition} \tag{4.135}$$

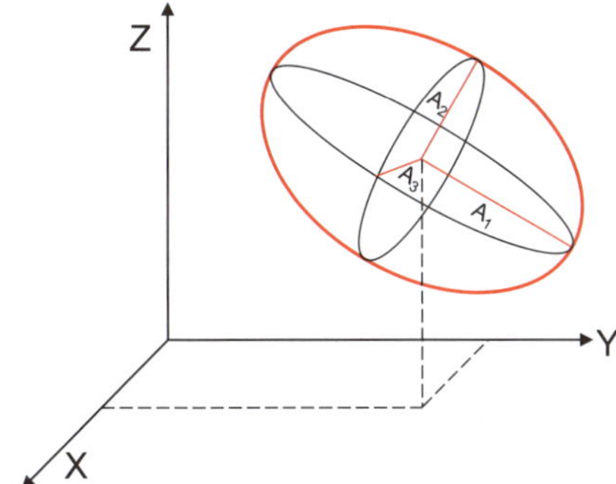

Fig. 4.76: Confidence ellipsoid

The directions of the semi-axes of the ellipsoid are defined by the eigenvectors and their length is given by the eigenvalues (Fig. 4.76):

$$A_i = \sqrt{\lambda_i \cdot \chi^2_{3,1-\alpha}} \quad \text{: length of semi-axis } A_i,\ i=1,2,3 \tag{4.136}$$

where $\chi^2_{3,1-\alpha}$: quantile of χ^2-distribution

In contrast to the standard bundle adjustment, possible correlations between adjusted point coordinates and orientation parameters are not taken into account in the spatial intersection. Assuming image coordinates of equal accuracy in all images, a stretched error ellipsoid indicates a weak intersection of homologous rays (Fig. 4.77).

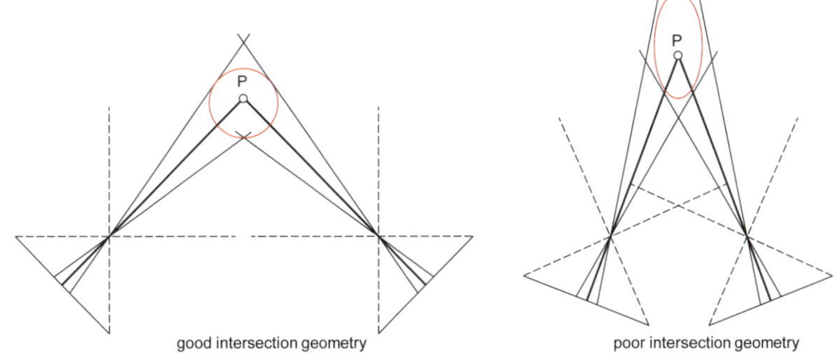

Fig. 4.77: On the geometry of spatial intersection

4.4.7.2 Direct determination of geometric elements

Geometric 3D elements (straight line, circle, sphere, cylinder etc.) can be determined in two ways:

- calculation of best-fit elements from measured 3D coordinates
- best-fit adjustment of elements to measured 2D contours in multiple images

The first method is preferred in conventional coordinate metrology. The required 3D coordinates are delivered by coordinate measuring machines (CMM) from direct probing of the object surface. The calculation of best-fit elements, and the intersection of these elements, is of major importance because most industrial objects offered for inspection are composed of regular geometric elements. The most important algorithms are discussed in section 2.3.

The second method[1], contour measurement, is based on the idea that the imaged edges of geometric elements generate unique grey level contours which can be extracted by suitable edge operators or manual measurements (see sections 5.2.4, 5.4.3). Normally, there are no discrete or identifiable points along the imaged edge which would enable individual 3D points on the object surface to be measured, for example by intersection. However, if sufficient edge image points belonging to a common object surface can be detected in a well-configured set of images, an adjustment can be formulated for estimating the parameters of the unknown element. Assuming known parameters of interior and exterior orientation, each image point defines a light ray in space which, in principle, touches or is tangent to the surface of the element. For this purpose, a distance offset is defined between the light ray and the spatial element which is minimised by adjustment of the element's parameters.

The method can be used to calculate, for instance, straight lines, 3D circles, cylinders or other elements which generate an appropriate contour in the image. The number of images is unlimited. However, for a reliable determination usually more than two images are required.

Straight line in space

Straight lines often occur in real objects, for example the intersection lines of walls, rectangular patches defined by windows and the linear textures defined by bricks and tiles. In a central perspective projection, straight lines are projected as straight lines[2].

For example, according to section 2.3.2.1 a straight line in space is defined by

$$\mathbf{x}_G = \mathbf{a} + \lambda \cdot \mathbf{b} \tag{4.137}$$

---

[1] The following derivations are due to Andresen (1991).

[2] In the following it is assumed that the image coordinates have been corrected for distortion.

If the distance vector **a** is perpendicular to the direction vector **b**, which has unit length, then the additional conditions

$$\begin{aligned} |\mathbf{b}| &= 1 \\ \mathbf{a} \cdot \mathbf{b} &= 0 \end{aligned} \tag{4.138}$$

ensure that four independent parameters remain for the definition of a straight line in space.

With known interior and exterior orientation, each image point $\mathbf{x'} = (x',y',-c)$ corresponds to a point in the object coordinate system XYZ:

$$\mathbf{X}_B = \mathbf{X}_0 + \mathbf{R}^{-1}\mathbf{x'} \tag{4.139}$$

The image point's spatial ray through the perspective centre then gives a line in parametric form

$$\mathbf{x}_S = \mathbf{X}_0 + \lambda \cdot (\mathbf{X}_B - \mathbf{X}_0) = \mathbf{X}_0 + \lambda \cdot \mathbf{c} \tag{4.140}$$

The shortest distance $d_{i,k}$ according to (2.123) between the image rays $\mathbf{x}_S$ and the desired spatial straight line $\mathbf{x}_G$ is minimised for all image points i of image k:

$$\sum_k \sum_i d_{i,k}^2 = \min. \tag{4.141}$$

The linearised correction equations

$$\frac{\partial d_{i,k}}{\partial a_x}\Delta a_x + \frac{\partial d_{i,k}}{\partial a_y}\Delta a_y + \frac{\partial d_{i,k}}{\partial a_z}\Delta a_z + \frac{\partial d_{i,k}}{\partial b_x}\Delta b_x + \frac{\partial d_{i,k}}{\partial b_y}\Delta b_y + \frac{\partial d_{i,k}}{\partial b_z}\Delta b_z = -d_{i,k} \tag{4.142}$$

and the conditions (4.138) build an extended normal system of equations according to (2.183). For each program run the approximate parameters of the straight line are iteratively corrected by the adjusted increments Δ until there is no significant change.

Approximate values of the straight line can be derived from the intersection of the two planes E_1 and E_2, derived from the image contour points (Fig. 4.78). It is obvious that the calculation of approximate values, as well as the solution of the normal system of equations, will fail in the case of a dual image configuration where the baseline and required straight line are parallel. The addition of at least one image with a perspective centre outside their plane will provide the data for a unique solution.

Fig. 4.79 shows an application where edges are determined on the interior trim of a car door. The irregular contour line is caused by the rough surface texture of the trim. The computed lines can be used to calculate additional elements such as intersecting points or quantities such as distance.

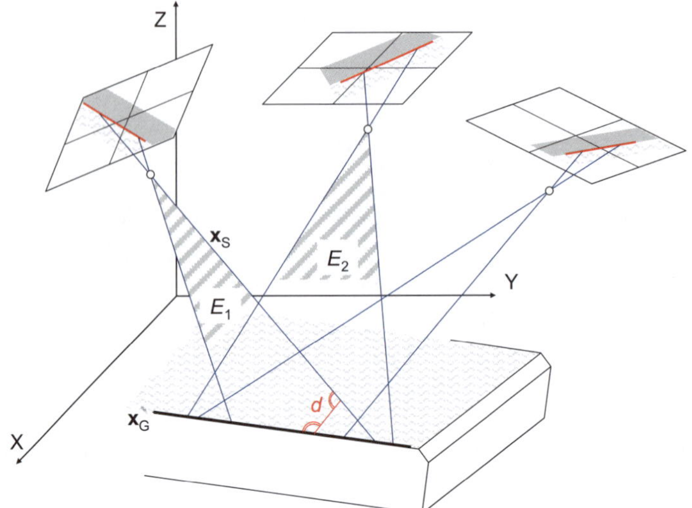

Fig. 4.78: Contour method for the determination of a spatial straight line

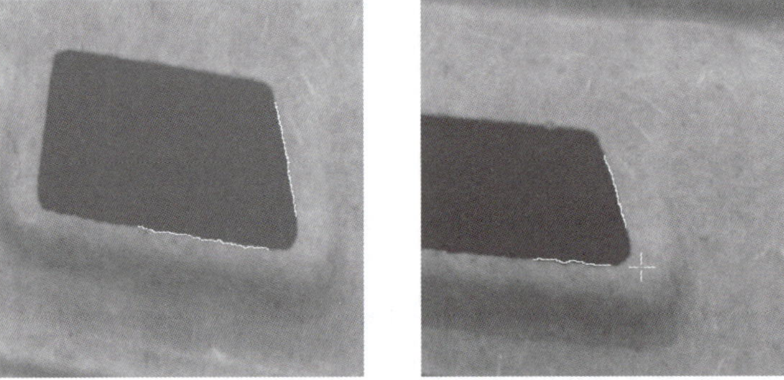

Fig. 4.79: Edge measurements on car door trim

3D circle

Circles occur on industrial objects mainly as drill holes, end sections of cylindrical features or disks. In general the circles are imaged as ellipses.

According to section 2.3.2.3, a 3D circle is defined by the centre $\mathbf{X}_M = (X_M, Y_M, Z_M)$, the vector normal to the plane of the circle $\mathbf{n} = (a,b,c)$ and the radius r. Taking into account $|\mathbf{n}| = 1$, the circle has 6 independent parameters.

The radial distance d between the circumference and the point of intersection of an image ray with the plane of the circle is taken as the quantity to be minimised (Fig. 4.80). It corresponds to the distance e in Fig. 2.28. The solution follows the adjustment principle explained above for the straight line in space.

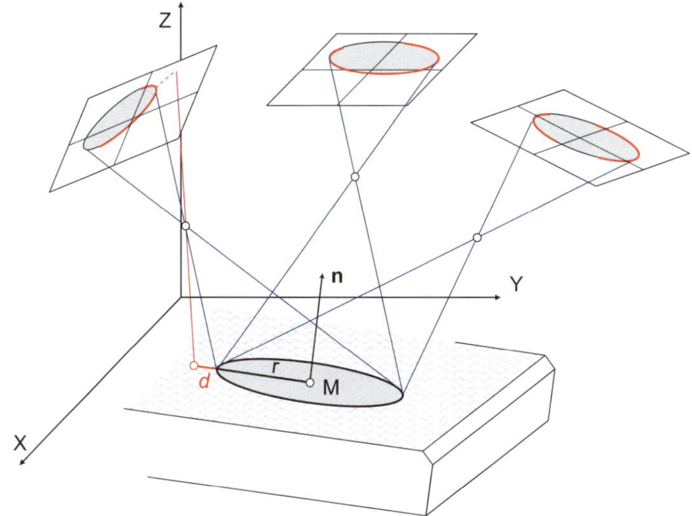

Fig. 4.80: Contour method for determining a 3D circle

The following procedure is used for determining approximate values of the circle parameters:

1. Pseudo orientation parameters are calculated for two images such that the perspective centres of each image are preserved, but the principal points of the cameras are shifted into the centre of the imaged ellipse. The centre of the 3D circle in space is computed by spatial intersection of the rays corresponding to the ellipse centres after transformation into the new pseudo image planes.
2. In each of the two pseudo images, the major axis of the ellipse is now approximately parallel to a diameter of the circle. The approximate normal vector to the circle plane can therefore be derived by the vector product of these two major ellipse axes after transforming them from image to object space.
3. The circle radius is approximately given by multiplying either major axis by the corresponding image scale number.

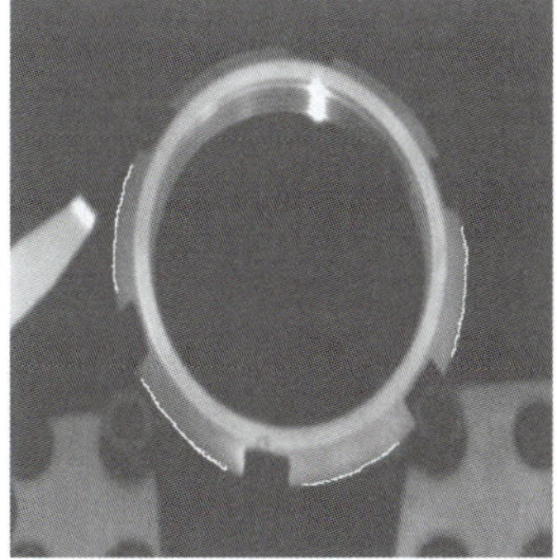

Fig. 4.81: Contour measurement of a 3D circle created by a coupling ring

For reasonable parameter estimation, more than half of the circle should be visible in at least one image. The procedure fails if the perspective centres are located in or close to the plane of circle, in which case the imaged ellipse becomes a straight line.

Fig. 4.81 shows an application where the position and radius of a coupling ring is determined from measurements of two partial contours in two images.

Right circular cylinder in space

Cylindrical objects often occur as pipes, tubes or columns. A cylinder appears in an image as straight line contours of the cylinder surface which result from the intersection of the image plane with the plane through the perspective centre and tangential to the cylinder surface (Fig. 4.82).

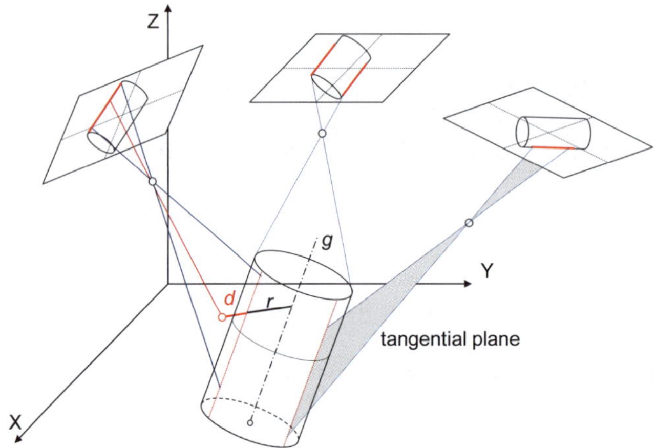

Fig. 4.82: Contour method for determining a right-circular cylinder

A right circular cylinder is defined by its axis g and radius r, hence by five independent parameters. The distance value defined in section 2.3.2.3 between an arbitrary point in 3D space and the cylinder surface cannot be used in this case, since no unique point can be defined on the image ray. Instead the distance d between the cylinder surface and the intersection point of image ray and radius vector (vector perpendicular to the axis of the cylinder) is used.

Alternatively the shortest distance between cylinder axis and image ray, reduced by the radius r, can be used as a distance measure.

In a similar way to the straight line in space, approximate parameters of the cylinder axis can be generated using two imaged lines, each of which are the average of opposite cylinder edges. The approximate radius can be calculated as for the 3D circle above, if a cylinder end is visible. If not, many applications use standardized pipe diameters which are the cylinders under evaluation. Alternatively, a radius can be calculated as the shortest distance between cylinder axis and tangential contour plane. The adjustment process operates in a similar way to the contour methods explained above.

In some simple applications only the cylinder axis is in fact required. The best-fit procedure is an optimized intersection of planes on which the axis lies. This assumes that each image sees both sides of the cylinder. This provides an averaged line in the image which approximates to the image of the axis.

There is no solution for the cylinder calculation if the cylinder axis is parallel to the baseline between the images (Fig. 4.83). As indicated above, a unique solution can then be found if at least one additional image is used whose perspective centre is not located on the baseline between the first two images.

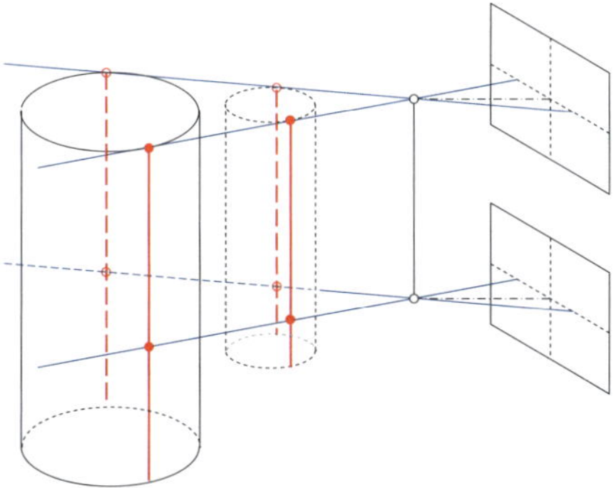

Fig. 4.83: Ambiguity of cylinder determination

Fig. 4.84 shows the measurement of the cylindrical section of a coolant hose.

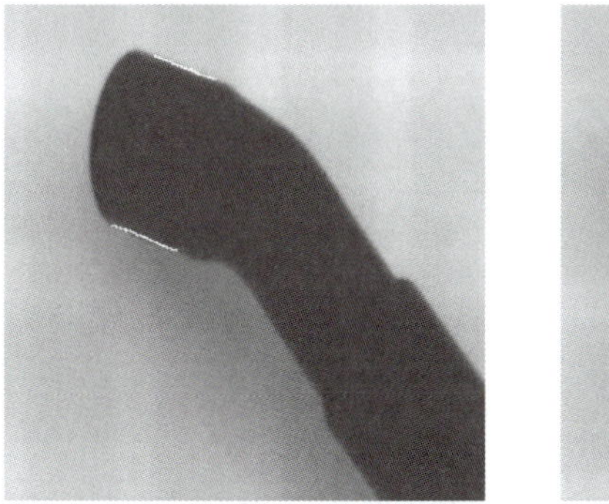

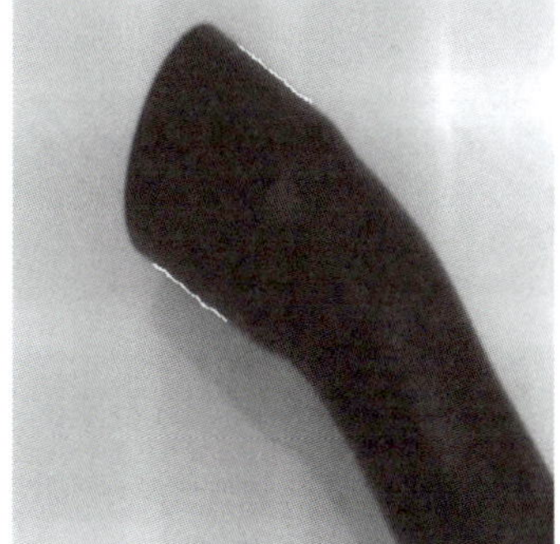

Fig. 4.84: Contour measurement of the cylindrical section of a coolant hose

Accuracy of the contour method

The accuracy of the contour method depends on the following factors:

- Consistency of optical and physical object edges:

  The main problem for the contour method is correct illumination which must ensure that the physical edge of the object corresponds to the optical edge in the image, regardless of viewing direction, surface orientation and material characteristics. This requirement applies to all images involved in the measurement.

  Simpler objects such as flat components, which are imaged by multiple photos, can usually be made visible by a uniform diffuse illumination. More complex objects whose surfaces have varying spatial orientations, and possibly different materials, must in principle be individually illuminated for each camera position and component (see section 3.6.2.4).

- Roughness and sharpness of the object edge:

  The physical properties of the object edge have a major influence on the quality of contour measurement.

  A rough object edge leads to higher standard deviations in the image measurements. In this case, the least-squares approach calculates an average geometric element. In principle, the adjustment model could incorporate circumscribed or inscribed elements as used to determine form deviations in industrial metrology (see section 2.3.4). Fig. 4.79 shows automatic measurement of rough edges.

  The sharpness of an object edge is also significant, i.e. the shape of the interface between one object surface and another. Although the element to be measured, and other derived quantities, are often related to manufactured features, the contour element can still only measure the optically generated edge. Fig. 4.81 shows an object with sharp edges. Whilst sharp edges can normally be uniquely measured, round or chamfered edges are much more difficult to measure and, even with sophisticated illumination techniques, often cannot be exactly defined (Fig. 4.85).

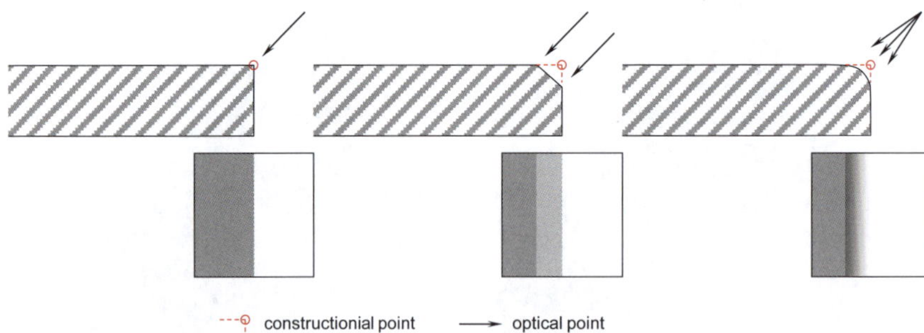

Fig. 4.85: Contour points for different types of edge

- Elimination of gross errors:

  False measurements along the contour can be eliminated by robust error detection methods (see section 2.4.4).

- Number of measured contour points:

  Each contour point delivers one observation for the adjustment, which usually leads to high redundancies. As an example, the measurement of a straight line, with an average length of 30 pixels and measured in three images, leads to 90 observations (1 pixel per contour point) which are then used to solve for only four unknowns.

- Accuracy of image contour points:

  Aside from the camera quality, the accuracy of measured image coordinates depends on the quality of the measuring instrument or method. Using digital image edge extraction, an accuracy of about 1/10 to 1/20 of a pixel can be expected, depending on the image operator and the edge sharpness.

- Distribution of contour points:

  Since the method does not require homologous points, any parts of the contour can be used for calculation. The reliability of calculation increases if contour points are uniformly distributed over the complete element. Straight line contours from straight edges and cylinders should at least be defined by points taken from both ends of the element in all images. For circular objects, more than half of the circle edge should be measured in at least one image.

- Image configuration:

  As for other photogrammetric point measurements, the quality of the result strongly depends on the configuration of camera stations and viewing directions. Ideally at least three images with good intersection angles should be included in the calculation of the element.

4.4.7.3 Determination of spatial curves (snakes)

Arbitrary curves in space can be approximated by suitable polynomial functions, in particular spatial Bézier or B-spline functions (see section 2.3.3.2). These functions provide smoothing effects which can be adjusted by an appropriate choice of parameters (degree of polynomial, curvature constraints). The final interpolation curve is defined by a set of coefficients which are determined by a system of equations.

The curve approximations discussed in section 2.3.3.2 are based on specific 3D points on the object surface. However, in a similar way to the contour method for direct measurement of elements (section 4.4.7.2), if no discrete 3D points are available on the object contour, the parameters of the curve must be calculated by a multi-image process based on 2D image points. Starting with approximate values of the curve, the related coefficients are computed iteratively by least-squares adjustment. If the adjustment converges, the calculated curve gradually conforms to the spatial object curve. If B-spline functions are used, *snakes* are computed which iteratively wind around the real object curve.

The extension of the plane form of (2.116) to three-dimensional B-splines gives:

$$P(t) = \begin{cases} x(t) = \sum_{i=0}^{k} x_i B_{i,n}(t) \\ y(t) = \sum_{i=0}^{k} y_i B_{i,n}(t) \\ z(t) = \sum_{i=0}^{k} z_i B_{i,n}(t) \end{cases} \qquad 0 \le t \le k-n+1 \tag{4.143}$$

In the adjustment process, the unknown B-spline functions B are calculated as functions of the unknown object coordinates x,y,z, which are themselves functions of the measured image coordinates x',y'.

The method can be extended by a photometric model which uses the grey values of imaged contour points as observations, in a similar way to least-squares matching (see section 5.5.4). For known parameters of interior and exterior orientation, additional geometric constraints (epipolar geometry) can be utilized. The approach has been published as *LSB-snakes*. Fig. 4.86 shows the spatial contours of a component calculated by this method.

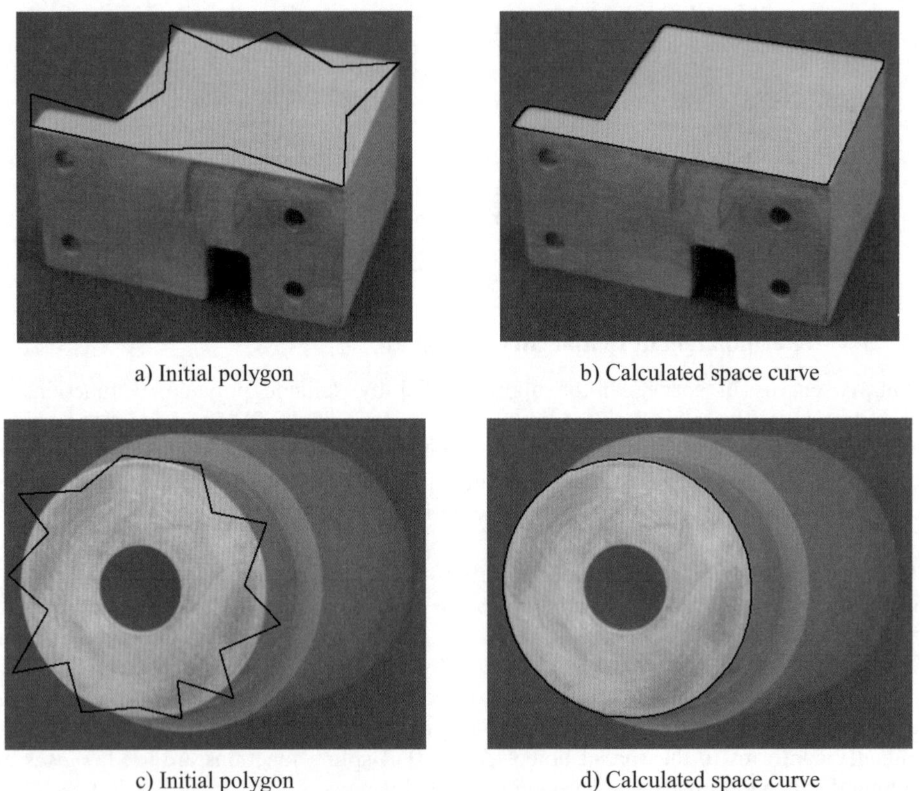

a) Initial polygon

b) Calculated space curve

c) Initial polygon

d) Calculated space curve

Fig. 4.86: Contour measurement by LSB-snakes (Li & Gruen 1997)

4.5 Panoramic photogrammetry

Panoramic photogrammetry is a special branch of close-range photogrammetry that uses panoramic images instead of conventional perspective imagery. Panoramic imagery is created either by digitally stitching together multiple images from the same position (left/right, up/down) or by rotating a camera with conventional optics, and an area or line sensor, in a specially designed mounting fixture (see section 3.5.6). This section gives an overview of the basic panoramic imaging model, orientation methods and algorithms for 3D reconstruction.

4.5.1 Cylindrical panoramic imaging model

The most common method of panoramic photogrammetry is based on a cylindrical imaging model, as generated by numerous analogue and digital panoramic cameras or by a computational fusion of individual central perspective images. Assuming the camera rotation corresponds to a horizontal scan, the resulting panoramic image has central perspective imaging properties in the vertical direction only.

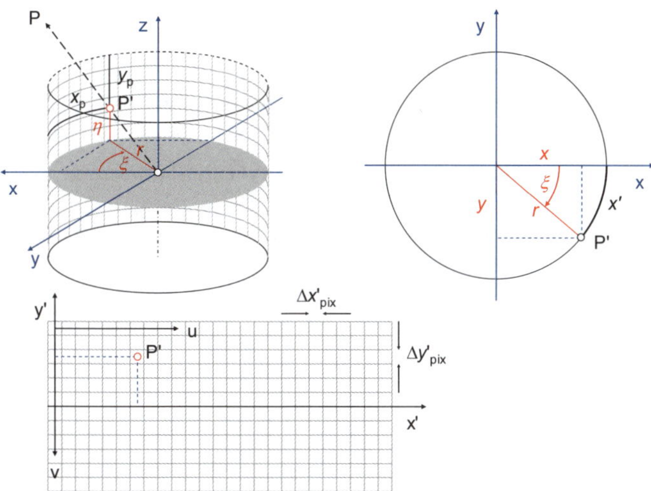

Fig. 4.87: Coordinate systems defining a cylindrical panorama

An image point P' can be defined either by the cylindrical coordinates r,ξ,η or by the Cartesian panoramic coordinates x,y,z (Fig. 4.87). The panorama is assumed to be created by a clockwise rotation when viewed from above. The metric image coordinates x',y' and the pixel coordinates u,v are defined within the cylindrical surface of the panorama, which is a plane when the cylinder is unrolled.

$$\begin{bmatrix} x \\ y \\ z \end{bmatrix} = \begin{bmatrix} r \cdot \cos\xi \\ -r \cdot \sin\xi \\ \eta \end{bmatrix} \tag{4.144}$$

$$\begin{bmatrix} x' \\ y' \end{bmatrix} = \begin{bmatrix} r \cdot \xi \\ z \end{bmatrix} \tag{4.145}$$

The digital plane panoramic image has n_C columns and n_R rows with a pixel resolution $\Delta x'_P$ and $\Delta y'_P$. The circumference of the panorama is therefore $n_C \cdot \Delta x'_P$ and its height is $n_R \cdot \Delta y'_P$. The radius of the digital image is, in the ideal case, equal to the principal distance of the camera. It can be calculated from the circumference or the horizontal angular resolution $\Delta\xi$:

$$r = \frac{n_C \cdot \Delta x'_P}{2\pi} = \frac{\Delta x'_P}{\Delta \xi} \tag{4.146}$$

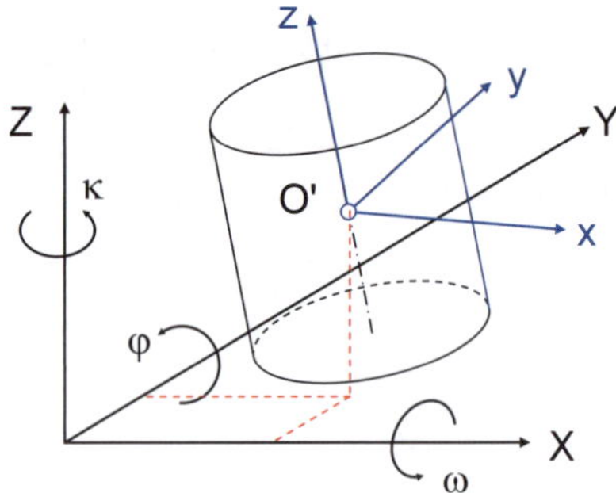

Fig. 4.88: Exterior orientation of cylindrical panorama

By introducing the parameters of exterior orientation, the transformation between object coordinate system XYZ and panoramic coordinate system xyz (Fig. 4.88) is given by:

$$\mathbf{X} = \mathbf{X}_0 + \frac{1}{\lambda}\mathbf{R} \cdot \mathbf{x} \tag{4.147}$$

Rearrangement gives

$$\begin{aligned} \mathbf{x} &= \lambda \cdot \mathbf{R}^{-1} \cdot (\mathbf{X} - \mathbf{X}_0) \\ x &= \lambda \cdot [r_{11}(X - X_0) + r_{21}(Y - Y_0) + r_{31}(Z - Z_0)] = \lambda \cdot \overline{X} \\ y &= \lambda \cdot [r_{12}(X - X_0) + r_{22}(Y - Y_0) + r_{32}(Z - Z_0)] = \lambda \cdot \overline{Y} \\ z &= \lambda \cdot [r_{13}(X - X_0) + r_{23}(Y - Y_0) + r_{33}(Z - Z_0)] = \lambda \cdot \overline{Z} \end{aligned} \tag{4.148}$$

where $\overline{X}, \overline{Y}, \overline{Z}$ define a temporary coordinate system that is parallel to the panorama system. With the scale factor

$$\lambda = \frac{r}{\sqrt{\overline{X}^2 + \overline{Y}^2}} \tag{4.149}$$

the image coordinates in the unrolled (plane) panorama are given by

$$x' = r \cdot \arctan\left(\frac{\overline{X}}{\overline{Y}}\right) - \Delta x'$$
$$y' = y'_0 + \lambda \cdot \overline{Z} - \Delta y' \tag{4.150}$$

or directly as collinearity equations between object coordinates and image coordinates:

$$x' = r \cdot \arctan\left(\frac{r_{12}(X - X_0) + r_{22}(Y - Y_0) + r_{32}(Z - Z_0)}{r_{11}(X - X_0) + r_{21}(Y - Y_0) + r_{31}(Z - Z_0)}\right) - \Delta x'$$
$$y' = y'_0 + \lambda \cdot \left[r_{13}(X - X_0) + r_{23}(Y - Y_0) + r_{33}(Z - Z_0)\right] - \Delta y' \tag{4.151}$$

Here y'_0 denotes a shift of the principal point in the y direction and $\Delta x', \Delta y'$ are correction parameters for potential imaging errors in the camera.

The pixel coordinates u,v of a digital panoramic image can be derived as:

$$u = \frac{x'}{\Delta x'_{pix}}$$
$$v = \frac{n_R}{2} - \frac{y'}{\Delta y'_{pix}} \tag{4.152}$$

4.5.2 Orientation of panoramic imagery

The method of determining the exterior orientations of one or more panoramic images is analogous to that for central perspective imagery; that is, by application of the panoramic collinearity equations to space resection and/or bundle adjustment. Both methods require suitable approximate values.

4.5.2.1 Approximate values

For the usual case of panoramas with an approximate vertical rotation axis, initial values for exterior orientation can easily be derived from reference points. The centre coordinates of the panorama X_0,Y_0 can be calculated by plane resection using three control points P_1, P_2, P_3 and the angles $\Delta\xi_{12}$, $\Delta\xi_{23}$, which can be derived from the corresponding image points P'_1, P'_2 and P'_3 (Fig. 4.89).

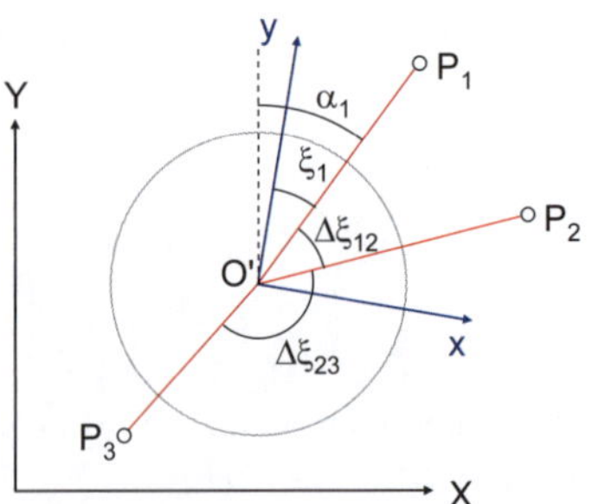

Fig. 4.89: Plane resection

The Z_0 coordinate can, for example, be calculated from reference point P_1:

$$Z_0 = Z_1 - d_1 \cdot \frac{z_1}{r} \tag{4.153}$$

where $d_1 = \sqrt{(X_1 - X_0)^2 + (Y_1 - Y_0)^2}$

The approximate rotation angles of exterior orientation can be given to sufficient accuracy by

$$\begin{aligned} \omega &= 0 \\ \varphi &= 0 \\ \kappa &= \xi_1 - \alpha_1 \end{aligned} \tag{4.154}$$

4.5.2.2 Space resection

Space resection for a panoramic image can be formulated as an over-determined adjustment problem. For this purpose the panoramic imaging equations (4.151) can be linearised at the approximate values given above. However, the use of Cartesian panoramic coordinates from (4.151) is much more convenient since they can be directly used as virtual three-dimensional image coordinates for the observation equations of a standard space resection (see section 4.2.3). Each image point provides three individual coordinates *x,y,z*, while for central perspective images the principal distance $z' = -c$ is constant for all image points.

As for the standard space resection, a minimum of three reference points is required, although this can generate up to four possible solutions. Alternatively, using four reference points a unique solution for the six parameters of exterior orientation is always obtained. If the reference points are distributed over the whole horizon, then even with a small number of points (greater than 4) a very reliable solution is achieved.

4.5.2.3 Bundle adjustment

Bundle adjustment of panoramic images is based on the same general approach as a bundle adjustment of central perspective images, including unknown object points and self-calibration of the camera if required. Some programs permit the simultaneous processing of both panoramic and central perspective images.

One advantage of panoramic images is that a stably oriented set of images can be obtained using a comparatively small number of object points. The example in Fig. 4.90 shows a set of four panoramas taken for the interior survey of a room. In this example, and using only four of the unknown object points, the following number of unknowns must be calculated:

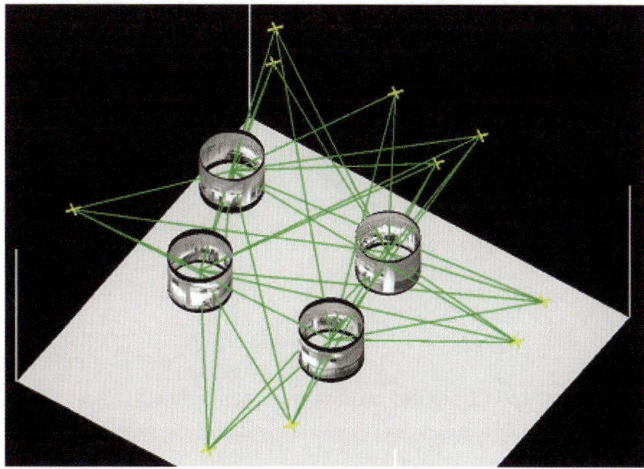

Fig. 4.90: Distribution of 4 panoramas in 3D space

Parameter group	unknowns	number	total
exterior orientations	6	4	24
object points	3	4	12
datum definition	-7	1	-7
sum			**29**

The 29 unknowns can be determined by measuring the object points in all four images, so providing 4 x 4 x 2 = 32 observations. Since this minimum solution is very sensitive to noise and outliers, the number of object points should be increased to at least 8 in this case. Nevertheless, compared to standard image blocks the number of required object points is much smaller. This is mainly due to the stable geometry of a cylindrical image which can be oriented in 3D space with very little object information.

The bundle adjustment can be extended with parameters for the correction of imaging errors. When a panoramic image results from stitching together single images calibrated using the original camera parameters, no additional corrections are necessary. In contrast, if panoramic images from a rotating line scanner are to be adjusted, parameters specific to the scanner must be introduced (see section 3.5.6.1).

4.5.3 Epipolar geometry

Given two oriented panoramas, epipolar plane and epipolar lines can be constructed for an imaged object point analogously to a stereo image pair (Fig. 4.91). The epipolar plane *K* is defined by the object point P and either both image points P' and P" or the projection centres O' and O". The epipolar plane intersects the two arbitrarily oriented panoramic cylinders in the elliptical epipolar lines *k*' and *k*".

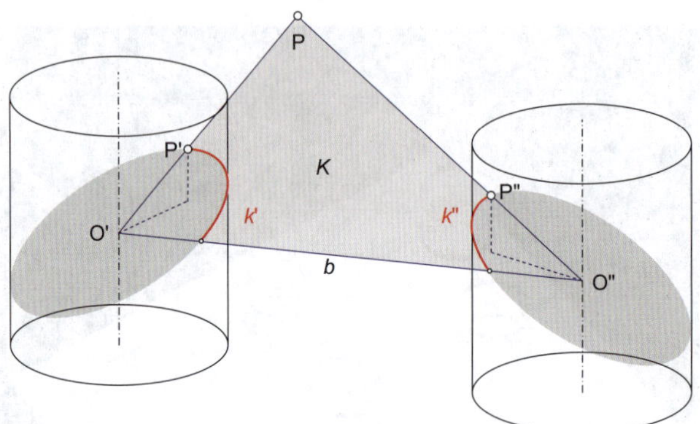

Fig. 4.91: Epipolar geometry for panoramic images

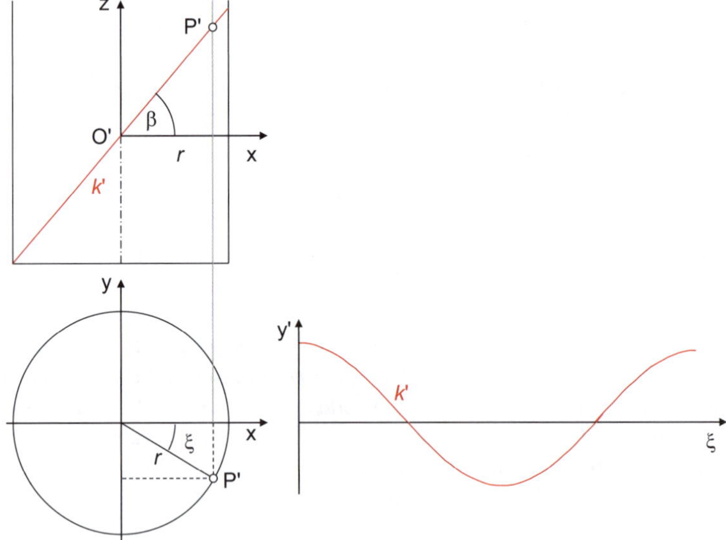

Fig. 4.92: Sine form of panoramic epipolar lines

On the unrolled panoramic plane, the epipolar lines are sinusoidal. For an image point P' measured in the left-hand image, an arbitrary object point P can be defined. The corresponding image point P" in the right-hand panorama can be calculated according to

(4.151). If the epipolar plane intersects the panoramic cylinder at angle β, corresponding to slope $m = \tan\beta$, the intersection straight line

$$z = m \cdot x \tag{4.155}$$

corresponding to Fig. 4.92 is obtained. With $y' = z$ and $x = r \cdot \cos\xi$ the equation for the epipolar line is given by

$$y' = m \cdot r \cdot \cos\xi \tag{4.156}$$

Fig. 4.93 shows an example of a panoramic image with the epipolar line superimposed.

Fig. 4.93: Measured point (top) and sine form of corresponding panoramic epipolar line (bottom)

4.5.4 Spatial intersection

As for space resection and bundle adjustment for panoramic images, a general spatial intersection using three-dimensional panoramic coordinates can also be calculated. Spatial intersection fails if the object point lies on the baseline b (see Fig. 4.91).

If intersection is formulated on the basis of equations (4.151), the derivatives of the observation equations with respect to the unknown object coordinates are required. Using (4.148) they are given by

$$\begin{aligned}
\frac{\partial x'}{\partial X} &= \frac{r}{x^2+y^2}(r_{12}x - r_{11}y) & \frac{\partial y'}{\partial X} &= \frac{\lambda}{z}\left(\frac{r_{11}x - r_{12}y}{x^2+y^2} + \frac{r_{13}}{z}\right) \\
\frac{\partial x'}{\partial Y} &= \frac{r}{x^2+y^2}(r_{22}x - r_{21}y) & \frac{\partial y'}{\partial Y} &= \frac{\lambda}{z}\left(\frac{r_{21}x - r_{22}y}{x^2+y^2} + \frac{r_{23}}{z}\right) \\
\frac{\partial x'}{\partial Z} &= \frac{r}{x^2+y^2}(r_{32}x - r_{31}y) & \frac{\partial y'}{\partial Z} &= \frac{\lambda}{z}\left(\frac{r_{31}x - r_{32}y}{x^2+y^2} + \frac{r_{33}}{z}\right)
\end{aligned} \tag{4.157}$$

As usual, the unknown object coordinates *X,Y,Z* are calculated by iterative adjustment until corrections to the unknowns become insignificant.

4.5.5 Rectification of panoramic images

4.5.5.1 Orthogonal rectification

As with central perspective images, panoramic images can also be rectified if the geometric relation between object coordinate system and image coordinate system is known. The panoramic images can then be rectified onto any chosen reference plane.

Fig. 4.94 shows the rectification of the panoramic image from Fig. 4.93 onto an interior side wall (XZ plane, upper image) as well as onto the floor and ceiling (XY planes with different Z values, lower images). The resulting images are true to scale with respect to the chosen reference plane but objects outside this plane are distorted. The black circles visible in the XY rectifications represent the areas in the vertical direction which are not covered by the original images.

a) Rectification onto XZ plane

b) Rectification onto XY plane (floor and ceiling)

Fig. 4.94: Rectification of the panoramic images from Fig. 4.93

4.5.5.2 Tangential images

Central perspective images can be generated from panoramic images by defining a new central projection image plane as a tangential plane to the panoramic cylinder. Then according to Fig. 4.95, every image point P' in that part of the panoramic image which faces the tangential plane, can be projected onto the tangential plane. The result is a central perspective image in which object lines again appear as image lines. The perspective centre is a point on the axis of the panoramic cylinder and the principal distance corresponds to the cylinder radius. The exterior orientation of the tangential image can be directly derived from the exterior orientation of the panorama. The derived central perspective images can be further processed like other conventional photogrammetric images.

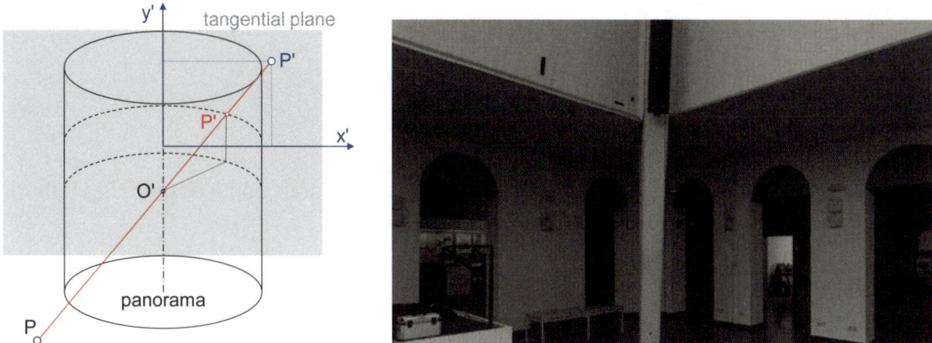

Fig. 4.95: Generation of central perspective tangential images from panoramic images

4.6 Multi-media photogrammetry

4.6.1 Light refraction at media interfaces

4.6.1.1 Media interfaces

The standard photogrammetric imaging model (see sections 1.2.3, 4.2.1) assumes collinearity of object point, perspective centre and image point. Deviations caused by the lens or sensor are modelled by image-based correction functions. This approach is useful for most image configurations and applications.

If light rays in image or object space pass through optical media with differing refractive indices, they no longer follow a straight line. Using extended functional models for multi-media photogrammetry it is possible to calculate the optical path of the rays through additional media interfaces and take this into account in object reconstruction. For example, interfaces exist if the optical path intersects the following media:

- walls made of glass in object space (glass container, window panes),
- water (under, through),
- inhomogeneous atmosphere (refraction),
- filter glasses in front of the lens,

- individual lenses within a compound lens,
- glass covers on CCD sensors,
- réseau plates.

For a rigorous model of the optical path, a geometric description of the media interface must be available; for example:

- plane in space,
- second order surface (sphere, ellipsoid),
- wave-shaped surfaces.

Usually the transmission media are assumed to be homogenous and isotropic, i.e. light rays propagate uniformly in all directions inside the media.

Applications of multi-media photogrammetry can be found in fluid flow measurement (recording through glass, water, gas), underwater photogrammetry (underwater archaeology or underwater platform measurement) or hydrology (river bed recording).

4.6.1.2 Plane parallel media interfaces

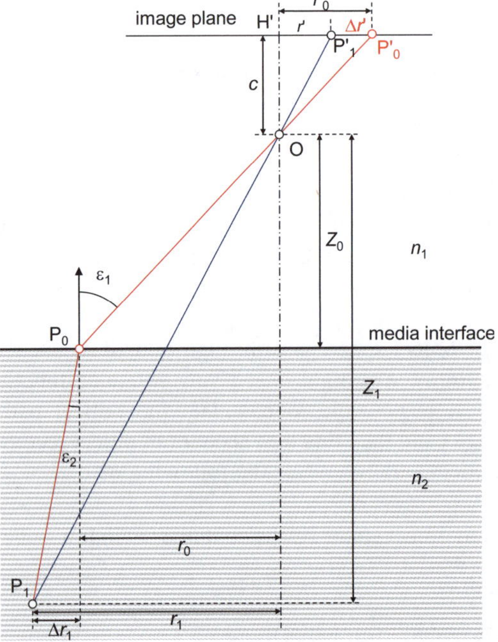

Fig. 4.96: Planar media interface parallel to the image plane

The simplest multi-media case occurs if there is only one planar interface located parallel to the image plane (Fig. 4.96). An object point P_1 is then projected onto point P'_0, passing through intermediate point P_0 which lies on the interface. As a result, a radial shift $\Delta r'$ occurs with respect to the image point P'_1, which corresponds to a straight line projection without refraction.

According to Fig. 4.96, the radial shift $\Delta r'$, taking account of (3.8), is given by[1]:

$$\Delta r' = r'_0 \cdot Z_{rel}\left(1 - \frac{1}{\sqrt{n^2 + (n^2 - 1)\tan^2 \varepsilon_1}}\right) \tag{4.158}$$

$$\text{where } Z_{rel} = \frac{Z_i - Z_0}{Z_i} \quad \text{and } r'_0 = c \cdot \tan \varepsilon_1$$

Eqn. 4.158 shows that the effect of refraction is a function of distance. The radial shift is zero when:

1. $\varepsilon_1 = 0$: the light ray is perpendicular to the interface and is therefore not refracted;
2. $n = 1$: both media have equal refractive indices;
3. $Z_i = Z_0$: the object point is located on the interface ($Z_{rel} = 0$).

The radial shift $\Delta r'$ can become significantly large:

Example 4.11:

A submerged underwater camera with air-filled lens and camera housing, $r'_0 = 21$ mm (small-format 35mm camera), $Z_{rel} = 1$ (media interface located in perspective centre) and $c = 28$mm (wide angle lens) has a maximum radial shift of $\Delta r' = 6.9$ mm.

Example 4.12:

A medium-format camera used for airborne measurement of the seabed, where $r'_0 = 38$ mm, $Z_i = 4$m, $Z_0 = 3$m ($Z_{rel} = 0.25$) (media interface is the water surface) and $c = 80$mm (standard angle lens) leads to a maximum radial shift of $\Delta r' = 2.8$mm.

It can be shown that (4.158) can be expressed as the following power series:

$$\Delta r' = Z_{rel} \cdot (A_0 \cdot r'_0 + A_1 \cdot r'^3_0 + A_2 \cdot r'^5_0 + \ldots) \tag{4.159}$$

This power series expansion is similar to the standard functions for correction of simple radial distortion in (3.54), and also to the correction of distance-dependent distortion (3.67). For applications where Z_{rel} is close to 1, complete compensation for the effect of a plane parallel media interface can be done by the correction function for distance-dependent radial distortion.

---

[1] The following derivations and illustrations are due to Kotowski (1987).

The model of (4.158) can be extended to an unlimited number of plane parallel media interfaces, for instance for the modelling of plane parallel plates inside the camera (filter, réseau), or glass panes in object space (Fig. 4.97). For p interfaces, the radial shift becomes:

$$\Delta r' = \frac{r'_0}{Z_i} \cdot \left[(Z_i - Z_{01}) - \sum_{l=1}^{p} \frac{d_l}{\sqrt{N_l^2 + (N_l^2 - 1)\tan^2 \varepsilon_1}} \right] \tag{4.160}$$

where

$d_l = Z_{l+1} - Z_l$: distance between two adjacent interfaces

$N_l = \frac{n_{l+1}}{n_l}$: relative refractive index

If differential distances between interfaces are used in (4.160), multi-layered media can be modelled. For example, this approach can be used to set up an atmospheric model for the description of atmospheric refraction.

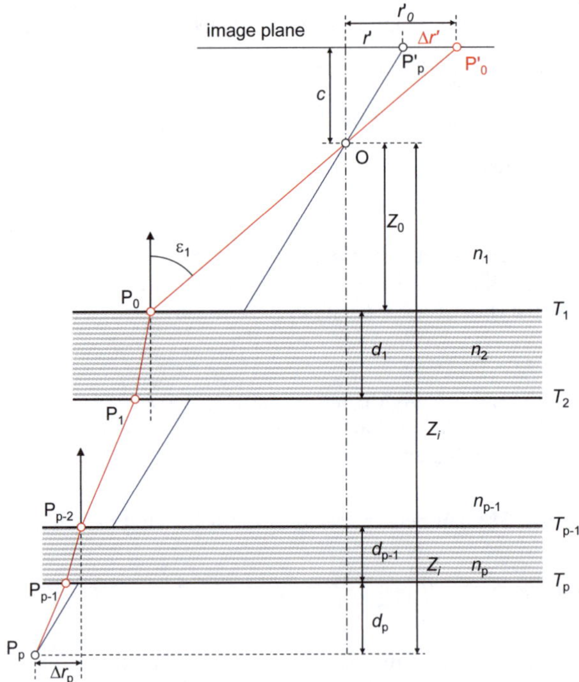

Fig. 4.97: Multiple plane parallel interfaces

4.6.1.3 Ray tracing through refracting interfaces

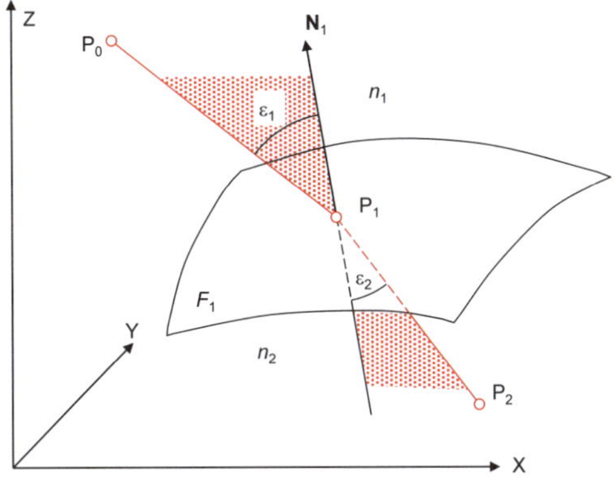

Fig. 4.98: Ray tracing through an optical interface

If arbitrary interfaces must be taken into account in the imaging model, each light ray must be traced through all contributing media by the successive application of the law of refraction (*ray tracing,* see section 5.3.3.1). For this purpose a set of three constraint equations is set up for each refracting point of a media interface. These provide the 3D coordinates of the point of refraction and the path of the ray between the point P_0, the interface point P_1 and the point P_2 (Fig. 4.98):

1. P_1 is located on surface F_1:

   This condition is fulfilled by

   $$F_1(X_1, Y_1, Z_1) = 0$$

   where for a general second order surface:

   $$\begin{aligned} F_1 &= \mathbf{X}^T \mathbf{A} \mathbf{X} + 2\mathbf{a}^T \mathbf{X} + a = 0 \\ &= f_S(\mathbf{X}, \mathbf{A}, \mathbf{a}, a) \end{aligned} \tag{4.161}$$

   $$\text{where } \mathbf{X} = \begin{bmatrix} X \\ Y \\ Z \end{bmatrix}, \ \mathbf{a} = \begin{bmatrix} a_1 \\ a_2 \\ a_3 \end{bmatrix} \text{ and } \mathbf{A} = \begin{bmatrix} a_{11} & a_{12} & a_{13} \\ a_{21} & a_{22} & a_{23} \\ a_{31} & a_{32} & a_{33} \end{bmatrix}, \ a_{ij} = a_{ji} \text{ for } i \neq j$$

   The surface can be parameterised with respect to the temporary coordinate system $\overline{X}, \overline{Y}, \overline{Z}$ according to section 4.4.2.3, e.g. as rotationally symmetric surfaces.

2. Compliance with the law of refraction:

   $$F_2 = n_1 \cdot \sin\varepsilon_1 - n_2 \cdot \sin\varepsilon_2 = 0$$

   The angles of incidence and refraction are introduced as a function of the object coordinates and the normal vector $\mathbf{N}_1$, e.g. for ε_1:

$$\cos\varepsilon_1 = \frac{\mathbf{N}_1^T \cdot (\mathbf{X}_0 - \mathbf{X}_1)}{\left|\mathbf{N}_1^T\right| \cdot \left|\mathbf{X}_0 - \mathbf{X}_1\right|}$$

3. Surface normal at P_1, and the projection rays, lie in a common plane:

   This is implemented by applying the coplanarity constraint to the three vectors:

$$F_3 = \begin{vmatrix} X_0 - X_1 & Y_0 - Y_1 & Z_0 - Z_1 \\ X_1 - X_2 & Y_1 - Y_2 & Z_1 - Z_2 \\ N_X & N_Y & N_Z \end{vmatrix} = 0$$

The three constraint equations are linearised at approximate values. Solving the system of equations results in the object coordinates X_1,Y_1,Z_1 of the refraction point.

For an imaging system of p media interfaces, the system of equations is set up for each refraction point P_l, $l = 1 \ldots p$. This principle is used in optics for the calculation of lens systems. This could be used in photogrammetry for a rigorous determination of distortion which took account of all optical elements. To date this approach has not, however, been used.

4.6.2 Extended model of bundle triangulation

Using the fundamentals of optical interfaces and ray tracing discussed above, the functional model for bundle adjustment can be extended. For this purpose, the ray tracing algorithm with arbitrary interfaces is integrated into the imaging model of the collinearity equations (4.9).

It is natural to distinguish two major imaging configurations:

- constant (object invariant) position of interfaces relative to the measured object
- constant (bundle invariant) location of interfaces relative to the camera system

4.6.2.1 Object-invariant interfaces

Object-invariant interfaces occur, for instance, if an object is imaged through a glass window or through water (Fig. 4.99).

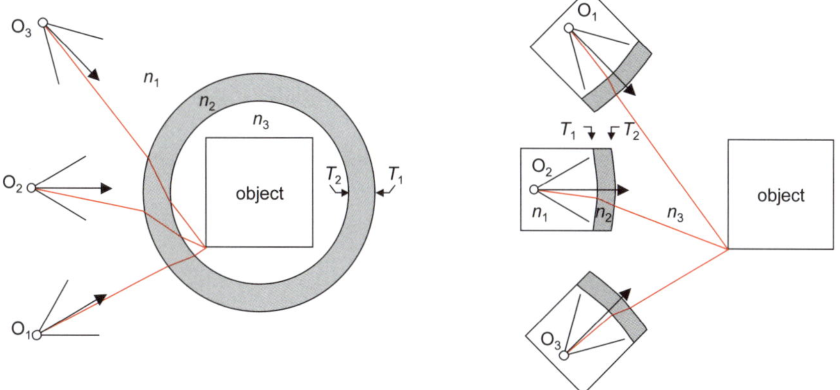

Fig. 4.99: Object-invariant interfaces

Fig. 4.100: Bundle-invariant interfaces

The extended observation equations can be derived in three steps:

1. Ray tracing according to equation (4.161) with p interfaces:

$$\overline{\mathbf{X}}_i^l = f_S(\mathbf{X}_i, \mathbf{X}_{0j}, \mathbf{A}^l, \mathbf{a}^l, a^l, n^l) \tag{4.162}$$

where

$\mathbf{A}^l$, $\mathbf{a}^l$,a^l:	parameters of interface
n^l:	relative refractive indices
$l = 1 \ldots p$:	index of interface
i:	point index
j:	image index
k:	camera index

2. Spatial rotation and translation:

$$\mathbf{X}_{ij}^* = \mathbf{R}_j \cdot \overline{\mathbf{X}}_i^1 - \mathbf{X}_{0j}$$

$$\begin{bmatrix} X_{ij}^* \\ Y_{ij}^* \\ Z_{ij}^* \end{bmatrix} = \mathbf{R}_j \cdot \begin{bmatrix} \overline{X}_i^1 - X_{0j} \\ \overline{Y}_i^1 - Y_{0j} \\ \overline{Z}_i^1 - Z_{0j} \end{bmatrix} \tag{4.163}$$

3. Extended collinearity equations:

$$\begin{bmatrix} x'_{ij} \\ y'_{ij} \end{bmatrix} = -\frac{c}{Z_{ij}^*} \cdot \begin{bmatrix} X_{ij}^* \\ Y_{ij}^* \end{bmatrix} + \begin{bmatrix} x'_{0k} \\ y'_{0k} \end{bmatrix} + \begin{bmatrix} \Delta x'_k \\ \Delta y'_k \end{bmatrix} \tag{4.164}$$

4.6.2.2 Bundle-invariant interfaces

Bundle-invariant interfaces are given by additional optical refracting interfaces which are part of the image acquisition system, e.g. underwater housing, add-on front filter, conversion lenses, parallel plate cover on sensor or réseau plate (Fig. 4.100).

In contrast to the object-invariant approach, the calculation is performed in reverse order:

1. Spatial translation and rotation:

$$\mathbf{X}_{ij}^{*} = \mathbf{R}_{j} \cdot \mathbf{X}_{i}^{1} - \mathbf{X}_{0j} \tag{4.165}$$

2. Ray tracing:

$$\overline{\mathbf{X}}_{ij}^{l} = f_S(\mathbf{X}_{ij}^{*}, \mathbf{A}^{l}, \mathbf{a}^{l}, a^{l}, n^{l}) \tag{4.166}$$

3. Extended collinearity equations:

$$\begin{bmatrix} x'_{ij} \\ y'_{ij} \end{bmatrix} = -\frac{c}{\overline{Z}_{ij}^{1}} \cdot \begin{bmatrix} \overline{X}_{ij}^{1} \\ \overline{Y}_{ij}^{1} \end{bmatrix} + \begin{bmatrix} x'_{0k} \\ y'_{0k} \end{bmatrix} + \begin{bmatrix} dx'_{k} \\ dy'_{k} \end{bmatrix} \tag{4.167}$$

The image shifts caused by bundle-invariant interfaces are usually independent of the distance (see example 4.8). For rotationally symmetric elements (lenses) or plane parallel plates set parallel to the image plane, these shifts are radially symmetric. With suitable image configurations they can be corrected by camera calibration which employs standard functions for distortion.

5 Digital image processing

5.1 Fundamentals

5.1.1 Image processing procedure

Since digital image processing techniques have become available, the possibilities for photogrammetric image measurement have changed significantly. As in other fields, digital images not only enable new methods for the acquisition, storage, archiving and output of images, but most importantly for the automated processing of the images themselves.

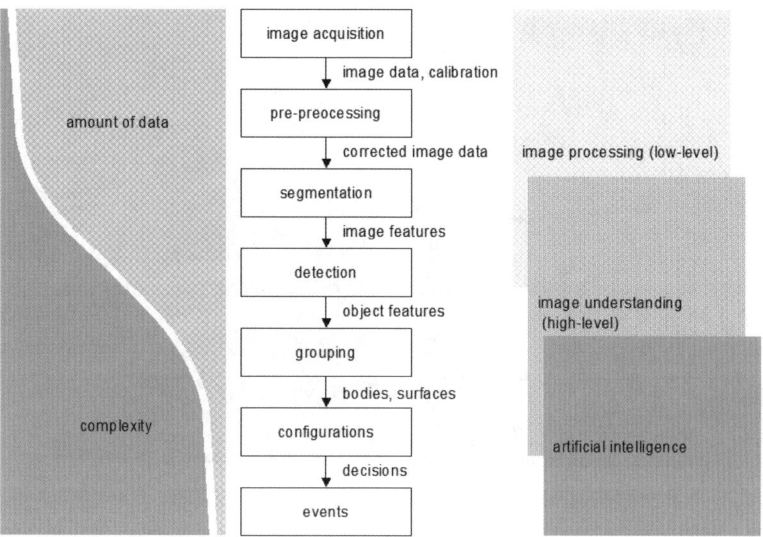

Fig. 5.1: Image processing sequence

Fig. 5.1 illustrates a typical image-processing workflow, starting with image acquisition and ending with some intelligent initiation of events. As the sequence proceeds from top to bottom, the volume of data is reduced whilst the complexity of processing increases.

Photogrammetric image processing methods are primarily developed and applied in the fields of image acquisition (sensor technology, calibration), pre-processing and segmentation (image measuring, line following, image matching). Major considerations for these methods are the automation of relatively simple measuring tasks and the achievement of a suitable accuracy. Everyday methods which apply standard tasks such as digital point measurement (fiducial marks, réseau arrays, targets) or stereo photogrammetry (automated orientation, surface reconstruction) are well developed. In contrast, methods appropriate to high-level image processing (object recognition) and image understanding are still under development.

In close-range photogrammetry two characteristic features can be identified which clearly illustrate the possibilities and limitations of automatic image processing:

- The consistent use of object targeting composed of retro-reflective marks, some of which are coded, combined with suitable illumination and exposure techniques, results in quasi-binary images that can be processed fully automatically (Fig. 5.2 left).
- Arbitrary image configurations which result in large variations in image scale, occlusions, incomplete object imaging etc. (Fig. 5.2 right). In contrast to simple stereo configurations such as those found in aerial photogrammetry, close-range applications are often characterised by complex object surfaces and convergent multi-image network configurations which require a large amount of interactive processing.

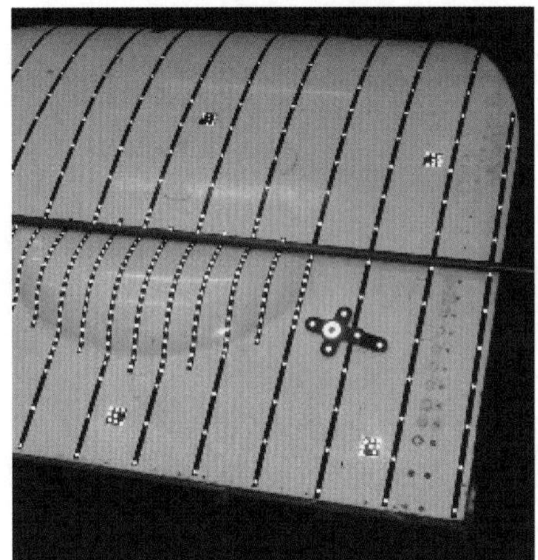

Fig. 5.2: Targeted and non-targeted object scenes

This chapter concentrates on those image processing methods which have been successfully used in practical digital close-range photogrammetry. The emphasis is on methods for geometric image processing and image measurement. Extensions to basic principles and specialised algorithms can be found in the literature on digital image processing and in computer vision.

5.1.2 Pixel coordinate system

The definition of the pixel coordinate system is fundamental to image processing methods used for image measurement[1]. For an image

$$S = s(x, y) \tag{5.1}$$

[1] Here the expression "pixel coordinate system" is used to define the coordinate system of a digital image. In contrast, the image coordinate system defines the reference system physically defined in the camera. The systems can be identical in digital cameras.

conventional processing usually adopts a left-handed xy system of rows and columns which is related to the display of the image on the computer monitor and where x denotes the row and y the column direction.

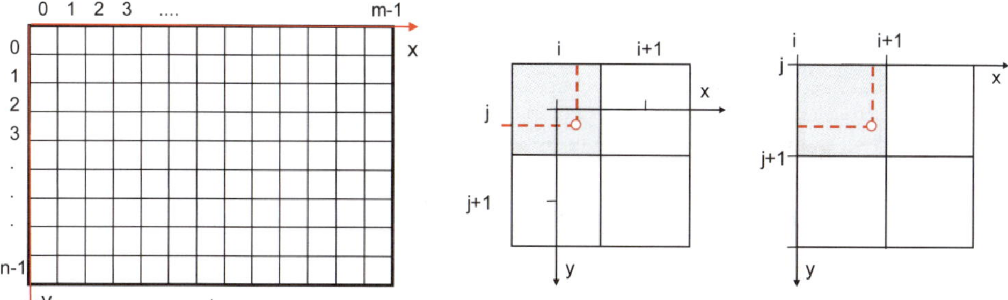

Fig. 5.3: Definition of the pixel coordinate system

Fig. 5.4: Position of a pixel in the xy coordinate system

The origin of this system is located in the upper left-hand corner and the first image element has the row and column numbers (0,0). m is the number of columns, n the number of rows. The last image element has the coordinates (m–1,n–1). Width and height of a pixel are equal to 1 (Fig. 5.3).

In this discrete grid, each pixel has whole integer coordinate values. When an object is imaged by a sensor characterised by such a grid, each pixel acquires a grey value corresponding to the local image brightness across its area. Grey values are usually quantised with an 8 bit depth to provide 256 grey levels ranging from 0 (black) to 255 (white). Since human vision can only distinguish about 60 shades of grey, this grey level depth is sufficient for a visual representation of images. However, machine systems can handle a much higher information content and quantisations of 10 bits (1024 grey levels), 12 bits or 16 bits can be used. True colour images are usually stored with 24 bits per pixel, 8 bits for each red, green and blue (RGB) colour channel.

Due to the pixel dimensions and optical transfer characteristics (MTF, PSF) of the image acquisition system (see section 3.1.5), an imaged object can cover more than one pixel. This leads to a possible sub-pixel position for the coordinates of the imaged object. Measurement to sub-pixel level is only possible if the position of an imaged object can be interpolated over several pixels. Here it is assumed that a small shift of an object edge leads to a corresponding sub-pixel change in the imaged grey values. When sub-pixel coordinates are employed, it is conventional to consider the integer xy-coordinate (i,j) as applying to either the upper left corner or the centre of a pixel (Fig. 5.4).

Imaged objects must usually cover more than one pixel in order to be detected or processed. Adjacent pixels belonging to one object are characterised by grey values that have uniform properties within a limited region (connectivity). Within a discrete image raster, each pixel possesses a fixed number of neighbours, with the exception of the image border. In defining connectivity, neighbouring pixels are classified according to the N4 or the N8 scheme (Fig. 5.5). In the following example three objects A, B and C are imaged. If N4 connectivity is assumed, then object B decomposes into individual pixels and objects A and C are separated. In contrast, N8 connectivity leads to a single integral object B. However, A and C merge together due to their corner connection. Extended algorithms for connectivity must therefore consider the distribution of grey values within certain regions by using, for example, appropriate filters (see section 5.2.3).

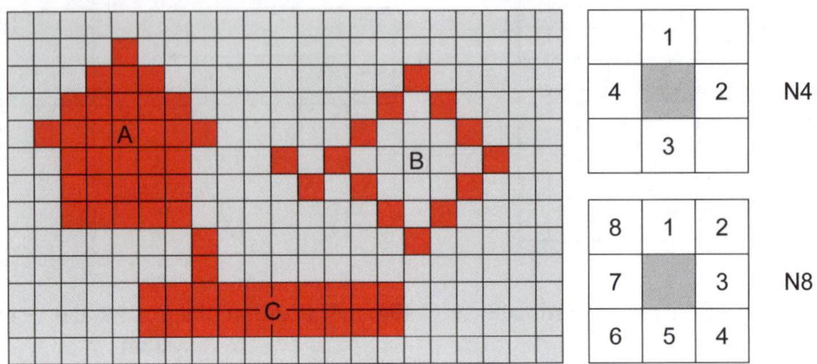

Fig. 5.5: Objects and connectivity

5.1.3 Handling image data

5.1.3.1 Image pyramids

Image or resolution pyramids describe sequences where successive images are reductions of the previous image, usually by a factor of 2 (Fig. 5.6). Prior to image reduction the image can be smoothed, for example using Gaussian filters (see section 5.2.3.2). As resolution is reduced, smaller mage structures disappear, i.e. the information content decreases (Fig. 5.7). The total amount of data required to store the pyramid is approximately only 30% more than in the original image. Image pyramids are typically used in hierarchical pattern recognition problems which start with a search of coarse features in the lowest resolution image (pyramid top). The search is refined with increasing resolution, working progressively down through the pyramid layers, each time making use of the results of the previous resolution stage.

5.1.3.2 Data formats

There are many ways to organise digital image data. Numerous data formats have been developed for digital image processing and raster graphics that, in addition to the actual image data, allow the storage of additional information such as image descriptions, colour tables, overview images etc. For photogrammetric purposes, the unique reproducibility of the original image is of major importance. Loss of information can occur not only in certain data compression methods but also by using an insufficient depth of grey values (bits per pixel).

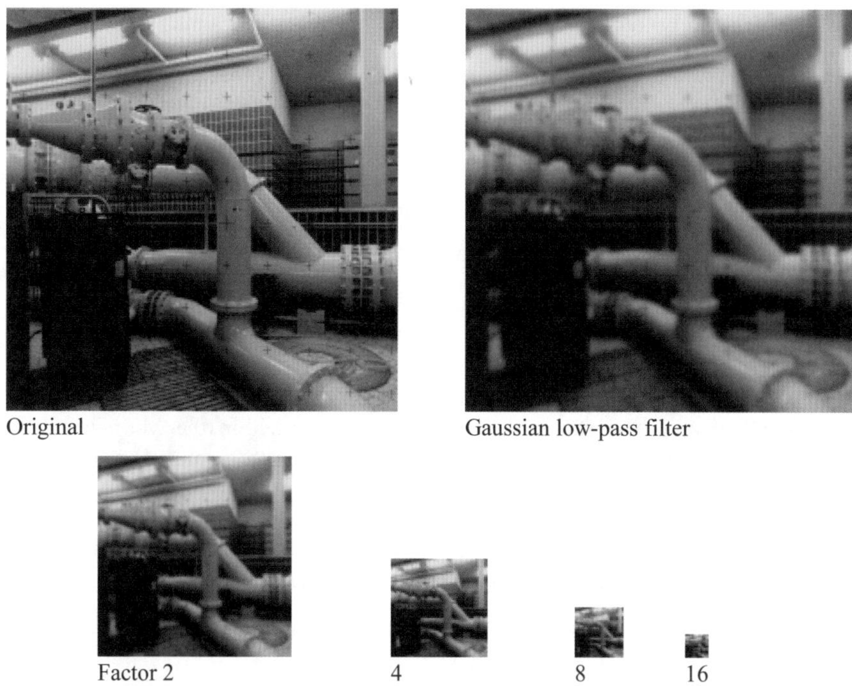

Fig. 5.6: Image pyramids with 5 steps

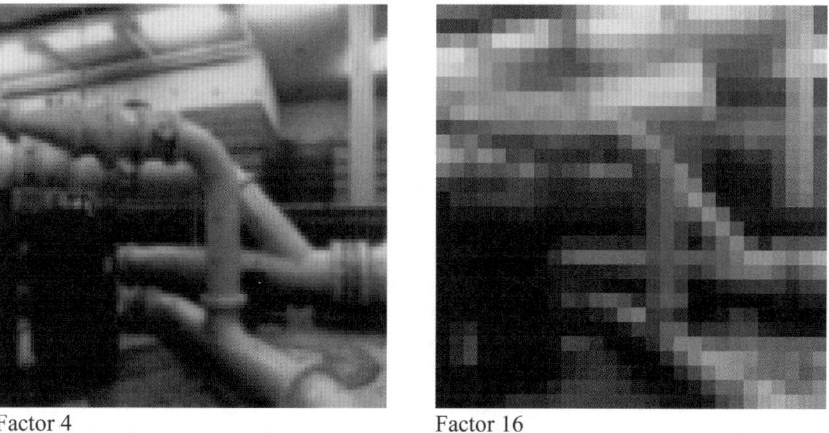

Fig. 5.7: Information content at reduction factors 4 and 16

From the large number of different image formats the following are in common use:

- Direct storage of raw data:

  Here the original grey values of the image are stored in a binary file without compression. Using one byte per pixel, the resulting file size in bytes is exactly equal to the total number of pixels in the image. The original image format in rows and columns must be separately recorded since the raw image data cannot otherwise be read correctly. Multi-channel images can usually be stored in the order of their spectral bands, either pixel by pixel (pixel interleaved), line by line (line interleaved) or channel by channel (band interleaved).

- TIFF - Tagged Image File Format:

  The TIFF format is widely used due to its universal applicability. It is based on a directory of pointers to critical image information such as image size, colour depth, palettes, resolution etc. which must be interpreted by the import program. Numerous variants of TIFF exist and this can sometimes lead to problems with image transfer. The format permits different methods of image compression (LZW, Huffman, JPEG).

- BMP - Windows Bitmaps:

  The BMP format is the standard format for images within the Microsoft Windows environment. It enables the storage of arbitrary halftone and colour images (up to 24 bit) with varying numbers of grey levels or colours.

- GIF - Graphic Interchange Format:

  Images stored in GIF format are compressed without loss of information. Grey level and colour images are limited to 8 bits per pixel.

- PNG - Portable Network Graphics:

  A replacement of GIF free of patents, which provides an alpha channel (variable transparency), RGB storage and a higher compression level.

- JPEG - Joint Photographic Expert Group:

  JPEG is a format which allows compression levels up to a factor of 100 and in which a certain loss of information is accepted (see below). The baseline JPEG format utilises discrete cosine representation and a table of coefficients to store a tiled version of the image, whilst an updated version (JPEG2000) utilises wavelet compression (see section 5.1.3.3)

Table 5.1 provides a comparison between different image storage formats applied to two images with different structures. The first image 'pipes' is the pipe measurement image shown in Fig. 5.2. The second image 'testfield' is taken from a set of images which have been taken for camera calibration purposes (see Fig. 7.22). The resulting file sizes show that the 'pipes' image with its inhomogeneous structures can be compressed up to only 80% of the original size without loss of information. In contrast, the more homogeneous 'testfield' image can be reduced to less than 30% of its original size. However, compression methods with information loss, such as JPEG or wavelet compression, permit much higher levels of data reduction.

Table 5.1: File size in bytes for different image data formats (see also to Fig. 5.9 and Fig. 5.10)

Data format	Zero-loss	Image1 'pipes'	Compression level	Image2 'testfield'	Compression level
image size [pixel]		440 x 440		1000 x 1000	
raw data	x	193 600	100%	1 000 000	100%
BMP	x	194 678	101%	874 698	100%
TIFF	x	194 762	101%	1 001 722	101%
TIFF LZW-compression	x	141 088	73%	305 022	25%
GIF	x	174 266	90%	284 770	26%
PNG	x	111 492	58%	268 528	22%
JPG comp. factor 1	x	114 408	59%	214 616	21%
JPG comp. factor 10		50 796	26%	68 097	7%
JPG comp. factor 20		34 570	18%	33 299	3%
JPG comp. factor 99		3660	2%	13 005	1%
JPEG2000, factor 1	x	10 1607	52%	197 734	20%
JPEG2000, factor 10		93 281	48%	180 180	18%
JPEG2000, factor 20		82 851	43%	160 160	16%
JPEG2000, factor 100		1996	1%	2015	0.2%

5.1.3.3 Image compression

Image compression is of major practical importance to digital photogrammetry, due to the large amounts of data which are handled. For example, a monochromatic set of 50 images each of 3000 x 2000 pixels represents some 300 MB of raw data. The same number of digitised aerial images (pixel size 15 µm) requires more than 11 GB of disk space.

Run-length encoded compression methods count the number of identical grey values within a line or a region and code the corresponding image area by its grey value and a repetition factor. This method is useful for images with extensive homogeneous regions, but for images of natural scenes this often leads to an increased amount of data.

Frequency-based compression methods apply a spectral analysis to the image (Fourier, cosine or wavelet transformations, see section 5.2.3.1) and store the coefficients of the related functions. Eliminating coefficients of low significance compresses data with loss of information, often called lossy compression. Particular attention is drawn to wavelet compression which permits higher compression levels than the original JPEG method (see below) whilst maintaining visual image quality (see Fig. 5.9 and Fig. 5.10). For natural image scenes an increasing smoothing effect can be observed at high compression levels (Fig. 5.9c).

The basic JPEG image format is also based on a compression method with loss of information. The goal is to preserve the essential image content without a significant loss of visual quality, even at high compression levels. Due to the high compression performance, JPEG is widely used in many graphical and technical fields and the procedure is standardised to ensure consistent and appropriate image quality for the given application.

The compression algorithm is based on the 3-stage *Baseline Sequential* method (Fig. 5.8). A discrete cosine transformation (DCT) is calculated in disjoint 8x8 pixel patches. The resulting coefficients are weighted using a selectable quantisation table and stored using run-length encoding. Data decompression is performed in the reverse order. For colour images an additional IHS (intensity, hue, saturation) colour transformation is performed.

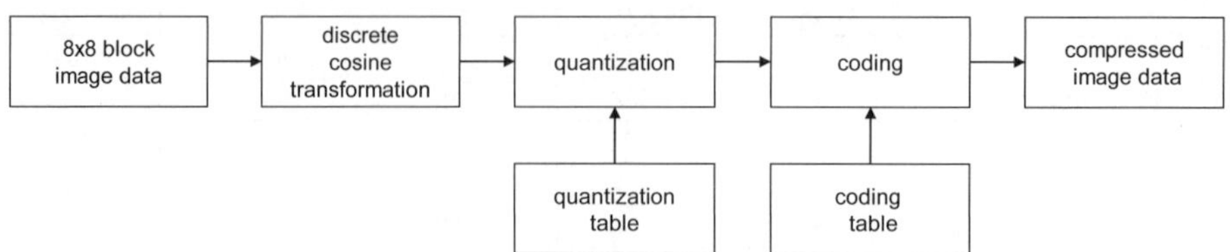

Fig. 5.8: JPEG image compression

The actual loss of information is controlled by the choice of intervals in the quantisation table. Usually the table is designed such that no significant loss of image quality can be visually observed (Fig. 5.9 and Fig. 5.10).

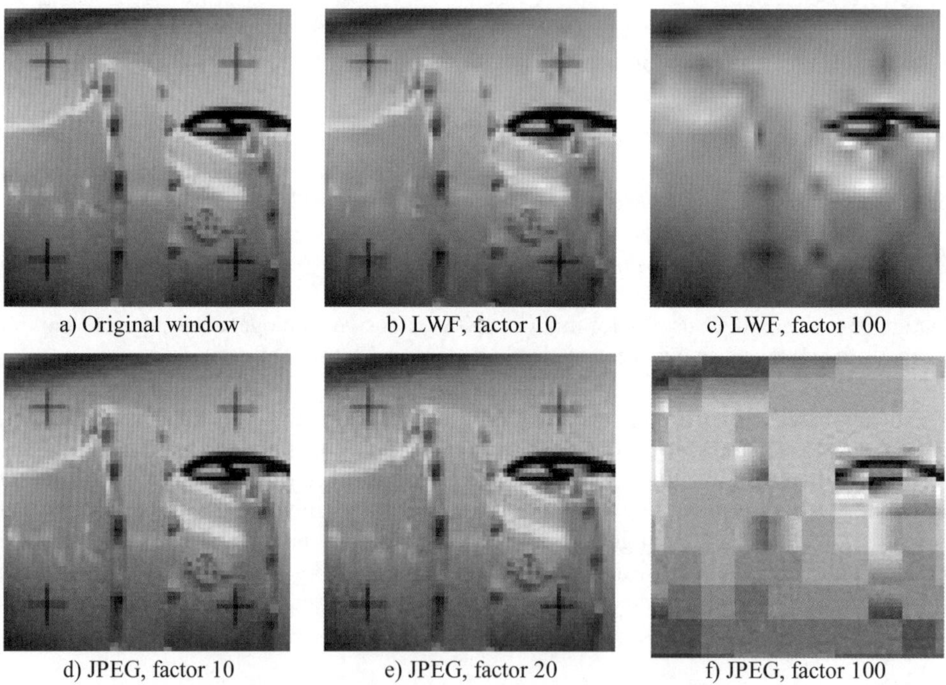

Fig. 5.9: Effect on quality of compression losses in image 'pipes' (see Table 5.1)

Extensions to the JPEG format are available with JPEG2000. Here the cosine transformation is replaced by a wavelet transformation. In addition, different areas of the image can be compressed to different levels. JPEG2000 also supports transparent images.

The effect of JPEG compression on photogrammetric measurement mainly depends on the image content. In general, JPEG compression can give rise to localised image displacements of the order of 0.1 to 1 pixel. This significantly exceeds the accuracy potential of automatic point measurement which lies around 0.02–0.05 pixel (see section 5.2.4.7). In addition the use of 8x8 pixel tiles within the JPEG process can cause undesirable edge effects.

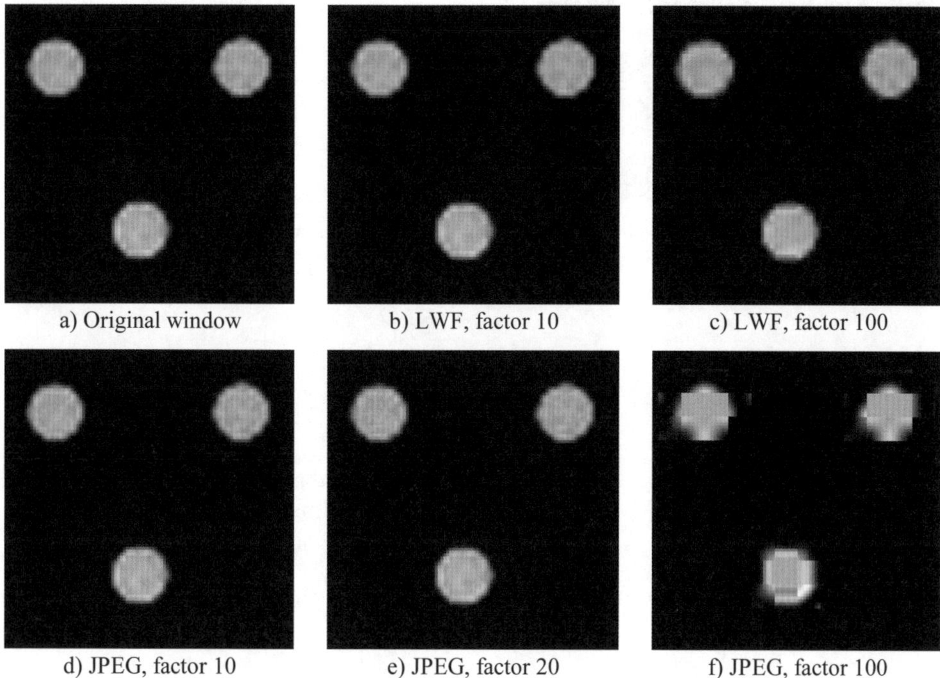

Fig. 5.10: Effect on quality of compression losses in image 'testfield' (see Table 5.1)

5.2 Image preprocessing

5.2.1 Point operations

5.2.1.1 Histogram

The histogram provides the frequency distribution of the grey values in the image. It displays the absolute or relative frequency of each grey value either in tabular or graphical form (Fig. 5.11)

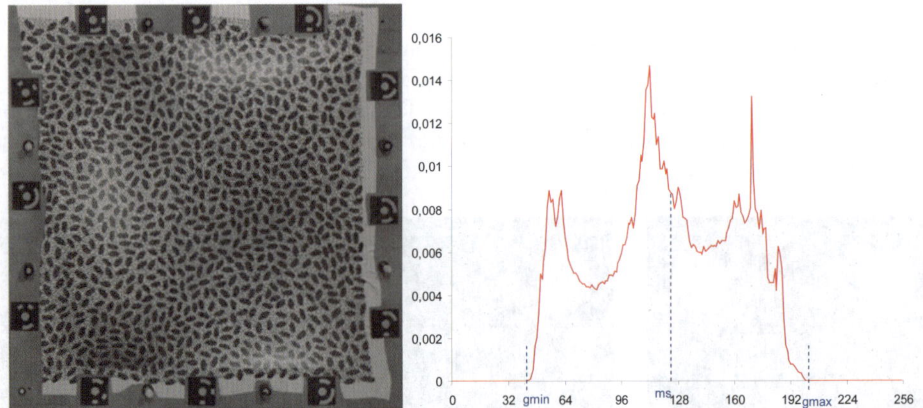

Fig. 5.11: Grey level image with corresponding histogram and parameters

The most important parameters of a histogram are:

$$p_S(g) = \frac{a_S(g)}{M}$$: relative frequency $p_S(g)$ and absolute frequency $a_S(g)$ where $M = m \cdot n$ and $0 \le p_S(g) \le 1$ (5.2)

g_{min}, g_{max} : minimum and maximum grey value of the image

$$K = \frac{g_{max} - g_{min}}{g_{max} + g_{min}}$$: contrast (5.3)

$$m_S = \frac{1}{M}\sum_{u=0}^{m-1}\sum_{v=0}^{n-1} s(u,v) = \sum_{g=0}^{255} g \cdot p_S(g)$$: mean of grey values (5.4)

$$q_S = \frac{1}{M}\sum_{u=0}^{m-1}\sum_{v=0}^{n-1} [s(u,v) - m_S]^2 = \sum_{g=0}^{255} (g - m_S)^2 \cdot p_S(g)$$: variance of grey values (5.5)

$$H = \sum_{g=0}^{255} [p_S(g) \cdot \log_2 p_S(g)]$$: entropy (5.6)

$$\alpha = \frac{-\sum_{g=0}^{k}\left[p_S(g)\cdot \log_2 p_S(g)\right]}{H} \quad : \text{symmetry (anisotropic coefficient)} \tag{5.7}$$

k: minimum grey value where

$$\sum_{g=0}^{k} p_S(g) \geq 0.5$$

Whilst minimum and maximum grey values define the image contrast, the mean is a measure of the average intensity (brightness) of the image. For statistical image processing, variance or standard deviation is also calculated but both are of minor interest in metrology applications.

The information content in an image can be measured by its entropy. It corresponds to the average number of bits necessary to quantise the grey values. Entropy can also be used to calculate an appropriate factor for image compression (see section 5.1.3.3).

The degree of symmetry of a histogram is determined by the anisotropic coefficient. Symmetrical histograms have a coefficient $\alpha = 0.5$. This coefficient can also be used to determine a threshold for bimodal histograms (see section 5.2.1.4).

Example 5.1:

The histogram of the image in Fig. 5.11 has the following parameters:

Minimum grey value:	$g_{min} = 42$
Maximum grey value:	$g_{min} = 204$
Contrast:	$K = 0.658$
Mean value:	$m_s = 120.4$
Variance:	$q = 1506.7$
Standard deviation:	$q^{1/2} = 38.8$
Entropy:	$H = -7.1$
Symmetry:	$\alpha = 0.49$

5.2.1.2 Lookup tables

Lookup tables (LUT, colour tables, palettes) are simple tools for the global manipulation of grey values. Each grey value of an input image is assigned a unique grey value in an output image. This method of processing grey values is simple to implement and can be found in almost all graphics or image processing programs. LUTs are easily displayed in diagrammatic form (Fig. 5.12). LUT operations are unique but usually non reversible.

Given the lookup table

$$LUT(g) \qquad g = 0,1,2,...,255 \tag{5.8}$$

the grey values of the output image are calculated:

$$g' = LUT(g) \tag{5.9}$$

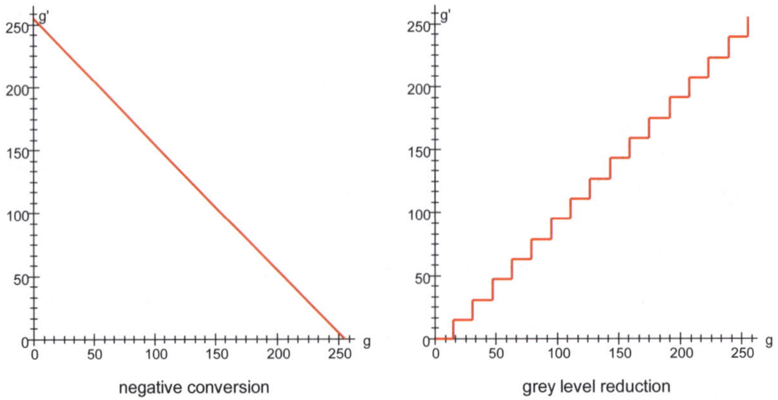

Fig. 5.12: Examples of lookup tables

5.2.1.3 Contrast enhancement

Manipulating the brightness and contrast of an original image results in a change of the grey value distribution within certain regions, for example along an image edge. Variations of grey value are not only a function of object intensity but are also influenced by the relative position of camera and object. In general a non-linear manipulation of grey values can affect the geometry of an object's image and should therefore be avoided if possible. However, contrast enhancement can provide a better visual interpretation for interactive image processing[1].

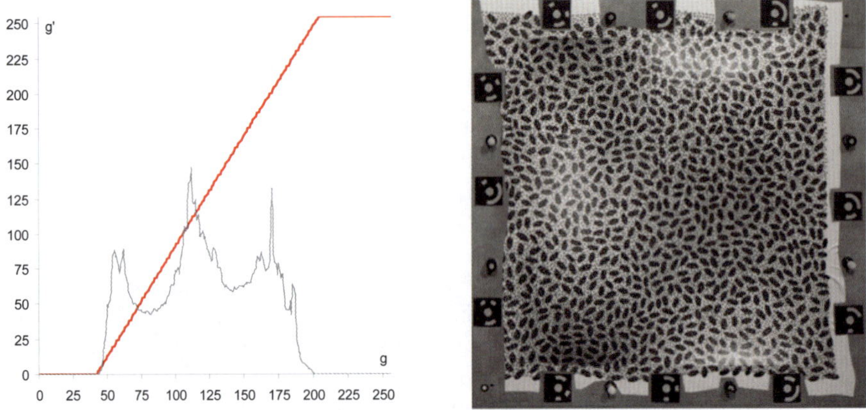

Fig. 5.13: Linear contrast enhancement (original image in Fig. 5.11)

Image contrast changes are easily applied by a lookup table. Table values can be defined interactively, pre-calculated or derived from the image content itself. Common methods for adjusting brightness and contrast are:

[1] Various digital image measuring methods are independent of contrast or calculate radiometric correction parameters (see section 5.4.2.4).

- Linear contrast stretching:

  The LUT is a linear interpolation between g_{min} and g_{max} (Fig. 5.13). Minimum and maximum grey values can be derived from the histogram, or defined interactively. The calculation of the LUT is derived from a shift (offset) r_0 and a scale factor (gain) r_1 as follows:

$$LUT(g) = r_0 + r_1 g \tag{5.10}$$

where

$$r_0 = -\frac{255 \cdot g_{\min}}{g_{\max} - g_{\min}} \qquad r_1 = \frac{255}{g_{\max} - g_{\min}}$$

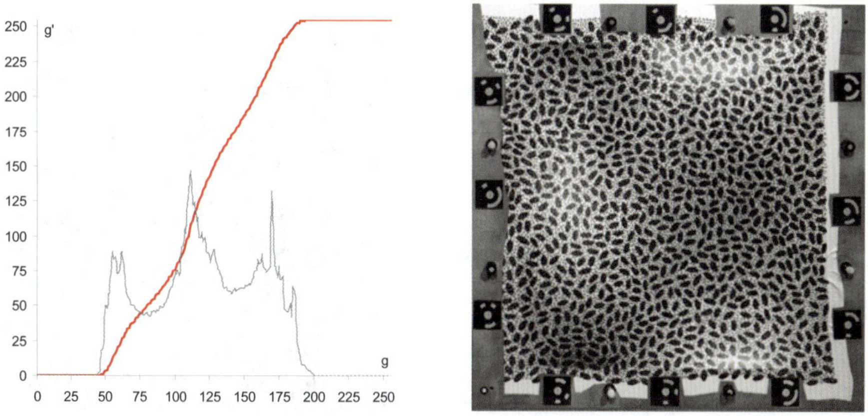

Fig. 5.14: Contrast stretching by histogram equalisation (original image in Fig. 5.11)

- Histogram equalisation:

  The cumulative frequency function is calculated from the histogram of the original image:

$$h_S(g) = \sum_{k=0}^{g} p_S(k) \qquad g = 0,1,2,...,255$$
$$\text{mit } 0 \leq h_S(g) \leq 1 \tag{5.11}$$

The LUT values are given by:

$$\text{LUT}(g) = 255 \cdot h_s(g)$$

The function is dependent on the histogram since the slope of the LUT is proportional to the frequency of the corresponding grey value (Fig. 5.14). Image contrast is consequently strongly enhanced in areas where grey values have high frequencies. The output image S' therefore has a histogram $p_{S'}(g')$ with relative cumulative frequencies which are constant for each grey value according to the definition $h_{S'}(g) = 1/255 \cdot g'$.

- Gamma correction:

  Many interactive image processing programs permit a Gamma correction where, in analogy with the gamma characteristic in the photographic process, the slope of the LUT is adjusted logarithmically. This essentially results in an increase or decrease of the mean grey value (Fig. 5.15).

  The LUT of a gamma correction is given by:

$$\mathrm{LUT}(g) = 255 \cdot \left(\frac{g}{255} \right)^{\frac{1}{\gamma}} \tag{5.12}$$

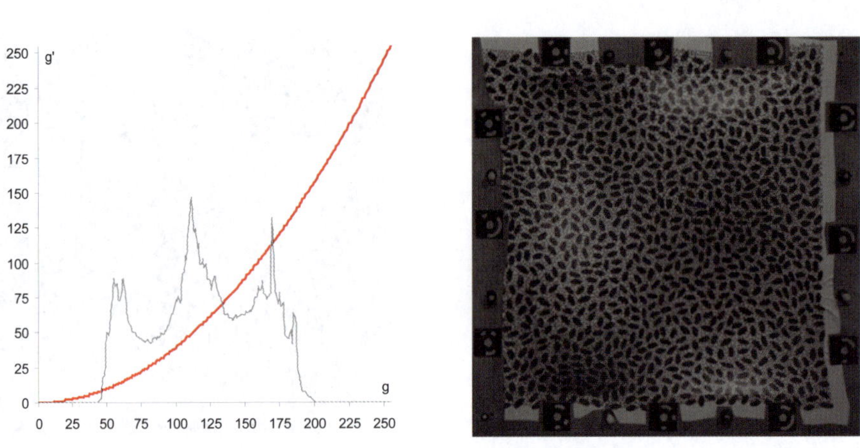

Fig. 5.15: Gamma correction with $\gamma = 0.5$ (original image in Fig. 5.11)

- Local contrast adjustment:

  In principle, all the above image processing methods can be applied to any image detail (window, area of interest). Using the Wallis filter (section 5.2.3.4) the image is modified so that the local contrast in a filter window is optimised.

5.2.1.4 Thresholding

In general, thresholding is used to clearly differentiate grey values which belong to different object classes, e.g. to separate objects and background. Thresholding is often a pre-processing stage prior to segmentation.

Consider a simple case where the image consists only of two classes, i.e.

- class K_1: background (e.g. dark)
- class K_2: objects (e.g. bright)

The corresponding histogram can be expected to be bimodal having two significant groupings of data, each represented by a maximum (peak), which are separated by a minimum (valley). Clearly both classes can be separated by a single threshold t (bimodal thresholding) located within the minimum region between the class maxima, e.g. by defining

$$T = \frac{m_2 - m_1}{2} \tag{5.13}$$

where m_1, m_2 are the mean grey values of classes K_1, K_2

Applying the lookup table

$$LUT(g) = \begin{cases} g_1 & \text{for } g \le T \\ g_2 & \text{for } g > T \end{cases} \tag{5.14}$$

where g_1, g_2 are the new grey values for classes K_1, K_2

results in a binary image (two-level image) consisting only of grey values g_1 (e.g. 0) and g_2 (e.g. 1 or 255). Fig. 5.16 shows the histogram of the image in Fig. 5.17 which has two significant primary maxima ($m_1 \approx 18$, $m_2 \approx 163$), representing background and object. The secondary maximum ($m_3 \approx 254$) is caused by the imaged retro-reflective targets. The binary image of Fig. 5.18 is the result of thresholding with $T = 90$ (between the primary maxima).

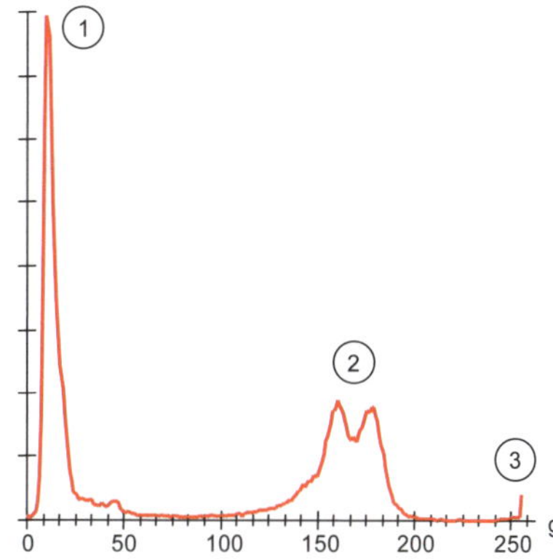

Fig. 5.16: Histogram of Fig. 5.17 with two primary maxima (1, 2) and one secondary maximum (3)

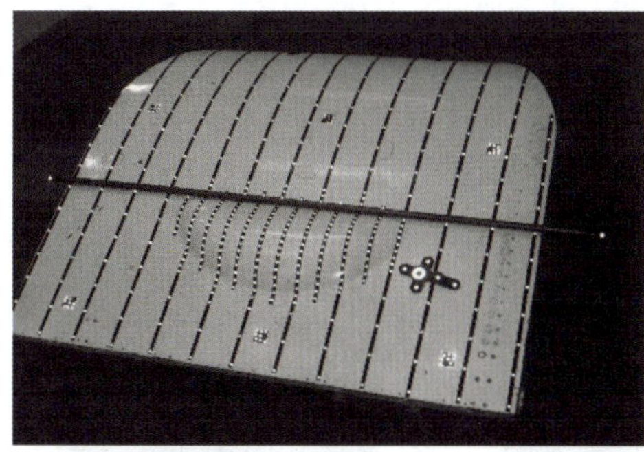

Fig. 5.17: Metric image with retro-reflective targets

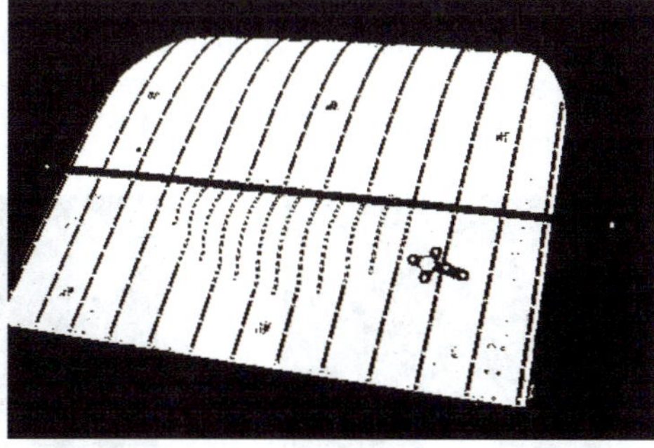

Fig. 5.18: Result after thresholding with $T = 90$

Fig. 5.19 shows the result with threshold value $T = 192$, located near maximum 2. This preserves some image information in addition to the targets. With $T = 230$ (Fig. 5.20) almost all targets are separated or segmented out from the background (but see Fig. 5.34).

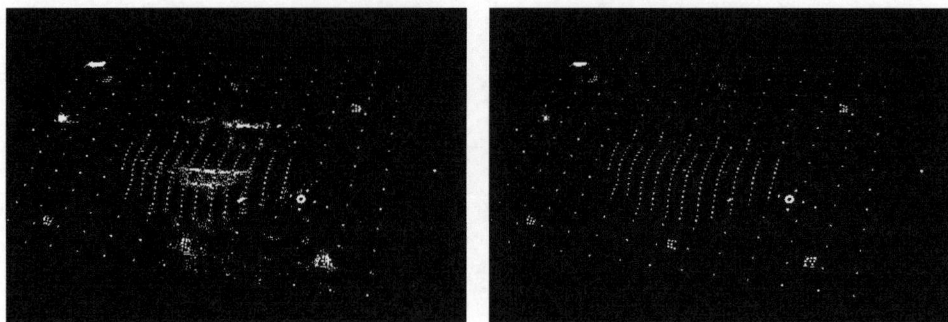

Fig. 5.19: Result after thresholding with $T = 192$ **Fig. 5.20:** Result after thresholding with $T = 230$

For more complex images the problem of thresholding lies in both the calculation of representative class averages and with the subsequent definition of the threshold value itself. Natural image scenes usually have more than two grey-value classes and this requires a much more complex thresholding procedure (multi-modal or dynamic thresholding).

5.2.1.5 Image arithmetic

Two or more images or image subsets can be combined numerically:

- arithmetic: addition, subtraction, division, multiplication

  The grey values of both images are combined arithmetically, e.g. subtracted:

$$s'(x,y) = s_2(x,y) - s_1(x,y) \tag{5.15}$$

  The results of this image operation (difference image) show the differences between both input images (example in Fig. 5.21). If necessary the grey values of the output image, where negative values are possible, must be transformed into the positive range [0…255], or alternatively stored as 16 bit signed integers.

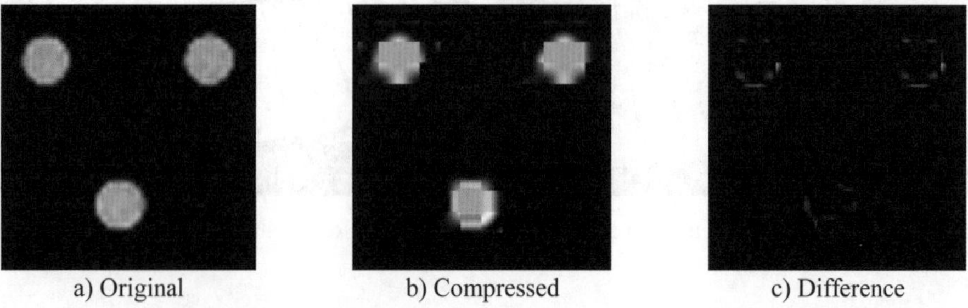

a) Original b) Compressed c) Difference

Fig. 5.21: Example of a difference image to illustrate information loss after JPEG image compression

- logical: $=, <, >, \leq, \geq, \neq$

  The grey values of both images are compared logically, resulting in Boolean values 1 (true) or 0 (false).

- bitwise: AND, OR, NOT, XOR

  The grey values of both input images are combined bit by bit. The XOR operation has practical use for the temporary superimposition of a moving cursor on the image. The original grey value can be recovered without the use of temporary storage by executing two sequential XOR operations with the value 255. Example: 38 XOR 255 = 217; 217 XOR 255 = 38.

5.2.2 Colour operations

5.2.2.1 Colour spaces

True-colour images are composed of three spectral (colour) image channels which store their respective intensity or colour distributions. Storage normally requires 8 bits per channel so that RGB images have a colour depth of 24 bits (Fig. 5.22). An optional alpha channel can be stored to control the transparency of an image (the proportion of the image displayed with respect to the background). RGB images with an alpha channel have a storage depth of 32 bits per pixel.

An image with n colour or spectral channels can be defined as an image vector **S:**

$$\mathbf{S} = \begin{bmatrix} s_0(x,y) \\ s_1(x,y) \\ \vdots \\ s_{n-1}(x,y) \end{bmatrix} \tag{5.16}$$

For an RGB image ($n = 3$), this corresponds to:

$$\mathbf{S}_{RGB} = \begin{bmatrix} s_R(x,y) \\ s_G(x,y) \\ s_B(x,y) \end{bmatrix} = \begin{bmatrix} \mathbf{R} \\ \mathbf{G} \\ \mathbf{B} \end{bmatrix} \tag{5.17}$$

The most important colour models in photogrammetry are the RGB colour space and the IHS colour space (also called the HSL colour space, see below and Fig. 5.23). In the RGB colour space, colour is defined by three-dimensional Cartesian coordinates R,G,B. The origin of the coordinate system ($R = 0,\ B = 0,\ G = 0$) defines black and the maximum position ($R = 1,\ B = 1,\ G = 1$) defines white. Values on the principal diagonal, with equal RGB components, define grey values. The additive primary colours, red, green and blue, as well as the subtractive colours (complementary colours) yellow, magenta and cyan (Y,M,C)

lie at the corners of the RGB cube. It is simple to convert between additive and subtractive primary colours:

$$\begin{bmatrix} R \\ G \\ B \end{bmatrix} = \begin{bmatrix} 1-C \\ 1-M \\ 1-Y \end{bmatrix} \qquad \begin{bmatrix} C \\ M \\ Y \end{bmatrix} = \begin{bmatrix} 1-R \\ 1-G \\ 1-B \end{bmatrix} \tag{5.18}$$

a) Original true-colour image b) Red channel

c) Green channel d) Blue channel

Fig. 5.22: True-colour image with separate RGB channels

The conversion of an RGB image to a grey-value image is done for every pixel as follows:

$$s'(x,y) = [s_R(x,y) + s_G(x,y) + s_B(x,y)]/3 \tag{5.19}$$

or more simply written as:

$$I = \frac{R+G+B}{3} \tag{5.20}$$

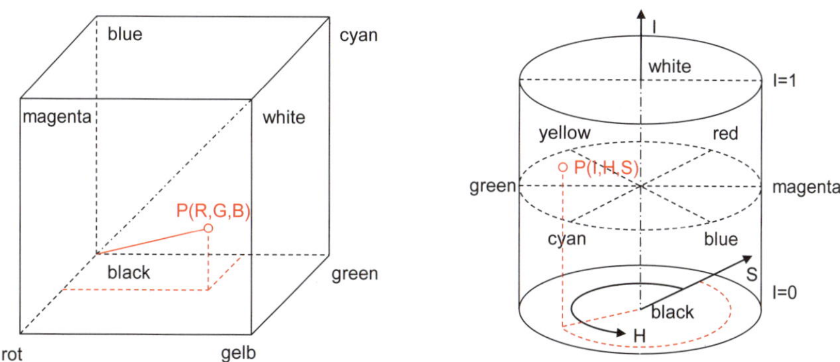

Fig. 5.23: RGB and IHS colour spaces

In the IHS colour space, colours have components of intensity (I) or luminance (L), hue (H) and saturation (S). The IHS values can be interpreted as cylindrical coordinates as shown in Fig. 5.23. Grey values between black and white lie on the cylinder axis which also represents the intensity axis. Primary colours are represented at angle H on the colour circle and their saturation is given by the distance from the cylinder axis.

The CIE XYZ colour model is a standardised representation of colour which relates physical colours to how the average human eye perceives them. Colour representation is in an XYZ system where each colour coordinate is defined in the interval [0,1]. An extension to the CIE model is the CIE *lab* model, where *l*, *a* and *b* represent colour axes. Both models are based on standardised illumination and reflection conditions.

5.2.2.2 Colour transformations

RGB ↔ IHS:

A unique transformation is possible between RGB and IHS colour spaces. However, the formulas used in practice are not consistent. The following is a widely used transformation:

with $I_{max} = \max(R,G,B)$ and $I_{min} = \min(R,G,B)$ then

$$I = \frac{I_{max} + I_{min}}{2}$$

$$H = \begin{cases} 0° & \text{for } I_{max} = I_{min} \\ 60° \cdot \left(0 + \dfrac{G-B}{I_{max} - I_{min}}\right) & \text{for } I_{max} = R \\ 60° \cdot \left(2 + \dfrac{B-R}{I_{max} - I_{min}}\right) & \text{for } I_{max} = G \\ 60° \cdot \left(4 + \dfrac{R-G}{I_{max} - I_{min}}\right) & \text{for } I_{max} = B \end{cases} \tag{5.21}$$

$$S = \begin{cases} 0 & \text{for } I_{max} = I_{min} \\ \dfrac{I_{max} - I_{min}}{I_{max} + I_{min}} & \text{for } I < 0.5 \\ \dfrac{I_{max} - I_{min}}{2 - I_{max} - I_{min}} & \text{for } I \geq 0.5 \end{cases}$$

The calculated values lie in the intervals $I[0\ldots1]$, $H[0\ldots360]$ und $S[0\ldots1]$ and can subsequently be transformed into the grey-scale range [0…255].

Using the substitutions

$$q = \begin{cases} I(1+S) & \text{for } I < 0.5 \\ I + S - (I \cdot S) & \text{for } I \geq 0.5 \end{cases} \qquad p = 2I - q \qquad H' = H/360$$

$$t_R = H'+1/3 \qquad t_G = H' \qquad t_B = H'-1/3$$

then for every colour $C \in (\mathrm{R,G,B})$

$$C = \begin{cases} p + 6(q-p)t_C & \text{for } t_C < 1/6 \\ q & \text{for } 1/6 \leq t_C < 1/2 \\ p + 6(q-p)(2/3 - t_C) & \text{for } 1/2 \leq t_C < 2/3 \\ p & \text{else} \end{cases} \qquad \text{where } t_C = t_C \bmod 1 \tag{5.22}$$

Fig. 5.24 shows the IHS channels which result from the transformation of a true-colour image.

RGB ↔ XYZ:

The transformation of RGB values into the CIE XYZ model is achieved with the following operation:

$$\begin{bmatrix} X \\ Y \\ Z \end{bmatrix} = \mathbf{M} \cdot \begin{bmatrix} R \\ G \\ B \end{bmatrix} \qquad \begin{bmatrix} R \\ G \\ B \end{bmatrix} = \mathbf{M}^{-1} \cdot \begin{bmatrix} X \\ Y \\ Z \end{bmatrix} \tag{5.23}$$

where

$$\mathbf{M} = \begin{bmatrix} 0.4887180 & 0.3106803 & 0.2006017 \\ 0.1762044 & 0.8129847 & 0.0108109 \\ 0 & 0.0102048 & 0.9897952 \end{bmatrix}$$

The transformation matrix **M** is defined by the CIE standard. There are deviations from this matrix which are optimised for printing and computer graphics.

a) Original true-colour image

b) Intensity channel

c) Hue channel

d) Saturation channel

Fig. 5.24: IHS transformation

Example 5.2:

The following table shows the transformation of RGB colour values to IHS and XYZ.

Colour	RGB image	RGB value	IHS value	IHS image	XYZ from RGB
Red	255,0,0	1.0,0.0,0.0	0.5,0.0,1.0	128,0,255	0.49,0.18,0.00
Green	0,255,0	0.0,1.0,0.0	0.5,120.0,1.0	128,85,255	0.31,0.81,0.01
Yellow	255,255,0	1.0,1.0,0.0	0.5,60.0,1.0	128,43,255	0.80,0.99,0.01
Khaki	195,176,145	0.76, 0.69,0	0.64,28.8,0.67	163,20,172	0.70,0.67,0.39
White	255,255,255	1.0,1.0,1.0	1.0,0.0,0.0	255,0,0	0.94,0.98,1.00
Black	0,0,0	0.0,0.0,0.0	0.0,0.0,0.0	0,0,0	0.0,0.0,0.0

5.2.2.3 Colour combinations

The following procedures are commonly used to create new colour assignments in grey-scale and colour images:

Pseudo colour

Pseudo colours are obtained when RGB colour values are assigned to the values in a single-channel, grey-scale image using a colour palette (lookup table, Fig. 5.25), and the result displayed as a colour image. Pseudo colours are used, for example, in the colouring of thermal images so that the intensity values (representing temperature) can be visualised using a colour scale (see example in Fig. 5.26).

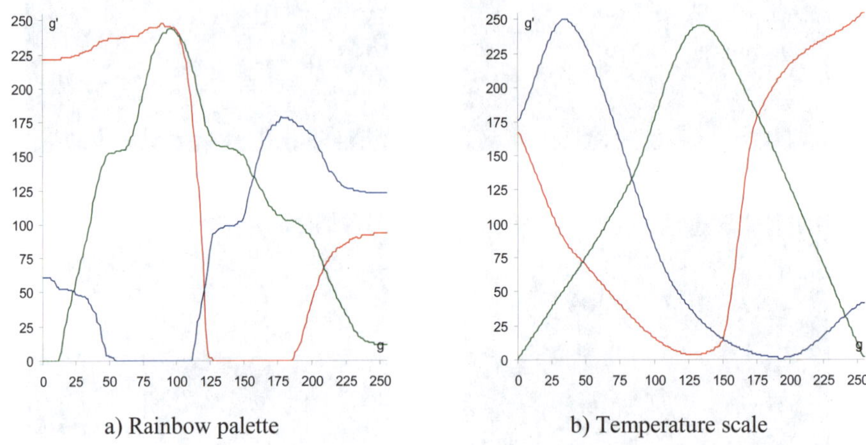

a) Rainbow palette b) Temperature scale

Fig. 5.25: Lookup tables for creating pseudo colour images

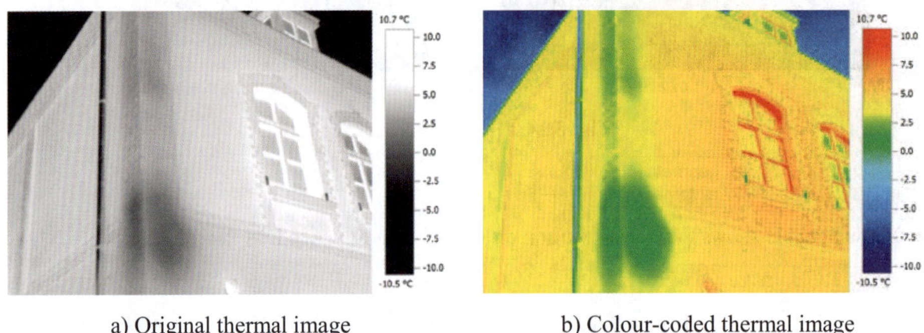

a) Original thermal image b) Colour-coded thermal image

Fig. 5.26: Pseudo colour for thermal image

False colour

False colour images are obtained from multi-channel images (e.g. RGB images) in which every input channel is freely assigned to the channel of an RGB output image. For example, using the following assignments:

$$\text{a)}\begin{bmatrix} s_R(x,y) \\ s_G(x,y) \\ s_B(x,y) \end{bmatrix} = \begin{bmatrix} s_B(x,y) \\ s_R(x,y) \\ s_G(x,y) \end{bmatrix} \qquad \text{b)}\begin{bmatrix} s_R(x,y) \\ s_G(x,y) \\ s_B(x,y) \end{bmatrix} = \begin{bmatrix} s_G(x,y) \\ s_R(x,y) \\ s_{IR}(x,y) \end{bmatrix} \tag{5.24}$$

a false colour image with the channels exchanged (a) or a false colour infrared image (b) is obtained.

Pan sharpening

Pan sharpening (also called resolution merging) is the combination of a sensor with high resolution in the panchromatic region (P channel) with the lower resolution channels of a colour sensor (RGB channels). The concept stems from human vision where the retina has a higher number of rods (detecting brightness) and a significantly lower number of cones (detecting colour) which together give the impression of a high-resolution image.

According to Fig. 5.27, the RGB image and the P channel are geometrically aligned so that all input images represent the same object at the same resolution. The RGB image is subsequently transformed into HIS format. The resulting I channel is then replaced by the P channel and a reverse transformation into RGB space is calculated. The final image then has the geometric resolution of the high-resolution P channel as well as the colour information from the original RGB image.

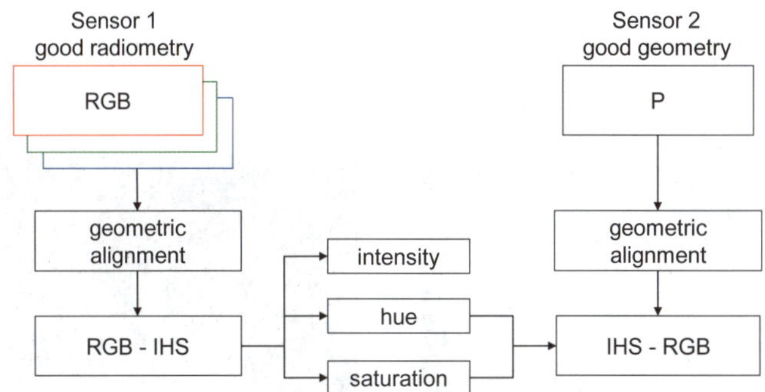

Fig. 5.27: Principle of pan sharpening

5.2.3 Filter operations

5.2.3.1 Spatial domain and frequency domain

The theory of digital filtering is based on the fundamentals of digital signal processing (communication engineering, electronics). Its fundamental method is the Fourier transform which represents arbitrary signals (series of discrete values, waves) as linear combinations of trigonometric functions. The discrete one-dimensional Fourier transformation for n samples of an input signal $s(x)$ is given by:

$$F(u) = \frac{1}{n}\sum_{x=0}^{n-1} s(x)\cdot\exp\left(-i\frac{2\pi}{n}ux\right) \\ = \mathrm{Re}(F(u)) + i\cdot\mathrm{Im}(F(u)) \tag{5.25}$$

Here u denotes the spatial frequency and $i = \sqrt{-1}$. The inverse Fourier transformation is given by:

$$s(x) = \frac{1}{n}\sum_{u=0}^{n-1} F(u)\cdot\exp\left(+i\frac{2\pi}{n}ux\right) \tag{5.26}$$

i.e. the original signal can be exactly reconstructed from its Fourier transform.

The Euler formulas show the connection with the underlying trigonometric functions:

$$e^{-i2\pi ux} = \cos(2\pi ux) - i(2\pi ux) \\ e^{i2\pi ux} = \cos(2\pi ux) + i(2\pi ux) \tag{5.27}$$

The power spectrum of $s(x)$ is defined by

$$P(u) = |F(u)|^2 = \mathrm{Re}^2(u) + \mathrm{Im}^2(u) \tag{5.28}$$

The 1D Fourier transformation can easily be extended to the discrete 2D Fourier transformation.

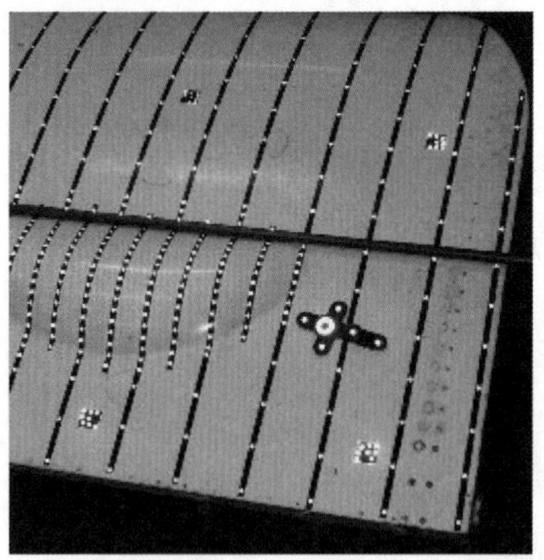
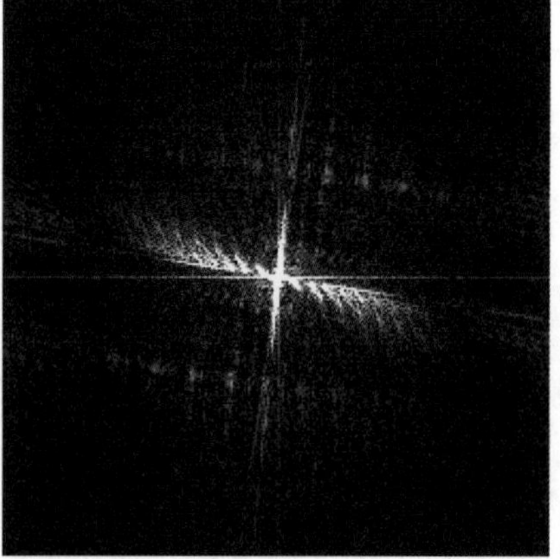

Fig. 5.28: Grey-value image (left) and corresponding power spectrum (right)

When applied to an image, the discrete Fourier transformation transforms it from the spatial domain S=$s(x,y)$ into the frequency domain $F(u,v)$. A visual evaluation of the spatial frequencies (wave numbers) in the image can be made through the power spectrum. For example, edges generate high frequencies. A power spectrum example is illustrated in Fig. 5.28, where bright points in the spectrum correspond to large amplitudes. This example shows high amplitudes which are perpendicular to the significant edges in the original image (spatial domain). The two horizontal and vertical lines in the spectrum are caused by the image borders.

Different basic functions from the Fourier transform result in alternative image transformations. For example the *discrete cosine transformation* (DCT) uses only cosine terms. The *wavelet transformation* uses various basic functions such as the Haar function to transform the original signal not only into the frequency domain but also into a scale domain of different resolutions. Wavelets are especially useful for image operations which must simultaneously account for coarse (smoothed) structures and detailed features having high information content. One application of the wavelet transformation is in image compression with information loss (see section 5.1.3.3).

Filters can be used to select or suppress certain spatial frequencies in the original image (high-pass filter, band-pass filter, low-pass filter). In the frequency domain the desired frequencies are multiplied by a filter function which defines the filter characteristics. Fig. 5.29 (left) shows the spectrum after applying a ring-shaped filter in the frequency domain. In this example the spectrum within the inner circle and beyond the outer circle is multiplied by 0 and within the ring zone it is multiplied by 1. After transformation back into the spatial domain this band-pass filter produces edge enhancement (Fig. 5.29 right).

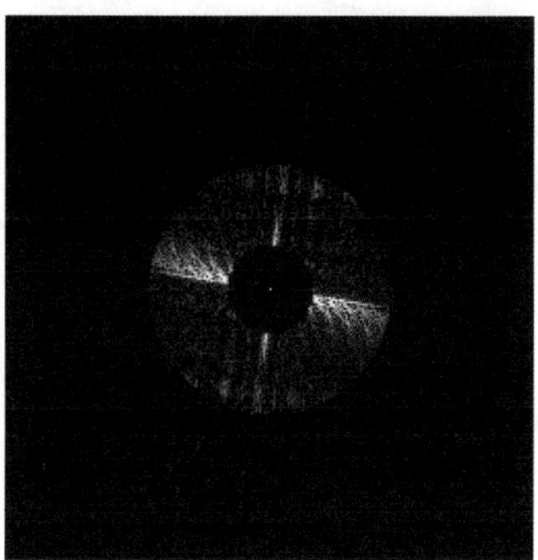

Fig. 5.29: Power spectrum of band-pass filtered image (left) and resulting image (right)

In the spatial domain, filters are applied by convolution with a filter operator. It can be shown that both approaches have identical results (Fig. 5.30).

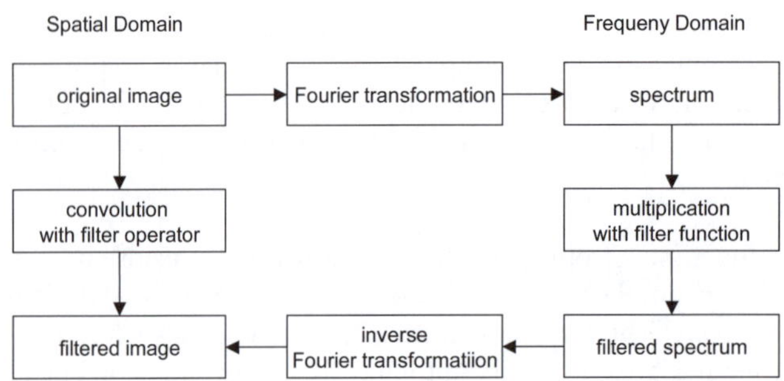

Fig. 5.30: Filtering in the spatial and frequency domains

Filter methods based on a one or two-dimensional convolution calculate a weighted sum of grey values in a given pixel region of the input image *S*. The result is assigned to the output image *S'* at the position of the region's central pixel.

$$s'(x,y) = \frac{1}{f}\sum_{u=-k}^{+k}\sum_{v=-l}^{+l} s(x-u, y-v)\cdot h(u,v) \qquad (5.29)$$
$$S' = S \otimes H$$

Here $H = h(u,v)$ denotes the filter matrix (filter operator) with $p \times q$ elements, where $k = (p-1)/2$ and $l = (q-1)/2$. Usually p and q are odd numbers and often $p = q$. The factor f is used for normalisation to the range [0…255]. The operator ⊗ symbolises the convolution operation.

In order to filter the complete image, the filter operator is shifted over the image in rows and columns as shown in Fig. 5.31. At each *x,y* position the convolution is calculated and the resulting grey value stored in the output image. The number of computational instructions amounts to $(2k+1)^2$ multiplications and $(2k+1)–1$ additions. For example, an image with 1024 x 1024 pixels, $p = 5$, $k = 2$ requires around $26 \cdot 10^6$ multiplications and $25 \cdot 10^6$ additions. Some filter masks can be split into one-dimensional convolutions which can be separately computed in the x and y directions. In this case only $(4k+2)$ multiplications and $4k$ additions are required. In the example this results in around $10 \cdot 10^6$ multiplications and $8 \cdot 10^6$ additions.

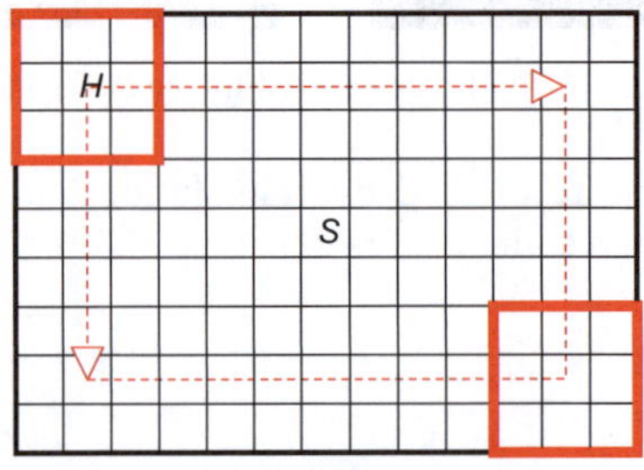

Fig. 5.31: Scheme for image filtering with $p=q=3$

5.2.3.2 Smoothing filters

Smoothing filters (low-pass filters) are mainly used for the suppression of grey-level noise, such as the quantisation noise associated with digitisation. These types of filter principally divide into linear smoothing filters, based on convolution, and non-linear smoothing filters based on rank orders (median filter). Table 5.2 shows one- and two-dimensional examples of typical filter operators.

Table 5.2: Examples of filter operators for image smoothing

Filter method	1D	2D
smoothing filter (moving average)	$\underset{3,1}{H}=\frac{1}{3}\cdot[1\;\;1\;\;1]=[1/3\;\;1/3\;\;1/3]$	$\underset{3,3}{H}=\frac{1}{9}\cdot\begin{bmatrix}1&1&1\\1&1&1\\1&1&1\end{bmatrix}=\begin{bmatrix}1/9&1/9&1/9\\1/9&1/9&1/9\\1/9&1/9&1/9\end{bmatrix}$
smoothing filter (binomial filter)	$\underset{3,1}{H}=\frac{1}{4}\cdot[1\;\;2\;\;1]=[1/4\;\;1/2\;\;1/4]$ $\underset{5,1}{H}=\frac{1}{16}\cdot[1\;\;4\;\;6\;\;4\;\;1]$	$\underset{3,3}{H}=\frac{1}{4}(1\;\;2\;\;1)\cdot\frac{1}{4}\begin{pmatrix}1\\2\\1\end{pmatrix}=\frac{1}{16}\begin{pmatrix}1&2&1\\2&4&2\\1&2&1\end{pmatrix}$ $\underset{5,5}{H}=\frac{1}{256}\begin{pmatrix}1&4&6&4&1\\4&16&24&16&4\\6&24&36&24&6\\4&16&24&16&4\\1&4&6&4&1\end{pmatrix}$

Linear, low-pass filters smooth the image but, depending on choice of filter coefficients, also smear image edges. The smoothing effect increases with larger filter sizes.

Gaussian filters possess mathematically optimal smoothing properties. The coefficients of the filter are derived from the two-dimensional Gaussian function:

$$f(x,y)=\frac{1}{2\pi\sigma^2}\cdot\exp\left(-\frac{x^2+y^2}{2\sigma^2}\right) \tag{5.30}$$

A suitable filter matrix of size p is usually chosen empirically. As an example, given $\sigma = 1$ a filter size of $p=7$ is appropriate. The Gaussian filter coefficients can be well approximated by the binomial coefficients of Pascal's triangle (see Table 5.2).

The non-linear median filter performs good smoothing whilst retaining sharp edges (Fig. 5.32d). The median filter is not based on convolution. Instead, it calculates the median value (as opposed to the mean) of a sorted list of grey values in the filter matrix, and uses this as the output grey value. The output image therefore consists only of grey values which exist in the input image. This property is essential for the filtering of images where, instead of intensities, the image elements store attributes or other data. The median filter is a member of the group of rank-order filters.

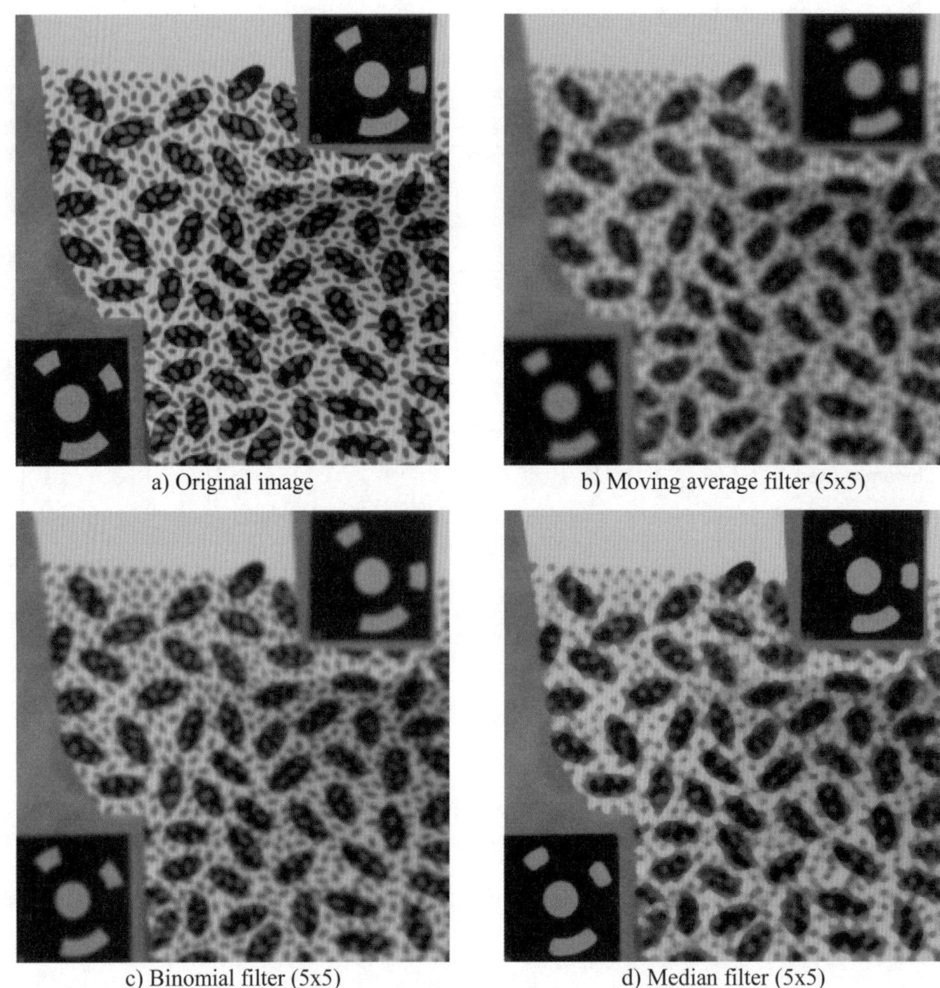

a) Original image b) Moving average filter (5x5)

c) Binomial filter (5x5) d) Median filter (5x5)

Fig. 5.32: Effect of different smoothing filters

5.2.3.3 Morphological operations

Morphological operations form their own class of image processing methods. The basic idea is the application of non-linear filters (see median filter, section 5.2.3.2) for the enhancement or suppression of black and white image regions with (known) shape properties, e.g. for the segmentation of point or circular features. Filtering is performed with special *structuring elements* which are tuned to the feature type to be detected and are successively stepped across the whole image. It is necessary to ensure that the focus or active point for the structuring element is carefully defined in order to avoid offsets in the identified locations of the features being detected.

Two fundamental functions based on Boolean operations for binary images are defined for morphological image processing:

- Erosion:

  Erosion leads to the shrinking of regions. The value 1 is set in the output image if all pixels in the filter region (e.g. 3x3 elements) correspond to the structuring element, i.e. the structuring element is in complete agreement with the image region. Otherwise the value 0 is set.

- Dilation:

  Dilation leads to the enlargement of connected regions. The number 1 is set if the structuring element includes at least one matching pixel within the image filter region.

Sequential application of dilation and erosion can be used to close gaps or to separate connected regions in the image. The following combinations of basic operations are useful:

- Opening:

  Opening is achieved by an erosion followed by dilation (Fig. 5.33 top). Small objects are removed.

- Closing:

  The reverse process is referred to as closing. Dilation is followed by erosion in order to close gaps between objects (Fig. 5.33 bottom).

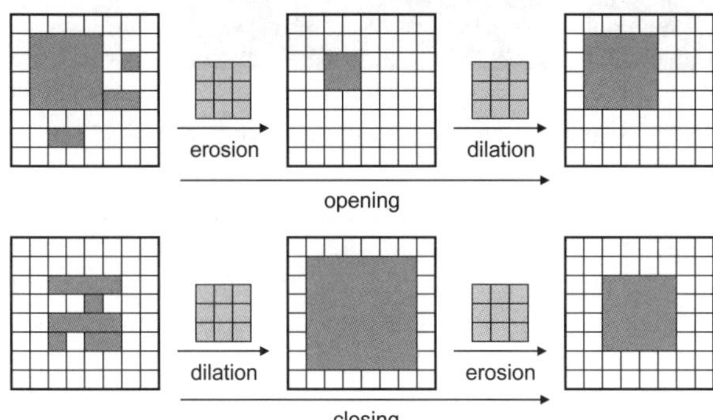

Fig. 5.33: Morphological operations with 3x3 structure element (after Bässmann & Kreyss 1998)

An application of opening is the segmentation of bright targets in a photogrammetric image. Fig. 5.34 shows a problematic situation for an image region with a number of targets lying close together. After simple thresholding, the intermediate result shows several small circular features which do not correspond to target points, as well as connected regions which actually represent separate targets. Segmentation, based on the procedure of section 5.2.1.4 with a minimum point size of 15 pixel, results in nine objects of which two have joined together features from separate adjacent objects.

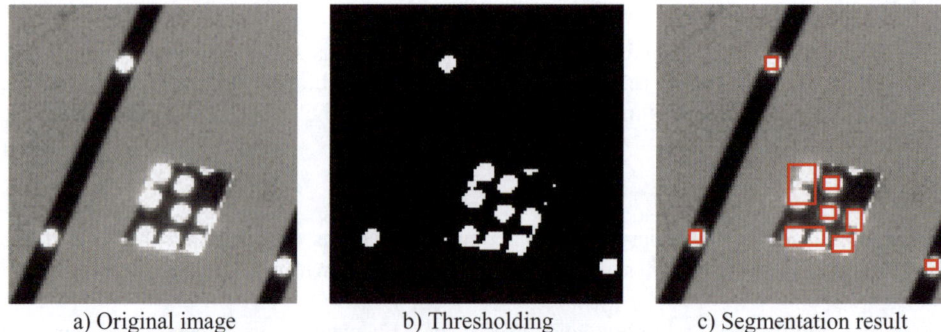

a) Original image b) Thresholding c) Segmentation result

Fig. 5.34: Simple segmentation after thresholding (9 objects)

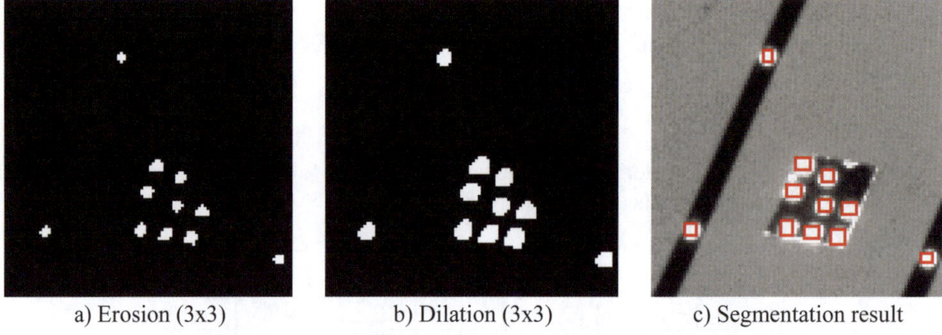

a) Erosion (3x3) b) Dilation (3x3) c) Segmentation result

Fig. 5.35: Segmentation after thresholding and opening (11 objects)

Fig. 5.35 shows the same image region processed using the opening operator. Starting with the same binary image, erosion eliminates objects smaller than the applied 3x3 structure element. In addition the targets shrink. Subsequent application of dilation enlarges the remaining objects to their original size. This type of segmentation correctly extracts all 11 objects.

Morphological operators must be extended for grey-level images. Instead of the Boolean decision for each pixel of the structuring element, the minimum grey-value difference (grey-value erosion) or the maximum grey-value sum (grey-value dilation), between structure element and original image, is used as the result.

5.2.3.4 Wallis filter

The objective of the Wallis filter is a local optimisation of contrast. The grey values in the output image are calculated as follows:

$$s'(x,y) = s(x,y)r_1 + r_0 \tag{5.31}$$

Here the parameters r_0,r_1 of a linear contrast stretching (eqn. 5.10) are determined in a local filter window of size n x n pixel. The calculation of the filter parameters for each pixel is based on the following values:

$$r_1 = \frac{\sigma' c}{\sigma c + \frac{\sigma'}{c}} \tag{5.32}$$

where σ: standard deviation in input window
σ': standard deviation in output window
c: contrast control factor $0 < c < 1.3$

$$r_0 = b m_{S'} + (1 - b - r_1) m_S \tag{5.33}$$

where m_S: average value in input window
$m_{S'}$: average value in output window
b: brightness control factor $0 \leq b \leq 1$

By fixing the average value $m_{S'}$ (e.g. 120) and standard deviation σ' (e.g. 60), the brightness and contrast ranges of the output window are defined for the entire image. The control factors b and c are determined empirically, e.g. $b = 0.8$ and $c = 1.3$. The filter size n is set according to the further processing of the image and lies, for example, between $n = 3$ and $n = 25$.

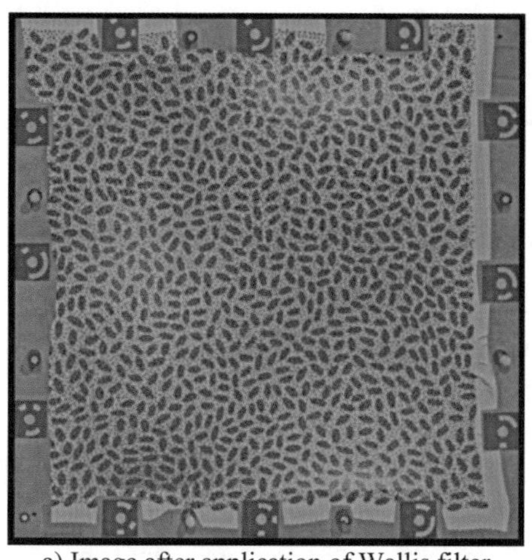

a) Image after application of Wallis filter (original image in Fig. 5.11)

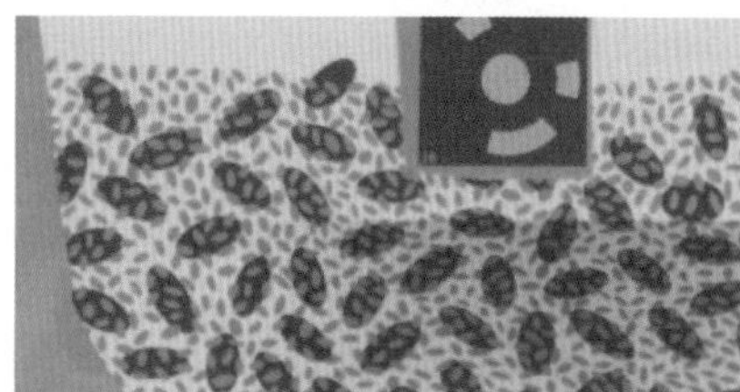

b) Detail from original image

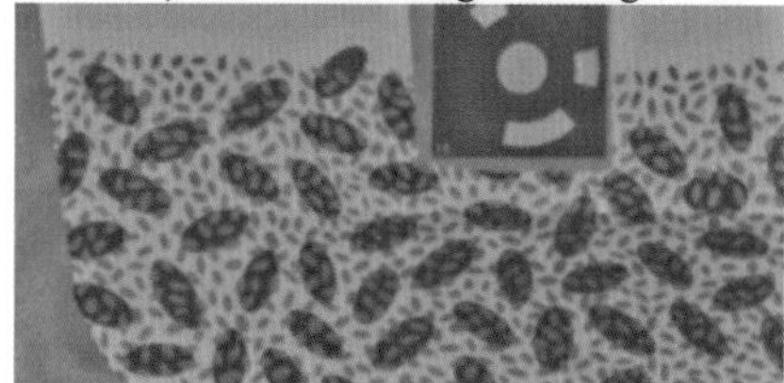

c) Detail from Wallis image

Fig. 5.36: Local contrast improvement with Wallis filter ($n = 25$, $b = 0.6$, $c = 1.6$, $m_{S'} = 120$, $\sigma' = 60$)

Amongst other applications, the Wallis filter is used as a preparation for various image-matching techniques (see section 5.4.2.4 and section 5.5) in order to improve the quality of matching independently of image contrast. Fig. 5.36 shows an example of Wallis filtering. In this case, the selected filter parameters lead to a lower contrast image, but also one with a more consistent contrast across the whole image.

5.2.4 Edge extraction

Grey-value edges are the primary image structures used by the human visual system for object recognition. They are therefore of fundamental interest for digital image measurement and pattern recognition. Every object stands out from the background on the basis of a characteristic change in the relevant image structures (Fig. 5.37).

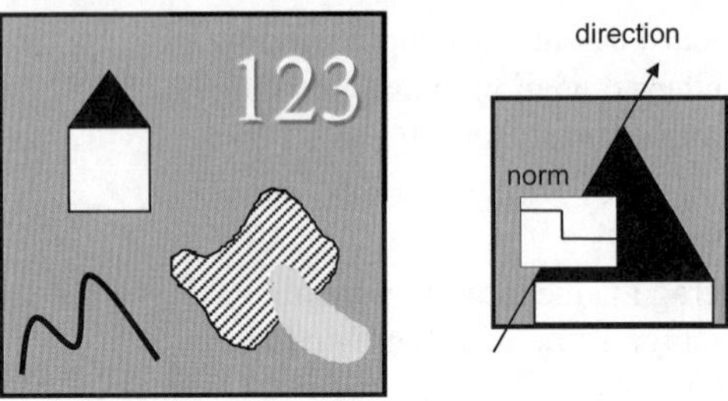

Fig. 5.37: Objects and edges

This change in image structure can be due to:

- significant change in grey values (grey-value edge) along the physical object edge
  → edge extraction
- change in colour values (colour edge)
  → multi-spectral edge extraction
- change in surface texture (e.g. hatched vs. point pattern)
  → texture analysis

The following discussion concerns only the extraction of grey-level edges which can be characterised by the following properties:

- a significant change in adjacent grey values perpendicular to the edge direction
- edges have a direction and magnitude
- edges are formed by small image structures, i.e. the region of grey value change may not be too large
- small image structures are equivalent to high frequencies in the frequency domain

If edges are to be extracted by means of filters then, in contrast to smoothing filters, high frequencies must be enhanced and low frequencies suppressed (= high-pass filter or band-pass filter).

An edge, or more precisely ramp, is a significant change of intensity between two grey value areas of a particular size. In contrast, a line is a thin grey value image area which cannot be resolved into two opposite edges.

5.2.4.1 First order differential filters

The first derivative of a continuous function $s(x)$ is given by

$$s'(x) = \frac{ds}{dx} = \lim_{\Delta x \to 0} \frac{s(x+\Delta x) - s(x)}{\Delta x} \tag{5.34}$$

and for a discrete function by:

$$s'(x) = \frac{s(x+1) - s(x)}{1} = s(x+1) - s(x) \tag{5.35}$$

A filter mask $H_2 = [-1 \quad +1]$, known as a Roberts gradient, can be derived from (5.35). Larger filter masks, which also have a smearing effect, are often used, e.g.

$H_3 = [-1\ 0\ +1]$

$H_5 = [-1\ \ 0\ \ 0\ \ 0\ \ +1]$ or $[-2\ \ -1\ \ 0\ \ +1\ \ +2]$

These filters approximate the discrete first derivative of the image function by (weighted) differences. They evaluate local extremes in grey value distribution from gradients calculated separately in the x and y directions. The zero crossing of the first derivative is assumed to give the position of a grey value line and its maximum or minimum gives the position of an edge (ramp), see Fig. 5.38.

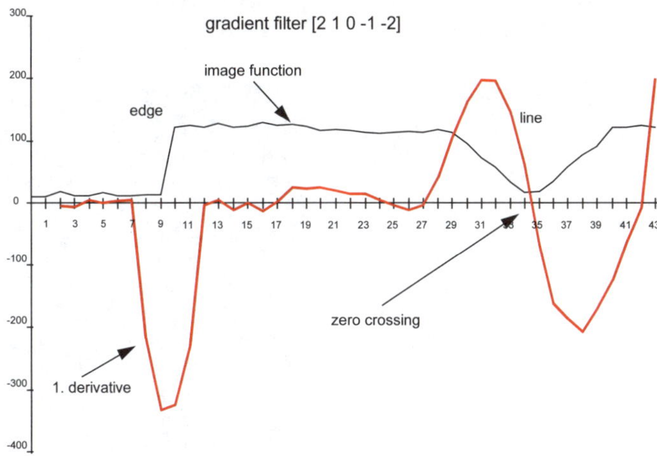

Fig. 5.38: Convolution of an image row with gradient filter $H = [2\ 1\ 0\ -1\ -2]$

From the gradient

$$grad(s) = \left(\frac{ds(x,y)}{dx}, \frac{ds(x,y)}{dy} \right)^{\mathrm{T}} \tag{5.36}$$

magnitude and direction of an edge can be derived:

$$\sqrt{\left(\frac{ds(x,y)}{dx}\right)^2+\left(\frac{ds(x,y)}{dy}\right)^2} \quad \text{: magnitude of gradient} \tag{5.37}$$

$$\left(\frac{ds(x,y)}{dy}\right) \Big/ \left(\frac{ds(x,y)}{dx}\right) \quad \text{: direction of gradient} \tag{5.38}$$

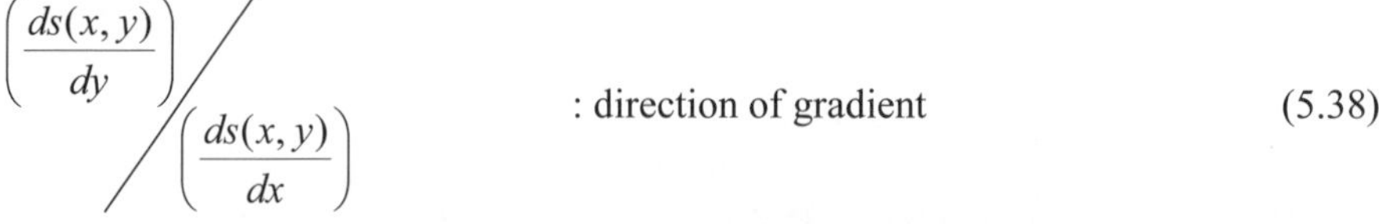

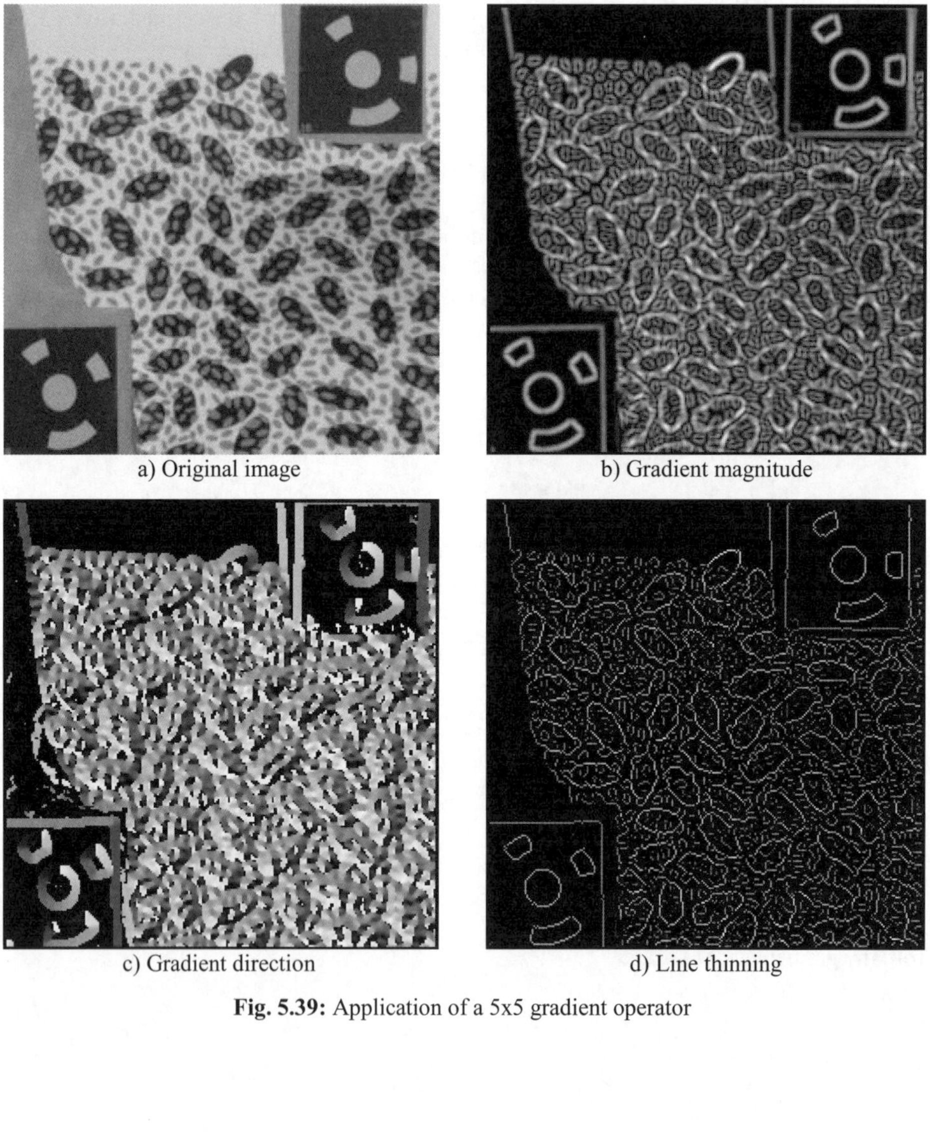

a) Original image

b) Gradient magnitude

c) Gradient direction

d) Line thinning

Fig. 5.39: Application of a 5x5 gradient operator

A well-known two-dimensional gradient filter is the Sobel operator. It approximates the first derivatives in x and y by separated convolution with the filter masks

$$H_x = \begin{bmatrix} 1 & 0 & -1 \\ 2 & 0 & -2 \\ 1 & 0 & -1 \end{bmatrix} \qquad H_y = \begin{bmatrix} 1 & 2 & 1 \\ 0 & 0 & 0 \\ -1 & -2 & -1 \end{bmatrix}$$

The operator can be extended to larger filter masks. The magnitude of the gradient is calculated from the intermediate convolution results and stored as a grey value. The direction of the gradient can also be stored as a coded grey value image. Magnitude and direction images can be used for further line and edge extraction, e.g. for line thinning and chaining. Fig. 5.39 shows the application of a 5x5 gradient operator followed by line thinning.

5.2.4.2 Second order differential filters

For a continuous function $s(x)$ the second derivative is given by

$$s''(x) = \frac{d^2 s}{dx^2} = \lim_{\Delta x \to 0} \frac{s(x+\Delta x) - s(x) - [s(x) - s(x-\Delta x)]}{(\Delta x)^2} \tag{5.39}$$

and for a discrete function:

$$s''(x) = \frac{s(x+1) - s(x) - [s(x) - s(x-1)]}{(1)^2} = s(x+1) - 2s(x) + s(x-1) \tag{5.40}$$

Here a filter mask $H_3 = [+1 \ \ -2 \ \ +1]$ can be derived. The second derivative can also be generated by double application of the first derivative.

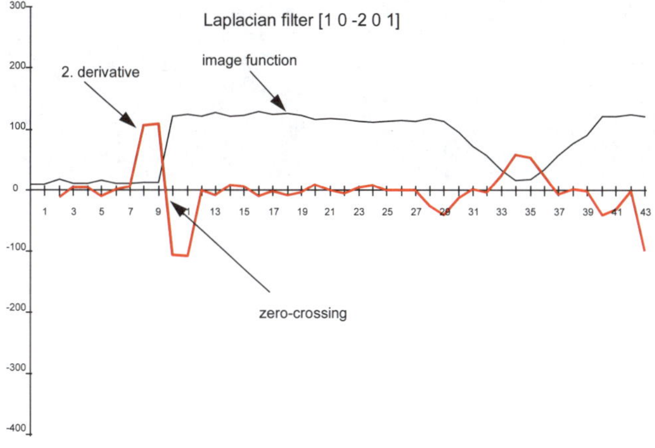

Fig. 5.40: Convolution of an image row by a Laplacian filter $H_5 = [1 \ 0 \ -2 \ 0 \ 1]$

A grey value edge is formed by a ramp change in grey values. The position of the edge is given by the zero crossing of the second derivative (Fig. 5.40).

The second derivative of a two-dimensional function is given by the total second order differential:

$$s''(x,y) = \nabla^2 s = \frac{d^2 s}{dx^2} + \frac{d^2 s}{dy^2} \tag{5.41}$$

For a discrete function $s''(x,y)$ the second derivative can therefore be formed by addition of the partial second derivatives in the x and y directions.

$$\begin{bmatrix} 0 & 1 & 0 \\ 0 & -2 & 0 \\ 0 & 1 & 0 \end{bmatrix} + \begin{bmatrix} 0 & 0 & 0 \\ 1 & -2 & 1 \\ 0 & 0 & 0 \end{bmatrix} = \begin{bmatrix} 0 & 1 & 0 \\ 1 & -4 & 1 \\ 0 & 1 & 0 \end{bmatrix}$$

The resulting convolution mask is called a *Laplace operator*. Its main properties are:

- edges are detected in all directions and so it is invariant to rotations
- light-dark changes produce an opposite sign to dark-light changes

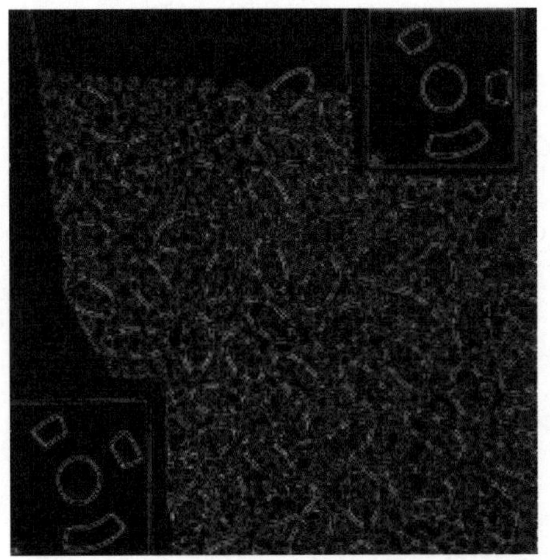
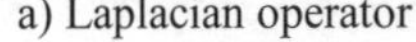
a) Laplacian operator

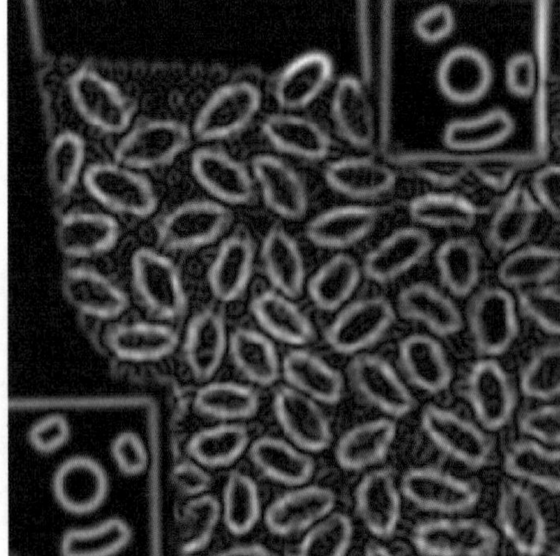
b) Laplacian of Gaussian operator (σ=3.0)

Fig. 5.41: Edge extraction with differential operators

5.2.4.3 Laplacian of Gaussian filter

Fig. 5.41 illustrates the sensitivity to noise of the Laplace operator, hence minor intensity changes are interpreted as edges. A better result could be expected if the image were smoothed in advance. As the Gaussian filter is an optimal smoothing filter (see section

5.2.3.2), the second derivative of the Gaussian function is regarded as an optimal edge filter which combines smoothing properties with edge extraction capabilities.

The second derivative of the Gaussian function (5.30) is given by:

$$f''(r,\sigma) = \nabla^2 GAUSS = -\left(1 - \frac{r}{\sigma^2}\right) \cdot \exp\left(-\frac{r^2}{2\sigma^2}\right) \tag{5.42}$$

where $r^2 = x^2 + y^2$

The filter based on this function is known as the *Laplacian of Gaussian* or LoG filter. Due to its shape it is also called a *Mexican hat* (Fig. 5.42). Empirical analysis shows that a filter size of 11–13 corresponds to a ±3σ interval. Fig. 5.41 shows the result of LoG filtering with σ = 3.0.

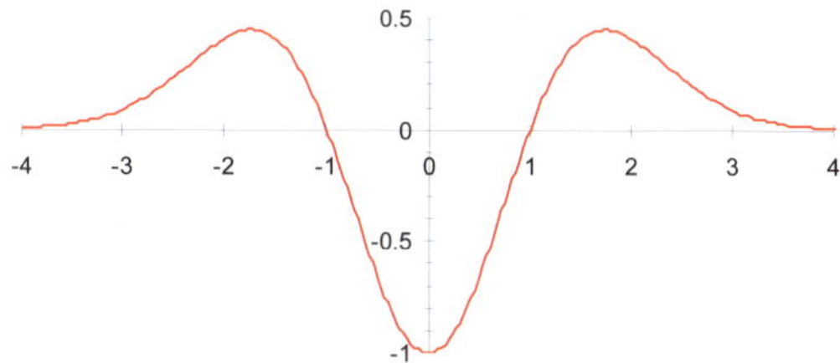

Fig. 5.42: Laplacian of Gaussian (LoG) with σ = 1

5.2.4.4 Image sharpening

Image sharpening is used to enhance the sharpness of an image. An optically defocused image cannot subsequently be focused without further information, but the visual impression can be improved. Out-of-focus images are characterised by reduced contrast and blurred edges and filters can be defined which reduce these effects. A possible sharpening filter is given by the following definition:

$$\begin{aligned} s'(x,y) &= 2 \cdot s(x,y) - [s(x,y) \otimes h(u,v)] \\ &= s(x,y) + \left(s(x,y) - [s(x,y) \otimes h(u,v)]\right) \end{aligned} \tag{5.43}$$

Here $h(u,v)$ represents a smoothing filter, e.g. the binomial filter listed in Table 5.2. The subtraction of the smoothed image from the original corresponds to a high-pass filtering. The enhancement in the output image therefore results from adding the high frequencies in the original image. Fig. 5.43 shows an example of the effect of a sharpening filter based on a 5x5 binomial filter.

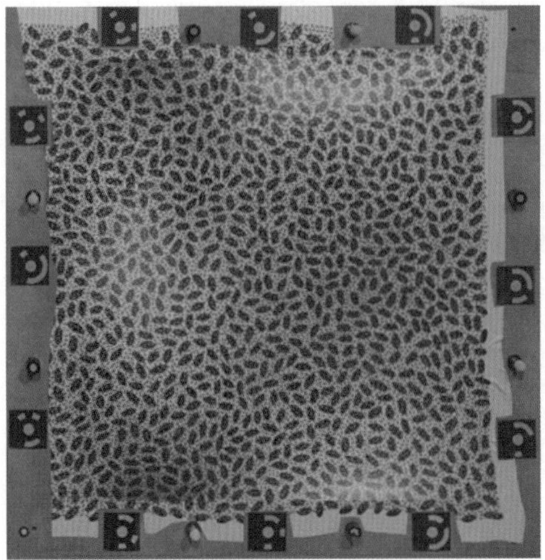

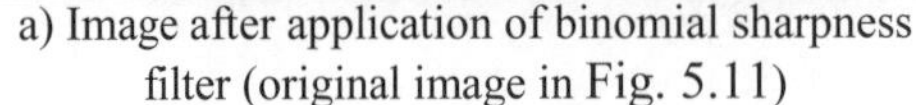

a) Image after application of binomial sharpness filter (original image in Fig. 5.11)

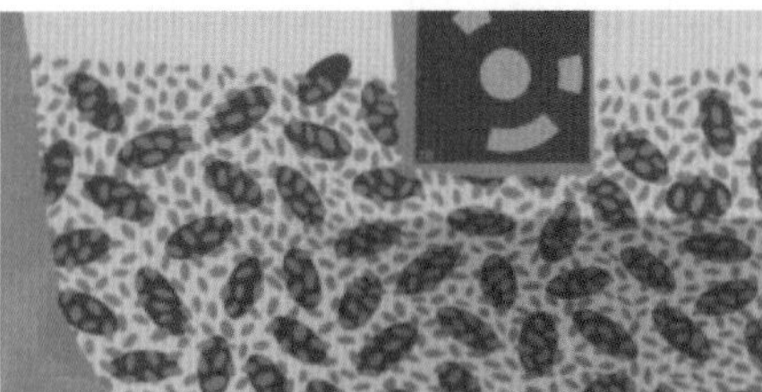

b) Detail from original image

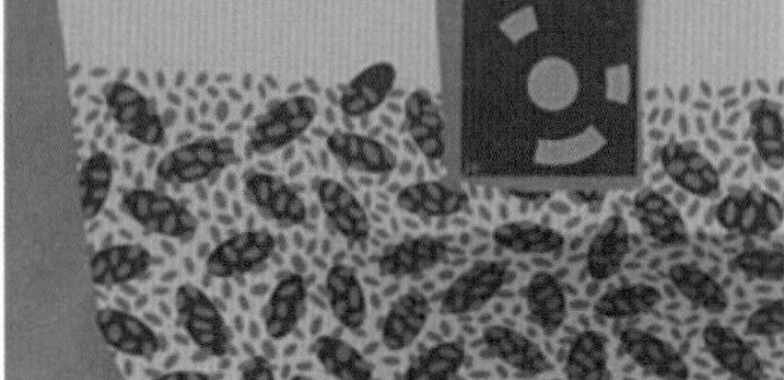

c) Detail in sharpened image

Fig. 5.43: Enhanced sharpness

5.2.4.5 Hough transform

The Hough transform is based on the condition that all points on an analytical curve can be defined by one common set of parameters. Whilst a Hough transform can be applied to the detection of a wide variety of imaged shapes, one of the simplest solutions is the case of imaged straight lines where all points must fulfil the line equation in Hesse's normal form

$$r = x\cos\varphi + y\sin\varphi \tag{5.44}$$

In order to determine the parameters r and φ, a discrete two-dimensional parameter space (Hough space) is spanned with elements initialised to zero. For each edge point at position x,y in the image, the direction of the gradient $\varphi+90°$ is known and r can therefore be determined. At position r,φ the corresponding value in Hough space is increased by 1, i.e. each point on the line accumulates at the same position in Hough space (due to rounding errors and noise it is actually a small local area). Hence, straight lines can be detected by searching the Hough accumulator for local maxima.

Fig. 5.44 illustrates the application of the Hough transform to an image with several well-structured edges (a). Edge extraction is performed by a 5x5 gradient filter (in analogy to Fig. 5.39), which delivers a magnitude and direction image (b). Several maxima (clusters) can be recognised in Hough space (c). Clusters which are arranged in one column of the Hough accumulator represent parallel edges in the original image. The value pairs in Hough space can be transformed back into the spatial domain x,y. Analytical lines are determined as a result, although their start and end points cannot be reconstructed (d).

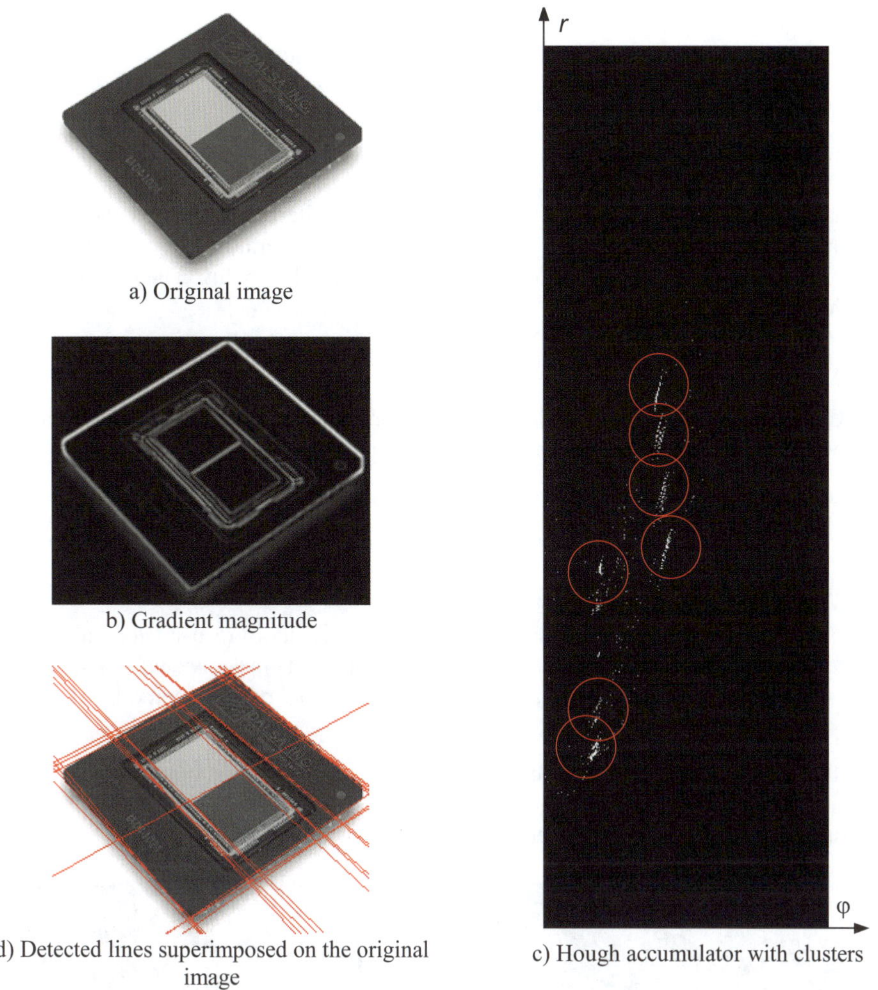

a) Original image

b) Gradient magnitude

c) Hough accumulator with clusters

d) Detected lines superimposed on the original image

Fig. 5.44: Hough transform

Such a Hough transform is most relevant for the detection of objects predominantly formed by straight lines. The method can be expanded to curves of higher order (e.g. circles of unknown radius) although the dimension of the Hough accumulator is then no longer two-dimensional.

5.2.4.6 Enhanced edge operators

The simple methods of edge extraction discussed in the previous sections often do not deliver satisfactory results. An edge filter suitable for measurement tasks should have the following properties:

- complete extraction of all relevant edges (robustness)
- simple parametrisation (preferably without interactive input)
- high sub-pixel accuracy
- minimum computional effort

Numerous methods for edge extraction are given in the literature, such as the following well-established methods. In contrast to simple convolution operators, they are extended by pre and post processing as well as adaptive parameter adjustment. They provide good results even for complex images.

- Canny operator and Deriche operator:

  The Canny operator belongs to the class of operators based on LoG. It optimises the following quality criteria in edge measurement:

  - sensitivity to true edges (uniqueness of edge)
  - robustness to noise (maximum signal-to-noise ratio)
  - accuracy of edge position

  A function is defined for each criterion whose parameters are used to build a non-recursive linear filter. The Canny operator delivers a list of connected contour points with sub-pixel resolution and can be classified as one of the methods of contour following (see section 5.4.3).

  A further development is the Deriche operator which achieves the quality criteria above by recursive filtering. Fig. 5.45 and Fig. 5.46 show the application of both filters, where almost identical results are obtained by appropriate parameter settings.

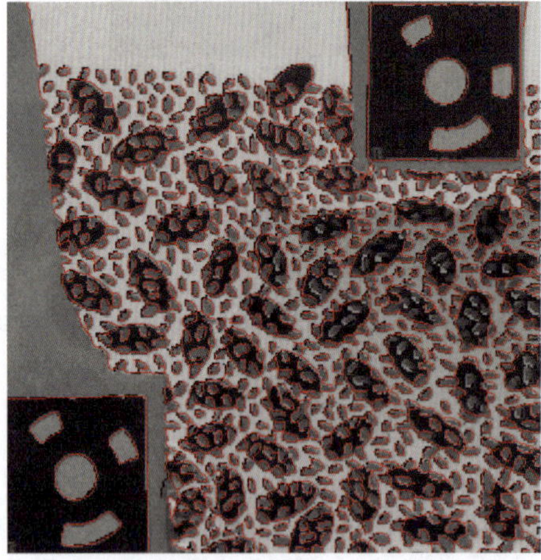

Fig. 5.45: Canny operator with $\sigma = 1.5$

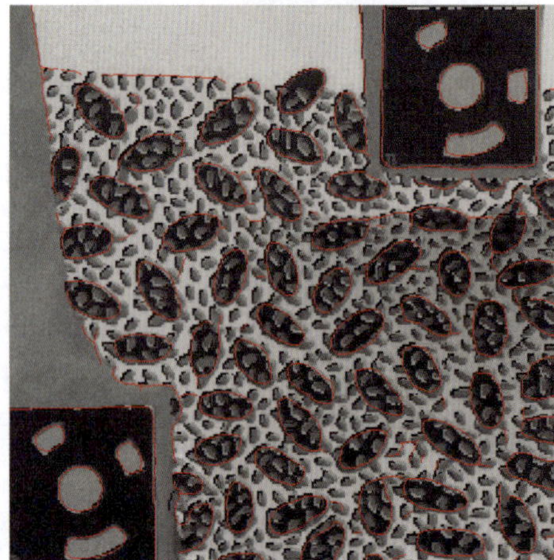

Fig. 5.46: Deriche operator with $\sigma = 1.5$

- Edge extraction in image pyramids:

  Image pyramids (see section 5.1.3.1) represent the image content at different spatial resolutions. Since the ability to detect edges in natural images varies with image resolution (image scale), an approach is sought which determines the optimal scale for each edge pixel. LoG filters, or morphological operators, can be used as edge operators.

- Least-squares edge operators:

  A model describing the geometric and radiometric properties of the edge can be determined by least-squares parameter estimation. By adjusting their initial parameter values, a priori edge models (templates) can be fitted to the actual region defining an edge (see section 5.4.2.4, *least-squares template matching*). An optimal fit could be the least squares estimate in which the differentials have the least entropy.

  A global approach to edge extraction is possible if the energy function

  $$E = E_{int} + E_{grey} + E_{ext} \tag{5.45}$$

  where E_{int}: curve energy function
  E_{grey}: grey-value conditions
  E_{ext}: geometric constraints

  is minimised in a least-squares solution. The curve energy function E_{int} describes the behaviour of the curvature along the edge or the sensitivity with respect to possible changes in direction. Grey-value conditions E_{grey} along the edge can be defined by requiring, for example, maximum gradients. Additional geometric constraints such as straight lines or epipolar geometries are specified by E_{ext}.

5.2.4.7 Sub-pixel interpolation

Sub-pixel resolution

In digital photogrammetry, line and edge filters are used for the measurement of geometric elements (points, lines) which are described by their contours (see section 5.4.2.5). The objective is to locate these patterns to the highest accuracy. As discussed in section 5.1.2, object structures covering several pixels can be measured by interpolation to the sub-pixel level.

The theoretical resolution of the position of a digitised grey value edge is, in the first instance, a function of the slope of the grey values along the edge and quantisation depth (number of bits per grey value). It is defined by a parameter, sometimes called the slack value, *d,* where the position of the imaged edge can be varied without changing the related grey values (Fig. 5.47). For a step change ($\beta = 0$ in diagram) the slack is maximised and amounts to 1.

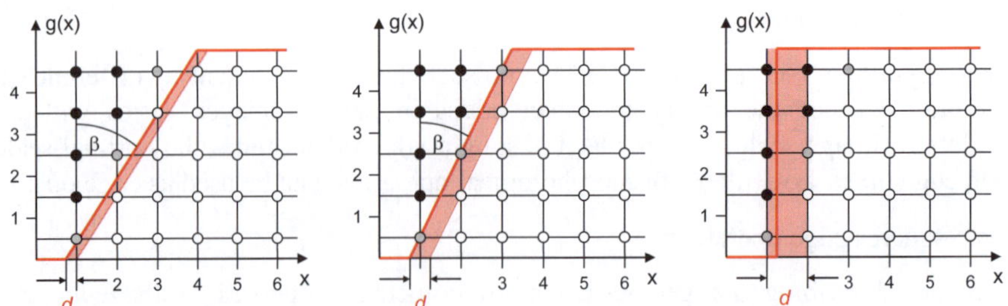

Fig. 5.47: Slack in a digitised edge (according to Förstner 1985)

It can be shown that, for N quantisation steps, the uncertainty of edge positioning is at least $1/(N–1)$. This corresponds to 0.004 pixels for $N = 256$ grey levels. In this case the average deviation of a *single* edge point can be estimated, independently of the slope of the edge, as

$$\overline{\sigma}_d = 0.015 \text{ pixel} \tag{5.46}$$

This theoretical quantity will be higher in practice due to optical limitations and noise but it defines a lower limit for the estimation of positional accuracy of image measurements. Hence, the accuracy of an edge of length n pixels can be estimated as

$$\overline{\sigma}_K = \frac{\overline{\sigma}_d}{\sqrt{n}} \text{ pixel} \tag{5.47}$$

This is also approximately true for the centre of a circle measured by n edge points. For a circular target of diameter 6 pixels there are approximately $n = 19$ edge points. Hence, in the ideal case, the centre of such a target can be determined to an accuracy of about 0.004 pixel (4/1000 pixel). Further investigations are discussed in section 5.4.2.6.

A selection of edge extraction methods providing sub-pixel interpolation, together with their principal parameters, are summarised in Table 5.3:

Table 5.3: Methods for sub-pixel interpolation of edge points

Method	Model	Intermediate result	Sub-pixel interpolation
differential filter	deflection point	gradient image	linear interpolation
moment preservation	Grey-value plateaux	1st, 2nd and 3rd moments	solution of 3 equations
feature correlation (*template matching*)	edge template, cross-correlation	correlation coefficients	2nd order interpolation
least-squares matching	edge template, geometric and radiometric transformation	up to 8 transformation parameters	shift parameters

Zero-crossing interpolation

As shown in Fig. 5.38 and Fig. 5.40 the derivative functions do not pass zero at an integer pixel coordinate. The sub-pixel position of an edge can be determined by first or second order interpolation in the neighbourhood of the zero crossing (Fig. 5.48a). This method is used, for example, for the edge measurement of patterns based on points (see section 5.4.2.5). Interpolation of zero crossings provides edge location with a precision of up to 1/100 pixel.

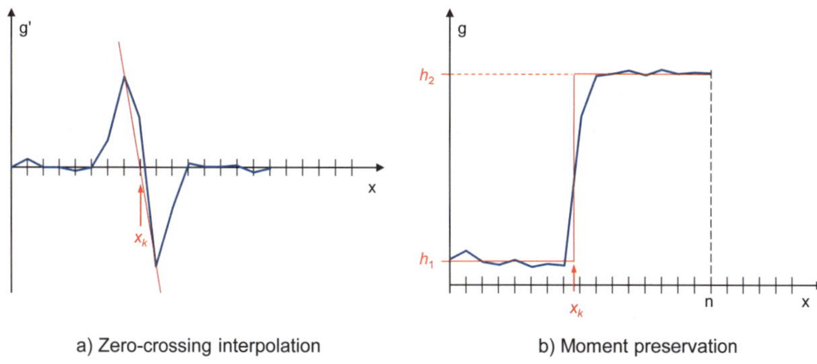

Fig. 5.48: Edge interpolation methods

Moment preservation

The basic idea of the moment preservation method is based on the assumption that an edge within a one-dimensional image function (e.g. an image row) of window size n can be described by three parameters. These define the left grey value plateau h_1, the right grey value plateau h_2 and the coordinate of the grey value step x_k (Fig. 5.48b). The three required equations are formed by the 1st, 2nd and 3rd moments:

$$m_i = \frac{1}{n}\sum_{j=1}^{n} g_j^i \quad i = 1,2,3 \tag{5.48}$$

The following parameters are determined (without derivation):

$$\begin{aligned} h_1 &= m_1 - \bar{\sigma}\sqrt{\frac{p_2}{p_1}} \\ h_2 &= m_1 + \bar{\sigma}\sqrt{\frac{p_1}{p_2}} \\ p_1 &= \frac{1}{2}\left[1 + \bar{s}\sqrt{\frac{1}{4+\bar{s}^2}}\right] \end{aligned} \tag{5.49}$$

where

$$\bar{s} = \frac{m_3 + 2m_1^3 - 3m_1 m_2}{\bar{\sigma}^3}$$
$$\bar{\sigma}^2 = m_2 - m_1^2 \tag{5.50}$$

The desired edge position is given by:

$$x_K = n \cdot p_1 \tag{5.51}$$

The moment preservation method is easy to implement and it delivers the sub-pixel edge location without any further interpolation. It is used, for example, with the Zhou operator for the edge extraction of elliptically shaped target images (see section 5.4.2.5).

Correlation methods

Correlation methods determine the position in a search image which has the highest similarity with a reference pattern (*template*). The reference pattern can be a subset of a natural image, or a synthetically created image. For example, when searching for a vertical edge which switches from dark to light, a reference pattern similar to Fig. 5.49 can be used.

The similarity between two patterns can be measured by the normalised cross-correlation coefficient r (see section 5.4.2.3). If r is plotted as a function of x, the position of maximum correlation is most likely to be the true position of the reference pattern. If the curve around the maximum is approximated by a quadratic function (parabola), the desired position can be determined to sub-pixel precision (Fig. 5.49).

Least-squares matching (see section 5.4.2.4) determines a transformation which describes both the change of contrast and the geometric projection between reference pattern and search image. The approach requires reasonably good initial values of the unknown parameters, in particular the required shift parameters. The adjustment process directly delivers the sub-pixel positions of the pattern (edge).

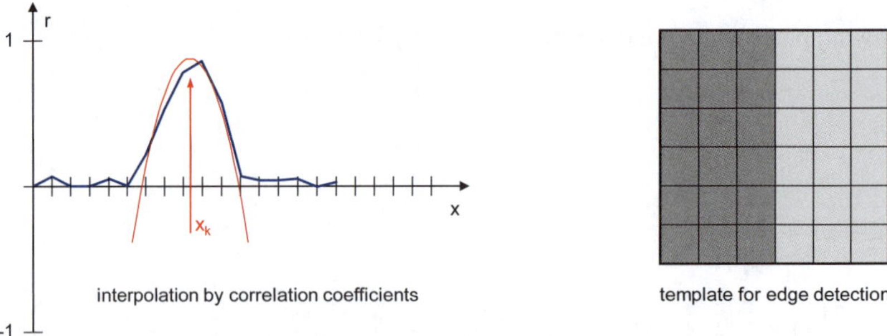

Fig. 5.49: Interpolation by correlation coefficients

5.3 Geometric image transformation

The process of modifying the geometric projection of a digital image is here referred to as a geometric image transformation. Related methods are required in photogrammetry for photo rectification and orthophoto production (section 4.2.6), for combining images with CAD models and for template matching procedures. Rendering and morphing methods also belong to this category. Fig. 5.50 shows an example of the projective rectification of a façade (see also Fig. 4.16).

Fig. 5.50: Geometric rectification of a façade

The term *rectification* denotes a general modification of pixel coordinates, e.g. for

- translation and rotation
- change of scale or size (magnification, reduction)
- correction of distortion effects
- projective rectification (from central perspective to parallel projection)
- orthophoto production (differential rectification)
- rectification of one image with respect to another
- superimposition of natural structures onto a surface, e.g. a CAD model (*texture mapping*)

Geometric image transformations are generally performed in two stages:

- transformation of pixel coordinates (image coordinates) into the target system (rectification) – this transformation is the reverse of the imaging process
- calculation (interpolation) of output grey values

5.3.1 Fundamentals of rectification

Rectification is founded on the geometric transformation of pixel coordinates from an original image to an output image:

$$s'(x',y') = G(s(x,y)) = g'$$
where
$$x' = f_1(x,y) \quad\quad (5.37)$$
$$y' = f_2(x,y)$$

Here the grey value g, at position (x,y) in the original image, appears in the output image, after grey value interpolation G, as grey value g' at position (x',y').

The geometric transformations f_1 and f_2 can be almost arbitrary coordinate transformations. The affine transformation (2.8) is often used for standard modifications such as translation, scaling, rotation or shearing. The projective transformation (8 parameters, see eqn. 2.17) is suited to the rectification of images of planar objects. Where a digital surface model is available, arbitrary free-form surfaces can be transformed into orthophotos by use of the collinearity equations (4.9) (see section 4.2.2).

In all cases there is a transformation between the pixel coordinates (x,y) of the input image and the pixel coordinates (x',y') of the output image. The grey values of the input image in the region of (x,y) must be stored in the output image at position (x',y'). For this purpose the indirect rectification method is usually applied in which the output image is processed pixel by pixel. By reversing the geometric transformation, the grey value of the input image is interpolated at the reverse-transformed position (x,y) and then stored in the output image. This algorithm is easy to implement and avoids gaps or overlapping regions in the output image (Fig. 5.51).

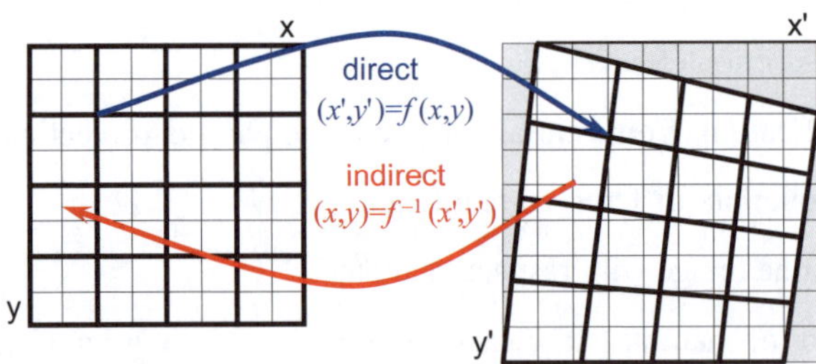

Fig. 5.51: Rectification methods

5.3.2 Grey-value interpolation

The second step in rectification consists of the interpolation of a suitable grey value from the local neighbourhood using an arbitrary non-integer pixel position, and then storing this quantity in the output image (resampling). The following methods are normally used for grey value interpolation:

- zero order interpolation (nearest neighbour)
- first order interpolation (bilinear interpolation)
- second order interpolation (bicubic convolution, Lagrange polynomials)

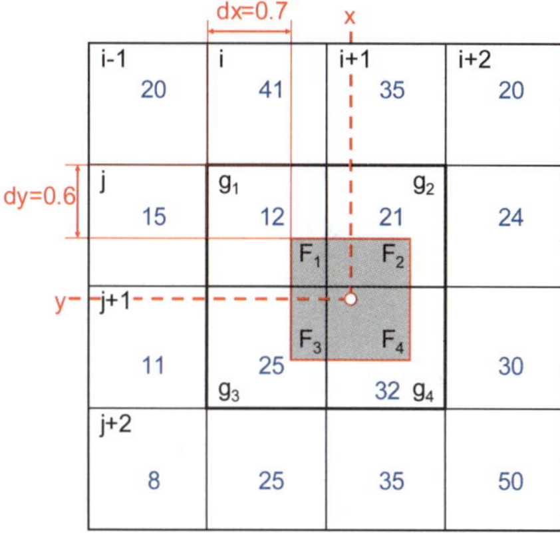

Fig. 5.52: Grey-value interpolation in 2x2 and 4x4 neighbourhood

For the method of nearest neighbour, the grey value at the rounded or truncated real pixel coordinate is used in the output image. The interpolation rule is given by:

$$g' = s(\mathrm{round}(x), \mathrm{round}(y)) \tag{5.38}$$

Nearest-neighbour grey value in the example of Fig. 5.52: $g' = 32 = g_4$

This approach leads to the visually worst rectification result. However, the computational effort is small and the output image consists only of grey values which also exist in the input image.

The bilinear or biquadratic interpolation takes into account the 2x2 adjacent grey values of the computed pixel position. The interpolated grey value is the result of the weighted average of adjacent grey values in which the weight is given by the relative coverage of the current pixel. The interpolation rule is given by:

$$g' = F_1 \cdot s(i,j) + F_2 \cdot s(i+1,j) + F_3 \cdot s(i,j+1) + F_4 \cdot s(i+1,j+1) \tag{5.39}$$

where $F_1 + F_2 + F_3 + F_4 = 1$

or analogously to equation (2.15):

$$\begin{aligned} g' = s(i,j) &+ dx \cdot [s(i+1,j) - s(i,j)] + dy \cdot [s(i,j+1) - s(i,j)] \\ &+ dx \cdot dy \cdot [s(i+1,j+1) - s(i+1,j) - s(i,j+1) + s(i,j)] \end{aligned} \tag{5.40}$$

Bilinear grey value in the example of Fig. 5.52: $g' = 25.3 \approx 25$

With modest computational effort, bilinear interpolation generates slightly smoothed rectifications of good quality.

Bicubic convolution and Lagrange interpolation are usually applied only in output rectifications where the highest image quality is required. These methods use a 4x4 environment for interpolation which results in computation times up to 10 times higher. The algorithm for bicubic convolution is as follows:

$$\begin{aligned} df(x) &= |x|^3 - 2\cdot|x|^2 + 1 \\ df(x) &= -|x|^3 + 5\cdot|x|^2 - 8\cdot|x| + 4 \\ df(x) &= 0 \end{aligned} \quad \begin{cases} \text{for } |x| < 1 \\ \text{for } 1 \le |x| < 2 \\ \text{other cases} \end{cases}$$

$$\begin{aligned} a(n) = {} & s(i-1, j+n-2)df(dx+1) \\ & + s(i, j+n-2)\cdot df(dx) \\ & + s(i+1, j+n-2)\cdot df(dx-1) \\ & + s(i+2, j+n-2)\cdot df(dx-2) \end{aligned} \quad \text{for } n = 1,2,3,4 \tag{5.41}$$

$$g' = a(1)\cdot df(dy+1) + a(2)\cdot df(dy) + a(3)\cdot df(dy-1) + a(4)\cdot df(dy-2)$$

Bicubic grey value for the example of Fig. 5.52: $g' = 21.7 \approx 22$

The arbitrarily chosen example of Fig. 5.52 shows clearly that the three different interpolation methods generate quite different values for the interpolated grey level. Fig. 5.53 shows an enlarged image region after rotation, generated by the three methods. The nearest neighbour approach gives rise to clearly visible steps along sloping edges. The other two interpolation methods yield results which are visually very similar.

The problem of grey value interpolation at non-integer positions also occurs in least-squares matching (see section 5.4.2.4). Bilinear interpolation is normally used as an efficient compromise between computation time and image quality. If edge detection is required in the output image, it is better to apply it prior to rectification. This avoids interpolation bias and less processing power is required to rectify edges than to rectify large pixel arrays.

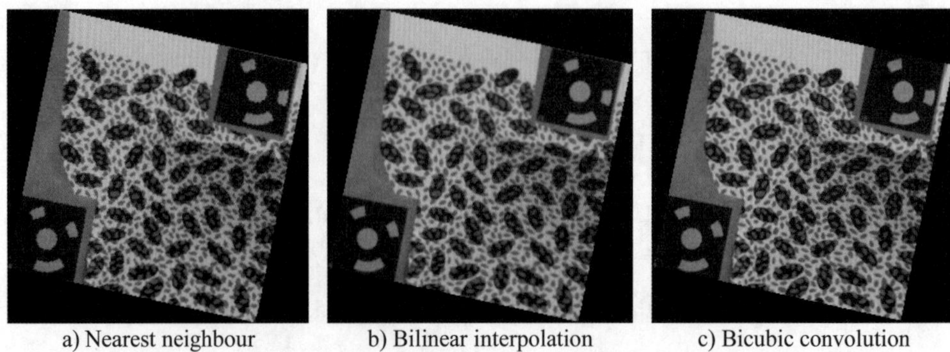

a) Nearest neighbour b) Bilinear interpolation c) Bicubic convolution

Fig. 5.53: Rectification results with different methods of grey-value interpolation

5.3.3 3D visualisation

5.3.3.1 Overview

The field of computer graphics offers a number of methods for visualizing three-dimensional objects in the form of a photo-realistic image. Visualisation of 3D objects is increasingly important, e.g. for the following applications:

- analysis of manufactured CAD models
- design studies
- quality analysis of manufactured components
- illustration of volume models, e.g. computer tomograms
- animation of movies, TV scenes and advertisements
- design of virtual-reality scenes
- visualisation of dynamic processes, e.g. simulated movements
- combination of natural and artificial objects, e.g. for the planning of new buildings in existing residential districts

In close-range photogrammetry, the main focus of 3D visualisation is the representation of photogrammetrically reconstructed, real objects. In general, 3D graphics or CAD programs are used here to provide additional elements and graphical editing of the 3D data.

The mathematical and, in particular, physical imaging models central to many visualisation methods also help in understanding object-based image processing methods (see section 5.5.5). These algorithms model the object surface with geometric and radiometric parameters determined from a least-squares adjustment to observed image coordinates and grey values.

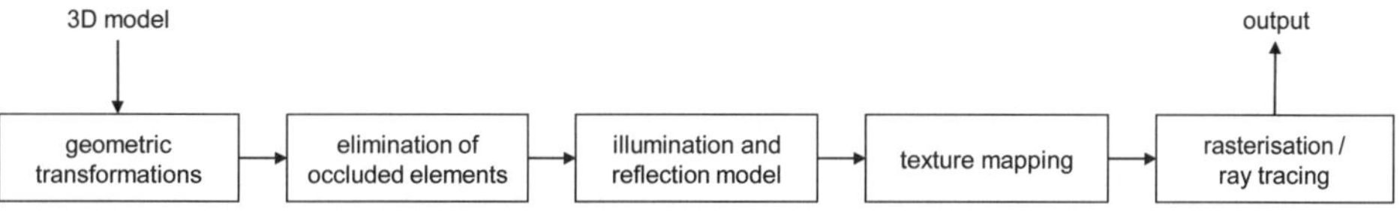

Fig. 5.54: Visualisation chain

Fig. 5.54 illustrates the typical stages in the graphical visualisation of a 3D scene (visualisation chain):

- 3D model:

  The object to be visualised is defined by a discrete 3D object surface model (e.g. by triangulation) and/or as a set of geometric elements (lines, planes, cylinders etc.). A 3D wire-frame model is not in itself sufficient because the single surface points must be linked topologically and combined to form (opaque) surface patches.

- Geometric transformations:

  After selecting the spatial observation position, object points can be projected onto a (virtual) display or projection plane using the transformations described in section 2.2.2.3. Object areas outside the visible volume can be excluded from further processing.

- Elimination of occluded elements:

  Reverse-facing object regions, which are not visible, can be eliminated by simple comparison of the observation direction with the surface normal vector (*backface culling*). Subsequently any remaining object elements must be checked for partial or complete occlusion by other objects (*hidden surface removal*) (Fig. 5.55). For rasterisation this is carried out using a Z buffer which stores the range to each object in the scene.

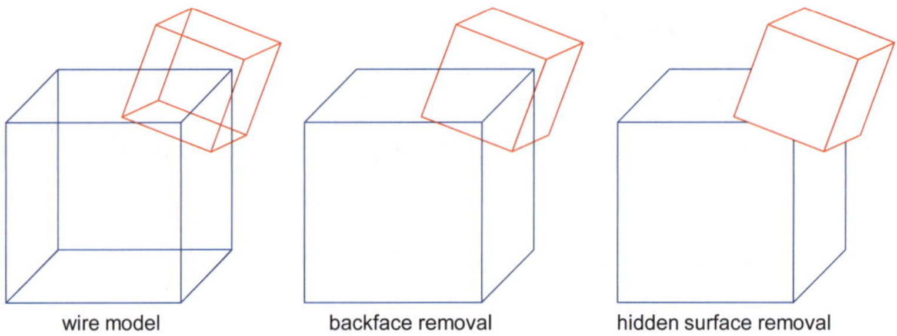

Fig. 5.55: Elimination of occluded elements

- Illumination and reflection model:

  Light sources with various characteristics (illumination power, direction, colour) can be positioned in the 3D object space. In addition, individual reflection parameters, dependent on the properties of the surface materials, can be assigned to each partial surface element.

- Texture mapping:

  In texture mapping, an artificial or natural structure is assigned to the visualised surface element. Structure or texture can be generated in different ways:

  - by assignment of a colour value (if no spectral reflection model is available),
  - by superimposition of an artificial texture (pattern),
  - by superimposition of a natural texture derived from existing photos of the object.

- Ray tracing:

  A virtual camera position (observer's position) and the required geometric projection is defined. The light ray associated with a new output image pixel is traced back into object space (*ray tracing*). If the ray strikes a surface element the corresponding pixel value is modified according to the local reflection function. If appropriate, the pixel's output intensity or colour value takes into account multiple reflections and the effects of shadowing.

Rasterisation can be used as a faster alternative. After the corner points of each polygonal surface element (esp. triangles) are projected onto the projection plane (see above), the inner part of the polygon is filled row by row in the projection plane's discrete pixel space. This is also known as scan conversion. Rasterisation is the most widely used technique for 3D visualisation. It is implemented in major industry standards, such as OpenGL and DirectX and is supported by hardware acceleration. Rasterisation lends itself well to real-time and interactive visualisation. While the basic form of rasterisation does not incorporate any means to compute the colour and intensity of the single pixels, the addition of shaders, especially programmable shaders, can perform complex lighting calculations.

5.3.3.2 Reflection and illumination

Illumination and reflection models describe the light paths which start from one or more light sources and strike one or more object surfaces before registration in the (virtual) camera. An intensity or colour value is then registered at the corresponding position in the camera image. The visualised intensity value is therefore a function of:

- spectral characteristics of light sources
- spatial position of light sources
- properties of the atmosphere (media)
- spectral characteristics of surface materials (reflection, absorption, transmission)
- surface structure (roughness)
- location and orientation of surfaces
- optical and radiometric characteristics of the camera (sensor)
- location and orientation of the camera (observer)

The extent to which these properties are applied depends on the complexity of the model. Most cases are restricted to simple models which use only one light source, do not permit multiple reflections off surfaces and impose only limited properties of materials. Here it is more important for the observer to visualise a three-dimensional situation rather than view a photo-realistic image.

Reflection types

The *ambient* reflection model is based on the idea that a certain proportion k_a of the incident light I_i reflects uniformly in all directions. The reflected radiation I_r is given by:

$$I_r(\lambda) = k_a \cdot I_i(\lambda) \quad (5.42)$$
$$\text{where } 0 \le k_a \le 1$$

This model does not consider the spatial orientation of the surface. The factor k_a results in a uniform intensity variation of the whole scene.

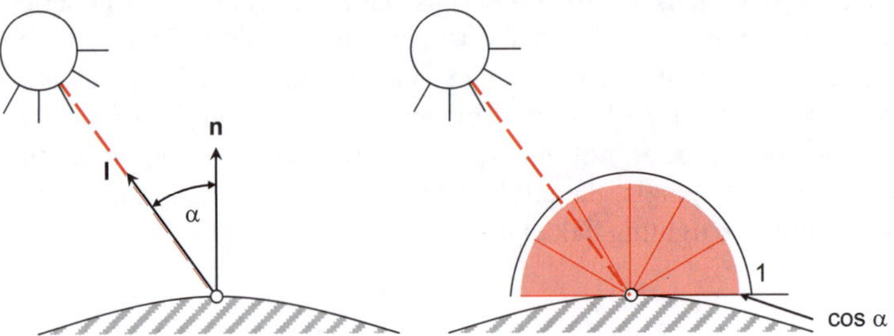

Fig. 5.56: Diffuse reflection

Diffuse reflection (Fig. 5.56) is based on Lambert's law. The intensity of the reflected light reduces as a function of the cosine of the angle α between the surface normal **n** and the direction to the light source **l** and is expressed by the scalar triple product of the two vectors:

$$I_r(\lambda) = k_d(\lambda) \cdot \max((\mathbf{n} \cdot \mathbf{l}), 0) \cdot I_i(\lambda) \quad (5.43)$$

The term $k_d(\lambda)$ is dependent on material and defined as a function of wavelength. It results in the perception of colour. The reflected component is independent of the observer's position.

The basic idea of *mirrored* reflection (Fig. 5.57) is the principle that angle of incidence = angle of reflection. For a perfect mirror the light is reflected in direction **r**. Consequently, in viewing direction **b** the intensity is given by:

$$I_r(\lambda) = \begin{cases} I_i(\lambda) & \text{if } \mathbf{b} \cdot \mathbf{r} = 0 \\ 0 & \text{otherwise} \end{cases} \quad (5.44)$$

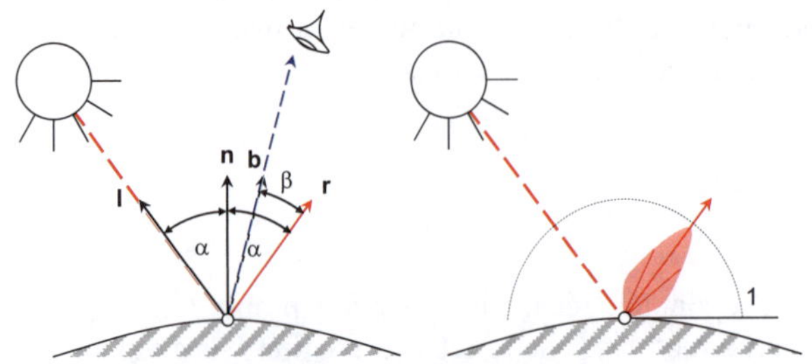

Fig. 5.57: Mirrored reflection

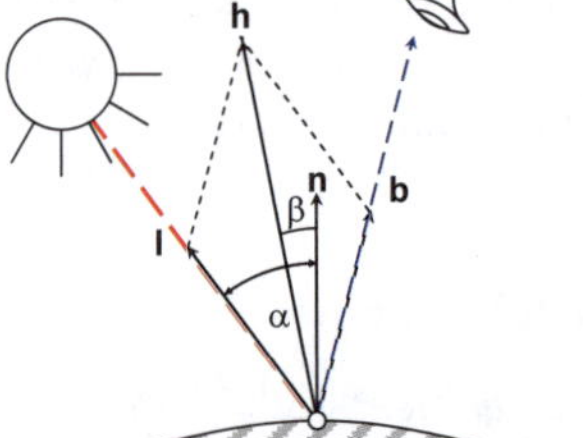

Fig. 5.58: Phong's illumination model

Since perfect mirrors rarely exist in reality, and as they would generate a signal only in one direction **b** = **r** (β = 0), some spread around this direction is permitted:

$$I_r(\lambda) = k_s(\lambda) \cdot \max((\mathbf{b} \cdot \mathbf{r}), 0)^m \cdot I_i(\lambda) \tag{5.45}$$

Here the exponent m defines the spread characteristic which is dependent on material and term $k_s(\lambda)$ specifies the spectral reflective properties. Large values of m describe polished surfaces (metals, mirrors), small values specify matt, non-metallic surfaces.

Phong's illumination model

The illumination model according to Phong (Fig. 5.58) assumes parallel rays of light between light source, surface and observer for each point on the surface. With the highlight vector **h** defined as:

$$\mathbf{h} = \frac{\mathbf{b} + \mathbf{l}}{|\mathbf{b} + \mathbf{l}|} \tag{5.46}$$

a complete model which includes all reflection types above can be defined:

$$I_r(\lambda) = I_i(\lambda) \cdot [k_a + k_d(\lambda) \cdot \max(0, (\mathbf{n} \cdot \mathbf{l})) + k_s(\alpha, \lambda) \cdot \max(0, (\mathbf{n} \cdot \mathbf{h}))^m] \tag{5.47}$$

In the special case where light source and observer have the same position then **l** = **b** = **h**. If this position is in the negative z direction[1], it follows that

$$(\mathbf{n} \cdot \mathbf{l}) = \begin{bmatrix} n_x \\ n_y \\ n_z \end{bmatrix} \cdot \begin{bmatrix} 0 & 0 & -1 \end{bmatrix} = -n_z \tag{5.48}$$

and therefore

$$I_r(\lambda) = I_i(\lambda) \cdot [k_a + k_d(\lambda) \cdot \max(0, -n_z) + k_s(\alpha, \lambda) \cdot \max(0, -n_z)^m] \tag{5.49}$$

Phong's illumination model generates acceptable results with moderate computational effort (examples in Fig. 5.59).

[1] In computer graphics the position of the viewer is often the point (0,0,–z).

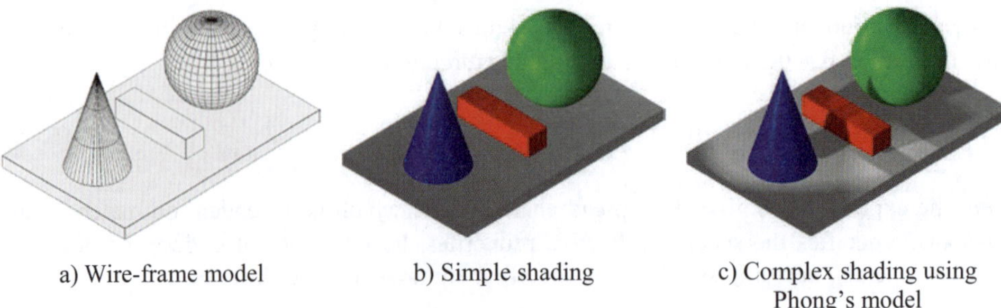

a) Wire-frame model b) Simple shading c) Complex shading using Phong's model

Fig. 5.59: Visualisation of illuminated 3D scene

Complex illumination models

In order to generate realistic scenes, Phong's model must be extended. The various approaches cannot be discussed in sufficient detail here, but they are based on the following features:

- introduction of surface roughness by means of micro-facets,
- introduction of absorption which varies with material and incident angle as a function of the refractive index n at the interface between different media (e.g. air/metal),
- introduction of a visual colour perception model (CIE model),
- introduction of a bi-directional reflection distribution function (BRDF) which is a function of the specified material for each individual wavelength λ of the spectrum,
- introduction of stochastic models defining the probability of a certain reflection.

These functions are firstly used to model the physical reflection at the surface. In addition, a global illumination model requires a description, using *ray tracing*, of the complete path taken by all light rays (emitted by light sources and/or surfaces) including transmission, multiple reflection and shadowing. Since this procedure would require an unreasonable computational effort, the process is instigated in a similar way to indirect rectification (see section 5.3.1) where the algorithm starts at the output pixel position and computes the corresponding intensity by tracing backwards into object space.

The principle of calculation is as follows. The light ray defined by the image coordinates and the perspective centre strikes the first surface. This ray is divided into a transmitted and a reflected ray and both are recursively traced until:

- the corresponding intensity is less than a threshold,
- a certain recursion depth (e.g. 5) is reached or
- a light source (which can also include a window) is intersected.

This procedure defines a tree structure which is processed from the leaves down to the root. At each division, the registered intensity values are accumulated and stored in the output image.

An extended model is provided by the *radiosity* approach in which every surface point acts as an emitter, either as a light source itself or by reflected radiation. The unknown emission is determined in a system of equations which must be solved for every surface point and for every wavelength. The approach requires high computational power and leads to realistic visualisations.

5.3.3.3 Texture mapping

The pattern or structure on a surface is referred to as texture. It results from a variation of small surface elements within a limited neighbourhood. Textures are physically generated by the different reflection properties and geometric characteristics of surface particles which, in the limit, can be as small as molecules. With the aid of textures, visualised objects achieve their realistic appearance.

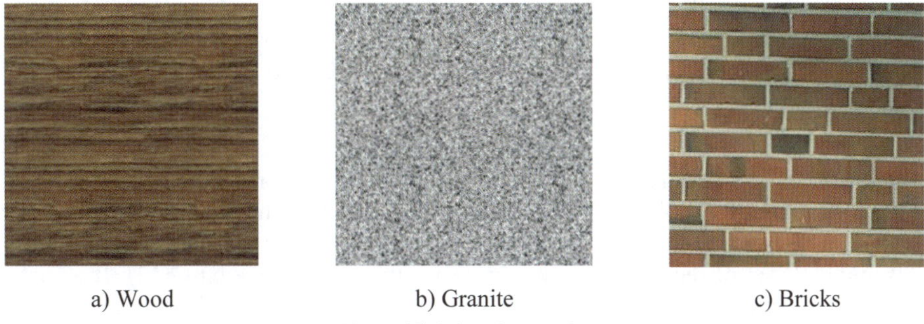

a) Wood b) Granite c) Bricks

Fig. 5.60: Artificial and natural textures

The generation of texture in a visualised scene can be generated most simply by projection of a texture image onto the surface (texture mapping). A texture image is a digital image whose pattern is superimposed on the object in the manner of a slide projection. The texture image can be a real photograph or can consist of artificial patterns (examples in Fig. 5.60).

Texture mapping includes the following principal techniques:

- 2D texture mapping onto plane object surfaces
- 3D texture mapping onto plane object surfaces
- 3D texture mapping onto arbitrarily shaped object surfaces
- 3D texture mapping onto arbitrarily shaped object surfaces in combination with an illumination model

For 2D texture mapping, a plane transformation between a region in the visualised image and the texture image is calculated. According to the principle of plane rectification (see section 4.2.6.4), corresponding image patches are defined in both images and their corner points are mapped by affine, bilinear or projective transformation. There is no direct relationship to the 3D object model, i.e. for each new visualisation the texture regions must be re-defined, if necessary by manual interaction.

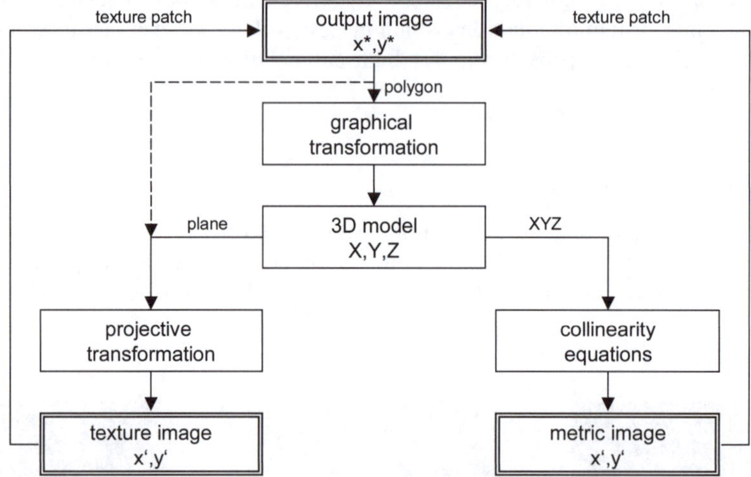

Fig. 5.61: Texture mapping

Texture mapping is more flexible if the known geometric relationship between visualisation and 3D model is utilised. Assuming an object composed of plane surface elements defined by polygons (e.g. triangles or quadrangles), each polygon can be mapped onto a corresponding area of the texture image by projective transformation (2.17). If a spatially oriented metric image provides the natural surface textures, the photogrammetric projection equations can map the individual surface elements onto the metric image texture to define a texture patch (Fig. 5.61). These texture patches can then be transferred into the output image by interpolation within the transformed polygon.

Fig. 5.62: Texture mapping with natural patterns

Fig. 5.62 shows an example of texture mapping in which terrestrial photographs are used for the façades, and aerial photographs for the ground. The trees are artificial, graphics objects.

For texture mapping of curved surfaces, the workflow of Fig. 5.61 must be implemented pixel by pixel in the output image. By incorporating a (global) illumination model it is possible to generate a photo-realistic scene consisting of lights, shadows and natural surface structures.

5.4 Digital processing of single images

This section deals with methods for locating objects in single digital images. It distinguishes between algorithms for the determination of single point features (pattern centres) and those for the detection of lines and edges. The common aim of these methods is the accurate and reliable measurement, to sub-pixel resolution, of image coordinates for use in analytical object reconstruction. Stereo and multi-image processing methods are discussed in section 5.5.

5.4.1 Approximate values

5.4.1.1 Possibilities

The image processing methods discussed here require initial approximations for the image position which is to be accurately determined. These approximate values can be found in different ways:

- by pre-determined (calculated) image coordinates, e.g. from (approximately) known object coordinates and known image orientation parameters
- by manually setting an on-screen cursor
- by pattern recognition (segmentation) during image pre-processing, e.g. by searching for certain grey value patterns in the image
- using interest operators which detect regions of significant image structure (see section 5.5.2.1)

In close-range photogrammetry, and especially in the industrial field, simple image structures can often be engineered through specific targeting and illumination techniques so that, for example, only (bright) object points on a homogeneous (dark) background exist in the image (examples in Fig. 5.2, Fig. 7.22). In this case, the generation of approximate values reduces to the location of simple image patterns and can, in many cases, be fully automated (see section 5.4.1.2).

The use of coded targets in industrial applications has been particularly successful. These encode a point identification which can be detected and decoded automatically (see section 3.6.1.5).

5.4.1.2 Segmentation of point features

The measurement (segmentation) of bright targets is an important special case in practical photogrammetric image acquisition. If no information about the position of target points is available, the image must be searched for potential candidates. Since measurement tasks in practice involve different image scales, perspective distortion, extraneous lighting, occlusions etc., the following hierarchical process of point segmentation has proven effective:

1. adaptive binarisation by thresholding (see section 5.2.1.4)
2. detection of connected image regions exceeding a threshold (connected components)
3. analysis of detected regions with respect to size (number of pixels) and shape
4. storage of image positions which meet appropriate conditions of size and shape

Thresholding is relatively simple for artificial targets. If retro-reflective targets with flash illumination are used, or LED targets, these generate significant peaks in the upper region of the grey value histogram (see Fig. 5.16). For non-reflective targets (e.g. printed on paper) the contrast against the image background is often weaker. In some circumstances, thresholding cannot then be performed globally but must be adapted to different parts of the image.

Connected regions are detected by a neighbourhood or connectivity analysis. Using the sequential process outlined in Fig. 5.63, the left and the upper three neighbouring pixels are analysed at the current pixel position. If the grey value of one of these neighbours exceeds the threshold, the neighbour belongs to a region already detected. Otherwise a new region is created. Pixels in connected regions are marked in list or label images. V-shaped objects which are temporarily assigned to different regions are recombined by a contact analysis.

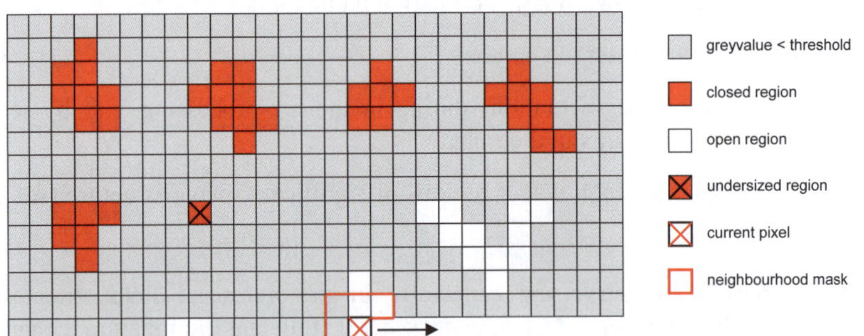

Fig. 5.63: Sequential connectivity analysis (according to Maas 1992)

In a non-sequential connectivity analysis, adjacent pixels are traced recursively in all directions until one or more of the following termination criteria are met:

- the corresponding grey value is less than the threshold
- the maximum number of pixels permitted in a region is exceeded
- the pixel already belongs to a known region

The resulting connected regions can now be analysed with respect to their shape and size. Here the number of pixels is only a coarse indicator of suitable feature size, which is possibly known a priori. In the case of circular targets, shape parameters describing the expected elliptical shape of the feature can be derived from the region's pixel distribution. A possible test criterion is given by the following shape parameter f_E:

$$f_E = \frac{U^2}{4\pi A} \tag{5.50}$$

where

U: periphery

A: area

The value of the shape parameter is 1 for a circle and greater than 1 for all other shapes. For an ellipse a suitable threshold for f_E can be estimated by the ratio of semi-axes b/a which depends on the imaging angle α between image plane and circle plane in object space (Fig. 5.64). As imaging angles of more than 70° result in highly elongated ellipses, a threshold of 1.5 is very suitable.

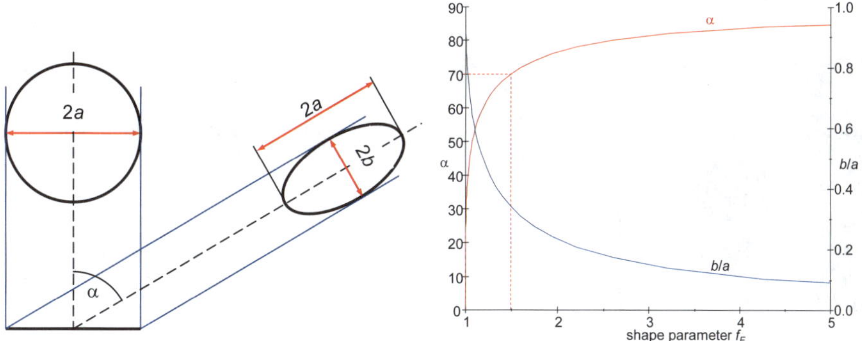

Fig. 5.64: Shape parameter of ellipse (according to Ahn 1997)

The regions which remain after binarisation and shape analysis are then finally indexed and stored sequentially.

Fully-automated segmentation is particularly successful if the following criteria are met:

- good contrast in brightness between target and background through suitable targeting and illumination techniques (retro-targets, flash)
- no extraneous reflections from the object surface (avoidance of secondary light sources)
- no occlusions due to other object parts

- good separation of individual targets in image space
- minimum target size in image space (5 pixel diameter)
- no excessive differences size between imaged targets

The problems of target imperfections and false detections are illustrated in the examples of Fig. 5.17ff, Fig. 5.34, Fig. 5.35 and Fig. 5.65a. In some cases the segmentation steps above must be extended by additional pre-processing (e.g. morphological operations) and further analysis.

5.4.2 Measurement of single point features

Here single point features are taken to mean image patterns where the centre of the pattern is the reference point. Examples are shown in Fig. 5.65.

5.4.2.1 On-screen measurement

Arbitrary image features can be measured manually by positioning a digital floating mark (cursor) on the computer screen. For this purpose the cursor should be displayed as a cross or circle. The minimum movement of the floating mark is 1 pixel in screen coordinates. The average measurement accuracy of non-targeted points (e.g. building corners) is around 0.3-0.5 pixel. If the image is zoomed the measurement accuracy can be improved to approximately 0.2 pixel.

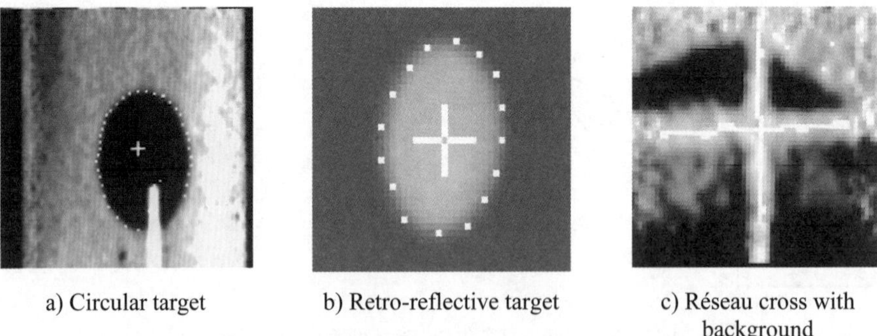

a) Circular target b) Retro-reflective target c) Réseau cross with background

Fig. 5.65: Examples of single point features with detected contour points

5.4.2.2 Centroid methods

If the feature to be measured consists of a symmetrical distribution of grey values, the local centroid can be used to determine the centre. The centroid is effectively a weighted mean of the pixel coordinates in the processing window:

$$x_M = \frac{\sum_{i=1}^{n}(x_i \cdot T \cdot g_i)}{\sum_{i=1}^{n}(T \cdot g_i)} \qquad y_M = \frac{\sum_{i=1}^{n}(y_i \cdot T \cdot g_i)}{\sum_{i=1}^{n}(T \cdot g_i)} \tag{5.51}$$

Here n is the number of processed pixels in the window, g_i is the grey value at the pixel position (x_i, y_i). The decision function T is used to decide whether a pixel is used for calculation. T can be defined by an (adaptive) grey value threshold t, for example:

$$T = \begin{cases} 0 & for \quad g < t \\ 1 & for \quad g \geq t \end{cases}$$

For features whose structure is defined by grey value edges, such as the circumference of a circle, it is reasonable to include edge information in the centroid calculation. For this purpose a weighting function based on gradients is employed:

$$x_M = \frac{\sum_{i=1}^{n} x_i \cdot grad^2(g_{x,i})}{\sum_{i=1}^{n} grad^2(g_{x,i})} \quad \text{or} \quad y_M = \frac{\sum_{i=1}^{n} y_i \cdot grad^2(g_{y,i})}{\sum_{i=1}^{n} grad^2(g_{y,i})} \tag{5.52}$$

Centroid operators are computationally fast and easy to implement. In general they also work for very small features (∅ < 5 pixel) as well as slightly defocused points. However, the result depends directly on the grey value distribution of the environment so they are only suitable for symmetrical homogenous patterns as shown in Fig. 5.65b. Defective pixels within the processing window will negatively affect the calculation of centre coordinates.

The theoretical accuracy of the centroid can be estimated by applying error propagation to (5.51):

$$\begin{aligned} \sigma_{x_M} &= \frac{1}{\sum g_i}\sqrt{\sum (x_i - x_M)^2} \cdot \sigma_g \\ \sigma_{y_M} &= \frac{1}{\sum g_i}\sqrt{\sum (y_i - y_M)^2} \cdot \sigma_g \end{aligned} \tag{5.53}$$

The standard deviation of the centroid is clearly a linear function of the grey value noise σ_g, and the distance of a pixel from the centre. It is therefore dependent on the size of the feature.

Example 5.3:

A targeted point with the parameters

point diameter:	6 pixel
operator window size:	13 x 13 pixel
target grey value:	200
background grey value:	20
grey-value noise:	0.5 grey level

results in a theoretical standard deviation of the centroid of $\sigma x_M = \sigma y_M = 0.003$ pixel.

In practice, centroid operators can reach an accuracy 0.03–0.05 pixel if circular or elliptical white targets on dark backgrounds are used (see section 5.2.4.7).

5.4.2.3 Correlation methods

In image processing, correlation methods are procedures which calculate a similarity measure between a reference pattern $f(x,y)$ and a target image patch extracted from a larger search area within the acquired image $g(x,y)$. The position of best agreement is assumed to be the location of the reference pattern in the image.

A common similarity value is the normalised cross-correlation coefficient. It is based on the following covariance and standard deviations (see section 2.4.3.3):

$$\rho_{fg} = \frac{\sigma_{fg}}{\sigma_f \cdot \sigma_g} \qquad \text{: correlation coefficient} \tag{5.54}$$

where

$$\sigma_{fg} = \frac{\sum\left[(f_i - \bar{f})(g_i - \bar{g})\right]}{n}$$

$$\sigma_f = \sqrt{\frac{\sum (f_i - \bar{f})^2}{n}} \qquad \sigma_g = \sqrt{\frac{\sum (g_i - \bar{g})^2}{n}}$$

and

$\bar{f}, \bar{g}$: arithmetic mean of grey values

n : number of pixels in the reference pattern

For pattern recognition, the reference pattern is successively shifted across a window of the search image according to Fig. 5.31, with the correlation coefficient calculated at each position.

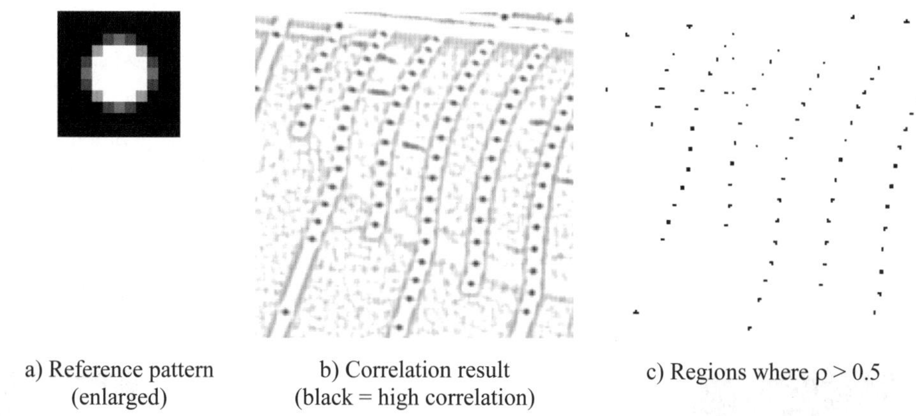

a) Reference pattern (enlarged) b) Correlation result (black = high correlation) c) Regions where $\rho > 0.5$

Fig. 5.66: Cross correlation (window of search image of Fig. 5.2a)

Fig. 5.66 shows the reference image and correlation result of the search for circular targets in Fig. 5.2. Dark spots in the correlation image indicate high correlation results. As expected, correlation maxima occur at bright targets, but medium correlation values are also caused by background noise and edges.

To identify pattern positions in the search image, all x,y positions with correlation coefficient ρ greater than a threshold t are stored. The choice of a suitable threshold t depends on the image content. For correlating stereo images of similar appearance, the threshold t can be derived from the auto-correlation function. A suitable threshold lies in the range $t = 0.5$ to 0.7. Where synthetic patterns (templates) are used for image correlation, optimal correlation values may be lower if the corresponding patch in the search image deviates from the reference image in terms of background or other disturbances.

In the regions where correlation maxima are detected, a further interpolating function applied to the neighbouring correlation values can determine the feature position to sub-pixel coordinates (see section 5.2.4.7).

The calculation process can be accelerated by prior calculation of the values which are constant in the reference image (σ_f in eqn. 5.54) and by reducing the image resolution. The pattern matrix can then be shifted across the image in larger steps but this effectively leads to a loss of information. However, a hierarchical calculation based on image pyramids (see section 5.1.3.1) can be performed in which the search results of one stage are used as prior knowledge for the next higher resolution stage.

Cross correlation is a robust method, independent of contrast but requiring a high computational effort. Target patterns can have an almost arbitrary structure. However, differences in scale and rotation, or any other distortions between reference image and target image, are not readily modelled and lead directly to a reduction in similarity value. An image measuring accuracy of about 0.1 pixel can be achieved.

5.4.2.4 Least-squares matching

Principle

The method of least-squares matching (LSM) employs an iterative geometric and radiometric transformation between reference image and search image in order to minimise the least-squares sum of grey value differences between both images. The reference image can be a window in a real image which must be matched in a corresponding image (e.g. stereo partner). For a known grey value structure, the reference image can be generated synthetically and used as a template for all similar points in the search image (*least squares template matching*).

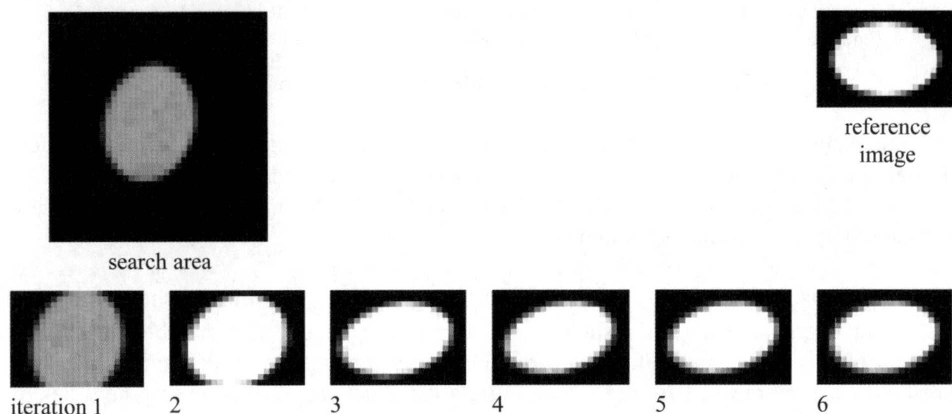

Fig. 5.67: Iterative transformation of an image patch

The geometric fit assumes that both image patches correspond to a plane area of the object. The mapping between two central perspective images can then be described by the projective transformation (2.17). For sufficiently small image patches where the 3D surface giving rise to the imaged area can be assumed to be planar, the 8-parameter projective transformation can be replaced by a 6-parameter affine transformation (2.8). The six parameters are estimated by least-squares adjustment using the grey values of both image patches as observations. The radiometric fit is performed by a linear grey value transformation with two parameters. Fig. 5.67 shows the iterative geometric and radiometric transformation of a reference pattern to the target pattern, which has been extracted from a larger search area.

The formulation as a least-squares problem has the following implications:

- An optimal solution is obtained if the mathematical model is a reasonably good description of the optical imaging process.
- The mathematical model can be extended if additional information or conditions are available, e.g. geometric constraints between images or on the object.
- The approach can be adapted to simultaneous point matching in an unlimited number of images.

- The observation equations are non-linear and must therefore be linearised at given initial values.
- The adjustment is normally highly redundant because all grey values in the image patch are used as observations to solve for only eight unknowns.
- Accuracy estimates of the unknowns can be derived from the covariance matrix. Internal quality measures are therefore available which can be used for blunder (gross error) detection (see section 2.4.4), quality analysis and post-processing.

Least-squares matching was developed in the mid-eighties for digital stereo image analysis and is now established as a universal method of image analysis. In two-dimensional image processing it can, in addition to single point measurement, be applied to edge extraction and line following. For three-dimensional object reconstruction it can be configured as a spatial intersection or bundle triangulation, and also integrated into the determination of object surface models and geometric elements.

Mathematical model

Given two image patches $f(x,y)$ and $g(x,y)$, identical apart from a noise component $e(x,y)$[1]:

$$f(x,y) - e(x,y) = g(x,y) \tag{5.55}$$

For a radiometric and geometric fit, every grey value at position (x,y) in the reference image f_i is expressed as the corresponding radiometrically and geometrically transformed grey value g_i at position (x',y') in the search image as follows:

$$\begin{aligned} f_i(x,y) - e_i(x,y) &= r_0 + r_1 \cdot g_i(x',y') && i = 1,\ldots,n \\ x' &= a_0 + a_1 x + a_2 y && n = p \cdot q \quad \text{(window size)} \\ y' &= b_0 + b_1 x + b_2 y && n \geq 8 \end{aligned} \tag{5.56}$$

Both translation parameters a_0 and b_0 are of major importance as they define the relative shift between reference image and search image. Coordinates x',y' are non-integer values and so the corresponding grey values must be appropriately interpolated, e.g. using bilinear interpolation (see section 5.3.2).

The observation equation (5.56) must be linearised since the image function $g(x',y')$ is non-linear. In summary, the linearised correction equations are as follows (ignoring index i):

$$\begin{aligned} f(x,y) - e(x,y) = g^{\circ}(x,y) &+ g_x da_0 + g_x x da_1 + g_x y da_2 + \\ &+ g_y db_0 + g_y x db_1 + g_y y db_2 + dr_0 + dr_1 g^{\circ}(x,y) \end{aligned} \tag{5.57}$$

The partial differentials are given by the grey value gradients (see section 5.2.4.1):

[1] The notation is adapted from Grün (1996).

$$g_x = \frac{\partial g^o(x,y)}{\partial x} \qquad g_y = \frac{\partial g^o(x,y)}{\partial y} \tag{5.58}$$

It is convenient and sufficient for most purposes to set initial parameter values as follows:

$$\begin{aligned} a_0^o &= a_2^o = b_0^o = b_1^o = r_0^o = 0 \\ a_1^o &= b_2^o = r_1^o = 1 \end{aligned} \tag{5.59}$$

If the transformation parameters are written as the vector of unknowns $\hat{\mathbf{x}}$, the partial derivatives as the design matrix **A** and the grey-value differences between reference image and search image as the vector of observations **l**, then the linearised correction equation are given by:

$$\underset{n,1}{\mathbf{l}} + \underset{n,1}{\mathbf{v}} = \underset{n,u}{\mathbf{A}} \cdot \underset{u,1}{\hat{\mathbf{x}}} \tag{5.60}$$

where

$$\underset{u,1}{\hat{\mathbf{x}}^T} = [da_0, da_1, da_2, db_0, db_1, db_2, r_0, r_1]$$

$$\hat{\mathbf{x}} = (\mathbf{A}^T\mathbf{P}\mathbf{A})^{-1} \cdot (\mathbf{A}^T\mathbf{P}\mathbf{l})$$

$$\hat{s}_0 = \sqrt{\frac{\mathbf{v}^T\mathbf{P}\mathbf{v}}{n-u}}$$

n: number of observations = number of pixels in window
u: number of unknowns (8)

It is usual to give all observations equal weight by setting **P** = **I**. The adjustment equations must to solved iteratively. In every iteration the unknowns are corrected. This leads to new grey-value differences between search image and transformed (rectified) reference image, until the least-squares sum of the corrections is less than a threshold.

During computation the estimated parameters should be tested for significance. Depending on image content, i.e. the similarity between reference and search image and the accuracy of initial values, the chosen transformation model (5.56) may have to be simplified or extended in successive iterations. This effect can be demonstrated by the least-squares matching of an elliptical pattern to a circular target in the search image. For this purpose the affine transformation is over-parametrised because rotation and scaling can either be modelled by a shear angle β and different scales in x and y (Fig. 5.68 left), or equivalently by a global rotation α and a scale factor (Fig. 5.68 right). In this case a 5-parameter transformation without a parameter for shear should be used. In addition it is useful to compute the geometric parameters first, with radiometric coefficients included in the final iterations.

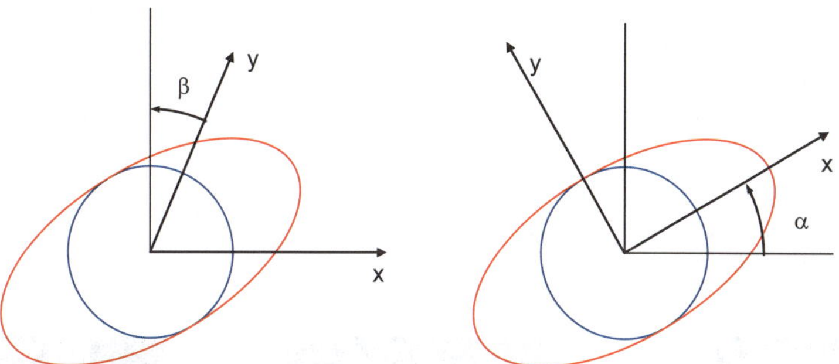

Fig. 5.68: Transformation of a circle into a rotated ellipse

Quality of least-squares matching

The adjustment equations for least-squares matching are usually highly redundant. For example, a window size of 21x21 pixels generates $n = 441$ observations for only $u = 8$ unknowns.

Grey-level gradients are used in the linearised correction equations (5.57), and a solution exists only if enough image structures (edges) are available in the matched windows. For homogeneous image patches the normal system of equations is singular.

The approach requires approximate initial values, especially for the shift coefficients a_0 and b_0. As a rule of thumb, the approximate window position should not be displaced more than half the window size from the desired point. Approximate values can be derived from a previous segmentation process or from known object geometry and orientation data.

After solving the system of adjustment equations (5.60), any residuals describe the remaining grey-value differences between reference image and adjusted search image which are not described by the mathematical model.

$$v_i = f_i(x, y) - \hat{g}_i(x, y) \tag{5.61}$$

They are a measure of the noise level in the image as well as the quality of the mathematical model.

The calculated cofactor matrix $\mathbf{Q}_{ll}$ can be used to judge the quality of parameter estimation. Similarly to (2.188), the a posteriori standard deviation of estimated parameter j is given by:

$$\hat{s}_j = \hat{s}_0 \cdot \sqrt{q_{jj}} \tag{5.62}$$

where

q_{jj}: diagonal elements of $\mathbf{Q}_{ll}$

The standard deviations of parameters a_0 and b_0 can reach high accuracies of the order of 0.01–0.04 pixel if there is good similarity between reference and search image. However, the standard deviation is only an analytical error estimate. The example in Fig. 5.69 shows the result of a least-squares matching with 5 geometric and 2 radiometric parameters. Standard deviations of shift parameters for the good-quality search image are less than 0.02 pixel. For low quality images, standard deviations can still be of the order of 0.08 pixel, although the centre coordinates are displaced by 0.29 pixel and –0.11 pixel with respect to the non-defective optimum point.

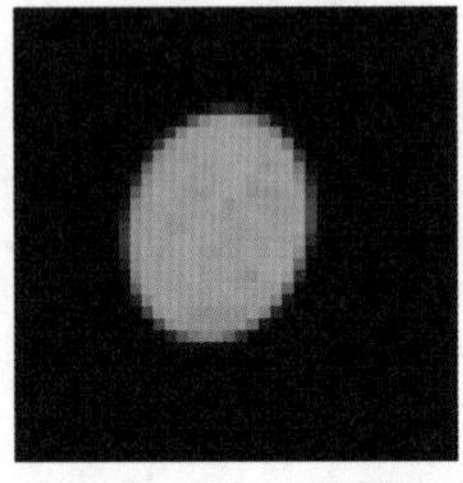

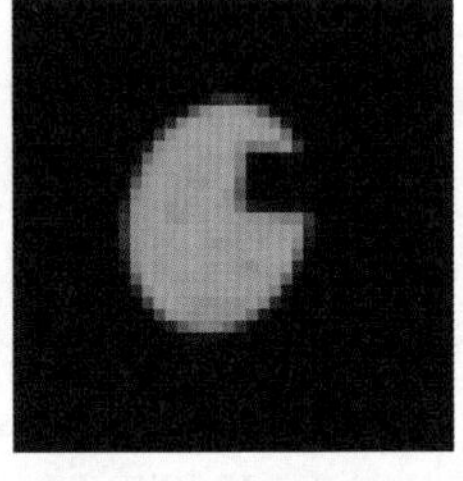

a) Reference image (template)

b) High quality search image
$a_0 = 17.099 \pm 0.020$
$b_0 = 17.829 \pm 0.012$

c) Defective search image
$a_0 = 16.806 \pm 0.088$
$b_0 = 17.940 \pm 0.069$

Fig. 5.69: Least-squares matching (5+2 parameters): shift parameters with standard deviations

Using blunder detection, as explained in section 2.4.4, it is possible to eliminate a limited number of gross errors in the observations. Here blunders refer to pixels whose grey values are caused, for example, by occlusions or other artefacts, and are not described by the functional model. If the adjustment additionally tests for the significance of parameters, and if non-significant parameters are eliminated automatically, then least-squares matching becomes an adaptive, self-controlled method of point measurement.

The least-squares matching algorithm described above can be extended by the integration of simultaneous processing of multiple images (multi-image matching) and by the introduction of geometric constraints (epipolar geometry), as described in section 5.5.4.2.

5.4.2.5 Structural measuring methods

Structural measuring methods extract edges in the image which are relevant to the object, and reconstruct its geometry with the aid of mathematically defined shapes.

Circular and elliptical features

In general, circular objects are imaged as ellipses. To a first approximation the ellipse centre corresponds to the projected circle centre (see section 3.6.1.2). For this purpose the star operator or the Zhou operator have been proven effective. The centre of the ellipse is determined in several steps:

1. Definition of a search window based on a given approximate position
2. Extraction of edge points (ellipse boundary)
3. Calculation of ellipse parameters

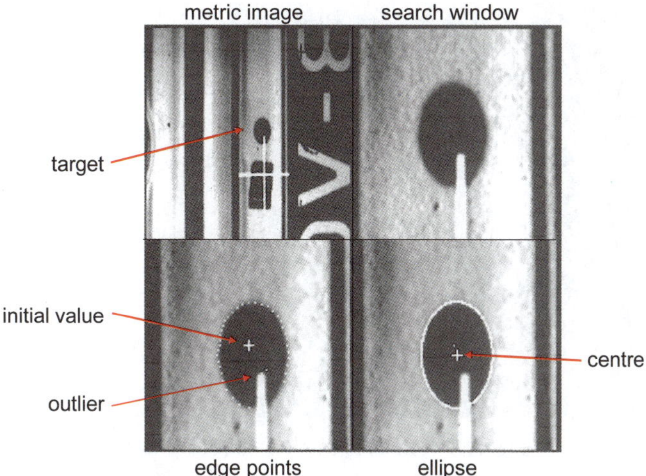

Fig. 5.70: Procedure for ellipse measurement

The *star operator* determines points on the ellipse by edge detection (e.g. according to section 5.2.4) along search lines radiating from an approximation position inside the ellipse (Fig. 5.70, Fig. 5.71). These search lines intersect the ellipse at favourable angles and grey values must be appropriately interpolated along the lines.

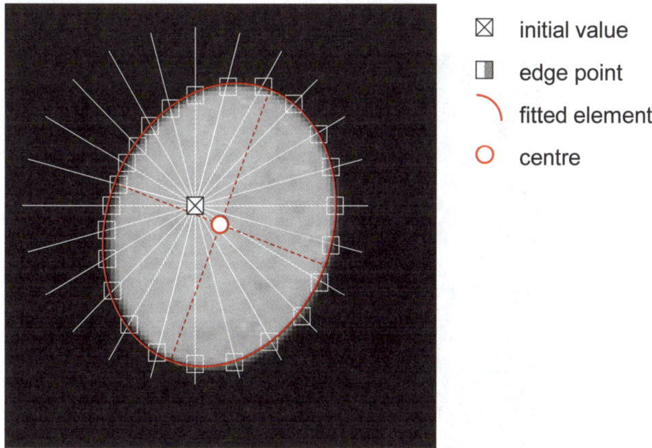

Fig. 5.71: Principle of the star operator

The coordinates of the extracted edge points are subsequently used as observations for calculating the parameters of a best-fit ellipse (see section 2.3.1.3). Individual false edge points can be eliminated by blunder detection. Of the five ellipse parameters the centre coordinates can be directly used. However, they depend on the initial approximate centre position and should therefore be used as improved starting values for a further iteration until changes in the centre coordinates are below a threshold value.

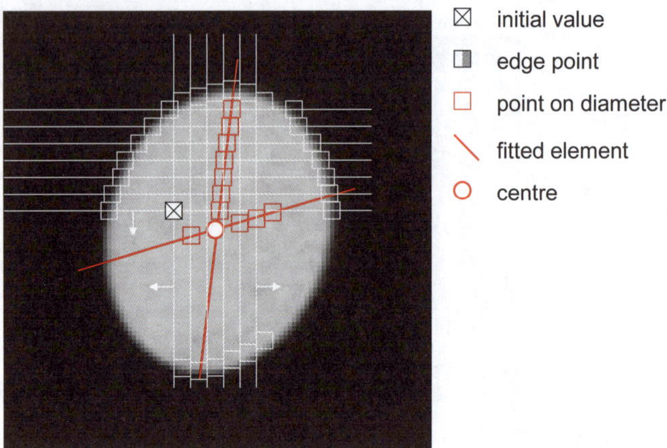

Fig. 5.72: Principle of Zhou ellipse operator

The Zhou operator makes use of conjugate ellipse diameters (see section 2.3.1.3). Ellipse diameters are straight lines connecting mid points of parallel chords. The intersection point of conjugate diameters corresponds to the ellipse centre. In the image the ellipse edge points are determined within rows and columns and the corresponding middle point is calculated. Two regression lines are determined for the middle points belonging to the diameters whose intersection point then corresponds to the desired ellipse centre (Fig. 5.72).

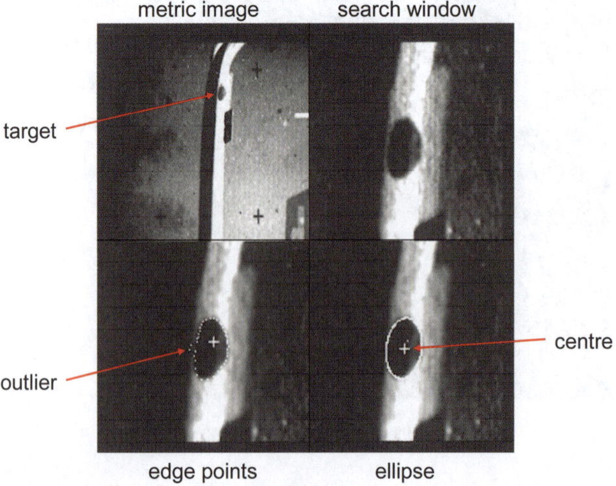

Fig. 5.73: Example of ellipse measurement on curved surface

This method of ellipse measurement largely requires good quality targets. A small number of defective edge points can be handled by robust blunder detection within the ellipse or line adjustment. There are limits to the measurement of damaged or occluded targets (e.g. Fig. 3.145) as the number of false edge points increases. Circular targets cannot be correctly measured by ellipse approximation if they are located on non-planar surfaces (Fig. 5.73).

Cross-shaped features

Cross-shaped features (e.g. réseau crosses, object corners) can be measured in a similar way to ellipse measurement, by edge detection. Here the objective is also to extract the relevant grey value edges which define the cross.

For upright crosses (see example in Fig. 5.65c) the centre points of the bars are extracted along rows and columns, with the central region ignored. As with the Zhou operator, a regression line is fitted to the centre points of each bar. The intersection point of both lines defines the centre of the cross.

Arbitrarily rotated crosses can only be reliably measured by the above algorithm if the extracted line points are analysed in order to correctly assign them to the appropriate cross bar. A rotation-invariant method is provided by the ring operator which extracts edge points within concentric rings around the approximate initial centre point (Fig. 5.74). Extracted edge points within a ring are initially defined by polar coordinates (radius and arc length). These are easily transformed into Cartesian coordinates to which regression lines can be fitted.

Measurements based on edge detection have the advantage that feature and background can be separated relatively easily and only those image points used which describe the actual shape of the feature. As an example, imaged réseau crosses are often disturbed by background structures. False edge points can be detected and eliminated a priori by a classification of image gradients (sign and magnitude). Any remaining blunders can be identified as outliers on the regression line.

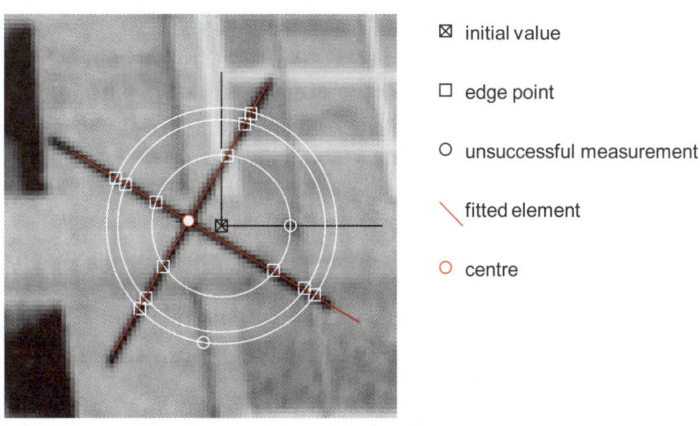

Fig. 5.74: Principle of the ring operator

5.4.2.6 Accuracy issues

The location accuracy of single point features can be assessed as follows:

- Comparison with nominal coordinates of synthetic reference features:

  Single point features with a regular geometric structure, variable centre coordinates, shape parameters and contrast, can be generated synthetically. These can be analysed after known arbitrary sub-pixel displacements are applied, for example, by geometric transformations and appropriate grey value interpolations. Accuracy analysis using synthetic features is particularly useful for testing individual effects and parameters of an algorithm.

- Analytical error analysis of adjustment:

  If the centre of a feature is calculated by adjustment (e.g. by least-squares matching or a best-fit ellipse) standard deviations and reliability figures can be computed for the centre coordinates. However, these only indicate the precision to which the chosen mathematical model fits the observations supplied, such as grey values or edge points (see section 5.4.2.4).

- Analysis of bundle adjustment:

  Multi-image bundle adjustment can offer the possibility of a more rigorous accuracy assessment which takes into account all influences in the measurement process (image acquisition system, digitisation, point detection operator, mathematical model). This can be achieved, for example, by calibration against a test field of high-accuracy reference points. The image residuals remaining after the adjustment can be interpreted as a quality measure for the target point accuracy, although these are also potentially influenced by systematic effects that are not accounted for in the functional model. Ultimately however, only independently measured reference points, distances or surface shape provide a strictly rigorous method of analysing point measurement quality (see sections 4.4.5 and 7.2).

Many comparative investigations of different measuring algorithms, applied to synthetic and real test images, have shown the promising potential of digital point measurement.

Measurement resolution is limited to about 1/1000–2/1000 pixel if adjustment methods (least-squares matching, ellipse operators) are applied to the localisation of appropriately sized and undamaged synthetic features. This result corresponds to the theoretical positioning accuracy of edge-based operators (see section 5.2.4.7).

For real imagery with well exposed and bright elliptical targets, accuracies of 2/100–5/100 pixel can be achieved if least-squares operators or adaptive centroid operators are applied to multi-image configurations. Ellipse measurement based on edge detection tends to be slightly more accurate than least-squares matching and centroid methods if there is increased image noise or distinct blunders are present.

A significant factor in determining point accuracy is the size (diameter) of imaged points. The optimum target size ranges between about 5 and 15 pixels in diameter. Smaller points do not provide enough object information, which limits the localisation accuracy of matching procedures or edge-oriented operators. Larger point diameters result in larger numbers of observations but the number of significant edge points increases only linearly whilst the number of pixels in the window increases quadratically. In addition, disturbances in the image are more likely with the result that the centre of the ellipse would be displaced with respect to the actual centre of the target circle as it increases in size (see section 3.6.1.2).

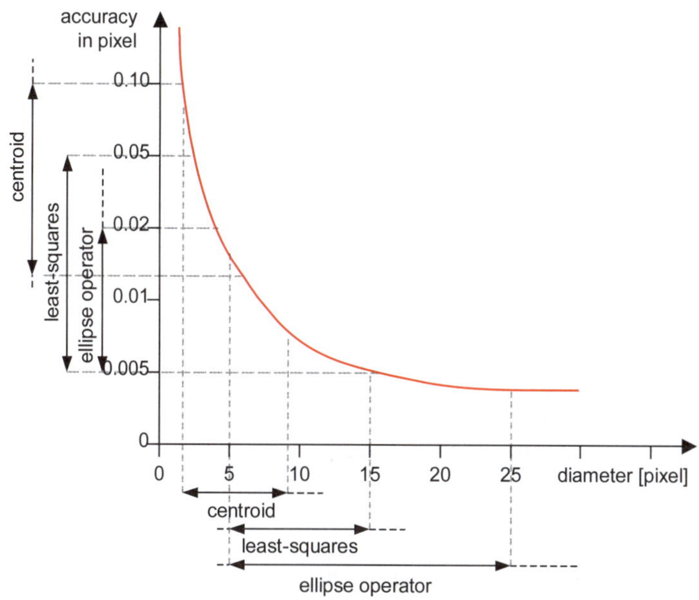

Fig. 5.75: Practical use and accuracy potential of different point measurement operators as a function of target diameter (y-axis not to scale)

Fig. 5.75 shows the empirical relationship between target size and achievable point accuracy (illustration not scaled), together with typical application areas for different operators. As target size increases the function converges to an accuracy of about 0.005 pixel.

5.4.3 Contour following

Raster-to-vector conversion not only leads to a significant reduction in data but also enables the description of symbolic and topological information. One method of conversion is contour following where the objective is to create vector data structures by combing connected edge or line points.

The initial result of contour following is a series of point coordinates. Integer contour coordinates can be stored, for example, using N8 connectivity (section 5.1.2) in which the direction code of subsequent pixels is recorded. Floating point coordinates with sub-pixel resolution can be stored as point lists. Start and end points, crossings and divisions can be described by additional attributes within the point list.

There are different approaches to contour following which differ mainly in terms of the required a priori knowledge about the contour. If information is available about start and end points, or the approximate contour path, the extraction of contour points by edge detection can be simplified and accelerated. Contour following without any a priori knowledge requires processing of the complete image, the analysis of connected pixels and the elimination of false image contours, for example caused by noise.

Two algorithms selected from a large variety of published methods are presented here for illustration.

5.4.3.1 Profile-driven contour following

Assume the starting point and direction of a single contour are known, e.g. by interactive input or by calculated image coordinates. Contour points are extracted within grey value profiles perpendicular to the current contour direction. Within these cross profiles, line or edge points are extracted to sub-pixel resolution according to the methods described in section 5.2.4 (Fig. 5.76).

From a minimum of two contour points extracted in this way, the approximate position of the next cross profile p_i is estimated. The sampling width s along the contour can be adjusted by evaluating the lateral displacement d of the next edge point. If the displacement exceeds a certain threshold, the width is reduced and a new cross profile p' is measured. This approach leads automatically to a denser point distribution along contours with higher curvatures.

Geometric and radiometric variations along the contour, e.g. contrast changes due to background differences, can be determined by analysing and comparing a profile's grey values with previously measured profiles. Profile-driven contour following can therefore adaptively determine the current contour path. Fig. 5.76 illustrates the principle using the example of a form line on a car body component.

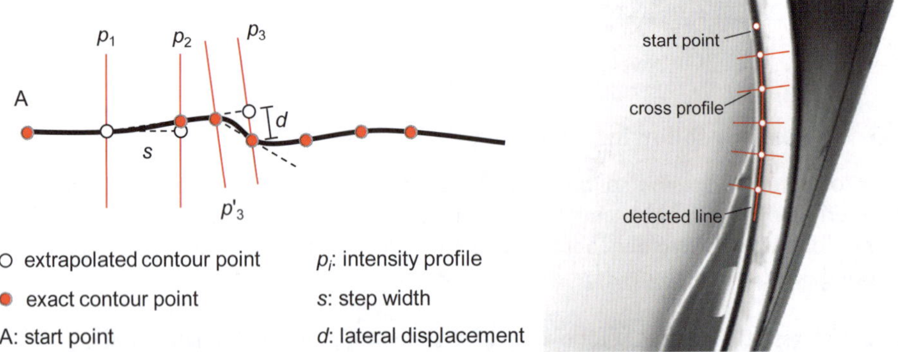

Fig. 5.76: Principle and application of profile-driven contour following

5.4.3.2 Contour following by gradient analysis

Given a gradient image (e.g. as created by the Sobel operator, section 5.2.4.1) the window at an arbitrary approximate position is analysed. If the magnitude of the gradient exceeds a threshold, the next window position along the current contour path is searched. This process is repeated until a termination criterion determines that there is no suitable further window to be analysed. Of the numerous possible strategies for contour following, the following two methods are outlined:

Analysis of gradient direction

Edge detection generates a grey value image of gradient magnitudes and a corresponding image of gradient directions (see example in Fig. 5.39). The following strategy can be used for contour following:

1.	if $grad(x,y) > t$, then	gradient magnitude > threshold
2.	mark the contour point $P(x,y)$	$P(x,y)$ is permitted only once as a contour point
3.	proceed with sample width s in gradient direction $\varphi+90°$ to $x+\Delta x, y+\Delta y$	contour must be perpendicular to the gradient direction
4.	if $grad(x,y)$ is similar to $grad(x+\Delta x, y+\Delta y)$, then set $x=x+\Delta x$ and $y=y+\Delta y$ and proceed with step 2; otherwise: reduce step width s or terminate contour	similarity measure controls integrity of contour

This algorithm follows the contour depending on the gradient direction. Its sensitivity can be controlled by the selection of a suitable similarity measure in step 4 and by geometric constraints, e.g. a minimum contour length or a maximum permitted change in direction. In step 4 topological termination criteria can also be introduced, for instance the detection of contours already measured or closed contour lines.

Analysis of gradient environment

If no information is available about gradient direction, the contour path can be followed by analysing the centroid of the gradients according to (5.52). Within the processed window, all gradients above a threshold are used for a centroid calculation. If the contour is characterised by suitable number of high gradients inside the window, the centroid is shifted in the direction of the contour.

The steps in the procedure are:

1.	calculate centroid x_s,y_s at current position x,y	using (5.52) for all pixels exceeding the threshold
2.	set gradient to 0 at contour point $P(x,y)$	$P(x,y)$ is permitted only once as a contour point and therefore does not participate in any further centroid calculation
3.	if $grad(x_s,y_s) > t$, then set $x=x_s$ and $y=y_s$ and proceed with step 2, otherwise terminate contour	the next contour point must again exceed the threshold

This algorithm is very simple and fast. It does not require any additional parameters except the gradient threshold t.

5.5 Image matching and 3D object reconstruction

5.5.1 Overview

Image matching methods are used to identify and uniquely match identical object features (points, patterns, edges) in two or more images of the object. Matching methods are also required for:

- identification of discrete (targeted) image points for 3D point measurement
- identification of homologous images features for 3D surface reconstruction
- identification and tracking of objects in image sequences

One of the earliest problems in computer vision, and still one of the most researched topics, was automatic matching of corresponding image features. Correspondence analysis is a fundamental requirement in understanding images of spatial scenes and is closely related to human visual perception. Whilst digital image matching of suitably structured object scenes (e.g. using targets or projected patterns) can surpass human performance in some areas (measuring accuracy, processing speed), the analysis of arbitrary object scenes is still an issue of intensive research.

Correspondence analysis can be classified as an ill-posed problem, i.e. it is uncertain if any solution exists which is unique and robust with respect to variations in the input data. In principle, the following problems may occur when matching arbitrary scenes:

- due to occlusion, an image point P_{ij} (point i in image j) does not have a homologous partner point P_{ik};
- due to ambiguous object structures or transparent surfaces, there are several candidates P_{ik} for image point P_{ij};
- for regions with poor texture, the solution becomes unstable or sensitive with respect to minor disturbances in the image (noise).

Solutions in practice assume the following preconditions for object and image acquisition:

- intensities in all images cover the same spectral regions
- constant illumination, atmospheric effects and media interfaces for the period of image acquisition
- stable object surface over the period of image acquisition
- macroscopically smooth object surface
- opaque object surface
- largely diffuse reflection off the surface
- known approximate values for orientation data (image overlap) and object data (geometric and radiometric parameters)

From the wide range of matching techniques, a selection of successful methods is discussed below. These are directly related to geometric surface reconstruction.

Image matching methods are based on a hierarchical strategy (Fig. 5.77). The necessary calibration and orientation data for the input images are usually calculated in advance. Preprocessing includes image enhancement (smoothing, noise reduction, contrast adjustment) and the reduction of resolution (image pyramids).

In the feature extraction step, image features such as distinct points or edges are extracted from the images independently of one another and a large percentage are assumed to be common to all images. For this purpose, not only interest operators (section 5.5.2.1) and edge operators (section 5.2.4) are used, but also segmentation methods for the coarse detection of object points (e.g. coded targets, section 3.6.1.5). In addition to the geometric parameters, the extracted features are characterised by additional attributes (e.g. topological relationships, point numbers).

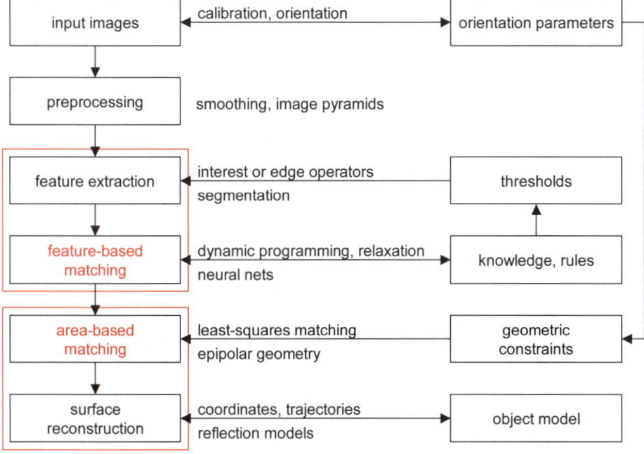

Fig. 5.77: Strategy for image matching

The subsequent step of *feature-based matching* attempts to identify as many corresponding features as possible in all images. Additional information in the form of knowledge or rules can be used here to limit the search space and to minimise mismatches. In general the matching of features is the most problematical step in correspondence analysis because the type and extent of additional knowledge may vary widely and success in matching depends directly on the shape and form of the object scene. A successful matching process results in approximate values for homologous image structures.

Area-based precise matching of original grey values then determines corresponding object elements to high accuracy. Correlation and least-squares methods are classified as area-based matching methods. In this step additional geometric information (e.g. epipolar geometry (see section 4.3.2), object constraints) can be used to improve accuracy and reliability. 3D object data can be derived from the calculated homologous image elements. Extended object-based matching methods use a geometric and a radiometric object model where the corresponding image regions are determined iteratively. The final result is a structured spatial object description in the form of coordinates, geometric elements or vector fields.

5.5.2 Feature-based matching procedures

5.5.2.1 Interest operators

Interest operators are algorithms for the extraction of distinctive image points which are potentially suitable candidates for image-to-image matching. Suitable candidates for homologous points are grey value patterns which, as far as possible, are unique within a limited region and are likely to have a similar appearance in the corresponding image. For each pixel, interest operators determine one or more parameters (interest value) which can be used for subsequent feature matching. They are mainly applied to the matching of digitised aerial images, but are also useful for determining approximate points for surface reconstruction in close-range imagery. However they are rarely applied when searching for artificial target points.

Criteria for such distinctive candidate features, and the requirements for an optimal interest operator, can be summarised as follows:

- individuality (locally unique, distinct from background)
- invariance in terms of geometric and radiometric distortions
- robustness (insensitivity to noise)
- rarity (globally unique, distinct from other candidates)
- applicability (interest values are suitable for further image analysis)

Within a local window, the following is a selection of possible criteria for determining the presence of readily identifiable structures:

- Local variance:

  The grey value variance in a window can be calculated by equation (5.5). Highly structured image patterns have high variances, homogeneous regions have zero variance. The variance does not have any geometric meaning, high numbers can also result from edges and therefore it is not suitable as an interest value.

- Auto-correlation function:

  The auto-correlation function results from calculation of the normalised cross-correlation (section 5.4.2.3) of an image patch with itself. If the function shows a sharp maximum it indicates a distinctive image structure which is not repeated locally. An example function is the Harris operator.

- Grey-value surface curvature:

  If the grey values in a local region are regarded as a spatial surface (grey value mountains), distinctive points have a high local curvature in all directions, declining rapidly in the near neighbourhood. Curvature can be calculated by differential operators (section 5.2.4.2) which approximate the second derivative.

- Gradient sums:

  This operator computes the squared gradient sums in the four principal directions of a window. If the smallest of the four sums exceeds a threshold a distinct feature is indicated. This approach eliminates single edges which show little structural variation along the edge. The remaining image regions are those which have significant intensity changes in all directions. Distinctive point features are recognisable by the fact that the gradient sums are significant in all directions. Individual edges, which have little structural change along the edge direction, are therefore eliminated. Examples of methods based on gradient sums are the Moravec, SURF and SIFT operators (more on SIFT below).

- Low self-similarity:

  The Förstner operator calculates the covariance matrix of the displacement of an image window. The corresponding error ellipse becomes small and circular for image features which are distinctive in all directions (more on the Förstner operator below).

- Local grey-value comparison:

  If the number of similar grey values in a window is below a threshold value, then this indicates a distinctive point feature (SUSAN and FAST operators, more information below).

Förstner operator

The Förstner operator is based on the assumption that the region around a point $f(x,y)$ is a shifted and noisy copy of the original image signal $g(x,y)$ (see eqn. 5.55):

$$f(x,y) = g(x + x_0, y + y_0) + e(x,y) \tag{5.63}$$

Linearisation at initial values $x_0 = 0$, $y_0 = 0$ gives

$$\begin{aligned} &dg(x,y) - e(x,y) = \frac{\partial g}{\partial x} x_0 + \frac{\partial g}{\partial y} y_0 = g_x x_0 + g_y y_0 \\ &\text{where} \\ &dg(x,y) = f(x,y) - g(x,y) \end{aligned} \tag{5.64}$$

Using the unknown shift parameters x_0,y_0 and the uncorrelated, equally weighted observations (grey value differences) $dg(x,y)$, the following normal system of equations for a least-squares estimation is obtained:

$$\begin{aligned} &\mathbf{N} \cdot \mathbf{x} = \mathbf{A}^T \cdot \mathbf{A} \cdot \mathbf{x} = \mathbf{A}^T \cdot \mathbf{dg} \\ &\text{or} \\ &\begin{bmatrix} \sum g_x^2 & \sum g_x g_y \\ \sum g_y g_x & \sum g_y^2 \end{bmatrix} \cdot \begin{bmatrix} x_0 \\ y_0 \end{bmatrix} = \begin{bmatrix} g_x dg \\ g_y dg \end{bmatrix} \end{aligned} \tag{5.65}$$

The normal equation matrix **N** contains the functional model of a displacement of the image window in x and y. Its inverse can be interpreted as a variance-covariance matrix whose eigenvalues λ_1,λ_2 indicate the semi-axes of an error ellipse. Features forming well-defined points are characterised by small circular error ellipses. In contrast, elongated ellipses are obtained for edge points. Unstructured or noisy image features result in large error ellipses.

Based on the parameters

$$\begin{aligned} &w = \frac{\det(\mathbf{N})}{\operatorname{trace}(\mathbf{N})} = \frac{1}{\lambda_1 + \lambda_2} \qquad \text{: measure of ellipse size} \\ &w > 0 \\ &q = \frac{4\det(\mathbf{N})}{\operatorname{trace}(\mathbf{N})^2} = 1 - \left(\frac{\lambda_1 - \lambda_2}{\lambda_1 + \lambda_2}\right)^2 \qquad \text{: measure of roundness of ellipse} \\ &0 \le q \le 1 \end{aligned} \tag{5.66}$$

a distinct point is observed if thresholds w_{min} and q_{min} are exceeded. Suitable windows are 5 to 7 pixels in size, appropriate thresholds are in the ranges

$w_{min} = (0.5 \dots 1.5) \cdot w_{mean}$, w_{mean} = mean of w for the complete image

$q_{min} = 0.5 \dots 0.75$

Fig. 5.78b shows the result of the Förstner operator applied to a stereo image after a thinning procedure. Distinct corners are detected but a slight displacement can also be observed here. A more precise point location can be obtained if the unknowns x_0,y_0 in (5.65) are calculated. As expected, several points are detected, whilst other similar features fall below the threshold.

The Förstner operator is in some respect similar to the independently developed Harris operator. Today their key idea is sometimes referred to as the Förstner-Harris approach.

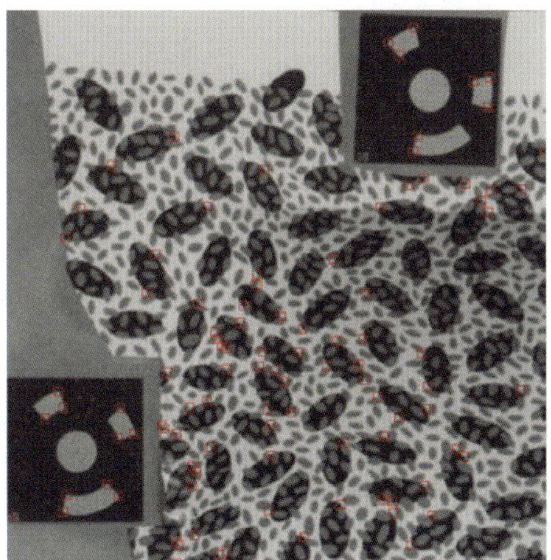
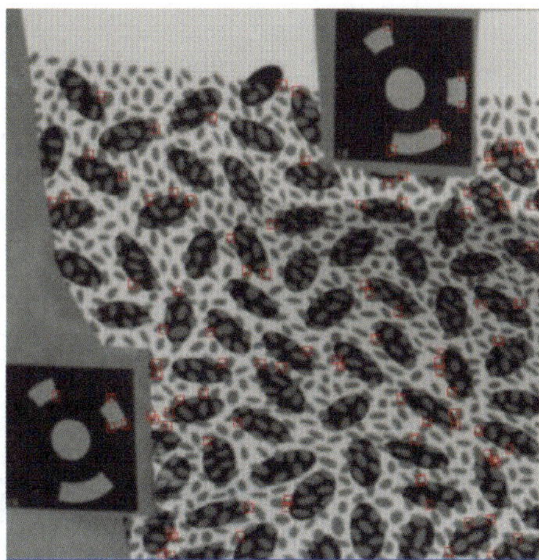

Fig. 5.78: Point detection in a stereo image with Förstner operator (filter size 5x5)

SUSAN operator

The SUSAN operator (*smallest univalue segment assimilating nucleus*) compares the intensities of the pixels in a circular window with the grey value of the central pixel, designated as the *nucleus*. A distinctive point has been found when the number of similar grey values in the window lies under a threshold value.

A decision value c is calculated for every pixel in a filter region, u,v:

$$c(u,v) = \begin{cases} 1 & \text{für} \quad |s(x+u,y+v) - s(x,y)| \\ 0 & \text{für} \quad |s(x+u,y+v) - s(x,y)| > t \end{cases} \tag{5.67}$$

where

t: threshold value for similarity of grey value to nucleus

A more stable decision function is given by:

$$c(u,v) = \exp\left[-\left(\frac{s(x+u,y+v) - s(x,y)}{t}\right)^6\right] \tag{5.68}$$

The sum of all values $c(u,v)$ in the window

$$n(x,y) = \sum c(u,v) \tag{5.69}$$

is compared with the geometry threshold T:

$$R(x,y) = \begin{cases} T - n(x,y) & \text{wenn } n(x,y) < T \\ 0 & \text{sonst} \end{cases} \tag{5.70}$$

In order to detect corner features a threshold value $T = n_{max}/2$ is set (n_{max} = number of pixels in the filter window). When less than half the grey values in the window are similar to one another, a distinctive point with the interest value R is generated at the position x,y.

Fig. 5.79 shows examples of points detected in a stereo pair by the SUSAN operator. Because the operator is affected directly by image brightness, in the area occupied by the random pattern many similar points are found in the left and right hand images but in the high-contrast pattern areas (coded target locations), very varied features are detected.

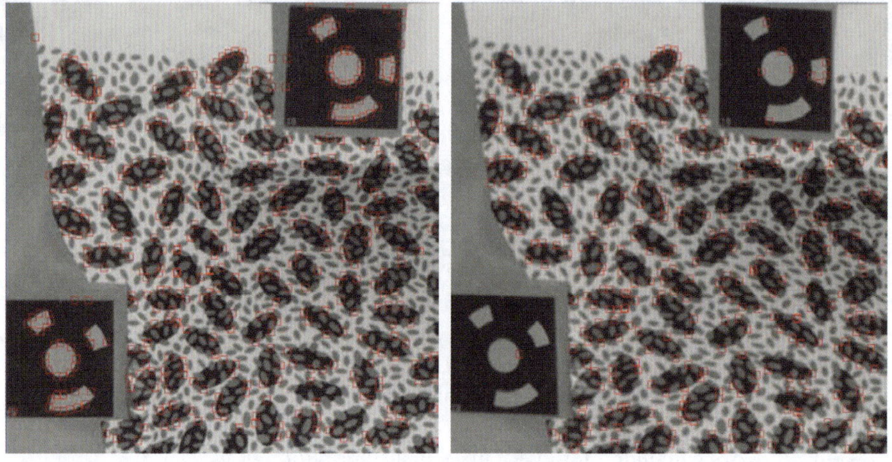

Fig. 5.79: Feature detection in stereo pair using SUSAN operator (search window 5x5)

FAST operator

Like the SUSAN operator, the FAST operator (*features from accelerated segment test*) analyses the intensities of the pixels in a circular window. A distinctive feature is assumed to be found when a number of connected pixels with similar grey values is found in a ring around the central pixel. An analysis based on a ring is largely invariant to changes in scale and rotation.

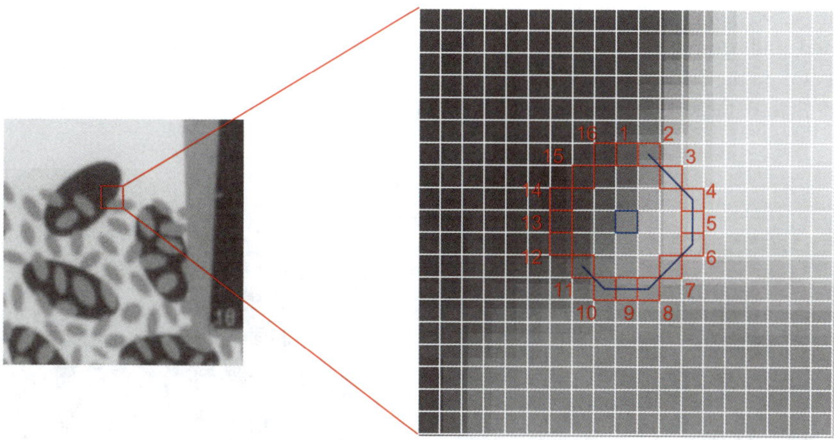

Fig. 5.80: Principle of the FAST operator

The principle behind the operator is shown in Fig. 5.80. In this example, within a ring which is 16 pixels in length, 10 neighbouring pixels lie below the threshold value. The grey values $s(u)$, $u = 1\ldots16$, are compared with the grey value of the central pixel and assigned a similarity value $c(u)$:

$$c(u) = \begin{cases} d & \text{for} \quad s(u) \le s(x,y) - t & \text{(darker)} \\ s & \text{for} \quad s(x,y) - t < s(u) < s(x,y) + t & \text{(similar)} \\ b & \text{for} \quad s(x,y) + t \le s(u) & \text{(brighter)} \end{cases} \tag{5.71}$$

where

t: threshold value for similarity of grey value to central pixel

In the simplest case a corner point is found when, for 12 neighbouring pixels, $c(u) = d$ or $c(u) = b$. For a more robust analysis of the area, the distribution of the grey values can be evaluated by sampling and different types of corner features identified.

5.5.2.2 Feature detectors

Feature detectors combine the detection of interest points (or key points) with the extraction of descriptors. Descriptors are compact and distinct representations of the area surrounding the key point. The descriptors serve as feature vectors for each interest point and can directly be used for correspondence analysis (see below). Generally any interest point detector can be combined with any feature descriptor. However, most successful approaches provide an integrated method of key point detection and descriptor extraction.

SIFT operator

The SIFT operator (*scale invariant feature transform*) initially performs edge detection by means of the DoG filter (section 5.2.4.3) at every level of an image pyramid. Key points are detected by identifying local extrema (maxima and minima) in the DoG filtered images (section 5.2.4.3). From the calculated gradients a histogram of the gradient directions is calculated and from this a 128-dimensional feature vector is derived. Since features of interest are related to the main gradient directions and image pyramid levels with the greatest differentiation, the operator is largely invariant to rotational and scale changes.

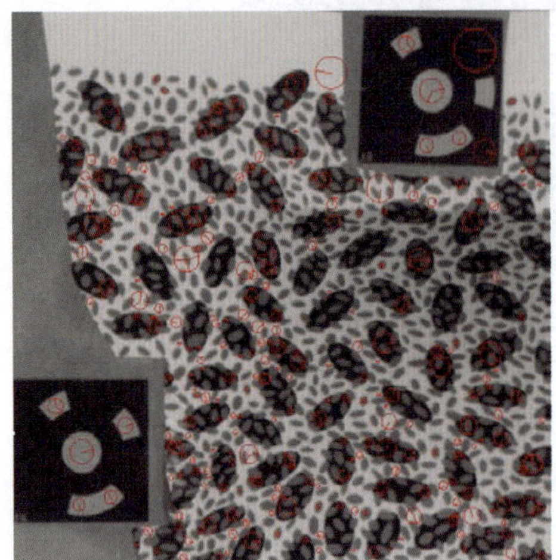
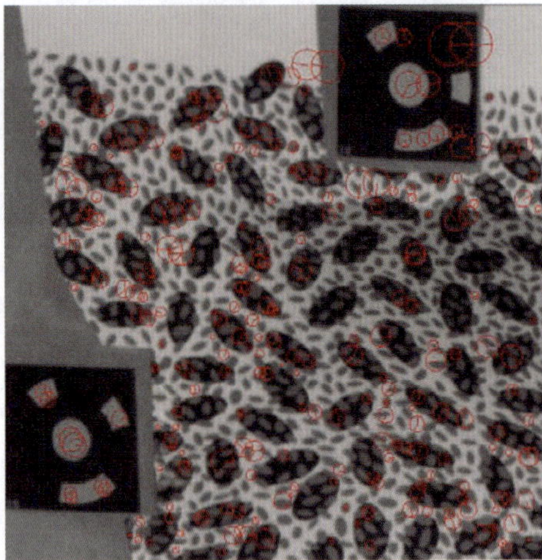

Fig. 5.81: Feature detection in stereo pair using SIFT operator. The size of the circle indicates the scale at which the feature was detected. The ticks indicate the direction of the dominant gradient.

The SIFT operator is particularly suitable for assignments between images which have significant differences with respect to scale, rotation and small differences in perspective. Typically the operator does not detect points with visual impact but features which have a similar form in neighbouring images (Fig. 5.81).

Affine invariant SIFT (ASIFT)

While the SIFT descriptor is invariant to shift, rotation and scale, there is no explicit provision for perspective distortion. Experimental evidence exists for the SIFT descriptor to be tolerant to view obliqueness of about 30 degrees.

In photogrammetry it has been long established that an affine transformation can be used to locally approximate a perspective distortion. This has been used for example in least-squares matching. The ASIFT feature descriptor attempts to combine the power of the SIFT operator with an affine transformation to make a feature descriptor which is invariant to affine and ultimately perspective distortions. The SIFT descriptor uses normalisation to eliminate rotation and translation effects. To deal with scale influences the scale space is sampled and multiple representations of the local feature area are stored. ASIFT takes the same approach to deal with affine distortions, namely sampling the affine transformation space. While the scale space is a one-dimensional space, the affine transformation has six parameters (section 2.2.1.2). A naive sampling in a six dimensional space is computationally prohibitive. A core idea of the ASIFT is therefore the observation that shift, rotation and scale of the SIFT feature already accounts for four of the affine distortions. Only two more parameters are required to fully compensate affine effects. These two parameters are described as longitude and latitude angles of a camera on a sphere pointing towards the sphere centre. In photogrammetry these parameters are often referred to as omega and phi of the exterior orientation in classical Euler notation. This space is sampled using six latitudes (including the pole) for example at 0°, 45°, 60°, 70°, 75° and 80°. Along these latitudes 1, 5, 6, 8, 10 and 13 samples are taken respectively. So besides the undistorted patch 42 affine distorted patches are extracted and the SIFT descriptors is computed for each of these patches. This gives an indication of the additional computational effort for the ASIFT descriptor.

SURF Operator

The *Speed-Up Robust Features* (SURF) approach uses the maxima of the determinant of the Hessian to detect interest points. The computation of the Hessian is accelerated by approximating the underlying Gaussian filter process with simple box filters. The SURF detector (much like the SIFT detector) uses image areas that resemble blobs as interest points.

The neighbourhood of the interest point is characterised by a descriptor using a histogram of Haar wavelet responses. This histogram is computed over a square region oriented along the dominant orientation to achieve invariance to rotation. As with the SIFT, scale space is exploited to gain invariance to scale. Scale space is computed implicitly by varying the size of the filters. Integral images are used to accelerate filtering particularly for large filter sizes. Fig. 5.82 gives an example of the interest points detected using the SURF approach.

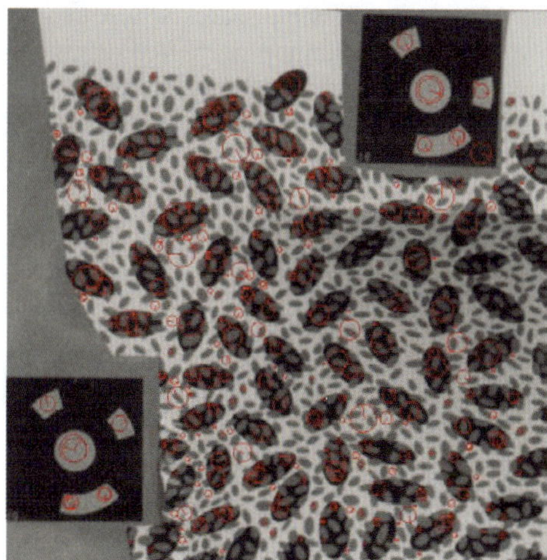
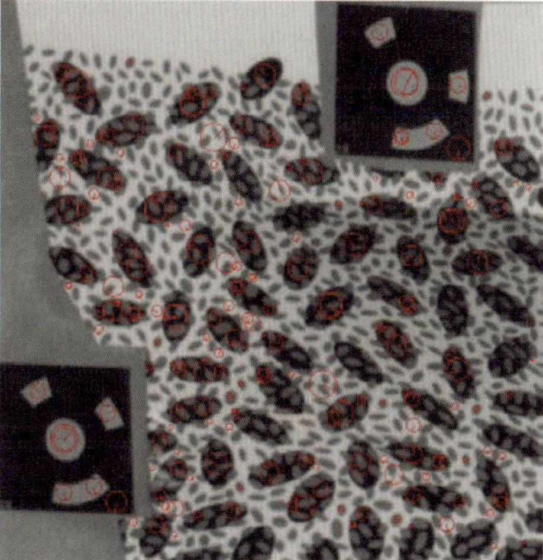

Fig. 5.82: Feature detection in stereo pair using SURF operator. The size of the circle indicates the scale at which the feature was detected. The ticks indicate the direction of the dominant gradient.

5.5.2.3 Correspondence analysis

As a result of applying an interest operator, an unstructured list of image points (coordinates and interest values) is generated for each image. Correspondence analysis then has the task of extracting a set of corresponding points based on the stored interest values and coordinates in both lists. The complexity of this process mainly depends on:

- configuration of images (stereo images, convergent multi-image configuration)
- available geometric information (initial values, orientation parameters)
- object's surface structure (projected patterns, variety of edges, occlusions)

The simplest case is given by normal stereo imaging of an object with a simply structured surface and no occlusions (example in Fig. 6.9). The interest value is a suitable matching criterion due to the high similarity between the content of both images. In addition, the search space in the partner image can be considerably reduced if the relative overlap between the images is known.

The matching process becomes much more complicated for convergent multiple images of complex objects with no artificial pattern projection. Point features are not then normally appropriate for image matching as numerous distinctive image locations are detected at non corresponding image features. In these cases edge extraction and matching is more useful.

A list of potential matches, with associated attributes, is derived from the extracted point or edge features. Either primitive image features (grey values, interest values, descriptors, edge parameters) or higher-level characteristics derived from them (e.g. relations) can be used as attributes. The list of candidates, and consequently the possible search space, can be significantly reduced by employing a priori knowledge, e.g. maximum permitted distances in image space (parallaxes, disparities). The subsequent correspondence analysis solves the matching of homologous features and the elimination of non-corresponding objects, as far as possible by generating a quality measure for each match.

The following basic procedures are derived from the numerous feature-based matching methods available:

- Best match:

  The simplest strategy to establish matching features is to select the best matching feature from the corresponding image. In the case of feature vectors associated with the interest points this involves computing the Euclidian distance of the two vectors is used. The smaller the distance is the better they match. This method can cause problems for repetitive structures in the images. The match to each occurrence of a repetitive pattern should be equally similar. This can easily lead to wrong matches. Hence if the best match is too similar to the second best match the correspondence is rejected. Only if the best match is significantly better to the second best it is accepted. Fig. 5.83 shows an example of this strategy. For this normal stereo pair, matches that run diagonally across the two images indicate false matches. Often this strategy is used in combination with a RANSAC filter.

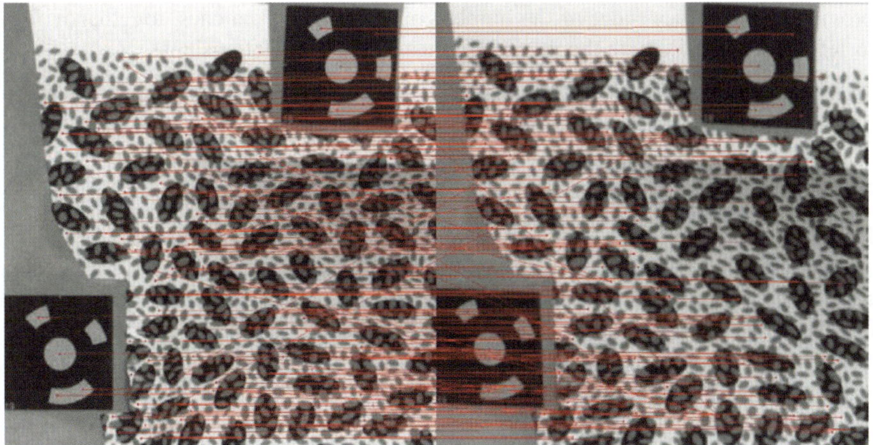

Fig. 5.83: Matching corresponding SIFT features in a normal stereo pair using best matches. Matching interest points are connected by lines.

- Relaxation:

  The imaged objects to be matched (e.g. points resulting from interest operators applied to two images) are associated with properties (e.g. their interest values from the Förstner operator) which, after appropriate normalisation, can be interpreted as probabilities. In an initialisation stage the probability of a match is computed for every possible pair of candidates. During an internal process these probabilities are altered step by step, taking the properties of adjacent objects into account using a compatibility function in order to arrive at the most probable set of matches. Properties have a physical expression, e.g. separation. After each pass in the analysis, the best (most probable) assignments are kept and the rest are evaluated in the next iteration or are excluded.

- Dynamic programming:

  This process is most commonly used where the number of possible combinations of candidate features to be matched exceed the ability of the system to make all comparisons. Instead of commencing with the first possible pair of matching features and testing them, dynamic programming starts by considering the properties of the final outcome and defining it with a cost function. The method then minimises the cost function to determine the best set of matches.

  Dynamic programming can be used, for example, in a one-dimensional search for corresponding points in epipolar images where image features are directly given by the grey value profile.

- Relational matching:

  Correspondence analysis by relational matching uses relationships between image objects (e.g. "to left of", above" etc.) in addition to the features themselves. A matching function is required which takes into account the extracted features and also applies the relationships in the left image to those in the right. An error value for matching is provided by an assessment of the number of left-to-right matches, together with right-to-left matches made with the inverse application of the relationships.

Relational matching is of major importance in high-level object recognition and scene description.

- Matching in image pyramids:

  Features are extracted at different resolutions in an image pyramid. At the lowest resolution few distinctive features are visible which, for instance, can be matched by one of the above methods. The matching result from a coarser resolution is used as approximate information for next higher resolution where additional features are extracted and matched. The approach combines robustness (coarse feature extraction) with precision (precise matching at the highest resolution).

- RANSAC filter:

  RANSAC (see 2.4.4.5) is a voting scheme to identify a set of observations which are consistent with a target function. To find a consistent set of matching points across a stereo pair, the coplanarity constraint (eqn. 4.53), which is used to compute the relative orientation or the essential matrix (eqn. 4.75), can be used as a target function for calibrated images. In the case of uncalibrated images, the fundamental matrix (eqn. 4.72) can be used. In the RANSAC scheme, the orientation of the image pair is calculated repeatedly using the minimum number of matching point pairs randomly selected from all correspondences. All remaining point pairs are tested against this orientation. All point pairs which are consistent with the calculated orientation are regarded as a consensus set. The largest set of matches is finally accepted as valid matches, all other point pairs are rejected as false matches.

5.5.3 Correspondence analysis based on epipolar geometry

The correspondence problem can be greatly simplified if the relative orientation of the images is known. This is the case for fixed stereo vision systems and in close-range applications where image orientation is solved in a separate process, e.g. by application of coded targets. The search space for a corresponding point in the neighbouring image can then be reduced to an epipolar line (see section 4.3.2 and Fig. 5.84).

5.5.3.1 Matching in image pairs

Consider first the matching process in an image pair B_1,B_2 (Fig. 5.84). Point P" corresponding to P' is located within a small band either side of epipolar line *k*". The band width ε depends on the uncertainty of the orientation parameters and the image measurement quality of P'. The search space is strictly reduced to the straight line *k*" only for perfect input data. The length of the search space l_{12} is a function of the maximum depth ΔZ in object space.

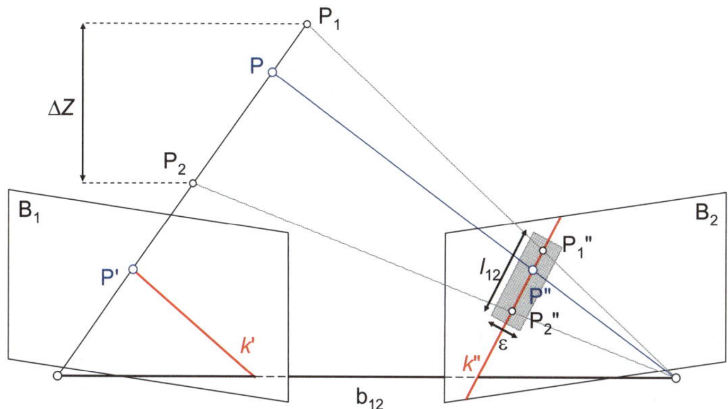

Fig. 5.84: Matching in an image pair based on epipolar lines (according to Maas 1992)

With increasing number of image points *n,* and area *f* of the search space, the probability P_a of ambiguous point matches also increases:

$$P_a = 1 - e^{-\frac{n \cdot f}{F}} \tag{5.72}$$

where
n: number of image points
F: image area
f: area of epipolar search space

The total number of ambiguities N_a for an image pair is given by:

$$N_a = (n^2 - n)\frac{2 \cdot c \cdot \varepsilon \cdot b_{12} \cdot (Z_{max} - Z_{min})}{F \cdot Z_{max} \cdot Z_{min}} \tag{5.73}$$

where
c: principal distance
b_{12}: base length between image B_1 and image B_2
ε: width (tolerance) of search space
$Z_{min} < Z < Z_{max}$: depth of object

It therefore increases

- quadratically with the number of image points
- linearly with the length of the epipolar lines
- linearly with the base length
- approximately linearly with object depth
- linearly with the width of the search space

Example 5.4:

Consider an image pair (Kodak DCS 420) with parameters $c = 18$ mm, $F = 18 \times 28$ mm², $b_{12} = 1$ m, $Z_{max} = 3$ m, $Z_{min} = 1$ m and $\varepsilon = 0.02$ mm. Depending on the number of image points n the following ambiguities N_a result:

$n = 50$: $N_a = 2$
$n = 100$: $N_a = 10$
$n = 250$: $N_a = 60$
$n = 1000$: $N_a = 950$

The example above shows that epipolar geometry leads to a match if the number of image points is relatively small. It depends mainly on the application as to whether ambiguities can be reduced by an analysis of interest values, a reduction of object depth using available approximations or by a reduced search width.

5.5.3.2 Matching in image triples

Ambiguities can be considerably reduced if the number of images is increased. Fig. 5.85 illustrates the matching principle for a configuration of three images. Starting with an image point P' in image B_1, the corresponding epipolar lines k_{12} and k_{13} can be calculated for the other images. For both partner images, ambiguities are represented as candidates P_a'', P_b'', P_c'' in image B_2 and P_d'', P_e'', P_f'' in image B_3. The homologous points of P' cannot therefore be uniquely determined.

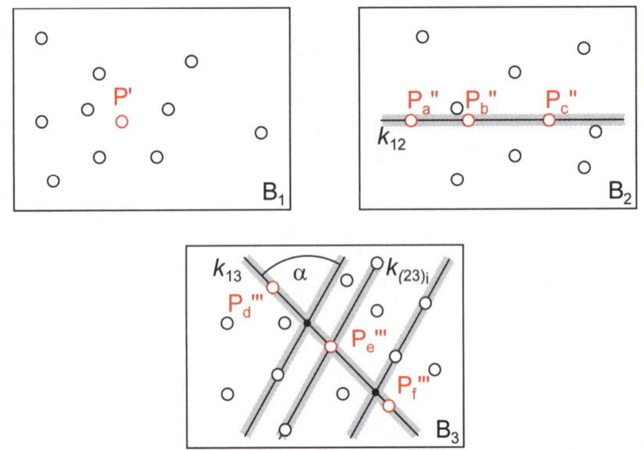

Fig. 5.85: Matching in image triples based on epipolar lines (according to Maas 1992)

If, in addition, the epipolar lines $k_{(23)i}$ are calculated in image B_3 for all candidates P_i in image B_2, it is most likely that only one intersection point with k_{13} is located close to a candidate in image B_3, in this case P_e'''. The search space is therefore restricted to the tolerance region of the intersection points. The number of possible ambiguities for three images is given by:

$$N_a = \frac{4(n^2 - n)\cdot \varepsilon^2}{F \cdot \sin\alpha}\left(1 + \frac{b_{12}}{b_{23}} + \frac{b_{12}}{b_{13}}\right) \tag{5.74}$$

where
n: number of image points
F: image area
ε: width (tolerance) of the search space
α: angle between epipolar lines in image B_3
b_{jk}: base length between image B_j and image B_k

Ambiguities are minimised if the three cameras are arranged in an equilateral triangle such that $b_{12} = b_{13} = b_{23}$ and $\alpha = 60°$. In the numerical example above ($n = 1000$) the number of ambiguities is then reduced to $N_a = 10$.

5.5.3.3 Matching in an unlimited number of images

The method can be extended to a virtually unlimited number of images. In order to search for homologous points of P' in all other images, a combinatorial approach must be applied which investigates all likely combinations of epipolar line intersections. The following matching strategy represents a practical solution for limiting the required computational effort:

1. Selection of an image point in image B_i
2. Search for candidates on epipolar lines in images B_{i+j}, until at least 1 candidate is found
3. Verification of all candidates by calculating their object coordinates using spatial intersection followed by back projection into all images $B_{i+j+1} \ldots B_n$
4. Counting all successful verifications, i.e. the calculated (back projected) image position must contain an image point
5. Acceptance of the candidate possessing the significantly largest number of successful verifications.

This approach offers some major advantages:

- an arbitrary number of images can be processed
- an object point need not be visible in every image, e.g. due to occlusions or limited image format
- interest values can optionally be used in order to reduce the number of candidates
- approximate values of image points are not required, i.e. there are no pre-conditions for the continuity (smoothness) of the object surface nor the spatial distribution of object points.

5.5.4 Area-based multi-image matching

The least-squares matching approach introduced in section 5.4.2.4 for two image patches (reference and search image) can be extended by the following features:

- simultaneous matching of one point in multiple images (*multi-image matching*)
- simultaneous matching of multiple points in multiple images (*multi-point matching*)
- introduction of geometric conditions in image space and object space (*multi-image geometrically constrained matching*)
- introduction of object models (object-space matching)

5.5.4.1 Multi-image matching

Consider a reference image $f(x,y)$ and m search images $g_i(x,y)$, which are to be matched to the reference image. Equation (5.55) can be extended to multiple images:

$$f(x,y) - e_i(x,y) = g_i(x,y) \qquad i = 1,\ldots,m \tag{5.75}$$

Here $e_i(x,y)$ indicates random noise in image i. In a similar way to least-squares matching, the following adjustment system results:

$$\underset{n,1}{\mathbf{v}} = \underset{n,u}{\mathbf{A}} \cdot \underset{u,1}{\hat{\mathbf{x}}} - \underset{n,1}{\mathbf{l}} \tag{5.76}$$

where

$$\hat{\mathbf{x}}_i^T = [da_0, da_1, da_2, db_0, db_1, db_2, r_0, r_1]_i \qquad i = 1,\ldots,m$$

m: number of search images

n: total number of observations
$n = n_1+n_2+\ldots+n_m$; n_i = number of observations (pixel) in image $g_i(x,y)$

u: total number of unknowns, $u = 8 \cdot m$

The m parameter vectors $\hat{\mathbf{x}}_i^T$ can be determined independently within the system of equations (5.76) because they have no cross connections within the design matrix **A**.

5.5.4.2 Geometric constraints

Functional model

The multi-image approach of (5.75) enables the simultaneous determination of all matches of a point. However, it does not consider the constraint that all homologous points must correspond to one common object point. This constraint can be formulated by the condition that all homologous image rays, taking account of image orientation parameters, must intersect optimally at the object point.

This constraint is also the basis for the bundle adjustment model (section 4.4.2) where the collinearity equations (4.9) are used. The three-dimensional coordinates of point P in image k are given by[1]:

$$\mathbf{x'}_{pk} = \frac{1}{m_{pk}} \cdot \mathbf{R}_k^{-1} \cdot (\mathbf{X}_p - \mathbf{X}_{0k})$$

$$\begin{bmatrix} x'_p \\ y'_p \\ -c \end{bmatrix}_k = \frac{1}{m_{pk}} \cdot \begin{bmatrix} r_{11} & r_{21} & r_{31} \\ r_{12} & r_{22} & r_{32} \\ r_{13} & r_{23} & r_{33} \end{bmatrix}_k \cdot \begin{bmatrix} X_p - X_{0k} \\ Y_p - Y_{0k} \\ Z_p - Z_{0k} \end{bmatrix} \tag{5.77}$$

In simplified notation, the image coordinates x'_p, y'_p are given by

$$\begin{aligned} x'_{pk} &= -c_k \cdot \left(\frac{k_X}{N}\right)_{pk} = -F_{pk}^X \\ y'_{pk} &= -c_k \cdot \left(\frac{k_Y}{N}\right)_{pk} = -F_{pk}^Y \end{aligned} \tag{5.78}$$

Based on initial values $(x'_p, y'_p)^0$, the following non-linear observation equations for the required shifts $\Delta x'_p, \Delta y'_p$ result:

$$\begin{aligned} \Delta x'_{pk} + F_{pk}^X + x'^0_{pk} &= 0 \\ \Delta y'_{pk} + F_{pk}^Y + y'^0_{pk} &= 0 \end{aligned} \tag{5.79}$$

Here $\Delta x'_p, \Delta y'_p$ correspond to the shift coefficients da_0, db_0 in (5.76). The equations (5.79) establish the relationship between image space and object space. In contrast to the collinearity equations, which primarily only establishes the functional relationship between observed image coordinates and unknown point and orientation data, this approach uses as observations the original grey values, in combination with least-squares matching. Additional observation equations can be set up using the parameters in terms F_{pk}, e.g. for the simultaneous calculation of object coordinates XYZ or for the formulation of geometric constraints (e.g. Z = const.).

Object restrictions

It is first assumed that the interior and exterior orientation parameters of all images are known, e.g. by a prior bundle triangulation. The system of equations (5.79) must be linearised at initial values of the remaining unknowns X,Y,Z:

[1] Image coordinates have their origin at the principal point and are corrected for distortion.

$$\Delta x'_{pk} + \frac{\partial F^X_{pk}}{\partial X_p} dX_p + \frac{\partial F^X_{pk}}{\partial Y_p} dY_p + \frac{\partial F^X_{pk}}{\partial Z_p} dZ_p + F^{X\,0}_{pk} + x'^{\,0}_{pk} = 0$$
$$\Delta y'_{pk} + \frac{\partial F^Y_{pk}}{\partial X_p} dX_p + \frac{\partial F^Y_{pk}}{\partial Y_p} dY_p + \frac{\partial F^Y_{pk}}{\partial Z_p} dZ_p + F^{Y\,0}_{pk} + y'^{\,0}_{pk} = 0 \tag{5.80}$$

In summary the system of additional observation equations is as follows:

$$\underset{m',1}{\mathbf{w}} = \underset{m',3}{\mathbf{B}} \cdot \underset{3,1}{\hat{\mathbf{y}}} - \underset{m',1}{\mathbf{t}} \tag{5.81}$$

where

$\hat{\mathbf{y}}^T = [dX, dY, dZ]$

m': number of images:
$m' = m$, if no transformation of the reference image is permitted
$m' = m+1$, if the reference image is also to be transformed, e.g. with respect to an artificial template

For the extended system of (5.79) and (5.81) the complete vector of unknowns is given by

$$\bar{\mathbf{x}} = (\mathbf{A}^T\mathbf{PA} + \mathbf{B}^T\mathbf{P}_t\mathbf{B})^{-1} \cdot (\mathbf{A}^T\mathbf{Pl} + \mathbf{B}^T\mathbf{P}_t\mathbf{t}) \tag{5.82}$$

The parameter vector $\bar{\mathbf{x}}$ consists of the simultaneously calculated displacements of the image patches in $g_i(x,y)$ as well as the adjusted corrections of the object coordinates. Because of their relationship to the XYZ object coordinates, the image shifts cannot take arbitrary values but are constrained to follow an epipolar line (Fig. 5.86). The influence of this geometric condition can be controlled by the weight matrix $\mathbf{P}_t$.

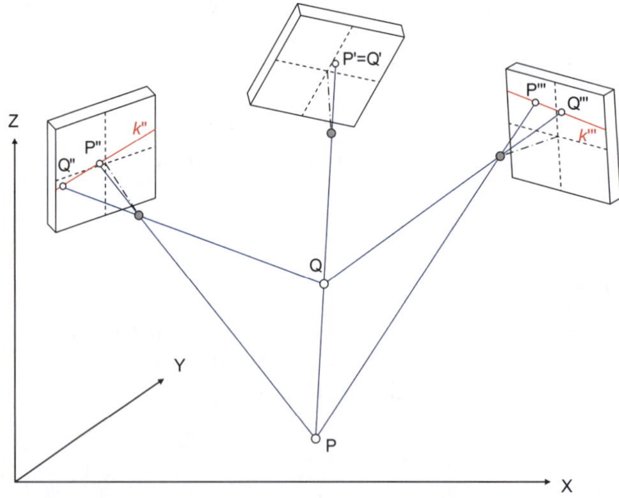

Fig. 5.86: Geometric condition of ray intersection

The model in (5.80) is appropriate for determining arbitrary 3D object points. In order to measure surface models, some of the coordinate components of the surface points can be fixed:

- Constant *X,Y*:

  If the surface points are located on a particular grid of *XY* coordinates (e.g. where $\Delta X = \Delta Y$=const.), corresponding terms can be eliminated from (5.80) and only *Z*-value adjustments remain:

$$\begin{aligned} \Delta x'_{pk} + \frac{\partial F^X_{pk}}{\partial Z_p} dZ_p + F^{X\,0}_{pk} + x'^{\,0}_{pk} = 0 \\ \Delta y'_{pk} + \frac{\partial F^Y_{pk}}{\partial Z_p} dZ_p + F^{Y\,0}_{pk} + y'^{\,0}_{pk} = 0 \end{aligned} \qquad (5.83)$$

  In consequence, the normal system of equations becomes simpler. The approach corresponds to the manual measurement of terrain models in stereoscopic plotters. In this case the image shifts do not take place within epipolar lines but on the *vertical line locus* instead (see section 4.3.6.3).

- Constant *Z*:

  In order to measure contour lines at a given height Z = const., only coordinate displacements in the *X* and *Y* directions are permitted and (5.80) reduces to:

$$\begin{aligned} \Delta x'_{pk} + \frac{\partial F^X_{pk}}{\partial X_p} dX_p + \frac{\partial F^X_{pk}}{\partial Y_p} dY_p + F^{X\,0}_{pk} + x'^{\,0}_{pk} = 0 \\ \Delta y'_{pk} + \frac{\partial F^Y_{pk}}{\partial X_p} dX_p + \frac{\partial F^Y_{pk}}{\partial Y_p} dY_p + F^{Y\,0}_{pk} + y'^{\,0}_{pk} = 0 \end{aligned} \qquad (5.84)$$

The possibility of restricting particular coordinate components enables surface models to be recorded with respect to specific reference planes or along specific sections through the surface. The (topological) structure of the surface model is therefore defined directly during measurement and not by a later analysis of an unstructured point cloud which could be the result, for instance, of an active projection method (section 3.7.2 and also see section 5.5.6.1).

Additional extensions

The concept of additional observation equations, described above, permits the introduction of further constraints on the adjustment system The influence of additional observations can be controlled by appropriate weighting from $p_i = 0$ where the constraint has no effect through to $p_i = \infty$ where the constraint is strictly enforced.

As examples, additional constraints can be formulated for the following tasks:

- Edge extraction by least-squares matching:

  Single edge points can be determined by a suitable edge template (see example in Fig. 5.49). As edges are linear features, an additional constraint can be introduced which forces the template to move in the direction of the gradient, i.e. perpendicular to the edge.

- Determination of spatial object contours:

  Spatial object contours can be described by geometric elements (straight line, circle etc.) or spline functions. The corresponding analytical parameters can be included as unknowns in a least-squares fit. An example of this approach is the method of LSB snakes (section 4.4.7.3) which combines edge extraction and determination of a B-spline spatial curve (*snake*) in a single analysis.

- Measurement of point grids:

  The least-squares matching concept for a single object point can easily be extended to an unlimited number of points. However, if no geometric relationship between the points exists then the result is identical to single point matching. Geometric relations can be formulated either by associating points with a common geometric element (e.g. all point belong to a plane or a cylinder), or by defining neighbourhood relationships.

  The latter approach can be compared to the interpolation of digital surface models where points on a grid are connected by additional constraints such as minimum surface curvature (section 3.3.4.5). The approach of *multi-patch matching* utilises this idea by defining constraints between adjacent grid elements (*patches*).

- Bundle concept:

  If the interior and exterior orientation parameters of images are known only approximately, they can be included as unknowns in the least-squares matching. Equation (5.80) is then extended by corresponding differential coefficients of the additional orientation parameters.

5.5.5 Semi-global matching

Correlation methods, and other area-based methods, match the often rectangular-shaped areas of two stereo images onto each other. This area or window acts as a low-pass filter and blurs depth discontinuities. The result is a depth image with washed-out edges. The larger the window, the stronger the blurring effect. The smaller the window, the lesser the blurring is. Ideally one would match only individual pixels not areas. However, the intensity information of a single pixel does not yield enough information for meaningful matches and additional constraints are needed to enforce consistency, e.g. a smoothness constraint. In the case of matching the left image I_l of a stereo pair onto the right image I_r, this can be expressed as an energy function to be minimised

$$E(D) = \sum_{p} C(p, d_p) + \sum_{q \in N_p} P \left[\left| d_p - d_q \right| \geq 1 \right] \tag{5.85}$$

The cost function C expresses the cost for matching pixel p of the left image I_l with a parallax or disparity of d_p, i.e. matching it to a corresponding pixel of the right image I_r at a distance d_p along the epipolar line in image space. The first term summarises these costs for all pixels p of the image I_r. Since individual pixels are compared, only the intensities $I_l(p)$ and $I_r(q)$ are used to calculate cost. Simple approaches use the absolute difference of intensities to calculate cost, whereas better results are obtained by using mutual information. The second term enforces local smoothness over the disparities. The penalty term P penalises disparities of all pixels q from the neighbourhood N_p of p, which are larger than 1 pixel. The penalty term can lead to a match being rejected, even if the intensity values $I_l(p)$ and $I_r(q)$ are very similar but their disparity is very different to the neighbours. In such a case a match can be preferred where the intensity values are less similar but the disparity agrees better.

The disparities d_p for all pixels p are typically stored in an image D of the same dimension as I_l. To solve the matching problem, all disparities d_p must be determined while ensuring $E(D)$ is minimal. The smoothness penalty term P links the disparity at p to that of its neighbours q, who in turn are linked to their neighbours and so on. This effectively links all pixels of the image to each other. Hence this approach is referred to as *global matching*. Unfortunately, the computational effort to solve this problem in practice grows exponentially with the size of the input data. This makes it unsuitable for many applications.

Semi-global matching compromises the global optimum of the solution to gain computational efficiency. Rather than considering all pixels of the image simultaneously while minimizing for $E(D)$, only eight linear paths through the image are considered. The paths run radially symmetrically from the image boundaries and meet at pixel p. (For a graphical depiction of such paths see Fig. 5.71.) The individual costs for each path are summed up and the disparity d_p with the least cost is chosen.

Semi-global matching is much faster than global matching. Implementations have been presented where the computational effort for a solution grows only linearly with the number of pixels and the allowed range of disparities. Semi-global matching delivers a dense disparity map D and subsequently a dense range map (ideally one range value for each pixel of I_r), see Fig. 5.87 for an example. It should be noted that semi-global matching only matches stereo pairs. It cannot directly use the information of multi-view photogrammetry where multiple rays meet at an object point. Instead, it is recommended to fuse the depth maps obtained from multiple stereo pairs.

Fig. 5.87: Left and right stereo pair and dense depth map from semi-global matching

5.5.6 Matching methods with object models

The matching methods described in the previous sections are mainly based on geometric relationships between images, or between images and object. Although contrast differences are modelled in the least-squares matching by two radiometric parameters, they are completely independent of the reflection characteristics of the surface.

In order to create a complete object model, it is necessary to combine the geometric object properties (position and shape) with the reflection properties of the surface. This idea is used in various 3D visualisation methods (see section 5.3.3) and can also be used for object reconstruction. The features of such a complete reconstruction method are:

- introduction of a surface reflection model (material characteristics, illumination)
- ray tracing through different media (as a minimum through the atmosphere)
- multi-image adjustment based on least-squares matching
- topological structuring of the surface by surface grid and shape lines

The global objective is the 3D surface reconstruction by simultaneous calculation of all orientation parameters, object point coordinates and geometric element parameters, as well as illumination and reflection parameters. Grey values in multiple images are available as observations, as are initial values of unknowns. The number of parameters is extremely high and the resulting system of equations is correspondingly complex.

5.5.6.1 Object-based multi-image matching

Object-based multi-image matching is based on a relationship between the intensity (colour) value of a surface element G_i and the grey values g_{ij} of the associated images (Fig. 5.88). The grey values are a function of the orientation parameters $\mathbf{O}_j$ of an image j and the surface parameters $\mathbf{Z}_i$ of an element i:

$$G_i = g_{ij}(\mathbf{Z}_i, \mathbf{O}_j) \qquad \begin{array}{l} i = 1,\ldots,m \\ j = 1,\ldots,n \end{array} \tag{5.86}$$

where
m: number of surface elements
n: number of images

In order to solve this system of equations, initial values for the unknown surface and the orientation parameters are required. The difference between the grey values calculated from the initial values can be regarded as a stochastic value which gives

$$\Delta g_{ij} = G_i^0 - g_{ij}(\mathbf{Z}_i^0, \mathbf{O}_j^0) \tag{5.87}$$

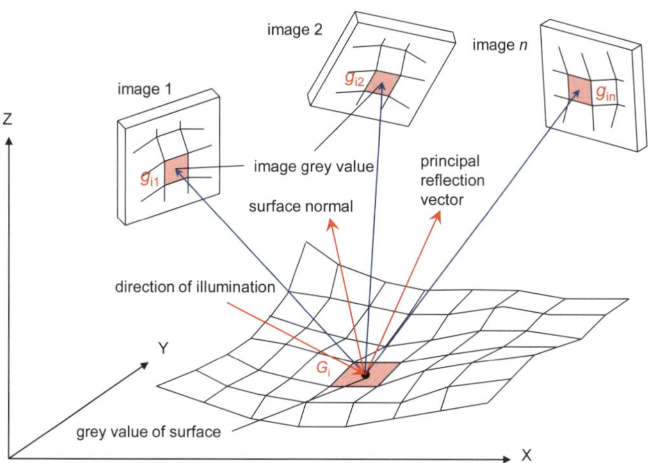

Fig. 5.88: Relationship between intensity value of surface and grey value of image (Schneider 1991)

The imaged light intensity depends on the properties of the surface material and of the geometric orientation of surface element, light source and imaging sensor (Fig. 5.88) as, for example, described by Phong's illumination model (section 5.3.3.2). Similarly to least-squares matching, the grey values in the individual images are matched using two radiometric parameters which are dependent on material:

$$\Delta g_{ij} = G_i^0 - \left(r_{1,j}^0 + r_{2,j}^0 \cdot g'_{ij}(\mathbf{Z}_i^0, \mathbf{O}_j^0)\right) \tag{5.88}$$

where

$r_{1,j}^0, r_{2,j}^0$: approximate radiometric correction parameters

g'_{ij}: observed image grey value: $g_{ij} = r_{1,j}^0 + r_{2,j}^0 \cdot g'_{ij}$

Equation (5.88) can form the observation equation for each grey value in all images where a specific surface element is visible. Using the substitution

$$r_{2,j} = 1 + r'_{2,j}$$

and re-arranging the linearised correction equations, the following is obtained:

$$\begin{aligned} v_{ij} = G_i^0 - dG_i - (r_{1,j}^0 + dr_{1,j}) - (r'^0_{2,j} + dr'_{2,j}) \cdot g'_{ij}(\mathbf{Z}_i^0, \mathbf{O}_j^0) \\ - \frac{\partial g'}{\partial \mathbf{Z}} d\mathbf{Z} - \frac{\partial g'}{\partial \mathbf{O}} d\mathbf{O} - g'_{ij}(\mathbf{Z}_i^0, \mathbf{O}_j^0) \end{aligned} \tag{5.89}$$

The differential coefficients

$$\frac{\partial g'}{\partial \mathbf{Z}} = \frac{\partial g'}{\partial x'} \cdot \frac{\partial x'}{\partial \mathbf{Z}} + \frac{\partial g'}{\partial y'} \cdot \frac{\partial y'}{\partial \mathbf{Z}} \qquad \frac{\partial g'}{\partial \mathbf{O}} = \frac{\partial g'}{\partial x'} \cdot \frac{\partial x'}{\partial \mathbf{O}} + \frac{\partial g'}{\partial y'} \cdot \frac{\partial y'}{\partial \mathbf{O}} \tag{5.90}$$

contain the grey value gradients $\partial g'/\partial x'$ and $\partial g'/\partial y'$. This approach can therefore only be applied when an appropriate number of edges or structures exist in the images. The remaining differential coefficients $\partial x'/\partial \mathbf{Z}$ etc. correspond to the derivatives from the space resection and bundle triangulation models (see section 4.4.2).

Now all the relevant parameters are available for an iterative adjustment

$$\mathbf{v} = \mathbf{A} \cdot \hat{\mathbf{x}} - \mathbf{l}$$

with observations

$$\mathbf{l} = \mathbf{g'}(\mathbf{Z}^0, \mathbf{O}^0)$$

which leads to a vector of unknowns

$$\hat{\mathbf{x}}^T = (d\mathbf{Z}^T, d\mathbf{O}^T, d\mathbf{r}_1^T, d\mathbf{r'}_2^T, d\mathbf{G}^T) \tag{5.91}$$

The importance of object-based multi-image matching has two aspects. One is the simultaneous determination of all parameters influencing the formation of the image, including the object itself. The other is that the method can be extensively modified by altering object parameters relevant to the application. In this way the following tasks can be solved by use of appropriate functions:

- Measurement of artificial targets:

  Circular targets can be defined as a circle in space (see section 2.3.2.3) whose centre must be determined. An additional transformation converts the coordinates of the plane of the circle into the object coordinate system. The target's grey values have a radiometric model which is a function of the radius and consists, for example, of a white circle on a black background (Fig. 5.89).

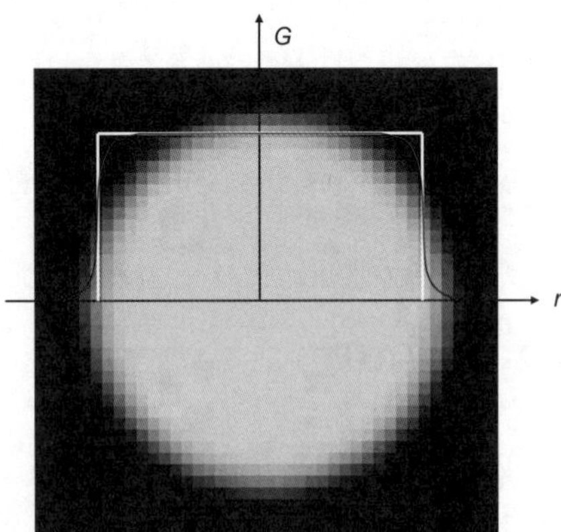

Fig. 5.89: Distribution of smoothed object grey values for a circular target

The surface element size should approximately correspond to the size of a pixel in image space. The circle centre in object space is determined ignoring any eccentricities (see section 3.6.1.2).

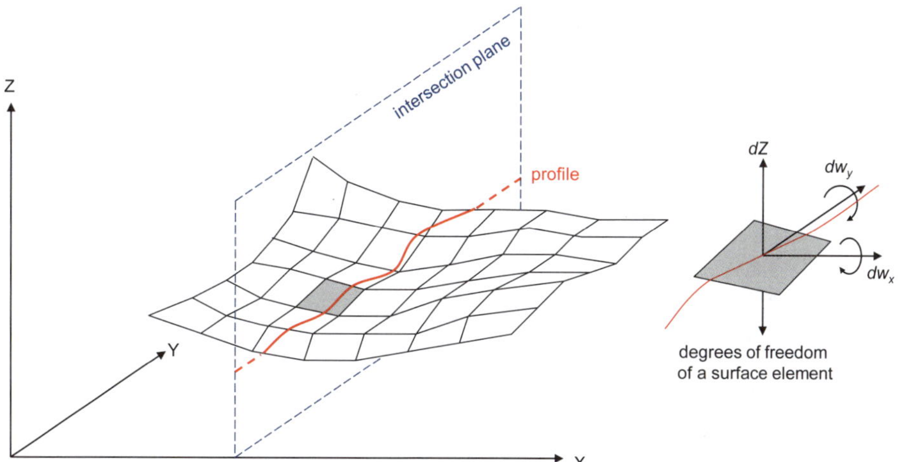

Fig. 5.90: Resulting profile from intersection of section plane and surface

- Measurement of surfaces:

  Surfaces are recorded as profiles, surface models or free-form surfaces. If the measurement is based on profiles, an arbitrarily oriented section is defined in which the profile is represented by a two-dimensional curve (Fig. 5.90). Within the section a surface element is defined by a tangential plane whose degrees-of-freedom are reduced to two rotations and a shift within the section plane.

- Bundle triangulation:

  The system of equations (5.88), extended to determine 3D circles and additional parameters of interior orientation (simultaneous calibration), leads logically to an object-based bundle triangulation for photogrammetric images with circular targets.

- Orthophoto production:

  Object-based multi-image matching can also be applied to the production of orthophotos where the calculated grey values of the object are here used for image generation. The required 3D object model can be estimated within the multi-image matching process itself, or it can be generated from other sources. In principle, this approach enables orthophotos to be generated from an arbitrary number of images, with occlusions and radiometric variations no longer affecting the result.

Object-based multi-image matching, briefly described here, has considerable potential for wide application and makes use only of the original image grey values. Feasibility in principle, and some test applications, have been demonstrated but so far practical use has been limited by the high computational effort needed to solve the large systems of normal equations. The generation of appropriate initial values for the unknown parameters is, in addition, a non-trivial problem.

5.5.6.2 Multi-image matching with surface grids

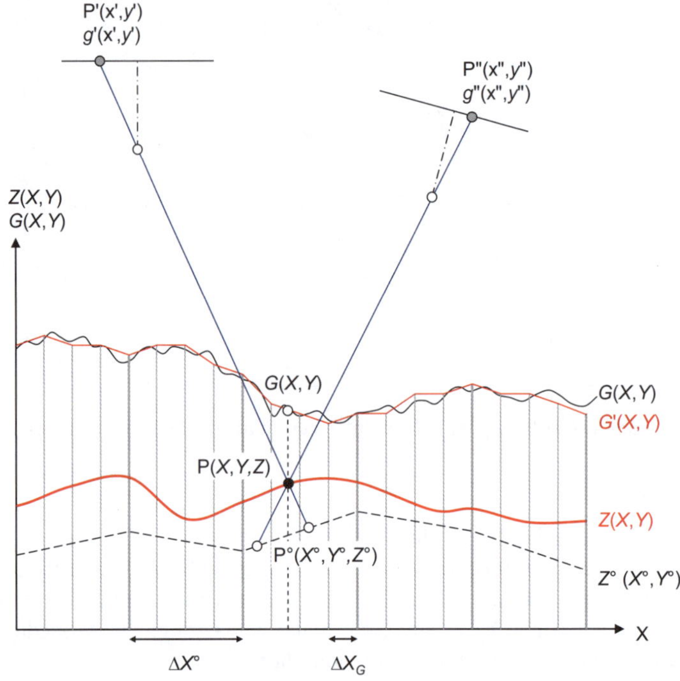

Fig. 5.91: Interpolation and projection of surface and reflection model (after Wrobel 1987)

Surface elements calculated by object-based multi-image matching (section 5.5.6.1) are independent of their adjacent elements in terms of geometry and radiometry. However, this assumption is invalid for piecewise smooth surfaces. With the exception of discontinuities (breaklines), adjacent object elements can be connected by radiometric and geometric interpolation functions. The above approach can therefore be extended by coefficients of piecewise linear functions for both the radiometric model and the geometric model (terrain model). For this purpose the surface is represented by triangular or quadrilateral patches (facets) within which linear or bilinear interpolation can be applied.

Fig. 5.91 illustrates the principle. The required surface $Z(X,Y)$ consists of grey values $G(X,Y)$. Photographic recording of surface points $P(X,Y,Z)$ results in image points $P'(x',y')$ with grey values $g'(x',y')$. Based on an approximately known height model $Z^0(X^0,Y^0)$, surface points are interpolated, e.g. on a fixed grid with separations $\Delta X^0 = \Delta Y^0$=const.. The surface reflection model $G'(X,Y)$ is calculated on a dense grid width $\Delta X_G,\Delta Y_G$ (Fig. 5.92).

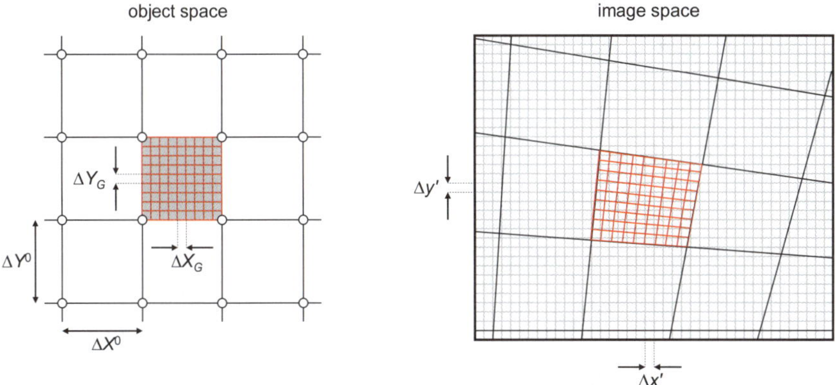

Fig. 5.92: Projection of object facets into image space

The grid width of the reflection model should be adjusted to the maximum spatial frequency in object space. If this information is missing, the grid width should be chosen with respect to the mean pixel size $\Delta x'$,$\Delta y'$ in the images:

$$\Delta X_G \geq m \cdot \Delta x' \qquad \text{where } m\text{: image scale number} \tag{5.92}$$
$$\Delta Y_G \geq m \cdot \Delta y'$$

For the height model, the grid width should adapt to the local object shape in order to achieve a reasonable result. For example, grid width can be reduced on object breaklines. Usually a square grid is chosen where

$$\Delta X^0 = \Delta Y^0 > \Delta X_G = \Delta Y_G \tag{5.93}$$

The introduction of radiometric facets with corner point grey values $G(X_k,Y_l)$ leads to:

$$G(X,Y) = \sum a_{kl}(X,Y) \cdot G(X_k,Y_l) \tag{5.94}$$
$$\text{where } \sum_{kl} a_{kl} = 1$$

Correspondingly, coefficients for the height model's geometric facets $Z(X_r,Y_s)$ are introduced to give:

$$Z(X,Y) = \sum b_{rs}(X,Y) \cdot Z(X_r,Y_s) \tag{5.95}$$
$$\text{where } \sum_{rs} b_{rs} = 1$$

The system of equations (5.88) is extended by (5.94) and (5.95). Coefficients a_{kl} and b_{rs} are determined by the nodal point values of each object grid or triangle facet.

Object-based multi-image matching based on object facets offers additional benefits:

- By choosing suitable weights for the coefficients, the smoothness of a surface can be controlled so that image matching can bridge critical areas containing intensity distributions where texture is not present.
- Occlusions and discontinuities can be handled by correspondingly detailed object surfaces or 3D models.

5.6 Range imaging and point clouds

At the start of this chapter an image was introduced as a discrete two dimensional grid, where each grid cell (or pixel) stores the local brightness. For a range image this concept changes only slightly. A range image is the same two dimensional grid but, in addition, each grid cell holds the distance (or range) from camera to object.

5.6.1 Data representations

A *range image* in its most basic form is, analogously to eqn. 5.1:

$$R = r(x, y) \tag{5.96}$$

where the function r contains the distance from the camera to the object. While an intensity image normally contains quantised integer values, floating point numbers are usually used to represent range. In the photogrammetric context presented here, the range is defined as the distance from the projection centre of the camera to the object's surface. It should be noted that other definitions are possible, e.g. where distance is defined relative to the housing of the camera.

As with any image, range images can contain additional channels. Additional channels frequently include

- intensity
- measurement uncertainty
- colour

Where additional colour channels are used, the image is commonly called an *RGBD image*, to indicate the three colour channels Red, Green, Blue and the Distance (range).

Each pixel in the range image uniquely identifies a point in three-dimensional space. The two integer coordinate values and the range measured at this pixel form a triplet $(x, y, r(x,y))$. In order to obtain three-dimensional Cartesian coordinates (X,Y,Z), it is necessary to apply a projection model to the triplet, typically the pin-hole camera model (Fig. 1.7). This is very similar to transforming polar coordinates to Cartesian coordinates. To account for any geometric errors in the projection (see section 3.1.3), a full lens correction model can be applied. This is consistent with the normal workflow in photogrammetry.

Since a projection model with additional parameters can be quite complex, it can be desirable to store a representation of a range images with the projection model already applied. In this case a three-channel image is stored where the three channels contain the X, Y and Z coordinates. This representation is, for example, used in the popular ASCII exchange format PTX.

As every pixel of the image in this case holds a Cartesian coordinate triplet (X,Y,Z), this is one form of representation for a point cloud. The grid still contains the information about pixel connectivity and thus provides the neighbourhood relationship for each pixel, such as a N8 or N4 neighbourhood relationship (see section 5.1.2). This representation is therefore referred to as an *organised point cloud.* If all coordinate triplets are stored only as a list, without regard to grid structure, it is known as an *unorganised point cloud.* In this case, establishing neighbourhood relationships between points is computationally expensive and typically requires the use of a spatial search structure such as a kd-tree..

a) Colour image from triangulation sensor

b) Range image from triangulation sensor

c) Colour image from TOF camera

d) Range image from TOF camera

Fig. 5.93: Examples images from low cost range cameras (UCL London)

Fig. 5.93 shows two example range images with associated RGB images. The images were captured with low-cost consumer-grade range cameras. The top row (a, b) shows the output of a triangulation sensor similar to the Microsoft Kinect. The bottom row (c, d) shows the output of a time-of-flight range camera intended for gesture recognition. The intensity values of the range image indicate the distance of the object from the sensor. The brighter the value of the pixel, the closer the object. In the two range images on the right, the different heights of the three boxes can be assessed. This is not possible using the 2D RGB images on the left.

5.6.2 Registration

Since range images can be converted to XYZ coordinate triplets, registering two range images requires the computation of the *rigid-body transformation* which brings the two three-dimensional coordinate systems into alignment. The rigid-body transformation is a special case of the spatial similarity transformation (see section 2.2.2.2), where the scale factor m is kept to 1. Closed-form solutions exist to compute the transformation from a minimum of three corresponding points. The orientation or registration of range images is thus much simpler than the orientation of 2D images which incorporate a perspective projection.

To compute the transformation, correspondences between the range images must be established in a way very similar to image matching. Correspondences can come from a variety of sources:

- 3D geometries imaged in the scene (natural or artificial)
- 2D intensity features imaged in the scene (natural or artificial)
- geometry of the whole scene

5.6.2.1 3D target recognition

The best accuracy can be achieved when placing artificial objects of known geometry into the scene. For three-dimensional registration, spheres have proven very reliable. They can be imaged from any direction and their centre provides a stable 3D point (Fig. 5.94a). To estimate the sphere parameters from a given point cloud, the partial derivatives given in eqn. 2.142 are used to perform a least-squares adjustment.

To detect spheres in a scene automatically, the RANSAC robust estimation scheme (section 2.4.4.5) is commonly used. In eqn. 2.138 a linear form of the sphere equation indirectly estimates the sphere parameters from a minimum of 4 points. In each iteration of the RANSAC algorithm, four points are randomly selected from the point cloud and the sphere parameters are determined. All remaining points are tested to see if their distance from the sphere centre corresponds to the radius of the sphere. The largest consensus set is accepted as representing points on the surface of the sphere and the corresponding centre, derived from them, can be used as a tie point for registration (Fig. 5.94b).

a) Spherical target

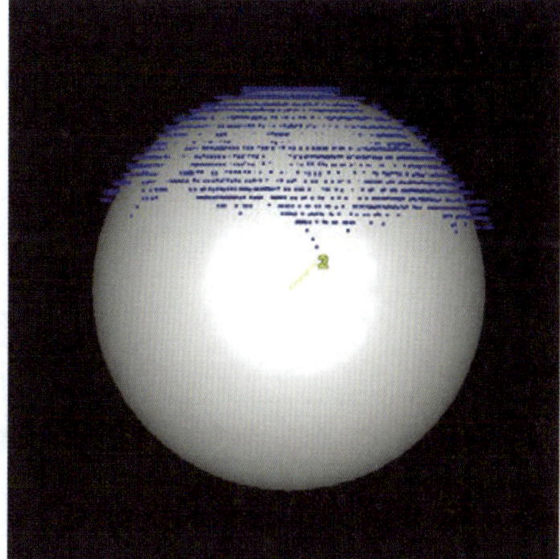

b) Measured point cloud and best-fit sphere

Fig. 5.94: Measurement and analysis of a spherical target

5.6.2.2 2D target recognition

For 2D target recognition it is necessary to have a perfectly registered intensity or RGB channel, i.e. every pixel in the intensity or RGB channel must correspond to the same pixel in the range or XYZ channel. In this case, any of the target extraction methods in close-range photogrammetry can be employed. This is particularly necessary for the extraction of artificial targets (circular shapes or checkerboard patterns) and the use of feature detectors (e.g. SIFT, SURF, see section 5.5.2.1). As target detectors in 2D image processing typically result in sub-pixel coordinates, the distance value must also be interpolated at sub-pixel level from the range or XYZ channel. If the target is known to be planar, accuracy can be improved if a plane is derived from the points corresponding to the target surface. The least-squares estimation process will reduce the noise of the range measurement contained in individual pixels.

5.6.2.3 Automated correspondence analysis

For the automated correspondence analysis of features extracted from two intensity images, a RANSAC filter approach was introduced in section 5.5.2.3. The same principle can be used to establish automatically the correspondence of targets extracted from two range images. However, the coplanarity constraint is here replaced by the rigid-body transformation. A minimum set of three target points is sampled in each iteration from both images. The transformation is computed and all remaining target points are tested against this transformation.

5.6.2.4 Point-cloud registration - iterative closest-point algorithm

If the separate 3D point clouds do not have common 3D or 2D targets, a transformation into a common coordinate system can still be made by using common surfaces in overlapping areas. The overlapping areas must have surfaces with at least three linearly independent normal vectors to provide enough information for a unique registration. Point clouds which represent a single plane, a set of parallel planes or a single slightly curved surface (e.g. the roof of a car or a section of an aircraft wing) do not provide sufficient geometric information. Likewise, if the area contains only a single rotationally symmetric surface such as a sphere or cone, registration is not possible.

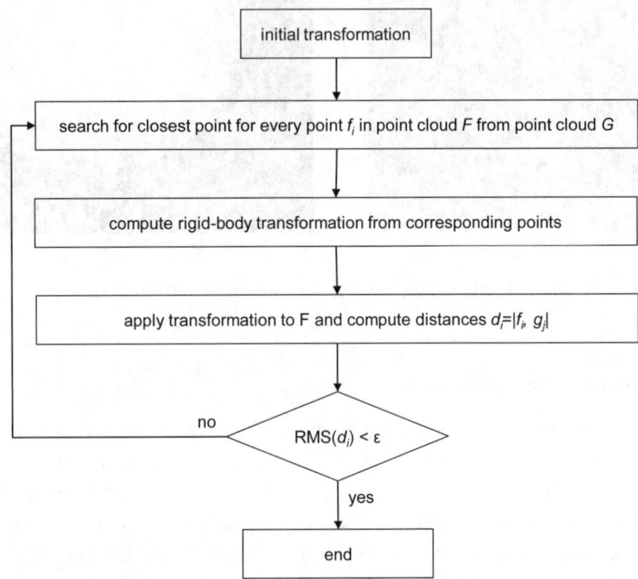

Fig. 5.95: Flow diagram for iterative closest-point algorithm

The most well-known algorithm for registering two point clouds based on overlapping surfaces is the *iterative closest-point* algorithm (ICP). As the name suggests, it is an iterative algorithm which requires a starting value for the transformation, i.e. the two point clouds must be roughly aligned. The following discussion considers the case of aligning point cloud F to point cloud G, where F consists of n points f_i and G of m points g_j. The ICP starts by searching for the closest point for every point f_i among all points in G. The closest point is the point g_j which has the shortest Euclidean distance to f_i. All closest point pairs (f_i, g_j) are assumed to be corresponding points. From this set of corresponding points a rigid body transformation is computed as described above. This transformation is applied to point cloud F. The Euclidean distance $d_i(f_i, g_j)$ for all corresponding point pairs is computed after transformation. If the overall error, e.g. the RMS of all d_i, is below a pre-set threshold, the iteration is aborted. Otherwise the iteration is continued by searching again for the closest points and establishing new corresponding points. See Fig. 5.95 for a flow diagram of the algorithm.

The computationally most expensive step of the algorithm is the search for the closest point. This problem is also known as the *nearest neighbour search*. A naïve implementation would compute distances $d_i(f_i, g_j)$ for every point g_j of G and select the point with smallest d_i. This would require m computations. Well known spatial search structures such as the kd-tree can reduce the search time to *log m*. If a surface meshing is available for point cloud *G*, the method can be adapted to consider distances between points and surfaces. This typically improves the accuracy of the alignment.

5.6.3 Range-image processing

Many of the image processing operators described above can also be used for range-image processing. The fact that the data is stored in a two-dimensional grid makes it possible to apply many of the well-known two-dimensional filters to range images. A typical pre-processing step, which is often required for range images, is smoothing (section 5.2.3.2). This is particularly important when working with consumer-grade sensors as they often deliver noisy range data. A standard moving-average filter can be used to reduce the noise level. More complex edge-preserving smoothing filters, such as the bilateral filter, can also be used. Fig. 5.96 shows an example of different smoothing techniques applied to a range image. The large averaging mask (15x15) shrinks the object from the boundaries.

Other operators can also be adapted for use on range images. However, they do not always produce the desired result. The application of standard edge-extraction operators (section 5.2.4) to a range image results in "jump edges", which are edges in the range image where the object in the foreground "jumps" to the background. This is, for example, useful for detecting the silhouette of a foreground object. However, standard image edge extraction cannot detect crease edges, which are continuous changes in surface direction. An example of this type of three-dimensional edge is the ridge of a roof.

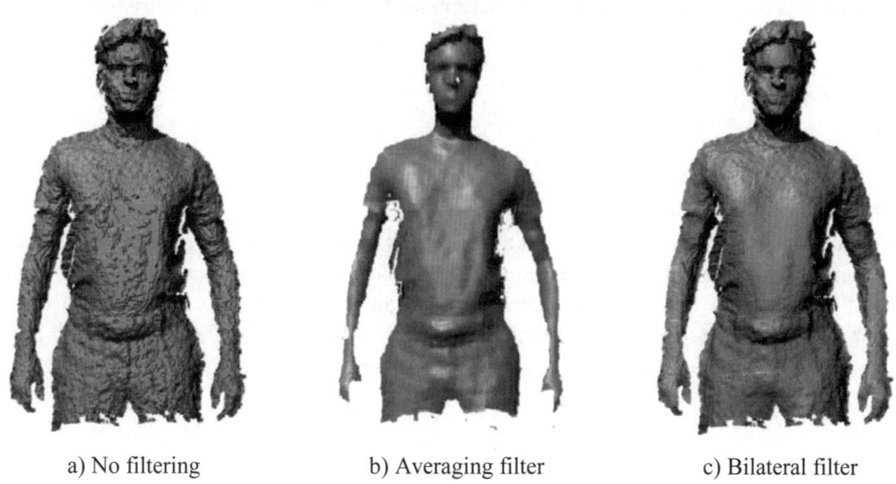

a) No filtering b) Averaging filter c) Bilateral filter

Fig. 5.96: Rendered range image with different filters applied

Work on feature extraction for range images therefore often concentrates on surface curvature for which local derivatives act as a building block in their computation. For the computation of curvature the first and second derivatives are required. This can be done either numerically using derivative filters (sections 5.2.4.1 and 5.2.4.2) or analytically by fitting polynomial surfaces. Since second order derivative filters are highly sensitive to noise, they need to be combined with smoothing filters. Local curvature can be used both for surface segmentation, edge extraction and feature-point detection. Either the two principal curvatures k_1 and k_2 are used directly or mean curvature H and Gaussian curvature K are derived from them:

$$H = \frac{k_1 + k_2}{2} \qquad K = (k_1 \cdot k_2) \tag{5.97}$$

Using H and K local surface types can be classified into nine categories shown in Table 5.4.

Table 5.4: Surface types classified by mean curvature H and Gaussian curvature K

	$K<0$	$K=0$	$K>0$
$H<0$	saddle surface (negative)	cylinder (negative)	ellipsoid (negative)
$H=0$	minimal surface	plane	-
$H>0$	saddle surface (positive)	cylinder (positive)	ellipsoid (positive)

Cylindrical surface types indicate the location of an edge. An alternative description of local surface shape is the shape parameter. Again using the principal curvatures k_1 and k_2, the parameter S to describe shape and the parameter C to describe strength are derived.

$$S = \arctan\left(\frac{k_2}{k_1}\right) \qquad C = \sqrt{k_1^2 + k_2^2} \tag{5.98}$$

This separation of shape and strength can be seen as an analogy to the separation of colour and intensity used in alternative colour spaces. The parameter C allows for the easy detection of planes (where C is smaller than a certain threshold). The parameter S can be used for the simple detection of umbilical points ($k_1 = k_2$) and minimal points ($k_1 = -k_2$).

6 Measuring tasks and systems

6.1 Overview

This chapter presents practical, working systems, often commercially available, which utilise photogrammetric principles to solve a range of measuring tasks. The very wide variety of potential application areas and system solutions in close-range photogrammetry makes it difficult to classify typical system concepts. A start is therefore made with a classification according to the number of images or cameras used. A common feature of these systems is that they primarily permit the acquisition of individual object points. Further sections continue with the presentation of surface scanning systems for measuring free-form surfaces and systems for measuring dynamic processes. The chapter concludes with a particular group of measuring systems which makes use of mobile platforms.

In contrast to traditional opto-mechanical instruments, digital processing systems consist of hardware and software components which are subject to rapid technical change. Hence, the solutions and products presented here represent only a snapshot of current technical development. This should, however, also provide clear pointers to the concepts and possibilities of future systems. Note in the following that any stated technical details and measurement accuracies are based on the manufacturer's claims and specifications.

6.2 Single-camera systems

6.2.1 Camera with hand-held probe

Based on the principle of (inverse) space resection (single image, 6DOF calculation, see section 4.2.5), a single camera can determine the current spatial position of a measurement probe in the camera's coordinate system or another reference system (Fig. 6.1). For this purpose the probe (locator) must be provided with at least three reference targets who 3D coordinates, together with the 3D coordinates of the touch point, are known in a local probe coordinate system. If there are more than three reference targets visible on the probe, over-determined measurements are possible which contribute to an increase of accuracy and reliability in the measurement. The reference targets can be distributed spatially, in a plane or simply arranged on a straight line. Depending on the chosen arrangement, the number of degrees of freedom which can be calculated drops from 6 (spatial) to 5 (straight line). The 3D coordinates of the touch point are generated according to eqn. 4.40.

It is not possible with a single-camera system to identify the probe targets on the basis of their measured 3D coordinates, as is possible with stereo or multi-camera systems (section 6.4.2.2). Point identification must therefore be achieved by using coded targets (section 3.6.1.5) or unique target separations (e.g. enabling the use of cross ratios in the image, see section 2.2.1.5).

Fig. 6.2 shows the SOLO system from Norwegian company Metronor. It consists of a factory-calibrated camera and a hand-held touch probe with 5 LED targets arranged on a plane. The operator records a measurement by depressing a switch on the probe. The accuracy of a point-pair separation (distance measurement) in a measuring volume of 1.5 x 1.5 x 1.5 m^3 is 0.12 mm (2 sigma).

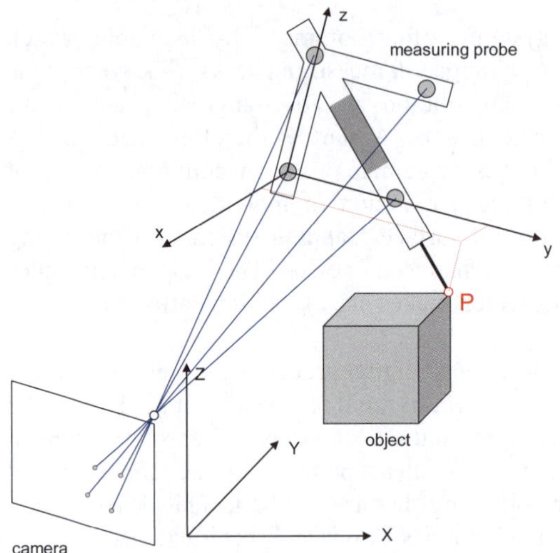

Fig. 6.1: Principle of a single-camera system with a touch probe for object measurement

Fig. 6.2: Example of a single-camera system with 3D touch probe (Metronor Solo)

6.2.2 Probing system with integrated camera

Self-locating cameras automatically determine their own orientation with respect to a fixed or moving set of reference targets.

The AICON ProCam system is a touch probe which incorporates such a self-locating camera. In detail the probe consists of a CCD camera, a recording button and a bayonet connector for attaching a range of exchangeable probe tips which are calibrated with respect to each other in a local xyz system (Fig. 6.3, Fig. 6.4). A system of reference points with known positions in a superior XYZ coordinate system are arranged on the borders of the measuring volume. If the measuring probe is positioned in such a way that the camera images a minimum of three reference points, the exterior orientation of the probe can be calculated by space resection. Coded targets provide automatic identification of the reference points.

The measuring probe is connected to the control computer by wire in such a way that a synchronised image is acquired when the operator records a measurement. The images of the reference points are found and measured automatically. A measuring accuracy of about 0.1mm can be achieved in a variable measuring volume that is limited only by the dimensions and accuracy of the reference point field. The reference field can be mounted in a fixed position, or it can be mobile. By adapting the size of the reference points, the distance to the reference field can be freely chosen.

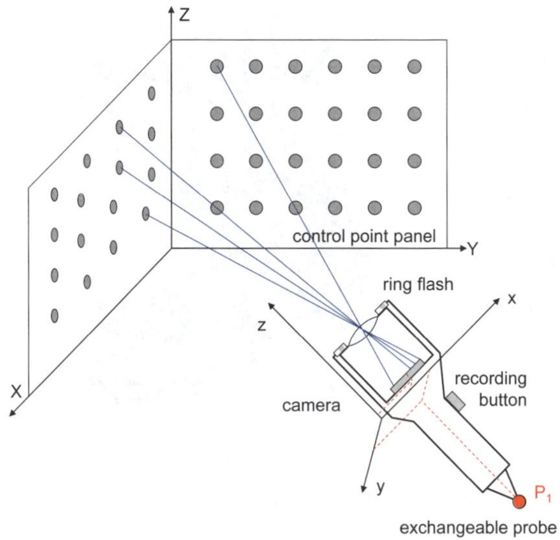

Fig. 6.3: Principle of a probing device with integrated camera

Fig. 6.4: Probe with integrated camera (AICON ProCam)

Compared to the usual online systems, the principle of the self-locating measuring probe has the advantage that the measuring accuracy is independent of the intersection angles of image rays. In addition, any number of object points can be measured without re-arranging the camera configuration, provided a sufficient number of reference points are visible.

6.2.3 Camera system for robot calibration

In general, robot calibration is used to check the robot arm position with respect to its nominal position, and to derive correction data for angle and distance adjustments. This task can be solved by online or offline photogrammetry.

One possible solution is to have digital cameras fixed in such a way that they observe the workspace of the robot. The robot arm is fitted with a calibrated reference object which is moved to various positions in space. By space resection, the position and orientation of the calibrated reference object with respect to the cameras and the robot arm are determined.

Fig. 6.5a, shows the CAROS$^+$ system (dating from around 1997) which is an online photogrammetric system using one (or optionally two) digital cameras. The single-camera version is used for adjustment of industrial robots, the measured poses (robot position and orientation) being determined by space resection with respect to the test field plate which is automatically tracked and measured within the workspace. Fig. 6.5b shows a robot calibration system in current use. This employs a high-frequency, three-line camera (compare with Fig. 3.125) which records a set of LED targets on the reference. The typical workspace (measuring volume) is about 3 x 3 x 2m^3 within which an object accuracy of 0.1 mm can be achieved. The OptoPose pose estimation system (metronom, I3 Mainz) is an example of a multi-camera system that also can be applied to robot calibration.

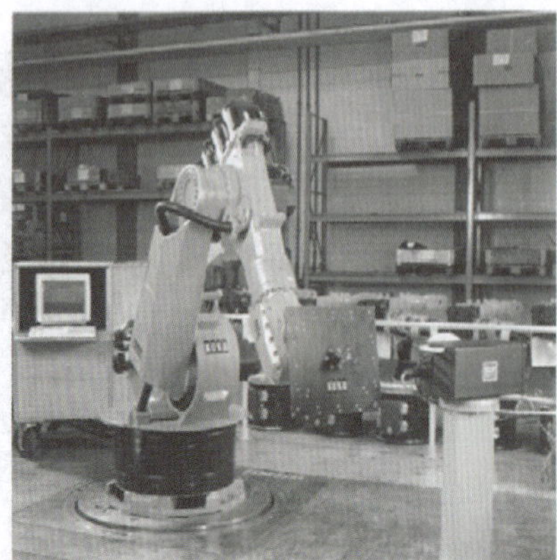

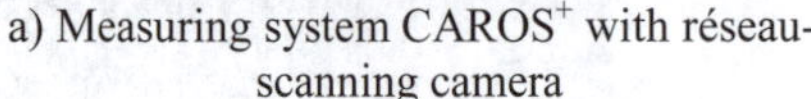

a) Measuring system CAROS$^+$ with réseau-scanning camera

b) Robot calibration with three-line camera (Nikon Metrology)

Fig. 6.5: Photogrammetric robot calibration and adjustment

6.2.4 High-speed 6 DOF system

Fig. 3.119 shows a high-speed camera system used to record the dynamic changes in position and angular orientation of a moving object. The example is the AICON WheelWatch, used to monitor wheel movements on a moving car.

The principle of operation is again based on a space resection, as described in section 4.2.3. By monitoring two separate sets of targets, one on the car body around the wheel arch and one on the wheel itself, the movements of the wheel relative to the body can be determined.

This particular system has a processing unit integrated into the camera which enables real-time target detection, and hence real-time calculation of the relative 6DOF between wheel and car body. This means that a dynamic measurement sequence can be of unlimited duration since there is no image recording but instead a continuous output of 6DOF values.

6.3 Stereoscopic systems

6.3.1 Digital stereo plotters

6.3.1.1 Principle of stereoplotting

For the strictly visual analysis of stereo imagery, today's dominant tool of choice is the digital stereo workstation which is the successor to the analytical stereoplotter. Both systems are based on the same principles, although the mechanical image carriers of an analytical plotter (Fig. 6.6) are replaced by image projection on a stereo monitor in the digital workstation. Although analytical plotters are now seldom used in practice, they are presented in this section in order to illustrate the principles. Design features relevant to digital workstations appear in later sections.

Using suitable input devices (3D cursor/mouse, hand and foot wheels), a floating mark is moved in 3D model space in coordinate increments $\Delta X,\Delta Y,\Delta Z$. Using the collinearity equations (4.9) and the parameters of interior and exterior orientation, the current XYZ position is transformed into corresponding image positions x',y' and x'',y''. Using servo control or digital panning, the images are also incrementally shifted in real time so that these image positions match the corresponding floating mark locations. The operator stereoscopically controls the 3D position of the floating mark in model space and, by interactive corrections to the floating mark's XYZ coordinates, ensures that it appears to touch the model surface ("setting the floating mark"). Only when the two measuring marks lie on homologous image points does the operator have a convincing visual impression that the floating mark lies on the model surface. At that stage the computed XYZ values may be stored. Hence, the analytical plotter is based not on direct measurement of image coordinates but on the computation of coordinates in 3D space. Fig. 1.32 shows an example of an analytical plotter.

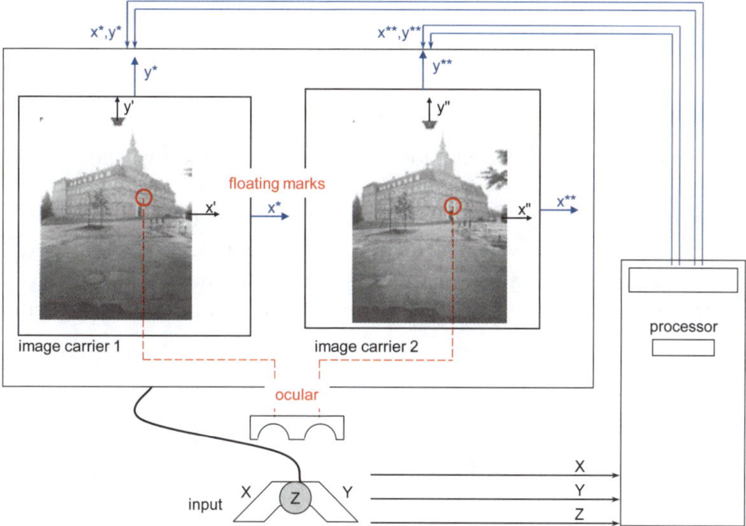

Fig. 6.6: Principle of the analytical plotter

Visual stereoscopic measurement is restricted to the normal case in which the camera axes are approximately parallel and perpendicular to the base. It is possible to compensate for rotations, κ, about the optical axis by optical image rotation (dove prism). However, large differences in image scale or rotation angles φ,ω between the two images cannot be accommodated by a human operator. For this reason, strongly convergent stereo images are often first transformed into normal-case images (see section 4.3.3.5).

Instrument control (image carriers, illumination, input devices), and transformation of the input XYZ values, is performed by a processor which delivers continuous 3D object coordinates. Hence the analytical plotter can be regarded as a 3D digitiser. Using a suitable interface, these 3D coordinates can be transferred to any post-processing system, for example a 3D CAD system or a geographic information system (GIS) which can be located on a separate computer system.

6.3.1.2 Orientation procedures

The orientation parameters required for stereoscopic measurement are determined according to the orientation procedures explained in section 4.3. In the first instance, the image coordinate system must be established. For digital images, this is inherently defined by the pixel array on the sensor. For analogue images, machine coordinates of fiducial marks or réseau crosses, defining the camera reference system, are measured monocularly, their calibrated values being known. Parameters are computed for a plane transformation which will bring the machine coordinates into their calibrated values.

Typically, the parameters of exterior orientation are then determined by one of the following procedures:

- Two step orientation:

  Following conventional practice, parameters are determined in two separate steps of relative orientation and absolute orientation. It will already be possible to acquire 3D model data after a successful relative orientation.

- Transfer of known orientation parameters:

  If the parameters of exterior orientation are known from previous calculations, they can be loaded either for single images or for stereo models. Consequently the system will be ready for use as soon as the image coordinate system has been established. Usually the exterior orientation will have been determined by bundle triangulation (section 4.4). When using existing orientation parameters, care must be taken to apply the rotation angles in the correct order, or calculate a transformation for a different rotation order.

6.3.1.3 Object reconstruction

Analytical and digital stereo instruments essentially only allow the 3D measurement of object points. Polygons are recorded as lines, areas are represented by dense point grids (digital surface models).

In order to measure linear object structures, an analytical plotter can be switched into a continuous measuring mode. In this mode, object point coordinates are continuously registered at fixed time or spatial increments as the operator moves the floating mark along the linear feature.

To a large extent, the measurement of digital surface models can be automated. For this purpose the control program calculates a fixed grid of points on the reference plane (for example $\Delta X = \Delta Y =$ const.). The stereo plotter automatically moves to these points, at each of which the operator (or a correlation algorithm) only has to set the floating point on the surface in order to determine its Z coordinate.

6.3.2 Digital stereo viewing systems

Digital stereo workstations have no mechanical image carriers. Instead, digitised images are displayed and moved digitally on a suitable stereo monitor. Each image may be moved relative to a superimposed measuring mark. The measuring marks, the images, or both simultaneously, can thereby be moved.

Stereoscopic viewing is achieved through either optical or electronic image separation. Essentially, the following techniques are used:

- Split screen:

  Both images are displayed side by side on a standard monitor and are viewed stereoscopically using a simple optical system similar to that of a mirror stereoscope. The images move which requires a fixed floating mark. The operator is forced to sit in a fixed position in front of the monitor and the system can only be used by one observer.

- Anaglyphs:

  Usually left and right images are coloured red and green, and viewed through glasses that have corresponding red and green filters. This principle is only useful for grey level images. Stereoscopic viewing is possible over large viewing distances and angles, and for a number of people simultaneously.

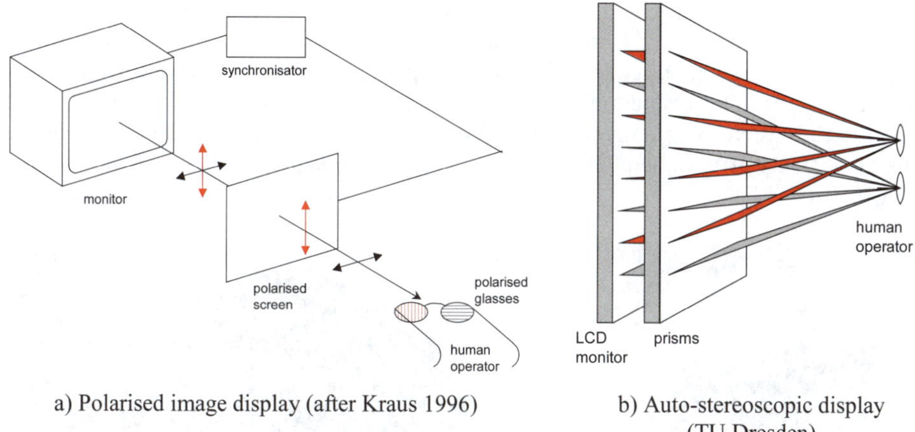

a) Polarised image display (after Kraus 1996)

b) Auto-stereoscopic display (TU Dresden)

Fig. 6.7: Possible designs of stereo monitors

- Alternating image display:

  Left and right images are alternately displayed at high frequency (>100 Hz). The operator wears glasses which have synchronised shutters so that only the currently displayed image is visible to the corresponding eye. Stereoscopic viewing is possible within a large workspace in front of the monitor, and for a number of people simultaneously.

- Polarised image display:

  An LCD filter is mounted in front of the monitor. The filter switches polarisation direction as the left and right images are alternately displayed. The operator wears glasses with correspondingly oriented polarising filters in order to present the correct image to the correct eye. Electronic control of these glasses is not required (Fig. 6.7a).

- Beam splitting display:

  A beam splitter (semi-silvered mirror) is located at the bisecting angle between the two displays (Fig. 6.8b). Polarised light emitted from the bottom monitor is transmitted through the mirror while polarised light emitted from the top monitor is laterally inverted upon reflection off the beam splitter. Polarised glasses are used for stereoscopic viewing (Fig. 6.7a).

- Stereo projection:

  Two digital projection devices (see section 3.6.2.2) can be used to display a stereo image on a large screen. Stereo viewing is usually enabled by shutter glasses that are synchronised with the projectors. It is mostly applications in virtual and augmented reality, where observers interact with a virtual scene, which are addressed. The position of the observer is normally measured by 3D tracking systems (see section 6.6.3).

- Auto-stereoscopic display (micro-prisms):

  In this case the stereoscopic image is displayed in separated image columns (even columns for the left image and odd columns for the right image). A vertical micro-prism system is mounted over the monitor surface and deflects the column images in the desired spatial viewing direction. The stereoscopic effect may be observed without any additional aids. The resolution of the stereoscopic image corresponds to half of the monitor resolution. An automatic image and prism adjustment (eye finder) compensates for movements of the operator, so creating a large workspace in front of the monitor (Fig. 6.7b).

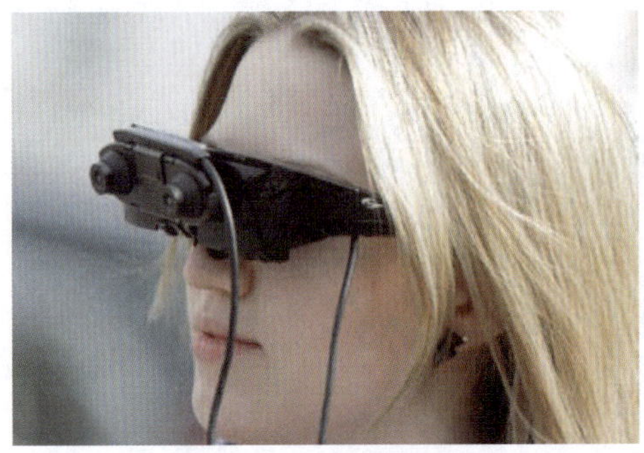

a) Stereo glasses (Vuzix)

b) Digital stereo monitor (Planar)

Fig. 6.8: Examples of digital stereo viewing systems

- LCD monitor glasses:

  The operator wears glasses with two separate LCD mini-monitors. The monitors can either display digital stereo images or can be connected to two mini cameras which observe the scene instead of the human eye. This kind of stereo viewer is used mostly for augmented reality applications (example in Fig. 6.8a).

Compared with analogue systems, digital stereo workstations offer a variety of benefits:

- simultaneous display of image (raster) and graphic (vector) information is possible;
- several workstations can be networked together in order to have parallel access to the same image material;
- the functions of comparator, analytical plotter, rectifier and graphical workstation are integrated in a single system;
- orientation procedures, as well as the measurement of points, lines and surfaces, can be automated.

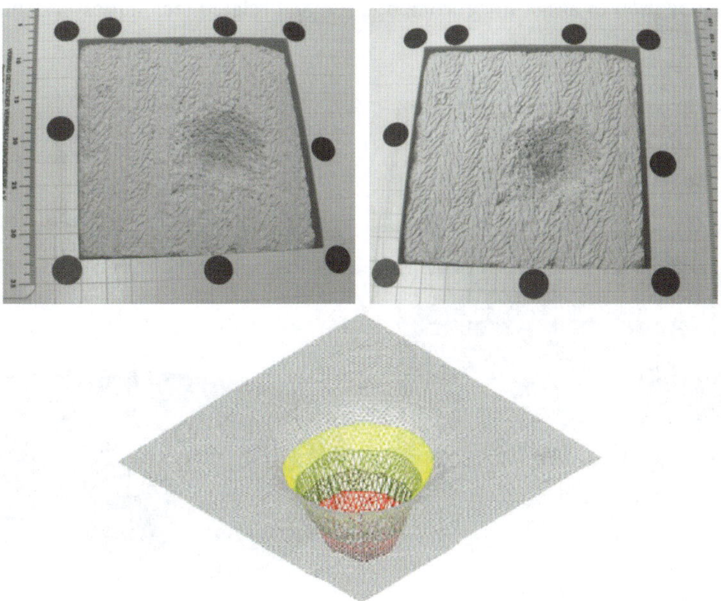

Fig. 6.9: Digital stereo-image pair and height measurement using BAE SocetSet (camera: Kodak DCS420, f = 15 mm, m = 15) (IAPG Oldenburg)

Digital stereoplotters can in principle be used for the processing of any stereo pairs. There are no technical restrictions on the imaging sensors used. However, the standard integrated software is frequently adapted to aerial imagery. As an example of a close-range application, Fig. 6.9 shows a digital stereo-image and the automatically measured height model of a crater in a cellular concrete block, used to demonstrate the properties of high-pressure injection nozzles.

6.3.3 Stereo vision systems

Stereo vision systems are dual-camera systems which provide an online 3D measurement space for applications such as the following:

- navigation of autonomous vehicles (detection of obstacles)
- 3D navigation for computer-controlled surgical operations
- control of manufacturing robots (location of workpieces)
- recording the internal 3D shape of pipes
- examination using stereo endoscopy (medicine, material testing)

In general these systems consists of two identical digital video cameras, mounted with their axes parallel. Their images are processed on line, or transmitted for visual observation. Fig. 6.10 shows a robot with stereo cameras. Fig. 6.11 shows a stereo camera which has a base of 60 mm and is used to assist in neuronavigation and related techniques in dental and aural surgery.

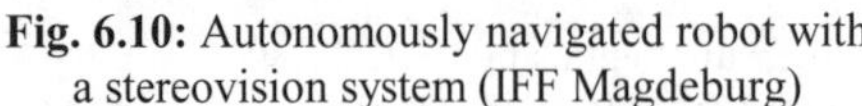

Fig. 6.10: Autonomously navigated robot with a stereovision system (IFF Magdeburg)

Fig. 6.11: Stereo camera system with 60mm base for small measurement volumes (AXIOS 3D)

A stereo camera's measurement volume is determined by its base, focal length and direction of camera axes. It is clear from Fig. 6.12 that the height-to-base ratio h/b can be improved by a suitable choice of convergence angle between the camera axes. It is therefore possible to reduce the object distance and improve the accuracy in both XY and Z (towards object). See also section 4.3.6.2.

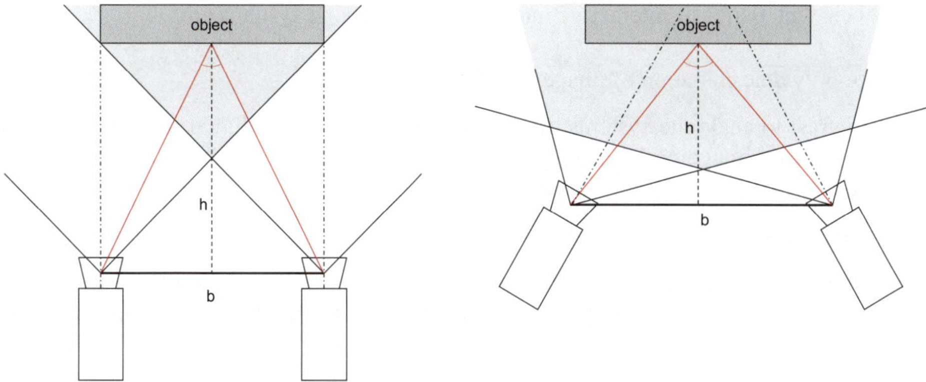

Fig. 6.12: Measurement volume and height-to-base ratio for different stereo configurations

Example 6.1:

A stereo camera system ($c = 12$ mm) is used in two configuration according to Fig. 6.12. For an image or parallax measuring accuracy of 0.5 µm, the following accuracies are achieved in object space:

Imaging parameters:	$b = 600$ mm	$b = 700$ mm
	$h = 600$ mm	$h = 420$ mm
	$h/b = 1$	$h/b = 0.7$
	$m = 50$	$m = 35$
Measurement accuracy:	$s_X = 0.025$ mm	$s_X = 0.017$ mm
	$s_Z = 0.025$ mm	$s_Z = 0.012$ mm

With convergent geometry a significantly higher measurement accuracy can be achieved, by as much as a factor of 2 in the viewing direction.

Based on the properties of digital video cameras, the measurement frequency of stereo cameras lies between 10 and 50 Hz. Higher frequencies can be achieved by the use of high-speed cameras which can be configured to measure in the same way as video cameras. Using a stereo mirror attachment (section 3.4.3.7), a stereo system can be created with a single camera.

6.4 Multi-image systems

6.4.1 Interactive processing systems

Interactive, multi-image processing systems permit the measurement of image points in more than two images. Checking of point identification and correspondence is performed by the operator. As well as pure coordinate measurement, these systems are particularly used for the reconstruction of graphical features, such as lines, or the production of virtual 3D models. The data acquired are usually transferred to a CAD program for further processing.

Typical characteristics of interactive multi-image processing systems are:

- Once-only determination of image coordinate system:

  In contrast to analogue systems, the transformation from the digital pixel system to the camera-based image coordinate system need be established once only as its parameters can then be stored within the image file.

- Image processing:

  The contrast, sharpness and brightness of the images can be enhanced[1] for ease of use. Image rectification or orthophoto production can be integrated into the processing system.

- Automation:

  Functions for the detection and measurement of point features not only allow automated interior orientation but they also expand interactive systems to photogrammetric online or offline systems.

- Superimposition of graphical information:

  Digital images and graphical information (vector data) can easily be superimposed permitting improved interactive control of the measurement.

- Integration into CAD environment:

  Display and measurement of images and 3D CAD processing of the resulting object data can take place within one closed system.

The PHIDIAS (Phocad) interactive photogrammetric system (Fig. 6.13) demonstrates the integration of digital photogrammetry with the 3D CAD environment. The system is embedded in the Microstation CAD program (Bentley) which therefore makes available all the standard CAD functions for the photogrammetric reconstruction of the object. The system is well suited to the complex graphical reconstruction of objects such as occur in architectural applications or in the as-built documentation of industrial plants. Integrated bundle adjustment with self-calibration, as well as function such as mono and stereo plotting, are included within the scope of the system.

Low-cost systems, with a basic range of construction tools for the simple generation of 3D models, have a much more restricted scope and performance. Examples of such systems, which become widespread in recent years, include Photomodeler (EOS Systems) and iWitness (Photometrix). Here the level of user guidance and automation in the orientation process ensure that even non-specialist users can deliver usable results after a short learning period.

Although the iWitness system (Fig. 6.14) is mainly designed for accident and crime-scene recording (see section 8.5.1.1), it can essentially be used to record any scene of choice. It supports a database of commercially available, off-the-shelf digital cameras which is constantly updated. Notable features are the integrated camera calibration which makes use

[1] Manipulating images can alter the imaging geometry.

of colour-coded targets (see section 3.6.1.5), as well as robust orientation algorithms which can handle the often unfavourable imaging configurations typical in forensic measurement.

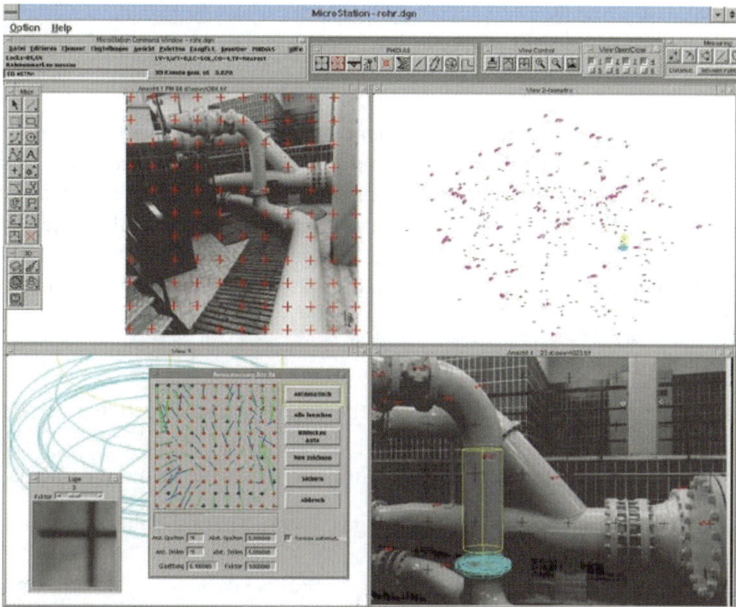

Fig. 6.13: PHIDIAS-MS interactive multi-image system (Phocad)

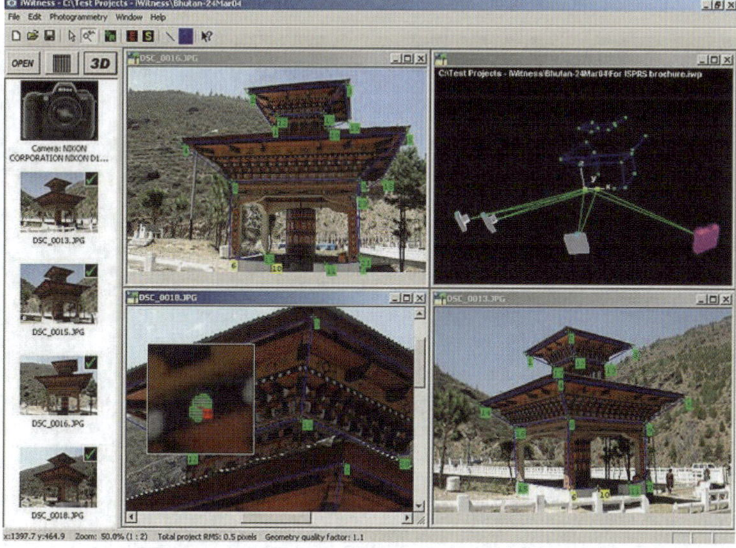

Fig. 6.14: iWitness interactive multi-image system (Photometrix)

6.4.2 Mobile industrial point measuring-systems

With the introduction and use of digital cameras and digital image processing, the customarily separate procedures of image acquisition on-site followed by image processing in the office may be combined. The complete measurement of an object can be performed on-site.

For point-based photogrammetric measurement of objects, a distinction is made between online and offline systems. In both cases discrete object points are measured, in some cases by optical setting on distinctive object features or targets and in other cases by use of a mechanical probe. In the latter case the point of the probe is set on the object point while targets on the probe are imaged. From the computed positions of these targets and a knowledge of the geometry of the probe the coordinates of the probed point are found.

6.4.2.1 Offline photogrammetric systems

An offline photogrammetric system has these characteristics (see section 3.2.1.1):

- photography of the object with at least two images from one or more cameras;
- subsequent orientation of the set of images, simultaneous calibration and 3D point determination by bundle triangulation.

The above two steps are separated in time and, possibly, in space (offline, see Fig. 3.34). As in much of photogrammetry, there are no restrictions in terms of imaging sensors, object targeting and image configuration (number and position of images).

Fig. 6.15: Object recording using a photogrammetric offline system (GSI) (lower left: detail enlargement showing coded targets)

Offline photogrammetry is very common in industrial applications where the use of digital cameras and simple or coded retro-reflective targets is an integral part of the process. As a result, automatic orientation and image measurement are possible.

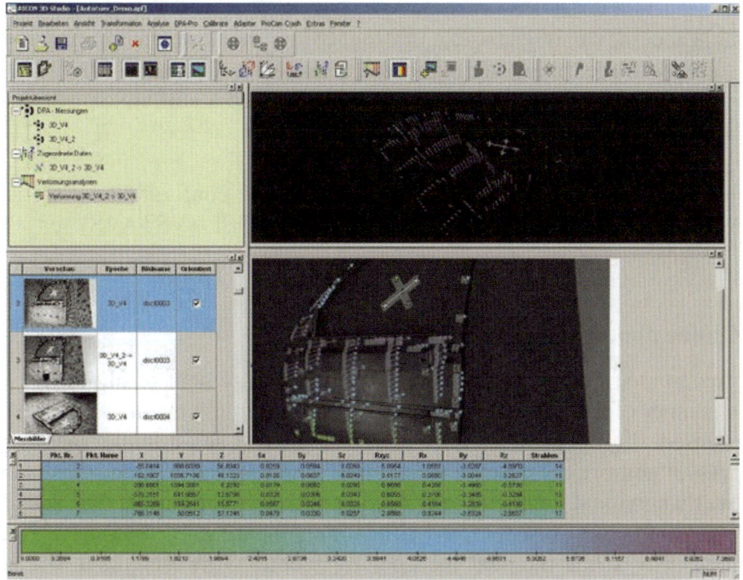

Fig. 6.16: User-interface of the system AICON 3D Studio

Fig. 6.15 shows the recording of an object which is fitted with a number of simple targets (tie points) and coded targets. The use of coded targets ensures that all target types can be reliably identified and correlated across all images and this, in turn, enables fully automatic image orientation (compare with section 4.4.4.2). Fig. 6.16 shows the user-interface of a typical multi-image system. Object measurement with industrial offline systems has the following characteristics:

- sequential image acquisition with high-resolution digital cameras with image storage and processing within the camera; optional image transfer via WLAN and ring flash
- object marking with circular retro-reflective targets
- coded targets for automated generation of approximate values and image orientation
- calibrated reference tools which establish a local 3D object coordinate system
- bundle triangulation with self-calibration
- digital point measurement with sub-pixel accuracy (0.02–0.05 pixel)
- typical number of images between 10 and 100 (no limit in principle)
- typical number of images per object point between 6 and 20 (no limit in principle)
- typical object dimensions between 1 m and 15 m (no limit in principle)

- typical duration for object recording and processing between about 10 min and 60 min
- achievable relative accuracy about 1:100 000 to 1:250 000 (1-sigma RMS) or 1:50 000 to 1:100 000 (length measurement error, see section 7.2.2)

The achievable accuracy in object space is strongly dependent on the imaging configuration and on the stability of the camera during image acquisition. The values quoted above can be achieved with good and multiple ray intersection angles and a uniform distribution of points over the available image format.

As examples, commercial offline systems in different configurations are offered by: AICON (MoveInspect/3D Studio), GOM (TriTop), GSI (VSTARS) or Photometrix (Australis).

6.4.2.2 Online photogrammetric systems

Online photogrammetric systems enable the direct measurement of 3D object coordinates much in the manner of a coordinate measuring machine. In the majority of systems, at least two synchronised digital cameras are used for image acquisition, each with known calibration values and pre-determined orientation with respect to an established coordinate system[1]. Usually a manual touch probe is used to make measurements (see section 3.6.1.6). Its spatial position and orientation within the measuring volume are determined photogrammetrically at the moment the operator records a measurement, the positions of the reference points on the probe then being calculated by intersection with respect to the globally defined XYZ coordinate system. Since the position of the probe tip P is known relative to the reference points in the probe's local xyz coordinate system, the tip coordinates are readily transformed into the global coordinate system (Fig. 6.17).

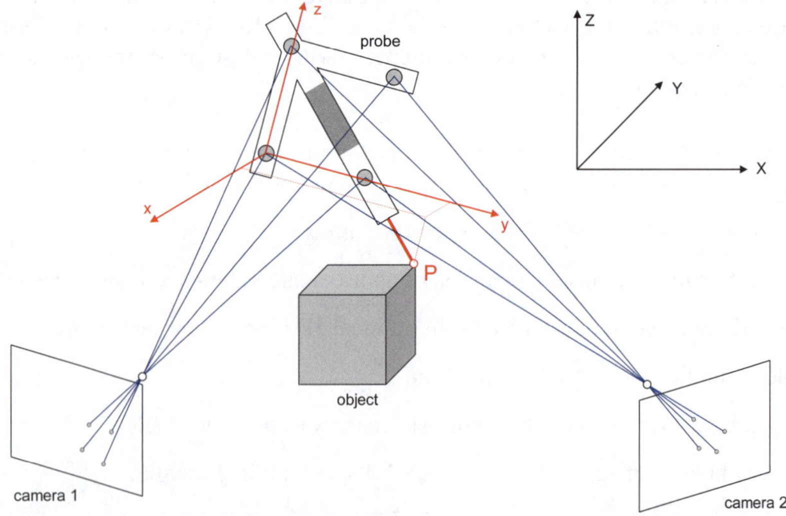

Fig. 6.17: Measurement principle of an online system with tactile probing

[1] See section 6.2.1 for manual touch-probe systems with a single camera.

Depending on probe type, measurement recording is made by one of the following methods:

- by depressing a button or activating a switch on the probe which remotely triggers image recording and measurement,
- by the use of a touch-trigger probe, commonly seen on coordinate measuring machines, which remotely triggers an image recording and measurement when the probe tip is touched lightly against the object surface,
- by mechanical displacement of one of the probe's reference points, which is then detected by the cameras and used to record a measurement (see section 3.6.1.6),
- by operation of an external switch, mouse click or keyboard command.

Camera operation, illumination (flash) and image transfer are controlled by a networked computer which also provides image processing and coordinate calculations. The computer can be located externally (laptop, computer trolley) or within the camera. In the latter case, image processing (point measurement) is performed directly within the camera, so that only a small number of image coordinates are transferred to the post-processing computer (example in Fig. 3.116).

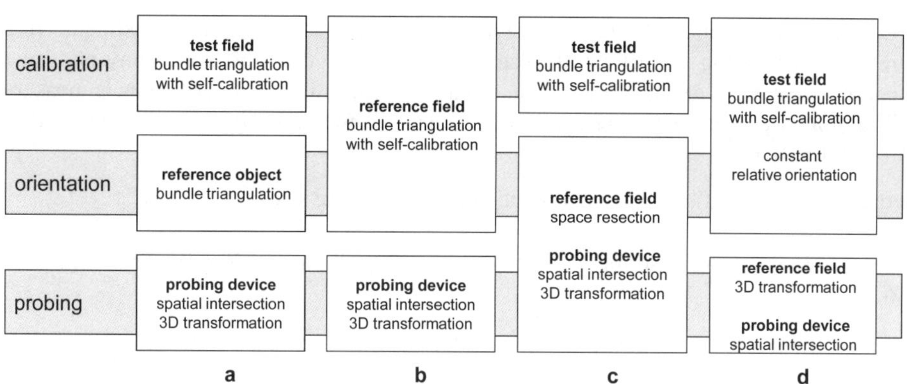

Fig. 6.18: Strategies for calibration, orientation and point measurement for online systems

Calibration and orientation of online systems can be performed in various ways according to the following schemes (Fig. 6.18):

(a) Separate pre-calibration, orientation and point measurement:

The parameters of interior orientation of the cameras are determined separately from the process of object measurement, for example by test field calibration. All cameras are oriented by simultaneous imaging of a reference field that defines a local 3D coordinate system. Subsequently the object is measured point by point, by intersection and 3D transformation of the target points and the probing tip.

This approach requires stable interior orientation of the cameras for the complete period between calibration and completion of the measurement. In addition, the exterior orientation of the cameras must be kept constant, for example by the use of fixed mechanical camera mountings or stable tripods.

(b) Calibration and orientation using a reference object added to the measuring object or measurement space:

A calibrated reference field is positioned on or around the object to be measured and provides local 3D control points or reference lengths. After setting up the cameras, their positions and orientations can be calculated from the reference data by space resection (separately for each camera) or by bundle triangulation. Under some circumstances, the interior orientation can be determined simultaneously if the reference field provides a suitable sufficient number and distribution of control points, or if it is moved and photographed in a number of different spatial positions. Point measurement follows as in (a).

(c) Integrated orientation and point measurement:

The orientation of the cameras can be integrated with the point measurement itself. For each new measurement of the probe position, the cameras also record a reference field which is used to check or recalculate the exterior orientations. Such a method is essential if stability of the cameras cannot be ensured.

(d) Stable relative orientation and 3D transformation:

Camera calibration and relative orientation and done by bundle adjustment. If the relative orientation is sufficiently stable, the probe is then located in any convenient coordinate system at the same time as a stable set of reference coordinates is measured. The probe tip coordinates are then transformed into the target system by a 3D similarity transformation.

There are various configurations and applications for online systems, such as:

- dual-camera system with manually tracked touch probing (as described above, see Fig. 6.19),
- dual-camera system for navigation of medical instruments or other sensors (Fig. 8.58),
- dual-camera system for tracking dynamic processes, for example the spatial movement of a robot,
- multi-camera system for analysing human motion (MoCap - motion capture), typically for applications in medical science and entertainment industries (Fig. 6.40, Fig. 6.41),
- multi-camera system for deformation monitoring by repeated measurements of a set of reference points (example application in Fig. 6.20).

Fig. 6.19: Online system with flexible camera set-up and manual touch probing (Metronor)

Fig. 6.20: Online system with fixed camera set-up for machine control (GOM PONTOS)

Assuming that the system components are the same in each case, the measuring accuracy of online systems will be less than that of offline systems, mainly for the following reasons:

- small number of images for each point measurement (default: 2);
- possible different measuring accuracies in X, Y and Z (see section 4.3.6.2);
- the strict stability requirements for interior and, especially, exterior orientation are difficult to meet;
- contact of the probe with the surface, and the triggering of the cameras, must occur simultaneously; unless a self-triggering type of probe is being used, this is dependent on the operator's skill.

With these points in mind, commercial online systems offer an accuracy in the object space of about 0.1–0.2 mm for a measuring distance up to 2 m (assumptions: image scale $m = 100$, base length $b = 2$ m). This corresponds to a relative accuracy of about 1:10 000 to 1:20 000, or an image measuring accuracy of 1 µm.

Available systems differ mainly in terms of the operating procedure and the strategies for orientation and error detection. Mobile online systems for point-by-point measurement are available, for example, from GOM (PONTOS), Metronor (DUO), AICON (MoveInspect), AXIOS 3D (CamBar) or Nikon Metrology (K-Series CMM).

6.4.3 Static industrial online measuring systems

Some specialised and repetitive measuring tasks are well suited to the development of specialised solutions with a high degree of automation. In contrast to mobile systems, these are static systems which occupy a semi-permanent position within a factory. They can be designed so that the photogrammetric measuring components (such as cameras, projectors, devices for rotating the object) can remain calibrated for a long period and the cameras can be oriented automatically using a fixed and known reference point field. Such a system is in a permanent state of readiness for repeating the same or similar measurements within a fixed, limited volume and with a considerable degree of automation.

Multi-camera photogrammetric systems can also be integrated directly into production lines for purposes of quality control. The measuring problem usually concerns a limited variety of parts, so that a fixed arrangement of cameras, light sources and control points is possible. The function of such a system is real-time measurement of the objects followed by quality analysis and the transfer of results to the manufacturing control system.

6.4.3.1 Tube inspection system

AICON's TubeInspect offers a solution for the automatic 3D measurement of formed tubes, rods or wires, a typical example being hydraulic brake lines for the automotive industries. This non-contact approach has replaced the previous use of gauges.

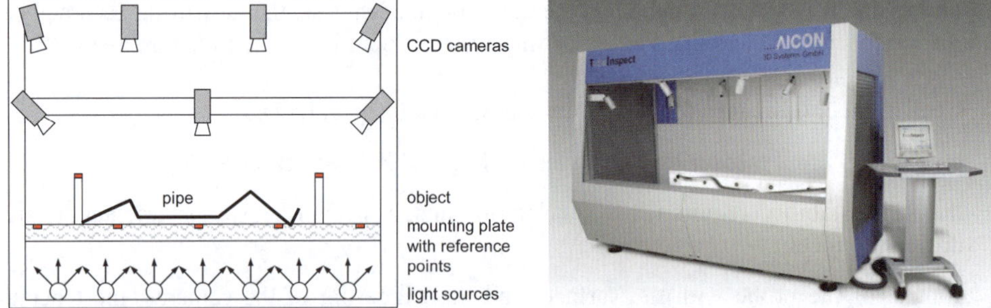

Fig. 6.21: Online system for check measurement of tubes, e.g. hydraulic brake lines (AICON)

The system is arranged on several horizontal planes distributed vertically (Fig. 6.21). The upper two planes contain up to 16 CCD cameras. The lower two planes consist of an illumination array of light sources and a transparent mounting plate with spatially distributed reference points which are used for camera orientation. The tube to be measured is placed on the mounting plate in an arbitrary position.

In each image, the tube appears as an easily identified black contour against a bright background (Fig. 6.22). Bending points, bending angles, straight line sections and arcs can be determined fully automatically by digital multi-image processing. The system also calculates correction data with respect to given nominal values that are directly transferred to the bending machine for adjustment of the manufacturing process (Fig. 6.23). The measuring accuracy has been reported as 0.3–0.5 mm within a measuring volume of about 2.5 x 1.0 x 0.7 m^3.

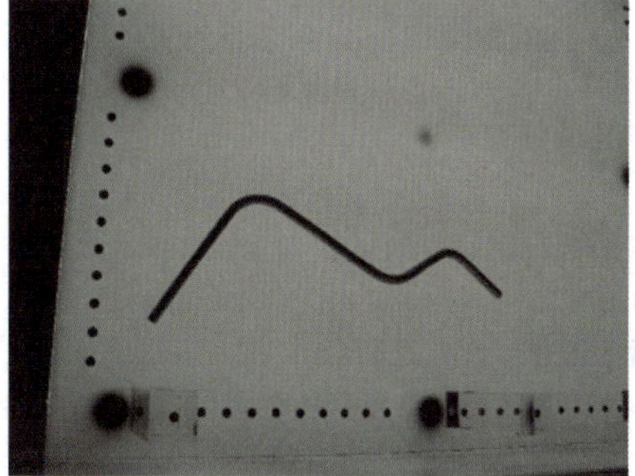

Fig. 6.22: Backlit image of hydraulic brake line (AICON)

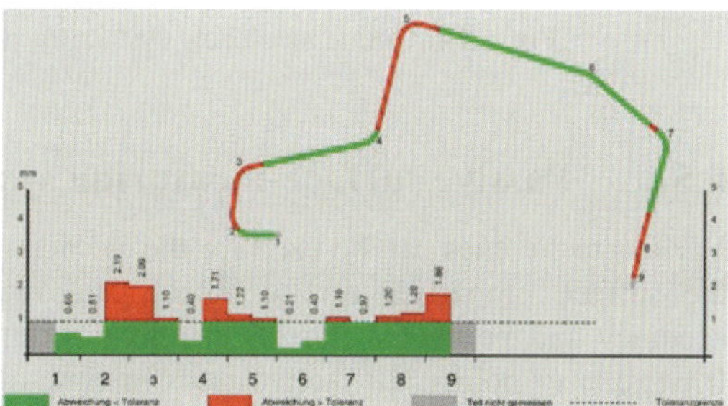

Fig. 6.23: Comparison with nominal values in hydraulic brake line measurement (AICON)

6.4.3.2 Steel-plate positioning system

A stereo camera system can be applied to the task of three-dimensional online positioning of components in manufacturing and assembly plants. This is illustrated by the following example of an online stereo system (AXIOS 3D CamBar B2) used to position steel plates for ship construction, with dimensions up to 30 m x 30 m, on laser cutting machines.

The system itself only has a measurement volume of 280 mm x 360 mm and an offset distance of 600 mm. It is mounted on a coarse positioning device which places it over an area to be measured on the plate. Cross-shaped marks are etched on the surface of the plate and have known (nominal) positions in the component's local coordinate system. The stereo system can identify the marks automatically, even under highly variable illumination conditions, and can determine their 3D locations relative to known reference drill holes. Depending on the quality of the target marks, measurement accuracy can be as high as a few hundredths of a millimetre. The entire plate can be positioned to an accuracy of 0.1 mm in the machine coordinate system. Fig. 6.24 shows the stereo system in place in the factory.

a) Stereo camera

b) Camera attached to positioning system

Fig. 6.24: Online measuring system for positioning steel plates (AXIOS 3D)

6.5 Passive surface-measuring systems

Various techniques can be used for the 3D acquisition of free-form surfaces, depending on the surface properties of the object, and requirements for duration of measurement, accuracy and point density. 3D imaging systems with active projection techniques, e.g. fringes, laser points, diffractive grids or time-of-flight principles, are presented in section 3.7. The optical systems described here all make use either of patterns projected onto the surface, or the existing surface texture itself imaged by two or more cameras. The measurement of 3D coordinates is based on correlation and matching methods, hence pure image information is used to reconstruct the three-dimensional object surface.

In general, area-based systems are used when measuring the 3D shape of arbitrary free-form surfaces. All relevant image-based techniques require sufficient structure in the surface texture in order to allow for accurate and reliable matching of homologous features. The resolution of the texture must match the geometric complexity of the surface and the desired resolution/density of the final point cloud. Sufficient contrast and gradients in the texture are necessary in order to achieve accurate matching results. The following approaches can be distinguished:

- Pattern projection:

  An array of regular or non-regular features is projected onto the object surface, e.g. by LCD devices or diffraction grids (section 3.6.2). No knowledge of calibration and orientation of the projector or of the pattern geometry is required. The pattern simply provides a visible structure defining the surface. Active pattern projection provides high contrast object textures for surfaces with few natural textures. The method is not suitable when using sequential images to measure deformation at specific object points (see section 6.5.2.1).

- Physical surface marking:

  In this case a pattern is physically applied to the object surface using adhesive film with a textured surface or power or paint. Adhesive film can be printed with any suitable pattern but it should be noted that film and adhesive thickness must be taken into account when analysing measurements. White power and similar can be applied to the object and brushed in order to generate a random pattern. Both methods have the disadvantage that the object surface is physically marked, but the advantage is that physically defined points can be tracked through image sequences, e.g. for deformation analysis or strain measurement.

- Natural surface texture:

  If the object surface has a suitable texture of its own, it can be measured directly without further preparation. The texture should consist of unique patterns that enable a reliable matching of homologous areas (see Fig. 6.9 for an example).

The two approaches can be combined. For example, a fast active projection method can be applied for coarse surface measurement which is then refined using a number of cameras and passive illumination of texture on the surface. They can be combined with standard, point-by-point photogrammetric measuring methods.

6.5.1 Point and grid projection

6.5.1.1 Multi-camera system with projected point arrays

A non-targeted, free-form surface can be measured point by point, by projecting an array of target points onto the surface (section 3.6.2) and imaging the reflected points by a number of cameras synchronised with the projection device. The projector's orientation must not be known as the point coordinates are calculated by intersected rays from oriented cameras, as is the case with conventional targets. However, if the projector's relative orientation is known then the target points can be projected on pre-programmed locations.

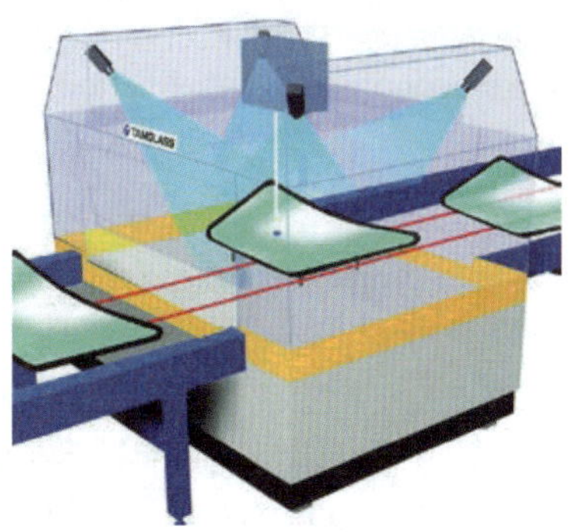

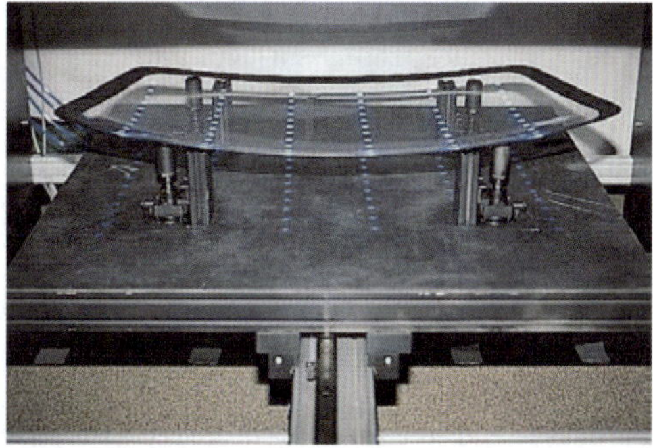

a) Schematic construction

b) Windscreen positioned for measurement

Fig. 6.25: Multi-image measuring system with laser point projection for the 3D measurement of windscreen surfaces (Mapvision)

Fig. 6.25 shows a multi-camera measuring system with target point projection, used to measure car windscreens (Mapvision). The scanner's projector emits ultraviolet light rays which cause fluorescence when they penetrate the glass. The resulting light spots on the surface of the glass are imaged by CCD cameras sensitive to the relevant wavelength. The measuring accuracy of the system has been reported as 0.2 mm.

6.5.1.2 Multi-camera systems with target grid projection

Grid projection methods are used mainly in materials testing for mapping displacement and strain. The surface shape can be reconstructed by analysing the deformation of the raster with respect to a reference position. The reference grid can be generated in different ways:

- by a grid that is physically created on the object surface,
- by projection of a grid from a position to the side of the viewing camera (Fig. 6.26 left),
- by reflection of a grid (Fig. 6.26 right).

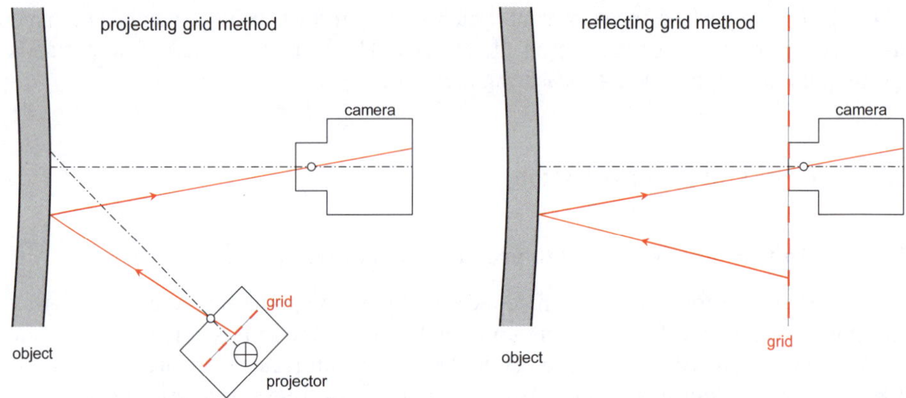

Fig. 6.26: Target grid measuring methods (after Ritter 1995)

The configuration for grid projection corresponds to that of stationary fringe projection (section 3.7.2.1) and, therefore, can be extended to multiple cameras. The reflection method assumes virtually specular surfaces and only allows measurement of surface inclination.

6.5.1.3 Multi-camera system with grid projection

The AICON ProSurf system (see Fig. 6.27) used an analogue projector to place a regular grid structure on the object surface. This structure is imaged with at least three cameras, all oriented to one another and arranged around the object.

Measurement of the object surface was done by measuring the grid intersection points in the images and locating them in object space by multiple ray intersection. Epipolar geometry was used for the correct assignment of corresponding rays. With a minimum of three images this leads to a unique and correct assignment (see section 5.5.3.2). To measure the entire grid a recursive search strategy was implemented which ensured that all measurable points were located, even when points were obscured or otherwise missing. Object point location accuracy was around 0.1 mm in a measurement volume of approximately 1 m x 1 m x 0.5 m.

In a similar way, the GSI Pro-Spot system presented in section 3.6.2.2 projects a regular array of target points onto an object surface. If projector and object have no relative movement, then the targeted object surface can also be measured by a standard offline photogrammetric system (section 6.4.2.1).

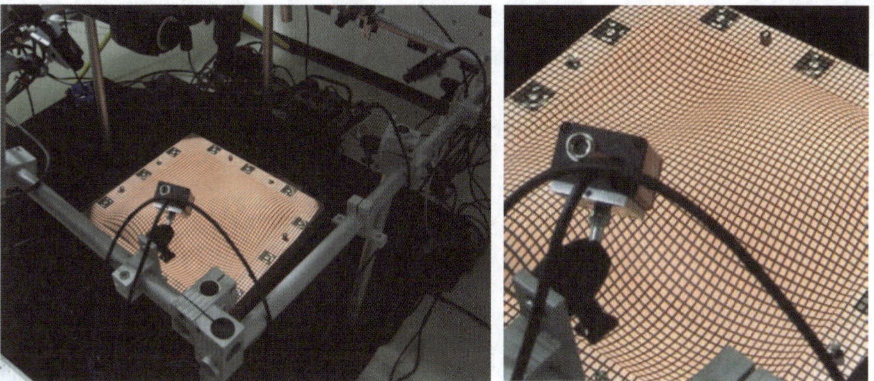

Fig. 6.27: Multi-camera system with grid projection (AICON)

6.5.2 Digital image correlation with random surface-texture patterns

6.5.2.1 Techniques for texture generation

Digital image correlation (DIC) and matching can be used to measure free-form surfaces if there is enough surface texture information. Typical image measuring accuracy is of the order of 0.1 pixels. Lateral and depth accuracy further depend on the image configuration, e.g. the base-to-height ratio of cameras. The approach has already been used in the early stages of digital photogrammetry, e.g. by the Zeiss Indusurf system (see section 1.4).

The texture must have a structure that allows for reliable and accurate image matching in two or more images. Texture resolution, structure and contrast must be designed for the required 3D result and take into account object curvature, camera resolution and image scale, illumination conditions and object reflectance properties.

Fig. 6.28 shows examples of suitable textures. For stereo systems conforming to the normal case of stereo photogrammetry, the surface texture should consist of image edges that are perpendicular to the (horizontal) epipolar lines (example in Fig. 6.28a). If the texture is designed for hierarchical image matching, e.g. in different image pyramids or scales, multiple pattern resolutions can be overlayed (example in Fig. 6.28b). High resolution textures are nessecary if high resolution is required for surface reconstruction (example in Fig. 6.28c).

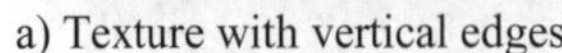
a) Texture with vertical edges

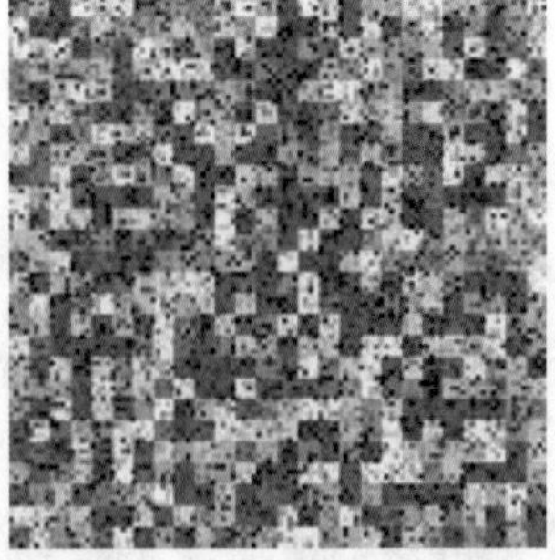
b) Texture suitable for image pyramids

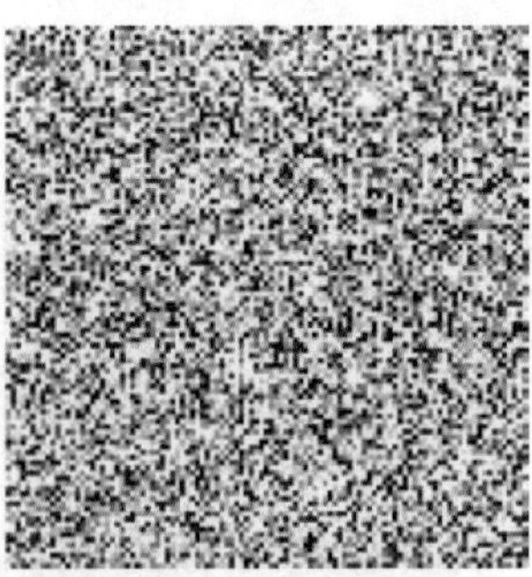
c) High resolution texture

Fig. 6.28: Different texture patterns suitable for digital image correlation (DIC)

If the natural surface texture does not support image matching, artificial textures can be provided by the following techniques. (Note that any physical marking of a surface may not always be permitted.)

- Pattern projection:

  An analogue slide projector, or digital projector (see section 3.6.2.2) can project any required texture pattern onto an object surface. However, a projected pattern requires a stable imaging configuration, hence projector, cameras and object must not move with respect to each other. The technique is therefore best suited for static applications.

- Adhesive patterned sheets:

  Thin adhesive sheets with a printed surface pattern can be attached to the target surface. Fig. 4.33 shows a test object to which such a textured target film has been added for purposes of evaluating surface measurement accuracy. The adhesive target film has a thickness of around 0.1 mm and can be printed with any convenient random pattern. Depending on application, the film thickness has to be taken into account for surface analysis.

- Paint and powder sprays:

  It is very common to spray white powder or paint onto the surface and apply brush strokes to provide a random pattern. The object surface then has a structure useful for image correlation. However, manual surface preparation may lead to non-uniform textures, inhomogeneous structures or different thickness of applied material. The texture layer can be removed from the surface after measurement.

- Etching:

  For applications such as strain measurement of metal sheets, a physical texture can be applied by chemically or laser etching random features into the surface. In this case the texture is of uniform thickness, predefined contrast and resolution and can be optimised for the latter analysis of the measurement, e.g. strain analysis. Since the texture is burned into the object material, it cannot be removed later.

6.5.2.1 Data processing

Recording object surface points with respect to the object or component coordinate system can be done in one of the following ways:

- Regular XY grid:

  Changes in surface position can be measured using a regular XY grid of points, i.e. at every XY location, and for every deformation state (epoch), a new Z value is determined. A suitable measurement technique for this purpose is the principle of the *vertical line locus* (VLL, see section 4.3.6.3).

- Irregular distribution of points:

  Based on result of feature detection algorithms (see section 5.5.2), the location of irregularly distributed surface points is calculated for every epoch by spatial intersection in which new XYZ values are determined every time. The result is an irregular 3D point cloud.

- Target tracking:

  Surface points which are determined at the start, e.g. distinctive features found by interest operators, are tracked through the deformation epochs. The result is a spatial curve for every measured surface point. Tracking requires prediction of the location of the surface points in the next epoch, which can be done by Kalman filtering.

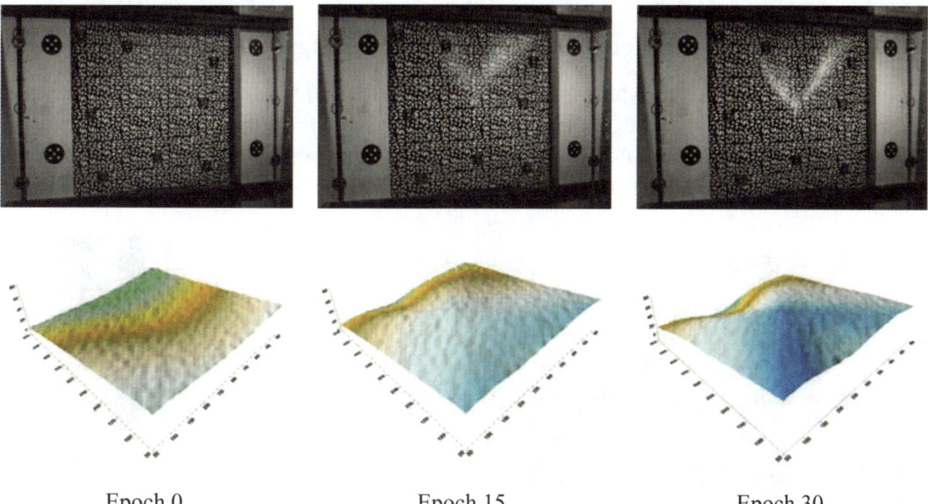

Epoch 0 Epoch 15 Epoch 30

Fig. 6.29: Image sequence (left hand image only) and the corresponding model of surface deformation (IAPG Oldenburg)

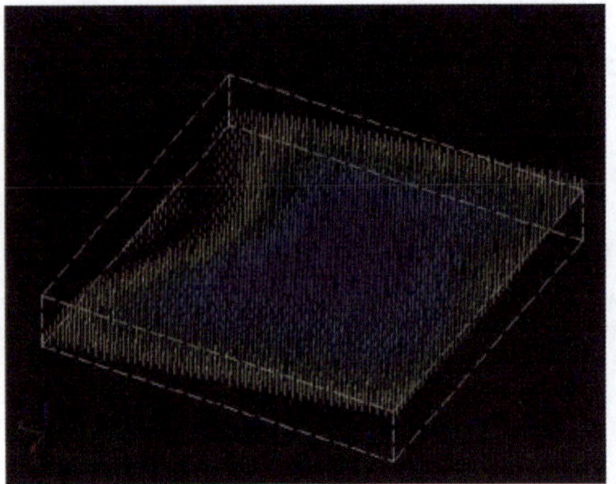

a) Deformation vectors on a regular XY grid

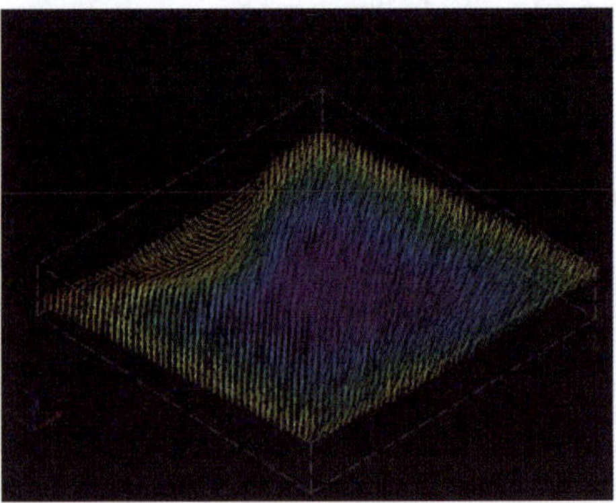

b) Trajectories of discrete surface points

Fig. 6.30: Deformation vectors between epochs 1 and 25 (IAPG Oldenburg)

Fig. 6.29 illustrates the recording and evaluation of a surface deformation. The surface was marked with a random pattern and its deformation recorded by a stereo camera. The images are processed in a multi-level image pyramid. This starts with a coarse measurement using normalised cross correlation (section 5.4.2.3) and finishes with a fine measurement using a least-squares adjustment (section 5.4.2.4). The derived deformation vectors are shown in Fig. 6.30, in one case on a fixed XY grid and in another as space trajectories of discrete surface points.

6.5.2.2 Multi-camera system for dynamic surface changes

It is only possible to model dynamic surface changes in 3D if the measurement frequency is higher than the object's rate of deformation. Usually, therefore, methods based on surface scanning or sequential pattern illumination (fringe projection) are unsuitable and optical surface capture with simultaneous multi-image recording is required.

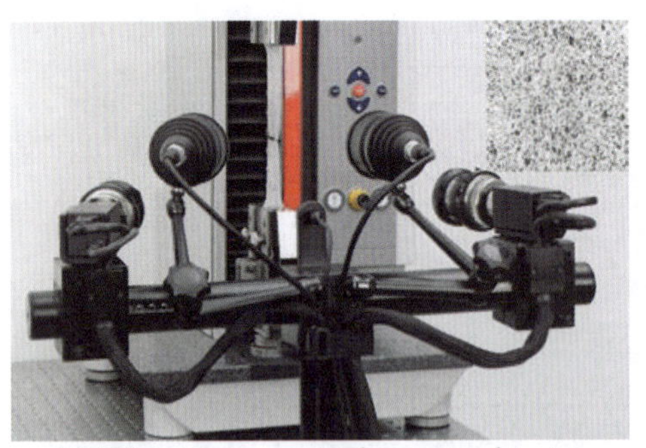

a) Strain measurement using sprayed texture

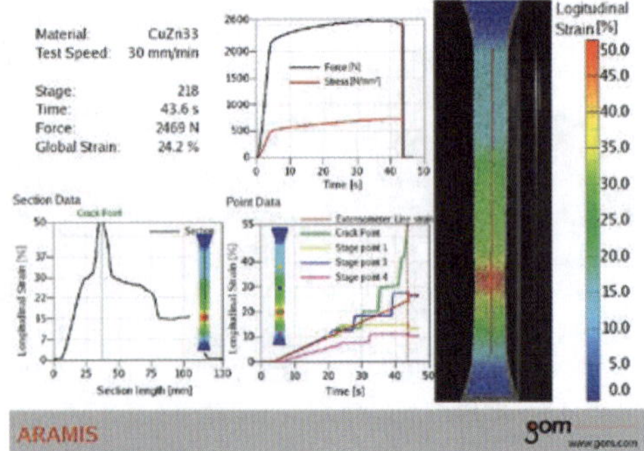

b) Result of measurement

Fig. 6.31: Digital correlation system (DIC) for strain analysis (GOM ARAMIS)

Fig. 6.31a shows an example application of a digital image correlation (DIC) system for strain analysis (GOM ARAMIS). In this case a tensile test specimen, whose length in the field of view is about 125 mm, is mounted on a tensile testing machine which applies controlled strain to the object. The metal surface has been textured by spraying with a random pattern (see upper right corner in Fig. 6.31a). The three-dimensional deformation of the test material is observed by two synchronised cameras and physical surface points are tracked through the image sequence by digital correlation techniques (Fig. 6.31b). The achieved accuracy is reported to 2 µm in XY and 4 µm in Z (= height).

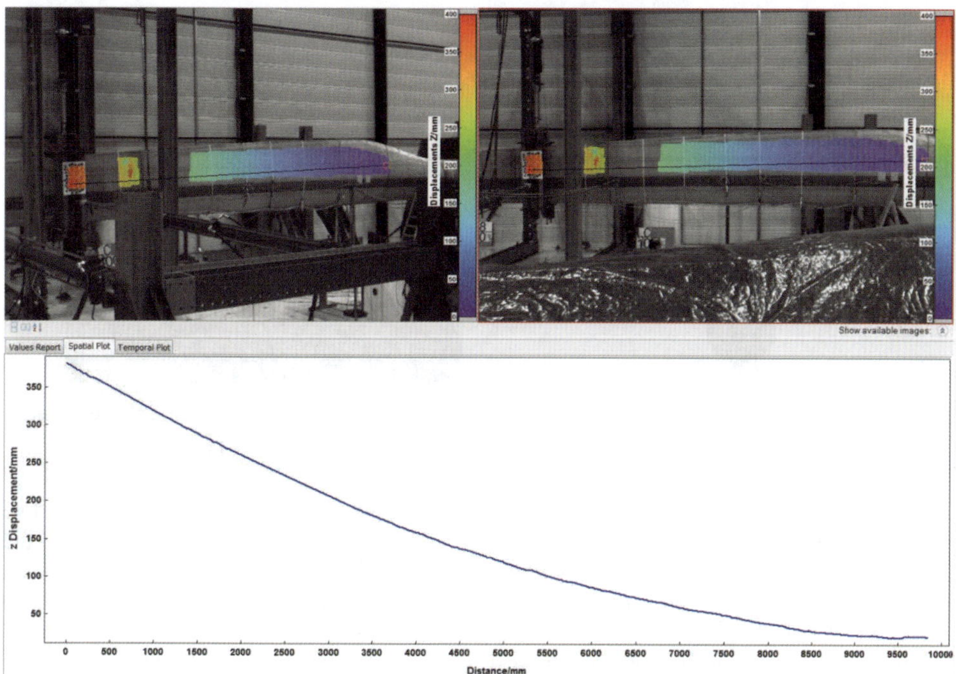

Fig. 6.32: Rotor blade deformation measurement with 3D DIC system (LIMESS)

Another application based on digital image correlation is shown in Fig. 6.32 (upper image). A wind-turbine rotor blade is measured over a length of 10 m. Here the natural texture of the unfinished surface provides sufficient structure for stereo correlation. Fig. 6.32 (lower image) shows the result of a bending test where the blade deformation in Z direction is measured. The achieved accuracy is of the order of 0.1 mm in Z (=out of plane).

6.5.3 Measurement of complex surfaces

When the size or form of the surface to be measured is such that it cannot be covered in a single measurement stage, then the surface must be measured in parts which are subsequently transformed into a common coordinate system to create a single 3D representation of the object. When 3D point clouds are transformed in this way, the process is also known as *registration*. The following section describes optical or mechanical positioning methods for sensor systems while registration methods based on pure point cloud information are explained in section 5.6.2.

6.5.3.1 Self-locating scanners - orientation with object points

In a technique called *self-location*, targets are attached to the object which are suitable for photogrammetric measurement in a common coordinate system. Cameras in surface scanners use these initially unknown points to orient themselves and their scan data relative to the object. Local 3D point clouds, scanned from different locations around the object, can then be transformed into a complete model via these points.

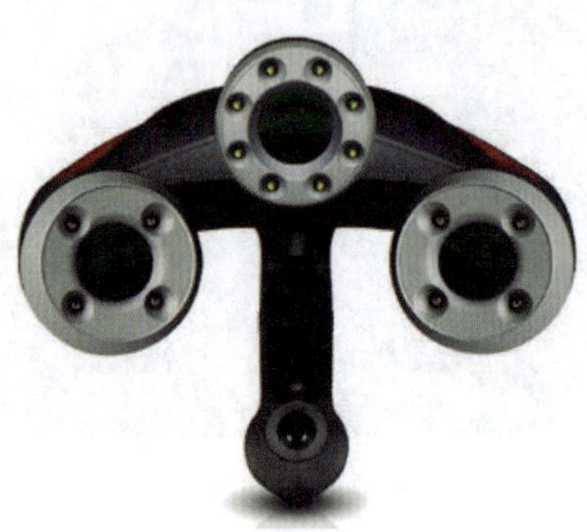

a) Handyscan 3D: MaxScan

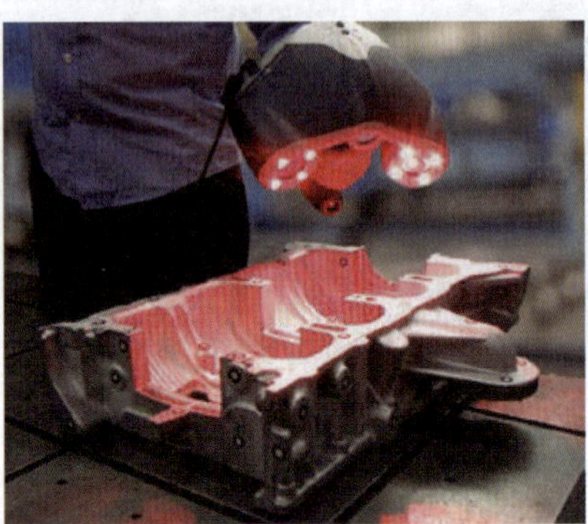

b) 3D surface capture using object reference targets

Fig. 6.33: Orientation of a surface scanner using passive reference points (Creaform)

The Handyscan 3D family of scanners (Creaform), which is built around a laser line projector and a stereo camera, operates in this way (Fig. 6.33). The stereo camera in the scanner locates the initially unknown reference targets in the camera's coordinate system by triangulation (space intersection). Simultaneously it locates the projected laser line to give a profile of 3D points along the object surface. When the system is moved to another position, in order to create another laser profile on the surface and so build up a full surface model, the current reference targets in the field of view are again located by the same procedure as before. Although the current stereo camera position is different from the preceding one, the current reference target positions, and laser profile points, can be added into the previous measurement by a spatial transformation, provided at least three common points have been measured in both camera positions. As new reference targets come into view, they are transformed into the common reference system and hence extend the spatial reference field.

This is a very flexible procedure in use. For example, if the object is small and light, both object and scanner can be freely moved during scanning, e.g. the object held in one hand and the scanner in the other The accuracy of surface point measurement is quoted at around a few tenths of millimetres, but is dependent on the quality and distribution of the reference targets. In fact, since subsequent data is built on preceding data, this process can lead to a stackup of measurement and orientation errors.

Accuracies can potentially be improved if the targets on the object are first located in a unified coordinate system using a single camera imaging network plus bundle adjustment (section 4.4 and Fig. 6.33b). During subsequent surface scanning, at every scan position the stereo camera can transform each scan profile into this network, provided it can triangulate at least three reference points in its field of view. Each scan is thereby locked directly into the common network and not onto the previous scan, hence the potential for improvement. However, the disadvantage of this procedure is that the usability of self-location is removed.

The accuracy of surface point measurement is quoted at around a few tenths of millimetres, but this is highly dependent on the quality and distribution of the reference targets.

6.5.3.2 Scanner location by optical tracking

The surface scanning system, typically a laser line scanner, has a number of targets (minimum three) attached to its body and these are tracked by an external camera system. Using 6DOF measurement principles (section 4.2.5) the position and angular orientation of the surface scanner can be determined. The measurements made by both scanner and tracker must be synchronised and the ability to identify and measure the locating targets must be guaranteed throughout the measurement process. Errors in 6DOF measurement directly affect the quality of the surface measurement.

Fig. 6.34a illustrates the concept by showing a three-line camera (section 3.5.4) tracking a laser line scanner. The scanner has a number of LED targets, suitable for this type of camera tracker, which are calibrated in a local scanner coordinate system. These are tracked by the camera which transforms the local LED and laser profile coordinates into its own coordinate system in real time. See also a similar concept in which a surface scanner is tracked by a laser tracker (section 3.7.1.3).

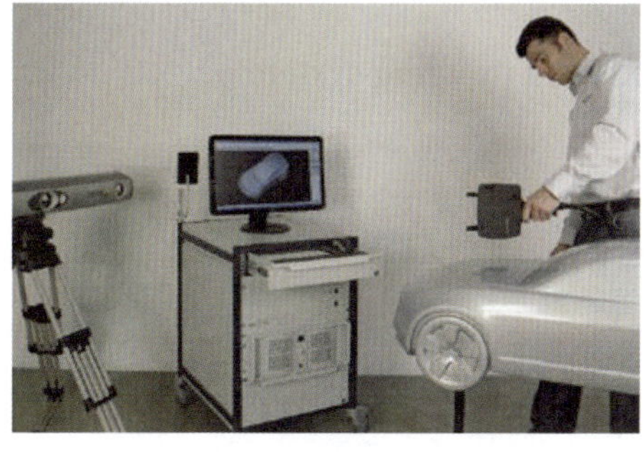

a) Optically tracked surface scanner (Steinbichler, NDI)

b) Complete 3D model generated from a number of individual surface scans (Steinbichler)

Fig. 6.34: Optically tracked surface scanner

6.5.3.3 Mechanical location of scanners

If the surface scanner is connected directly to a mechanical positioning system, then its spatial location and orientation can be determined by the positioning system provided it has the appropriate angle and displacement sensors. Relevant examples of such systems are robots and CMM arms (Fig. 6.35).

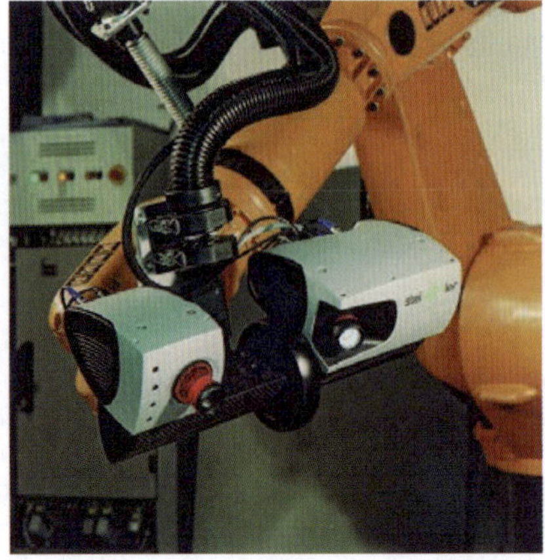

a) Area scanner attached to robot arm (Steinbichler)

b) Line scanner attached to CMM arm (API)

Fig. 6.35: Orientation of surface scanners by mechanical means

To connect the scanner's measurement data with those of the positioning platform, a corresponding calibration is required so that both systems operate in the same coordinate system.

6.6 Dynamic photogrammetry

6.6.1 Relative movement between object and imaging system

6.6.1.1 Static object

Relative movement between a static object and moving camera occurs in a number of applications:

- hand-held photography
- photography from an airborne vehicle
- image acquisition from a moving car
- image acquisition on unstable ground (oscillations, vibrations)

Stationary objects can be recorded in an offline process by sequential imaging with only one camera. Movements of the camera, during exposure, cause an image blur $\Delta s'$, dependent on velocity, exposure time and image scale:

$$\Delta s' = \frac{\Delta t \cdot v}{m} \tag{6.1}$$

where

Δt: exposure time
v: speed of moving camera
m: image scale factor

Blurring due to image motion results in a decreased modulation transfer in the direction of movement. The maximum permitted image motion can therefore be expressed as a function of resolving power. Investigations of aerial cameras have shown a maximum tolerable image motion of:

$$\Delta s'_{\max} = 1.5 \cdot RP^{-1} \tag{6.2}$$

where

RP: resolving power of the sensor [L/mm]

For applications in close-range photogrammetry, a maximum image blur of 1 pixel might, for example, be tolerated.

The maximum permitted exposure time in a given situation can be derived from (6.1) as:

$$\Delta t_{\max} = \frac{\Delta s'_{\max} \cdot m}{v} \tag{6.3}$$

Example 6.2:

A row of houses is imaged from a moving car ($v = 30$ km/h $= 8.33$ m/s) at image scale factor $m = 2000$. The maximum image blur should not exceed 1 pixel (e.g. 6 µm). It is necessary to find the maximum exposure time Δt and the resulting blur in object space:

Solution:

1. Permitted image motion: $\Delta s'_{\max} = 6\,\mu\text{m}$
2. Maximum exposure time: $\Delta t_{\max} = \frac{0.006 \cdot 10^{-3} \cdot 2000}{8.33} = 0.0014\,s \approx 1/700\,s$
3. Blurring in object space: $\Delta S = 2000 \cdot 0.006 = 12\,\text{mm}$

In practice, exposure times must also be selected on the basis of illumination conditions, available lens aperture, required depth of field and film or sensor sensitivity.

6.6.1.2 Moving object

The relationships discussed above for blurring due to camera motion are also valid for a moving object. However 3D measurement of a moving object requires at least two synchronised cameras.

Depending on object velocity, synchronisation errors lead to positional errors $\Delta s'$ which are proportional to the corresponding distance moved ΔS.

$$\Delta s' = \frac{\Delta t \cdot v}{m} = \frac{\Delta S}{m} \tag{6.4}$$

where

Δt: synchronisation error
v: object velocity
ΔS: distance moved
m: image scale factor

If the object is moving parallel to the baseline between two cameras, the positional error is effectively an x-parallax error $\Delta px'$. According to the standard case of stereo photogrammetry shown in Fig. 6.36, it is clear that the movement ΔS between times t_0 and t_1 results in a measurement of virtual point P^*. The corresponding error ΔZ in the viewing direction is given by the following expression (compare with eqn. 4.84):

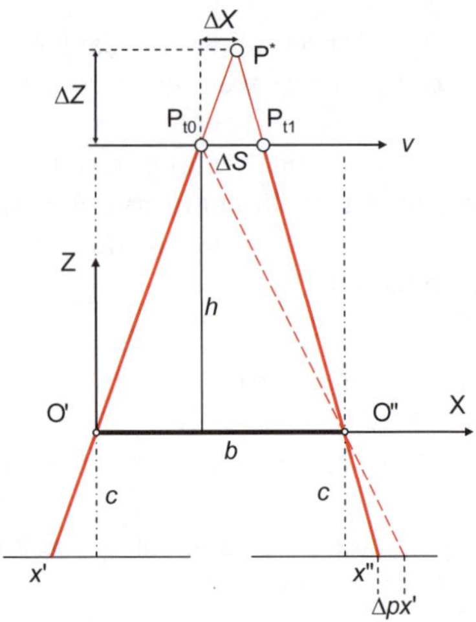

Fig. 6.36: Lateral and range errors caused by synchronisation error

$$\Delta Z = \frac{h}{b}\Delta S = \frac{h}{b} \cdot m \cdot \Delta px' \tag{6.5}$$

where

h: object distance
b: stereo base
$\Delta px'$: x-parallax error
m: image scale factor

In the direction of movement, the lateral error ΔX is as follows:

$$\Delta X = \frac{x'}{c}\Delta Z \tag{6.6}$$

where

x': image coordinate in left image
c: principal distance

Example 6.3:

For the application in example 6.2 ($v = 30$ km/h = 8.33 m/s; $m = 2000$) stereo images with two CCD video cameras are also to be recorded ($b = 1.5$ m, $c = 8$ mm, $h = 16$ m, $x' = 4$ mm). Technical limitations in the installation require a synchronisation error of $\Delta t = 1/500$ s to be taken into account. (In fact, suitable frame grabbers permit exactly synchronous image input, see below):
Solution:

1. Distance moved: $\Delta S = 1/500 \cdot 8.33 = 0.017 \text{ m}$

2. Lateral image error: $\Delta px' = \Delta s' = \dfrac{0.017}{2000} = 83 \cdot 10^{-4} \text{ m} = 8.3\,\mu\text{m}$

3. Error in viewing direction: $\Delta Z = \dfrac{16}{1.5} \cdot 2000 \cdot 8.3 \cdot 10^{-3} = 178 \text{ mm}$

4. Error in direction of movement: $\Delta X = \dfrac{4}{8} \cdot 178 = 89 \text{ mm}$

The example demonstrates the serious effect of a synchronisation error on the quality of object coordinates. The lateral image error of 8.3 µm is of the order of 1–2 pixels.

Conventional cameras can be synchronised by an electrical pulse linked to the shutter release. Progressive scan cameras (see section 3.5.1) work in non-interlaced mode and are therefore particularly suitable for dynamic imaging. For the synchronous image input from multiple video cameras there are frame grabbers with multiple parallel A/D converters (e.g. RGB inputs), whose synchronisation signal is used to control the cameras. The synchronisation pulse can also be generated by one camera in master/slave mode.

6.6.2 Recording dynamic sequences

Image sequences can provide chronological records of an object's spatial movements (deformations, trajectories, velocity and acceleration curves). In addition to recording time, suitable measures must be taken to identify and image discrete object points. Examples of dynamic applications are:

- recording crash tests in the automotive industry
- investigating turbulent flow in fluids or gases
- surveying structural deformations (buildings, bridges, …)
- material testing under mechanical or thermal stress
- vibration analysis
- calibrating and evaluating robot movement
- human motion analysis

Image sequences usually imply a sequential series of multiple images which record an object movement at an appropriate frequency. In object space, a stationary set of reference points is required to which the object movements can be related. It can also be used for image orientation, e.g. with moving camera stations.

Slow movements, such as the deformation of a cooling tower due to changes in direction of the sun's illumination, can be recorded as a sequential set of 3D reconstructions, each made using the offline photogrammetric process i.e. using a single conventional camera followed by a standard photogrammetric object reconstruction.

The recording of high-speed image sequences can be achieved using video cameras (camcorders) and digital high-speed cameras (section 3.5.3). Dynamic spatial modelling can be achieved by using multi-camera systems or a single camera with stereo image splitting (section 3.4.3.7). Fig. 6.37 shows part of a high-speed video sequence to determine the path of glass fragments in the simulation of a head striking a car windscreen. The sequence was recorded with two synchronised cameras (Weinberger Speedcam Visario, 1024 x 768 pixels, 2000 frames per second) in order to determine the three-dimensional paths of the glass particles.

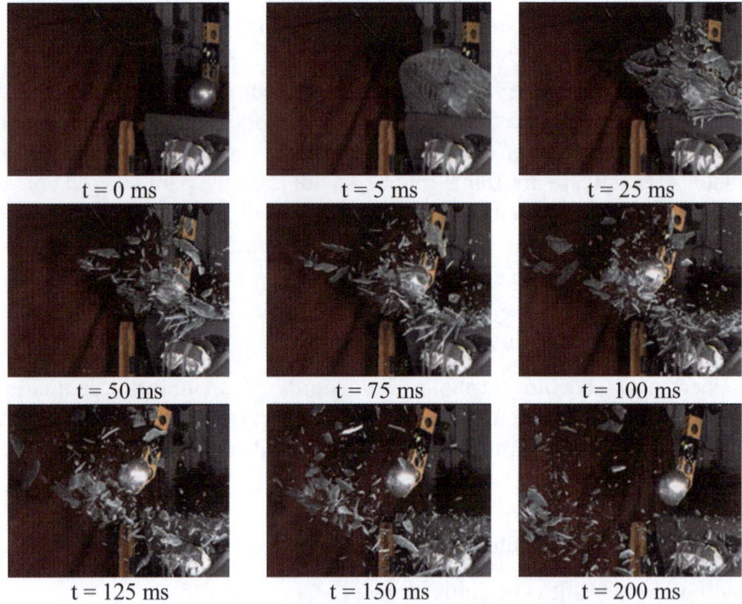

Fig. 6.37: High-speed video sequence (Porsche)

Fig. 6.38 illustrates another experimental set-up for recording spatial trajectories by a multi-camera image sequence. Three CCD video cameras are used to observe particle flow in a water tank. A light-section projector creates a "box" of light in which particles are visible. The image sequence is stored at video rate on a video recorder. The multi-media path of the imaging rays through the transparent wall of the tank is taken into account by the photogrammetric model (section 4.6).

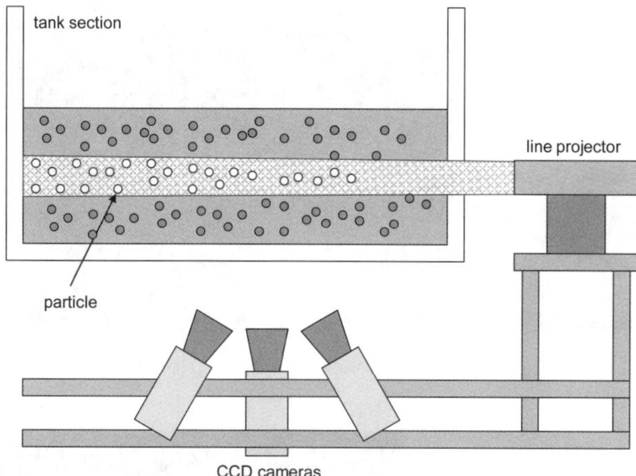

Fig. 6.38: Experimental set-up for multi-image recording of particle flow (after Maas 1992)

For this application the key problem lies in solving the correspondence, at a given point in time, between the three images of a large number of particles which are recorded as bright points (Fig. 6.39). The correspondence problem is solved by an image matching process based on epipolar geometry as outlined in section 5.5.3.

As a result of the photogrammetric analysis, the 3D coordinates of all particles throughout the complete image sequence can be determined. Particle trajectories can subsequently be derived, although this requires solution of the correspondence problem between consecutive images in the sequence.

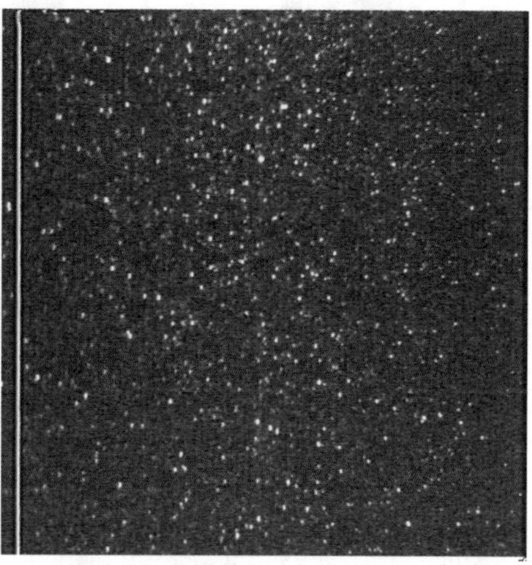

Fig. 6.39: Particle flow image (Maas 1992)

6.6.3 Motion capture (MoCap)

A number of commercial photogrammetric systems (motion capture systems) exist for the 3D recording and analysis of human movement (e.g. for applications in sport, medicine, ergonomics and entertainment). They are based either on off-the-shelf digital video cameras or specialised measurement cameras which integrated processors for the online acquisition of target marker locations. The cameras can be freely positioned (on tripods), attached to a mobile base (Fig. 6.40) or mounted in fixed locations around a measurement space (Fig. 6.41). System calibration and orientation is generally made using local reference point arrays or targeted scale bars which are moved around the measurement area.

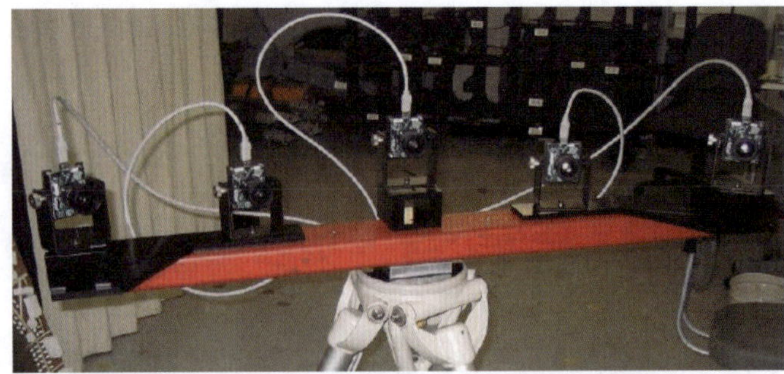

Fig. 6.40: Mobile tracking system (Hometrica)

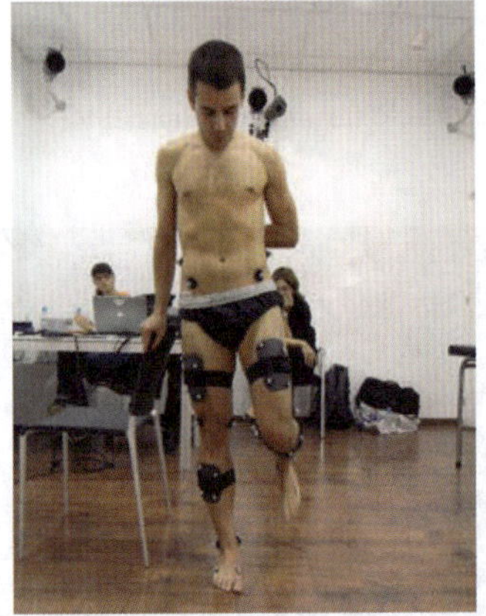

a) Cameras attached to wall

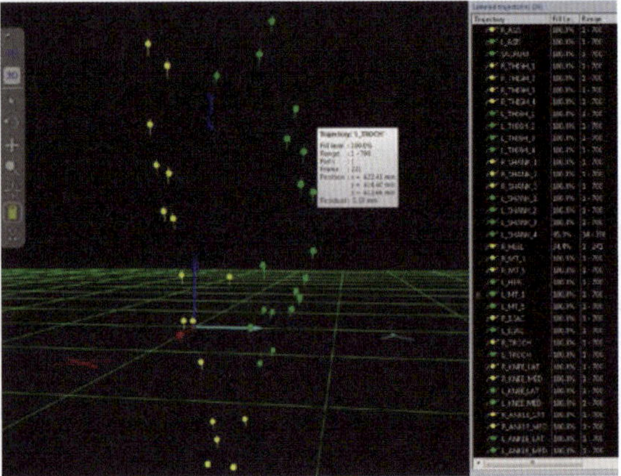

b) Example of a 3D motion analysis

Fig. 6.41: Stationary tracking system (Qualisys)

Retro-reflective targets placed on the body are normally used to record movement. They are illuminated from the camera locations and provide reliable target identification and tracking. The results of a measurement are the spatial trajectories of the target points which typically support a motion analysis or computer animation.

6.7 Mobile measurement platforms

6.7.1 Mobile mapping systems

Mobile mapping systems (Fig. 6.42) are moving platforms equipped with a range of sensors (cameras, laser scanners, radar, positioning systems) and used to measure transport routes (roads, railways, waterways), and both open and built environments. The objective is a continuous sensing of the scene with simultaneous registration of exterior sensor orientation using GPS and INS.

Mobile mapping systems for railway modelling are used to determine rail geometry and areas of damage, as well as bridge and tunnel measurement. To record and check road surfaces and edges, laser and radar systems are used (see example in Fig. 6.43). In populated areas it is mostly building façades and 3D city models which are generated. These are used for both tourism and technical purposes, in geoinformation systems and internet portals.

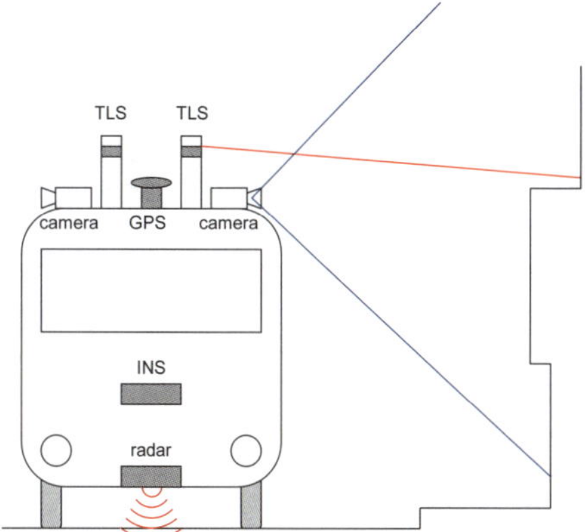

Fig. 6.42: Schematic mobile mapping system

Fig. 6.43: Mobile mapping system for road modelling (gispro)

A particular challenge for mobile mapping systems is the calibration and orientation of the individual sensors in a common coordinate reference system. For this purpose, large area reference fields must normally be created. These should provide a large number of reference points which can be recorded by the various sensors (e.g. image and laser systems).

6.7.2 Close-range aerial imagery

Images from low flying aerial platforms are used, for example, in the following applications:

- archaeological surveys
- volumetric measurement of spoil heaps and waste disposal sites
- roofscape mapping
- large-scale mapping of inaccessible areas
- test site mapping for geographical investigations
- biotope monitoring
- accident recording

- water pollution control
- reconnaissance

The acquisition of low-altitude aerial images is, in the first instance, a problem of flight technology (choice of sensor platform, navigation, flight authorisation). Photogrammetric processing is based on standard methods.

Manned aircraft are suitable for large-scale aerial images if image blur at their minimum speed is tolerable. For example, with an exposure time of $\Delta t = 1/1000$ s, flying speed $v = 180$ km/h and image scale factor $m = 1000$, image blur is approximately 50 µm according to (6.1). Compensation for image blur is widely implemented in aerial cameras by *forward motion compensation* (FMC). In digital cameras a corresponding shift of the image on the sensor can be achieved (*time delayed integration*). Low-speed ultra-light aircraft (microlights, $v_{min} = 40$ km/h) provide advantageous flight properties but lower payloads (<100 kg). Manned helicopters generate large vibrations whose negative influence on image quality must be corrected by a special camera mounting or hand-held photography.

For large scale mapping of smaller areas, pilotless aircraft and helicopters (*remoted piloted vehicle* RPV, *unmanned aerial vehicle* UAV, *unmanned aerial systems* UAS), balloons, blimps and kites are all usable (Table 6.1). A recent development has been remotely controlled helicopters, with automatic flight stabilisation and navigation, which can carry a payload of up to 18 kg. This permits, for example, the simultaneous use of digital cameras and airborne laser scanning systems. This development extends to multi-copters with up to 8 rotors (e.g. quadrocopters, octocopters). These provide a high level of flight stability although with a significantly lower payload. Significantly more powerful systems exist for military applications.

Table 6.1: Unmanned platforms for low-altitude aerial imagery

Type	Typical (max.) altitude [1] [m]	Payload [kg]	Range [2] [m]	Number of operators	Max. flying time [3] [min]
fixed-wing aircraft	50 – 100 (150)	3 – 18	1000	1 – 2	30 – 90
helicopter	10 – 100 (150)	1 – 25	500	1 – 2	10 – 120
multi-copter	10 – 100 (150)	0.5 – 2	500	1	up to 60
hot-air balloon, blimp	10 – 100 (200)	10 – 50	1000	3 – 5	60
gas balloon, blimp	10 – 100 (200)	10	stationary	2 – 3	unlimited [4]
kite	50 – 100 (300)	10 – 50	stationary	2	unlimited [4]

1) limited to 100 m in Germany; 2) usually required to stay in view; depending on sender/receiver, 3) payload dependent, 4) weather dependent

In addition to the photogrammetric camera, the flying platform must carry additional components such as remote control, navigation devices, video camera or camera housing. Even with skilled operators, navigation remains a practical problem. If uniform aerial blocks are to be recorded, it is useful to target the required image centres on the ground so that these can be observed by an airborne video camera. Alternatively, the platform can be observed by using the live image of a vertically directed terrestrial camera if the UAV is visible in the camera's field of view. Altimeters, or optical aids based on the principle of geodetic range measurement by base length, can be used in order to ensure a constant altitude. UAVs with GPS receivers can, in principle, be navigated autonomously, i.e. within the GPS accuracy they can fly a predefined course.

Fig. 6.44: Quadrocopter (Microdrones)

Fig. 6.45: Helicopter (Aeroscout)

Fig. 6.44 shows a remotely controlled quadrocopter which carries a light digital camera. Quadrocopters are very stable and can be flown by non-experts with some knowledge of flight. Use of the helicopter shown in Fig. 6.45 requires pilot training but it can carry a significantly greater payload (see Table 6.1). Fig. 6.46 illustrates an orthophoto and cadastral plan created from UAV images. Fig. 6.47 shows the use of a quadrocopter to acquire imagery of a church tower. Use of a balloon for archaeological recording of an oasis is shown in Fig. 6.48.

Fig. 6.46: Orthophoto (left) generated from UAV images and used to create a cadastral plan (right) (Land Surveying Office, City of Winterthur, Switzerland)

Fig. 6.47: Use of a quadrocopter to record a church tower

Fig. 6.48: Use of a balloon to photograph the Siwa oasis, Egypt (German Mining Museum)

7 Measurement design and quality

This chapter considers photogrammetric projects from the perspective of planning, accuracy and optimisation. In most practical cases the required measurement accuracy in object space is critical to the design and configuration of a measurement process. Generally speaking, accuracy is therefore always closely connected with costs and commercial viability of a system solution.

The planning of a photogrammetric configuration and strategy is a complex process requiring due regard for a wide range of issues. Here the question of camera calibration plays a particular role and requires careful consideration, depending on the selected imaging technology and required accuracy. The measurement network must be so configured that constraints imposed by local conditions, the specified accuracy, camera calibration and technical effort required are all taken fully into account.

7.1 Project planning

7.1.1 Planning criteria

The planning of a photogrammetric project includes the description of the actual measuring task, the concept for a solution und the presentation of results. It should be carried out in close co-operation with the client as planning errors are often detected at a late stage when they are then difficult to correct. Planning the imaging configuration is one aspect of the complete project plan which should include, in addition to metrology issues, economic aspects such as staff and time management, use of instruments, cost management etc.

The initial project plan should specify the following features of the measuring task:

- number and type of object areas to be measured, including a description of the object, its situation and the measuring task requirements
- dimensions of the object
- specified accuracy in object space (tolerances, measurement uncertainty)
- smallest object feature (resolution of fine detail)
- environmental conditions (variations in temperature, humidity, pressure and the presence of any vibration)
- options for object targeting
- definition and implementation of the object coordinate system
- determination of scale and reference points (geodetic measurements)
- alternative or supplementary measuring methods
- online or offline measurement
- acceptance test procedure or verification of accuracy

- available times for on-site work
- maximum permitted time for analysis and evaluation
- output of results (numerical and graphical, interfaces)

In the subsequent detailed planning, an appropriate concept for solution can be developed. In particular the image acquisition system and imaging configuration, as well as the type of image and data processing, must be defined. Apart from purely technical considerations (e.g. accuracy), the choice of components used also depends on the availability of instruments and personnel.

The following criteria should be defined in detail:

- estimation of average image scale
- selected processing system (analogue/digital, monoscopic, stereoscopic, multi-image)
- camera stations (number of images, network design, ray intersection geometry)
- required image measuring accuracy
- selected imaging system (image format, focal lengths)
- method of camera calibration (in advance, simultaneously)
- optical parameters (depth of focus, resolution, exposure)
- amount of memory for image data (type and cost of archiving)

7.1.2 Accuracy issues

Besides economic aspects, meeting accuracy requirements is generally of highest priority in practical project planning. Approximate accuracy estimates can based on the relationships given in section 3.3.1 which depend on three primary parameters:

- image measurement accuracy
- image scale
- design factor of the imaging configuration

Image measurement accuracy depends on the performance of the camera (stability and calibration), the accuracy of the image processing system (target image quality, measurement algorithm, instrumental precision) and the positioning capability (identification, targeting). These criteria must be balanced with respect to each other and given appropriate weight in the planning process. The camera selected for any given project is particularly important as it not only defines the quality of image acquisition and processing but, through choice of lens, also defines image scale and configuration.

Digital imaging systems can reach accuracies of 0.2–1 µm (1/50–1/10 pixel) depending on the mechanical stability and signal transfer type (A/D conversation). Digital processing systems providing sub-pixel operators for signalised targets can yield image accuracies between 1/100–5/100 pixels.

As explained in section 3.3.1, large image or sensor formats are advantageous for photogrammetric accuracy. It is essentially also the case that a larger number of pixels on a sensor will, in the first instance, improve resolution and therefore indirectly also the image measuring accuracy. However, achievable accuracy at the object will really only be improved by higher pixel numbers if the quality of the optical imaging components (resolving power of the lens), mechanical stability of the camera and image noise are consistent with the properties of the imaging sensor. It is particularly the case with sensors which have very small pixels (<2.5 µm), that an image measurement accuracy of better than 1/10 pixel (< 0.25 µm) cannot be expected due to limits in construction (camera mechanics, sensor mounting) and optics (diffraction).

In applications where there is essentially no limitation on the number of images, the achievable accuracy can be increased by taking additional images from locations which improve imaging geometry. The accuracy enhancement due to an increase of images at every station *k*, as expressed by eqn. 3.47, essentially only results in an improvement in precision. In this case, the increased numbers of observations going into the bundle adjustment improve the standard deviation of unit weight σ_0 and thereby the statistical precision of the calculated object coordinates. If object targeting permits fully automatic image measurement, then an increase in number of images has no significant disadvantages.

Specification of measurement accuracy should be done in close consultation with the end user or client. It must be made clear which accuracy parameters are to be used, the justification for the specified limits, and the method of verifying that the accuracy has been achieved. Quality measures obtained from various adjustment methods are explained in section 2.4.3 and 7.2.1. Metrologically defined parameters of accuracy (measurement uncertainty) are presented in section 7.2.1.

7.1.3 Restrictions on imaging configuration

A generally applicable geometric configuration for photogrammetric measurement cannot be defined because it always depends on circumstances specific to the object. A compromise must normally be found between different and partly incompatible restrictions:

- Image scale:

  The image scale is influenced by object distance, focal length (principle distance) and usable image format (eqn. 1.1, see Fig. 3.37). A larger image format enables shorter object distances for the same imaged object area. It not only leads to a larger scale (and higher accuracy) but also to a smaller number of images (economic benefits in reduced data processing and storage). It must be remembered that for complex object structures and highly convergent images, image scale can vary significantly within an image or from image to image.

- Image quality and resolution:

  The ability to detect and measure object details is again a function of image scale. The size of imaged object structures should lie between certain typical limits as follows:

  - visual analogue processing: one to three times the diameter of the floating mark
  - visual digital processing: 2–10 pixel
  - automatic digital point measurement: 6–10 pixel
  - automatic digital surface measurement: 11–25 pixel window size

- Object environment:

  The selection of suitable camera stations is often restricted by inaccessible object areas. It is therefore often necessary to use either additional lenses, or to increase the number of images or to dispense with optimal ray intersections. Additional camera/lens combinations may also increase the effort required for camera calibration. This is particularly applicable to zoom lenses with adjustable focal lengths (see section 3.4.3.4).

- Depth of field:

  The available depth of field (section 3.1.2.3) is mainly a function of image scale and f-stop. It restricts the choice of camera stations, especially for large image scales and under difficult lighting conditions. If applied targets are measured automatically, a slightly defocused image can be accepted if image contrast and point diameter are sufficiently large.

- Imaging angle:

  The imaging angle β at which an object is photographed should not be less than 20° for critical object features, and not less than 45° for retro-reflective targets, in order to achieve suitable image sizes and contrasts (Fig. 7.1). Furthermore, extremely oblique views have an eccentricity effect on the centre of circular targets as explained in section 3.6.1.2. Spherical targets permit all-round imaging but note that they are also elliptical in shape when imaged off axis (see Fig. 3.148).

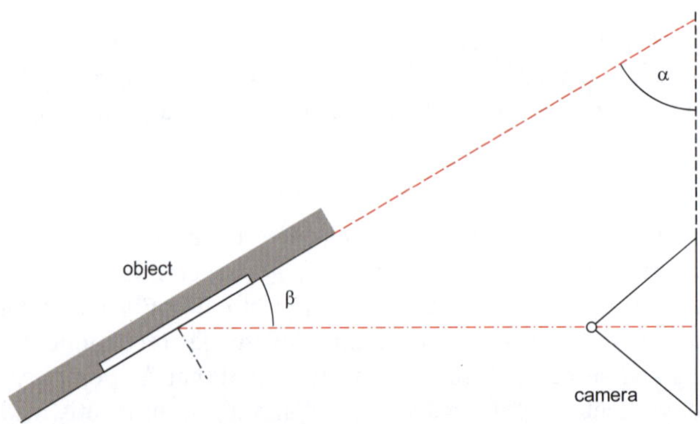

Fig. 7.1: Imaging angle

- Number and distribution of image points:

  The total number of object points has little effect on the total redundancy of the bundle adjustment. The quality of bundle triangulation depends more on the number and configuration of camera stations which, in addition to creating a reasonable intersection geometry (see below), should ideally utilise the full image format (see Fig. 7.28).

- Intersection angle:

  Good intersection angles are critical to the accuracy of point measurement (see Fig. 4.77). For the (graphical) reconstruction of approximately flat object surfaces (building facades) a reduced accuracy in the viewing direction can often be tolerated and in these cases it is possible to work with poor intersection angles (or insufficient height-to-base ratios).

  Engineering or industrial projects often require an object accuracy which is equal in all directions. Optimal ray intersection angles are around 90°–100°. In practice, intersection angles between 45° and 120° are sufficient if at least 4 to 6 images contribute to the measurement.

- Field of view and visibility:

  The field of view of a camera is defined by the format angle (section 3.4.3.2). If all object points can be imaged in all photos then the imaging configuration is simplified and the number of images required is reduced. At the same time, the total redundancy of the bundle triangulation increases. Since this assumption is only valid for simple objects (e.g. test fields for calibration purposes, see Fig. 7.20), objects with occluded areas must be recorded with additional photos which only include a small portion of all object points. To ensure network stability, these additional images should contain a reasonable number of well distributed tie points. For simultaneous calibration of distortion it is advantageous to have an irregular distribution of object points across the image format, with many occupying the image corners. In different images, it is an additional advantage to have the same points appear in different image areas.

- Image analysis and use:

  Depending on subsequent image analysis and presentation of results, additional images in diverse arrangements may be required. For image rectification and texture projection onto 3D models, useful images have few occlusions and view the object surfaces perpendicularly. For visual stereoscopic viewing, images with a good height-to-base ratio and parallel camera axes are required.

7.1.4 Monte Carlo simulation

A Monte Carlo simulation is a computational method for statistically analysing results from complex systems of calculation. The input data is randomised or altered in a structured way and its effect on the calculated output data observed. Every input parameter can be introduced with its own statistical distribution and variance, as well as a function to describe systematic deviations. For every calculation the input data is modified with the aid of a random number generator and the calculation is repeated many times until a statistically meaningful output sample has been obtained (Fig. 7.2).

The number of repeated simulations should be in the thousands in order to guarantee a consistent spread of random deviations within the possible distribution of values. It is usual to assume that the random variations follow a normal distribution. They can, for example, be generated by applying the Box-Müller method to uniformly distributed random numbers. Depending on application, it may be sensible to limit the noise on the input data, for example to 2 or 3-sigma, in order to avoid outliers in the data set.

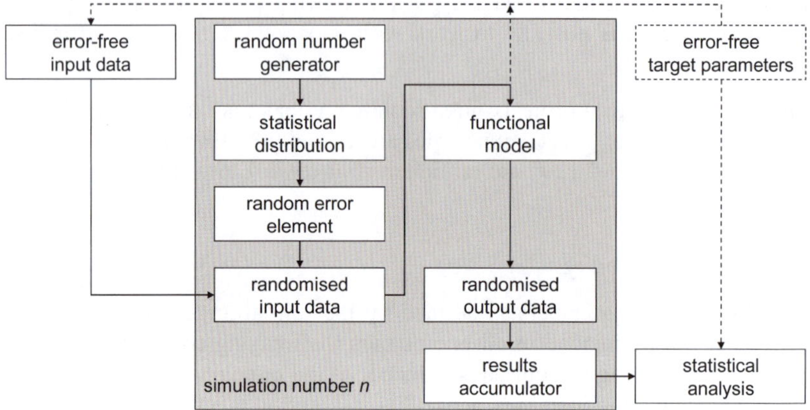

Fig. 7.2: Monte Carlo simulation

Results from all simulations are accumulated and statistically analysed at the end. This enables the generation of a histogram of the output parameters from which their standard deviations can be derived. It is also possible to make a direct comparison with defined, error-free target parameter values, from which error-free input data can be simulated using the functional model (dotted line connection in Fig. 7.2).

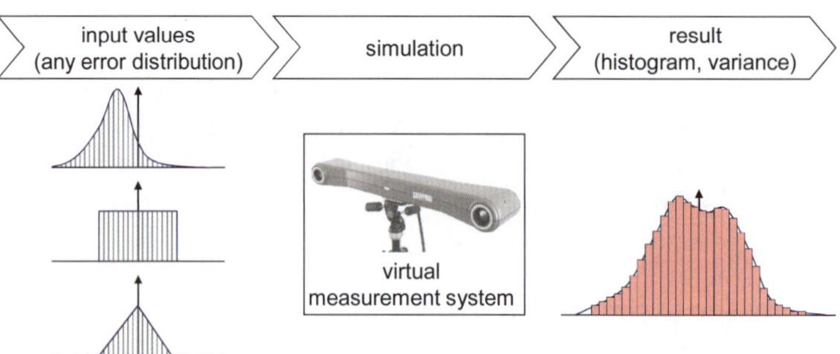

Fig. 7.3: Principle of a virtual measuring system by Monte Carlo simulation (after Schmidt et al. 2008)

The Monte Carlo method permits the computational modelling of a measuring system which, in this form, can also be regarded as a virtual measuring system. A necessary condition is accurate knowledge of the sources of error in a system and their mathematical description. It is then possible to create a realistic simulation of complex measuring tasks, and various influences on the measurement method, without the need to have a system physically available (Fig. 7.3).

Fig. 7.4 shows the result from a Monte Carlo simulation of the expected measurement accuracy of a stereo camera. Using defined, error-free object coordinates, as well as parameters for the interior and exterior orientation of each camera, error-free image coordinates are calculated according to eqn. 4.9. The defined object points are representative of the complete measurement volume of the system. In the subsequent Monte Carlo simulation, positive and negative random deviations, according to their respective expected measurement uncertainties, are added to the error-free image coordinates, camera parameters and orientation parameters. Using this randomised data, object point intersections are repeatedly calculated and their spread analysed.

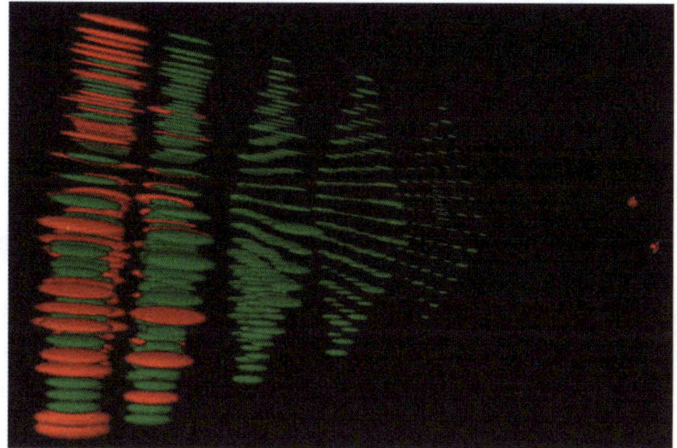

Fig. 7.4: Result from a Monte Carlo simulation of a stereo camera

In addition to numerical values for the calculated for the standard deviation and maximum value of the deviations, Fig. 7.4 also shows, at every object point, an error ellipsoid to illustrate the spatial distribution of the expected measurement uncertainty.

7.1.5 Computer-aided design of the imaging network

The restrictions mentioned in section 7.1.3 limit the choice of camera stations and viewing directions. In practice, a well configured network which surrounds the object, as indicated in Fig. 3.36, is impossible for many applications because of object restrictions or economic circumstances (e.g. maximum number of images). Image scales and intersection angles can therefore vary widely within a project, so that eqn. 3.47 is only valid for average accuracy estimations. Weak areas in an imaging network cannot be detected by this approach.

The image configuration can be simulated by bundle adjustment (section 4.4.6.1) if there are sufficient a priori object coordinates to represent object geometry. In this process the principal distance and exterior orientation of all cameras are iteratively varied until the following criteria are optimised:

- Maximum accuracy (minimum standard deviations of object coordinates):

  In general, object point accuracy and associated derived quantities (lengths, distances etc.) are of greatest importance in practice. In order to assess the expected accuracy it is helpful to split up the vector of unknowns in the bundle adjustment and express eqn. 4.99 in the form:

$$\begin{bmatrix} \hat{\mathbf{x}}_1 \\ \hat{\mathbf{x}}_2 \end{bmatrix} = \begin{bmatrix} \mathbf{A}_1^T\mathbf{P}\mathbf{A}_1 & \mathbf{A}_1^T\mathbf{P}\mathbf{A}_2 \\ \mathbf{A}_2^T\mathbf{P}\mathbf{A}_1 & \mathbf{A}_2^T\mathbf{P}\mathbf{A}_2 \end{bmatrix}^{-1} \cdot \begin{bmatrix} \mathbf{A}_1^T\mathbf{P}\mathbf{l} \\ \mathbf{A}_2^T\mathbf{P}\mathbf{l} \end{bmatrix} \tag{7.1}$$

  and

$$\mathbf{Q} = \begin{bmatrix} \mathbf{Q}_1 & \mathbf{Q}_{12} \\ \mathbf{Q}_{21} & \mathbf{Q}_2 \end{bmatrix}$$

  where

  $\hat{\mathbf{x}}_1$: exterior orientation and additional parameters

  $\hat{\mathbf{x}}_2$: object point coordinates

  In order to optimise the object point accuracy, only $\mathbf{Q}_2$ need be analysed according to the following considerations:

- Maximum reliability (ability to control and detect outliers):

  Statistical measures for reliability and robustness of an adjustment can be derived from the cofactor matrix of corrections $\mathbf{Q}_{vv}$ (section 2.4.3.4, eqn. 2.198).

- Maximum economic efficiency:

  The objective is to achieve the specified accuracy and reliability for minimum instrumental and personal effort.

The optimisation of photogrammetric imaging configurations (network optimisation, network design) can be considered in four stages:

1. Zero order design: definition of datum (object coordinate system):

   Standard deviations of object coordinates are directly influenced by the definition of the object coordinate system (section 4.4.3). Potential network deformation can be avoided by a datum definition without constraints. Optimal standard deviations are obtained by a free net adjustment which minimises the trace of the covariance matrix. In most cases the covariances of the object coordinates of a point i can be estimated by

$$\mathbf{Q}_{2i} \approx s_0^2 (\mathbf{A}_2^T\mathbf{P}\mathbf{A}_2)_i^{-1} \tag{7.2}$$

2. First order design: optimisation of the observation configuration:

   The purpose of optimising the configuration is to define an observation network whose design matrix $\mathbf{A}$, given a predefined weight matrix $\mathbf{P}$, results in a covariance matrix $\mathbf{Q}_2$ corresponding to the specified accuracy. This is primarily a question of achieving good intersection angles at the object points. Assuming an appropriate minimum configuration which does this, $\mathbf{Q}_2$ can be estimated by:

$$\mathbf{Q}_2 = \frac{s^2_{x'y'}}{k}(\mathbf{A}^T_{2B}\mathbf{P}\mathbf{A}_{2B})^{-1} \tag{7.3}$$

where

$\mathbf{A}_{2B}$: basic configuration design matrix

$s^2_{x'y'}$: standard deviations of image coordinates

k: number of additional, symmetrically arranged camera stations

The design factor q and the image scale number of eqn. 3.47 are reflected here in the matrix product.

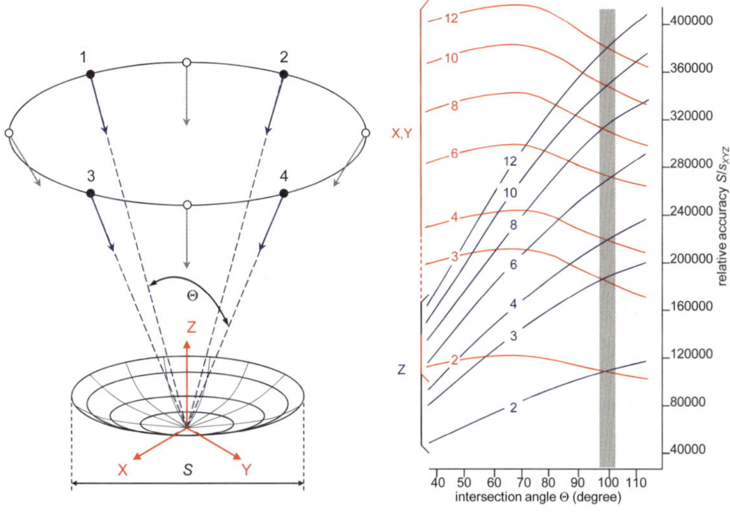

Fig. 7.5: Imaging configuration for antenna measurement and resulting object accuracies after variation of intersecting angles and number of camera stations (after Fraser 1996)

Fig. 7.5 illustrates the imaging configuration for an antenna measurement. The basic configuration consists of four camera stations. Additional images are added symmetrically at equal object distances. The intersection angle Θ is defined between the viewing directions of diametrically opposite camera stations. The diagram shows the resulting relative object accuracies S/s_{XYZ} as a function of the intersection angle and the number of camera stations. The relative accuracy increases with an increasing number of images. Equal values in all coordinate axes are obtained at $\Theta \approx 100°$.

3. Second order design: definition of observation weights (image measuring accuracy):

   The normal system of equations can be readily manipulated by appropriate choice of observation weights. Generally, the a priori standard deviations of image measurements are defined according to the precision of the image measuring device. Normally all image observations are given equal weights. Unequal weights are only justified if different measuring devices or algorithms are employed. In such cases an analysis of variances enables groups of observations of similar weights to be defined (see section 2.4.4.3).

4. Third order design: optimisation of point density (object points):

   Where self-calibration of the camera is not required, around 20–50 object points are sufficient to achieve a stable network geometry, optimised according to the above criteria. The quality of point measurement does not significantly improve if the number of object points is further increased. However, if the camera must be calibrated (self-calibration), targets must have a good distribution and density in the images in order to determine distortion parameters reliably across the entire image format (Fig. 7.28).

A relatively high effort is required for realistic simulations because, in many cases, object coordinates are either not available or must be generated manually. In addition, the selection of camera stations and viewing directions is usually performed interactively. Photogrammetric multi-image processing systems connected to 3D CAD systems offer an efficient basis for the simulation of imaging configurations if the object can be represented as a CAD model and if the camera stations can be edited graphically.

7.2 Quality measures and performance testing

The parameters and methods summarised in the following sub-sections are important criteria to use in planning and optimizing a measurement task, as well as for the subsequent verification of the accuracy achieved in object space.

7.2.1 Quality parameters

7.2.1.1 Measurement uncertainty

According to the International Dictionary of Metrology (VIM - Vocabulaire international de métrologie), and related publications such as DIN 1319-1, *measurement uncertainty* describes the range within which the *true value* of a measurand[1] lies. The true value is itself never known.

Measurement uncertainty is a characteristic value obtained from measurements, i.e. external conditions, influences due to the system and application and empirical values also contribute to this. Measurement and measurement uncertainty are estimated values and a complete measurement result only exists when both are specified. The measurement and its uncertainty must be traceable back to a reference standard (e.g. the SI unit of the metre).

The statement of measurement value and measurement uncertainty is typically given in the following form:

$$L = 1533.162 \text{ mm} \pm 0.015 \text{ mm}$$

[1] The quantity to be determined, for example, the length of a scale bar at 20°C. Effects such as correction for measurement at a different temperature would influence the uncertainty.

The measurement uncertainty encompasses all the unknown systematic and random error contributions to the measured value (Fig. 7.6). The extent defined by the uncertainty is itself uncertain. To deal with this in practice, a *coverage factor* k is introduced which is used as a multiplier of the coverage in the form:

$$L = (1533.162 \pm 0.030) \text{ mm}, k = 2$$

For a factor $k = 1$, the uncertainty value is a *standard uncertainty*, designated u and expressed in the form $\pm u$. In the example above it is ±0.015 mm. When a coverage factor greater than 1 is applied, the uncertainty is then known as the *expanded uncertainty* and designated U where $U = k \cdot u$ and is expressed in the form $\pm U$. In the example above with $k = 2$, $U = \pm 0.030$ mm.

If the probability density function which characterises the measurand is a normal distribution, then k defines the confidence interval (e.g. a 95 % confidence interval for $k = 2$) and the standard uncertainty limits correspond to the standard deviation.

7.2.1.2 Reference value

Reference values are frequently used in practice in order to provide a comparison for measurements and hence an estimate of their quality. A typical example is a reference length defined by the end targets on a reference bar which is measured by a system under test.

Reference values are themselves derived from measurements. Values measured by a system of higher order accuracy, or supplied with an officially recognised calibration certificate, are acceptable as reference values if their own measurement uncertainties, within the context of the application, are sufficiently small that they can effectively be regarded as true or error-free values. A reference uncertainty 5–10 times smaller than achievable by a system under test would generally be regarded as good.

7.2.1.3 Measurement error

The term *measurement error* defines the departure of an assigned value, derived from measurements of a quantity (e.g. length, coordinate), from a reference value.

Fig. 7.6 illustrates that measurement errors have systematic and random components. If the systematic errors are known, they can be determined by means of calibration and largely removed by applying corrections to the measurements. Unknown systematic errors in the measurements add to the random components to give the resulting measurement uncertainty.

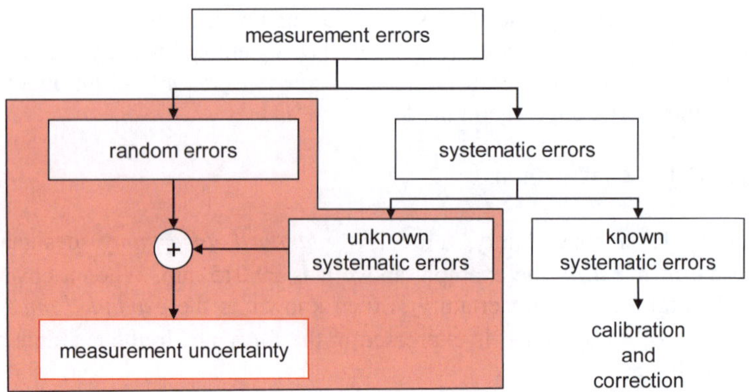

Fig. 7.6: Measurement error and measurement uncertainty

7.2.1.4 Accuracy

Accuracy describes the closeness of agreement between a measurement result and a measurement standard or accepted reference. Higher accuracy implies closer agreement but the term is strictly only qualitative, not quantitative. A statement about accuracy can only be made after a comparison is made with an independent, higher order reference value.

When estimating the accuracy of a photogrammetric process, reference objects and points used for comparison purposes should not also be included in the camera calibration or network adjustment procedures required to configure the photogrammetric system for object measurement.

Typical procedures for determining accuracy include:

- Checks of independent reference lengths:

  From the measured coordinates of end targets on reference lengths, distances between them can be calculated and compared with the calibrated lengths. Reference lengths for close-range testing are relatively easy to manufacture and their use follows normal practice in technical applications where coordinates are the source of derived elements (distances, deformations, surfaces, etc.). Furthermore, reference lengths permit the calculation of standardised length comparisons and the associated traceability back to the SI unit of the metre. The precondition is that at least one known length (e.g. scale bar) is included in the photogrammetric network. The test procedure is an industrial standard defined by the German guidelines VDI/VDE 2634/1 and can also be used as an evaluation test for camera calibration (see section 7.2.2).

- Comparison with independent reference points:

  If some of the object points used in the bundle adjustment have independent reference coordinates of higher accuracy, then a direct comparison between bundle coordinates and reference coordinates can be made.

This situation is difficult to achieve in practice. Firstly, comparison points which can be accurately identified, and have higher accuracy reference coordinates, are difficult to create in industrial situations. Secondly, independence of the data sets will not be achieved if the same measurement method (e.g. photogrammetry) is separately used to create both (bundle and reference) as this will result in correlations between them.

- Comparison with independent intersections:

  Following a bundle adjustment, it is more effective to use the separate measurement (by intersection) of additional image points which correspond to object reference points or which define object reference lengths. These intersections utilise the interior and exterior orientation parameters derived from the bundle adjustment and, as the target points are not included in the actual bundle adjustment process, their measurements are independent of it. They are therefore suitable for estimating the accuracy of a photogrammetric process which includes all uncertainties in the bundle adjustment. It also corresponds to the situation in practice where computed orientation data (from a bundle adjustment) are often used to make further object measurements.

7.2.1.5 Precision

Precision describes the statistical spread of a measured quantity as derived from repeated measurements or an adjustment process. It is normally expressed as a standard deviation or RMS value. Precision is a measure of relative accuracy. If a quantity is measured multiple times under repeatable conditions, it indicates the internal spread of a measurement result. In an adjustment calculation, precision is calculated as a standard deviation. Estimates of precision should always be provided with the coverage factor, e.g. 1 sigma.

7.2.1.6 Precision and accuracy parameters from a bundle adjustment

The quality parameters, which can be derived from the statistics generated by an adjustment process, reflect how well the measured values (observations) fit the functional model according to the chosen geometrical configuration (design matrix). If there are no systematic errors in the measurements, the parameters of precision describe the random variations in the measured values. In this case they also represent estimates of the measurement accuracies. Parameters of precision include the following (compare with section 2.4.3 and section 4.4.5):

- Standard deviation of unit weight:

  The a posteriori standard deviation $\hat{s}_0$ (sigma 0) is derived from the observation residuals and the redundancy r (eqn. 2.199). By increasing the redundancy, $\hat{s}_0$ can be reduced to almost any level required.

- Average residuals of image coordinates:

  The averages of residuals of all image coordinates are a measure of the quality of image measurement. In addition to the quality of point recognition, the quality of camera calibration (interior orientation) and camera stability are also covered by this measure.

- Standard deviation of object coordinates:

  Following an adjustment process, every unknown is assigned a standard deviation according to eqn. 2.188 (see section 2.4.3). If the unknowns relate to 3D coordinates, the average standard error of measured points (average point error) is given by:

$$s_{XYZ} = \sqrt{s_X^2 + s_Y^2 + s_Z^2} \tag{7.4}$$

  This, in turn, depends on the standard deviation of unit weight s_0. The average standard deviation of n adjusted object coordinates is given by the RMS values:

$$RMS_X = \sqrt{\frac{1}{n}\sum_{i=1}^{n} s_{X_i}^2} \qquad RMS_Y = \sqrt{\frac{1}{n}\sum_{i=1}^{n} s_{Y_i}^2} \qquad RMS_Z = \sqrt{\frac{1}{n}\sum_{i=1}^{n} s_{Z_i}^2} \tag{7.5}$$

  The standard deviations of object coordinates depend on the chosen datum and, depending on the positions of points in the object coordinate system, they can provide a false picture of the accuracy achieved. A more homogeneous distribution of standard deviations is achieved by using a free-net adjustment (section 4.4.3.3).

- Intersection residuals:

  The residuals resulting from ray intersections (section 4.4.7.1) is a measure of the precision of object point determination. In the stereo case (e.g. using a stereo camera as shown in Fig. 3.123), the intersection residual (y-parallax) can be zero, even when intersection error is present, due to the error being directed along the epipolar line.

7.2.1.7 Relative accuracy

The term relative accuracy, introduced in section 3.3.1, is used in photogrammetry to represent the performance of a measurement system independently of the dimension of its measurement volume. It is a dimensionless number, e.g. in the form 1:100 000, 10^{-6} or 10 ppm. It is usual in close-range photogrammetry to quote the achieved object measurement accuracy in relation to the maximum extent of the object.

If precision measures are used, e.g. RMS value of the object coordinates, then the derived relative value still remains a precision value. Since they are often based on a simple standard deviation, they then also only relate to a 68% coverage of the measurements.

If length comparisons are used to estimate the accuracy across an object, derived either as part of the bundle adjustment or by subsequent measurement, then lower relative accuracies may be calculated but they will be more practice-oriented and reproducible.

7.2.1.8 Tolerance

Tolerance is a parameter used in manufacturing to define the permissible limits to a feature's dimensions. Relative to a *nominal* value it can have different positive or negative values.

In a typical industrial measurement process it is necessary to measure critical dimensions on a part and decide if it is within tolerance, and therefore accepted, or out of tolerance and therefore rejected. Rejection is costly as parts must then either be discarded, recycled or reworked to bring them within tolerance.

Fig. 7.7 shows the relationship between the accuracy of a measuring device or procedure, and the extent to which it can decide if critical part dimensions are in tolerance (green) or out of tolerance (red). Depending on the ratio between tolerance and measuring uncertainty (e.g. 10:1), there is an area (yellow) where no clear decision can be made.

The device's accuracy is indicated by an agreed expanded measurement uncertainty $\pm U$ which corresponds, for example, to a 95% confidence level ($k = 2$). The diagram shows a measurement made within the tolerance limits (dotted lines). On the extreme left, a perfect measuring device with zero uncertainty can reliably determine if a feature lies anywhere within the entire tolerance band T. As the device uncertainty increases towards the right, a feature measured at the tolerance limits could, due to device uncertainty indicated by the yellow areas, be either in tolerance or out of tolerance. To be sure (within the confidence level) that the part is good, only measurements within the green area are acceptable. In effect, an increasing measurement uncertainty reduces the effective size of the tolerance band. In the extreme case (right hand side, not shown) where $U = T/2$, the device cannot determine if a part is in or out of tolerance.

The problem in practice is to select a compromise between measuring device or system with high accuracy (increased cost) giving rise to reduced rejections (lower cost).

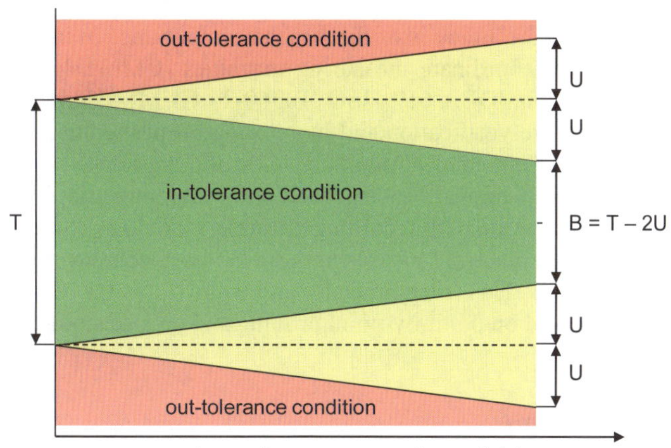

Fig. 7.7: Suitability of testing device or procedure for in-tolerance evaluation

7.2.1.9 Resolution

The *resolution* of a measurement system is the smallest increment which it can display or store (e.g. as represented by the most significant decimal place in a digital display). Resolution therefore defines the smallest change in the quantity to be measured which produces a significant change in the measurement signal, i.e. one which is above the noise level of the measuring system. In contrast, the *resolving power* (section 3.1.4) defined in optics and photogrammetry denotes the capability of an optical system to transmit a threshold frequency with sufficient contrast or modulation.

Fig. 7.8 illustrates the relationship between reference value, resolution, precision and measurement error, defined above. The measured or displayed values are spread according to their precision or repeatability, but may depart significantly (measurement error) from the reference value.

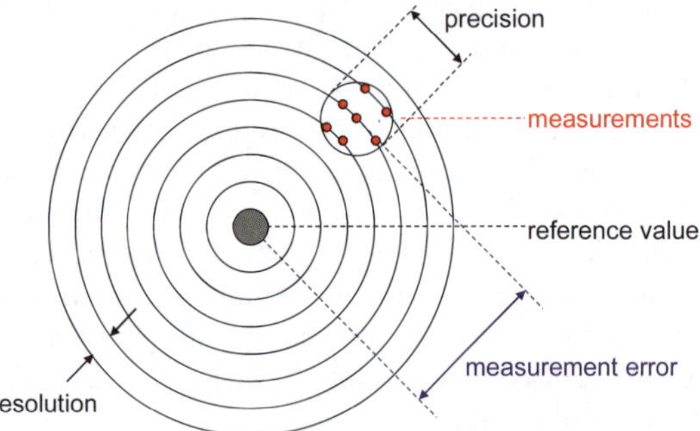

Fig. 7.8: Resolution, precision and measurement error (after Hennes, 2007)

7.2.2 Acceptance and re-verification of measuring systems

The checking of achievable accuracy is of fundamental importance in industrial metrology. In the field of mechanical coordinate measuring machines (CMM), long-established and standardised methods (VDI/VDE 2617, ISO 10360-2, GUM) define parameters and procedures for acceptance, re-verification and monitoring of measuring accuracy which are generally accepted and implemented in practice. They are applicable to both CMMs with touch probes and those with optical sensing heads. Photogrammetric systems employing touch probing can also be evaluated according to these guidelines. Since the year 2000, evaluation of optical non-contact 3D measuring systems has been covered by the German guideline VDI/VDE 2634. This recommends procedures for the acceptance and re-verification of systems based on point-by-point probing and area scanning.

7.2.2.1 Definition of terms

- Acceptance test:

  An *acceptance test* is the procedure for acceptance and approval of a measuring system after installation on the customer's site. The selected acceptance procedure is usually incorporated in the delivery contract. The goal of the test is the proof of the specified measuring accuracy under defined conditions. The acceptance test is usually performed jointly by system supplier and customer.

- Re-verification test:

  Re-verification or monitoring is the periodical checking of the measuring system after commissioning to ensure that it conforms with specifications. Compared with an acceptance test, re-verification can be simpler. As a rule it is carried out by the user, who also defines the time interval between such checks.

- Traceability:

  Traceability is the establishment of a link between the measured quantities, with their uncertainties, and a measurement standard, for example the standard metre. For this purpose a continuous chain of comparative measurements up to the national representation of the standard must be provided (Fig. 7.9). The standard used, for example a reference scale bar, must be calibrated and certified by a recognised calibration service.

- Characteristic parameter:

  A *characteristic parameter* is a measured or calculated value (threshold, maximum permitted value) which characterises the performance of a system or its individual components (for example probing error, see below).

- Probing error:

  Probing error is the value that describes the precision of probing of a single measured point. This parameter is used mainly with respect to CMMs in which the active probe, as distinct from the length measuring system, contributes to the total system accuracy. For example, the probing deviation can be determined by repeated measurement of geometrically known reference objects (for example a sphere). Residuals of single measurements with respect to the surface of the reference object indicate the probing error. Determination of probing error for area scanning systems is discussed in section 7.2.2.5.

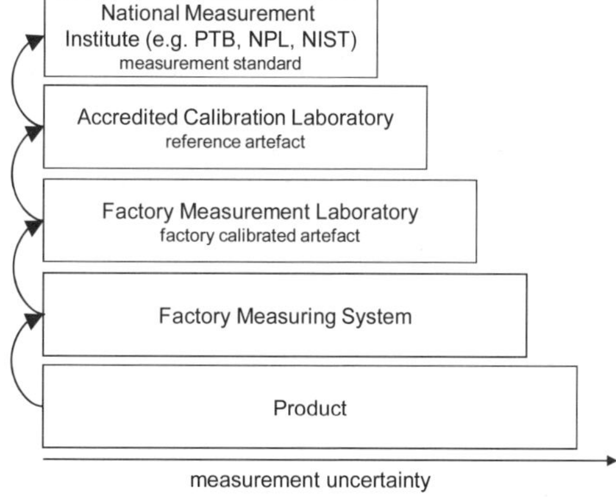

Fig. 7.9: Traceability to national standards

- Length error:

  The three-dimensional length error E (also called *length measurement error LME*) is defined as the difference between a measured (displayed) length L_m and the calibrated reference length L_r:

  $$E = L_m - L_r \tag{7.6}$$

  The length measurement error is usually derived from the measurement of two single probings (for example on a gauge block). Alternatively it can be determined from the distance measured between two spheres if the probing error can be eliminated.

  The length measurement error is used to analyse the accuracy of length measurement. Calibrated reference lengths can easily be established (for exceptions see below), and they can be traced back to a standard. Uncertainty of length measurement implicitly includes the probing uncertainty. The maximum permitted positive and negative limit of length error E is the maximum permissible error (*MPE*). It is defined as a length-dependent value that may not be exceeded in an acceptance or re-verification test. The *MPE* of length error E is shown graphically in Fig. 7.13 and expressed analytically as:

  $$MPE(E) = A + K \cdot L \leq B \tag{7.7}$$

  E: length error
  A, K: machine-specific constants
  L: measured length
  B: maximum permitted deviation of length measurement

  Since the error of photogrammetric length measurement does not necessarily depend on the length itself, the constant K may be zero.

- Sphere-spacing error:

  Sphere-spacing error indicates the capability of a system to measure the separation between the centres of two spheres which are derived from measurements on the spherical surfaces. The probing error is not explicitly included in the sphere-spacing error as multiple sampling of the surfaces averages it out. This parameter is particularly applicable to area scanning systems which cannot always directly determine a length measurement error. The length measurement error can be estimated from the sphere-spacing error if the probing error is known.

7.2.2.2 Differentiation from coordinate measuring machines (CMMs)

With respect to mechanical CMMs, photogrammetric 3D measuring systems have fundamentally different properties which are apparent in the procedures and parameters which characterise them.

- Image-based measurement of a large number of points:

  Optical 3D measuring systems, being image-based, enable the simultaneous registration of large numbers of object points (in the limit, each pixel). In contrast, touch-probe CMMs measure only one point per probe, although optical and line-scanning probing is also possible.

- Triangulation principle:

  Photogrammetric and fringe projection systems are based on triangulation, which leads to accuracies of object points which are dependent on scale and configuration. A homogeneous accuracy cannot, therefore, be expected within a specified measuring volume.

- Mobility:

  Optical 3D measuring systems are mobile and can be brought to the object. Consequently, from time to time their calibration data may change, or they operate under changing environmental conditions, or their imaging configuration may vary.

- Flexible configurations:

  Non-stationary photogrammetric systems allow a free choice of camera stations. Users themselves therefore determine the number and distribution of the images, the selection of cameras and lenses and the type of object targeting or probing.

- Unlimited measuring volume:

  In principle, the measuring volume of photogrammetric systems is unlimited in so far as depth of focus and field of view allow. If scale-dependent resolution of object details is taken into account, arbitrary object dimensions can be measured. Conversely, the measuring volume of mechanical CMMs is always limited.

7.2.2.3 Reference artefacts

A *reference artefact* is a physical object with known (calibrated) geometrical parameters. It should be economical to manufacture and easy to handle. As indicated in section 7.2.1.2, the calibration accuracy for a reference object should be approximately 5–10 times higher than that of the measuring system to be checked. Acceptance and re-verification procedures usually require a calibration certificate for the reference object.

If the measurement uncertainty of a system under test is determined using an artefact, the result is not only affected by random and systematic measurement error by also by the artefact's own calibration uncertainty. The measurement uncertainty cannot therefore be better than the calibration uncertainty.

The type of artefact used depends mainly on measurement volume, method of probing, availability and cost. Tactile probing implemented in coordinate measurement technology makes use of gauge blocks, step gauges and ball plates (Fig. 7.10) designed for a measurement volume with a typical diagonal length of 2 m. These are only suitable for photogrammetric systems using touch probing (example in Fig. 6.2).

a) Step gauge

b) Ball plate

Fig. 7.10: Standard reference objects for 3D coordinate measuring machines

a) Arrangement of reference scale bars

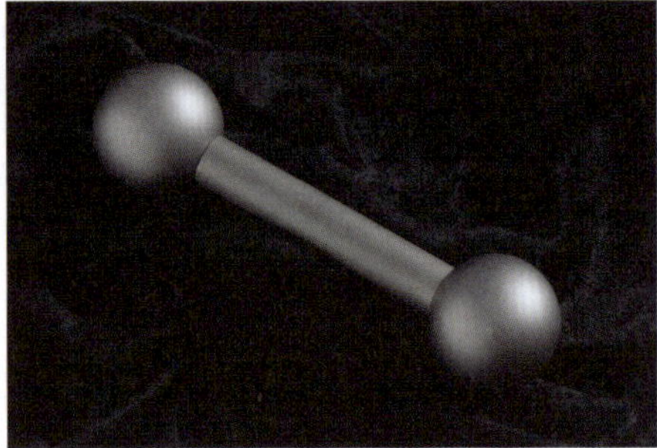

b) Dumb-bell artefact

Fig. 7.11: Test artefacts for 3D measurement systems

Fig. 7.11a shows an arrangement of reference scale bars for testing optical 3D measuring systems according to VDI/VDE 2634. The scale bars are provided with the type of targets which are otherwise used by the system under test. Fig. 7.11b shows a dumb-bell target which has a calibrated separation of its ball centres. This type of artefact can be used to evaluate the sphere-spacing error of area scanning systems.

Supplying reference objects for acceptance and re-verification tests for photogrammetric systems can be problematic. It is difficult to find scale bars suitable for testing large measuring volumes (dimensions >3 m). For smaller measuring volumes (<1 m^3) optical 3D systems achieve measuring accuracies of the order of 10 µm (1:100 000) requiring high-precision reference objects with suitable targets. For larger measuring volumes, such as required to measure parabolic antennas or ships, measuring accuracy is often checked using laser trackers (section 3.7.1.3). Differences in coordinates or in computed distances enable a comparison of accuracies. However, several points should be noted:

- high accuracy photogrammetric systems almost match the performance of laser trackers
- laser tracker measurements are costly in time and personnel
- laser trackers require geometric stability over the period of the observations
- multi-purpose targets are required which are measurable by both systems without additional significant loss of accuracy

7.2.2.4 Testing of point-by-point measuring systems

Part 1 of the VDI/VDE 2634 guidelines recommend parameters and methods for acceptance and re-verification testing of optical 3D systems which operate with point-by-point probing. The single parameter to be tested is the length measurement error and the testing is made by measurement of calibrated length artefacts.

Scale bars manufactured at appropriate lengths can serve as references. They should use the same type of targets as are used for the actual object measurement. The length can be calibrated, for example, by optical CMMs (for shorter lengths), high accuracy photogrammetry, by field survey (using total stations) or by laser interferometry.

In order to guarantee a sound analysis of the system, the arrangement of reference scale bars in object space should match the measuring task. If equal measuring accuracies are required for all coordinate axes, the scale bars must be arranged in such a way that direction-dependent length measurement deviations can be determined. A possible set-up is illustrated in Fig. 7.11a and Fig. 7.12. Of seven scale bars, three are arranged parallel to the coordinate axes and four along diagonals of a cuboid measuring volume. In order to increase the number of reference lengths of different sizes, individual scale bars can be divided into several sections. In this manner, the scale bar displayed in the foreground of Fig. 7.12 provides six partial distances which can be combined to give 21 different lengths.

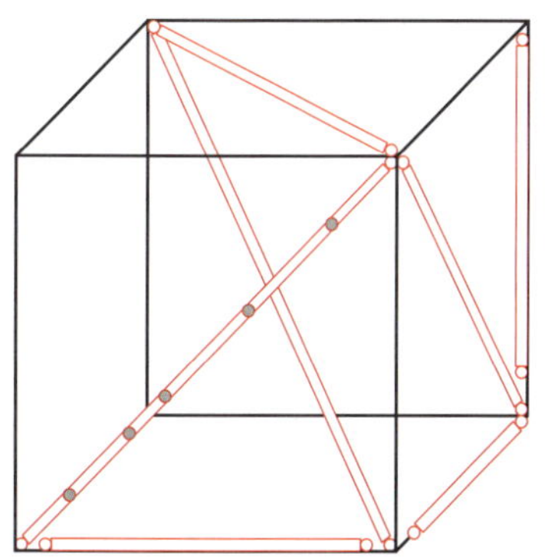

Fig. 7.12: Arrangement of scale bars in the measurement volume

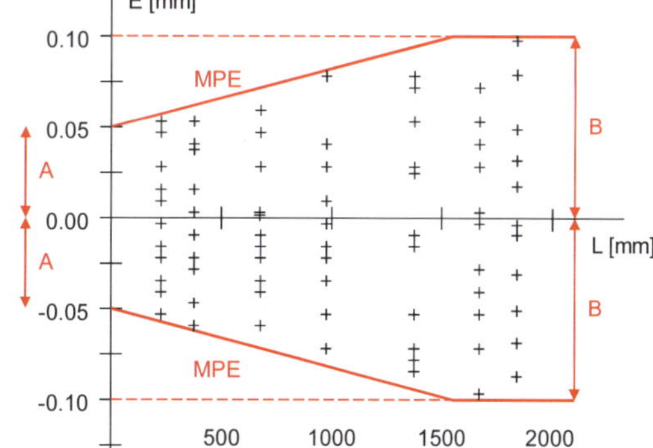

Fig. 7.13: Length measurement errors and limiting bounding box

As a practical tool for the display and analysis of length measurement errors, a diagram showing the measured differences with respect to the nominal distances can be used (Fig.

7.13). The maximum permitted limits shown by the red bounding box correspond to eqn. 7.7. A measuring system can be accepted as successful if all length errors lie within the box.

If the scale-bar arrangement illustrated in Fig. 7.12 is not possible, or the system under test can only view the measurement volume from one direction (e.g. the stereo camera system in Fig. 6.19), then a single scale bar can alternatively be moved to different positions in the volume for measurement.

If a photogrammetric calculation has provided standard deviations of object coordinates, then these can be used to estimate analytically the expected length measurement error. By applying error propagation to the equation for distance measurement:

$$L^2 = (X_2 - X_1)^2 + (Y_2 - Y_1)^2 + (Z_2 - Z_1)^2 \tag{7.8}$$

the variance of the calculated length can be obtained:

$$\begin{aligned} s_L^2 &= \left(\frac{\partial L}{\partial X_1}\right)^2 s_{X_1}^2 + \left(\frac{\partial L}{\partial X_2}\right)^2 s_{X_2}^2 + \left(\frac{\partial L}{\partial Y_1}\right)^2 s_{Y_1}^2 + \left(\frac{\partial L}{\partial Y_2}\right)^2 s_{Y_2}^2 \\ &+ \left(\frac{\partial L}{\partial Z_1}\right)^2 s_{Z_1}^2 + \left(\frac{\partial L}{\partial Z_2}\right)^2 s_{Z_2}^2 \\ &= \left(\frac{X_2 - X_1}{L}\right)^2 s_{X_1}^2 + \left(\frac{X_2 - X_1}{L}\right)^2 s_{X_2}^2 + \left(\frac{Y_2 - Y_1}{L}\right)^2 s_{Y_1}^2 + \left(\frac{Y_2 - Y_1}{L}\right)^2 s_{Y_2}^2 \\ &+ \left(\frac{Z_2 - Z_1}{L}\right)^2 s_{Z_1}^2 + \left(\frac{Z_2 - Z_1}{L}\right)^2 s_{Z_2}^2 \end{aligned} \tag{7.9}$$

Assuming coordinate standard errors are equal along the individual axes X, Y and Z ($s_{X1} = s_{X2}$ etc.), the following is obtained:

$$\begin{aligned} s_L^2 &= 2\frac{(X_2 - X_1)^2}{L^2} s_X^2 + 2\frac{(Y_2 - Y_1)^2}{L^2} s_Y^2 + 2\frac{(Z_2 - Z_1)^2}{L^2} s_Z^2 \\ &= \frac{2}{L^2}\left[(X_2 - X_1)^2 s_X^2 + (Y_2 - Y_1)^2 s_Y^2 + (Z_2 - Z_1)^2 s_Z^2\right] \end{aligned} \tag{7.10}$$

and if standard errors are equal in all directions ($s_X = s_Y = s_Z$):

$$\begin{aligned} s_L^2 &= \frac{2s_{XYZ}^2}{L^2}\left[(X_2 - X_1)^2 + (Y_2 - Y_1)^2 + (Z_2 - Z_1)^2\right] \\ &= \frac{2s_{XYZ}^2}{L^2} L^2 = 2s_{XYZ}^2 \\ s_L &= \sqrt{2} \cdot s_{XYZ} \end{aligned} \tag{7.11}$$

In order for the length measurement error to apply to all lengths measured by one system, a 3-sigma coverage factor (99 % confidence level) should be applied. The theoretical length measurement error is then given by:

$$E = 3\sqrt{2} \cdot s_{XYZ} = \sqrt{18} \cdot s_{XYZ} \tag{7.12}$$

Example 7.1:

After an on-site calibration using the reference artefact shown in Fig. 7.11a, the following RMS values for object coordinates were obtained:

RMS (1-sigma) X: 0.016 mm Y: 0.017 mm Z: 0.015 mm

According to eqn. 7.11, the standard error of length is given by $s_L = 0.023$ mm.

At a confidence level of 99%, corresponding to a 3-sigma coverage factor, the theoretical length measurement error according to eqn. 7.12 is given by $E = 0.068$ mm.

The length error is therefore approximately 4 times the simple standard error of object point location. This value corresponds well with the length error of 0.070 mm determined by testing against a reference.

7.2.2.5 Testing of area-scanning systems

Part 2 of the VDI/VDE 2634 guidelines recommend parameters and methods for acceptance and re-verification testing of optical 3D systems using, for example, fringe projection, tracked laser line scanners and image correlation methods.

The *probing error* parameter describes the error effects associated with surface point coordinates in a small measurement volume. It is derived from the measurement of a calibrated spherical surface to which a best-fitting sphere with variable radius is fitted. The range S_A of the measurement deviations from the best-fit sphere defines the probing error. The reference sphere is positioned at multiple locations within the measurement volume as illustrated in Fig. 7.14. The sphere's diameter should amount to around 10–20 % of the diagonal L_0 of the measurement space.

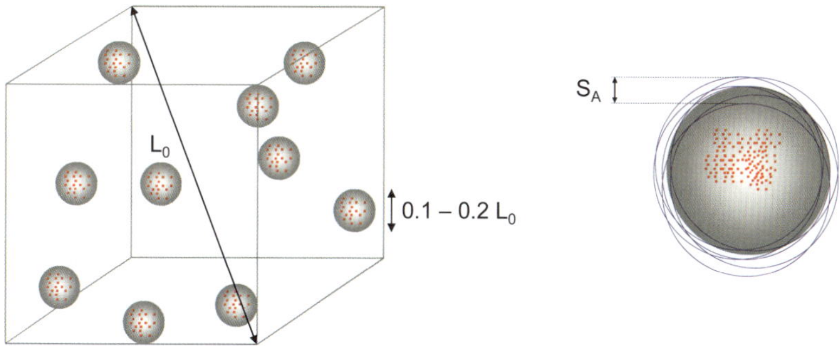

Fig. 7.14: Method of determining probing error

The *flatness error* parameter indicates the capability of the test system to measure a plane surface. For this purpose, a reference flat surface is measured at multiple locations within the test volume and the range of deviations of the measurements from a best-fit reference plane is determined at each (Fig. 7.15). The length of the reference flat should be around $0.5 \cdot L_0$.

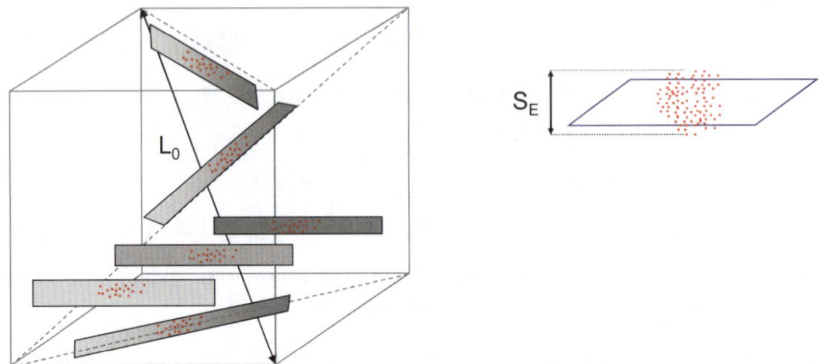

Fig. 7.15: Method of determining flatness error

The *sphere-spacing error* parameter indicates the deviation between the calibrated and measured separation of two spheres whose surfaces are scanned and centres found using a best-fit sphere with a given radius (Fig. 7.16). The sphere-spacing error must be representative of the entire measurement volume. The length of the dumb-bell artefact used for this purpose should be around $0.3 \cdot L_0$ with a sphere diameter of $0.1–0.2 \cdot L_0$. Since the probing error is not a part of the test parameter, due to multiple scanning and subsequent best fit of spheres, an ISO-compliant length measurement error is not generated (ISO 10360-2). Generally, the sphere-spacing error is always smaller than the length measurement error.

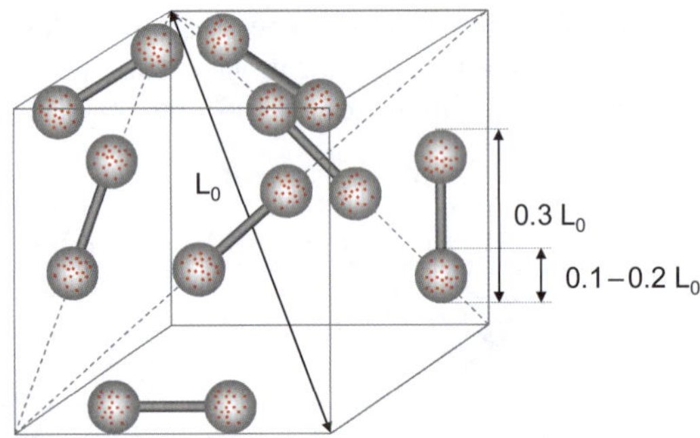

Fig. 7.16: Method of determining sphere-spacing error

When testing area scanning systems it is generally acceptable to eliminate up to 3 ‰ of the measured points from the raw data. This acknowledges the existence of unavoidable outliers in the data which are caused, for example, by reflective highlights off the object surface. It is also the case that many scanning methods acquire several million measurement points per scan and, as a matter of practicality, a thinning or filtering of the point cloud is necessary.

Part 3 of the VDI/VDE 2634 guidelines deals with systems that combine area scanning with multiple sensor orientations (multiple point clouds) in order to measure objects that are larger than the direct measuring volume of a surface measuring sensor. As examples, multiple point clouds can be created by moving the object (Fig. 7.17) and/or the sensor (Fig. 7.18), or by combining multiple sensors into one measurement system. Here it is necessary to deal with the task of merging individual point clouds into a single point cloud (registration), i.e. transforming them into a common coordinate system (compare with section 5.6.2).

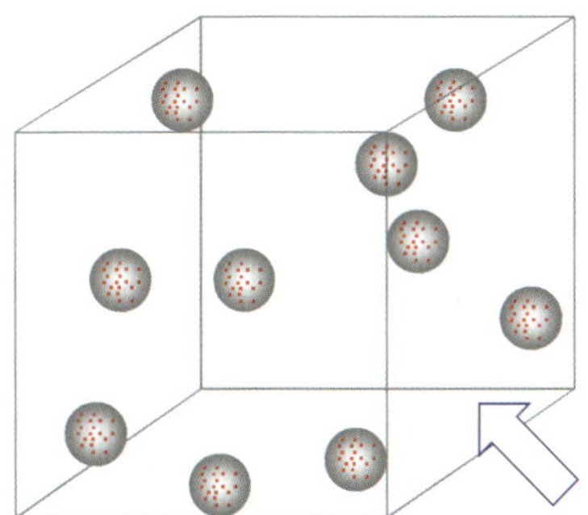

Fig. 7.17: Moving object with fixed sensor position

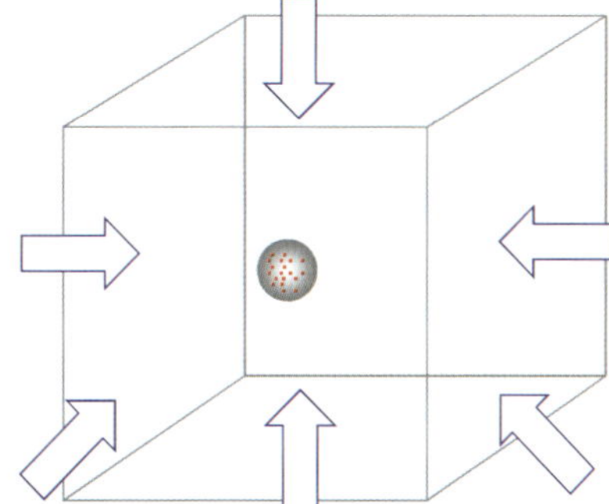

Fig. 7.18: Moving sensor with fixed object position

As with VDI/VDE 2634/2, the characteristics of probing error, sphere-spacing error and length measurement error are again evaluated. Practical constraints ensure that the recommended testing procedures of 2634/3 differ from those of 2634/2.

7.3 Strategies for camera calibration

7.3.1 Calibration methods

The purpose of camera calibration is to determine the geometric camera model described by the parameters of interior orientation (see section 3.3.2):

- spatial location of the perspective centre in the image coordinate system: principal distance and image coordinates of principal point
- parameters describing image errors: distortion and sensor corrections

In general, the interior orientation is assumed to be known and constant for metric cameras. The problem of camera calibration therefore mainly concerns those imaging systems (e.g. partial metric cameras, commercially available digital cameras) whose geometry is subject to variation over time. However, depending on the actual accuracy specifications, even metric cameras may have to be calibrated for the duration of image acquisition.

In general, imaging systems can be classified as follows:

- 1 un-calibrated, or approximately calibrated, camera for general network configuration
- 1 pre-calibrated, stable camera for general network configuration
- several calibrated and oriented cameras in a mechanically fixed configuration

If time and configuration conditions permit, then sufficient images can be taken during object measurement such that the camera can be calibrated simultaneously with the 3D object reconstruction (section 7.3.1.4). This procedure is often selected in offline photogrammetry and normally results in the highest measurement accuracy.

If only a restricted or weak camera network can be configured, then the camera can only be partly calibrated or not at all. In this case, if possible, the camera must be calibrated in a separate process directly before or after object measurement (e.g. using a test field, section 7.3.1.1). Here the validity of the camera parameters directly depends on its mechanical stability.

Today, camera calibration techniques involve a computational solution for camera parameters (camera model) which often cannot be separated from the actual object measurement. Consequently, an understanding of the different approaches to calibration requires a detailed knowledge of photogrammetric orientation and object reconstruction, especially bundle adjustment (see chapter 4.4).

Three calibration methods can effectively be distinguished. These are characterised by the reference object used and by the time and location of calibration:

- laboratory calibration (section 7.3.1.1)
- test-field calibration (section 7.3.1.1)
- self-calibration (section 7.3.1.5)

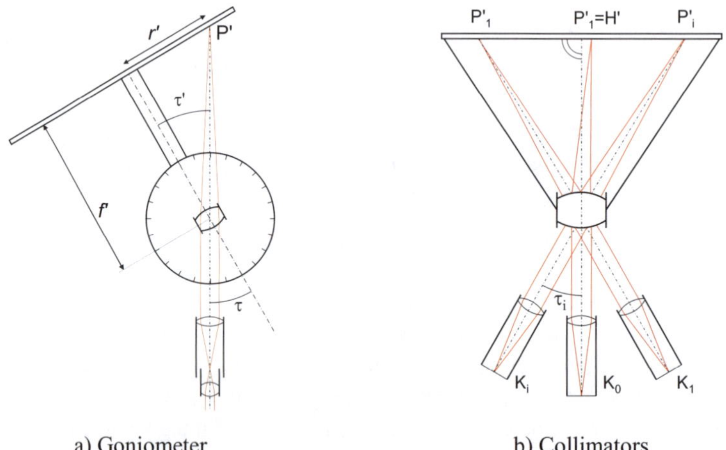

a) Goniometer b) Collimators

Fig. 7.19: Instruments for laboratory calibration (after Schwidefsky & Ackermann 1976)

7.3.1.1 Laboratory calibration

Laboratory calibration was used in the past for metric cameras. Interior orientation parameters are determined by goniometers, collimators or other optical alignment techniques where imaging direction or angles of light rays are measured through the lens of the camera (Fig. 7.19). Laboratory calibrations cannot normally be performed by a user or customer, and these methods have therefore seen little practical application in close-range photogrammetry.

7.3.1.2 Test-field calibration

Test-field calibration is based on a suitable targeted field of object points with optionally known coordinates or distances. This test field is imaged from several camera stations, ensuring good ray intersections and filling the image format. Test fields can be mobile (Fig. 7.20a), or stationary (e.g. building wall, Fig. 7.20b).

a) Mobile test field

b) Stationary test field on a wall

Fig. 7.20: Examples of photogrammetric test fields

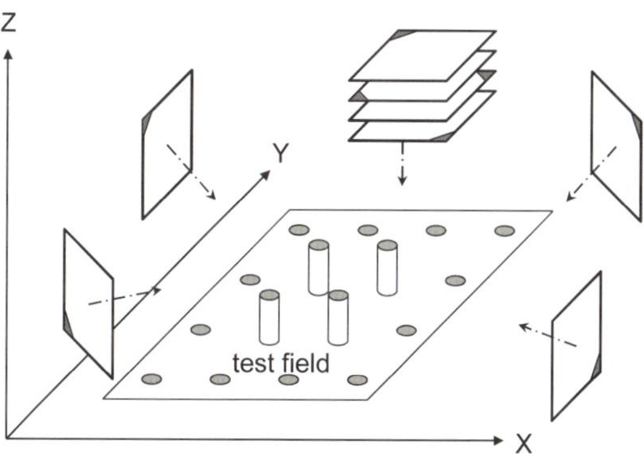

Fig. 7.21: Imaging configuration for test-field calibration

The parameters of the camera model are then calculated as part of a bundle adjustment in which normally the parameters of exterior orientation and the unknown 3D object coordinates are also calculated. Any known data (coordinates, distances) can be incorporated in different ways and are used to provide scale.

Fig. 7.21 shows a suitable image configuration for test-field calibration. In order to calibrate the camera, eight images which each image as many of the test field targets as possible, are sufficient. They should image the test field perpendicularly and obliquely and each image should have a relative rotation of 90° around the optical axis (see also Fig. 7.25). Fig. 7.22 shows a series of images acquired for test-field calibration. Further imaging configurations are discussed in section 7.3.2.

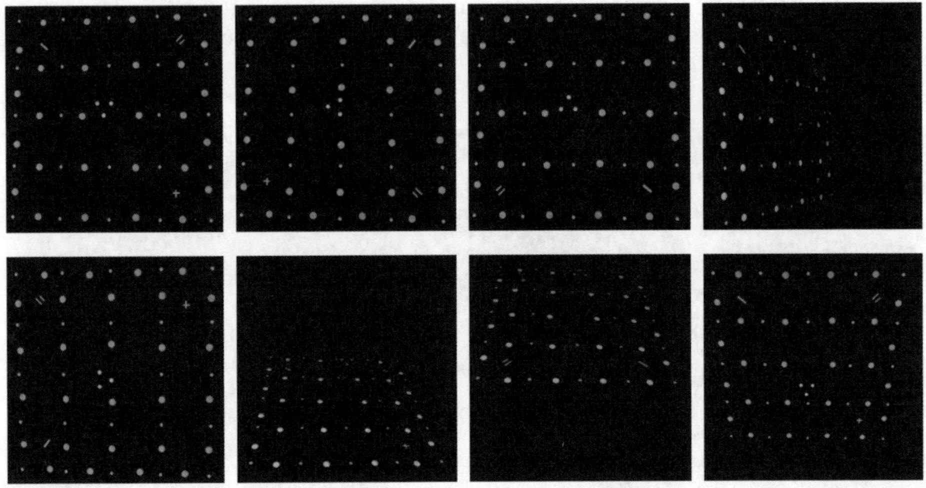

Fig. 7.22: Image series for test-field calibration

Measured image coordinates and approximately known object data are processed by bundle adjustment to give the parameters of the camera model (interior orientation) as well as the adjusted test-field coordinates and the parameters of exterior orientation.

For test-field calibration, the datum should be defined by an unconstrained technique (section 4.4.3) in order that possible inconsistencies between object point coordinates do not have a negative influence on the calculated parameters. An unconstrained datum can be created by a free net adjustment of the 3-2-1 method.

Numerical calculations can lead to unwanted correlations between the calculated parameters but these can largely be avoided by suitable imaging configurations. It is most important to provide at least one piece of scale information along the viewing direction in order to compute the principal distance. This can, for example, be achieved by a reference distance, by spatially distributed test-field points or by oblique images of a plane test field. Images rotated by 90° around the optical axis are used primarily to determine the principal point coordinates and affinity parameters. Three-dimensional test fields, e.g. with out-of-plane points, have the advantage over flat test fields that they can offer more points in depth which ensures that parameters are easier to determine and have smaller correlations.

Normally, test-field calibrations are done when a simultaneous calibration as part of an object measurement is not possible or an accuracy evaluation of a camera is required. In general, the design of the test field should represent the actual object to be measured. The number and distribution of image points are of major importance for an accurate determination of distortion parameters (see also section 7.3.3). In order to preserve the calibration parameters, there should never be any changes made to the camera (focusing, different lens) between test-field measurement and object reconstruction.

7.3.1.3 Plumb-line calibration

The plumb-line method uses a test field with several straight lines, created for example by vertically hanging wires (plumb lines, Fig. 7.23). Since in theory the projection of straight lines is invariant for perspective geometry, all departures from this condition must be caused by distortion effects. The deformed test field lines can only be used to determine distortion parameters and are insufficient to determine also principal distance and principal point. The calculated distortion parameters are not correlated with the further parameters of interior orientation or the exterior orientation parameters.

In a practical implementation of a plumb-line test field, targets can be added to the lines and these individually measured in the image. The lines can also be continuous features, such as thin white plastic cords set against a dark background for enhanced contrast. This arrangement facilitates automatic line following at high point densities (see section 5.4.3). Alternatively, natural straight line object features such as building edges can be used for calibration.

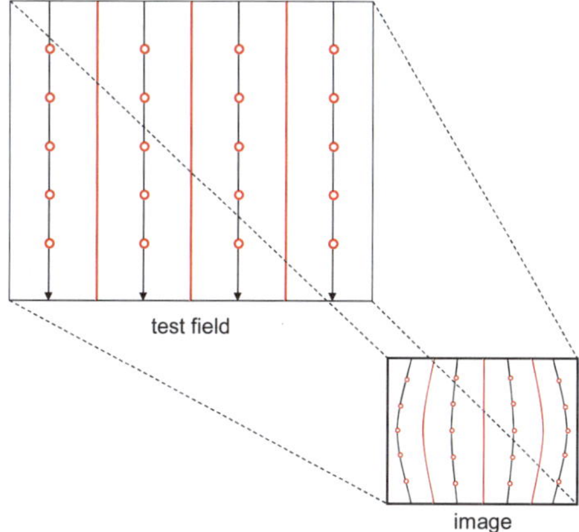

Fig. 7.23: Plumb-line method for test-field calibration

Plumb-line calibration can be sensibly applied in cases where a pre-calibration of distortion parameters is desired, e.g. if lenses with high distortions (e.g. fish-eye lenses) are used and the measured image coordinates are to be corrected for distortion prior to a system calibration (see section 7.3.1.6). Distance-dependent changes in distortion can also be determined by the plumb-line method.

7.3.1.4 On-the-job calibration

The term *on-the-job calibration* is often used where a test-field calibration (recording of a known point field) is combined with the actual object measurement. This approach is reasonable, for example, if the measuring object itself does not provide suitable geometry to enable self-calibration (see section 7.3.1.5).

A simple solution is provided by a portable frame consisting of several spatially distributed scale bars, positioned beside the measuring object and photographed simultaneously with it. The local coordinate system of the test field can be used as a three-dimensional object coordinate system and further reference points are not required.

7.3.1.5 Self-calibration

An extension to on-the-job calibration is *self-calibration* which simultaneously uses the images acquired for the actual object measurement. In effect, the test field is replaced by the object itself which must be imaged under conditions similar to those required for test-field calibration (spatial depth, tilted images and suitable ray intersections). Fig. 4.51 illustrates multiple imaging of a targeted car door which calibrates the camera as well as calculating the configurations of the ray bundles and the target coordinates.

The essential advantage of self-calibration is that the parameters of interior orientation are determined simultaneously with measurement of the object, so providing the highest of accuracies in object reconstruction.

Self-calibration does not require coordinates of known reference points. The parameters of interior orientation can be calculated solely by the photogrammetric determination of the object shape, i.e. by incorporating only image information and intersection conditions for unknown object points. If employed, reference points can be used to define a particular global coordinate system for the parameters of exterior orientation. In order to define scale it is sufficient to measure a single reference length in object space (although it is good practice to measure multiple reference lengths).

If the object to be measured does not permit a suitable image configuration, or if a multi-camera online system is used, then a test-field or on-the-job calibration must normally be performed.

7.3.1.6 System calibration

The expression *system calibration* is generally used for the determination of all geometric parameters of a complete measurement system, i.e. the interior and exterior orientation parameters of all the system components. System calibration is relevant to digital multi-camera systems that are either mobile and can be freely configured (e.g. dual camera online systems), or are mounted in a fixed position (e.g. 16-camera system for the inspection of brake pipes, see section 6.4.3.1).

For dual-camera online systems it is possible to calibrate each camera individually in advance. Alternatively, self-calibration can be applied to a set of images which have been acquired with both cameras simultaneously.

During operation, multi-camera systems on fixed mountings require particular care in monitoring and calibration. Such mechanical restrictions can cause problems, for instance, by not permitting convergent or tilted images. In general, exterior orientation parameters can be monitored on a regular basis by the use of reference points and, if necessary, can be recalculated by bundle adjustment or spatial resection. However, the interior orientation parameters can only be determined by object fields with a suitable distribution of object points. Fig. 1.39 shows an example of a multi-camera system which can be oriented and calibrated by a motor-driven rotating test field and where the orientation of the rotary table itself is simultaneously calculated.

7.3.2 Imaging configurations

The following imaging configurations are principally designed for self-calibration by bundle adjustment. The illustrated configurations of point fields and camera stations are a limited selection from many possibilities. Modifications and combinations are possible and often unavoidable. For stationary objects the camera is suitably positioned at locations around the object in order to obtain a good spatial distribution of images. For a stationary camera a mobile object is observed which is placed in different positions and angular orientations within the camera's field of view.

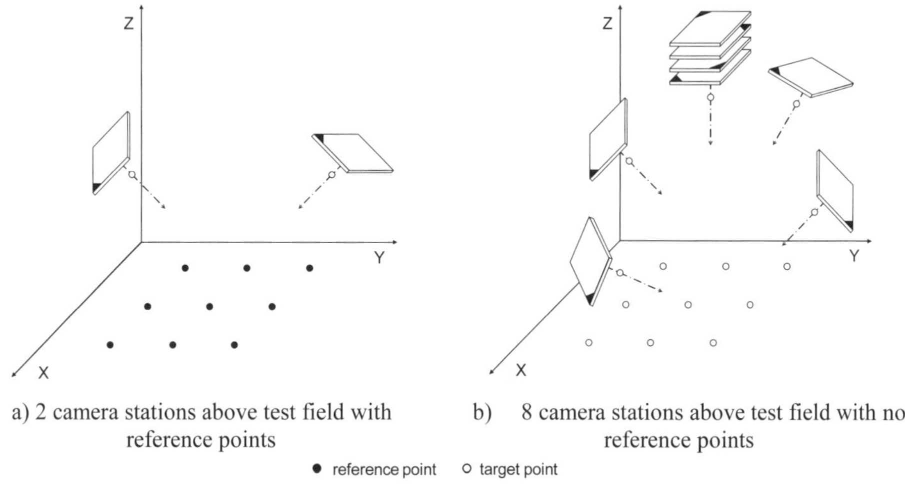

Fig. 7.24: Calibration configurations for plane test fields (after Wester-Ebbinghaus 1983, 1985)

7.3.2.1 Calibration using a plane point field

If only a plane point field (flat test field or measurement object) is available, several convergent images are necessary. The minimum number of images depends on the availability and distribution of reference points with known coordinates. If reference points are not provided, known distances (scale bars) in object space can also be used for calibration.

Fig. 7.24a illustrates a minimal image configuration for a plane test field with known reference points and an invariant camera interior orientation. The points are obliquely imaged with convergent camera axes and different roll angles (rotations about the camera axis). For a test field without control points, Fig. 7.24b shows a configuration of 8 images which generates better ray intersections, higher redundancy and improved use of the image format (more reliable determination of distortion).

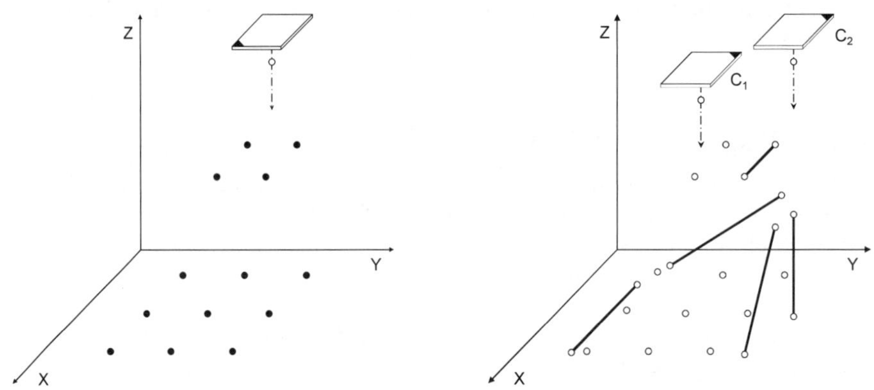

a) 1 image above spatial test field with reference points

b) 2 different cameras C_1,C_2 above test field with 5 scales and no reference points

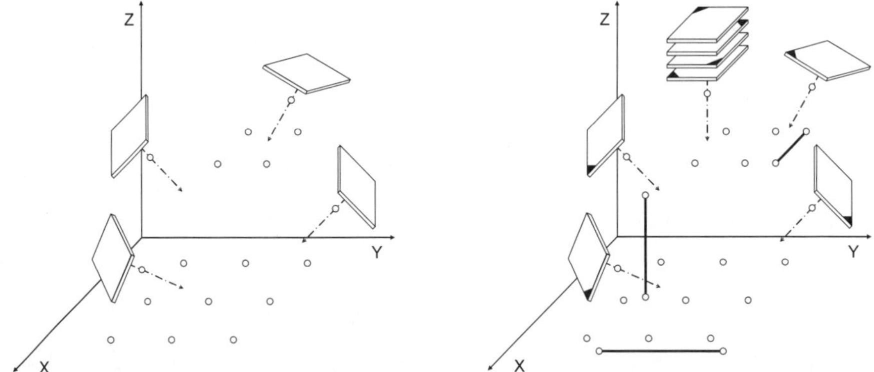

c) 4 images with arbitrary roll angles above spatial test field with no reference points

d) 8 images above spatial test field with 3 scales and no reference points

● reference point ○ target point ○——○ known distance

Fig. 7.25: Calibration configurations for spatial test fields (Wester-Ebbinghaus 1983, 1985)

7.3.2.2 Calibration using a spatial point field

Self-calibration is more reliable if object points are spatially distributed in three dimensions. Spatial point fields area preferable if the measuring task permits.

Fig. 7.25a shows an example of single image calibration using a known 3D test field (see section 4.2.3.1). The camera can be calibrated by means of an extended space resection or linear projective methods (see section 4.2.4). Fig. 7.25b illustrates the minimal configuration for a system composed of two different cameras (image-variant interior orientation), for example an online dual camera system. Explicitly rolled images are not necessary if at least four convergent images of a spatial point field are available (Fig. 7.25c). Finally, Fig. 7.25d displays the most demanding but also most reliable imaging configuration comprising eight images tilted with respect to each other and arranged above a spatial point field containing known distances. Similar configurations with around 16–25 images are recommended, although the total number is, in principle, unlimited.

In principle, all the imaging configurations shown above can be created by placing a portable test field in multiple positions with respect to the camera to be calibrated, such that the same perspective conditions between test field and camera are generated.

7.3.2.3 Calibration with moving scale bar

The principle of on-the-job calibration described in section 7.3.1.4 assumes that one or more cameras can be moved around the object in order to generate a sufficient number of convergent ray bundles. Alternatively, where cameras are in fixed locations, a test field can be moved into various positions in the object space. In this case, exterior orientations are initially based on the individual test field locations and must subsequently be transformed into a common coordinate system.

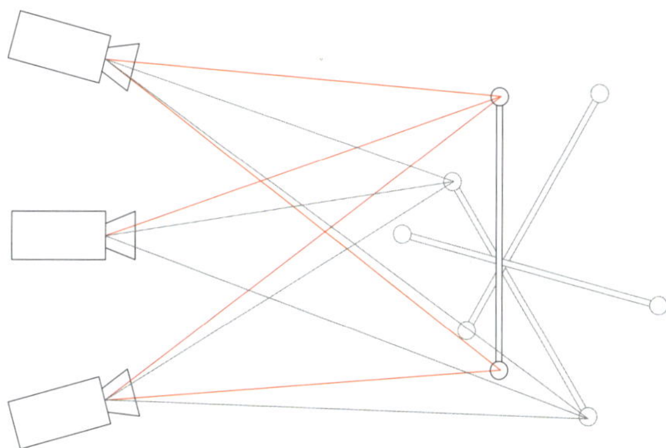

Fig. 7.26: Calibration of an imaging system with a moving scale bar

An alternative technique for calibrating and orienting sensors in fixed locations uses observations of a scale bar which is moved to multiple positions in the object space and is therefore measured from different directions (Fig. 7.26). This creates a set of unknown object points and every at scale bar position there is a length observation between a pair of points. Datum definition is achieved by a free net adjustment (six degrees of freedom) and scale information is derived from the length observations. In addition to the simple implementation, the method has the further advantage that the object space is completely defined by known lengths.

The method is particularly applicable for multi-camera systems which record image sequences and can automatically track and measure the targets on the scale bar. This type of system can be found in medical navigation (Fig. 8.58) or applications of motion capture (Fig. 6.41).

7.3.3 Problems with self-calibration

Practical problems with camera calibration typically arise in the following cases:

- Correlations between parameters:

  Using a bundle adjustment with self-calibration for the estimation of interior orientation parameters usually results in correlations between adjusted parameters. The presence of any significant correlations can be ascertained from analysis of the covariance matrix. High correlation values indicate linear dependencies between single parameters and should be avoided. Correlations often arise between the following parameters:

  - principal distance, principal point and exterior orientation
  - A_1, A_2 and A_3 will always be correlated to some extent as they are sequential terms in the radial lens polynomial model
  - principal point x'_0 and affine parameter C_1 or alternatively y'_0 and C_2

Ck	1.000									
Xh	0.017	1.000								
Yh	-0.071	0.011	1.000							
A1	-0.122	-0.001	0.010	1.000						
A2	0.037	-0.002	-0.010	-0.966	1.000					
A3	-0.006	0.005	0.010	0.905	-0.981	1.000				
B1	0.018	0.930	0.010	-0.003	-0.005	0.006	1.000			
B2	-0.034	0.010	0.905	0.004	0.002	-0.002	0.006	1.000		
C1	0.147	-0.010	-0.002	0.077	-0.106	0.108	-0.019	-0.006	1.000	
C2	0.002	-0.028	-0.028	0.007	-0.009	0.010	-0.014	-0.036	0.000	1.000
	Ck	Xh	Yh	A1	A2	A3	B1	B2	C1	C2

Fig. 7.27: Example of correlation between calibrated camera parameters

Fig. 7.27 illustrates the correlation matrix between the parameters of interior orientation. The corresponding measurement network consists of 140 images of the 3D test object shown in Fig. 7.11a. Larger correlation coefficients appear between the A parameters, as well as between the principal point coordinates (here given as *Xh*, *Yh*) and the B parameters of the tangential distortion.

Correlations between parameters can largely be neglected if object reconstruction and camera calibration are calculated in one simultaneous computation, as is the case for bundle adjustment with self-calibration. Nevertheless, parameters with no statistical significance can be detected by suitable test procedures and eliminated from the functional model. If individual interior orientation parameters are correlated and then used in subsequent, separate calculations, they no longer completely represent the chosen mathematical camera model. For example, if cameras in a fixed online measuring system are pre-calibrated using a different imaging configuration, then the subsequent online use of the resulting parameters can lead to errors in the computation of 3D coordinates. In general, calibration against a spatial test field results in lower correlations and more reliable camera parameters.

- Images without relative roll angles:

  Images with relative roll angles (rotations about the optical axis) are necessary for the determination of principal point coordinates, and possible affine transformation parameters, if the test field does not provide a suitable distribution of reference points, or if sufficient convergent images cannot be taken. The coordinates of the principal point are highly correlated with the parameters of exterior orientation if rolled images are not available.

- Incomplete use of the image format:

  The imaging sequence for camera calibration should be arranged in such a way that, within the full set of images, use of the complete image format is achieved. Only then is it possible to determine distortion parameters which are valid across this whole format (Fig. 7.28). It should also be noted that the optical axis should not, in all images, be directed at the centre of the object. The camera should be pointed in different directions, including if necessary, at only part of the object.

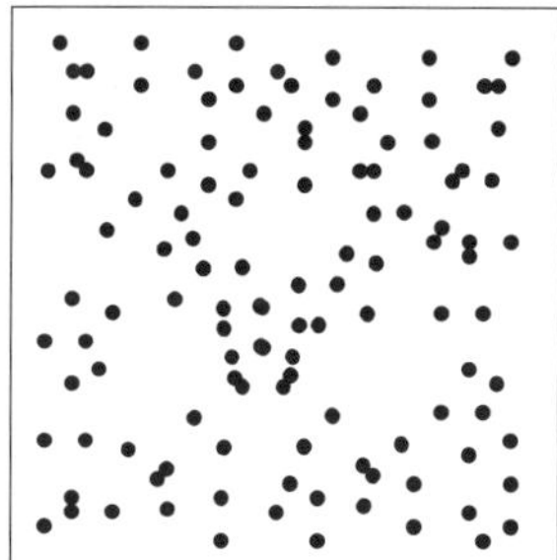

Fig. 7.28: Optimal (left) and poor (right) distribution of image points

- Use of high distortion lenses:

  Many wide-angle and super wide-angle lenses cause large distortions in the image corners. For these lenses standard distortion models are often insufficient and result in lower accuracies of points imaged in the corners. The problem becomes even more critical if the image format is incompletely used at the calibration stage. In cases where there is any doubt, it is good practice to ignore any image measurements made in the outer 10 % of the image format.

- Lack of camera stability:

  Determination of interior orientation becomes more uncertain if the camera geometry changes from image to image within a sequence, for example due to thermal effects, loose attachment of the lens or unstable mounting of the image sensor in the camera housing. Each variation in this case must be handled by defining a separate camera in the adjustment model. However, many imaging configurations only permit the simultaneously calibration of a small number of cameras. In section 4.4.2.4 an approach to the calibration of image-variant parameters is discussed.

- Missing scale information in the viewing direction:

  If scale information in the viewing direction is missing, as in the case of orthogonal images of plane test fields, principal distance and object distance cannot be uniquely determined. Just one known distance in the viewing direction (e.g. through the use of known reference points) or one known coordinate component (e.g. for convergent imagery of a plane test field) is sufficient for the calculation of principal distance. As a simple alternative, a network of convergent images of a planar target field will allow recovery of principal distance, but such a solution requires careful assessment. Another alternative is to set the principal distance to a fixed value, and not determine it in the bundle adjustment. Compensation for any potential scale error arising from this procedure in subsequent measurements can only be made using reference lengths in object space.

- Unknown pixel size in the image sensor:

  For many simple digital cameras (consumer cameras, mobile phone cameras) there is often insufficient technical data available. If there is missing information to determine the pixel size of the image sensor or the physical sensor format, then an arbitrary pixel size can be set. Although the image coordinate system is also arbitrarily scaled in this case, the calibrated parameters (in particular the principal distance) are determined with respect to the selected sensor scale. The form of the ray bundle defined by the interior orientation will, in fact, remain the same (Fig. 7.29).

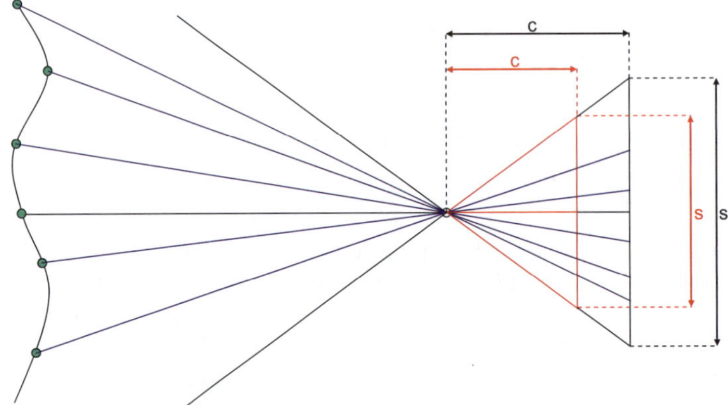

Fig. 7.29: Sensor format and principal distance

8 Example applications

The techniques of close-range photogrammetry provide universal methods and approaches to the geometric measurement of almost any kind of object. As a result there are a wide range of potential application areas. The following example applications represent only a small selection from the entire spectrum of possibilities. They are restricted to sample images, results and key technical specifications. Examples processed using film cameras can, conveniently and without restriction, be implemented using current digital cameras.

8.1 Architecture, archaeology and cultural heritage

8.1.1 Photogrammetric building records

Photogrammetric building records mostly aim to generate plan and elevation views for the following applications:

- preservation and restoration of the building
- art historical analysis
- documentation

The essential technical requirements for this field of architectural photogrammetry were already developed in the 19th century (see section 1.4). Photogrammetry offers a number of advantages compared with conventional manual methods of measured building surveys:

- non-contact measurement avoiding scaffolding on the facade
- reduced risk of accidents
- fast on-site image acquisition
- high accuracy
- three-dimensional coordinate measurement
- measurement of free-form contours and surfaces (ornamental details)
- combination of graphical output with rectified original images (photo maps)
- subsequent object measurement from archived metric images or historical photos

Targeted object points are mostly used as reference points and for image orientation. The actual object reconstruction is based on natural object features.

Analytical stereo instruments are still in use but digital multi-image systems are the systems normally used in current practice as they provide higher redundancy and direct superimposition of graphical information. In addition, digital processing systems are efficient in generating rectified images, e.g. for the production of facade mosaics (Fig. 8.1) or photo maps.

Fig. 8.1: Graphical elevation and rectified image of a terrace of buildings

8.1.1.1 Siena cathedral

The photogrammetric reconstruction of Siena Cathedral is a prime example of the measurement of a complex building in an environment with restricted access. The objective is to produce precise plans and detailed illustrations for research purposes and preservation of both the external facade and the interior.

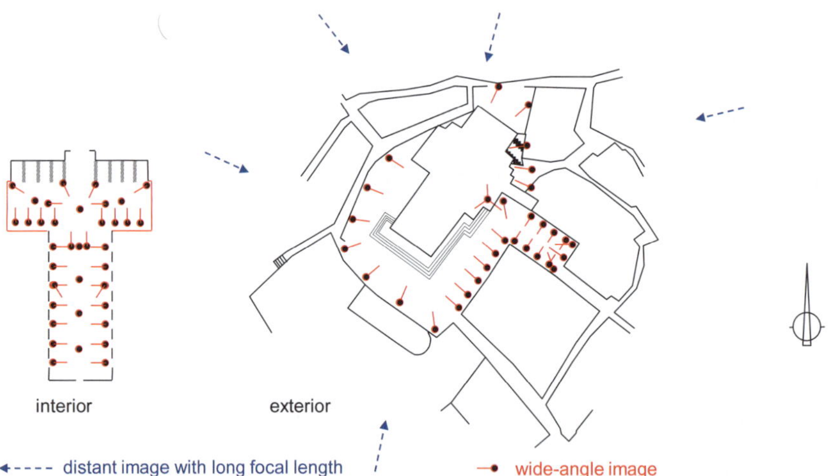

Fig. 8.2: Imaging configuration for Siena Cathedral

The complexity of the object demands a flexible multi-image configuration with a number of camera/lens combinations which are calibrated and oriented by bundle adjustment. For the actual reconstruction, more than 2200 images, taken with a semi-metric Rolleiflex SLX (Fig. 1.33) and Rolleiflex 6006, were processed in this way. However, it proved more efficient to record around 250 stereo images using the Jenoptik UMK analogue camera (image format 13 cm x 18 cm, Fig. 1.30). These could be stereoscopically analysed very simply, without significant calibration and orientation effort. Fig. 8.2 shows a plan view of the imaging configuration, Fig. 8.3 and Fig. 8.4 show sample metric images and results. The average object point accuracy was around 1–2 cm.

Fig. 8.3: Metric image (UMK) and stereoscopic mapping of façade (Messbildstelle)

Fig. 8.4: Metric image (Rollei) and elevation drawing (Fellbaum & Hau 1996)

8.1.1.2 Gunpowder tower, Oldenburg

The photogrammetric reconstruction of the historic Gunpowder Tower in Oldenburg provides an example of a multi-sided imaging configuration for the exterior and interior areas. The objective is to deliver stone-by-stone measurement data for the preparation of renovation work. The principal imaging configuration is shown in Fig. 4.53 although the internal dome is recorded by additional stereo imagery. Images taken by a Rollei 6006 were digitised and analysed using PHIDAS (Phocad), a CAD-based multi-imaging system. Image scales were between 1:100 und 1:300, and average object accuracy was around 1.5 mm. Fig. 8.5 shows metric images of the interior and exterior, and the cylindrical part of the reconstructed 3D CAD model. Superimposition of CAD data on a metric image is illustrated in Fig. 1.12, while Fig. 1.13 shows the cylindrical projection onto a plane.

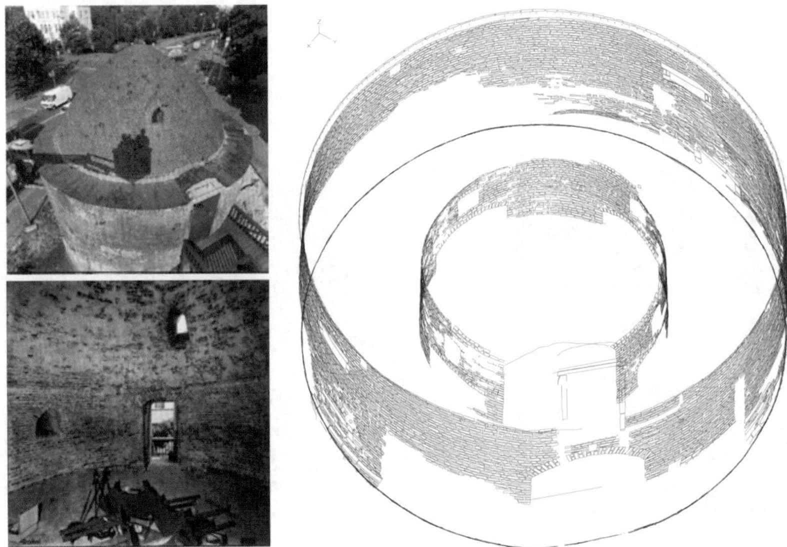

Fig. 8.5: Metric images and 3D CAD data for Oldenburg's Gunpowder Tower

8.1.1.3 Haderburg castle

The photogrammetric reconstruction of the Haderburg castle in Salurn (Salorno) demonstrates the advantages of close-range photogrammetry over alternative measurement methods based on laser scanners or total stations. Due to the inaccessible location, the necessary images could only be taken from a helicopter (Fig. 8.6). The 50 high-resolution, large-format images, taken with a calibrated Jenoptik UMK metric camera, provided a minimum number of images to ensure a sufficiently accurate mapping of the castle at a scale of 1:50 (example in Fig. 8.7).

Fig. 8.6: Recording images by helicopter using a UMK metric camera (Messbildstelle)

Fig. 8.7: Elevation drawing of the castle (Messbildstelle)

8.1.2 3D city and landscape models

8.1.2.1 Building visualisation

If a building's 3D CAD data is available in a topologically structured form (e.g. by a photogrammetric process), 3D visualisation methods (section 5.3.3: illumination models, texture mapping) can generate a photo-realistic representation. In addition to the purely aesthetic effect, these models also have practical application in building planning, facility management or building information management (BIM).

In contrast to the production of conventional plan and elevation drawings, 3D visualisation requires topological surfaces. Here points and lines (polygons) defining closed surfaces must be grouped together into logical surface patches, usually by interactive editing. For

CAD-based multi-image processing systems (see section 6.4.1) this can be implemented during the photogrammetric measurement stage, an approach also known as CAAD (*computer aided architectural design*). If techniques which produce unstructured point clouds are used (fringe projection, laser scanning) then surface modelling is achieved using triangle meshing of neighbouring 3D points (see section 2.3.3).

Data can be stored in standard 3D CAD formats such as DXF and DGN. Attention is increasingly focused on standardised 3D graphics languages such as VRML, X3D, 3DS or CityGML that include textures and which enable system-independent visualisations and interaction with the 3D model. Visualisation examples are shown in Fig. 1.6, Fig. 4.26 and Fig. 8.8, as well as in sections 8.1.2.3 und 8.1.3.2.

A more recent development is the use of simple solutions for the creation of 3D building models. These are based on simple interactive software solutions as offered, for example, by iWitness (Photometrix) or PhotoModeler (EOS Systems) . Fully automatic 3D modelling using any available imagery, including uncalibrated images, is possible with Internet-based solutions such as Photosynth (Microsoft). However, the achievable quality is generally insufficient for technical applications.

Fig. 8.8: Current state and reconstructed 3D model of the Monastery at Hude

8.1.2.2 City models

The application of 3D city models is becoming widespread, for example in:

- urban planning
- emissions analysis (sound, exhaust gases)
- planning mobile telephone networks
- setting up information systems (operational planning for rescue services, transport management)

- tourism (websites)
- three-dimensional city maps and navigation systems

These applications require the 3D model to be visually appealing, up-to-date and complete, rather than having high accuracy in the geometry and detail. Fast generation of city models can be achieved through:

- aerial photogrammetry (automatic extraction of buildings)
- airborne laser scanning (extraction of buildings from discontinuities in the height model; alignment with ground plan)
- video and laser scanning acquisition from moving vehicles (mobile mapping, section 6.7.1)

a) Addition of snowfall

b) Addition of trees

c) Complex building geometry

Fig. 8.9: Extracts from 3D city models (Oldenburg campus, IAPG)

Fig. 8.10: Texture image created from a number of individual images (IAPG)

Fig. 8.9 shows extracts from 3D city models whose components have been derived from aerial photography. Textures obtained from close-range imagery have been applied to the façades and create a realistic impression of street scenes. Different visual impressions can be further achieved by the addition of computer-generated graphics and animations (e.g. trees, snowfall).

Due to the large area covered by some of the façades, some of the textured images in this example have been created from a number of individual images. For this purpose the separate images are rectified in a common coordinate system and subsequently connected in an image mosaic (Fig. 8.10).

8.1.2.3 3D record of Pompeii

The complete recording of the excavations at Pompeii illustrates well the use of diverse data sources to create a full 3D documentation of a large, extended object. The approximately 350 finds distributed around the site, as well as the standing walls and their details, must be recorded in a single 3D model. This is required in order to prepare the data at different levels of resolution for purposes of digital conservation, animation (*virtual reality*) and connection to archaeological databases.

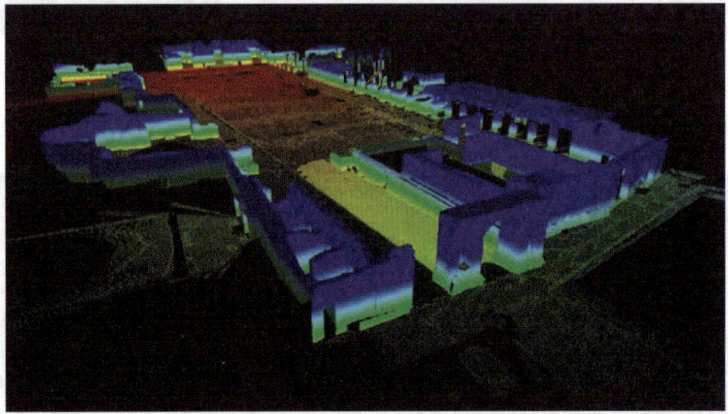

Fig. 8.11: Point cloud derived from terrestrial laser scanning (FBK Trento, Polytecnico di Milano)

Aerial and terrestrial imagery, as well as terrestrial laser scanning, are used for data acquisition and, by means of GPS measurements, transformed into a global coordinate system. Accuracy requirements are in the cm range for aerial images and the mm range for terrestrial measurements.

Fig. 8.11 shows a representative 3D point cloud created by merging multiple individual laser scans. Fig. 8.12 shows the complete model of the Forum in Pompeii and Fig. 8.13 shows detailed sections of the model with texture overlays.

Fig. 8.12: 3D model of the Forum in Pompeii (FBK Trento, Polytecnico di Milano)

Fig. 8.13: Detail sections of the 3D model of Pompeii (FBK Trento, Polytecnico di Milano)

8.1.3 Free-form surfaces

The art historical analysis and restoration of sculptures and ornamental building features requires the measurement of free-form surfaces. Conventional stereo measurement is suitable for line extraction if the object surface has distinct contours (Fig. 8.19). The surfaces of smaller objects can be measured by active pattern projection methods, as described in section 3.7.2. Larger objects are normally measured in parts which are merged into a single model using external tie points or other registration methods (see sections 5.6.2 and 6.5.3). Relevant measurement techniques include fringe projection, terrestrial laser scanning and photogrammetry. When measuring complex objects these are often used in combination.

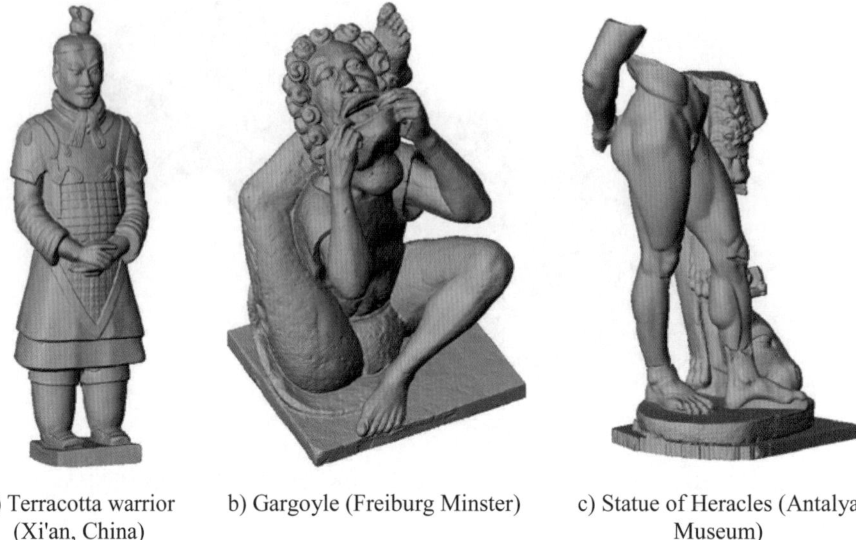

a) Terracotta warrior (Xi'an, China) b) Gargoyle (Freiburg Minster) c) Statue of Heracles (Antalya Museum)

Fig. 8.14: 3D digitizing of statues and sculptures (AICON/Breuckmann)

The raw measured 3D data are passed through the processing steps of triangle meshing, registration, smoothing and thinning after which Bézier or NURBS surfaces can be generated (see section 2.3.3). The derived 3D CAD data can be used, for example, to control NC milling machines or stereo lithography processing in order to generate facsimiles of the objects. They are also applicable in diverse aspects of 3D visualisation, e.g. in virtual museums or animations.

8.1.3.1 Statues and sculptures

Statues and sculptures mostly have complex surface shapes which cannot be interpreted as a 2½D surface and measured from only one side. If the objects of interest can be measured under controlled lighting conditions (e.g. in a laboratory or room with no external light), and they have reasonably bright, diffusely reflecting surfaces, then a fringe projection system is suitable for 3D recording.

For cultural heritage recording, the registration of individual 3D point clouds is generally achieved using the object's own geometry (natural points, distinctive surface features). As an alternative, or in combination, additional photogrammetrically measured control points can be used, either to bridge areas lacking in natural detail, or to enhance or control the overall accuracy. Smaller objects can make use of a servo positioning, e.g. rotary tables, to generate part scans in a common coordinate system.

Fig. 8.14 provides example results from digital, all-round surface measurement of diverse statues. Fig. 8.15 shows the measurement sequence in a fringe projection system.

Fig. 8.15: Projected pattern and image sequence in a fringe projection system (AICON/Breuckmann)

Fig. 8.16 shows a purely photogrammetric reconstruction of an Inca statue. This was achieved using an all-round imaging configuration with 25 images (Nikon D2XS, f=30 mm). These were oriented using a bundle adjustment (PhotoModeler) and the model created using multi-image correlation software (CLORAMA). The statue has a height of approximately 150 mm and was reconstructed using a point density of 5 points per mm which resulted in the calculation of more than 2 million object points. A comparison test with a fringe projection system showed an average difference of around 0.2 mm.

a) Statue b) Shaded 3D model c) Surface meshing

Fig. 8.16: Stereoscopic reconstruction of an Inca statue (FBK Trento)

8.1.3.2 Large free-form objects

The 3D reconstruction of Temple of Hadrian in Ephesus (Turkey) illustrates the measurement of a large, complex object with the aid of fringe projection and photogrammetric orientation of the resulting point clouds. Some 75 coded targets were attached to the object (Fig. 8.17a) and recorded in a multi-image network using a Nikon D3 digital camera. They were located in 3D using the AICON DPA Pro system. In addition, 35 of the points were intersected by theodolite which was used to reference them to the local geodetic coordinate system. The accuracy of the control points was around 1–2 mm.

a) Temple with photogrammetric targets

b) 3D model of the main entrance

Fig. 8.17: Recording the Temple of Hadrian using fringe projection and photogrammetry (AICON/Breuckmann, Austrian Archaeological Institute)

a) 3D model with natural texture

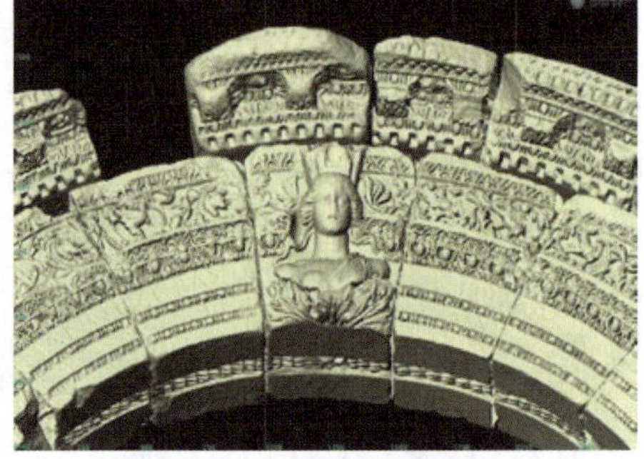

b) 3D model without texture

Fig. 8.18: Detail views of the Temple of Hadrian (AICON/Breuckmann)

Because fringe projection systems are sensitive to extraneous light, in particular sunlight, all scanning was performed at night. The temple extends across a volume of around 10 m x 10 m x 8 m and was recorded by around 1800 individual scans which generated a total of more than 1 million surface points. Fig. 8.17b shows the 3D model of the front view of the temple. Fig. 8.18 shows a detail view of the model, with and without texture, from which it can be seen that the model without texture gives a better 3D impression of the object.

8.1.3.3 Survey of the Bremen cog

The Hansa cog (a type of ship), found in Bremen harbour and dating from the 14th century, was completely measured by photogrammetry prior to water conservation lasting from 1982–1999. Recording and analysis were done using analogue cameras and analytical stereoplotters. Results were scaled plans with the principal contour lines of the object (Fig.

8.19). Following removal from the conservation tanks and a drying period of several years, a new digital measurement was made in 2003. This provided 3D profiles and models used to document changes in the cog during the conservation process and its subsequent presentation to the German Maritime Museum in Bremerhaven.

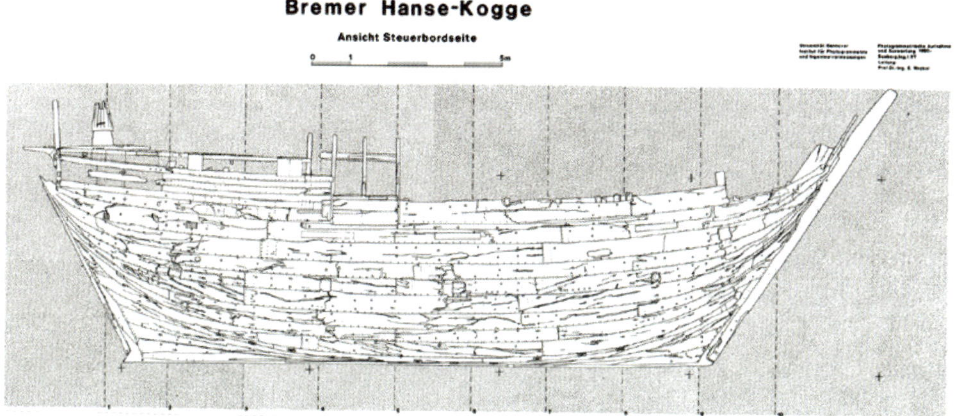

Fig. 8.19: Stereoscopic line extraction (Bremen Cog, IPI Hanover)

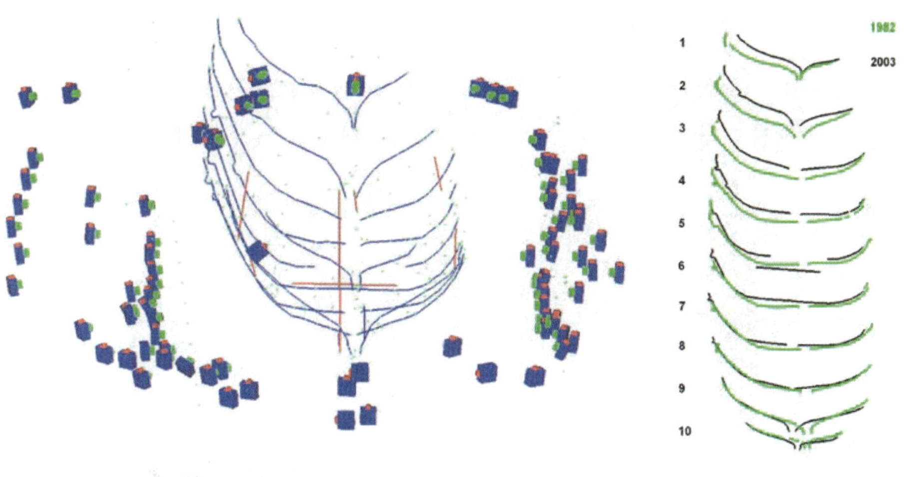

a) Digital image network configuration b) Comparison of profiles

Fig. 8.20: Digital survey of the Bremen Cog (IPI Hanover)

The entire set of targeted points and lines were digitally measured in the approximately 100 digital images. The image network was oriented and analysed using PhotoModeler (EOS System). At image scales between 1:100 and 1:900, object point accuracies of around 1–2 mm in XY and 8–10 mm in Z were achieved. Fig. 8.20 shows the digital imaging configuration and an example comparison of profiles.

8.1.4 Image mosaics

Digital archiving or restoration often requires pictorial documentation for the high resolution recording of large flat objects such as murals, floor mosaics and paintings. In some cases it is necessary to record the object in patches using a number of individual images, for example if the required overall resolutions, or the local conditions, do not permit recording with a single image.

The production of the corresponding image mosaics is based on the rectification of the individual images (see section 5.3) and a subsequent radiometric adjustment of the colours so that joins largely vanish between the original images. In a properly rectified (georeferenced) image mosaic, additional data overlays are possible (e.g. vector data) and geometric measurements can be made.

8.1.4.1 Image mosaics for mapping dinosaur tracks

The dinosaur tracks discovered in the Obernkirchen sandstone quarry are to be recorded photogrammetrically for the Hannover State Museum in case they are lost during further mining. More than 1300 digital images are to be recorded and, in the first instance, used to create 2D plans and image mosaics. Because of the built-in planning for image overlaps (see Fig. 8.21), three-dimensional reconstructions can subsequently be made in the future. Fig. 8.22 shows a true-scale image mosaic derived from rectified individual images. The measured dinosaur tracks appear in vector format.

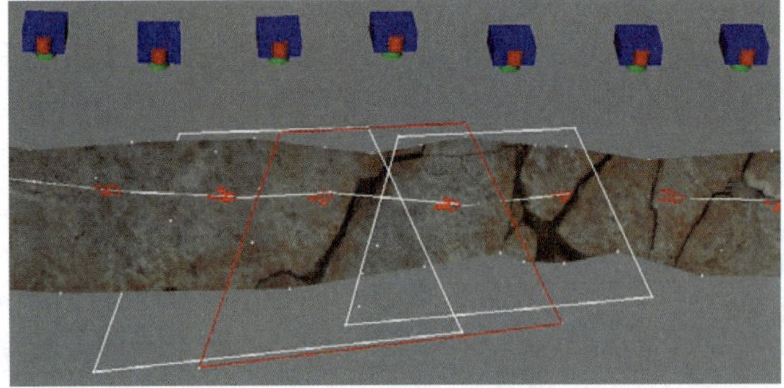

Fig. 8.21: Image configuration for generating an image mosaic (IPI Hannover)

Fig. 8.22: Mosaic based on 245 images with vector overlay of tracks (IPI Hannover)

8.1.4.2 Central perspective image mosaic

Using the example of the photogrammetric mapping of an archaeological site (Ihlow monastery), it is possible to demonstrate how an image mosaic with the property of central projection can be created from a number of overlapping images.

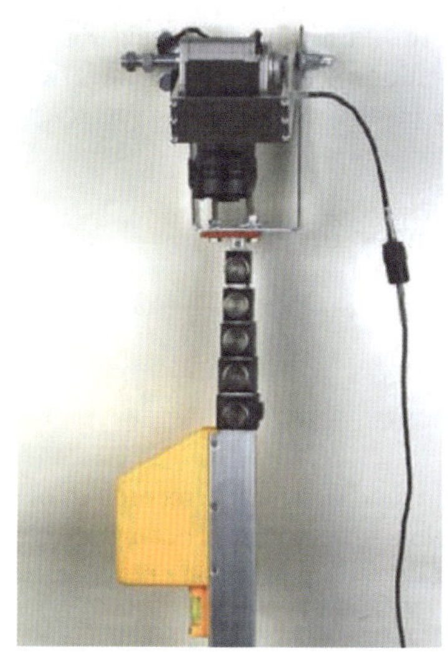

a) Telescopic stand with vertically pointed camera

b) Image mosaic generated from 12 individual images

Fig. 8.23: Production of a vertical, central perspective image mosaic (IAPG Oldenburg)

If the images are taken, to a good approximation, from the same location, e.g. using a telescopic stand supporting a camera pointed vertically down, then overlapping vertical and oblique images can be recorded. The images are then oriented using a bundle adjustment in which the translational components of the exterior orientations can be held fixed. With the calculated orientation data, a subsequent image mosaic can be created which, as a single entity, has the property of central projection and can therefore be used for further photogrammetric purposes (Fig. 8.23).

a) Individual overlapping images

b) Calculated image mosaic

Fig. 8.24: Generation of a central perspective image mosaic from 4 oblique images (IAPG Oldenburg)

Fig. 8.24a provides an example of multiple imaging from a hydraulic lift which is kept approximately in the same position for all images. The resulting image mosaic shown in Fig. 8.24b is used for stereoscopic mapping and calculation of an elevation model of an archaeological site for which a second image mosaic is used as a stereo partner image.

8.2 Engineering surveying and civil engineering

8.2.1 3D modelling of complex objects

8.2.1.1 As-built documentation

As-built documentation encompasses the measurement and inventory recording of an existing production facility. Most commonly, the complex structures and arrangements of pipes and machinery must be measured three-dimensionally for the following purposes:

- Generation of up-to-date construction plans and CAD models, to centimetre accuracy, for production planning and control, and plant information systems (facility management)
- Provision of precise geometric data (millimetre accuracy or better) to enable the replacement of large components which are manufactured off site to fit existing mechanical interfaces

For power plants, and especially for nuclear facilities, on-site measurement time must be minimised in order to reduce the danger to personnel and avoid interruption of processes in operation. Components such as pipe sections, heat exchangers and boilers are replaced during regular shut-down periods and the necessary geometric data must be available in advance. Using targets attached to the objects, accuracies are specified to about 0.5 mm for object dimensions of 10–20 m.

The documentation of complex pipework is of considerable importance. Relevant sites such as chemical plants, oil refineries and water works are characterised by difficult environmental conditions, bad visibility and complex geometries (Fig. 8.25). The objective of documentation is to produce a correct inventory and plans which in practice can only be processed by 3D CAD systems due the three-dimensional nature of the facility. The data acquired can be used for 2D and 3D views (Fig. 8.26), inventory analysis (e.g. location of pipes, parts lists), as well as the simulation of production processes.

Fig. 8.25: Metric images of pipework (INVERS)

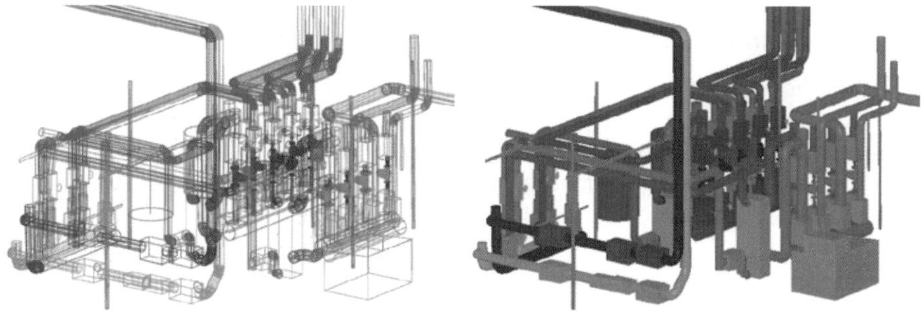

Fig. 8.26: CAD wire model and rendered view of pipework (INVERS)

Interesting opportunities are opened up by the combination of 3D laser scanning and photogrammetric measurement and analysis. Fast 3D acquisition by laser scanning requires relatively little effort on site. However, the relatively complex analysis of structural details is made easier if an oriented image is available from photogrammetric imagery recorded at the same time. Using the monoplotting method (section 4.2.7) details can be identified by interpretation of the image and interpolation in the point cloud (see Fig. 4.23).

8.2.1.2 Stairwell measurement

The three-dimensional recording of stairwells is a complex task due to the usual on-site conditions (accessibility, visibility). The problem can be solved photogrammetrically when the elements to be measured (edges, corners, steps, handrails) are suitably targeted and an appropriate imaging configuration is selected.

a) Targeting using special target adapters

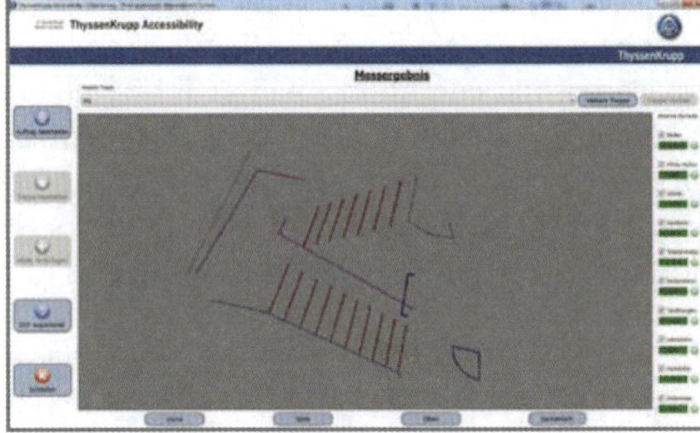

b) 3D analysis

Fig. 8.27: Photogrammetric stairwell measurement (AICON, ThyssenKrupp)

Fig. 8.27a shows a stairwell targeted with adapters whose measurement points define a unique relationship to the actual point of interest, e.g. the edge of a step (see section 3.6.1). The coding integrated into the adapters is designed so that the type of feature (corner, edge) is automatically identified. Measurement of features of interest is typically achieved to an accuracy of around 1 mm.

8.2.2 Deformation analysis

A key task in geodetic engineering surveys is to monitor deformations on buildings exposed to some particular mechanical or thermal loading. For these applications accuracy requirements are typically in the order of millimetres for object dimensions of more than 100 m (e.g. cooling towers, chimneys, wind turbines, dams, sluices, cranes, historical buildings etc.).

Photogrammetric deformation analysis is usually applied in cases where object or environmental conditions do not allow sufficient time on-site for extensive geodetic measurements, or a large number of object points are required. Image acquisition additionally provides an objective documentation of the object's state at the time of exposure and, if simultaneous measurements are made, can also record rapid object changes.

High precision photogrammetric object measurement requires high resolution cameras. Critical object points are targeted and a stable network of reference points is necessary for detecting possible object deformations or movements.

Measurement of the targeted points is done using digital image processing methods (section 5.4.2) with an image measurement accuracy in the region of 1/10 to 1/50 of a pixel. 3D object coordinates are calculated by bundle adjustment with simultaneous camera calibration and the inclusion of any additional geodetic measurements. Deformation analysis can also be performed within the bundle adjustment, or by separate 3D transformations of the object points in different measurement epochs.

8.2.2.1 Shape measurement of large steel converters

In the steel industry, raw iron is refined into to steel in a converter. Converters are pear-shaped vessels of around 10 m in height and 7 m in diameter (Fig. 8.28). After several years of operation, thermal and mechanical stresses lead to deformations which affect the exterior flow of cooling air. By measuring the wall of the vessel, critical areas can be detected and deformation monitored.

Camera stations are arranged in nine planes through the longitudinal axis of the converter. Viewing directions have a star-shaped form (Fig. 8.29). The required base length is given by the vertical overlap in adjacent images. Tie points, identified by white spots of paint, support image measurement. Inside the converter 19 reference points are arranged in two rings located close to its tilt axis. They are measured by total stations with respect to a nearby field of control points. At image scales between 1:50 and 1:85, achievable object point accuracy is around 1 mm with length errors <5 mm.

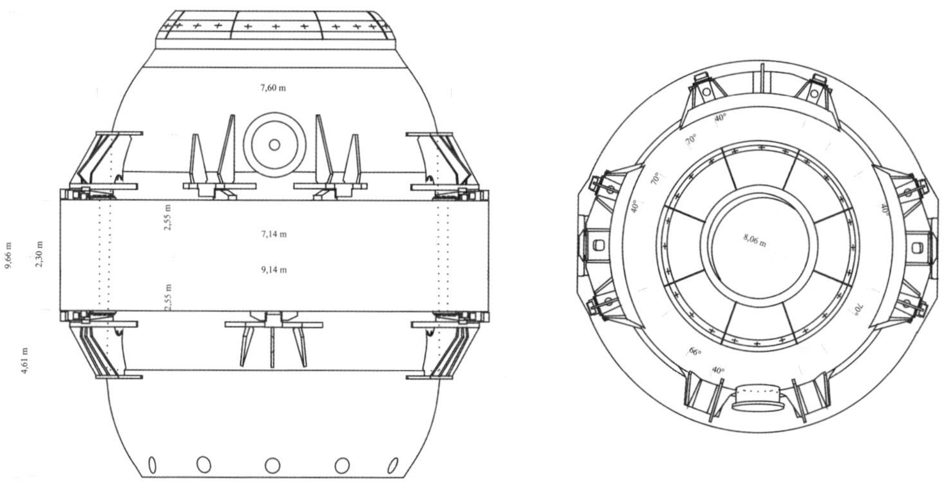

Fig. 8.28: Construction drawing of a steel converter (IPI Hannover)

Fig. 8.29: Steel converter and internal imaging configuration (IPI Hannover)

8.2.2.2 Deformation of concrete tanks

In this example, the deformation of concrete tanks used for galvanizing and electroplating must be measured under working conditions. The tanks are constructed from a special concrete and have dimensions of approximately 4 m x 1 m x 2 m. In operation they are slowly filled with liquid of total weight 7.5 tons. The tank walls are subject to critical deformations which must be observed photogrammetrically at 10 minute intervals. Around 325 points must be measured to an accuracy of less than 0.1 mm.

Due to the very confined object environment, the shortest focal length available for a Fuji FinePix S2 digital camera (f = 14 mm) must be used. Fig. 8.30 shows the measurement configuration which makes use of two reference lengths to define scale. However, points on the smaller object side could not be observed with optimal intersection angles, hence object point accuracy is weaker in those areas. The image scales achieved were between 1:70 and 1:120. The calculated deformations show a systematic behaviour (Fig. 8.31).

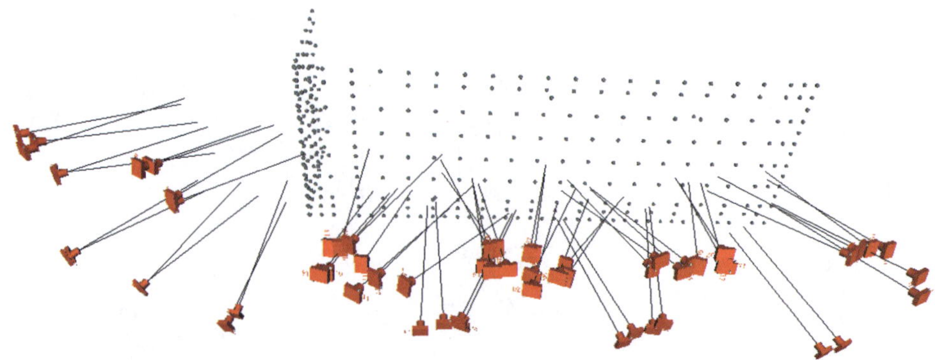

Fig. 8.30: Imaging configuration (IAPG Oldenburg)

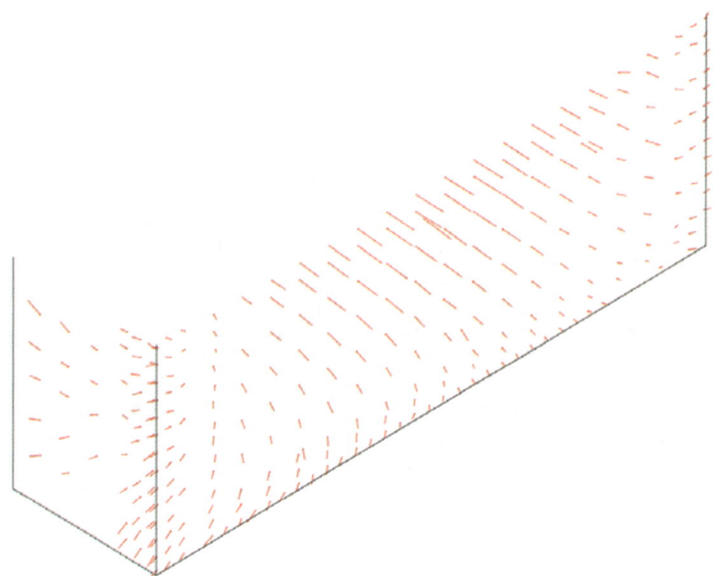

Fig. 8.31: Resulting deformation vectors between two sequential measurements (IAPG Oldenburg)

8.2.3 Material testing

8.2.3.1 Surface measurement of mortar joints in brickwork

The following application is an example of deformation analysis in building maintenance and is concerned with the measurement of erosion which is affecting the pointing (mortar joints) in brickwork. Over a period of some 10 years, the progress of erosion will be monitored on site every two years. The test sites are located on a church tower at a height of around 40 m (Fig. 8.32). Each test area is approximately 360 mm x 220 mm and is measured by stereo imagery. Due to the difficult lighting conditions, fringe projection systems cannot be used.

The accuracy is specified to about 0.1 mm. Four masonry bolts define a fixed object coordinate system. A separate reference frame containing calibrated reference points can be re-attached to a test area in order to deal with repeated measurement. A digital camera is used for image recording. Since the measurements must be made under difficult daylight conditions, a fringe projection system cannot be employed. The natural surface structure provides enough texture for image matching and 3D reconstruction. Results are presented in the form of contour lines and differential height models (Fig. 8.33). Typically around 100 000 surface points are measured.

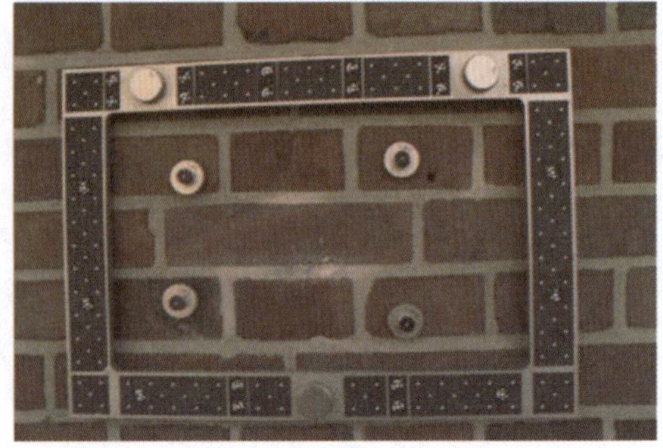

a) On-site location b) Sample image

Fig. 8.32: Data acquisition for erosion measurement (IAPG Oldenburg)

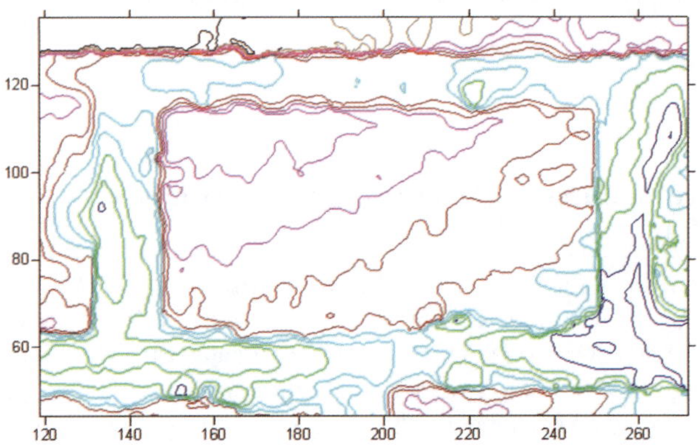

Fig. 8.33: Surface model derived from stereo matching (IAPG Oldenburg)

8.2.3.2 Structural loading tests

The following example describes the photogrammetric recording of deformations and cracks in fibre-reinforced concrete test objects (TU Dresden). The qualitative and quantitative development of cracks and deformations during the loading is of particular interest.

Fig. 8.34 shows the imaging configuration for the measurements. The test load object is recorded in three dimensions using digital stereo cameras. Mirrors positioned to the side provide side views of the test object. Only one camera views each side in order to record crack development in those areas. Camera calibration and orientation is done prior to testing using a multi-image network with bundle adjustment and self-calibration.

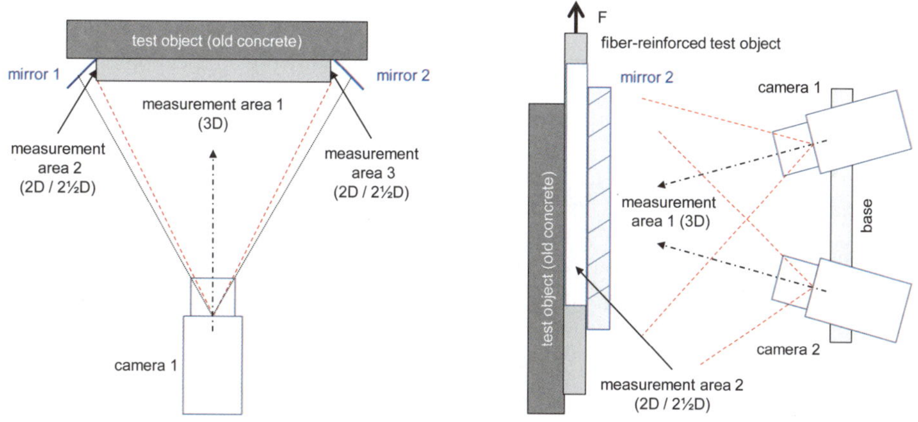

Fig. 8.34: Imaging configuration for displacement and crack analysis (IPF TU Dresden)

Cracks are detected by applying dense area-based matching to consecutive images and analysing the resulting shift of image patches for discrepancies. This way, cracks can be localised with pixel accuracy, and crack widths can be determined to an accuracy of about 1/20 pixel. Fig. 8.35a shows a greyscale-coded visualisation of the matching results, showing the position of cracks for one load stage. Cracks in the material are clearly distinguishable from the areas between them. Fig. 8.35b visualises measured crack patterns for single a load stage, where *dZ* denotes the width of the crack. A subsequent crack analysis (crack location and width) is made, for every load stage, along defined profiles shown in the image. The photogrammetric system permits continuous measurement of object deformation with an accuracy of up to 1 µm, and cracks with an accuracy of around 3–5 µm.

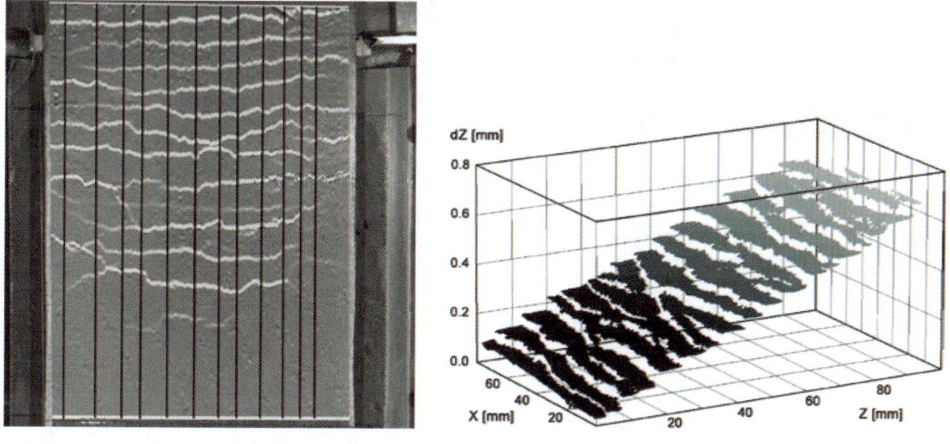

a) Crack pattern with marked profiles b) Surface evaluation (around 1.8 million points)

Fig. 8.35: Result of displacement and crack analysis (IPF TU Dresden)

8.2.4 Roof and façade measurement

The measurement of building surfaces is valuable for the following applications:

- measuring of façade areas for cleaning and painting
- measuring of façade areas for calculation quantities of construction and insulation material
- roof measurements for tiling
- simulation of building modifications and additions (e.g. colour, brickwork, tiles)
- measurement of windows and doors
- roof measurements to plan for solar panels

Fig. 8.36 shows the result of measuring the façade an apartment block. For this image the plane projective transformation parameters are known (section 4.2.6) so that the side lengths of the façade, and the location and areas of windows and other objects within the façade, can be determined. In this case, transformation coefficients are calculated using the four corner points of a window with known dimensions. This results in extrapolation errors at the edges of the image.

Fig. 8.36: Façade measurement and simulation of new building colour (IAPG Oldenburg)

Fig. 8.37a shows a measurement image with a graphical overlay of a planned solar panel array. Plane transformation parameters are obtained through the use of a calibrated reference cross whose target points can be automatically found in the image (compare with Fig. 4.17). Planning accuracy is around 5 cm. The required solar modules were inserted interactively into the image from a databank and, with texture projection, given a realistic appearance. Using an approximately computed exterior orientation of the uncalibrated image, it is possible to estimate the locations of points which are perpendicularly offset from the plane of the roof, for example in order to visualise the appearance of a raised module (Fig. 8.37b).

Fig. 8.37: Image analysis for planning the installation of a solar panel array on a roof (IAPG Oldenburg)

8.3 Industrial applications

8.3.1 Power stations and production plants

8.3.1.1 Wind power stations

There are a number of photogrammetric applications relevant to the construction and operation of wind power plants, for example:

- Measurement of rotor blades:

  Rotor blades have a complex laminated structure. For quality assurance during manufacture, shape and size are partly controlled by physical reference gauges but increasingly also by optical 3D measuring systems. During loading tests, deflections, deformations and strains are measured. Here conventional techniques (wires, strain gauges) are being replaced by optical methods. For example, the blade can be targeted and its deflection measured photogrammetrically against a fixed network of reference points (operational principle shown in Fig. 8.38).

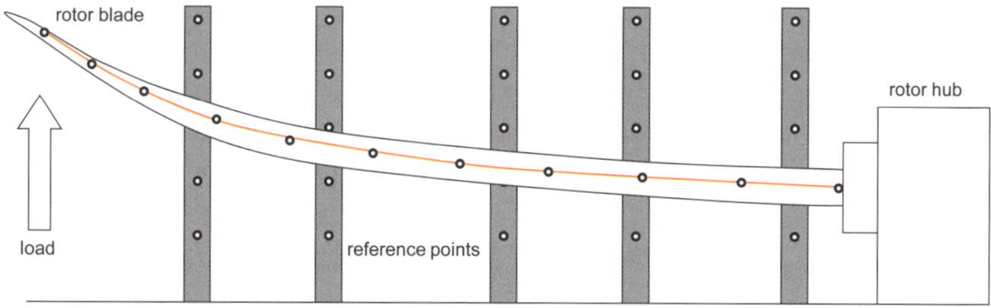

Fig. 8.38: Principle of photogrammetric measurement of load deflections

- Measurement of wind turbine tripods and rotor blade hubs

  Checking the flatness and roundness of the base of the turbine tripod (support tower) is done directly on site. This helps to ensure that the turbine is assembled according to plan. Measurement is by offline photogrammetry using a multi-image network which provides three-dimensional coordinates of targets attached to the surface of the tripod flange. Typical accuracy is around 0.1 mm for a base diameter of up to 4.5 m. Fig. 8.39 illustrates the tripod measurement and the results from analysis.

a) Targeting the flange of a wind turbine tripod

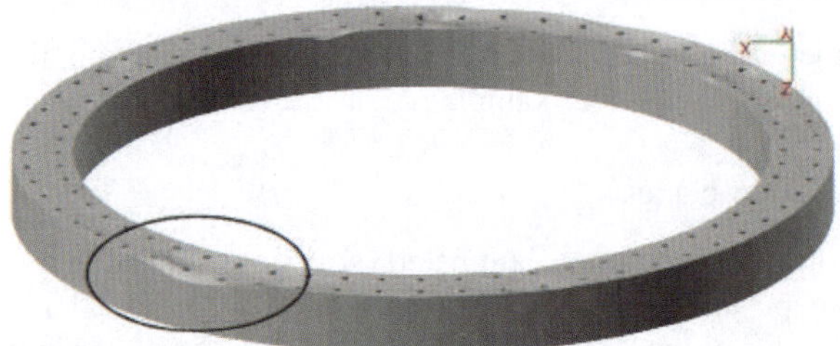

b) Form analysis of the tripod base showing areas of defects

Fig. 8.39: Photogrammetric recording of wind turbine tripods (AICON)

- Measurement of rotor blades in operation:

Deformation measurement in operation, and under actual wind conditions, requires a synchronously measuring, high-resolution, multi-camera system in order to achieve measurement accuracy in the centimetre region for objects of dimension up to 100 m. Fig. 8.40 illustrates the imaging configuration and targeting arrangement for measuring an operational wind turbine. The vibration and deformation of the rotor blades and tower are determined from 3D data sets acquired at measurement frequency of 100 Hz.

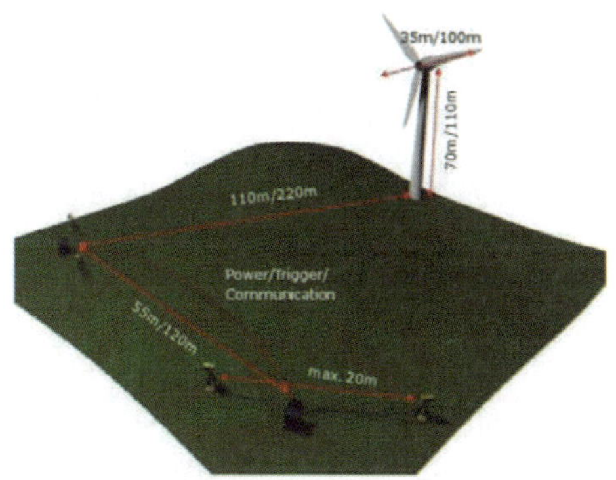

a) Imaging configuration with four synchronous cameras

b) Object targeting and online evaluation

Fig. 8.40: Photogrammetric measurement of a wind turbine in operation (GOM)

8.3.1.2 Particle accelerators

The photogrammetric measurement of detectors in a particle accelerator at CERN is an example of large object measurement requiring very high accuracy. Fig. 8.41a shows a sub-detector in the ATLAS experiment, part of the LHC (Large Hadron Collider), which has a diameter of 27 m. The detector has 12 sectors, each with 22 chambers and each marked with four targets. The objective of the measurement is the installation and determination of position of the chambers in the sub-detector system, as well as the alignment of the sub detector in the accelerator's global coordinate system.

a) Sub-detector (diameter 27 m)

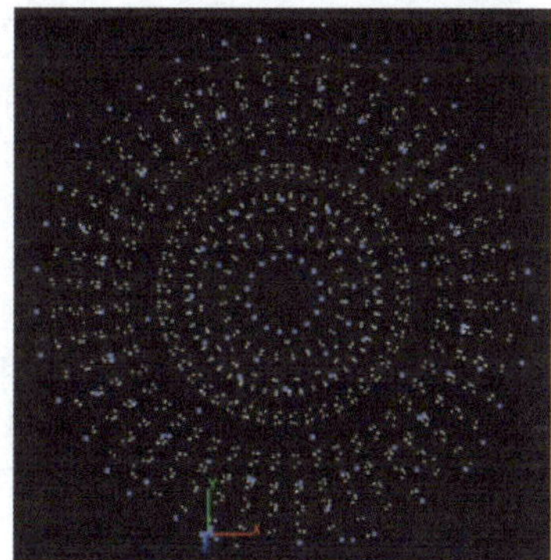

b) Photogrammetric analysis

Fig. 8.41: Sub-detector of particle accelerator LHC (CERN)

The photogrammetric task encompasses the recording of around 1200 object points using approximately 1000 images (Nikon D2XS, f = 17 mm und 24 mm) which ensures coverage of areas which are difficult to access. In addition, measurements with a total station provide scale information to an accuracy of around 0.3 mm, and enable the transformation of the photogrammetric data into the global coordinate system with an accuracy of around 0.5 mm. At average scales of 1:250–1:350, analysis with the AICON 3D Studio software provides an object coordinate precision of 0.022 mm in XY and 0.065 mm in Z (1-sigma RMS). Fig. 8.41b shows measured object points from part of the photogrammetric analysis. The absolute necessity for automatic outlier detection should be noted. Without this feature, a project like this, with some 89 000 observations and 9400 unknowns, could not be processed successfully.

8.3.2 Aircraft and space industries

Photogrammetric applications in the aerospace industry are distinguished by extremely high accuracy specifications for very large objects. Typical specifications for relative accuracies are between 1:100 000 and 1:250 000 or more. By the 1980s, large format, analogue cameras, in conjunction with réseau techniques and precise, digital comparators, were already in successful use (Fig. 8.42a). Since high-resolution digital cameras have become available, a number of applications for (automated) industrial photogrammetry have been developed. Examples of application areas include:

- measurement of parabolic antennas
- measurement of large tooling jigs and mechanical gauges
- production control of large components and assembly interfaces
- space simulations

a) Periodic checking of a tooling jig using Rollei LFC and GSI CRC-1 film cameras

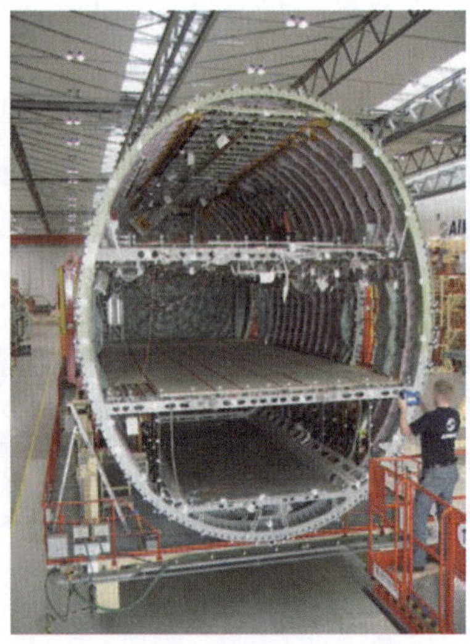

b) Measurement of A380 fuselage interface with GSI INCA digital metric camera

Fig. 8.42: Photogrammetric measurement of aircraft jigs and components (Airbus)

8.3.2.1 Inspection of tooling jigs

For the inspection of large tooling jigs in the aircraft industry (Fig. 8.42), accuracy is again specified to about 0.1 mm for object sizes up to 50 m (length of aircraft), i.e. a relative accuracy of up to 1:500 000 is required. This task can only be solved using highly redundant, multi-image networks. Verifying the achieved accuracy is problematic in measuring tasks of these dimensions. Reference scale bars longer than 3 m are difficult to handle in practice and significant effort is required to provide reference coordinates (see section 7.2.2). If time permits, laser trackers commonly offer a solution for reference measurement. However, target measurement is sequential and this option is expensive in terms of system and manpower resources required. There is a further need for targeting measurable by both systems in order to connect their measurements together.

8.3.2.2 Process control

Fig. 8.43 shows the application of online photogrammetry to process control in aircraft manufacture. Here two cameras are mounted on an unstable platform, i.e. their relative orientation is not constant. However, the exterior orientation is continuously updated by measurement of coded targets in fixed locations around the aircraft door, which is the component being measured. Simultaneous measurement of a manually placed touch probe at critical points around the door provides the 3D coordinates of the touch point in the same coordinate system. The achievable measurement accuracy is 10 µm + 10 µm/m.

a) Online system on unstable platform

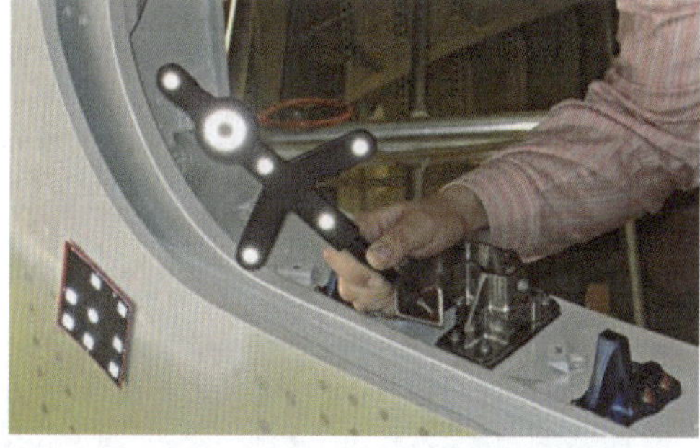

b) Manual measurement of points to be checked

Fig. 8.43: Online photogrammetric measurement in aircraft manufacture (GDV, Airbus)

The example in Fig. 8.44 shows a digital photogrammetric system employed for the online measurement of corner joints in aircraft wings. Tailored adhesive target sheets, with both standard and coded targets, are positioned on the interface between wing and aircraft body. Only 17 minutes, including system set-up, are allowed for the complete measurement of the corner connection. Image acquisition is an online procedure via a high-resolution video camera which, together with a light source, is mounted in a compact housing.

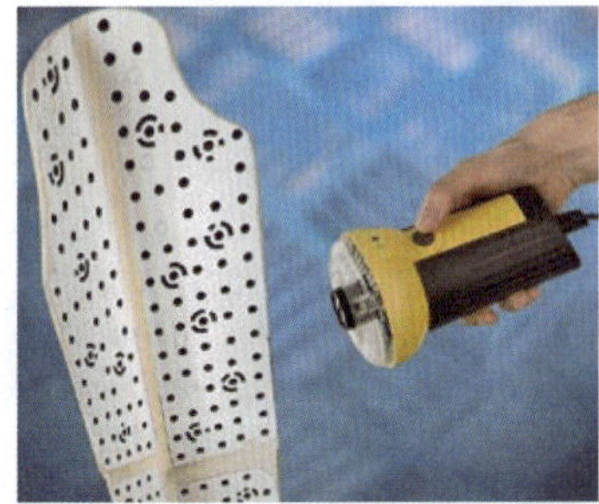

a) Targeting and image acquisition

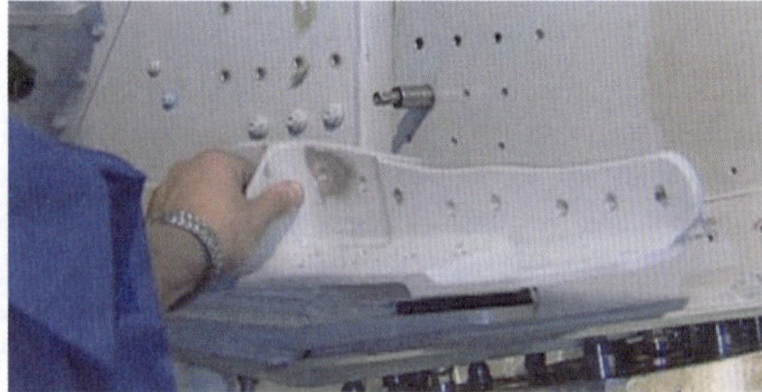

b) Installation of the specially machined corner fitting

Fig. 8.44: Photogrammetric measurement of corner fittings (AICON)

The imaged targets are measured fully automatically. An initial bundle adjustment provides object coordinates. If the resulting standard deviations exceed a certain threshold, the corresponding points are displayed in a different colour on the computer screen. The user can add more images until the specified accuracy of 0.02–0.1 mm is reached.

8.3.2.3 Antenna measurement

For the measurement of parabolic antennas and mirror telescopes, the shape of a hyperbolic surface must be checked. Antenna sizes range from 1 m to over 100 m. Applications cover a spectrum which includes size and shape control under varying thermal loads and the adjustment of large mirrors to the correct form. Typically the objects can only be measured from one side and the imaging configuration is designed so that every object point can be measured from as many convergent images as possible (see Fig. 7.5). The configuration is chosen to provide a homogeneous accuracy along all three coordinate axes. Retro-reflective targets are used to mark the object surface (Fig. 8.45). The size of the object, and access restrictions, typically demand the use of hydraulic lifting platforms (Fig. 8.46) which prevents the use of alternative techniques such as laser tracker measurement.

Fig. 8.45: Targeted parabolic reflector (GDV, Vertex Antennentechnik)

Fig. 8.46: Image acquisition from hydraulic platform (GDV, Vertex Antennentechnik)

The APEX telescope on the Chajnantor Plateau in the Chilean Andes (Fig. 8.45) provides a good example of the capabilities of photogrammetric measurement. Due to its position at 5100 m above sea level, only a limited time is available for measurement. The task is to measure the entire mirror surface (diameter 12 m) with a specified accuracy of 0.05 mm, in order to make adjustments to the component mirrors. Around 170 images are recorded using an INCA camera (Fig. 3.116) and evaluated using V-STARS (GSI). Measurement at approximately 1300 points determines the influence of gravity on the shape of the antenna in different positions. Measurement accuracy in object space is 10 µm (1-sigma RMS).

a) Imaging task in a thermal vacuum chamber

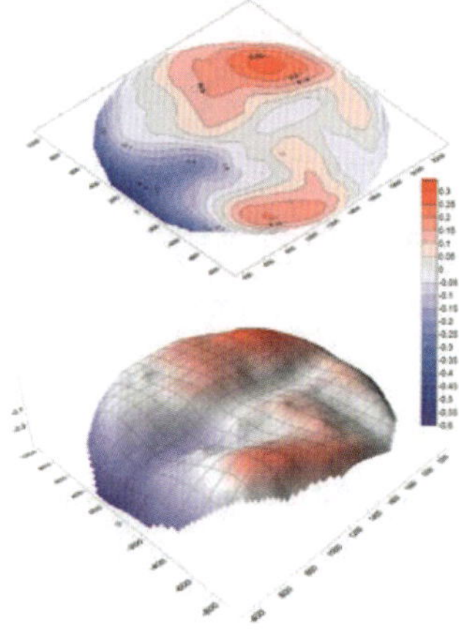

b) Results of shape analysis

Fig. 8.47: Measurement of a parabolic antenna in a space simulation chamber (GDV, IABG)

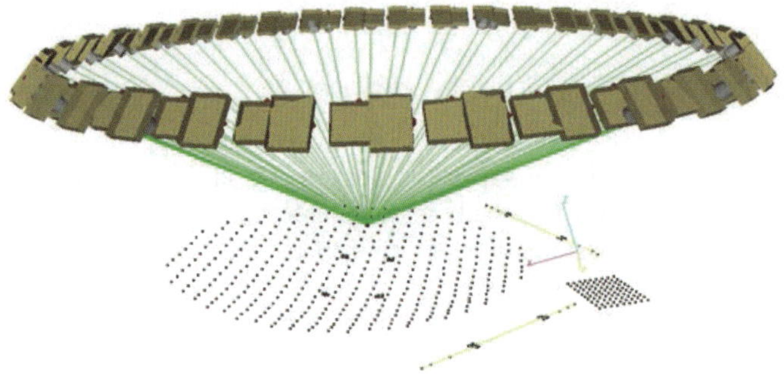

Fig. 8.48: Imaging positions and directions for antenna measurement (GDV)

If parabolic antennas are deployed in space, they are subject to extreme environmental conditions which can be re-created in a simulation chamber (Fig. 8.47a). Here antennas up to 4 m in diameter are exposed to temperatures between –120 C and +150 C. Either the camera or the antenna can be moved on a circular path in order to create the regular multi-image network illustrated in Fig. 8.48. Deformation analyses can then be derived from the measured object coordinates. Measurement accuracy is in the range 10–20 µm (1-sigma RMS).

8.3.3 Car industry

Three-dimensional measurement technology has, for some time, been one of the most important tools for quality control in the car industry. Mechanically probing coordinate measuring machines (CMMs) are mainly used for the high-precision measurement (1–10 µm) of small components. In contrast, optical 3D measuring methods, with typical accuracies of 0.05 to 0.2 mm, are mostly applied where an object cannot be transported to a stationary measuring system, cannot be probed mechanically or a very large number of object points must be measured within a short time. The following shows a selection of areas where photogrammetry can be applied:

- installation of production cells and assembly facilities
- surface measurement of design models (reverse engineering), (Fig. 6.184)
- car body measurement in a wind tunnel
- deformation measurement in torsion and car safety tests
- inspection of parts from third party suppliers (see windscreen measuring system, Fig. 6.25)
- driver assistance systems
- control of production machines (e.g. brake pipes, Fig. 6.21)

Measuring systems in production environments today almost exclusively utilise digital cameras, on or off line, in order to handle the necessary high data flows. In addition to the actual optical 3D measurement, these systems generally also provide data interfaces to CAD or CAM systems.

8.3.3.1 Rapid prototyping and reverse engineering

Prototypes are designed and manufactured prior to series production. If the corresponding parts contain free-form contours or surfaces, then the conventional generation of production data, e.g. for milling machines, is a costly and normally iterative process. Rapid prototyping methods can lead to much faster manufacturing of prototypes, permitting multiple passes through the production sequence shown in Fig. 8.49. The measuring task is essentially the complete 3D acquisition of the model in order to derive machine control data (reverse engineering).

Both single point and surface measurement systems are employed. They are portable, flexible and can be taken directly to the object. Since it is costly and time-consuming to place targets, it is common here to use online systems with manual touch probing (target adapters and touch probes, see also section 6.4.2.2). For surface measurement, area-based systems are used. Where large surfaces must be covered, additional photogrammetrically measured reference points are used to register individual scans (see Fig. 3.184 and Fig. 6.34).

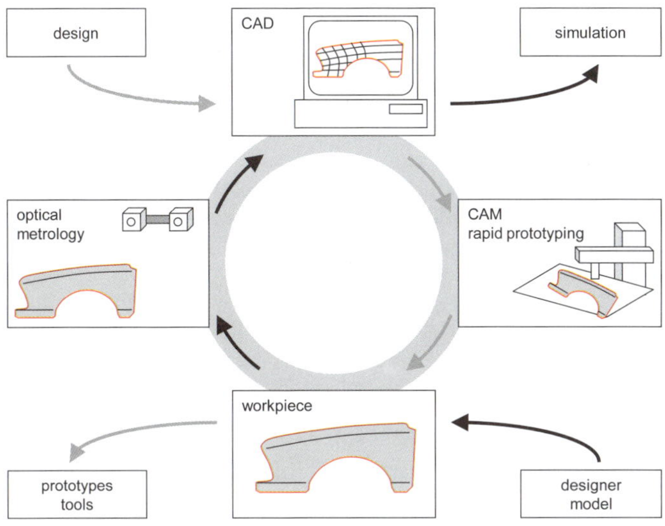

Fig. 8.49: Reverse engineering (after Bieder 1997)

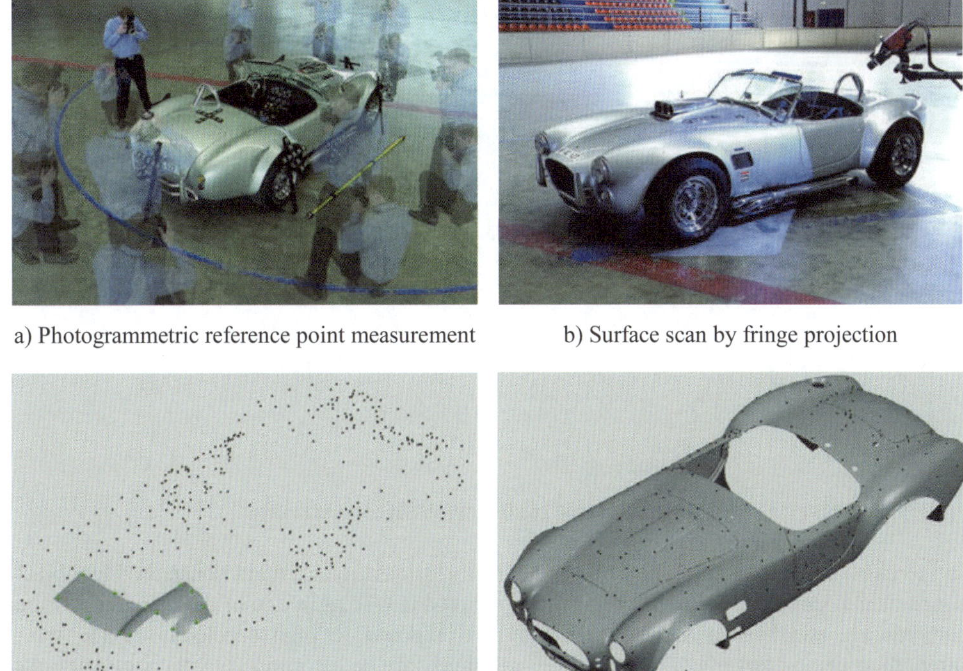

a) Photogrammetric reference point measurement b) Surface scan by fringe projection

c) Reference points and individual surface scans d) Complete model from merged individual scans

Fig. 8.50: 3D modelling of a car body (GOM)

Fig. 8.50 shows the photogrammetric measurement of local reference points placed on a car body. These are used to connect together individual surface point clouds generated by a fringe projection system. The final 3D model can be used for a variety of test and design tasks.

8.3.3.2 Car safety tests

In car safety tests (crash tests), photogrammetric evaluation of high-speed video is one of the tools used to the detailed displacements and changes in the car and dummies which represent the passengers (compare with section 3.5.3).

Typical investigations in car safety testing include:

- front and side crashes,
- protection of pedestrians,
- deformation analyses (engine compartment, roof and window structures, footwells).

Here imaging configurations can be classified into two types (Fig. 8.51):

- fixed camera(s) outside the car (A),
- moving camera(s) installed inside the car (B).

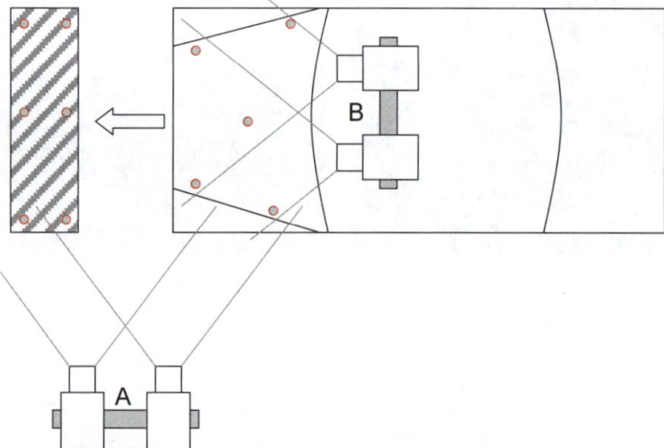

Fig. 8.51: Imaging configurations in car crash tests

In the case of fixed cameras, interior and exterior orientations remain constant. The object coordinate system can be defined on the car or outside it (e.g. in one of the cameras). If the cameras are attached to the car then they are exposed to high accelerations and their orientation parameters may change. These can then only be determined by the simultaneous measurement of fixed reference points.

To determine lateral displacements, image sequences are evaluated by 2D image analysis. To determine three-dimensional displacements and deformations, full photogrammetric techniques are employed. In crash tests, a typical accuracy is in the range 1–5 mm.

Fig. 8.52 shows extracts from an image sequence used for head-impact tests. They were made using a NAC HiDcam with a stereo mirror attachment (see section 3.5.3, Fig. 3.120). The objective is to measure the penetration depth of the head impact on the bonnet (hood). Trajectories and orientations of the impacting head can be determined in a global coordinate system from the sets of photogrammetrically measured 3D points, as illustrated in Fig. 8.53. By a combination of 3D image sequence analysis and electronic sensor technology, the forces and accelerations in pedestrian safety and frontal crash tests can be visualised.

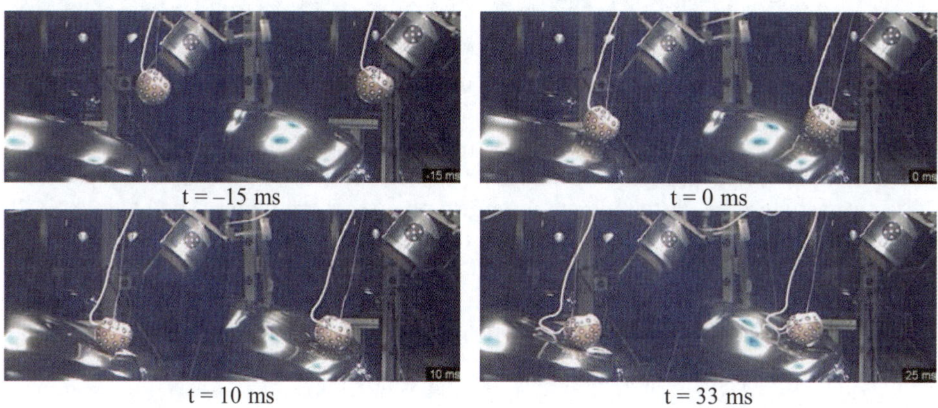

Fig. 8.52: Stereo image sequence of a head impact test (Volkswagen)

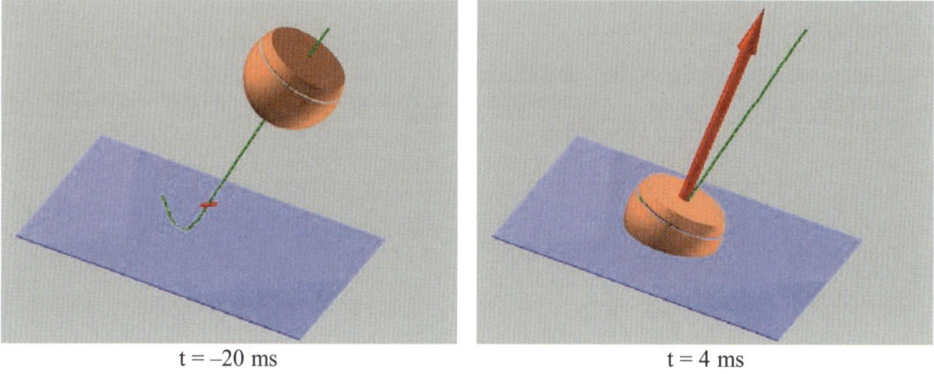

Fig. 8.53: Animated trajectory with overlay of acceleration vector (Volkswagen)

8.3.3.3 Car body deformations

To measure dynamic and thermal loading, car parts or entire cars are measured photogrammetrically point-by-point. If the individual deformation states are quasi static, that is they remain unchanged during a given period of time (e.g. before and after testing in a climate chamber), then the target points can be measured by a multi-image network using the offline photogrammetric method.

Fig. 8.54: Photogrammetric measurement of targets for a torsion experiment (Volkswagen)

Fig. 8.54 illustrates a car body torsion test. In this case strips of retro-reflective targets are placed along selected profiles and imaged from all sides using a digital camera. After interactive setting of a starting point, all targets on a strip can be measured automatically. An accuracy of 0.1 mm can be achieved here in object space.

8.3.4 Ship building industry

Photogrammetric tasks in shipbuilding include the following:

- measurement of steel plates and their orientation on metal cutting machines (see Fig. 6.24),
- measurement of ship sections,
- measurement of windows, hull and fittings,
- circularity checks on submarines.

Metrology applications in the ship building industry are characterised by:

- measurement of large objects (>30 m)
- restricted access
- vibrations and disadvantageous environmental conditions

An example of photogrammetric measurement in shipbuilding is the determination of shrinkage in the welding of large steel parts. For typical component sizes of 10–12 m, shrinkage during welding can amount to several mm and is taken into account in the construction plan. Fig. 8.55 illustrates the process.

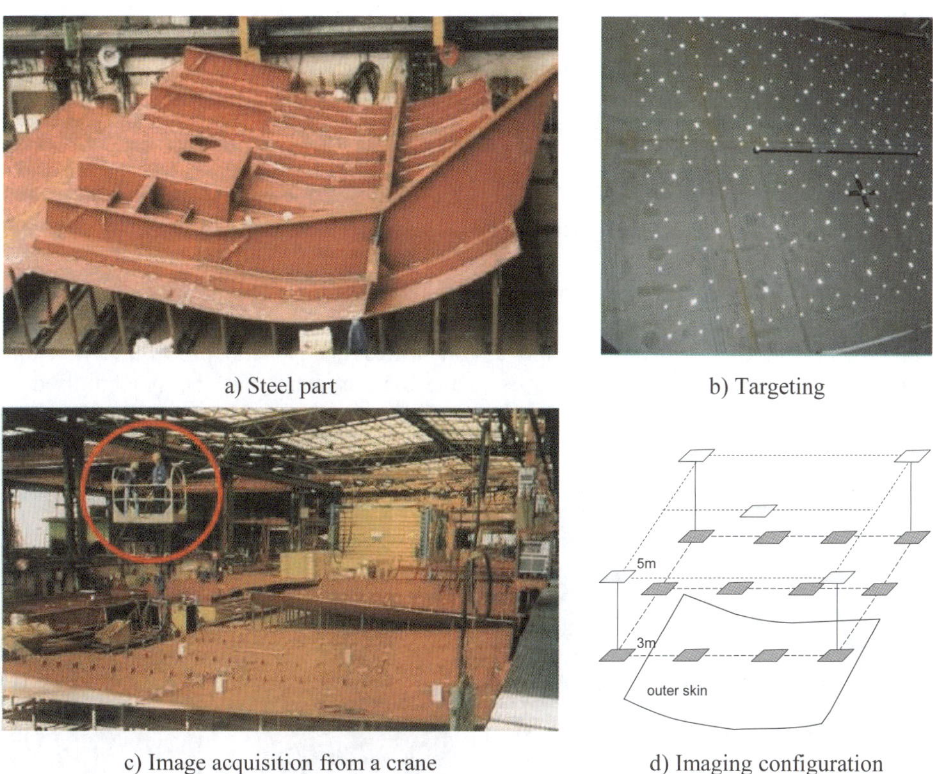

a) Steel part

b) Targeting

c) Image acquisition from a crane

d) Imaging configuration

Fig. 8.55: Photogrammetric acquisition of welding shrinkage (AICON)

Further typical applications are photogrammetric form measurement of fixtures (e.g. railings, windows) and hulls. In addition to photogrammetric methods, geodetic techniques and laser tracker measurements are also applied.

8.4 Medicine

In medicine, photogrammetry is mainly used to measure parts of the body, for example:

- to prepare and carry out operations (navigation)
- to construct prostheses and implants
- in plastic surgery
- in motion studies
- in the therapy for bone and spinal deformations
- to monitor growth

Photogrammetric solutions are applied in many areas of medicine such as orthopaedics, neurosurgery, dentistry and sports medicine.

8.4.1 Surface measurement

Medical surface measurements are mostly characterised by:

- the measurement of unstructured, soft, free-form surfaces
- the recording of subjects where are moving or not static
- the absence of permanent reference points on the subject

Poorly textured, free-form surfaces must be given an artificial structure, e.g. by fringe or pattern projection (see section 3.7.2). Stereoscopic configurations can be chosen for surface measurement and the term *biostereometrics* identifies this general procedure. Multiple images can also be used and in all cases image measurement must be synchronised to handle movement of the subject.

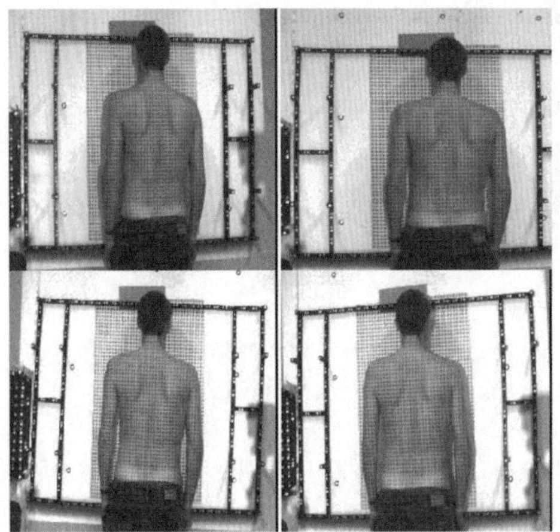
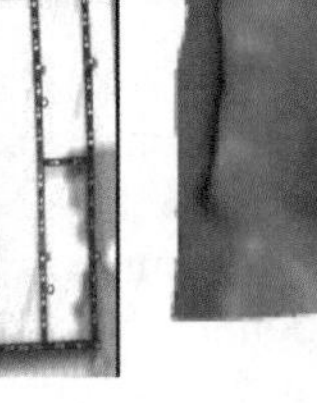

Fig. 8.56: Multiple image configuration for back measurement (Gäbel 1993)

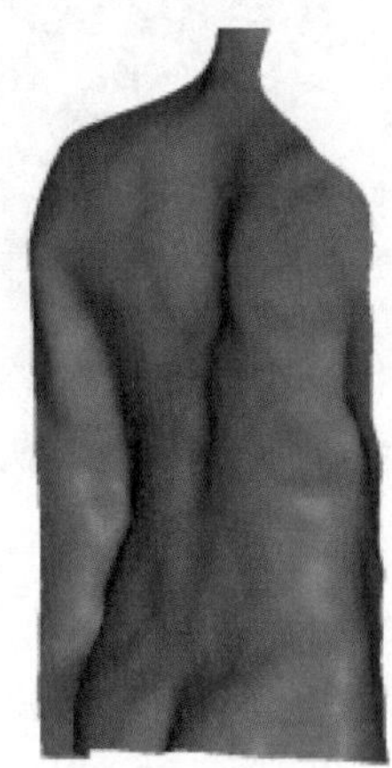

Fig. 8.57: 3D visualisation of human back measurement (Gäbel 1993)

Fig. 8.56 shows a set of four images for back measurement. The patient is positioned in front of a spatial field of reference points and a target grid is projected onto the surface of the patient's back. This is all imaged by four synchronised cameras. Least-squares matching is used for surface reconstruction. The resulting three-dimensional model can be analysed for asymmetries (Fig. 8.57). Features of the photogrammetric process are:

- non-contact measurement avoiding stress to the patient
- short measurement time (tenths of a second)
- no radiation exposure
- suitable also for infants and children
- subsequent measurement and diagnostic comparison over time

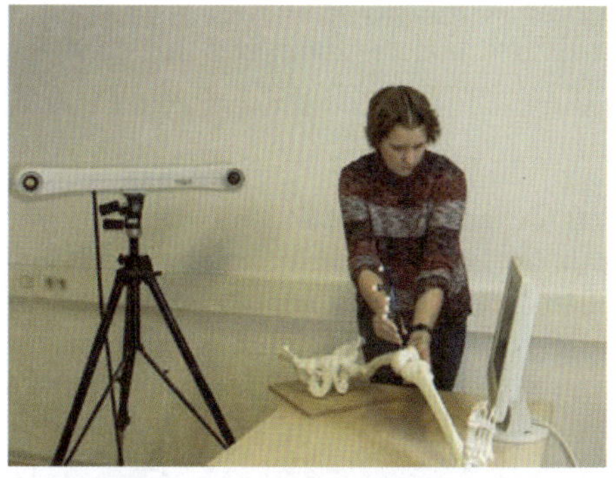

a) Point-by-point object measurement (AXIOS 3D)

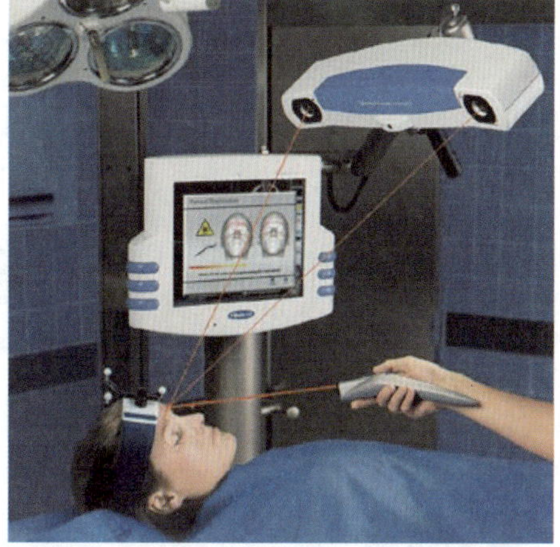

b) Projection and location of a laser spot (BrainLab)

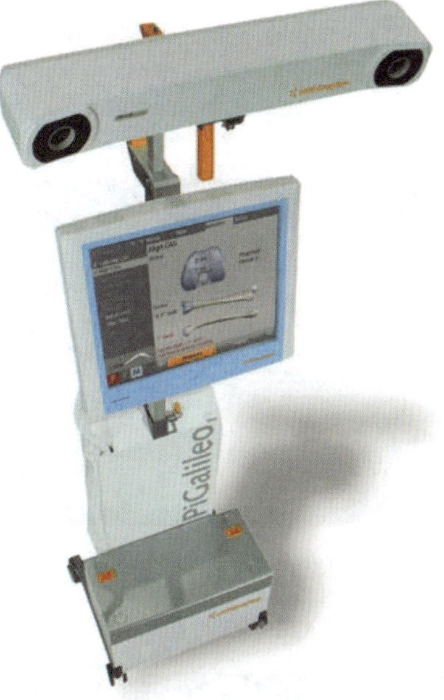

c) Online system for use in operating theatres (Smith & Nephews)

Fig. 8.58: Dual-camera systems for medical applications

8.4.2 Online navigation systems

Medicine increasingly uses digital, photogrammetric, online measuring systems with contact probing. They usually consist of two cameras fixed into a mobile housing. Fig. 8.58 shows examples of such dual camera, online systems with point-by-point tactile probing. These can not only use target adapters with probing tips (see section 3.6.1.6), but also the actual tools used in operations, if equipped with suitable targets (Fig. 8.59). They are a critical component in image-based planning and execution of operations (*image guided surgery*, IGS) where a spatial relationship between patient and surgical instrument must be established. In medicine this process is known as *navigation*. The key problem for navigation is the unstable position of the patient during the operation. It cannot be assumed that patient and measuring system have a constant geometric relationship with respect to one another. For this reason, local reference target arrays (locators) are attached to the patient whose potential motion can then be continuously monitored by the navigation system. The spatial position of surgical tools or predefined operation data (e.g. related computer tomograms) can be transformed into the coordinate system of the locator. Fig. 8.59b shows an example where a locator fixed to a bone is measured, together with a moving locator mounted on a surgical robot. In this way it is possible to compensate for any motion of the legs.

Accuracy requirements for navigation systems depend on application and range from around 0.1 mm for spinal surgery, around 0.5 mm for implanting knee and hip joints and up to 1 mm or so in brain surgery. These applications require measurement frequencies of 10 to 50 3D measurements per second, with up to 20 points or more simultaneously measured or tracked in 3D space. Motion analysis requires significantly lower accuracies but much higher measurement frequencies.

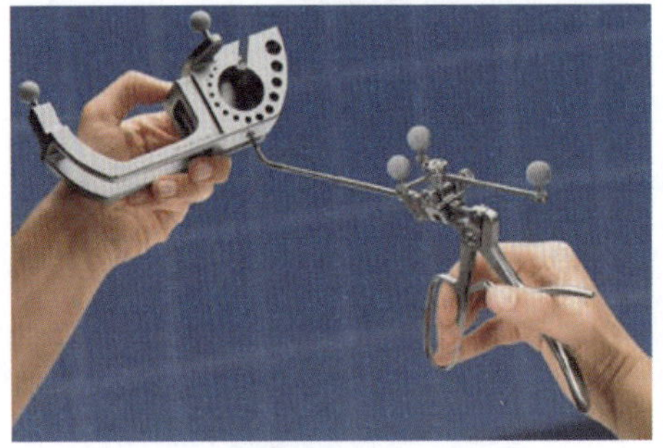

a) Targeted surgical tools (BrainLab)

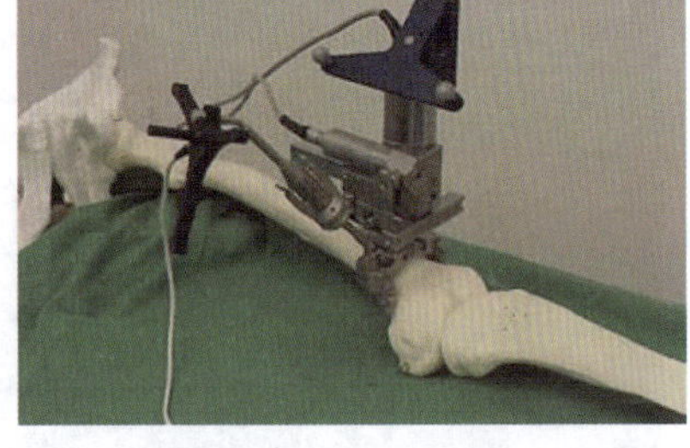

b) Surgical robot (Plus Orthopedics)

Fig. 8.59: Surgical tools with spatial reference points

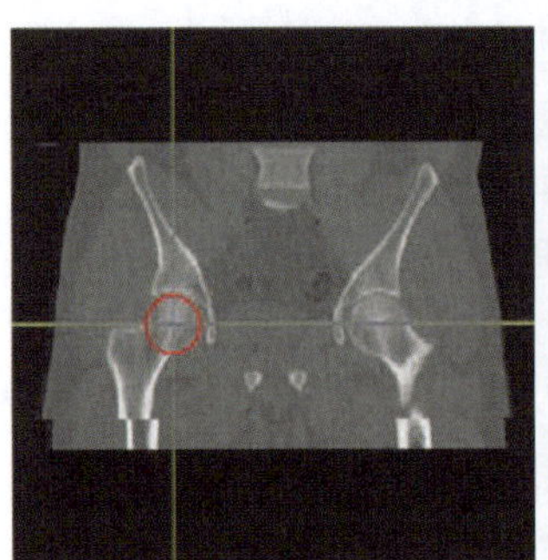

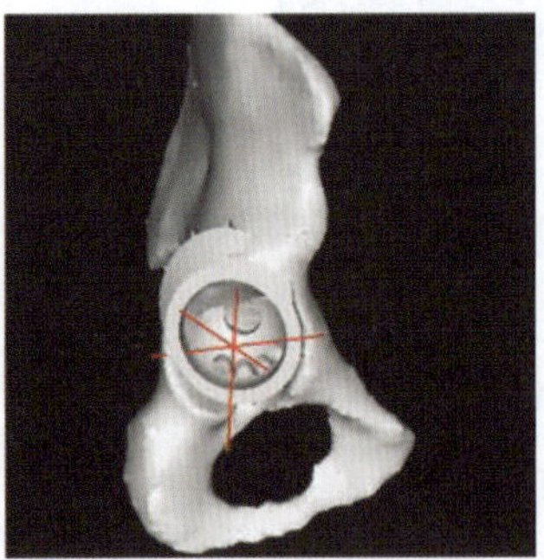

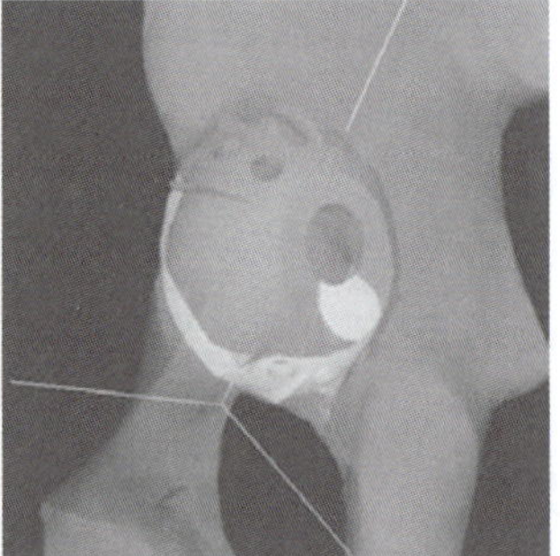

Fig. 8.60: Implementation plan for inserting a hip joint (Plus Orthopedics)

Photogrammetric navigation systems are also employed for pre-operative planning which is usually derived from computed tomographic (CT) images. Fig. 8.60 shows an example of the planned insertion of a CAD-designed hip joint in a CT image. Using anatomical navigation points (landmarks), the transformation between CT and real bone can be generated during the operation.

8.5 Miscellaneous applications

8.5.1 Forensic applications

Forensic photogrammetry covers applications such as:

- traffic accident recording
- recording and reconstruction of aircraft crashes
- scene-of-crime measurement
- estimating the height of criminals
- reconstructing bullet trajectories
- object reconstruction from amateur images
- detecting environmental pollution from aerial images

8.5.1.1 Accident recording

Fig. 8.61 shows metric images from a photogrammetric accident record. Some reference distances are measured in object space because control points cannot usually be provided due to the limited time available. The scene is recorded with a convergent set of images which can optionally include aerial images. Results can be output as scaled plans, CAD drawings or coordinates of the accident site (Fig. 8.62). Photogrammetric measurements are accepted as valid evidence in criminal trials and legal disputes.

a) Terrestrial image recording

b) Aerial photo from helicopter

Fig. 8.61: Photogrammetric images of an accident scene (Rolleimetric)

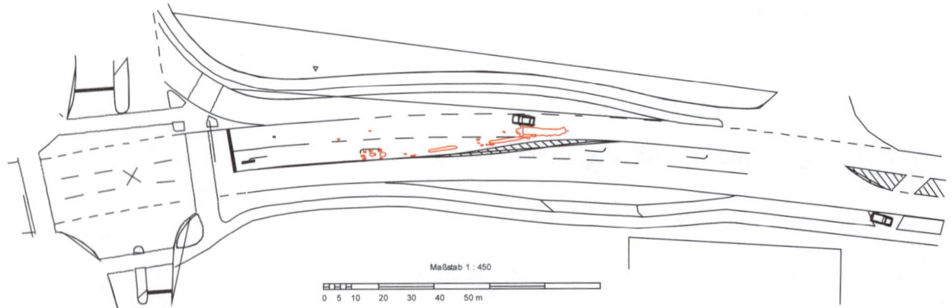

Fig. 8.62: Photogrammetric accident recording by a combination of aerial and terrestrial images

A typical measurement situation at an accident site is shown in Fig. 8.63. The accident scene was recorded with a 7 Megapixel amateur camera, the Olympus C7070WZ. The area covered by the camera network and object points have an extent of 80 m and maximum height variation of only 2 m. Imaging restrictions on site lead to largely horizontal and narrow ray intersections which demand particularly robust image orientation methods. The iWitness software (Photometrix) was used to orient the network of 22 images and produce drawings and CAD models. Accuracy in object space was around 3 cm.

a) Sample measurement images

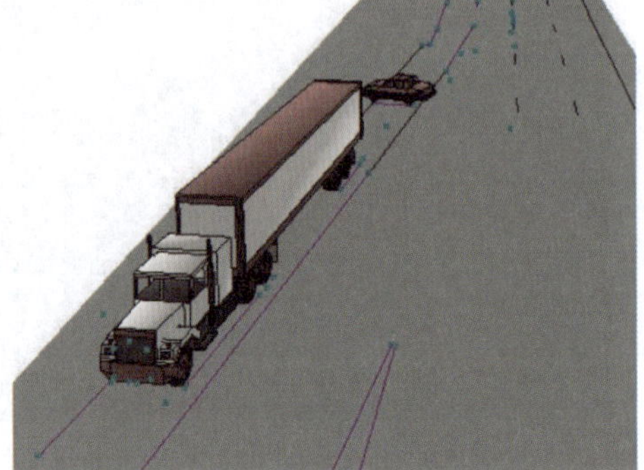

b) Imaging configuration c) Derived CAD model

Fig. 8.63: Photogrammetric accident recording (Photometrix)

8.5.1.2 Scene-of-crime recording

Crime scenes can be measured and documented by photogrammetry without disturbing the local environment or further police activities, such as the collection of evidence. Fig. 8.64 shows the result of a scene-of-crime reconstruction with full modelling of the local scene (a), overlay of CAD model and photo detail (b), blood spray trajectories (c) and the possible locations of the individuals involved (d). The scene in this case was recorded using a Nikon D700 digital camera. Elcovision 10 (PMS) and AutoCAD (Autodesk) were used for photogrammetric analysis and modelling. Measurement accuracy was in the mm region.

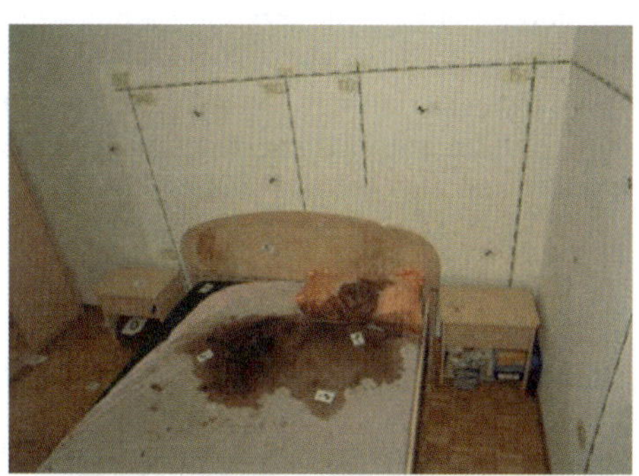

a) Measurement image from crime scene

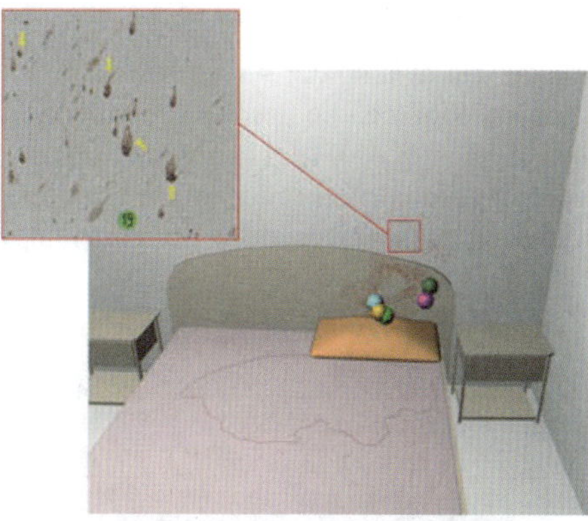

b) 3D model with image detail overlay

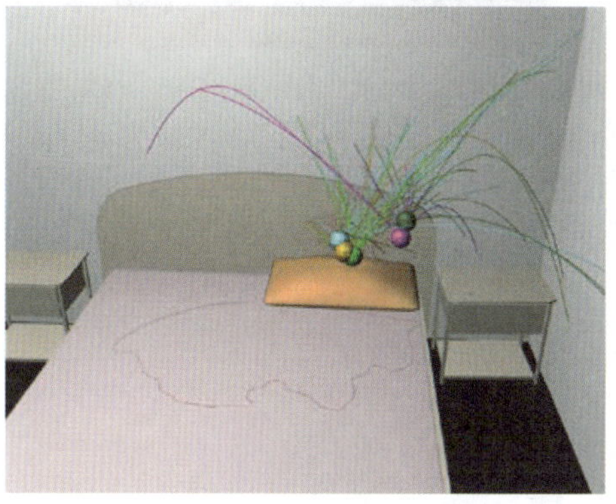

c) Reconstructed blood spray trajectories

d) Reconstructed positions of attacker and victim

Fig. 8.64: Photogrammetric crime record (Institut für Rechtsmedizin, University of Bern)

8.5.2 Scientific applications

Close-range photogrammetry can be advantageously applied in many scientific applications such as:

- recording physical phenomena,
- reconstructing biological processes (e.g. plant growth, spiders‘ webs),
- monitoring and modelling events for the earth sciences (e.g. landslides),
- static and dynamic measurement of glaciers,
- etc.

8.5.2.1 3D reconstruction of a spider‘s web

The example of a spider’s web shows how digital stereo photogrammetry, and a precision adjustment tool, can be used to record complex spatial structures, made visible by projected laser fans (Fig. 8.65). In this application, the connection points in the web are reconstructed three-dimensionally in order to use the derived model for an artificial construction. The 3D model can equally be used for fundamental investigations in biology and bionics.

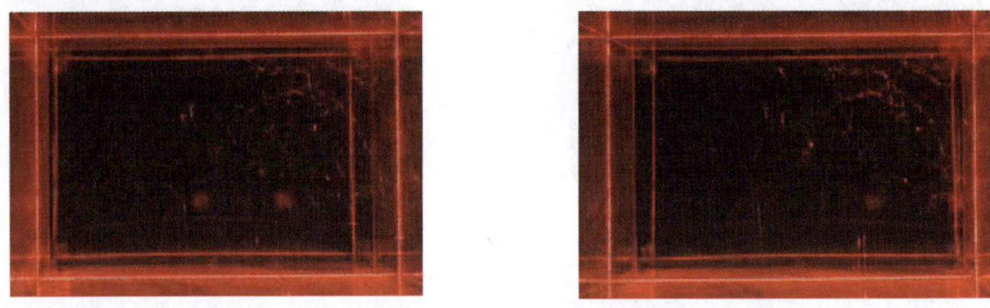

Fig. 8.65: Stereo images for photogrammetric recording of a spider’s web (TU Darmstadt)

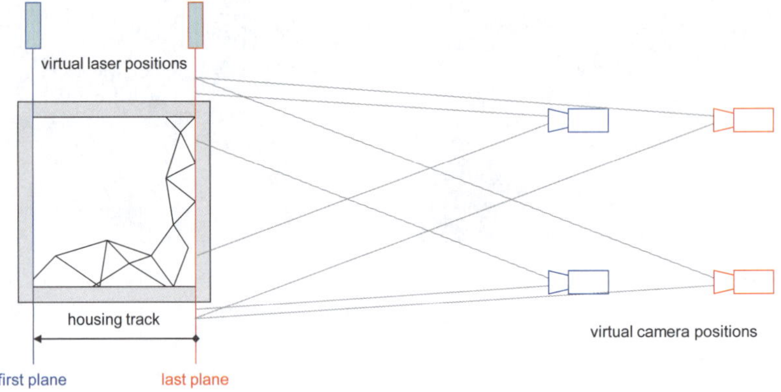

Fig. 8.66: Test configuration for measuring a spider’s web (TU Darmstadt)

The spider’s web is in a Plexiglas box which can be accurately positioned using a lathe (Fig. 8.66). In the start and end positions, two digital cameras (Canon EOS 5D M II) are relatively oriented using targets attached to the box. At intermediate positions, the known linear displacement imposed by the lathe enables the exterior orientations to be interpolated. A total of 110 stereo pairs are acquired in the process.

The full size of the web occupies a space of 53 x 50 x 30 cm^3, although the individual threads of the web have a thickness of around 50 µm. The cameras are equipped with tilt-shift lenses (see section 3.4.3.5) so that, by shifting the location of the principal point, an approximate normal stereo case with maximum object coverage is achieved. A laser generates a fan of light perpendicular to the viewing direction of the cameras in order to illuminate the connection points of the web.

Fig. 8.65 shows one of the stereo image pairs in which the spider's web is illuminated by the laser light plane. A stereo analysis using the ERDAS LPS software produces a 3D CAD model containing the points and connections making up the web (Fig. 8.67a). To produce the artistic model (created by artist Tomas Saraceno, Fig. 8.67b) it is not the absolute accuracy which is of importance but the completeness of the model.

a) CAD model (TU Darmstadt)

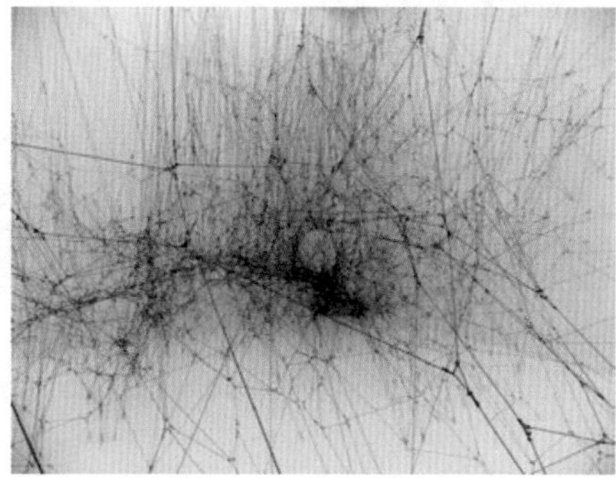

b) Artistic replica (Tomas Saraceno)

Fig. 8.67: Reconstruction of spider's web

8.5.2.2 Monitoring glacier movements

Another example of using photogrammetric analyses of image sequences is in monitoring glacier movements. The Jakobshavn Isbrae Glacier on the west coast of Greenland had, for many years, a constant speed of 20 m per day, but in more recent years this has increased to 40 m per day, combined with a dramatic retreat of the glacier front. Photogrammetric measurement campaigns were carried out in 2004, 2007 and 2010 with the objective of determining the glacier's spatial and temporal movement patterns using a sequence of terrestrial images taken with a digital camera from a hill.

The intention is to determine two-dimensional movement vectors showing the glacier movement in the direction of flow, as well as movements in a vertical direction which are induced by tidal effects. The recorded data was a monocular sequence of images made with a digital SLR camera. Images were taken at 30-minute intervals over a period of 12–36 hours. Scale was introduced with the aid of a hybrid geodetic and photogrammetric network. The application of correlation techniques enabled movement and trajectory vectors to be determined with a standard deviation of around 0.1–0.2 % of the trajectory length over a whole day. Accuracy was mainly limited by the difficult topography presented by the glacier surface and the effect of shadows.

Fig. 8.68 shows the camera in position and one of the images. The height changes over a 24-hour period at a point on the glacier near the glacier front, as calculated from the image sequence, are shown in Fig. 8.69. Here the glacier shows an almost perfect correlation with the tidal curve. This proves that the glacier front is floating on the fjord. The tidal influence fades away about 1 km behind the front. From this the position of the grounding line can be determined (Fig. 8.70). The results of this terrestrial photogrammetric measurement form the basis of glaciological studies of the Jakobshavn Isbrae Glacier.

a) Camera in position b) Example of a measurement image

Fig. 8.68: Photogrammetric recording of glacier movement (IPF TU Dresden)

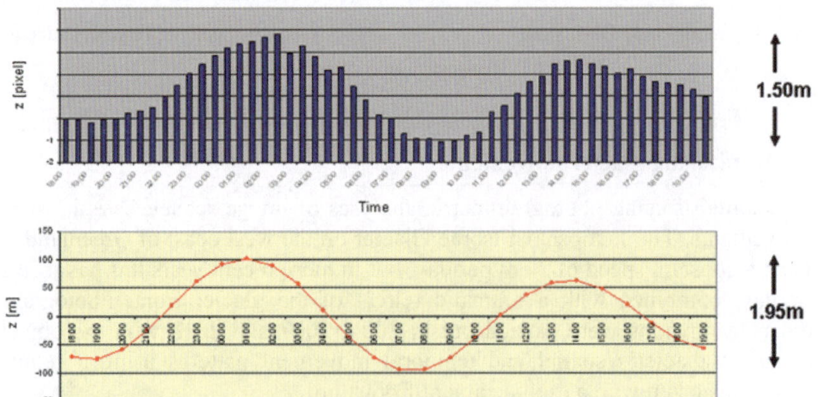

Fig. 8.69: Calculated height changes at a glacier location over 24 hours (upper diagram) and corresponding tidal curve (lower diagram) (IPF TU Dresden)

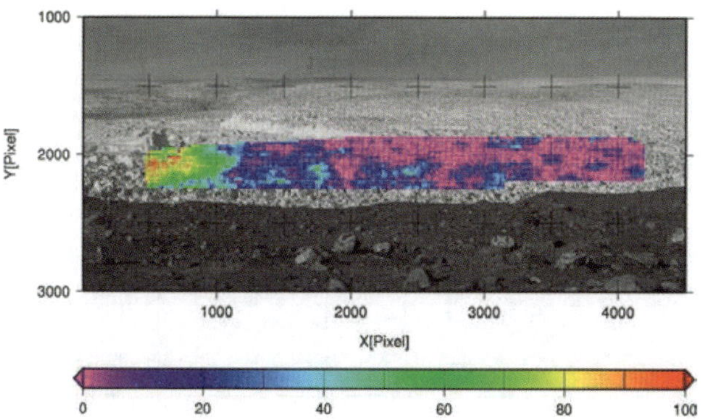

Fig. 8.70: Participation of the vertical glacier movement component with the tidal range (IPG/IPF TU Dresden)

8.5.2.3 Earth sciences

Recording and quantifying processes relevant to the earth sciences demands precise measurement methods. Digital photogrammetry is applied to a wide range of projects at both close and long ranges. Often surfaces must be modelled in order to compare them at different epochs. This may require the comparison of parameters such as roughness of the ground surface or an absolute comparison in order to quantify erosion and rates of material loss.

For example, photogrammetric methods can be used to monitor changes in ground surface or the evolution of rill network development due to rain. This can either be natural rain or rain simulated in a laboratory. In lab tests, ground samples with an extent of 1 x 2 m^2 up to 3.7 x 14.4 m^2 are measured photogrammetrically on a grid with 3 x 3 mm^2 resolution. For erosion research, comparison of surfaces at different epochs enables conclusions to be made concerning the change in surface roughness and the development and structure of drainage channels. Digital height models can be made in GIS software. Modelling such large areas to millimetre resolution places high demands on the design of the measurement network.

Fig. 8.71a shows the schematic arrangement of a rainfall test in the lab. Water drops (1) are formed on the end of capillary tubes from where they drop 1.5 m onto a net (2) in order to create a wider spectrum of drops. The drop from net to the upper edge of the sample box is approximately 6 m. This ensures that the drops reach a velocity corresponding to natural rainfall. The sample box (3) in this test has an area of 1 x 2 m^2 and is filled with a sample of earth or soil to be tested. Positioning struts (4) can be used to alter the tilt of the box. The surface water runoff is drained away at the lower end of the box (5). Sediment and water samples are taken here. The surface of the sample is recorded from two camera positions (Fig. 8.72b). The principal points of the cameras are displaced using lenses with fixed shift (see section 3.4.3.5) in order to obtain as large an area of stereo coverage as possible.

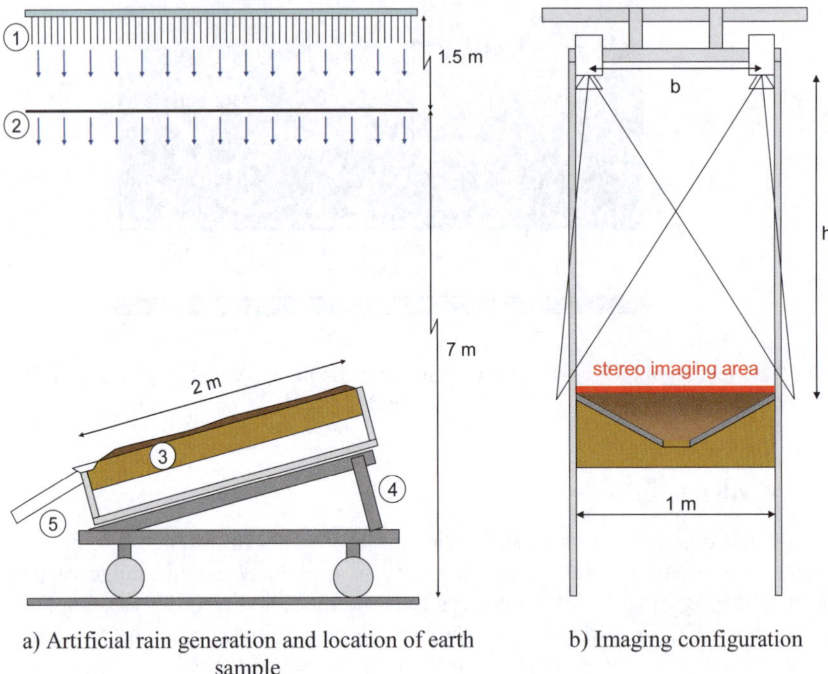

a) Artificial rain generation and location of earth sample

b) Imaging configuration

Fig. 8.71: Schematic arrangement of a rainfall test (University of Bern)

Fig. 8.72 shows the surface of the test sample before artificial rain (a), after 20 minutes of simulated rainfall (b) and after 65 minutes (c). It can be see that after 20 minutes the surface runoff has created a drainage pattern which becomes more established with increasing length of rainfall. Natural interaction makes the drainage pattern more efficient with time.

a) Before raining b) After 20 min c) After 65 min

Fig. 8.72: Photogrammetrically generated height model and drainage pattern (University of Bern)

9 Literature

9.1 Textbooks

9.1.1 Photogrammetry

Albertz, J. (2007): Einführung in die Fernerkundung – Grundlagen der Interpretation von Luft- und Satellitenbildern. 3. Aufl., Wissenschaftliche Buchgesellschaft, Darmstadt, 254 p.

Albertz, J., Wiggenhagen, M. (2009): Guide for Photogrammetry and Remote Sensing. 5th ed., Wichmann, Heidelberg, 334 p.

Atkinson, K.B. (ed.) (1980): Developments in Close Range Photogrammetry. Applied Science Publishers, London, UK.

Atkinson, K.B. (ed.) (1996/2001): Close Range Photogrammetry and Machine Vision. Whittles Publishing, Caithness, UK, 371 p.

Fryer, J.G., Mitchell, H.L., Chandler, J.H. (eds.) (2007): Applications of 3D Measurement from Images. Whittles Publishing, Caithness, UK, 304 p.

Karara, H.M. (ed.) (1989): Non-topographic Photogrammetry. 2nd ed., American Society for Photogrammetry and Remote Sensing, 445 p.

Kasser, M., Egels, Y. (2002): Digital Photogrammetry. Taylor & Francis, London, UK.

Kraus, K. (2007): Photogrammetry - Geometry from Images and Laserscans. 2nd ed., W. de Gruyter, Berlin, 459 p.

Luhmann, T. (ed.) (2002): Nahbereichsphotogrammetrie in der Praxis. Wichmann, Heidelberg, 318 p.

Luhmann, T., Robson, S., Kyle, S., Harley, I. (2006): Close Range Photogrammetry. Whittles Publishing, Caithness, UK, 510 p.

Luhmann, T. (2010): Nahbereichsphotogrammetrie. 3rd ed., Wichmann, Heidelberg, 668 p.

McGlone, J.C. (ed.) (2013): Manual of Photogrammetry. 6th ed., American Society for Photogrammetry and Remote Sensing, 1318 p.

Mikhail, E.M., Bethel, J.S., McGlone, J.C. (2001): Introduction to Modern Photogrammetry. John Wiley & Sons, New York, 479 p.

Schenk, T. (1999): Digital Photogrammetry. Vol. 1, TerraScience, Laurelville, USA, 428 p.

9.1.2 Optic, camera and imaging techniques

Arnold, C.R., Rolls, P.J., Stewart, J.C.J. (1971): Applied Photography, Focal Press, London, 510 p.

Born, M., Wolf, E., Bhatia, A.B. (1999): Principles of Optics. Cambridge University Press, 986 p.

Hecht, E. (2003): Optics. Addison Wesley, Longman, 698 p.

Holst, G.C. (1996): CCD Arrays, Cameras and Displays. SPIE Press, JCD Publishing, USA.

Inglis, A.F., Luther, A.C. (1996): Video Engineering. 2nd ed., McGraw-Hill, New York.

Jacobson, R.E., Ray, S.F., Attridge, G.G., Axford, N.B. (1983): The Manual of Photography. 7th ed., Focal Press, London, 628 p.

Langford, M.J. (1982): Basic Photography. 4th ed., Focal Press, London. 397 p.

Langford, M.J. (1983): Advanced Photography. 4th ed., Focal Press, London, 355 p.

Reinhard, E., Ward, G., Pattanaik, S., Debevec, P. (2006): High Dynamic Range Imaging. Elsevier, Amsterdam, 502 p.

Ray, S.F. (1994): Applied photographic optics. 2nd ed., Focal Press, Oxford, 586 p.

9.1.3 Digital image processing, computer vision and pattern recognition

Demant, C., Streicher-Abel, B., Waszkewitz, P. (1999): Industrial Image Processing – Visual Quality Control in Manufacturing. Springer, 353 p.

Förstner, W., Ruwiedel, S. (eds.) (1992): Robust Computer Vision. Wichmann, Heidelberg, 395 p.

Gonzales, R.C., Wintz, P. (1987): Digital Image Processing. 2nd ed., Addison Wesley, Reading, Ma., USA, 503 p.

Gonzales, R.C., Woods, R.E. (2003): Digital Image Processing. Prentice Hall, 813 p.

Graham, R. (1998): Digital Imaging. Whittles Publishing, Caithness, Scotland, UK.

Haralick, R.M., Shapiro, L.G. (1992): Computer and Robot Vision. Vol. I+II, Addison-Wesley, Reading, Ma., USA.

Hartley, R., Zisserman, A. (2004): Multiple View Geometry. 2nd ed., Cambridge University Press, Cambridge, UK, 672 p.

Jähne, B., Haußecker, H., Geißler, P. (1999): Handbook of Computer Vision and Applications. Vol. 1 Sensors and Imaging; Vol. 2 Signal Processing and Pattern Recognition; Vol. 3 Systems and Applications. Academic Press.

Jiang, X., Bunke, H. (1996): Dreidimensionales Computersehen: Gewinnung und Analyse von Tiefenbildern. Springer, 361 p.

Marr, D. (1982): Vision. Freeman, San Francisco.

Parker, J.R. (1996): Algorithms for Image Processing and Computer Vision. John Wiley & Sons, 432 p.

Rosenfeld, A., Kak, A.C. (1982): Digital Picture Processing, 2nd ed., Academic Press.

Russ, J.C. (2002): The Image Processing Handbook. CRC Press, 752 p.

9.1.4 Mathematics and 3D computer graphics

Foley, J.D., VanDam, A., Feiner, S.K. (1995): Computer Graphics. Addison-Wesley Professional, 1200 p.

Hunt, R.A. (1988): Calculus with Analytic Geometry. Harper & Row, New York. 1080 p.

Kuipers, J.B. (2002): Quaternions and Rotation Sequences. Princeton University Press, 371 p.

Laszlo, M. J. (1996): Computational Geometry and Computer Graphics in C++, Prentice-Hall, Upper Saddle River, NJ, 266 p.

Shirley, P. (2002): Fundamentals of Computer Graphics. A K Peters, 500 p.

Stroud, K.A. (2001): Engineering Mathematics. 5th ed., Industrial Press, 1236 p.

Watt, A. (1999): 3D Computer Graphics. Addison Wesley Publishing, 592 p.

9.1.5 Least-squares adjustment and statistics

Mikhail, E.M. (1976): Observations and least squares. IEP, New York, 497 p.

Mikhail, E.M, Gracie, G. (1981): Analysis & Adjustment of Survey Measurements. Van Nostrand Reinhold Company, 368 p.

Robert, C. P., Casella, G. (2002): Monte Carlo Statistical Methods. Springer, Berlin.

Wolf, P.R., Ghilani, C.D. (1997): Adjustment Computations: Statistics and Least Squares in Surveying and GIS (Surveying & Boundary Control). John Wiley & Sons.

9.1.6 Industrial and optical 3D metrology

Brinkmann, B. (2012): International Vocabulary of Metrology, German-English version, 4th ed., Beuth, Berlin, 76 p.

Gruen, A., Kahmen, H. (eds.) (1989-2009): Optical 3D Measurement Techniques. Wichmann, Heidelberg, since 2001 published by ETH Zurich and TU Vienna.

Luhmann, T., Müller, C. (2002-2013) (eds.): Photogrammetrie-Laserscanning-Optische 3D-Messtechnik – Beiträge der Oldenburger 3D-Tage, Wichmann, Heidelberg.

Vosselman, G., Maas, H.-G. (eds.) (2010): Airborne and Terrestrial Laser Scanning. Whittles Publishing, Caithness, UK, 336 p.

9.2 Introduction and history

Ackermann, F., Ebner, H., Klein, H. (1970): Ein Rechenprogramm für die Streifentriangulation mit unabhängigen Modellen. Bildmessung und Luftbildwesen, Heft 4, pp. 206-217.

Adams, L.P. (2001): Fourcade: The centenary of a stereoscopic method of photogrammetric surveying. Photogrammetric Record, 17 (99), pp. 225-242.

Albertz, J. (2009): 100 Jahre Deutsche Gesellschaft für Photogrammetrie, Fernerkundung und Geoinformation e.V. Photogrammetrie-Fernerkundung-Geoinformation, Heft 6, pp. 487-560.

Arthur, D.W.G. (1960): An automatic recording stereocomparator. Photogrammetric Record, 3(16), 298-319

Atkinson, K.B. (1980): Vivian Thompson (1880-1917): not only an officer in the Royal Engineers. Photogrammetric Record, 10 (55), pp. 5-38.

Atkinson, K.B. (2002): Fourcade: The Centenary – Response to Professor H.-K. Meier. Correspondence, Photogrammetric Record, 17 (99), pp. 555-556.

Brown, D.C. (1958): A solution to the general problem of multiple station analytical stereotriangulation. RCA Data Reduction Technical Report No. 43, Aberdeen.

Brown, D.C. (1976): The bundle adjustment – progress and prospectives. International Archives of Photogrammetry, 21 (3), ISP Congress, Helsinki, pp. 1-33.

Deville, E. (1895): Photographic Surveying. Government Printing Bureau, Ottawa. 232 p.

Deville, E. (1902): On the use of the Wheatstone Stereoscope in Photographic Surveying. Transactions of the Royal Society of Canada, Ottawa, 8, pp. 63-69.

Förstner, W. (1982): On the geometric precision of digital correlation. International Archives for Photogrammetry and Remote Sensing, Vol. 26/3, pp. 176-189.

Fourcade, H.G. (1903): On a stereoscopic method of photographic surveying. Transactions of the South African Philosophical Society, 14 (1), pp. 28-35.

Fraser, C.S., Brown, D.C. (1986): Industrial photogrammetry – new developments and recent applications. Photogrammetric Record, 12 (68), pp. 256-281.

Fraser, C.S. (1993): A resume of some industrial applications of photogrammetry. ISPRS Journal of Photogrammetry & Remote Sensing, 48(3), pp. 12-23.

Fraser, C.S. (1997): Innovations in automation for vision metrology systems. Photogrammetric Record, 15(90), pp. 901-911.

von Gruber, O., (ed.), McCaw, G.T., Cazalet, F.A., (trans) (1932): Photogrammetry, Collected Lectures and Essays. Chapman & Hall, London.

Gruen, A. (1985): Adaptive least squares correlation – a powerful image matching technique. South African Journal of Photogrammetry, Remote Sensing and Cartography, 14 (3), pp. 175-187.

Gruen, A. (1994): Digital close-range photogrammetry: progress through automation. International Archives of Photogrammetry & Remote Sensing, Melbourne, 30(5), pp. 122-135.

Haggrén, H. (1987): Real-time photogrammetry as used for machine vision applications. Canadian Surveyor, 41 (2), pp. 210-208.

Harley, I.A. (1967): The non-topographical uses of photogrammetry. The Australian Surveyor, 21 (7), pp. 237-263.

Helava, U.V. (1957): New principle for analytical plotters. Photogrammetria, 14, pp. 89-96.

Hinsken, L. (1989): CAP: Ein Programm zur kombinierten Bündelausgleichung auf Personalcomputern. Zeitschrift für Photogrammetrie und Fernerkundung, Heft 3, pp. 92-95.

Kruck, E. (1983): Lösung großer Gleichungssysteme für photogrammetrische Blockausgleichungen mit erweitertem funktionalem Modell. Dissertation, Wiss. Arbeiten der Fachrichtung Vermessungswesen der Universität Hannover, Nr. 128.

Laussedat, A. (1898, 1901, 1903): Recherches sur les instruments, les méthodes et le dessin topographiques. Gauthier-Villars, Paris.

Luhmann, T., Wester-Ebbinghaus, W. (1986): Rolleimetric RS - A New System for Digital Image Processing. Symposium ISPRS Commission II, Baltimore.

Masry, S.E., Faig, W. (1977): The analytical plotter in close-range applications. Photogrammetric Engineering and Remote Sensing, January 1977.

Poivilliers, G. (1961): Address delivered at the opening of the Historical Exhibition., International Archives of Photogrammetry, XIII/1, London.

Sander, W. (1932): The development of photogrammetry in the light of invention, with special reference to plotting from two photographs. In: von Gruber, O. (ed.): McCaw, G. T. & Cazalet, F. A., (Trans.): Photogrammetry, Collected Lectures and Essays. Chapman & Hall, London, pp. 148-246.

Schmid, H. (1956-57): An analytical treatment of the problem of triangulation by stereo-photogrammetry. Photogrammetria, XIII, Nr. 2 and 3.

Thompson, E.H. (1962): Photogrammetry. The Royal Engineers Journal, 76 (4), pp. 432-444.

Thompson, V.F. (1908): Stereo-photo-surveying. The Geographical Journal, 31, pp. 534ff.

Torlegård, K. (1967): On the determination of interior orientation of close-up cameras under operational conditions using three-dimensional test objects. Doctoral Thesis, Kungl, Tekniska Höskolan.

Wester-Ebbinghaus, W. (1981): Zur Verfahrensentwicklung in der Nahbereichsphotogrammetrie. Dissertation, Universität Bonn.

Wheatstone, C. (1838): Contribution to the physiology of vision – Part the first. On some remarkable, and hitherto unobserved, phenomena of binocular vision. Philosophical Transactions of the Royal Society of London for the year MDCCCXXXVIII, Part II, pp. 371-94.

9.3 Mathematical fundamentals

9.3.1 Transformations and geometry

Beder, C., Förstner, W. (2006): Direct solutions for computing cylinders from minimal sets of 3D points. In Leonardis et al. (eds.): Proceedings of the European Conference on Computer Vision, Springer, pp. 135-146.

Forbes, A.B. (1989): Least-squares best-fit geometric elements. National Physical Laboratory, Report DITC 140/89, Teddington, United Kingdom.

Förstner, W. (2000): Moderne Orientierungsverfahren. Photogrammetrie-Fernerkundung-Geoinformation, Heft 3, pp. 163-176.

Förstner, W., Wrobel, B. (2013): Mathematical concepts in photogrammetry. In McGlone (ed.): Manual of Photogrammetry, 6th ed., pp. 63-233.

Hinsken, L. (1987): Algorithmen zur Beschaffung von Näherungswerten für die Orientierung von beliebig im Raum angeordneten Strahlenbündeln. Deutsche Geodätische Kommission, Reihe C, Nr. 333.

Lösler, M., Nitschke, M. (2010): Bestimmung der Parameter einer Regressionsellipse in allgemeiner Raumlage. Allgemeine Vermessungs-Nachrichten, Heft 3, pp. 113-117.

Weckenmann, A. (1993): Auswerteverfahren und Antaststrategien in der Koordinatenmeßtechnik. In Neumann (ed.): Koordinatenmeßtechnik – Neue Aspekte und Anwendungen, Expert Verlag, Ehningen, pp. 133-169.

Späth, H. (2009): Alternative Modelle zum Ausgleich mit einer Geraden. Allgemeine Vermessungs-Nachrichten, Heft 11-12, pp. 388-390.

9.3.2 Adjustment techniques

Ackermann, F., Förstner, W., Klein, H., Schrot, R., van Mierlo, J. (1980): Grobe Datenfehler und die Zuverlässigkeit der photogrammetrischen Punktbestimmung. Seminar, Universität Stuttgart.

Baarda, W. (1968): A testing procedure for use in geodetic networks. Niederländische Geodätische Kommission, New Series 2, Nr. 5, Delft.

Fellbaum, M. (1996): Robuste Bildorientierung in der Nahbereichsphotogrammetrie. Dissertation, Technische Universität Clausthal, Geodätische Schriftenreihe der Technischen Universität Braunschweig, Nr. 13.

Förstner, W., Wrobel, B. (2013): Mathematical concepts in photogrammetry. In McGlone (ed.): Manual of Photogrammetry, 6th ed., pp. 63-233.

Förstner, W., Wrobel, B., Paderes, F., Fraser, C.S., Dolloff, J., Mikhail, E.M., Rujikietgumjorn, W. (2013): Analytical photogrammetric operations. In McGlone (ed.): Manual of Photogrammetry, 6th ed., pp. 785-955.

Gruen, A., Kersten, T. (1995): Sequential estimation in robot vision. Photogrammetric Engineering & Remote Sensing, Vol. 61, No. 1, pp. 75-82.

Hinsken, L. (1989): CAP: Ein Programm zur kombinierten Bündelausgleichung auf Personalcomputern. Zeitschrift für Photogrammetrie und Fernerkundung, Heft 3, pp. 92-95.

Kampmann, G. (1986): Robuster Ausreißertest mit Hilfe der L1-Norm-Methode. Allgemeine Vermessungs-Nachrichten, 93, pp. 139-147.

Kersten, T., Baltsavias, E. (1994): Sequential estimation of sensor orientation for stereo images sequences. International Archives of Photogrammetry and Remote Sensing, Vol. 30, Part 5, Commission V, pp. 206-213.

Klein, H. (1984): Automatische Elimination grober Datenfehler im erweiterten Blockausgleichungsprogramm PAT-M. Bildmessung und Luftbildwesen, Heft 6, pp. 273-280.

Kruck, E. (1983): Lösung großer Gleichungssysteme für photogrammetrische Blockausgleichungen mit erweitertem funktionalem Modell. Dissertation, Wiss. Arbeiten der Fachrichtung Vermessungswesen der Universität Hannover, Nr. 128.

Kruck, E. (1995): Balanced least squares adjustment for relative orientation. In Gruen/Kahmen (ed.): Optical 3-D Measurement Techniques III, Wichmann, Heidelberg, pp. 486-495.

Lehmann, R. (2010): Normierte Verbesserungen – wie groß ist groß? Allgemeine Vermessungs-Nachrichten, Heft 2, pp. 53-61.

9.4 Imaging technology

9.4.1 Optics and sampling theory

Bähr, H.-P. (1992): Appropriate pixel size for orthophotography. International Archives of Photogrammetry and Remote Sensing, Vol. 29/1, Washington.

Schmidt, F. (1993): 2D-Industriemeßtechnik mit CCD-Sensoren. Zeitschrift für Photogrammetrie und Fernerkundung, Heft 2, pp. 62-70.

Schwarte, R. (1997): Überblick und Vergleich aktueller Verfahren der optischen Formerfassung. GMA-Bericht 30, Optische Formerfassung, Langen, pp. 1-12.

9.4.2 Camera modelling and calibration

Barazzetti, L., Mussio, L., Remondino, F., Scaioni, M. (2011): Targetless camera calibration. International Archives of the Photogrammetry, Remote Sensing and Spatial Information Sciences, Vol. 38 (5/W16), on CD-ROM.

Beyer, H. (1992): Geometric and radiometric analysis of a CCD-camera based photogrammetric close-range system. Dissertation, Mitteilungen Nr. 51, Institut für Geodäsie und Photogrammetrie, ETH Zürich.

Brown, D.C. (1971): Close-range camera calibration. Photogrammetric Engineering, 37(8), pp. 855-866.

Clarke, T.A., Fryer, J.G. (1998): The development of camera calibration methods and models. Photogrammetric Record, 16(91), pp. 51-66.

Cronk, S., Fraser, C.S. Hanley, H.B. (2006): Automatic calibration of colour digital cameras. Photogrammetric Record, 21(116), pp. 355-372.

Förstner, W., Wrobel, B., Paderes, F., Fraser, C.S., Dolloff, J., Mikhail, E.M., Rujikietgumjorn, W. (2013): Analytical photogrammetric operations. In McGlone (ed.): Manual of Photogrammetry, 6th ed., pp. 785-955.

Fraser, C.S. (1980): Multiple focal setting self-calibration of close-range metric cameras. Photogrammetric Engineering and Remote Sensing, 46(9), pp. 1161-1171.

Fraser, C.S., Brown, D.C. (1986): Industrial photogrammetry – new developments and recent applications. The Photogrammetric Record, 12(68), pp. 197-216.

Fraser, C.S., Shortis, M. (1992): Variation of distortion within the photographic field. Photogrammetric Engineering and Remote Sensing, 58(6), pp. 851-855.

Fraser, C.S. (1997): Digital camera self-calibration. ISPRS International Journal of Photogrammetry & Remote Sensing, Vol. 52, pp. 149-159.

Fraser, C.S. (2013): Automatic camera calibration in close-range photogrammetry. Photogrammetric Engineering & Remote Sensing, 79(4), pp. 381-388.

Fryer, J.G. (1996): Camera calibration. In Atkinson (ed.): Close Range Photogrammetry and Machine Vision, Whittles Publishing, Caithness, UK, pp. 156-179.

Fryer, J.G., Brown, D.C. (1986): Lens distortion for close-range photogrammetry. Photogrammetric Engineering & Remote Sensing, 52 (1), pp. 51-58.

Goldschmidt, R., Peipe, J. (1994): Investigation of a HDTV camera for photogrammetric application. International Archives of Photogrammetry and Remote Sensing, Vol. 30, Part 1, pp. 122-124.

Gruen, A., Huang, T.S. (2001): Calibration and Orientation of Cameras in Computer Vision. Springer, Berlin/Heidelberg.

Kenefick, J.F., Gyer, M.S., Harp, B.F. (1972): Analytical self-calibration. Photogrammetric Engineering and Remote Sensing, 38(11), pp. 1117-1126.

Kotowski, R. Weber, B. (1984): A procedure for on-line correction of systematic errors. International Archives of Photogrammetry and Remote Sensing, Vol. 25(3a), pp. 553-560.

Li, W., Schulte, M., Bothe, T., von Kopylow, C., Köpp, N., Jüptner, W. (2007): Beam based calibration for optical imaging devices. Proceedings of 3DTV-CON.

Luhmann, T. (2005): Zum photogrammetrischen Einsatz von Einzelkameras mit optischer Stereostrahlteilung. Photogrammetrie-Fernerkundung-Geoinformation, Heft 2, pp. 101-110.

Maas, H.-G. (1999): Ein Ansatz zur Selbstkalibrierung von Kameras mit instabiler innerer Orientierung. Publikationen der DGPF, Band 7, München 1998.

Menna, F., Nocerino, E., Remondino, F., Shortis, M. (2013): Investigation of a consumer-grade digital stereo camera. Proc. of Videometrics, Range Imaging and Applications XII, SPIE Optical Metrology, Vol. 8791.

Peipe, J., Tecklenburg, W. (2009): Zur Bestimmung des Bildhauptpunktes durch Simultankalibrierung. In Luhmann/Müller (ed.): Photogrammetrie-Laserscanning-

Optische 3D-Messtechnik, Oldenburger 3D-Tage 2009, Wichmann, Heidelberg, pp. 340-347.

Piechel, J., Jansen, D., Luhmann, T. (2010): Kalibrierung von Zoom- und Shift-/Tilt-Objektiven. In Luhmann/Müller (ed.): Photogrammetrie-Laserscanning-Optische 3D-Messtechnik, Oldenburger 3D-Tage 2010, Wichmann, Heidelberg, pp. 284-291.

Pollefeys, M., Koch, R., Van Gool, L. (1999): Self-calibration and metric reconstruction inspite of varying and unknown internal camera parameters. International Journal of Computer Vision, 32 (1), pp. 7-25.

Remondino, F., Fraser, C.S. (2006): Digital camera calibration methods: considerations and comparisons. International Archives of the Photogrammetry, Remote Sensing and Spatial Information Sciences, Vol. 36/5, pp. 266-272.

Rieke-Zapp, D., Tecklenburg, W., Peipe, J., Hastedt, H., Haig, C. (2009): Evaluation of the geometric stability and the accuracy potential of digital cameras – Comparing mechanical stabilisation versus parametrisation. ISPRS Journal of Photogrammetry and Remote Sensing, Vol. 64/3, pp. 248-258.

Robson, S., Shortis, M.R. (1998): Practical influences of geometric and radiometric image quality provided by different digital camera systems. Photogrammetric Record, 16 (92), pp. 225-248.

Robson, S., Shortis, M.R., Ray, S.F. (1999): Vision metrology with super wide angle and fisheye optics. Videometrics VI, SPIE Volume 3641, pp. 199-206.

Schneider, D. (2008): Geometrische und stochastische Modelle für die integrierte Auswertung terrestrischer Laserscannerdaten und photogrammetrischer Bilddaten. Dissertation, Technische Universität Dresden.

Schneider, D., Maas, H.-G. (2003): Geometric modelling and calibration of a high resolution panoramic camera. In Grün/Kahmen (eds.): Optical 3D Measurement Techniques VI. Vol. II, Wichmann, Heidelberg, pp. 122-129.

Schwalbe, E. (2005): Geometric modelling and calibration of fisheye lens camera systems. International Archives of the Photogrammetry, Remote Sensing and Spatial Information Sciences, Vol. 36, Part 5/W8.

Shortis, M.R., Bellman, C.J., Robson, S., Johnston, G.J., Johnson, G.W. (2006): Stability of zoom and fixed lenses used with digital SLR cameras. International Archives of the Photogrammetry, Remote Sensing and Spatial Information Sciences, Vol. 36/5, pp. 285-290.

Stamatopoulos, C., Fraser, C.S. (2011): Calibration of long focal length cameras in close-range photogrammetry. Photogrammetric Record, 26(135), pp. 339-360.

Tecklenburg, W., Luhmann, T., Hastedt, H. (2001): Camera modelling with image-variant parameters and finite elements. In Grün/Kahmen (eds.): Optical 3D Measurement Techniques V, Wichmann, Heidelberg.

Tsai, R.Y. (1986): An efficient and accurate camera calibration technique for 3-D machine vision. Proc. International Conference on Computer Vision and Pattern Recognition, Miami Beach, USA, pp. 364-374.

van den Heuvel, F., Verwaal, R., Beers, B. (2007): Automated calibration of fisheye camera systems and the reduction of chromatic aberration. Photogrammetrie-Fernerkundung-Geoinformation, Heft 3, pp. 157-165.

Wester-Ebbinghaus, W. (1981): Zur Verfahrensentwicklung in der Nahbereichsphotogrammetrie. Dissertation, Universität Bonn.

Wester-Ebbinghaus, W. (1985): Verfahren zur Feldkalibrierung von photogrammetrischen Aufnahmekammern im Nahbereich. In Kupfer/Wester-Ebbinghaus (eds.): Kammerkalibrierung in der photogrammetrischen Praxis, Deutsche Geodätische Kommission, Reihe B, Heft Nr. 275, pp. 106-114.

9.4.3 Sensors and cameras

Atkinson, K.B. (1989): Instrumentation for non-topographic photogrammetry. In Karara (ed.): Non-Topographic Photogrammetry, 2nd ed., American Society for Photogrammetry and Remote Sensing, pp. 15-35.

Beyer, H. (1992): Geometric and radiometric analysis of a CCD-camera based photogrammetric close-range system. Dissertation, Mitteilungen Nr. 51, Institut für Geodäsie und Photogrammetrie, ETH Zürich.

Beyer, H., Kersten, T., Streilein, A. (1992): Metric accuracy performance of solid-state camera systems. SPIE Vol. 1820, Videometrics, pp. 103-110.

Bobey, K., Brekerbohm, L. (2005): CMOS vs. CCD-Bildsensoren und Kameras. In Luhmann (ed.): Photogrammetrie-Laserscanning-Optische 3D-Messtechnik, Oldenburger 3D-Tage 2005, Wichmann, Heidelberg, pp. 172-182.

Boyle, W.S., Smith, G.E. (1970): Charge coupled semiconductor devices. Bell Systems Technical Journal, Vol. 49, pp. 587-593.

Brown, J., Dold, J. (1995): V-STARS – A system for digital industrial photogrammetry. In Gruen/Kahmen (eds.): Optical 3D Measurement Techniques III, Wichmann, Heidelberg, pp. 12-21.

Chapman, D., Deacon, A. (1997): The role of spatially indexed image archives for "As-Built" modelling of large process plant facilities. In Gruen/Kahmen (eds.): Optical 3D Measurement Techniques IV, Wichmann, Heidelberg, pp. 475-482.

Dähler, J. (1987): Problems in digital image acquisition with CCD cameras. Fast Processing of Photogrammetric Data, Interlaken, pp. 48-59.

Dold, J. (1997): Ein hybrides photogrammetrisches Industriemeßsystem höchster Genauigkeit und seine Überprüfung. Dissertation, Heft 54, Schriftenreihe Studiengang Vermessungswesen, Universität der Bundeswehr, München.

El-Hakim, S. (1986): Real-time image metrology with CCD cameras. Photogrammetric Engineering & Remote Sensing, 52 (11), pp. 1757-1766.

Ganci, G., Handley, H. (1998): Automation in videogrammetry. International Archives of Photogrammetry & Remote Sensing, Hakodate, 32 (5), pp. 53-58.

Hauschild, R. (1999): Integrierte CMOS-Kamerasysteme für die zweidimensionale Bildsensorik. Dissertation, Universität Duisburg.

Läbe, T., Förstner, W. (2004): Geometric stability of low-cost digital consumer cameras. International Archives of the Photogrammetry, Remote Sensing and Spatial Information Sciences, Vol. 35 (1), pp. 528-535.

Lenz, R., Fritsch, D. (1990): On the accuracy of videometry with CCD-sensors. International Journal for Photogrammetry and Remote Sensing (IJPRS), 45, pp. 90-110.

Lenz, R., Beutlhauser, R., Lenz, U. (1994): A microscan/macroscan 3x12 bit digital color CCD camera with programmable resolution up to 20992 x 20480 picture elements. International Archives of Photogrammetry and Remote Sensing, Vol. 30/5, Melbourne, pp. 225-230.

Luhmann, T., Tecklenburg, W. (2002): Bundle orientation and 3D object reconstruction from multiple-station panoramic imagery. ISPRS Symposium Comm. V, Korfu.

Lyon, R., Hubel, P. (2002): Eyeing the camera: Into the next century. IS&T/TSID 10th Color Imaging Conference Proceedings, Scottsdale, Az., USA, pp. 349-355.

Peipe, J. (1995): Photogrammetric investigation of a 3000 x 2000 pixel high-resolution still-video camera. ISPRS Intercommission Workshop, From Pixels to Sequences, Zurich.

Riechmann, W. (1990): The reseau-scanning camera – conception and first measurement results. ISPRS Symposium Commission V, Zurich.

Rieke-Zapp, D.H. (2006): Wenn's etwas mehr sein darf – Verschieben der Hauptpunktlage für eine optimale Stereoabdeckung. In Luhmann/Müller (eds.): Photogrammetrie-Laserscanning-Optische 3D-Messtechnik, Oldenburger 3D-Tage 2006, Wichmann, Heidelberg, pp. 32-39.

Robson, S., Shortis, M.R., Ray, S.F. (1999): Vision metrology with super wide angle and fisheye optics. Videometrics VI, SPIE Vol. 3641, pp. 199-206.

Robson, S., Shortis, M.R. (1998): Practical influences of geometric and radiometric image quality provided by different digital camera systems. Photogrammetric Record, 16(92), pp. 225-248.

Schmidt, F. (1993): 2D-Industriemeßtechnik mit CCD-Sensoren. Zeitschrift für Photogrammetrie und Fernerkundung, Heft 2, pp. 62-70.

Shortis, M.R., Beyer, H.A. (1996): Sensor technology for digital photogrammetry and machine vision. In Atkinson (ed.): Close Range Photogrammetry and Machine Vision, Whittles Publishing, Caithness, UK, pp. 106-155.

Wester-Ebbinghaus, W. (1983): Ein photogrammetrisches System für Sonderanwendungen. Bildmessung und Luftbildwesen, Heft 3, pp. 118-128.

Wester-Ebbinghaus, W. (1989): Das Réseau im photogrammetrischen Bildraum. Zeitschrift für Photogrammetrie und Fernerkundung, Heft 3, pp. 3-10.

9.4.4 Targeting and illumination

Ahn, S.J.; Kotowski, R. (1997): Geometric image measurement errors of circular object targets. In Gruen/Kahmen (eds.): Optical 3D Measurement Techniques IV, Wichmann, Heidelberg, pp. 463-471.

Ahn, S.J., Schultes, M. (1997): A new circular coded target for the automation of photogrammetric 3D surface measurements. In Gruen/Kahmen (eds.): Optical 3D Measurement Techniques IV. Wichmann, Heidelberg, pp. 225-234.

Clarke, T.A. (1994): An analysis of the properties of targets used in digital close range photogrammetric measurement. Videometrics III, SPIE Vol. 2350, Boston, pp. 251-262.

Fraser, C.S., Cronk, S. (2009): A hybrid measurement approach for close-range photogrammetry. ISPRS Journal of Photogrammetry & Remote Sensing, 64(3), pp. 328-333.

Hattori, S., Akimoto, K., Fraser, C.S., Imoto, H. (2002): Automated procedures with coded targets in industrial vision metrology. Photogrammetric Engineering & Remote Sensing, 68(5), pp. 441-446.

Kager, H. (1981): Bündeltriangulation mit indirekt beobachteten Kreiszentren. Geowissenschaftliche Mitteilungen der Studienrichtung Vermessungswesen der TU Wien, Heft 19.

Mulsow, C. (2007): Ein photogrammetrisches Verfahren zur Kalibrierung eines Beamers. In Luhmann/Müller (ed.): Photogrammetrie-Laserscanning-Optische 3D-Messtechnik, Oldenburger 3D-Tage 2007, Wichmann, Heidelberg, pp. 58-67.

Otepka, J.O., Fraser, C.S. (2004): Accuracy enhancement of vision metrology through automatic target plane determination. International Archives of Photogrammetry, Remote Sensing and Spatial Information Sciences, 35(B5), pp. 873-879.

Shortis, M.R., Seager, J.W., Robson, S., Harvey, E.S. (2003): Automatic recognition of coded targets based on a Hough transform and segment matching. Videometrics VII, SPIE Vol. 5013, pp. 202-208.

Wiora, G., Babrou, P., Männer, R. (2004): Real time high speed measurement of photogrammetric targets. In Rasmussen et al. (ed.): Pattern Recognition, Springer, Berlin, pp. 562-569.

9.4.5 Laser-based systems

Benning, W., Becker, R., Effkemann, C. (2004): Extraktion von Ecken, Kanten und Profilen aus Laserscannerdaten, gestützt durch photogrammetrische Aufnahmen. In Luhmann (ed.): Photogrammetrie-Laserscanning-Optische 3D-Messtechnik, Oldenburger 3D-Tage 2004, Wichmann, Heidelberg, pp. 213-220.

Beraldin, J.-A., Blais, F., Lohr, U. (2010): Laser scanning technology. In Vosselman/Maas (eds.): Airborne and Terrestrial Laser Scanning, Whittles Publishing, Caithness, UK, pp. 1-42.

Dold, J. (2004): Neue Laser-Technologien für die Industrievermessung. Photogrammetrie-Fernerkundung-Geoinformation, Heft 1, pp. 39-46.

Dold, C., Ripperda, N., Brenner, C. (2007): Vergleich verschiedener Methoden zur automatischen Registrierung von terrestrischen Laserscandaten. In Luhmann/Müller (eds.): Photogrammetrie-Laserscanning-Optische 3D-Messtechnik, Oldenburger 3D-Tage 2007, Wichmann, Heidelberg, pp. 196-205.

Hennes, M. (2009): Freiformflächenerfassung mit Lasertrackern – eine ergonomische Softwarelösung zur Reflektoroffsetkorrektur. Allgemeine Vermessungs-Nachrichten, Heft 5, pp. 188-194.

Kersten, T., Mechelke, K., Lindstaedt, M., Sternberg, H. (2009): Methods for geometric accuracy investigations of terrestrial laser scanning systems. Photogrammetrie-Fernerkundung-Geoinformation, Heft 4, pp. 301-316.

Lichti, D., Skaloud, J. (2010): Registration and calibration. In Vosselman/Maas (eds.): Airborne and Terrestrial Laser Scanning, Whittles Publishing, Caithness, UK, pp. 83-133.

Loser, R., Kyle, S. (2003): Concepts and components of a novel 6DOF tracking system for 3D metrology. In Gruen/Kahmen (eds.): Optical 3D Measurement Techniques VI, Vol. 2, pp. 55-62.

Mechelke, K., Kersten, T., Lindstaedt, M. (2007): Comparative investigations into the accuracy behaviour of the new generation of terrestrial laser scanning systems. In Gruen/Kahmen (eds.): Optical 3D Measurement Techniques VIII, Zurich, Vol. I, pp. 319-327.

Runne, H., Niemeier, W., Kern, F. (2001): Application of laser scanners to determine the geometry of buildings. In Grün/Kahmen (eds.): Optical 3D Measurement Techniques V, Wichmann, Heidelberg, pp. 41-48.

Staiger, R. (1992): Automatische und dynamische Koordinatenmessung mit mobilen Sensorsystemen. In Welsch et al. (eds.): Geodätische Meßverfahren im Maschinenbau, Schriftenreihe DVW, 1/1992, Wittwer, Stuttgart, pp. 81-96.

Studnicka, N., Riegl, J., Ullrich, A. (2003): Zusammenführung und Bearbeitung von Laserscandaten und hochauflösenden digitalen Bildern eines hybriden 3D-Laser-Sensorsystems. In Luhmann (ed.): Photogrammetrie-Laserscanning-Optische 3D-Messtechnik, Oldenburger 3D-Tage 2004, Wichmann, Heidelberg, pp. 183-190.

Wehr, A. (1998): Scannertechniken zur dimensionellen Oberflächenbestimmung. 44. DVW-Seminar „Hybride Vermessungssysteme", Schriftenreihe des DVW, Band 29, pp. 125-148.

Wendt, K., Schwenke, H., Bösemann, W., Dauke, M. (2003): Inspection of large CMMs and machine tools by sequential multilateration using a single laser tracker. Laser Metrology and Machine Performance, Vol. 5, pp. 121-130.

9.4.6 3D imaging systems

Böhm, J. (2012): Natural user interface sensors for human body measurement. International Archives of the Photogrammetry, Remote Sensing and Spatial Information Sciences, 39, B3.

Böhm, J., Pattinson, T. (2010): Accuracy of exterior orientation for a range camera. International Archives of the Photogrammetry, Remote Sensing and Spatial Information Sciences 38 (Part 5), pp. 103 – 108.

Kahlmann, T., Remondino, F., Ingensand, H. (2006): Calibration for increased accuracy of the range imaging camera SwissRanger. International Archives of Photogrammetry, Remote Sensing and Spatial Information Sciences, 36, Part 5, pp. 136-141.

Karel, W. (2008): Integrated range camera calibration using image sequences from hand-held operation. International Archives of Photogrammetry, Remote Sensing and Spatial Information Sciences, Vol. 37, Part 5, pp. 945-951.

Khoshelham, K., Elberink, S.O. (2012): Accuracy and resolution of Kinect depth data for indoor mapping applications. Sensors, 12, pp. 1437-1454.

Lichti, D.D., Kim, C. (2011): A comparison of three geometric self-calibration methods for range cameras. Remote Sensing, 3 (5), pp. 1014-1028.

Remondino, F., Stoppa, D. (eds.) (2013): TOF Range-Imaging Cameras. Springer.

9.4.7 Phase-based measurements

Andrä, P. (1998): Ein verallgemeinertes Geometriemodell für das Streifenprojektionsverfahren zur optischen 3D-Koordinatenmessung. Dissertation, Strahltechnik Band 9, Universität Bremen.

Gühring, J. (2002): 3D-Erfassung und Objektrekonstruktion mittels Streifenprojektion. Dissertation, Universität Stuttgart, Fakultät für Bauingenieur- und Vermessungswesen.

Stahs, T., Wahl, F.M. (1990): Oberflächenmessung mit einem 3D-Robotersensor. Zeitschrift für Photogrammetrie und Fernerkundung, Heft 6, pp. 190-202.

Strutz, T. (1993): Ein genaues aktives Bildtriangulationsverfahren zur Oberflächenvermessung. Dissertation, Fakultät für Maschinenbau, Universität Magdeburg.

Thesing, J., Behring, D., Haig, J. (2007): Freiformflächenmessung mit photogrammetrischer Streifenprojektion. VDI Berichte Nr. 1996, Optische Messung technischer Oberflächen in der Praxis.

Wahl, F.M. (1986): A coded light approach for depth map acquisition. In Hartmann (ed.): Mustererkennung 1986, Springer, Berlin, pp. 12-17.

Winter, D., Reich, C. (1997): Video-3D-Digitalisierung komplexer Objekte mit frei beweglichen, topometrischen Sensoren. GMA-Bericht 30, Optische Formerfassung, Langen, pp. 119-127.

Wiora, G. (2001): 3D-Erfassung und Objektrekonstruktion mittels Streifenprojektion. Dissertation, Ruprechts-Karl-Universität Heidelberg, Naturwissenschaftlich-Mathematische Gesamtfakultät.

9.5 Analytical methods

9.5.1 Analytical photogrammetry

Abdel-Aziz, Y., Karara, H.M. (1971): Direct linear transformation from comparator coordinates into object space coordinates in close range photogrammetry. ASP Symposium on Close-Range Photogrammetry.

Andresen, K. (1991): Ermittlung von Raumelementen aus Kanten im Bild. Zeitschrift für Photogrammetrie und Fernerkundung, Nr. 6, pp. 212-220.

Beder, C., Förstner, W. (2006): Direct solutions for computing cylinders from minimal sets of 3D points. In Leonardis et al. (eds.): Proceedings of the European Conference on Computer Vision, Springer, pp. 135-146.

Brandstätter, G. (1991): Zur relativen Orientierung projektiver Bündel. Zeitschrift für Photogrammetrie und Fernerkundung, Nr. 6, pp. 199-212.

Brown, D.C. (1958): A solution to the general problem of multiple station analytical stereotriangulation. RCA Data Reduction Technical Report No. 43, Aberdeen.

Cross, P. A. (1990): Working Paper No. 6, Advanced least squares applied to position-fixing. Department of Land Surveying, Polytechnic (now University) of East London, 205 p.

Forkert, G. (1994): Die Lösung photogrammetrischer Orientierungs- und Rekonstruktions-aufgaben mittels allgemeiner kurvenförmiger Elemente. Dissertation, TU Wien, Geowissenschaftliche Mitteilungen der Studienrichtung Vermessungswesen, Heft 41, 147 p.

Förstner, W., Wrobel, B. (2013): Mathematical concepts in photogrammetry. In McGlone (ed.): Manual of Photogrammetry, 6th ed., pp. 63-233.

Förstner, W., Wrobel, B., Paderes, F., Fraser, C.S., Dolloff, J., Mikhail, E.M., Rujikietgumjorn, W. (2013): Analytical photogrammetric operations. In McGlone (ed.): Manual of Photogrammetry, 6th ed., pp. 785-955.

Fraser, C.S., Brown, D.C. (1986): Industrial photogrammetry – new developments and recent applications. The Photogrammetric Record, 12 (68), pp. 197-216.

Fraser, C.S. (2006) Evolution of network orientation procedures. International Archives of Photogrammetry, Remote Sensing and Spatial Information Sciences, Dresden, 35(5), pp. 114-120.

Haralick, R.M., Lee, C., Ottenberg, K., Nolle, M. (1994): Review and analysis of solutions of the three point perspective pose estimation problem. International Journal of Computer Vision, 13 (3), pp. 331-356.

van den Heuvel, F.A. (1997): Exterior orientation using coplanar parallel lines. Proc. 10th Scandinavian Conference on Image Analysis, Lappeenranta, pp. 71-78.

Horn, B.K. (1987): Closed-form solution of absolute orientation using unit quaternions. Journal of the Optical Society of America A, 4(4), pp. 629-642.

Kyle, S.A. (1990): A modification to the space resection. Allgemeine Vermessungs-Nachrichten, International Edition, Heft 7, pp. 17-25.

Läbe, T., Förstner, W. (2005): Erfahrungen mit einem neuen vollautomatischen Verfahren zur Orientierung digitaler Bilder. Publikationen der DGPF, Band 14, pp. 271-278.

Li, H., Gruen, A. (1997): LSB-snakes for industrial measurement applications. In Gruen/Kahmen (eds.): Optical 3D Measurement Techniques IV, Wichmann, Heidelberg, pp. 169-178.

Loser, R., Luhmann, T. (1992): The programmable optical 3-D measuring system POM – applications and performance. International Archives of Photogrammetry and Remote Sensing, Vol. 29, B5, Washington, pp. 533-540.

Luhmann, T. (2009): Precision potential of photogrammetric 6 DOF pose estimation with a single camera. ISPRS Journal of Photogrammetry and Remote Sensing, 64(3), pp. 275-284.

Mayer, H. (2007): Automatische Orientierung mit und ohne Messmarken – Das Mögliche und das Unmögliche. Publikationen der DGPF, Band 16, pp. 457-464.

Rohrberg, K. (2009): Geschlossene Lösung für den räumlichen Rückwärtsschnitt mit minimalen Objektinformationen. In Luhmann/Müller (eds.): Photogrammetrie-Laserscanning-Optische 3D-Messtechnik, Oldenburger 3D-Tage 2009, Wichmann, Heidelberg, pp. 332-339.

Sanso, F. (1973): An exact solution of the roto-translation problem. Photogrammetria, 29(6), pp. 203-216.

Schmid, H.H. (1958): Eine allgemeine analytische Lösung für die Aufgabe der Photogrammetrie. Bildmessung und Luftbildwesen, 1958, Nr. 4, pp. 103-113, and 1959, Nr. 1, pp. 1-12.

Smith, A.D.N. (1965): The explicit solution of the single picture resection problem, with a least squares adjustment to redundant control. The Photogrammetric Record, 5 (26), pp. 113-122.

Thompson, E.H. (1959): A rational algebraic formulation of the problem of relative orientation. The Photogrammetric Record, 3 (14), pp. 152-159.

Thompson, E.H. (1966): Space resection: failure cases. The Photogrammetric Record, 5 (27), pp. 201-204.

Hemmleb, M., Wiedemann, A. (1997): Digital rectification and generation of orthoimages in architectural photogrammetry. International Archives for Photogrammetry and Remote Sensing, Vol. XXXII, Part 5C1B, Göteborg, pp. 261-267.

Wrobel, B. (1999): Minimum solutions for orientation. In Huang/Gruen (eds.): Calibration and Orientation of Cameras in Computer Vision, Washington D.C., Springer, Heidelberg.

9.5.2 Bundle adjustment

Brown, D.C. (1976): The bundle adjustment – progress and prospectives. International Archives of Photogrammetry, 21 (3), ISP Congress, Helsinki, pp. 1-33.

Case, J.B. (1961): The utilization of constraints in analytical photogrammetry. Photogrammetric Engineering, 27(5), pp. 766-778.

El-Hakim, S.F., Faig, W. (1981): A combined adjustment of geodetic and photogrammetric observations. Photogrammetric Engineering and Remote Sensing, 47(1), pp. 93-99.

Fellbaum, M. (1996): Robuste Bildorientierung in der Nahbereichsphotogrammetrie. Dissertation, Technische Universität Clausthal, Geodätische Schriftenreihe der Technischen Universität Braunschweig, Nr. 13.

Fraser, C.S. (1982): Optimization of precision in close-range photogrammetry. Photogrammetric Engineering and Remote Sensing, 48(4), pp. 561-570.

Fraser, C.S. (1996): Network design. In Atkinson (ed.): Close Range Photogrammetry and Machine Vision, Whittles Publishing, Caithness, UK, pp. 256-281.

Granshaw, S.I. (1980): Bundle adjustment methods in engineering photogrammetry. The Photogrammetric Record, 10(56), pp. 181-207.

Hastedt, H. (2004): Monte Carlo simulation in close-range photogrammetry. International Archives of the Photogrammetry, Remote Sensing and Spatial Information Sciences, Vol. 35, part B5, pp. 18-23.

Hinsken, L. (1987): Algorithmen zur Beschaffung von Näherungswerten für die Orientierung von beliebig im Raum angeordneten Strahlenbündeln. Dissertation, Deutsche Geodätische Kommission, Reihe C, Nr. 333.

Hinsken, L. (1989): CAP: Ein Programm zur kombinierten Bündelausgleichung auf Personalcomputern. Zeitschrift für Photogrammetrie und Fernerkundung, Heft 3, pp. 92-95.

Hunt, R.A. (1984). Estimation of initial values before bundle adjustment of close range data. International Archives of Photogrammetry and Remote Sensing, 25 (5), pp. 419-428.

Jacobsen, K. (1982): Selection of additional parameters by program. International Archives of Photogrammetry and Remote Sensing, Vol. 24-3, pp. 266-275.

Jorge, J.M. (1977): The Levenberg-Marquardt algorithm: implementation and theory. In Watson (ed.): Numerical Analysis, Lecture Notes Math. 630, pp. 105-116.

Kager, H. (1981): Bündeltriangulation mit indirekt beobachteten Kreiszentren. Geowissenschaftliche Mitteilungen der Studienrichtung Vermessungswesen der TU Wien, Heft 19.

Kager, H. (1989): Orient: a universal photogrammetric adjustment system. In Gruen/Kahmen (eds.): Optical 3D Measurement Techniques, Wichmann, Karlsruhe, pp. 447-455.

Kruck, E. (1983): Lösung großer Gleichungssysteme für photogrammetrische Blockausgleichungen mit erweitertem funktionalen Modell. Dissertation, Universität Hannover.

Levenberg, K. (1944): A method for the solution of certain problems in least squares. The Quarterly of Applied Mathematics, 2, pp. 164-168.

Marquardt, D (1963): An algorithm for least-squares estimation of nonlinear parameters. SIAM Journal on Applied Mathematics, 11, pp. 431-441.

Triggs, B., McLauchlan, P., Hartley, R., Fitzgibbon, A. (1999): Bundle adjustment – a modern synthesis. ICCV '99, Proceedings of the International Workshop on Vision Algorithms, Springer, Berlin, pp. 298-372.

Wester-Ebbinghaus, W. (1985): Bündeltriangulation mit gemeinsamer Ausgleichung photogrammetrischer und geodätischer Beobachtungen. Zeitschrift für Vermessungswesen, 110 (3), pp. 101-111.

Zinndorf, S. (1985): Freies Netz – Anwendungen in der Nahbereichsphotogrammetrie. Bildmessung und Luftbildwesen, Heft 4, pp. 109-114.

9.5.3 Camera calibration

Bräuer-Burchardt, C., Voss, K. (2000): Automatic lens distortion calibration using single views. Mustererkennung 2000, Springer, Berlin, pp. 187-194.

Brown, D.C. (1971): Close-range camera calibration. Photogrammetric Engineering, 37(8), pp. 855-866.

Cronk, S., Fraser, C.S., Hanley, H. (2006): Automated metric calibration of colour digital cameras. The Photogrammetric Record, 21(116), pp. 355-372.

Fraser, C.S., Shortis, M. (1992): Variation of distortion within the photographic field. Photogrammetric Engineering and Remote Sensing, 58(6), pp. 851-855.

Fraser, C.S., Cronk, S., Stamatopoulos, C. (2012): Implementation of zoom-dependent camera calibration in close-range photogrammetry. International Archives of the Photogrammetry, Remote Sensing and Spatial Information Sciences, Vol. 39(B5), pp. 15-19.

Fraser, C.S. (2013): Automatic camera calibration in close-range photogrammetry. Photogrammetric Engineering & Remote Sensing, 79(4), pp. 381-388.

Fryer, J.G. (1996): Camera calibration. In Atkinson (ed.): Close Range Photogrammetry and Machine Vision, Whittles Publishing, Caithness, UK, pp. 156-179.

Fryer, J.G., Brown, D.C. (1986): Lens distortion for close-range photogrammetry. Photogrammetric Engineering & Remote Sensing, 52(1), pp. 51-58.

Godding, R. (1999): Geometric calibration and orientation of digital imaging systems. In Jähne et al. (eds.): Handbook of Computer Vision and Applications, Vol. 1, Academic Press, pp. 441-461.

Gruen, A., Huang, T.S. (2001): Calibration and Orientation of Cameras in Computer Vision. Springer, Berlin/Heidelberg.

Kenefick, J.F., Gyer, M.S., Harp, B.F. (1972): Analytical self-calibration. Photogrammetric Engineering and Remote Sensing, 38(11), pp. 1117-1126.

Luhmann, T., Godding, R. (2004): Messgenauigkeit und Kameramodellierung – Kernfragen der Industriephotogrammetrie. Photogrammetrie-Fernerkundung-Geoinformation, Heft 1, pp. 13-21.

Luhmann, T., Piechel, J., Roelfs, T. (2013): Geometric calibration of thermographic cameras. In Kuenzer/Dech (eds.), Thermal Infrared Remote Sensing: Sensors, Methods, Applications, Springer, Berlin, pp. 27-42.

Maas, H.-G. (1999): Ein Ansatz zur Selbstkalibrierung von Kameras mit instabiler innerer Orientierung. Publikationen der DGPF, Band 7, München 1998, pp. 47-53.

Maas, H.-G. (1999): Image sequence based automatic multi-camera system calibration techniques. ISPRS Journal of Photogrammetry and Remote Sensing, 54(5-6), pp. 352-359.

Remondino, F., Fraser, C.S. (2006): Digital camera calibration methods: considerations and comparisons. International Archives of the Photogrammetry, Remote Sensing and Spatial Information Sciences, Vol. 36/5, pp. 266-272.

Shortis, M.R., Robson, S., Beyer, H.A. (1998): Principal point behaviour and calibration parameter models for Kodak DCS cameras. Photogrammetric Record, 16(92), pp. 165-186.

Stamatopoulos, C., Fraser, C.S. (2013): Target-free automated image orientation and camera calibration in close-range photogrammetry. ASPRS Annual Conference, Baltimore, Maryland, March 24-28, 8 p, (on CD-ROM).

Tecklenburg, W., Luhmann, T., Hastedt, H. (2001): Camera modelling with image-variant parameters and finite elements. In Gruen/Kahmen (eds.): Optical 3D Measurement Techniques V, Wichmann, Heidelberg.

Wester-Ebbinghaus, W. (1981): Zur Verfahrensentwicklung in der Nahbereichsphotogrammetrie. Dissertation, Universität Bonn.

Wester-Ebbinghaus, W. (1983): Einzelstandpunkt-Selbstkalibrierung – ein Beitrag zur Feldkalibrierung von Aufnahmekammern. Deutsche Geodätische Kommission, Reihe C, Nr. 289.

Wester-Ebbinghaus, W. (1985): Verfahren zur Feldkalibrierung von photogrammetrischen Aufnahmekammern im Nahbereich. In Kupfer/Wester-Ebbinghaus (eds.): Kammerkalibrierung in der photogrammetrischen Praxis, Deutsche Geodätische Kommission, Reihe B, Heft Nr. 275, pp. 106-114.

Ziemann, H., El-Hakim, S.F. (1983): On the definition of lens distortion reference data with odd-powered polynomials. Canadian Surveyor 37(3), pp. 135-143.

9.5.4 Multi-media photogrammetry

Fryer, J.G., Fraser, C.S. (1986): On the calibration of underwater cameras. The Photogrammetric Record, 12 (67), pp. 73-85.

Höhle, J. (1971): Zur Theorie und Praxis der Unterwasser-Photogrammetrie. Dissertation, Deutsche Geodätische Kommission, Reihe C, Heft 163.

Kotowski, R. (1988): Phototriangulation in multi-media photogrammetry. International Archives of Photogrammetry and Remote Sensing, Vol. 27, Kyoto.

Maas, H.-G. (1995): New developments in multimedia photogrammetry. In Gruen/Kahmen (eds.): Optical 3D Measurement Techniques III, Wichmann, Heidelberg, pp. 91-97.

Mulsow, C. (2010): A flexible multi-media bundle approach. International Archives of the Photogrammetry, Remote Sensing and Spatial Information Sciences, Vol. 38, Part 5, pp. 472-477.

Putze, T. (2008): Erweiterte Verfahren zur Mehrmedienphotogrammetrie komplexer Körper. In Luhmann/Müller (eds.): Photogrammetrie-Laserscanning-Optische 3D-Messtechnik, Oldenburger 3D-Tage 2008, Wichmann, Heidelberg, pp. 202-209.

Shortis, M.R., Harvey, E.S., Abdo, D.A. (2009): A review of underwater stereo-image measurement for marine biology and ecology applications. In Gibson et al. (eds.): Oceanography and Marine Biology: An Annual Review, Vol. 47, CRC Press, Boca Raton FL, USA, 342 p.

9.5.5 Panoramic photogrammetry

Chapman, D., Kotowski, R. (2000): Methodology for the construction of large image archives of complex industrial structures. Publikationen der DGPF, Band 8, Essen 1999.

Heikkinen, J. (2005): The circular imaging block in close-range photogrammetry. Dissertation, Helsinki University of Technology, 142 p.

Luhmann, T., Tecklenburg, W. (2004): 3-D object reconstruction from multiple-station panorama imagery. ISPRS Panoramic Photogrammetry Workshop, International Archives of the Photogrammetry, Remote Sensing and Spatial Information Sciences, Vol. 34, Part 5/W16, pp. 39-46.

Luhmann, T. (2010): Panorama photogrammetry for architectural applications. Mapping, ISSN 1131-9100, N° 139, pp. 40-45.

Schneider, D. (2008): Geometrische und stochastische Modelle für die integrierte Auswertung terrestrischer Laserscannerdaten und photogrammetrischer Bilddaten. Dissertation, Technische Universität Dresden.

Schneider, D., Maas, H.-G. (2006): A geometric model for linear-array-based terrestrial panoramic cameras. The Photogrammetric Record 21(115), Blackwell Publishing Ltd., Oxford, UK, pp. 198-210.

9.6 Digital image processing

9.6.1 Fundamentals

Canny, J. (1986): A computational approach to edge detection. IEEE Transactions on Pattern Analysis and Machine Intelligence, PAMI-8, Vol. 6, pp. 679-698.

Düppe, R.D., Weisensee, M. (1996): Auswirkungen der verlustbehafteten Bildkompression auf die Qualität photogrammetrischer Produkte. Allgemeine Vermessungs-Nachrichten, Heft 1, pp. 18-30.

Graps, A. (1995): An introduction to wavelets. IEEE Computational Science and Engineering, 2 (2), Los Alamitos, USA.

Lanser, S., Eckstein, W. (1991): Eine Modifikation des Deriche-Verfahrens zur Kantendetektion. 13. DAGM-Symposium, München, Informatik Fachberichte 290, Springer, pp. 151-158.

Pennebaker, W.B., Mitchell, J.L. (1993): JPEG still image data compression standard. Van Nostrad Reinhold, New York.

Phong, B.T. (1975): Illumination for computer generated pictures. Comm. ACM, 18 (8), pp. 311-317.

Tabatabai, A.J., Mitchell, O.R. (1984): Edge location to subpixel values in digital imagery. IEEE Transactions on Pattern Analysis and Machine Intelligence, PAMI-6, No. 2.

Weisensee, M. (1997): Wechselwirkungen zwischen Bildanalyse und Bildkompression. Publikationen der DGPF, Band 5, Oldenburg 1996, pp. 37-45.

9.6.2 Pattern recognition and image matching

Ackermann, F. (1983): High precision digital image correlation. 39. Photogrammetric Week, Stuttgart.

Baltsavias, E. (1991): Multiphoto geometrically constrained matching. Dissertation, ETH Zürich, Institut für Geodäsie und Photogrammetrie, Nr. 49.

Bay, H., Ess, A., Tuytelaars, T., Van Gool, L. (2008): Speeded-up robust features (SURF). Computer vision and image understanding, 110(3), pp. 346-359.

Bethmann, F., Luhmann, T. (2010): Least-squares matching with advanced geometric transformation models. International Archives of Photogrammetry, Remote Sensing and Spatial Information Sciences, Vol. 38, Part 5, pp. 86-91.

Ebner H., Heipke C. (1988): Integration of digital image matching and object surface reconstruction. International Archives for Photogrammetry and Remote Sensing, (27) B11, III, pp. 534-545.

El-Hakim, S.F. (1996): Vision-based automated measurement techniques. In Atkinson (ed.): Close Range Photogrammetry and Machine Vision, Whittles Publishing, Caithness, UK, pp. 180-216.

Förstner, W. (1982): On the geometric precision of digital correlation. International Archives of Photogrammetry and Remote Sensing, Vol. 26/3, pp. 150-166.

Förstner, W. (1985): Prinzip und Leistungsfähigkeit der Korrelation und Zuordnung digitaler Bilder. 40. Photogrammetric Week, Stuttgart.

Förstner, W., Dickscheid, T., Schindler, F. (2009): Detecting interpretable and accurate scale-invariant keypoints. IEEE 12th International Conference on Computer Vision, pp. 2256-2263.

Förstner, W., Gülch, E. (1987): A fast operator for detection and precise location of distinct points, corners and centres of circular features. ISPRS Intercommission Workshop on „Fast Processing of Photogrammetric Data", Interlaken, pp. 281-305.

Furukawa, Y., Ponce, J. (2010): Accurate, dense and robust multi-view stereopsis. IEEE Trans. PAMI, Vol. 32, pp. 1362-1376.

Gruen, A. (1985): Adaptive least squares correlation – a powerful image matching technique. South African Journal of Photogrammetry, Remote Sensing and Cartography, 14 (3), pp. 175-187.

Gruen, A. (1996): Least squares matching: a fundamental measurement algorithm. In Atkinson (ed.): Close Range Photogrammetry and Machine Vision, Whittles Publishing, Caithness, UK, pp. 217-255.

Gruen, A., Baltsavias, E. (1988): Geometrically constrained multiphoto matching. Photogrammetric Engineering and Remote Sensing, 54 (5), pp. 633-641.

Haala, N., Rothermel, M. (2012): Dense multi-stereo matching for high quality digital elevation models. Photogrammetrie-Fernerkundung-Geoinformation, Vol. 4, pp. 331-343.

Heipke C. (1992): A global approach for least squares image matching and surface reconstruction in object space. Photogrammetric Engineering & Remote Sensing (58) 3, pp. 317-323.

Hirschmuller, H. (2005): Accurate and efficient stereo processing by semi-global matching and mutual information. Computer Vision and Pattern Recognition, Vol. 2, pp. 807-814.

Hirschmüller, H. (2008): Stereo processing by semi-global matching and mutual information. IEEE Transactions on Pattern Analysis and Machine Intelligence, 30 (2), pp. 328-341.

Hosseininaveh, A., Robson, S., Boehm, J., Shortis, M., Wenzel, K. (2013): A comparison of dense matching algorithms for scaled surface reconstruction using stereo camera rigs. ISPRS Journal of Photogrammetry and Remote Sensing, Vol. 78, pp. 157-167.

Kersten, T., Lindstaedt, M. (2012): Automatic 3D object reconstruction from multiple images for architectural, cultural heritage and archaeological applications using open-source software and web services. Photogrammetrie–Fernerkundung–Geoinformation, Heft 6, pp. 727-740.

Jazayeri, I., Fraser, C.S. (2008): Interest operators in close-range object reconstruction. International Archives of the Photogrammetry, Remote Sensing and Spatial Information Sciences, Vol. 37, Part B5, pp. 69-74.

Lowe, D.G. (2004): Distinctive image features from scale-invariant keypoints. International Journal of Computer Vision, 60 (2), pp. 91-110.

Luhmann, T. (1986a): Automatic point determination in a réseau scanning system. Symposium ISPRS Commission V, Ottawa.

Luhmann, T. (1986b): Ein Verfahren zur rotationsinvarianten Punktbestimmung. Bildmessung und Luftbildwesen, Heft 4, pp. 147-154.

Maas, H.-G., Grün, A., Papantoniou, D. (1993): Particle tracking in threedimensional turbulent flows - Part I: Photogrammetric determination of particle coordinates. Experiments in Fluids Vol. 15, pp. 133-146.

Maas, H.-G. (1996): Automatic DEM generation by multi-image feature based matching. Int. Archives for Photogrammetry and Remote Sensing, Vol. 31, Part B3, pp. 484-489.

Morel, J. M., Yu, G. (2009): ASIFT: A new framework for fully affine invariant image comparison. SIAM Journal on Imaging Sciences, 2(2), pp. 438-469.

Piechel, J. (1991): Stereobild-Korrelation. In Bähr/Vögtle (eds.): Digitale Bildverarbeitung – Anwendung in Photogrammetrie, Kartographie und Fernerkundung, Wichmann, Heidelberg, pp. 96-132.

Pons, J.-P., Keriven, R., Faugeras, O. (2007): Multi-view stereo reconstruction and scene flow estimation with a global image-based matching score. International Journal of Computer Vision, Vol. 72(2), pp. 179-193.

Remondino, F., El-Hakim, S., Gruen, A., Zhang, L. (2008): Turning images into 3D models - development and performance analysis of image matching for detailed surface reconstruction of heritage objects. IEEE Signal Processing Magazine, Vol. 25(4), pp. 55-65.

Rosten, E., Drummond, T. (2006): Machine learning for high-speed corner detection. European Conference on Computer Vision, Vol. 1, pp. 430-443.

Scharstein D., Szeliski. R. (2002): A taxonomy and evaluation of dense two-frame stereo correspondence algorithms. International Journal of Computer Vision, Vol. 47(1/2/3), pp. 7-42.

Schlüter, M. (1999): Von der 2½D- zur 3D-Flächenmodellierung für die photogrammetrische Rekonstruktion im Objektraum. Dissertation, TU Darmstadt. Deutsche Geodätische Kommission, Reihe C, Nr. 506.

Schneider, C.-T. (1991): Objektgestützte Mehrbildzuordnung. Dissertation, Deutsche Geodätische Kommission, Reihe C, Nr. 375.

Seitz, S.M., Curless, B., Diebel, J., Scharstein, D., Szeliski, R. (2006): A comparison and evaluation of multi-view stereo reconstruction algorithms. Computer Vision and Pattern Recognition, Vol. 1, pp. 519-526.

Shortis, M.R., Seager, J.W., Robson, S., Harvey, E.S (2003): Automatic recognition of coded targets based on a Hough transform and segment matching. Videometrics VII, SPIE Vol. 5013, pp. 202-208.

Shortis, M.R., Clarke, T.A., Short, T. (1994): A comparison of some techniques for the subpixel location of discrete target images. Videometrics III, SPIE Vol. 2350, pp. 239-250.

Triggs, B. (2004): Detecting keypoints with stable position, orientation, and scale under illumination changes. Computer Vision-ECCV, Springer, Berlin, pp. 100-113.

Viola, P., Jones, M. (2001): Rapid object detection using a boosted cascade of simple features. Conference on Computer Vision and Pattern Recognition, IEEE Computer Society , Vol. 1, pp. I/511-I/518.

Wrobel, B, Weisensee, M. (1987): Implementation aspects of facets stereo vision with some applications. Fast Processing of Photogrammetric Data, Interlaken, pp. 259-272.

Wrobel, B. (1987): Facets stereo vision (FAST vision) – A new approach to computer stereo vision and to digital photogrammetry. Fast Processing of Photogrammetric Data, Interlaken, pp. 231-258.

Zhou, G. (1986): Accurate determination of ellipse centers in digital imagery. ASPRS Annual Convention, Vol. 4, March 1986, pp. 256-264.

9.6.3 Range image and point cloud processing

Besl, P.J. (1988). Surfaces in range image understanding. Vol. 27, Springer, New York.

Besl, P.J., McKay, N.D. (1992): A method for registration of 3D-shapes. IEEE Transactions on Pattern Analysis and Machine Intelligence, 14 (2), pp. 239-256.

Böhm, J., Becker, S. (2007): Automatic marker-free registration of terrestrial laser scans using reflectance features. In Grün/Kahmen (eds.): Optical 3D Measurement Techniques, pp. 338-344.

Koenderink, J.J., van Doorn, A.J. (1992): Surface shape and curvature scales. Image and vision computing, 10(8), pp. 557-564.

Newcombe, R.A., Davison, A.J., Izadi, S., Kohli, P., Hilliges, O., Shotton, J., Fitzgibbon, A. (2011): KinectFusion: Real-time dense surface mapping and tracking. 10th IEEE International Symposium on Mixed and Augmented Reality (ISMAR), pp. 127-136.

Remondino, F. (2003): From point cloud to surface: the modeling and visualization problem. International Archives of the Photogrammetry, Remote Sensing and Spatial Information Sciences, Vol. XXXIV, part 5/W10 (CD-ROM).

Wendt, A. (2004): On the automation of the registration of point clouds using the Metropolis algorithm. The International Archives of the Photogrammetry, Remote Sensing and Spatial Information Science, Vol. 35/3.

9.7 Measurement tasks and systems

9.7.1 Overviews

Fellbaum, M, Godding, R. (1995): Economic solutions in photogrammetry through a combination of digital systems and modern estimation techniques. In Gruen/Kahmen (eds.): Optical 3D Measurement Techniques III, Wichmann, Karlsruhe, pp. 362-372.

Heipke C., 1995: State-of-the-art of digital photogrammetric workstations for topographic applications. Photogrammetric Engineering & Remote Sensing (61) 1, pp. 49-56.

Luhmann, T. (2010): Close-range photogrammetry for industrial applications. ISPRS Journal of Photogrammetry and Remote Sensing, Vol. 64/3, pp. 558-569.

9.7.2 Measurement of points and contours

Benning, W., Schwermann, R. (1997): PHIDIAS-MS – Eine digitale Photogrammetrie-applikation unter MicroStation für Nahbereichsanwendungen. Allgemeine Vermessungs-Nachrichten, Heft 1, pp. 16-25.

Beyer, H.A., Uffenkamp, V., van der Vlugt, G. (1995): Quality control in industry with digital photogrammetry. In Gruen/Kahmen (eds.): Optical 3D Measurement Techniques III, Wichmann, Heidelberg, pp. 29-38.

Bösemann, W. (1996): The optical tube measurement system OLM – photogrammetric methods used for industrial automation and process control. International Archives of Photogrammetry and Remote Sensing, Vol. 31, B5, pp. 55-58.

Brown, J., Dold, J. (1995): V-STARS – a system for digital industrial photogrammetry. In Gruen/Kahmen (eds.): Optical 3-D Measurement Techniques III, Wichmann, Heidelberg, pp. 12-21.

Dold, J. (1997): Ein hybrides photogrammetrisches Industriemeßsystem höchster Genauigkeit und seine Überprüfung. Dissertation, Heft 54, Schriftenreihe Studiengang Vermessungswesen, Universität der Bundeswehr, München.

Dold, J. (2002): 3D non-contact metrology for antenna manufacturing and testing. Proceedings 25th ESA Antenna Workshop on Satellite Antenna Technology, ESTEC, Nordwijk, Netherlands

Godding, R., Lehmann, M., Rawiel, G. (1997): Robot adjustment and 3-D calibration – photogrammetric quality control in daily use. In Gruen/Kahmen (eds.): Optical 3-D Measurement Techniques IV, Wichmann, Heidelberg, pp. 158-165.

Godding, R., Luhmann, T. (1992): Calibration and accuracy assessment of a multi-sensor online-photogrammetric system. International Archives of Photogrammetry and Remote Sensing, Vol. 29/5, Washington, pp. 24-29.

Luhmann, T., Broers, H. (1998): An automatic system for the measurement of flat workpieces. ISPRS Conference, Commission V, Hakodate, Japan.

Pettersen, H. (1992): Metrology Norway System – an on-line industrial photogrammetry system. International Archives of Photogrammetry and Remote Sensing, Vol. 24, B5, pp. 43-49.

Schwermann, R., Effkemann, C. (2002): Kombiniertes Monoplotting in Laserscanner- und Bilddaten mit PHIDIAS. In Luhmann (ed.): Photogrammetrie und Laserscanning, Wichmann, Heidelberg, pp. 57-70.

Sinnreich, K., Bösemann, W. (1999): Der mobile 3D-Meßtaster von AICON – ein neues System für die digitale Industrie-Photogrammetrie. Publikationen der DGPF, Band 7, München 1998, pp. 175-181.

Zinndorf, S. (2004): Photogrammetrische Low-Cost-Systeme. Photogrammetrie-Fernerkundung-Geoinformation, Heft 1, pp. 47-52.

9.7.3 Measurement of surfaces

Akça, D., Gruen, A. (2005). Recent advances in least squares 3D surface matching. In Gruen/Kahmen (eds.): Optical 3D Measurement Techniques VII, Vol. II, TU Vienna, pp. 197-206.

Akça, D., Gruen, A. (2007): Generalized least squares multiple 3D surface matching. International Archives of the Photogrammetry, Remote Sensing and Spatial Information Sciences, Vol. 36, Part 3/W52.

Kirschner, V., Schreiber, W., Kowarschik, R., Notni, G. (1997): Selfcalibrating shape-measuring system based on fringe projection. Proc. SPIE 3102, pp. 5-13.

Kludas, T. (1995): Three-dimensional surface reconstruction with the Zeiss photogrammetric industrial measurement system InduSURF Digital. ISPRS Intercommission Workshop „From Pixels to Sequences", Zürich, pp. 285-291.

Lichtenstein, M, Benning, W. (2010): Registrierung von Punktwolken auf der Grundlage von Objektprimitiven. Allgemeine Vermessungs-Nachrichten, Heft 6, pp. 202-207.

Niini, I. (2003): 3-D glass shape measuring system. In Gruen/Kahmen (eds.): Optical 3D Measurement Techniques VI, Vol. II, ETH Zürich, pp. 176-181.

Pfeifer, N., Karel, W. (2008): Aufnahme von 3D-Punktwolken mit hoher zeitlicher Auflösung mittels aktiver Sensoren. In Luhmann/Müller (eds.): Photogrammetrie-Laserscanning-Optische 3D-Messtechnik, Oldenburger 3D-Tage 2008, Wichmann, Heidelberg, pp. 2-13.

Riechmann, W., Thielbeer, B. (1997): Hochaufgelöste Oberflächenerfassung durch Photogrammetrie und Streifenprojektion. Photogrammetrie-Fernerkundung-Geoinformation, Heft 3, pp. 155-164.

Ritter, R. (1995): Optische Feldmeßmethoden. In Schwarz (ed.): Vermessungsverfahren im Maschinen- und Anlagenbau, Wittwer, Stuttgart, pp. 217-234.

Scharsich, P., Pfeifer, T. (1998): Aktive Mehrbildphotogrammetrie für die flächenhafte dreidimensionale Formerfassung. Publikationen der DGPF, Band 6, pp. 279-286.

Schewe, H. (1988): Automatische photogrammetrische Karosserievermessung. Bildmessung und Luftbildwesen, Heft 1, pp. 16-24.

Schreiber W., Notni G. (2000): Theory and arrangements of self-calibrating whole-body three-dimensional measurement systems using fringe projection technique. Optical Engineering 39, pp. 159-169.

Staiger, R., Weber, M. (2007): Die passpunktlose Verknüpfung von Punktwolken – ein Erfahrungsbericht. Terrestrisches Laserscanning, Schriftenreihe des DVW, Band 53, pp. 91-107.

Strutz, T. (1993): Ein genaues aktives Bildtriangulationsverfahren zur Oberflächenvermessung. Dissertation, Fakultät für Maschinenbau, Universität Magdeburg.

Wendt, A. (2004): On the automation of the registration of point clouds using the Metropolis algorithm. International Archives of the Photogrammetry, Remote Sensing and Spatial Information Sciences, Vol. 35/3.

Winter, D., Reich, C. (1997): Video-3D-Digitalisierung komplexer Objekte mit frei beweglichen, topometrischen Sensoren. GMA-Bericht 30, Optische Formerfassung, Langen, pp. 119-127.

9.7.4 Dynamic and mobile systems

Bäumker, M., Brechtken, R., Heimes, F.-J., Richter, T. (1998): Hochgenaue Stabilisierung einer Sensorplattform mit den Daten eines (D)GPS-gestützten Inertialsystems. Zeitschrift für Photogrammetrie und Fernerkundung, Heft 1, pp. 15-22.

Boeder, V., Kersten, T., Hesse, C., Thies, Th., Sauer, A. (2010): Initial experience with the integration of a terrestrial laser scanner into the mobile hydrographic multi-sensor system on a ship. International Archives of the Photogrammetry, Remote Sensing and Spatial Information Sciences, Vol. 38, part 1/W17.

Colomina, I.,_Blázquez, M., Molina, P., Parés, M.E., Wis, M. (2008): Towards a new paradigm for high-resolution low-cost photogrammetry and remote sensing. International Archives of the Photogrammetry, Remote Sensing and Spatial Information Sciences, 37(B1), pp. 1201-1206.

Eisenbeiss, H. (2004). A mini unmanned aerial vehicle (UAV): system overview and image acquisition. International Archives of Photogrammetry. Remote Sensing and Spatial Information Sciences, 36(5/W1).

Eisenbeiss, H. (2009): UAV photogrammetry. Dissertation, ETH Zurich, Switzerland, Mitteilungen Nr.105, doi:10.3929/ethz-a-005939264, 235 p.

Grenzdörffer, G.J., Engel, A., Teichert, B. (2008): The photogrammetric potential of low-cost UAVs in forestry and agriculture. The International Archives of the Photogrammetry, Remote Sensing and Spatial Information Sciences, 31(B3), pp. 1207-1214.

Heckes, J., Mauelshagen, L. (eds.) (1987): Luftaufnahmen aus geringer Flughöhe. Veröffentlichungen aus dem Deutschen Bergbaumuseum, Nr. 41, 135 p.

Kersten, T., Büyüksalih, G., Baz, I., Jacobsen, K. (2009): Documentation of Istanbul historic peninsula by kinematic terrestrial laser scanning. The Photogrammetric Record, 24(126), pp. 122-138.

Kutterer, H. (2010): Mobile mapping. In Vosselman/Maas (eds.): Airborne and Terrestrial Laser Scanning, Whittles Publishing, Caithness, UK, pp. 293-311.

Maas, H.-G., Grün, A., Papantoniou, D. (1993): Particle tracking in threedimensional turbulent flows - Part I: Photogrammetric determination of particle coordinates. Experiments in Fluids Vol. 15, pp. 133-146.

Maas, H.-G. (1997): Dynamic photogrammetric calibration of industrial robots. Videometrics V, SPIE Proceedings Series, Vol. 3174.

Maas, H.-G. (2005): Werkzeuge und Anwendungen der photogrammetrischen 3D-Bewegungsanalyse. In Luhmann (ed.): Photogrammetrie-Laserscanning-Optische 3D-Messtechnik, Oldenburger 3D-Tage 2005, Wichmann, Heidelberg, pp. 207-213.

Raguse, K., Heipke, C. (2005): Photogrammetrische Auswertung asynchroner Bildsequenzen. In Luhmann (ed.): Photogrammetrie-Laserscanning-Optische 3D-Messtechnik, Oldenburger 3D-Tage 2005, Wichmann, Heidelberg, pp. 14-21.

Remondino, F., Barazzetti, L., Nex, F., Scaioni, M., Sarazzi, D. (2011): UAV photogrammetry for mapping and 3D modelling – current status and future perspectives. International Archives of the Photogrammetry, Remote Sensing and Spatial Information Sciences, 38(1).

Siebert, S., Klonowski, J., Neitzel, F. (2009): Unmanned Aerial Vehicles (UAV) – historische Entwicklung, rechtliche Rahmenbedingungen und Betriebskonzepte. In Luhmann/Müller (eds.): Photogrammetrie-Laserscanning-Optische 3D-Messtechnik, Oldenburger 3D-Tage 2009, Wichmann, Heidelberg, pp. 376-383.

van Blyenburgh, P. (1999): UAVs: an Overview. Air & Space Europe, I, 5/6, pp. 43-47.

9.8 Quality issues and optimization

9.8.1 Project planning and simulation

Abdullah, Q., Bethel, J., Hussain, M., Munjy, R. (2013): Photogrammetric project and mission planning. In McGlone (ed.): Manual of Photogrammetry, 6th ed., American Society of Photogrammetry & Remote Sensing, Bethesda, Maryland, pp. 1187-1219.

Alsadik, B.S., Gerke, M., Vosselman, G. (2012): Automated camera network design for 3D modeling of cultural heritage objects. Journal of Cultural Heritage.

Fraser, C.S. (1984): Network design considerations for non-topographic photogrammetry. Photogrammetric Engineering and Remote Sensing. 50(8), pp. 1115-1126.

Fraser, C.S. (1987): Limiting error propagation in network design. Photogrammetric Engineering and Remote Sensing, 53(5), pp. 487-493.

Fraser, C.S. (1996): Network design. In Atkinson (ed.): Close Range Photogrammetry and Machine Vision, Whittles Publishing, Caithness, UK, pp. 256-281.

Fraser, C. S., Woods, A., Brizzi, D. (2005): Hyper redundancy for accuracy enhancement in automated close range photogrammetry. The Photogrammetric Record, 20 (111), pp. 205-217.

Hastedt, H. (2004): Monte Carlo simulation in close-range photogrammetry. International Archives of the Photogrammetry, Remote Sensing and Spatial Information Sciences, Vol. 35, part B5, pp. 18-23.

Mason, S.O. (1994): Expert system-based design of photogrammetric networks. Dissertation, Mitteilungen Nr. 53, Institut für Geodäsie und Photogrammetrie, ETH Zürich.

Schmitt, R., Fritz, P., Jatzkowski, P., Lose, J., Koerfer, F., Wendt, K. (2008): Abschätzung der Messunsicherheit komplexer Messsysteme mittels statistischer Simulation durch den Hersteller. In: Messunsicherheit praxisgerecht bestimmen, 4. Fachtagung Messunsicherheit, VDI-Wissensforum, Düsseldorf.

Scott, W.R., Roth, G., Riverst, J.-F. (2003): View planning for automated three-dimensional object reconstruction and inspection. ACM Computing Surveys, Vol. 35(1), pp. 64-96.

Trummer, M., Munkelt, C., Denzler, J. (2010): Online next-best-view planning for accuracy optimization using an extended criterion. Proc. IEEE International Conference on Pattern Recognition (ICPR'10), pp. 1642–1645.

Zinndorf, S. (1986): Optimierung der photogrammetrischen Aufnahmeanordnung. Dissertation, Deutsche Geodätische Kommission, Reihe C, Heft Nr. 323.

9.8.2 Quality

Hennes, M. (2007): Konkurrierende Genauigkeitsmaße – Potential und Schwächen aus der Sicht des Anwenders. Allgemeine Vermessungs-Nachrichten, Heft 4, pp. 136-146.

Luhmann, T., Bethmann, F., Herd. B., Ohm, J. (2010): Experiences with 3D reference bodies for quality assessment of free-form surface measurements. International Archives of Photogrammetry, Remote Sensing and Spatial Information Sciences, Vol. 38, Part 5, pp. 405-410.

Luhmann, T. (2011): 3D imaging - how to achieve highest accuracy. In Remondino/Shortis (eds.): Videometrics, Range Imaging, and Applications XI, Proc. SPIE 8085, 808502 (2011); doi:10.1117/12.892070.

Rautenberg, U., Wiggenhagen, M. (2002): Abnahme und Überwachung photogrammetrischer Messsysteme nach VDI 2634, Blatt 1. Photogrammetrie-Fernerkundung-Geoinformation, Heft 2, pp. 117-124.

Schwenke, H., Wäldele, F., Wendt, K. (1997): Überwachung und Meßunsicherheit von optischen 3D-Meßsystemen. GMA-Bericht 30, Optische Formerfassung, Langen, pp. 271-280.

Staiger, R. (2004): Was ist eigentlich Metrologie? In Luhmann (ed.): Photogrammetrie-Laserscanning-Optische 3D-Messtechnik, Oldenburger 3D-Tage 2004, Wichmann, Heidelberg, pp. 2-11.

Wiggenhagen, M., Raguse, K. (2003): Entwicklung von Kenngrößen zur Qualitätsbeurteilung optischer Prozessketten. Photogrammetrie-Fernerkundung-Geoinformation, Heft 2, pp. 125-134.

9.9 Applications

9.9.1 Architecture, archaeology, city models

Akça, D., Grün, A., Breuckmann, B., Lahanier, C. (2007): High-definition 3D-scanning of arts objects and paintings. In Grün/Kahmen (eds.): Optical 3D Measurement Techniques VIII, Vol. II, ETH Zürich, pp. 50-58.

Albertz, J. (ed.) (2002): Surveying and documentation of historic buildings – monuments – sites. International Archives of Photogrammetry and Remote Sensing, Vol. 34, Part 5/C7.

Albertz, J., Wiedemann, A. (eds.) (1997): Architekturphotogrammetrie gestern-heute-morgen. Technische Universität Berlin.

Brenner, C., Haala, N. (1998): Fast production of virtual reality city models. International Archives of Photogrammetry and Remote Sensing, Vol. 32, Part B4, pp. 77-84.

Dallas, R.W.A. (1996): Architectural and archaeological photogrammetry. In Atkinson (ed.): Close Range Photogrammetry and Machine Vision, Whittles Publishing, Caithness, UK, pp. 283-302.

El-Hakim, S.F., Beraldin, J.-A. (2007): Sensor integration and visualisation. In Fryer et al. (eds.): Applications of 3D Measurement from Images, Whittles Publishing, Caithness, UK, pp. 259-298.

El-Hakim, S., Gonzo, L. Voltolini, F., Girardi, S., Rizzi, A., Remondino, F., Whiting, E. (2007): Detailed 3D modelling of castles. International Journal of Architectural Computing, Vol.5(2), pp. 199-220.

Fellbaum, M., Hau, T. (1996): Photogrammetrische Aufnahme und Darstellung des Doms von Siena. Zeitschrift für Photogrammetrie und Fernerkundung, Heft 2, pp. 61-67.

Gruen, A., Remondino, F., Zhang, L. (2004): Photogrammetric reconstruction of the Great Buddha of Bamiyan, Afghanistan. The Photogrammetric Record, Vol. 19(107), pp. 177-199.

Guidi, G., Remondino, F., Russo, M., Menna, F., Rizzi, A., Ercoli, S. (2009): A multi-resolution methodology for the 3D modeling of large and complex archaeological areas. International Journal of Architectural Computing, 7 (1), pp. 40-55.

Kersten, T., Lindstaedt, M., Vogt, B. (2009): Preserve the past for the future - terrestrial laser scanning for the documentation and deformation analysis of Easter Island's moai. Photogrammetrie-Fernerkundung-Geoinformation, Heft 1, pp. 79-90.

Kersten, T., Sternberg, H., Mechelke, K. (2009): Geometrical building inspection by terrestrial laser scanning. Civil Engineering Surveyor - The Journal of the Chartered Institution of Civil Engineering Surveyors, pp. 26-31.

Korduan, P., Förster, T., Obst, R. (2003): Unterwasser-Photogrammetrie zur 3D-Rekonstruktion des Schiffswracks „Darßer Kogge". Photogrammetrie-Fernerkundung-Geoinformation, Heft 5, pp. 373-381.

Kotowski, R., Meid, A., Peipe, J., Wester-Ebbinghaus, W. (1989): Photogrammetrische Bauaufnahme der „Kirchen von Siena" – Entwicklung eines Konzepts zur Vermessung von Großbauwerken. Allgemeine Vermessungs-Nachrichten, Heft 4, pp. 144-154.

Patias, P. (2007): Cultural heritage documentation. In Fryer et al. (eds.): Applications of 3D Measurement from Images, Whittles Publishing, Caithness, UK, pp. 225-257.

Remondino, F. (2007): Image-based detailed 3D geometric reconstruction of heritage objects. Publikationen der DGPF, Band 16, pp. 483-492.

Remondino, F., El-Hakim, S. (2006): Image-based 3D modelling: a review. The Photogrammetric Record, 21(115), pp. 269-291.

Remondino, F., Rizzi, A., Barazzetti, L., Scaioni, M., Fassi, F., Brumana, R., Pelagotti, A. (2011): Review of geometric and radiometric analyses of paintings. The Photogrammetric Record, Vol. 26(136), pp. 439-461.

Remondino, F. (2011): Heritage recording and 3D modeling with photogrammetry and 3D scanning. Remote Sensing, 3(6), pp. 1104-1138.

Streilein, A. (1998): Digitale Photogrammetrie und CAAD. Dissertation, Mitteilungen Nr. 68, Institut für Geodäsie und Photogrammetrie, ETH Zürich, 160 p.

Tecklenburg, W., Jantos, R., Luhmann, T. (2005): Untersuchungen zur Nutzung von Bildclustern für die 2D/3D-Auswertung von fast ebenen Aufnahmesituationen aus erhöhten Aufnahmestandpunkten. Publikationen der DGPF, Band 17, Rostock, pp. 293-300.

Wiedemann, A. (1997): Orthophototechnik in der Architekturphotogrammetrie – Möglichkeiten und Grenzen. In Albertz/Wiedemann (eds.): Architekturphotogrammetrie gestern-heute-morgen, Technische Universität Berlin, pp. 79-94.

Wiedemann, A., Tauch, R. (2005): Mosaikbildung in der Architekturphotogrammetrie. In Luhmann (ed.): Photogrammetrie-Laserscanning-Optische 3D-Messtechnik, Oldenburger 3D-Tage 2005, Wichmann, Heidelberg, pp. 116-121.

Wiggenhagen, M., Elmhorst, A., Wissmann, U. (2004): Digitale Nahbereichsphotogrammetrie zur Objektrekonstruktion der Bremer Hanse Kogge. Publikationen der DGPF, Band 13, Halle, pp. 383-390.

9.9.2 Engineering and industrial applications

Abraham, S., Wendt, A., Nobis, G., Uffenkamp, V., Schommer, S. (2010): Optische 3D-Messtechnik zur Fahrwerksmessung in der Kfz-Werkstatt. In Luhmann/Müller (eds.): Photogrammetrie-Laserscanning-Optische 3D-Messtechnik, Oldenburger 3D-Tage 2010, Wichmann, Heidelberg, pp. 176-183.

Beyer, H.A. (1995): Quality control in industry with digital photogrammetry. In Gruen/Kahmen (eds.): Optical 3D Measurement Techniques III, Wichmann, Heidelberg, pp. 29-38.

Bieder, H. (1997): Optische Formerfassung und Reverse Engineering – Anforderungen an die Software. GMA-Bericht 30, Optische Formerfassung, Langen, pp. 139-146.

Bösemann, W., Drohne, U., Schneider, C.-T. (1997): Application of an integrated photogrammetric system for the Airbus A319-A321 wing assembly. In Gruen/Kahmen (eds.): Optical 3D Measurement Techniques IV, Wichmann, Heidelberg, pp. 151-157.

Bösemann, W., Uffenkamp, V. (2004): Industrieanwendungen der Nahbereichs-photogrammetrie. Photogrammetrie-Fernerkundung-Geoinformation, Heft 1, pp. 29-38.

Chapman, D. (2013): As-built modelling of large and complex industrial facilities. In McGlone (ed.): Manual of Photogrammetry, 6th ed., American Society of Photogrammetry & Remote Sensing, Bethesda, Maryland, pp. 1082-1093.

Dold, J. (1999): Stand der Technik in der Industriephotogrammetrie. Photogrammetrie-Fernerkundung-Geoinformation, Heft 2, pp. 113-126.

Fraser, C.S. (1986): Microwave antenna measurement. Photogrammetric Engineering & Remote Sensing, 52(10), pp. 1627-1635.

Fraser, C.S. (1992): Photogrammetric measurement to one part in a million. Photogrammetric Engineering and Remote Sensing, 58(3), pp. 305-310.

Fraser, C.S., Mallison, J.A. (1992): Dimensional characterization of a large aircraft structure by photogrammetry. Photogrammetric Engineering and Remote Sensing, 58(5), pp. 539-543.

Fraser, C.S. (1996): Industrial measurement applications. In Atkinson (ed.): Close Range Photogrammetry and Machine Vision, Whittles Publishing, Caithness, UK, pp. 329-361.

Fraser, C.S., Riedel, B. (2000): Monitoring the thermal deformation of steel beams via vision metrology. ISPRS Journal of Photogrammetry & Remote Sensing, 55(4), pp. 268-276.

Fraser, C.S. (2001): Vision metrology for industrial and engineering measurement. Asian Journal of Geoinformatics, 2(1), pp. 29-37.

Fraser, C.S., Brizzi, D. (2002): Deformation monitoring of reinforced concrete bridge beams. Proceedings, 2nd International Symposium on Geodesy for Geotechnical and Structural Engineering, Berlin, 21-24 May, pp. 338-343.

Fraser, C.S., Brizzi, D., Hira, A. (2003): Vision-based multi-epoch deformation monitoring of the atrium of Federation Square. Proceedings, 11th International Symposium on Deformation Measurements, Santorini, Greece, 25-28 May, pp. 599-604.

Fraser, C.S. (2007): Structural monitoring. In Fryer/Mitchell/Chandler (eds.): Applications of 3D Measurement from Images, Whittles Publishing, Caithness, UK, pp. 37-64.

Fraser, C.S. (2013): Industrial and engineering measurement. In McGlone (ed.): Manual of Photogrammetry, 6th ed., American Society of Photogrammetry & Remote Sensing, Bethesda, Maryland, pp. 1075-1082.

Hampel, U., Maas, H.-G. (2009): Cascaded image analysis for dynamic crack detection in material testing. ISPRS Journal of Photogrammetry and Remote Sensing, Vol. 64, pp. 345-350.

Kahlmann, T., Bösemann, W., Godding, R. (2008): Erfassung dynamischer Prozesse im industriellen Umfeld. Publikationen der DGPF, Band 17, pp. 47-57.

Luhmann, T. (2010): Close range photogrammetry for industrial applications. ISPRS Journal for Photogrammetry and Remote Sensing, Vol. 64/3, pp. 558-569.

Luhmann, T., Tecklenburg, W. (1997): Hybride photogrammetrische Aufnahme großer Ingenieurobjekte. Photogrammetrie-Fernerkundung-Geoinformation, Heft 4, pp. 243-252.

Luhmann, T., Voigt, A. (2010): Automatische projektive Bildentzerrung am Beispiel der bildgestützten Planung von Solardachanlagen. Publikationen der DGPF, Band 19, S. 469-476.

Maas, H.-G., Kersten, T. (1994): Digital close-range photogrammetry in a shipyard for dimensional checking and control. SPIE Proceedings Videometrics III, Vol. 2350, pp. 108-114.

Miebach, R. (2002): Anwendungen im Schiffbau. In Löffler (ed.): Handbuch Ingenieurgeodäsie – Maschinen und Anlagenbau, Wichmann, Heidelberg, pp. 355-387.

Fraser, C.S. (2013): Industrial and engineering measurement. In McGlone (ed.): Manual of Photogrammetry, 6th ed., American Society of Photogrammetry & Remote Sensing, Bethesda, Maryland, pp. 1075-1082.

Przybilla, H.-J. (1999): Sensorvermessung im Industrie- und Anlagenbau. VDI-Berichte Nr. 1454, Moderne Sensorik für die Bauvermessung, VDI Verlag, Düsseldorf, pp. 173-183.

Przybilla, H.-J., Woytowicz, D. (2004): Dokumentation industrieller Anlagen: Vom 2D-Bestandsplan über das GIS zur virtuellen Realität – eine Standortbestimmung. Photogrammetrie-Fernerkundung-Geoinformation, Heft 1, pp. 53-58.

Riechmann, W., Ringel, H. (1995): Photogrammetrische Verformungsanalysen in der Karosserie-Entwicklung. Optische Meßmethoden in der modernen Automobilentwicklung, Haus der Technik, Essen.

Robson, S., Shortis, M. (2007): Engineering and manufacturing. In Fryer et al. (eds.): Applications of 3D Measurement from Images, Whittles Publishing, Caithness, UK, pp. 65-101.

Sternberg, H., Kersten, T., Jahn, I., Kinzel, R. (2004): Terrestrial 3D laser scanning - data acquisition and object modelling for industrial as-built documentation and architectural applications. International Archives of the Photogrammetry, Remote Sensing and Spatial Information Sciences, Vol. 35, Part B2, pp. 942-947.

9.9.3 Medicine, forensics, earth sciences

D'Appuzo, N. (2009): Recent advances in 3D full body scanning with applications to fashion and apparel. In Grün/Kahmen (eds.): Optical 3-D Measurement Techniques IX, Vol. II, TU Vienna, pp. 243-253.

Dietrich, R., Maas, H.-G., Baessler, M., Rülke, A., Richter, A., Schwalbe, E., Westfeld, P. (2007): Jakobshavn Isbrae, West Greenland: Flow velocities and tidal interaction of the front area from 2004 field observations. Journal of Geophysical Research, 112 (f3): F03S21.

Fraser, C.S., Hanley, H.B., Cronk, S. (2005): Close-range photogrammetry for accident reconstruction. In Grün/Kahmen (eds.): Optical 3D Measurement Techniques VII, Vol. II, TU Vienna, pp. 115-123.

Fryer, J.G. (2007): Forensic photogrammetry. In Fryer et al. (eds.): Applications of 3D Measurement from Images, Whittles Publishing, Caithness, UK, pp. 103-137.

Gäbel, H. (1993): Photogrammetrische Verfahren zur Erfassung von menschlichen Körperoberflächen. Dissertation, Deutsche Geodätische Kommission, Reihe C, Nr. 405.

Krüger, T. (2008): Optische 3D-Messtechnik in der dentalen Implantologie. In Luhmann/Müller (eds.): Photogrammetrie-Laserscanning-Optische 3D-Messtechnik, Oldenburger 3D-Tage 2008, Wichmann, Heidelberg, pp. 31-42.

Maas, H.-G., Schwalbe, E., Dietrich, R., Bässler, M., Ewert, H. (2008): Determination of spatio-temporal velocity fields on glaciers in West-Greenland by terrestrial image sequence analysis. International Archives of Photogrammetry, Remote Sensing and Spatial Information Science, Vol. 37, Part B8, pp. 1419-1424.

Mitchell, H. (2007): Medicine and sport: measurement of humans. In Fryer et al. (eds.): Applications of 3D Measurement from Images, Whittles Publishing, Caithness, UK, pp. 171-199.

Mugnier, C.J., Gillen, L., Lucas, S.P., Walford, A. (2013): Forensic photogrammetry. In McGlone (ed.): Manual of Photogrammetry, 6th ed., American Society of Photogrammetry & Remote Sensing, Bethesda, Maryland, pp. 1108-1130.

Näther, S. (2010): 3-D-Vermessungstechniken im Einsatz bei Polizei und Rechtsmedizin in der Schweiz. Forum, Zeitschrift des Bundes für Öffentlich bestellte Vermessungsingenieure, Heft 1, pp. 192-201.

Newton, I., Mitchell, H.L. (1996): Medical photogrammetry. In Atkinson (ed.): Close Range Photogrammetry and Machine Vision, Whittles Publishing, Caithness, UK, pp. 303-327.

Rieke-Zapp, D., Nearing, M.A. (2005): Digital close range photogrammetry for measurement of soil erosion. Photogrammetric Record, 20 (109), pp. 69-87.

Shortis, M.R., Harvey, E.S., Abdo, D.A. (2009): A review of underwater stereo-image measurement for marine biology and ecology applications. In Gibson et al. (eds.): Oceanography and Marine Biology: An Annual Review, Vol. 47, CRC Press, Boca Raton FL, USA, 342 p.

Wulff, C., Steineck, D., Krell. A., Saraceno, T. (2010): Zu Hause bei der schwarzen Witwe – Pilotprojekt zur stereoskopischen Vermessung eines dreidimensionalen Spinnennetzes. In Luhmann/Müller (eds.): Photogrammetrie-Laserscanning-Optische 3D-Messtechnik, Oldenburger 3D-Tage 2010, Wichmann, Heidelberg, pp. 268-271.

9.10 Other sources of information

9.10.1 Standards and guidelines

DIN 1319: Fundamentals of metrology. Beuth, Berlin.

DIN 18716: Photogrammetry and remote sensing. Beuth, Berlin.

DIN 18709: Concepts, abbreviations and symbols in surveying. Beuth, Berlin.

DIN 18710: Ingenieurvermessung. Teil 1-4. Beuth, Berlin.

DIN V ENV 13005 (1999): Leitfaden zur Angabe der Unsicherheit beim Messen. ENV 13005, Beuth, Berlin.

GUM (1995): ISO Guide to the Expression of Uncertainty in Measurement (GUM). International Bureau of Weights and Measures, ISBN 92 67 10188 9.

ISO 10360: Geometrical Product Specifications (GPS) - Acceptance and reverification tests for coordinate measuring machines (CMM), Beuth, Berlin.

VDI/VDE 2617: Accuracy of coordinate measuring machines; characteristics and their checking. VDI/VDE guide line, Part 1-9, Beuth, Berlin.

VDI/VDE 2634: Optical 3D measuring systems. VDI/VDE guide line, part 1-3, Beuth, Berlin.

VIM (2010): International dictionary of metrology. 3rd ed., Beuth, Berlin.

9.10.2 Working groups and conferences

ASPRS (The American Society for Photogrammetry and Remote Sensing, USA):
Publications: Photogrammetric Engineering and Remote Sensing
www.asprs.org

CIPA (Comité International de Photogrammétrie Architecturale):
Publications and conference proceedings
http://cipa.icomos.org/

CMCS (Coordinate Metrology Systems Conference):
Publications and conference proceedings
www.cmsc.org

DAGM (Deutsche Arbeitsgemeinschaft für Mustererkennung):
Publications: Mustererkennung, Springer (proceedings of annual conferences)
www.dagm.de

DGPF (Deutsche Gesellschaft für Photogrammetrie, Fernerkundung und Geoinformation):
Publications: Bildmessung und Luftbildwesen (until 1989), Zeitschrift für Photogrammetrie und Fernerkundung (until 1997), Photogrammetrie-Fernerkundung-Geoinformation (since 1997); Publikationen der DGPF (proceedings of annual conferences)
www.dgpf.de

DGZfP (Deutsche Gesellschaft für zerstörungsfreie Prüfung):
Publications: Technical guidelines, conference proceedings
www.dgzfp.de

ISPRS (International Society for Photogrammetry and Remote Sensing):
Commission III: Photogrammetric Computer Vision and Image Analysis
Commission V: Close-Range Imaging, Analysis and Applications
Publications: International Archives of the Photogrammetry, Remote Sensing and Spatial Information Sciences; ISPRS Journal of Photogrammetry and Remote Sensing
www.isprs.org

SPIE (The International Society for Optical Engineering):
Publications and conference proceedings
www.spie.org

The Remote Sensing and Photogrammetry Society (UK):
Publications: The Photogrammetric Record
www.rspsoc.org

VDI/VDE-GMA (VDI/VDE-Gesellschaft für Mess- und Automatisierungstechnik):
Publications: guidelines, conference proceedings.
www.vdi.de

Abbreviations

ADC	analogue-to-digital converter
AGC	automatic gain control
ASIFT	affine invariant SIFT
EP	entrance pupil
ASCII	American Standard Code for Information Interchange
AVI	audio video interleaved
BIM	building information management
BRDF	bidirectional reflection distribution function
CAAD	computer-aided architectural design
CAD	computer-aided design
CAM	computer-aided manufacturing
CCD	charge-coupled device
CCIR	International Radio Consultative Committee
CD-ROM	compact disk – read-only memory
CID	charge injection device
CIE	Commission Internationale de l'Éclairage (International Commission on Illumination)
CIPA	Comité International de Photogrammétrie Architecturale
CMM	coordinate measuring machine
CMOS	complementary metal oxide semi-conductor
CT	computer tomogram, -tomography
CTF	contrast transfer function
DAGM	Deutsche Arbeitsgemeinschaft für Mustererkennung e.V.
DCT	discrete cosine transform
DGPF	Deutsche Gesellschaft für Photogrammetrie, Fernerkundung und Geoinformation e.V.
DGPS	differential GPS
DGZfP	Deutsche Gesellschaft für zerstörungsfreie Prüfung e.V.
DIC	digital image correlation
DLT	direct linear transformation
DMD	digital mirror device

DOF	degree(s) of freedom
DOM	digital surface model
DRAM	dynamic random access memory
DTP	desktop publishing
DVD	digital versatile (video) disk
DXF	Autocad Data Exchange Format
EPS	encapsulated postscript
FAST	features from accelerated segment test
FFT	full frame transfer or fast Fourier transform
FMC	forward motion compensation
FOV	field of view
FPGA	field programmable gate array
FT	frame transfer
GIF	graphic interchange format
GIS	geo(graphic) information system
GMA	Gesellschaft für Mess- und Automatisierungstechnik e.V.
GPS	global positioning system
HDR	high dynamic range (imaging)
HDTV	high definition television
HSL	hue, saturation, lightness
ICP	iterative closest point
IEEE	Institute of Electrical and Electronical Engineers
IFOV	instantaneous field of view
IGS	image guided surgery
IHS	intensity, hue, saturation
IL	interline transfer
INS	inertial navigation system
IR	infrared
ISPRS	International Society for Photogrammetry and Remote Sensing
JPEG	Joint Photographic Expert Group
LCD	liquid crystal display
LCOS	liquid crystal on silicon

LED	light emitting diode
LMA	Levenberg-Marquardt algorithm
LME	length measurement error
LoG	Laplacian of Gaussian
LSM	least squares matching
LUT	lookup table
LZW	Lempel-Ziv-Welch (compression)
MoCap	motion capture
MOS	metal oxide semiconductor
MPE	maximum permitted error
MPEG	Motion Picture Expert Group
MR	magnetic resonance
MRT	magnetic resonance tomography
MTF	modulation transfer function
NC	numerically controlled (machine)
PLL	phase-locked loop
PLY	polygon file format
PNG	portable network graphics
PSF	point spread function
RANSAC	random sample consensus
REM	raster electron microscope
RGB	red, green, blue
RGBD	red, green, blue, distance
RMS	root mean square
RMSE	root mean square error
RP	resolving power
RPV	remote piloted vehicle
SCSI	small computer systems interface
SFM	structure from motion
SGM	semi-global matching
SIFT	scale invariant feature transform
SLR	single lens reflex

SMR	spherically mounted retro-reflector
SNR	signal-to-noise ratio
SPIE	The International Society for Optical Engineering
STL	stereo lithography
SURF	speed-up robust features
SUSAN	smallest univalue segment assimilating nucleus
SVD	single value decomposition
TIFF	tagged image file format
TLS	terrestrial laserscanning
TOF	time-of-flight
UAV	unmanned aerial vehicle (also: unpiloted)
USB	universal serial bus
UTM	universal transverse Mercator (projection)
UV	ultraviolet
VDE	Verband der Elektrotechnik, Elektronik, Informationstechnik e.V.
VDI	Verband Deutscher Ingenieure e.V.
VLL	vertical line locus
VR	virtual reality
VRML	virtual reality modelling language
WLAN	wireless local area network
WMV	Windows media video

Image sources

ABW Automatisierung + Bildverarbeitung Dr. Wolf GmbH, Frickenhausen, Germany
3.163d

Aeroscout GmbH, Hochdorf, Switzerland
6.45

AICON 3D Systems GmbH, Braunschweig, Germany
3.119, 3.123b, 3.159, 6.4, 6.16, 6.21, 6.22, 6.23, 6.27, 8.27, 8.39, 8.44, 8.55

Airbus Operations GmbH, Hamburg, Germany
8.42, 8.43

Automated Precisions Inc. (API), Rockville, MD, USA
6.35b

AXIOS 3D Services Gmbh, Oldenburg, Germany
3.123a, 3.153b, 3.162, 4.12, 6.11, 6.24, 8.58a

BrainLAB AG, Heimstetten, Germany
8.58b, 8.59a

Breuckmann GmbH, Meersburg, Germany
3.180, 8.14, 8.15, 8.17, 8.18

Casio Europe GmbH, Norderstedt, Germany
3.111a

Carl Zeiss, ZI Imaging, Oberkochen, Jena, Germany
1.26, 1.27, 1.28, 1.29, 1.30, 1.31, 1.32, 1.38

CERN, Switzerland
8.41 (courtesy: Claudia Marcelloni)

Creaform, Québec, Canada
6.33

Cyclomedia, Netherlands
3.136, 3.137

Dalsa Inc., Waterloo, Ontario, Canada
3.74a

DMT Deutsche Montan Technologie, German Mining Museum, Bochum, Germany
6.48

EOS Systems Inc., Vancouver, Canada
1.44

FBK Trento, Polytecnico di Milano, Italy
8.11, 8.12, 8.13, 8.16

Fokus GmbH Leipzig, Germany
4.25

Fraunhofer-Institut für Fabrikbetrieb und -automatisierung (IFF), Magdeburg, Germany
6.10

GDV Systems GmbH, Bad Schwartau, Germany
1.15, 8.43, 8.45, 8.46, 8.47, 8.48

gispro Sp. z o.o., Szczecin, Poland
6.43

GOM Gesellschaft für Optische Messtechnik mbH, Braunschweig, Germany
3.184, 6.20, 6.31, 8.40, 8.50

GSI Geodetic Services Inc., Melbourne, Florida, USA
1.35, 1.42, 3.116, 3.154, 3.163ab, 4.70, 5.2, 6.15, 8.42

Hasselblad Svenska AB, Göteborg, Sweden
3.115

Hexagon, Leica Geosystems (Wild, Kern, LH Systems, Cyra), Heerbrugg, Switzerland
1.39, 3.48, 3.172, 3.175, 3.178, 4.79, 4.81, 4.84

Hometrica Consulting, Zurich, Switzerland
6.40

High Speed Vision GmbH (hs vision), Karlsruhe, Germany
3.120

Industrieanlagen-Betriebsgesellschaft mbH (IABG), Ottobrunn, Germany
8.47a

Institut für Angewandte Photogrammetrie und Geoinformatik (IAPG), Jade Hochschule Oldenburg, Germany
1.1, 1.12, 1.13, 1.16, 1.17, 3.3, 3.6, 3.13, 3.15, 3.17, 3.22, 3.26, 3.40, 3.42, 3.43, 3.66, 3.79, 3.89, 3.96, 3.100, 3.101, 3.107, 3.133, 3.135, 3.143, 3.145, 3.146, 3.147, 3.154, 3.165, 3.166, 3.168, 3.174, 3.183, 4.6, 4.16, 4.17, 4.22, 4.24, 4.26, 4.27, 4.28, 4.33, 4.37, 4.51, 4.90, 4.93, 4.94, 4.95, 5.2b, 5.11, 5.22,5.26, 5.53, 5.59, 6.8, 6.9, 6.13, 6.28bc, 6.29, 6.30, 6.47, 7.4, 7.11, 7.20, 7.22, 8.1, 8.5, 8.8, 8.9, 8.10, 8.23, 8.24, 8.30, 8.31, 8.32, 8.33, 8.36, 8.37

Institut für Geodäsie und Photogrammetrie (IGP), ETH Zurich, Switzerland
1.22, 4.86, 6.39

Institut für Geologie, Institut für Rechtsmedizin, University of Bern, Switzerland
8.64, 8.71, 8.72

Institut für Photogrammetrie und Fernerkundung (IFP), TU Dresden, Germany
8.34, 8.35, 8.68, 8.69, 8.70

Institut für Photogrammetrie und GeoInformation (IPI), Leibniz University Hanover, Germany
8.19, 8.20, 8.21, 8.22, 8.28, 8.29

Institut für Photogrammetrie und Kartographie (IPK), TU Darmstadt, Germany
8.65, 8.67a

INVERS – Industrievermessung und -systeme, Essen, Germany
1.21, 8.25, 8.26

Jenoptik Laser-Optik-Systeme GmbH, Jena, Germany
3.126, 3.128

Kamera-Systemtechnik GmbH, Pirna, Germany
3.132

Kodak AG, Stuttgart, Germany
1.41

Leica Camera AG, Solms, Germany
3.113

Mamiya GmbH, München, Germany
3.114

Mapvision Ltd, Espoo, Finland
1.37, 6.25

Messbildstelle GmbH, Dresden, Germany
8.3, 8.4, 8.6, 8.7

Metronor AS, Nesbru, Norway
6.2, 6.19

Microdrones GmbH, Siegen, Germany
6.44

Nikon Metrology, Metris, Leuven, Belgium
3.112, 3.125, 6.5b

Northern Digital Inc., Waterloo, Ontario, Canada
6.34a

Österreichisches Archäologisches Institut, Austria
8.17

Pentacon GmbH, Dresden, Germany
3.129

PCO AG, Kehlheim, Germany
3.118

Phocad Ingenieurgesellschaft mbH, Aachen, Germany
1.43, 4.23, 6.13

Photometrix, Kew, Australia
1.19, 6.14, 8.63

Porsche AG, Stuttgart, Germany
6.37

Precision Instruments, Smith & Nephews, Aarau, Switzerland
8.58c, 8.59b, 8.60

Qualisys AB, Göteburg, Sweden
1.20, 6.41

Riegl Laser Measurement Systems GmbH, Horn, Austria
3.173, 4.23

Rollei Fototechnic GmbH, Braunschweig, Rolleimetric GmbH, Germany
1.33, 1.31, 1.36, 1.40, 3.117, 6.5a, 8.61, 8.62

Sinar Photography AG, Zürich, Switzerland
3.91

Sony Deutschland GmbH, Köln, Germany
3.111b

Steinbichler Optotechnik GmbH, Traunstein, Germany
6.34, 6.35a

ThyssenKrupp Accessibility BV, Krimpen aan den IJssel, Netherlands
8.27

Tomas Saraceno:
Fig. 8.67b

University College London, UK
5.93

Vertex Antennentechnik GmbH, Duisburg, Germany
1.15, 8.45, 8.46

VEW Vereinigte Elektronik-Werkstätten GmbH, Bremen, Germany
3.163c

Volkwagen AG, Wolfsburg, Germany
2.4, 3.122, 7.10, 8.52, 8.53, 8.54

Winterthus, Switzerland, Land Surveying Office
6.46

Zoller & Fröhlich GmbH, Wangen im Allgäu, Germany
3.171

Index

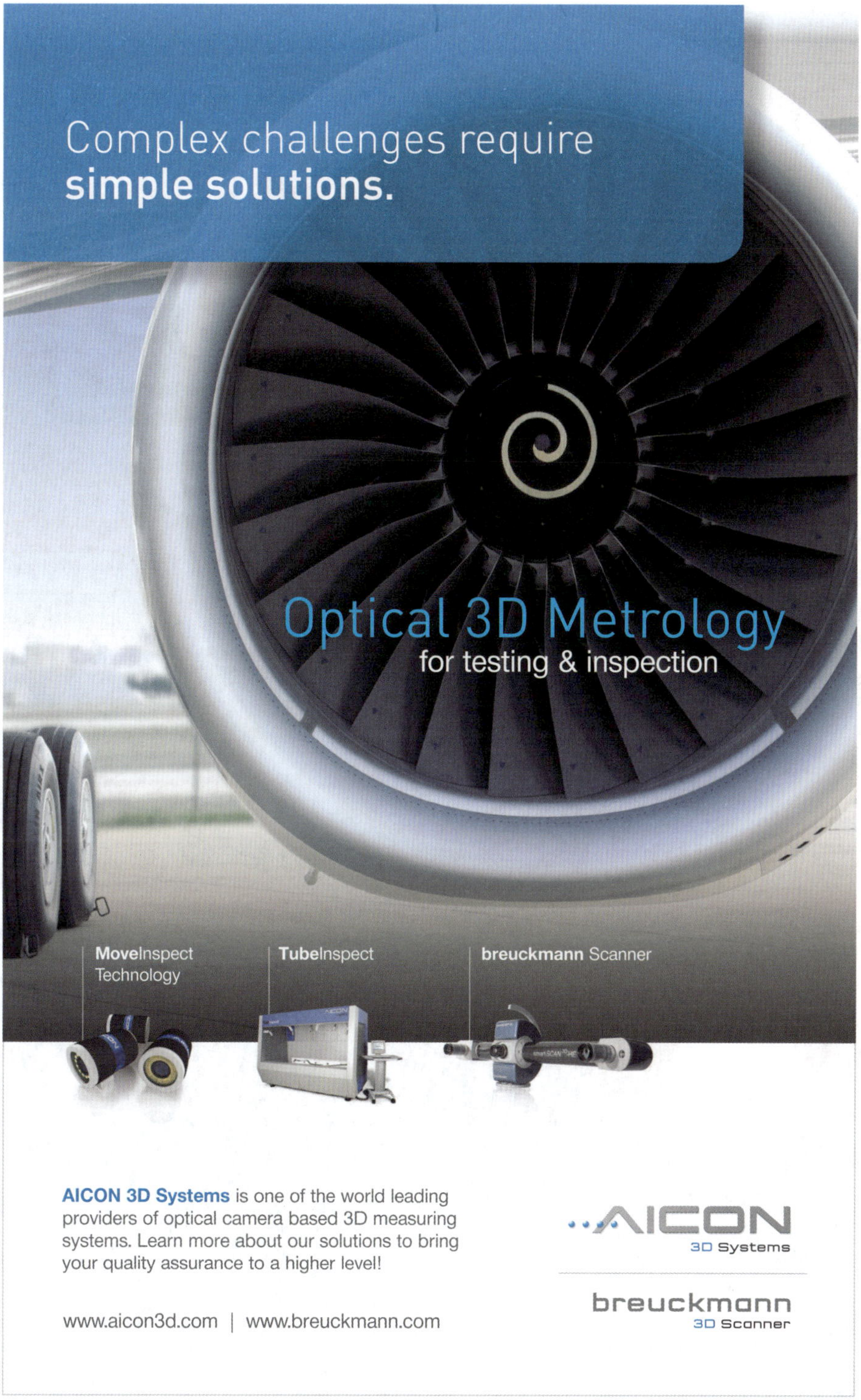

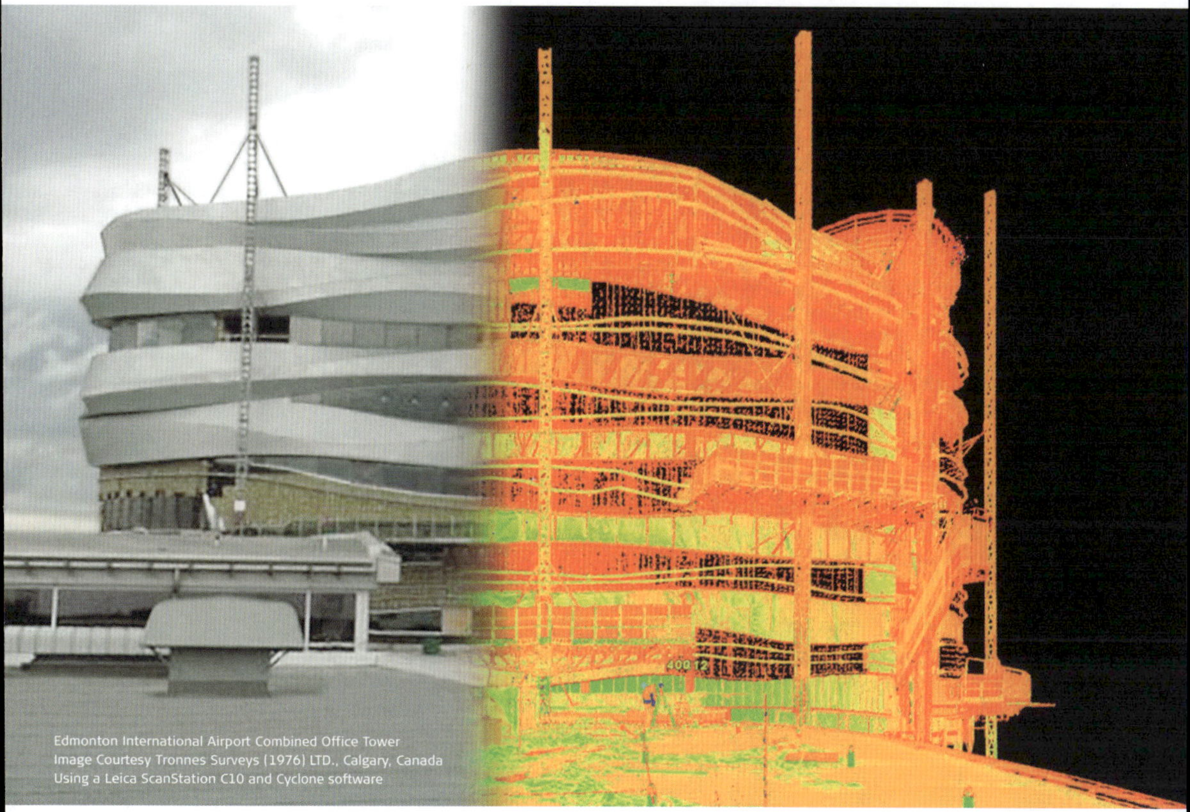